FLUORINATED HETEROCYCLIC COMPOUNDS

FLUORINATED HETEROCYCLIC COMPOUNDS

Synthesis, Chemistry, and Applications

Edited by

VIACHESLAV A. PETROV
DuPont Central Research and Development
Wilmington, DE, USA

WILEY
A JOHN WILEY & SONS, INC., PUBLICATION

Published by John Wiley & Sons, Inc., Hoboken, New Jersey
Published simultaneously in Canada

Library of Congress Cataloging-in-Publication Data:

Petrov, Viacheslav A.
Fluorinated heterocyclic compounds : synthesis, chemistry, and applications / Viacheslav A. Petrov.
p. cm.
Includes index.
ISBN 978-0-470-45211-0 (cloth)
1. Heterocyclic compounds–Synthesis. 2. Organofluorine compounds. I. Title.
QD400.5.S95P48 2009
547′.59–dc22

2009013357

Printed in the United States of America

10 9 8 7 6 5 4 3 2 1

CONTENTS

PREFACE

Heterocycles represent a larger group of organic compounds and play an important role in all aspects of pure and applied chemistry. The subgroup of this class called *fluorinated* heterocycles is relatively "young," since the intensive development of the synthetic chemistry of fluorinated heterocycles started only after World War II. Nowadays fluorinated heterocyclic compounds can be found among potent pharmaceuticals, crop protection agents, and products of technical importance. This merging area of organic, heterocyclic, and fluoroorganic chemistry is still rapidly growing and in the past six decades a large number of fluorinated heterocyclic materials have been discovered.

Several books containing sections on chemistry of fluorinated heterocyclic compounds have been published in last fifteen years. A short, nonexhaustive list includes *Organofluorine Chemistry* by T. Hiyama, Springer, 2000; *Modern Fluoroorganic Chemistry* by P. Kirsch, Wiley-VCH, 2004; *Fluorine in Organic Chemistry* by R. Chambers, Blackwell Publishing/CRC Press, 2004; and *Organofluorine Chemistry* by K. Uneyama, Blackwell Publishing, 2006. Some technical applications of selected fluorinated heterocycles were covered in Chapters 10, 11, 13, 15, 19, 21, and 24 of *Organofluorine Chemistry: Principles and Commercial Applications* by R.E. Banks, B.E. Smart, and J.C. Tatlow (Eds.), Plenum Press, published in 1994. Interestingly, so far only one book fully dedicated to the chemistry of fluorinated heterocyclic materials was published (Г. Г. Фурин, "Фторсодержащие Гетероциклические Соединения", Новосибирск, Наука, 2001 (G.G. Furin, *Fluorinated Heterocyclic Compounds*, Novosibirsk, Nauka, 2001)). Unfortunately, this monograph written in Russian language was not translated and due to its relatively small print run (550 prints), it is not readily available to the international chemical community. The list would not be complete without the recently published book *Fluorinated Heterocycles* by A. Gakh and K. Kirk (Eds.), ACS Symposium Series 1003, American Chemical Society, Washington, DC, 2009, dealing with different aspects of synthetic methodology for the preparation of selected classes of fluorinated heterocycles.

The diversity, complexity, and unique behavior of fluorinated heterocycles combined with a wide range of applications and the fact that a large body of experimental material has to be reviewed make the comprehensive coverage of the subject extremely difficult. As a compromise, it was decided to confine this book to the most representative routes, chemical transformations, and applications of fluorinated heterocycles containing oxygen, nitrogen, and sulfur (and to some extent other elements such as phosphorous and selenium) and also to limit a review of the literature to publications in open rather (than patent) scientific literature. The synthesis and transformations of three- and four-membered heterocycles containing oxygen, nitrogen, and sulfur are covered in Chapters 1 and 2, respectively. Due to a substantial number of publications on the synthesis and chemistry of five-membered heterocycles, it was decided to divide into two groups. The synthesis and chemistry of nitrogen-containing heterocycles are reviewed in Chapter 3; heterocycles containing oxygen, sulfur, and other elements are dealt with in Chapter 4. Data on the synthesis of fluorinated sugars are given in Chapter 5.

A similar approach was used in case of aromatic fluorinated heterocycles. Ring-fluorinated pyridines containing one, two, and three fluorine substituents are reviewed in Chapter 6, while Chapter 7 focuses on the synthesis and typical chemical transformations of aromatic heterocycles containing perfluoroalkyl groups.

Since *perfluorinated* heterocycles have distinct and often unique chemistry, this group was considered as a separate category and data on perfluorinated aromatic and nonaromatic compounds are given in Chapters 8 and 9, respectively. The information on seven-membered and larger ring heterocycles, including perfluorinated crown ethers and polyfluorinated macrocycles, is provided in Chapter 10, which concludes Part I of the book.

Part II contains information on different applications of fluorinated heterocycles. Chapter 11 focuses on the use of fluorinated heterocycles in agricultural products, Chapter 12 summarizes data on pharmaceuticals containing fluorinated heterocycles, and Chapter 13 reviews different aspects of technical applications of fluorinated heterocycles.

This book is intended for advanced students, graduates, and researchers from both academia and industry working in the area of organic, heterocyclic, and fluoroorganic chemistry and looking for a survey on the synthetic methods, chemistry, and applications of major classes of fluorinated heterocycles.

This book is written by an international team of world-recognized experts in the area of organic and industrial chemistry of fluorine. I would like to thank all contributors for their time and hard work, which made this first book on the chemistry and applications of fluorinated heterocycles possible. I am also indebted to Susan Farmer of Wiley-Blackwell, who came up with the idea of this book, for all her support and encouragement and also to the staff of the editorial office of Wiley-Blackwell for their cooperation and understanding.

This book would not have been possible without the encouragement of my family and I would like to thank Olga, Masha, Alexandra, and Andrew for all their help and continuous support.

February 2009 VIACHESLAV PETROV

CONTRIBUTORS

Anna V. Gromova, UNC Eshelman School of Pharmacy, Division of Medicinal Chemistry and Natural Products, The University of North Carolina, Chapel Hill, NC 27599-7360, USA

Wonpyo Hong, Crop Protection Products, E. I. DuPont de Nemours and Company, Stine Haskell Research Center, Newark, DE 19714, USA

Kenneth L. Kirk, Laboratory of Bioorganic Chemistry, National Institute of Diabetes and Digestive and Kidney Diseases, National Institutes of Health, DHHS, Bethesda, MD 20892, USA

Roman B. Larichev, DuPont Central Research and Development, Experimental Station, P.O. Box 80500, Wilmington, DE 19880-0500, USA

David M. Lemal, Department of Chemistry, Dartmouth College, 6128 Burke Laboratory, Hanover, NH 03755, USA

Viacheslav A. Petrov, DuPont Central Research and Development, Experimental Station, P.O. Box 80500, Wilmington, DE 19880-0500, USA

Feng-Ling Qing, Key Laboratory of Organofluorine Chemistry, Shanghai Institute of Organic Chemistry, Chinese Academy of Science, 354 Fenglin Lu, Shanghai 200032, China; College of Chemistry, Chemical Engineering and Biotechnology, Donghua University, 2999 North Renmin Lu, Shanghai 201620, China

Xiao-Long Qiu, Key Laboratory of Organofluorine Chemistry, Shanghai Institute of Organic Chemistry, Chinese Academy of Science, 354 Fenglin Lu, Shanghai 200032, China

Sudharsanam Ramanathan, Department of Chemistry, Dartmouth College, 6128 Burke Laboratory, Hanover, NH 03755, USA

Paul R. Resnick, FluoroScience LLC, 104 Rivergreen Court, Cary, NC 27518, USA

Lyudmila A. Rodinovskaya, N.D. Zelinsky Institute of Organic Chemistry, RAS, 47 Leninsky Prospect Moscow 119991, Russia

Graham Sandford, Department of Chemistry, Durham University, South Road, Durham, DH1 3LE, UK

Kenji Sasaki, Faculty of Pharmaceutical Science, Okayama University, 1-1-1, Tsushima-Naka, Okayama 700-8530, Japan

Yuriy Shermolovich, Institute of Organic Chemistry, NAS of Ukraine, Murmanskaya 5, Kiev-94, 02094, Ukraine

Alexander A. Shestopalov, Department of Chemistry, Duke University, Durham, NC 27708-0346, USA

Anatoliy M. Shestopalov, N.D. Zelinsky Institute of Organic Chemistry, RAS, 47 Leninsky Prospect Moscow 119991, Russia

Kenji Uneyama, Department of Applied Chemistry, Faculty of Engineering, Okayama University, 3-1-1, Tsushima-Naka, Okayama 700-8605, Japan

INTRODUCTION: NOMENCLATURE OF POLYFLUORINATED HETEROCYCLES

VIACHESLAV A. PETROV
DuPont Central Research and Development, Experimental Station, P.O. Box 80500, Wilmington, DE 19880-0500, USA

There are three types of nomenclature used for heterocyclic compounds.[1] Many heterocycles have *trivial* names, which are based on their occurrence, special properties, or historical reasons such as discovery of particular material. *Systematic* names of heterocyclic compounds derived from the structure of the compound are governed by IUPAC rules, which are divided into two groups: the Hantzsch–Widman and replacement nomenclatures.[1] In this book, we were trying to follow the guidelines for naming the heterocyclic compounds, which are summarized in Chapter 2 of the excellent book *The Chemistry of Heterocycles: Structures, Reactions, and Applications*.[1]

It should be pointed out that currently both trivial and systematic names are commonly used for naming the heterocyclic compounds. For example, an organic chemist will recognize without any difficulty the structures connected to names such as furane, pyrrole, pyrrolidine, pyrazole, imidazole, pyridine, or piperidine, despite the fact that all these names are trivial. On the other hand, the complex heterocycles require more sophisticated approaches in order to avoid ambiguity and correctly translate the chemical structure into the name. For these, compound names are often made using either trivial name (e.g., indazole for benzopyrazole, benzimidazole, indole, and isoindole) or the Hantzsch–Widman nomenclature, for example, 1,2,3- or 1,2,5-oxadiazoles, 1,3-dioxolane, 1,2- or 1,3-dithiolane, and 1,3- or 1,4-dioxane.[1] It should be noted that the Hantzsch–Widman nomenclature treats the unsaturated heterocycle with maximum number of conjugated double bonds as *parent* compound.[1] This adds another layer of complexity, giving rise to names such as tetrahydrofurane, tetrahydrothiophene, 2,3-dihydropyrrole, or 3,4-dihydrofurane. As it can be seen from the mentioned examples, names of heterocyclic compounds form a separate terminological group and the work with this terminology requires basic knowledge of the "language" used in this area of organic chemistry.

Heterocycles containing limited amount of fluorinated substituents (usually 1–3) can be named using trivial names or conventional nomenclature in combination with indication of the position of fluorinated substituents, for example, 2-fluoro-4-trifluoromethylpyridine. The situation becomes more complicated in case of polyfluorinated and completely fluorinated heterocycles. In case of heterocycles with relatively small number of fluorinated substituents and well-defined structures, Greek or Latin numeral roots can be used.[2] Names such as hexafluoropropene oxide, 2,2-bis(trifluoromethyl)oxirane, 2,2,3,3-tetrafluorooxetane, tetrakis(trifluoromethyl)furane, pentafluoropyridine, tetrafluoropyridazine, tetrafluoropyrimidine, and heptafluoroquinoline are unambiguous and commonly accepted (see Fig. 0.1).

It should be pointed out that for completely fluorinated materials, one could use the so-called perfluoro- or *F*-nomenclature.[2] The prefix perfluoro- or symbol *F*- have the same meaning and combined with trivial or standard systematic name of a heterocycle, it indicates that in parent compound all hydrogens *connected to carbons* were replaced by fluorines. Examples of different names for heterocycles, such as perfluoropropene oxide or *F*-propene oxide, perfluoro-(2,2-dimethyloxirane) or *F*-(2,2-dimethyloxirane), and so on, are shown in Fig. 0.2.

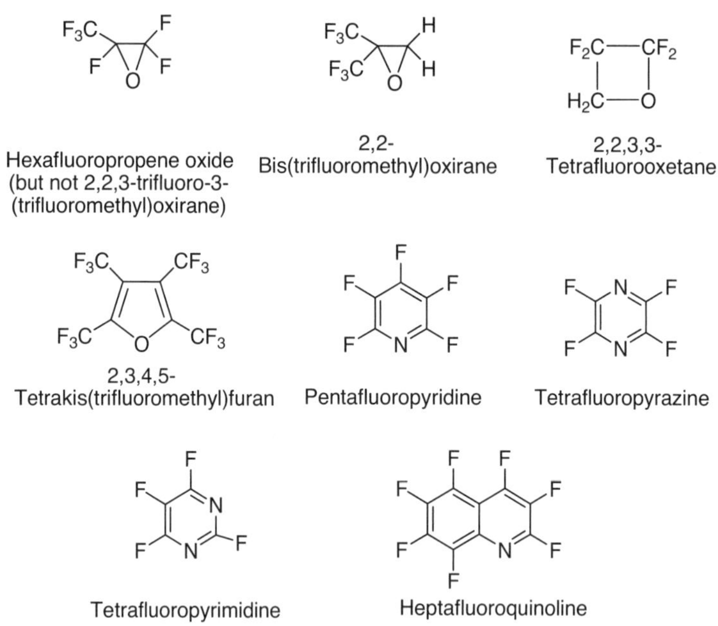

FIGURE 0.1 Nomenclature of polyfluorinated heterocycles.

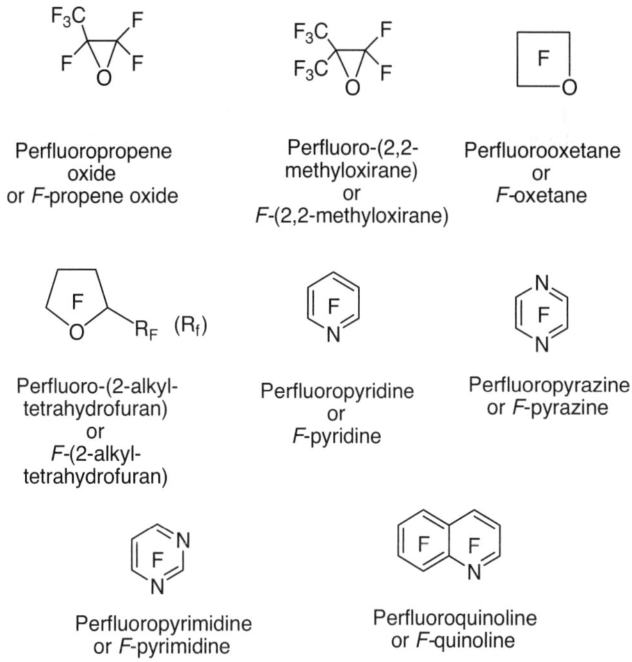

FIGURE 0.2 Examples of perfluoro- and *F*-nomenclatures of fluorinated heterocycles.

It should also be pointed out that the symbol "F" placed in the center of the heterocycle has the same meaning and denotes perfluorinated compound (see Fig. 0.2).[2]

Perfluoro- or *F*-nomenclatures are extremely convenient for the heterocyclic systems containing a large number of fluorines or perfluoroalkyl substituents. For example, listing all positions of 12 fluorine substituents in *F*-thiepane (see Fig. 0.3) makes systematic name of this compound long and cumbersome.

It should also be pointed out that the symbols R_F or Ar_F are also often used and they usually refer to monovalent perfluoroalkyl or perfluoroaryl group.[2] Despite the fact that R_F or Ar_F are recommended abbreviation of the corresponding groups, nowadays in scientific literature, symbols R_f or Ar_f are also used as equivalent of perfluoroalkyl or perfluoroaryl groups, respectively. In this book, both types of abbreviations can be found, although we were trying to adhere to original R_F or Ar_F abbreviations.

It is noteworthy that the case of cyclic polyfluorinated amines is special. Since the prefix perfluoro- is used to show the substitution of all hydrogens in the molecule with exception of those whose replacement affects the functionality,[2] the name perfluoropiperidine should be used for the compound containing NH group, but not for *N*-fluoroamine, which should be called perfluoro-*N*-fluoropiperidine (Fig. 0.3).

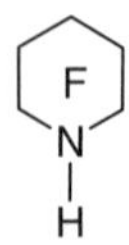

Perfluoropiperidine
or
F-piperidine

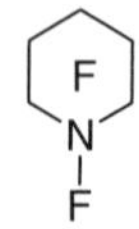

Perfluoro-N-fluoropiperidine
or *F*-N-fluoropiperidine

Perfluorothiepane
or F-thiepane
instead of
2,2,3,3,4,4,5,5,6,6,7,7-dodecafluoro-thiepane

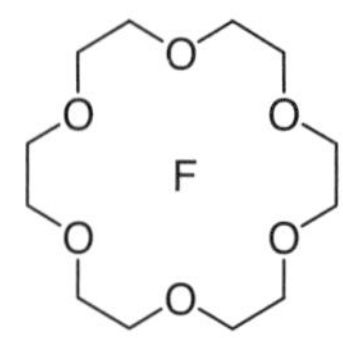

Perfluoro-18-crown-6
F-18-crown-6

FIGURE 0.3 Examples of nomenclature for perfluorinated heterocycles.

Due to overlap between perfluoro- and heterocycle nomenclatures, some names may be rather complicated and contradictory, such as *perfluoro*(*tetrahydro*furane), for the compound that has no single hydrogen in the molecule!

For the same reason naming of perfluorinated, nonaromatic derivatives may be difficult and in this case, replacement nomenclature is used, despite the fact that this it is recommended by IUPAC for use with larger ring heterocycles. The use of replacement nomenclature allows simplifying the process and shortening some names of heterocycles. For example, in case of two compounds shown in Fig. 0.4, perfluoro-1-azacyclopentene-1 and perfluoro-1-azacyclohexene-1 names are often used instead of perfluoro-3,4-dihydro-2*H*-pyrrole and perfluoro-2,3,4,5-tetrahydropyridine in scientific literature.

In this book, although we were trying to follow recommendations given in Refs 1 and 2, due to complexity of the subject it was difficult to keep it consistent, so the reader of the book should be prepared to find all types of nomenclatures and names of polyfluorinated heterocyclic compounds used in practice.

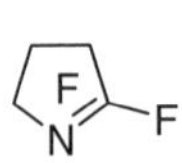

Perfluoro-1-azacyclopentene-1
or
Perfluoro-3,4-dihydro-2*H*-pyrrole

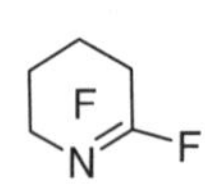

Perfluoro-1-azacyclohexene-1
or
Perfluoro-2,3,4,5-tetrahydropyridine

FIGURE 0.4 Examples of replacement and systematic nomenclatures for naming the perfluorinated nonaromatic compounds.

REFERENCES

1. Eicher, T.; Hauptmann, S. *The Chemistry of Heterocycles: Structures, Reactions, and Applications*, John Wiley & Sons: Chichester, UK, 2003; pp 5–16.
2. Banks, R.E.; Tatlow, J.C. *Organofluorine Chemistry: nomenclature and history*. In *Organofluorine Chemistry: Principles and Commercial Applications*; Banks, R.E.; Smart, B.E.; Tatlow, J.C., Eds.; Plenum Press: New York, **1994**; pp 2–4.

PART I

SYNTHESIS AND CHEMISTRY OF FLUORINATED HETEROCYCLES

1

FLUORINATED THREE-MEMBERED RING HETEROCYCLES

David M. Lemal and Sudharsanam Ramanathan

Department of Chemistry, Dartmouth College, 6128 Burke Laboratory, Hanover, NH 03755, USA

1.1 FLUOROOXIRANES[1–4]

1.1.1 Synthesis

Highly fluorinated oxiranes contrast sharply with their hydrocarbon counterparts with regard to both methods for their synthesis and chemical reactivity. Hydrocarbon-derived oxiranes are normally prepared by cyclization of an alcohol with a leaving group on the β-carbon or by electrophilic epoxidation of an alkene, but perfluoro- and halofluorooxiranes are usually synthesized via nucleophilic attack on the corresponding alkene. For example, hydrogen peroxide and aqueous alkali with a water-miscible cosolvent or phase-transfer agent have been used to prepare hexafluoropropylene oxide (HFPO, **1**).[5] Hydroperoxide anion adds to C1 of the alkene, and the resulting carbanionic center attacks at oxygen, breaking the weak O–O bond and expelling the hydroxide ion.

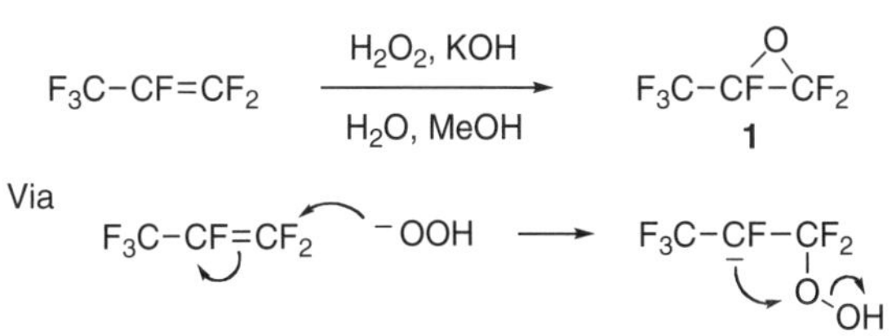

Fluorinated Heterocyclic Compounds: Synthesis, Chemistry, and Applications, Edited by Viacheslav A. Petrov

Yields are often higher with sodium hypochlorite, which reacts analogously to hydrogen peroxide, with expulsion of chloride ion. The oxidation of *cis*- and *trans*-perfluoroalkenes proceeds stereospecifically with retention of configuration.[2] 2,2-Bis (trifluoromethyl)ethylene is oxidized to **2** in 65–75% yield by sodium hypochlorite with phase-transfer catalysis (PTC) using Aliquat®-336.[7]

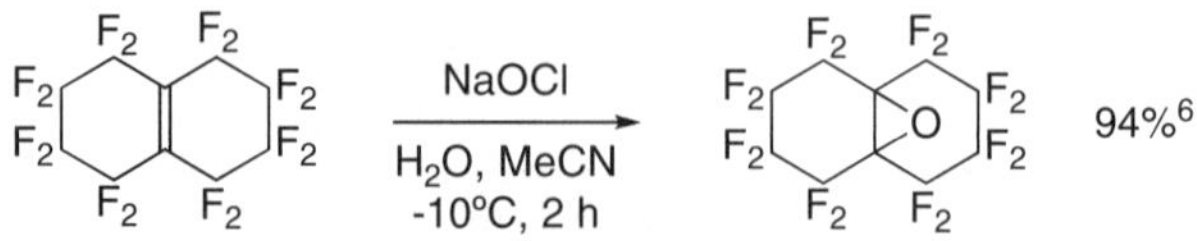

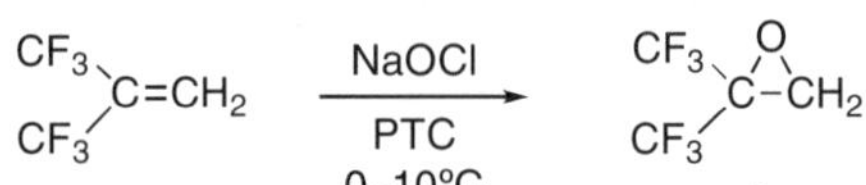

2

Another nucleophilic reagent that is effective with tri- and tetrasubstituted perfluoroalkenes is trimethylamine oxide. Attack by the oxygen atom is followed by elimination of trimethylamine. The amine oxide can be used catalytically as well as stoichiometrically, as *m*-chloroperbenzoic acid (MCPBA) or urea/hydrogen peroxide reoxidizes the amine.[8]

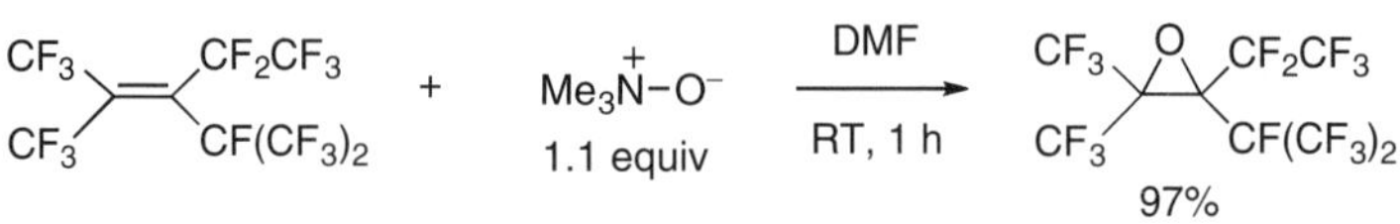

Though it requires vigorous conditions and is less common than nucleophilic epoxidation, electrophilic epoxidation of a perfluoroalkene is possible with the potent combination of chromic oxide and fluorosulfonic acid, providing another route to hexafluoropropylene oxide (**1**).[9] As a further example of electrophilic attack, hexafluoro Dewar benzene (**3**) is transformed into either a mono- or a diepoxide by the powerful hypofluorous acid–acetonitrile complex.[10] The fact that the much weaker electrophile MCPBA readily epoxides such electron-deficient alkenes as ethyl pentafluoromethacrylate (**4**)[11] suggests that it actually reacts via nucleophilic attack at the β-carbon.

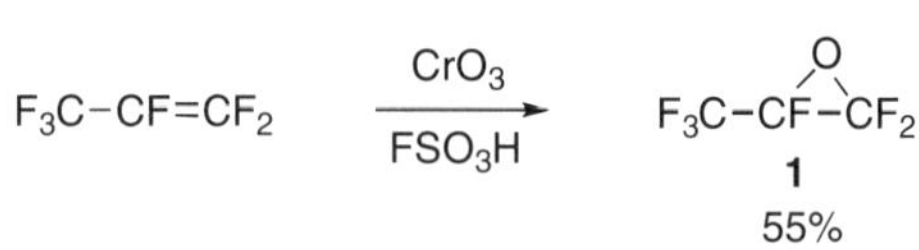

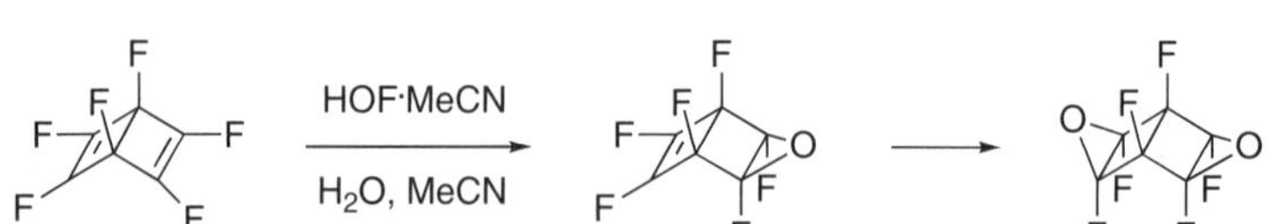

3

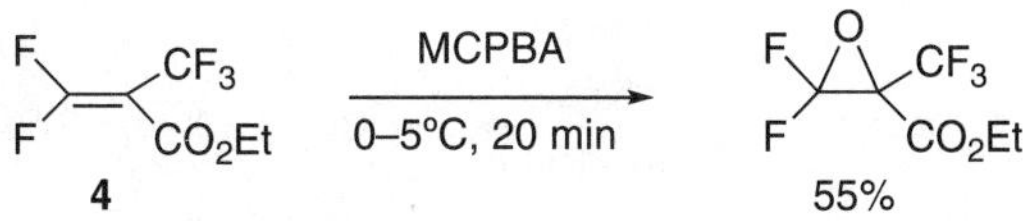

Oxirane formation can also occur via free radical mechanisms, as in the reaction of certain fluoroalkenes with oxygen. Under pressure at elevated temperatures, oxygen alone can suffice, but activation is frequently provided in the form of radical initiators (e.g., tribromofluoromethane) and ultraviolet light.[12] Thermolysis of dioxole **5**, comonomer from which DuPont's Teflon-AF® is made, offers an unusual route to an oxirane. Rearrangement of the heterocycle presumably takes place via a biradical intermediate.[13]

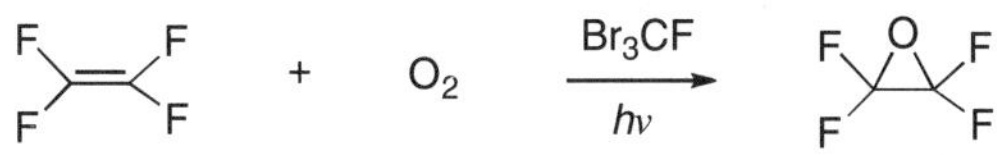

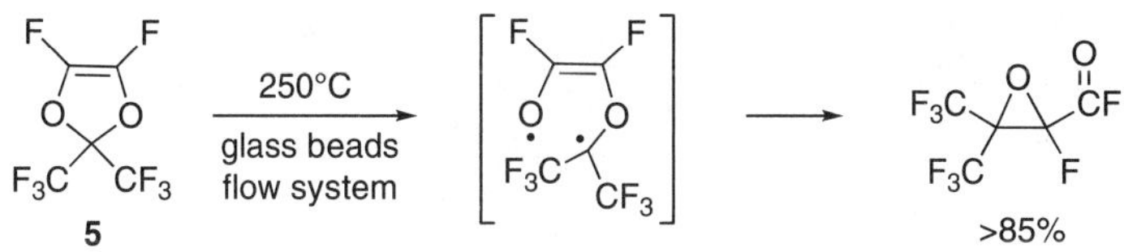

1.1.2 Reactions with Nucleophiles

Fluorooxiranes are easily ring opened by nucleophiles. Treatment of HFPO (**1**) with methanol, for example, affords methyl 2-methoxytetrafluoropropionate (96% yield) via the acid fluoride.[14] Nucleophilic attack on HFPO nearly always takes place at C3, the more hindered carbon, a surprising result for an S_N2 reaction.[3] This may be attributable at least in part to stabilization by negative hyperconjugation of the developing oxyanion in the ring-opening transition state. Both the X-ray structure and calculations show that this effect is very large in the trifluoromethoxide ion.[15] Passed over KF/activated carbon, HFPO is isomerized in excellent yield to perfluoropropionyl fluoride (**6**).[16] However, reaction of **1** with CsF in tetraglyme results in oligomerization.[17] The intermediate perfluoropropoxide ion (**7**) attacks another molecule of HFPO, and the process repeats itself to afford oligomers terminating as acyl fluorides (**8**). Fluorodecarbonylation of **8** produces the inert Krytox® fluids, which are useful as vacuum pump oils.[18] Thiourea behaves as a bifunctional nucleophile in its reaction with oxirane **9**, giving thiazolidinone **10**, with initial attack again at the more substituted carbon. 2-Aminophenol reacts with **9** in analogous fashion.[19]

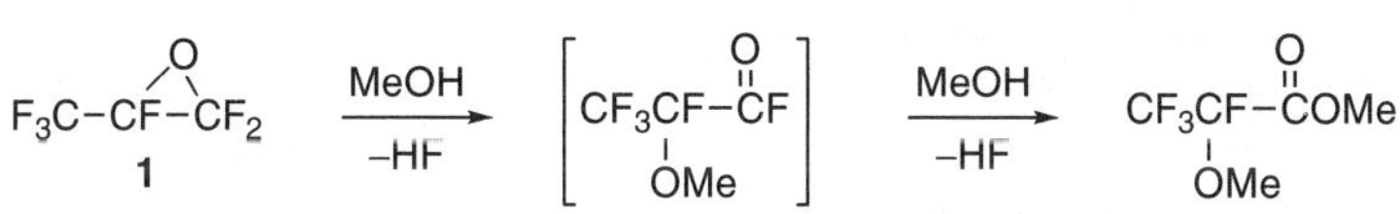

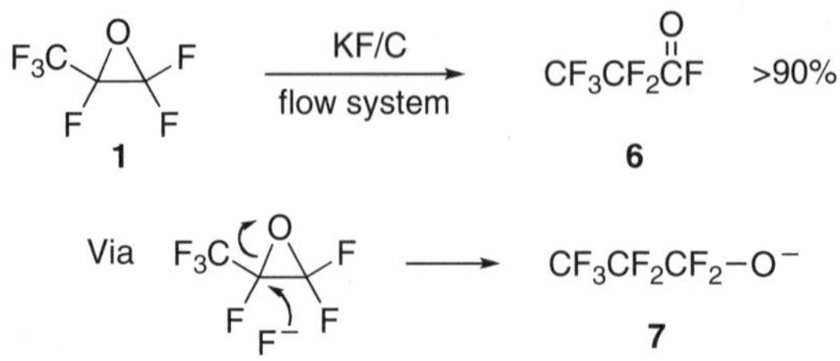

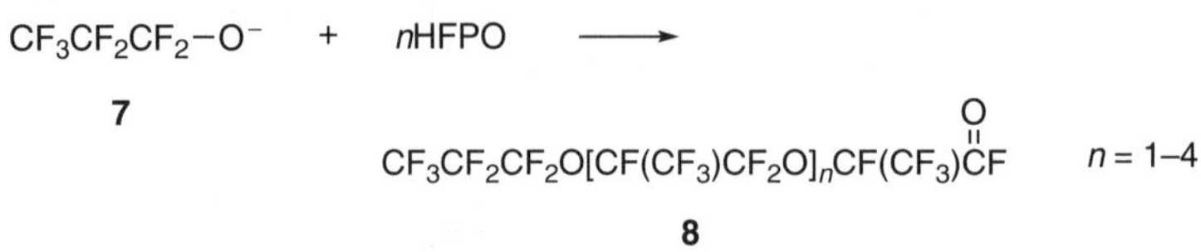

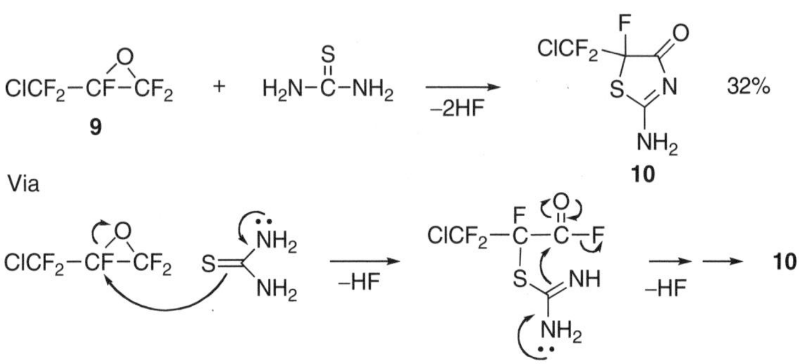

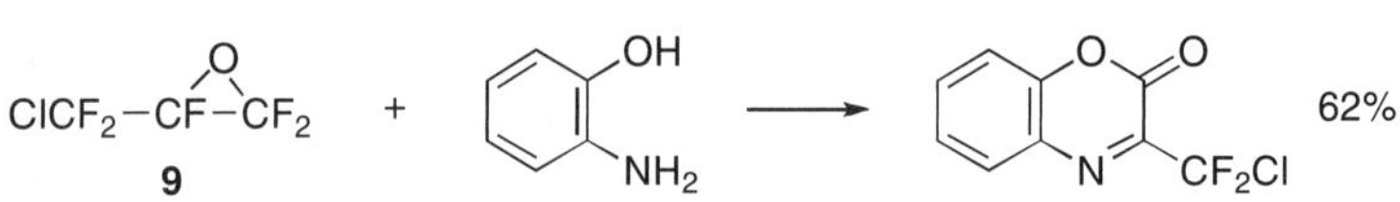

The primary amine ethylamine attacks at C3 of HFPO (**1**) to yield an acyl fluoride that reacts further to afford iminoamide **11**,[14] but the tertiary amine trimethylamine isomerizes **1** to perfluoropropionyl fluoride (**6**) almost quantitatively (30 h, 100°C).[20]

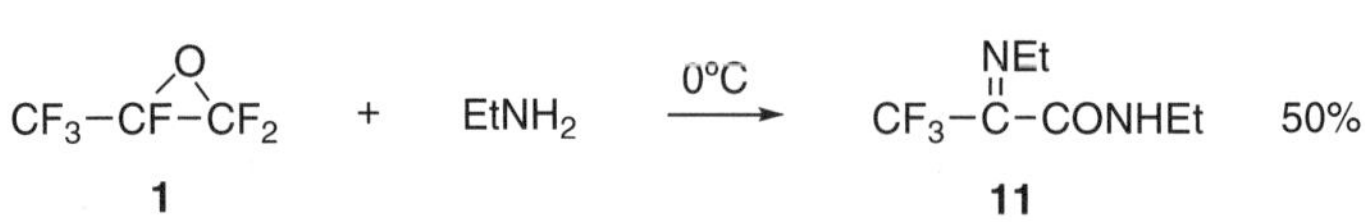

The mechanism shown below, in which the 3° amine just functions as an initiator, differs from others proposed in the literature[20,21] by avoiding unlikely substitution steps. Alternatively, fluoride migration may occur in the betaine **12** with expulsion of trimethylamine, yielding **6** directly.[14]

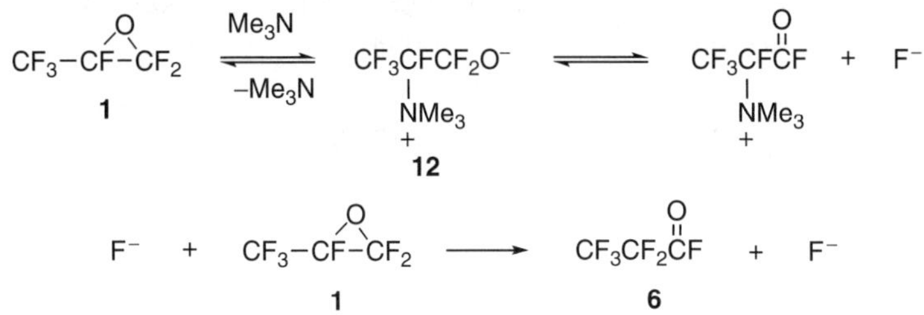

t-Butoxide reacts with HFPO (**1**) to give *t*-butyl perfluoropropionate (**13**).[22] Probably the reaction proceeds by butoxide attacking in the usual way to generate a little acyl fluoride and fluoride ion. Because the butoxide is too hindered to compete for **1** with fluoride, the fluoride ion then catalyzes isomerization of the rest of the HFPO to acyl fluoride **6**, which reacts with *t*-butoxide to give the ester **13**.

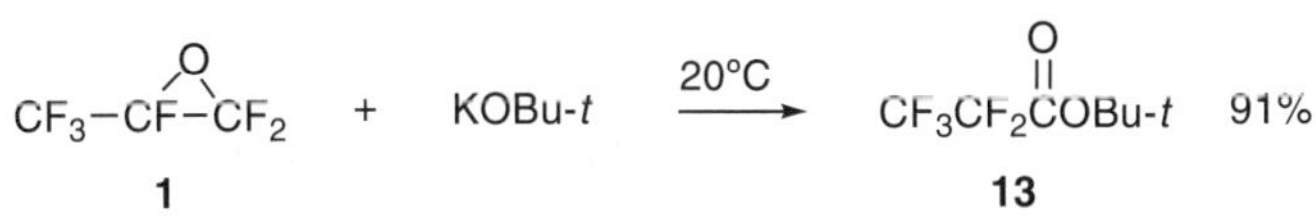

The reaction of **1** with an imine to give vinylogous amide **14**[23] can be interpreted similarly. Isomerization of **1** to **6** by fluoride ion is followed by attack on **6** of the enamine tautomer of the imine.

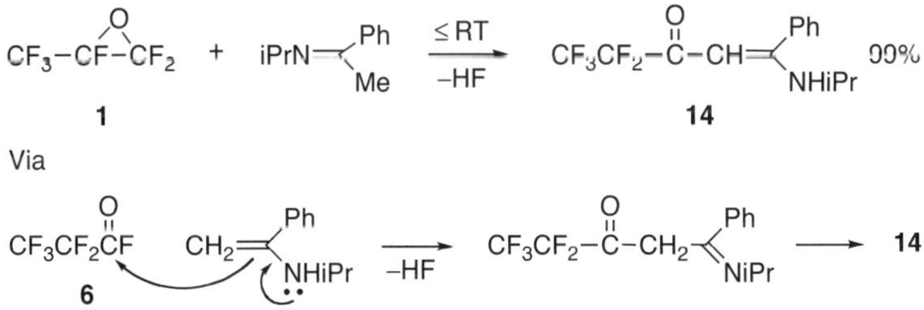

Finally, butyllithium is a rare example of a nucleophilic reagent that attacks at C2 of HFPO, leading after workup to tertiary alcohol **15**.[4]

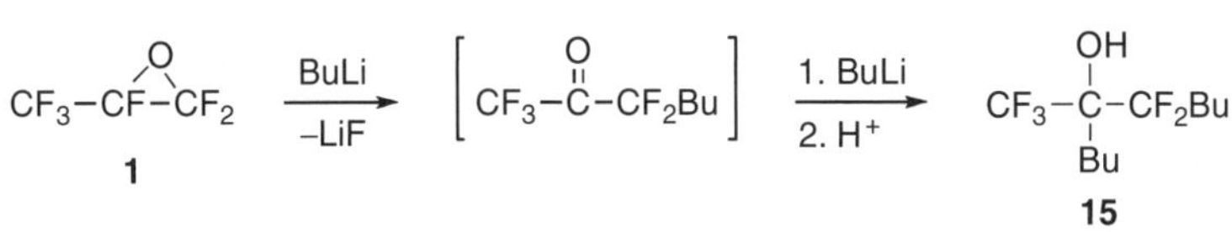

In contrast to HFPO, 2,2-bis(trifluoromethyl)oxirane (**2**) ring opens with oxygen, nitrogen, sulfur, and carbon nucleophiles at the less hindered carbon, yielding tertiary alcohols.[24,25] With diethylamine, for example, **2** affords an aminoalcohol in 83% yield. Oxirane **2** played an important role in the development of monomers from which to build highly transparent, yet readily alkali-soluble photoresist copolymers for use in semiconductor photolithography at 157 and 193 nm.[26] One such monomer was prepared by ring opening of **2** with a norbornene diol.

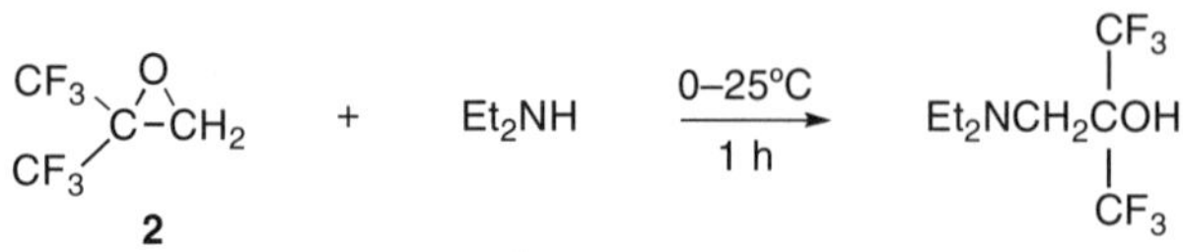

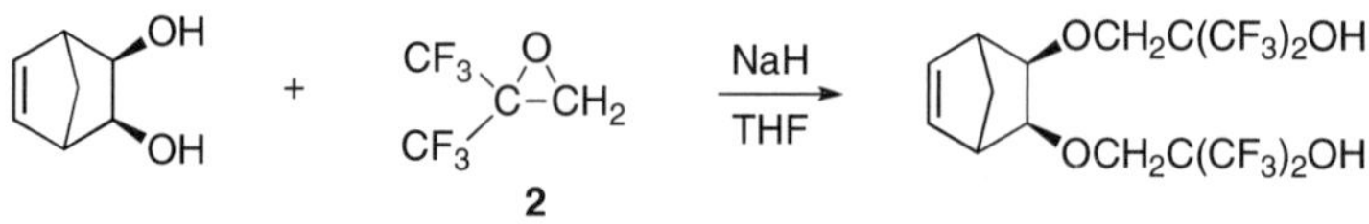

1.1.3 Reactions with Electrophiles

Hydrogen fluoride opens oxirane **16** under very vigorous conditions to give perfluoro-*t*-butanol (**17**).[27] However, Lewis acids isomerize perfluorooxiranes to carbonyl compounds, as illustrated by the transformation of perfluorocyclopentene oxide (**18**) into perfluorocyclopentanone (**19**) over alumina.[28]

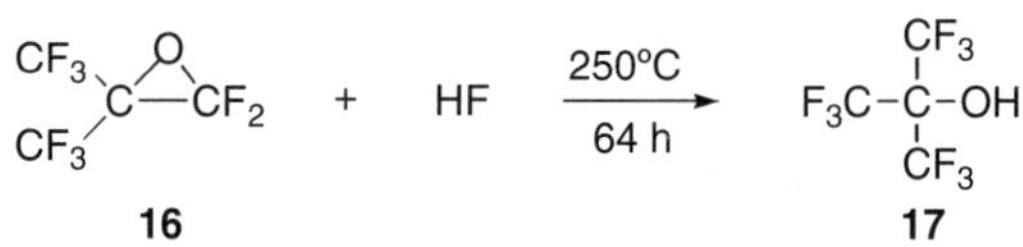

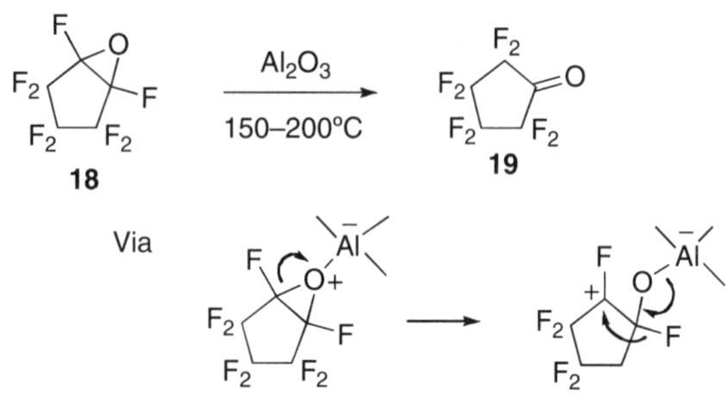

In similar fashion and in contrast to its isomerization by fluoride ion and other bases to perfluoropropionyl fluoride (**6**), HFPO is transformed by Lewis acids such as antimony pentafluoride[29,30] or aluminum chlorofluoride[31] into another isomer, hexafluoroacetone (**20**).

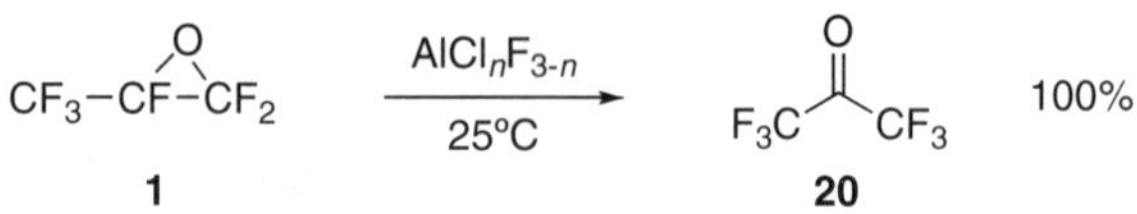

Treatment of the bicyclohexene oxide **21** with antimony pentafluoride yielded perfluorocyclopentene (**22**) instead of the expected bicyclic ketone.[10] Presumably ring strain is responsible for inducing rearrangement of the initially formed carbocation, then ring opening and decarbonylation ensue.

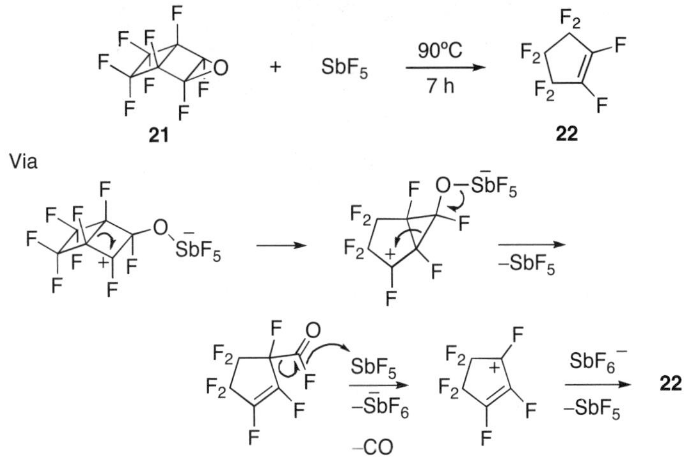

21 22

As would be expected, 2,2-bis(trifluoromethyl)oxirane (**2**) is more easily attacked by electrophiles than its perfluorinated counterpart.[24] It readily undergoes ring opening with concentrated hydrochloric acid to give a chlorohydrin in 88% yield.

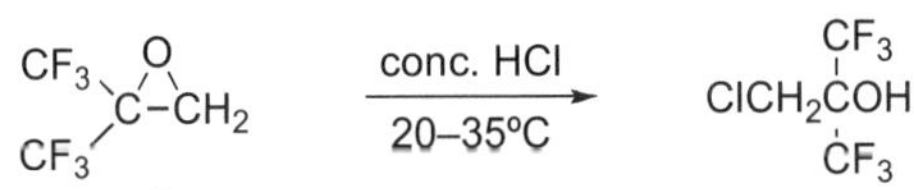

2

1.1.4 Difluorocarbene Chemistry

HFPO (**1**) is an excellent source of difluorocarbene, superior to hexafluorocyclopropane because it decomposes at lower temperatures.[3] It fragments into the carbene and trifluoroacetyl fluoride with a half-life of 169 min at 190°C, as determined by gas-phase NMR.[32] In the absence of a carbene trap, hexafluorocyclopropane and the acyl fluoride are the main products of HFPO decomposition at 200°C.[33]

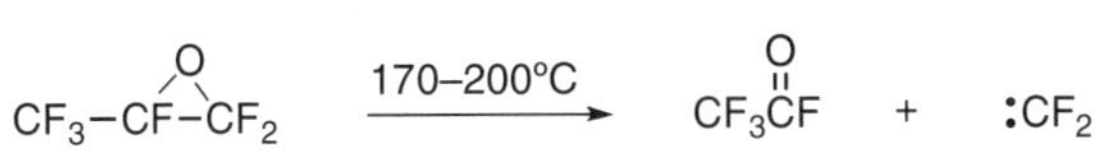

1

The reaction has been shown to be reversible. When trifluoroacetyl fluoride is heated at 130°C with difluorotris(trifluoromethyl)phosphorane (**23**), a lower temperature source of difluorocarbene, HFPO is formed.[34]

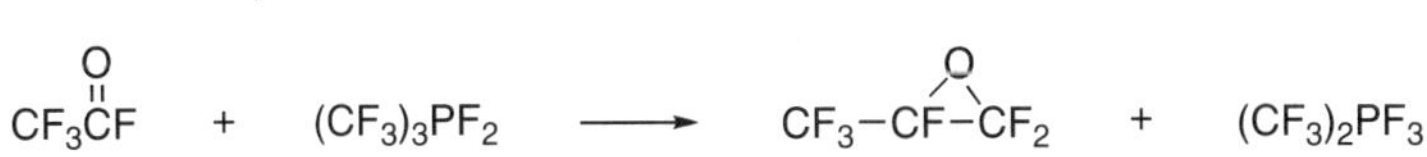

23 1

Generated from HFPO, difluorocarbene reacts with a wide variety of unsaturated compounds, both fluorinated and unfluorinated. It can react stereospecifically, as illustrated with the chloro-1,2-difluoroethylenes.[33]

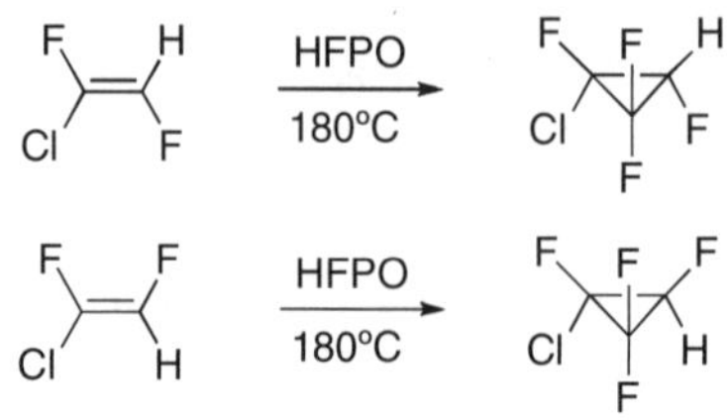

The highly strained perfluorospiropentane (**25**) has been prepared by the reaction of perfluoromethylenecyclopropane (**24**) with HFPO.[35] Cyclohexadiene **26** reacted with HFPO to give in 50% yield a norbornene (**27**) and stereoisomeric bicyclo[3.2.0] heptanes (**28**), all products of vinylcyclopropane rearrangement of the initial cyclopropane adducts.[36]

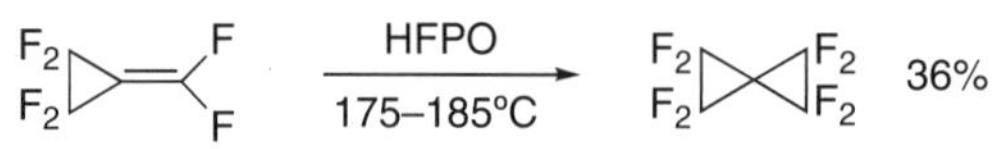

24 **25**

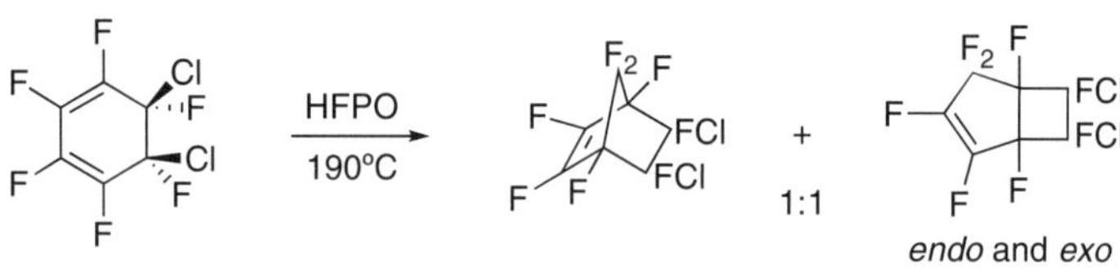

26 **27** **28**

The carbene is also capable of adding to some carbonyl groups to give oxiranes such as **29** from perfluorocyclobutanone.[37] Hexafluorothioacetone reacts analogously to yield thiirane **30**,[38] and an imine reacts with HFPO to afford aziridine **31**.[39]

29

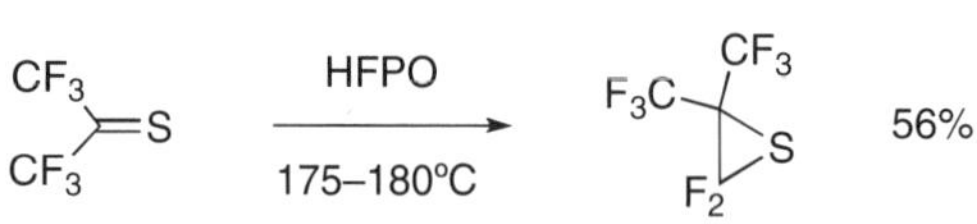

30

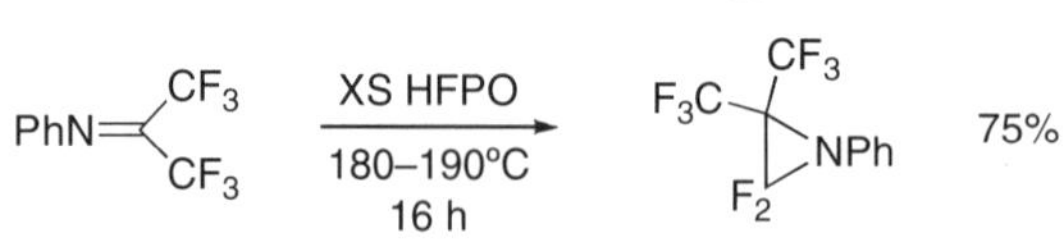

31

Perfluorooxiranes in which a different perfluoroalkyl group substitutes for the CF_3 of HFPO can also function as sources of difluorocarbene. In the presence of nickel powder, for example, oxirane **32** reacts with iodine to give difluorodiiodomethane (**33**) in high yield, accompanied by small amounts of oligomeric diiodides.[40] Yields are very low in the absence of nickel, and it is suggested that the reaction occurs on the surface of the metal with a nucleophilic nickel–carbene complex.

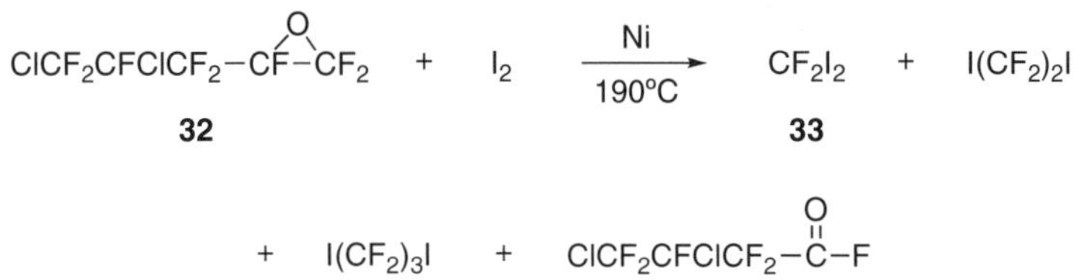

1.1.5 Other Oxirane Chemistry

Commercially available in 75% ee, trifluoromethyloxirane (**34**) can be obtained optically pure by enantioselective hydrolysis.[41,42] It has been elaborated via its lithium salt into a wide variety of derivatives incorporating quaternary chiral carbon centers. Treatment with butyllithium at about −100°C generates the trifluoromethyl-stabilized anion (**35**), which is stable for an hour at −78°C. It reacts with such electrophiles as aldehydes, ketones, and halides with retention of configuration, often in very good yield, to give products that are useful as synthetic intermediates. As an example, anion **35** reacts with a Weinreb amide to afford ketone **36**.[42]

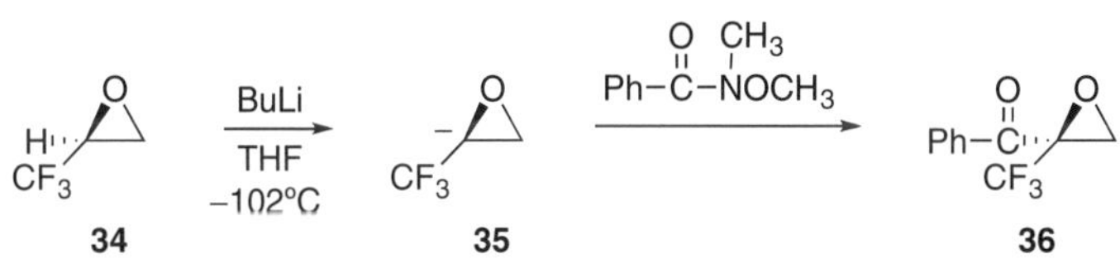

Ring opening of **34** with either acidic or nucleophilic reagents occurs at the unsubstituted carbon to give optically active alcohols, for example, hydroxysulfide **37** formed from phenylthiolate ion generated *in situ*. Oxidation of oxirane **34** yields trifluorolactic acid (**38**) without any loss of optical activity.[43]

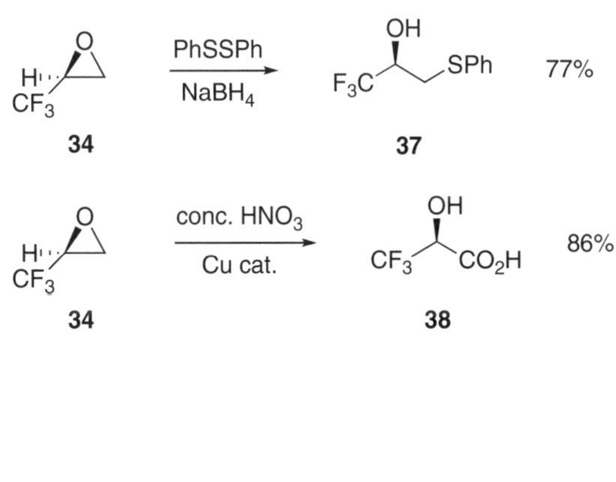

Hexafluorobenzene oxide (**39**) exists in rapid equilibrium with its valence isomer hexafluorooxepin (**40**) at RT, but the equilibrium lies far on the side of **39**.[44,45]

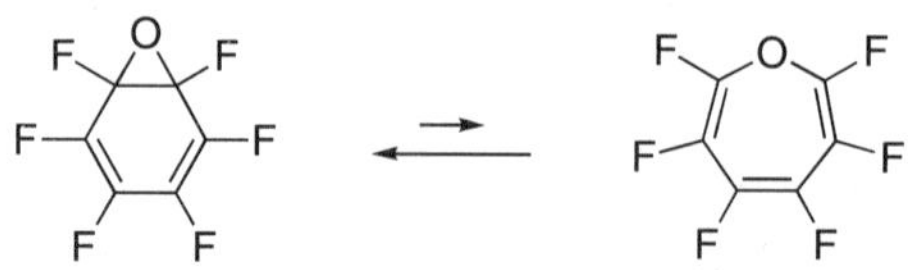

39 **40**

In contrast, the parent benzene oxide–oxepin equilibrium is quite evenly balanced.[46] In nonpolar solvents, **39** is rather stable at RT, but in acetonitrile or acetone it rearranges spontaneously to hexafluorocyclohexa-2,4-dienone (**42**), presumably via zwitterion **41**.

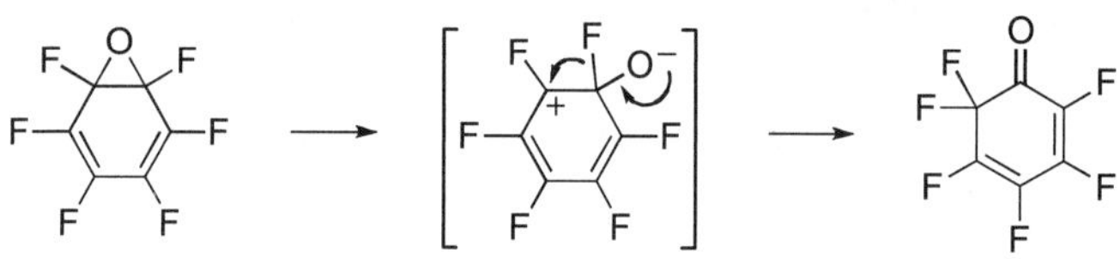

39 **41** **42**

Perfluorotetramethylcyclobutadiene oxide (**43**) rearranges with a half-life of about 20 min at 95°C.[47] As a Dewar furan, it might have been expected to aromatize to perfluorotetramethylfuran (**44**). That process is orbital topology forbidden,[48] however, and instead **43** is transformed cleanly into cyclopropenyl ketone **45**, apparently via ring opening to ketocarbene **46**.

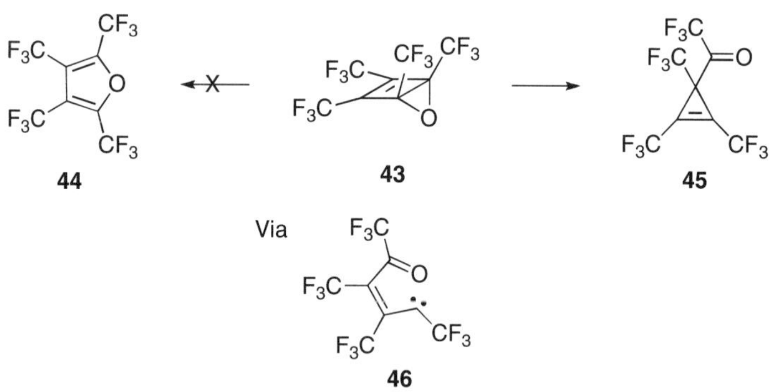

44 **43** **45**

Via

46

The unusually stable α-lactone bis(trifluoromethyl)acetolactone (**48**) has been synthesized by photolysis of the malonyl peroxide **47**.[49,50] It has a half-life in the gas phase of 8 h at 24°C, decarbonylating to give hexafluoroacetone (**20**). Its relatively high stability is another example of the perfluoroalkyl effect.[51] Treatment of **48** with ethanol opens the ring at the carbonyl carbon to yield ethyl α-hydroxyhexafluoroisobutyrate (**49**), thus showing that it reacts with the ring intact, not as a dipolar ion.[50]

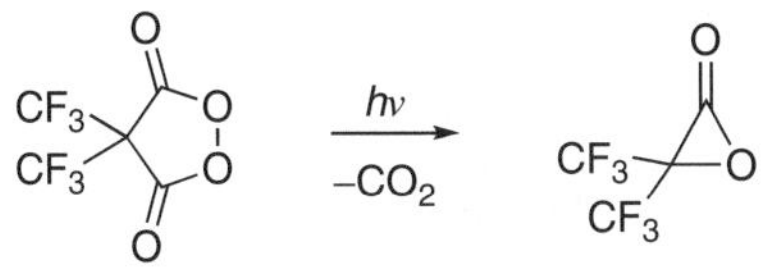

47 **48**

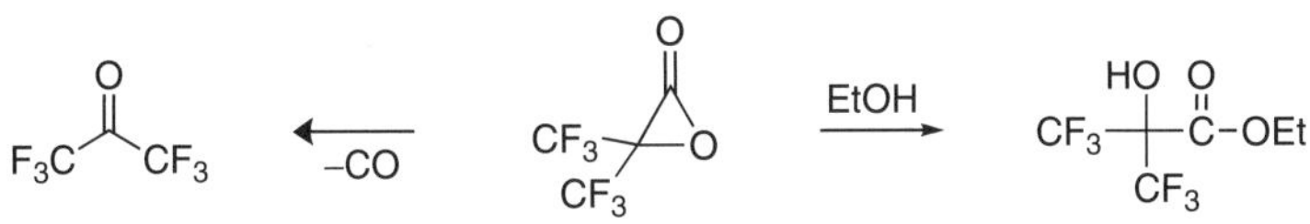

20 **48** **49**

1.2 FLUORODIOXIRANES

1.2.1 Difluorodioxirane (50)

This is the first dioxirane that is stable in the gas phase at RT. It was first postulated as an intermediate in the reaction with O_2 of triplet difluorocarbene, generated by attack of arc-produced carbon atoms on tetrafluoromethane.[52] When alkenes were present, they were stereospecifically epoxidized, and dioxirane **50** was proposed as the oxygen atom transfer reagent. Several years later, **50** was synthesized by CsF-catalyzed isomerization of fluoroformyl hypofluorite (**51**) in the presence of a halogen.[53] The procedure was improved by the use of KHF_2 as catalyst,[54] and the same transformation has been accomplished more recently by infrared photolysis of **51** using a pulsed CO_2 laser.[55] The unusual stability of **50** among dioxiranes is attributable primarily to the π donor ability of fluorine,[56] particularly to lone pair donation into the antisymmetric Walsh orbital of the ring.[57] Difluorodioxirane has the longest known O–O bond, 1.578 Å,[58] a consequence of that orbital interaction.

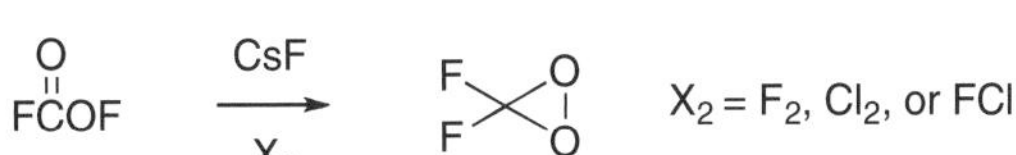

51 **50**

Reaction of **50** with *in situ* generated trifluoromethoxide ion yields fluoroformyl peroxides **52** and **53**.[54] It takes place by attack at oxygen, not carbon, as revealed by ^{13}C labeling.

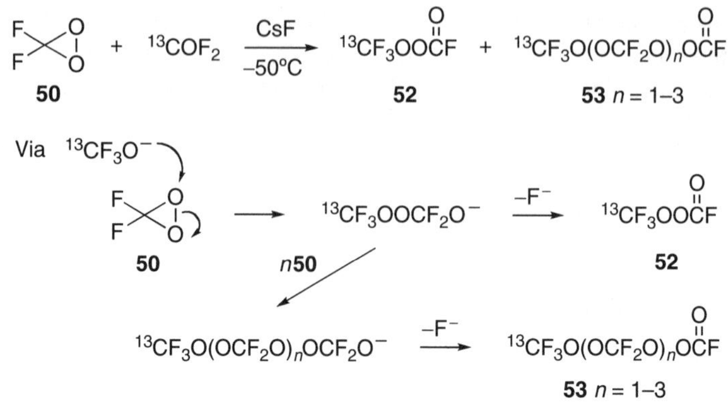

1.2.2 Methyl(trifluoromethyl)dioxirane (54)[59]

First synthesized and characterized in 1988,[60] this dioxirane (TMDO) is generally prepared by oxidation of 1,1,1-trifluoroacetone with potassium monoperoxosulfate (caroate). Yellow solutions of TMDO are quite stable at −20°C. It has also been generated in homogeneous organic solutions up to 1 M by oxidation with arenesulfonic peracids, formed *in situ* from the sulfonic acid, H_2O_2, and NaOH.[61]

Pyrolysis of **54** in the gas phase cleanly yields methyl trifluoroacetate (**55**), but thermal or photochemical decomposition in the liquid phase proceeds via a radical chain mechanism and gives all four possible esters (**55–58**).[62]

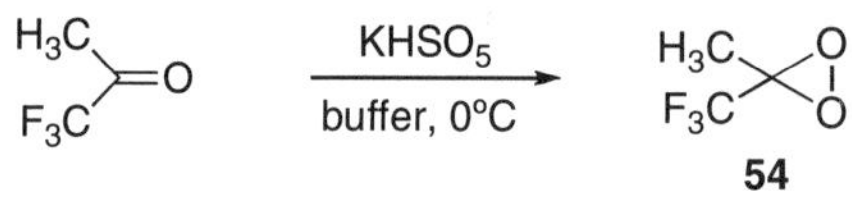

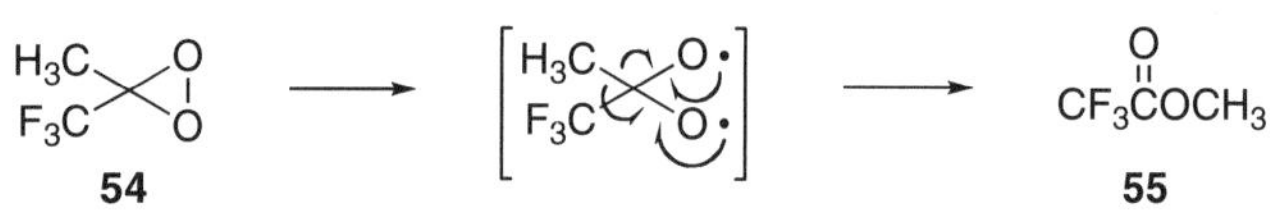

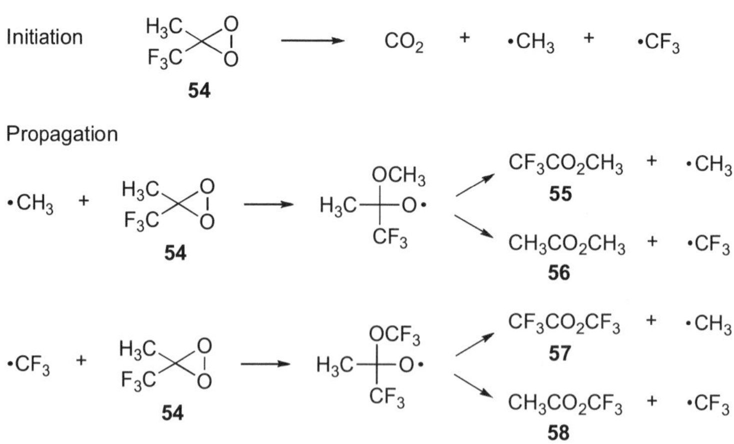

A very powerful oxidant, TMDO epoxidizes alkenes up to $\sim10^5$ times faster than the widely used dimethyldioxirane (DMDO), which in turn reacts $\sim10^2$ times faster than a peracid such as perbenzoic acid.[63] The electron-deficient enone C=C bond in anthracycline **59** resists attack by DMDO, but reacts with TMDO to give epoxide **60** in 95% isolated yield.[64] Naphthalene (**61**) is transformed by TMDO into dioxide **62** in 98% isolated yield (97% conversion).[65] TMDO introduces up to three epoxy functions into C_{60} at 0°C.[66]

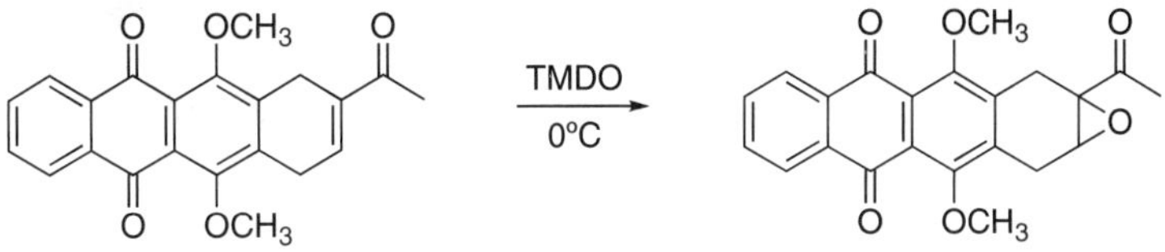

59 **60**

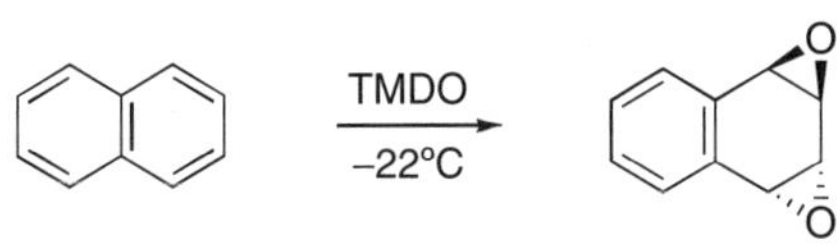

61 **62**

Acetylenes are oxidized by TMDO to diketones. Ynediol **63** affords dione **64** in 90% isolated yield at 95% conversion.[67]

63 **64**

TMDO inserts an oxygen atom into C–H bonds, with high selectivity in the order 3° > 2° > 1° bonds. The reaction is stereospecific with retention of configuration,[68] and apparently occurs concertedly.[69,70] As an example, *cis,cis*-1,3,5-trimethylcyclohexane (**65**) is oxidized cleanly to triaxial triol **66**,[71] and adamantane (**67**) is transformed to a mixture of bridgehead tetrol (**68**) and triol (**69**).[72] The reaction with adamantane is accompanied by chemiluminescence, with triplet trifluoroacetone ($E_T \sim 75$ kcal/mol) the apparent emitter.[73] This observation constitutes further evidence that the oxygen atom transfer takes place concertedly.

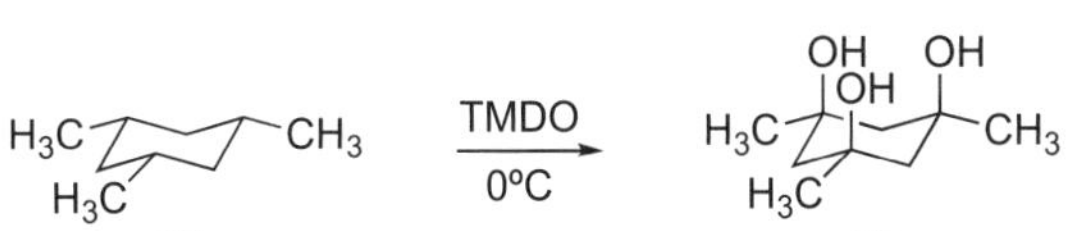

65 **66** 92%

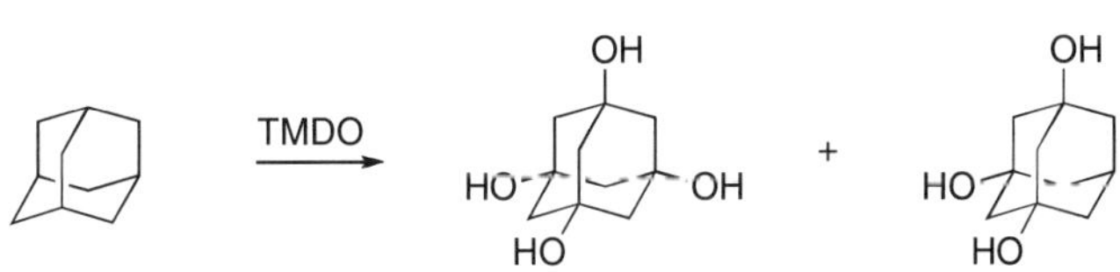

67 **68**, 73% **69**, 24%

TMDO also oxidizes secondary alcohols to ketones,[74,75] hydrazones to ketones,[76] sulfides to sulfoxides[77] and sulfones,[78] silanes to silanols,[79] amides to *N*-hydroxyamides,[80] and so on. An interesting application is the ring-opening cleavage of *p*-methoxybenzyl ethers, for example, **70**, to aldehydoester **71** (39% at 49% conversion).[81]

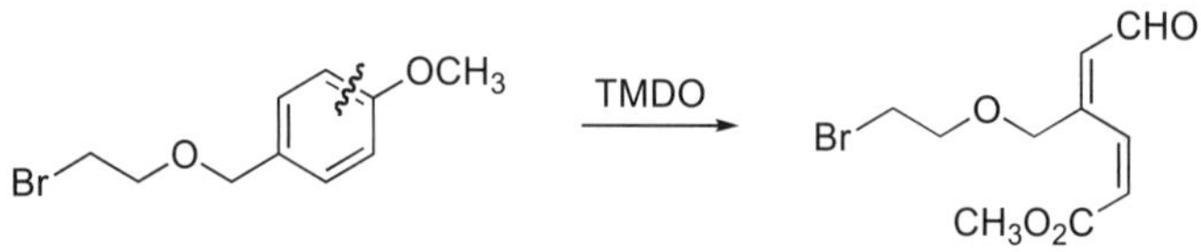

70 **71**

Treatment of TMDO (**54**) with trimethylamine oxide[82] or nucleophilic anions[83] results in the formation of singlet oxygen. Chloride ion, for example, is first oxidized to hypochlorite ion, which then attacks another molecule of the dioxirane. Fragmentation ensues, producing trifluoroacetone, singlet oxygen, and regenerated chloride ion.

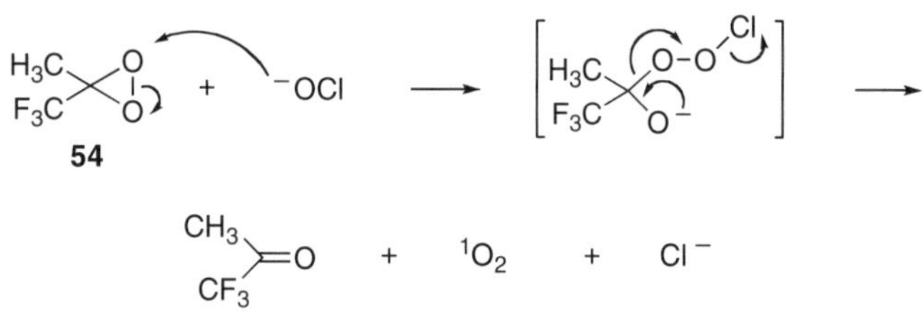

54

1.2.3 Other Fluorodioxiranes

Irradiation of trifluoromethylphenyldiazomethane (**72**) in an argon matrix at 10K afforded trifluoromethylphenylcarbene (**73**).[84] If the matrix was doped with oxygen, warming to 35K resulted in trapping of the carbene as trifluoroacetophenone oxide (**74**). Subsequent photolysis of **74** with long wavelength light gave the acetophenone (**75**) plus trifluoromethylphenyldioxirane (**76**). Photolyzed at shorter wavelengths, the dioxirane rearranged to phenyl trifluoroacetate (**77**).

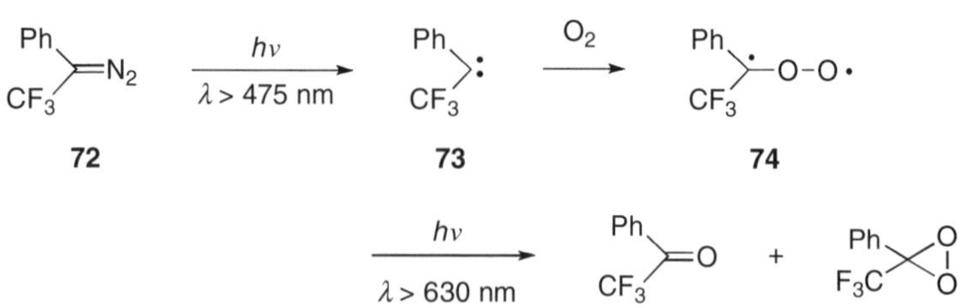

72 **73** **74**

75 **76**

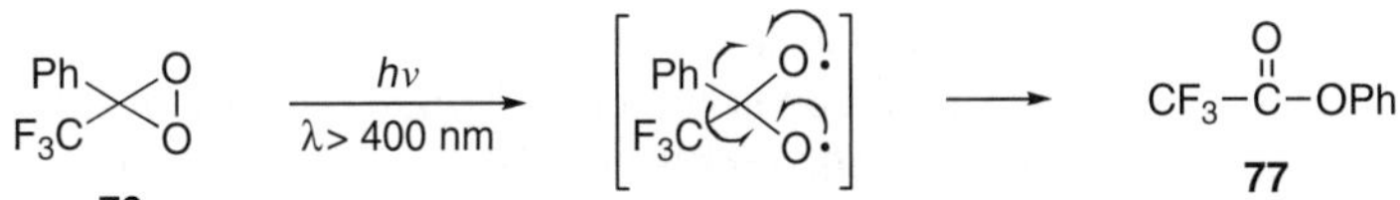

76 77

Based on its UV absorption, bis(trifluoromethyl)dioxirane (**79**) appears to have been generated by treatment of the mono- or dilithium salt of hexafluoroacetone hydrate (**78**) with elemental fluorine.[85]

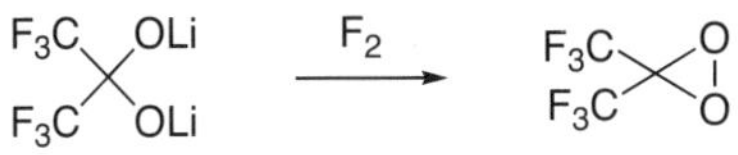

78 79

1.3 FLUOROTHIIRANES

1.3.1 Synthesis

Diazo compounds react with thiocarbonyl compounds to give thiiranes,[86] for example, diazopropionate **80** with thiobenzophenone (**81**) to afford thiirane **82**.[87]

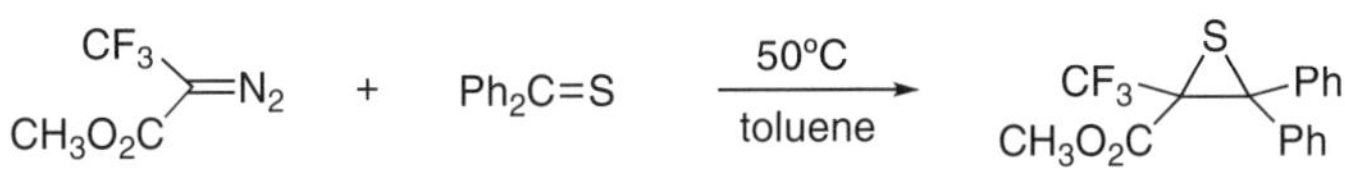

80 81 82

The assumption that such reactions occur via thiadiazoline intermediates received direct support when **84** was isolated in 92% yield from the reaction of bis(trifluoromethyl)diazomethane (**83**) with hexafluorothioacetone.[88] Refluxing **84** afforded perfluorotetramethylthiirane (**85**) in 95% yield.

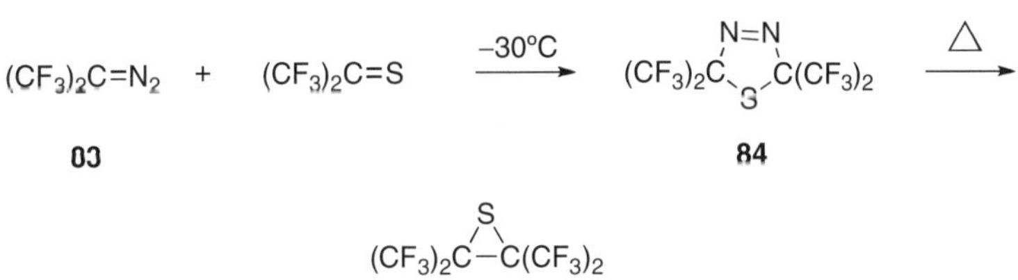

83 84 85

Difluorocarbene generated from HFPO (**1**) adds to thiocarbonyl compounds to give thiiranes, as illustrated in the section on oxiranes with the formation of thiirane **30**. Tetrafluorothiirane (**87**) has been made in 30–40% yield from thiocarbonyl fluoride (**86**) by this method.[89]

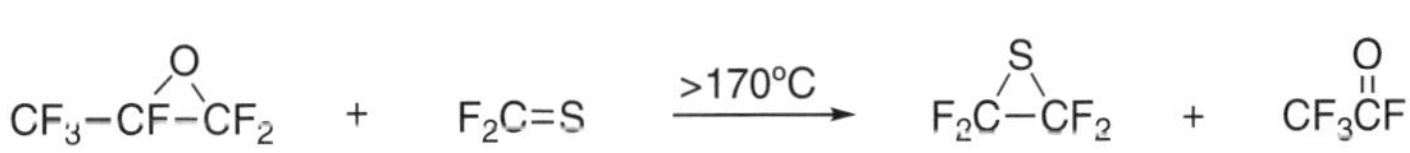

1 86 87

Dithietanes, obtained by dimerization of thiocarbonyl compounds, can be oxidized to monosulfones that yield thiiranes upon heating.[86] Trifluorothioacetyl chloride (**88**) dimerizes to give both stereoisomers of dithetane **89**, which yields sulfone **90** when oxidized with fuming nitric acid and chromium trioxide. Vacuum pyrolysis of **90** gives thiirane **91** as a *cis*/*trans* mixture.[90]

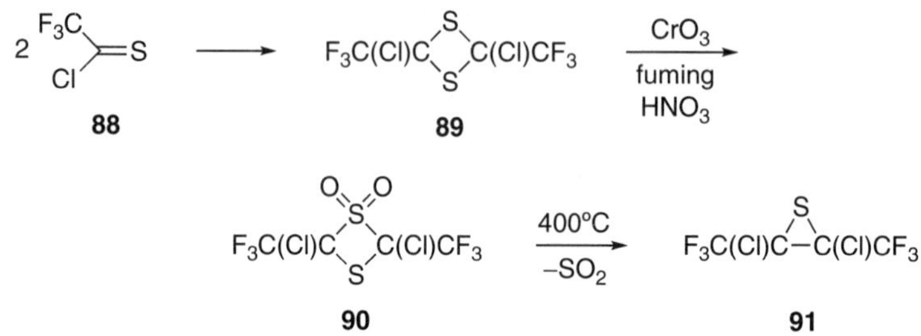

88 **89** **90** **91**

Fluorothiiranes are also prepared by cyclization of open-chain precursors. Treatment of sulfenyl chloride **92** with the triethylamine–boron trifluoride complex gave thiirane **30** in 68% yield.[91] Tris(diethylamino)phosphine transformed disulfide **94**, prepared from alkene **93** and sulfur monochloride, into thiirane **95**.[92] An indirect route from a chlorodisulfide that has the advantage of utilizing both halves of the molecule is exemplified by the transformation of **96** into thiirane **97**.[93]

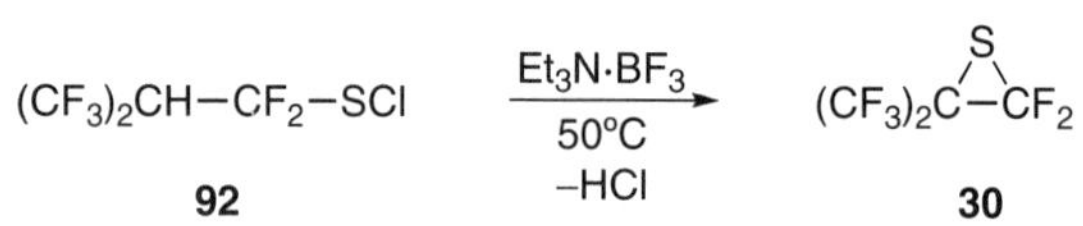

92 **30**

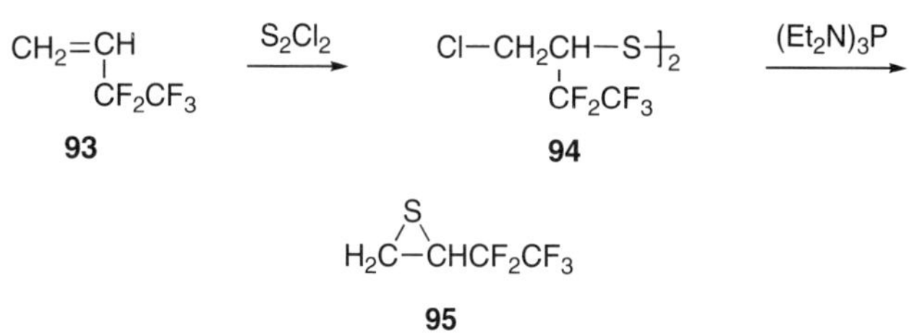

93 **94** **95**

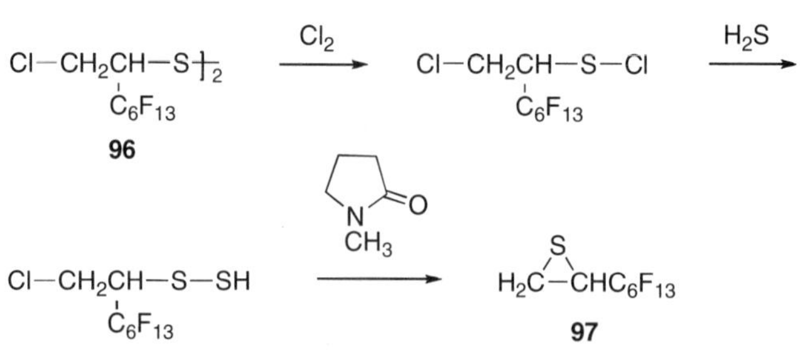

96 **97**

Valence isomerization provides another route to fluorothiiranes. Perfluorotetramethylthiophene (**98**) undergoes isomerization in the vapor phase to its extensively studied Dewar isomer (**99**).[94–96] Reaction of triplet sulfur atoms with fluoroalkenes[97] and fluoroalkynes[98] also yields thiiranes, among other products.

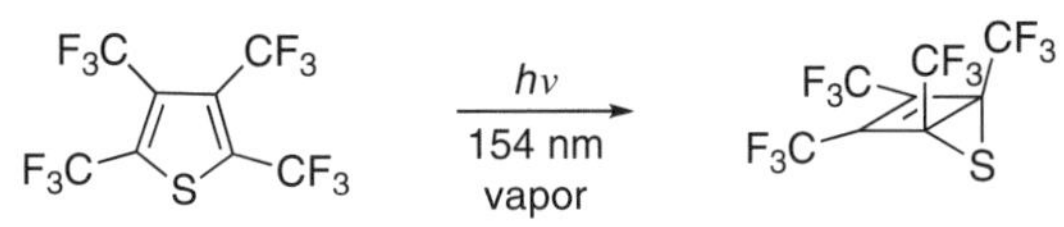

98 **99**

1.3.2 Reactions

Radical-initiated attack on tetrafluorothiirane (**87**) takes place at sulfur, as illustrated by the reaction with cyclohexane.[89] However, nucleophiles attack at carbon, and even a nucleophile as weak as benzophenone is capable of opening the ring to give oxathiolane **100**. Reaction with nucleophiles takes a different course with thiirane **30**, though, where again attack occurs at sulfur (e.g., formation of **101** by reaction with ethanol).[99] The ability of the *gem*-trifluoromethyl groups to stabilize developing negative charge in the transition state is presumably the reason for the contrasting behavior.

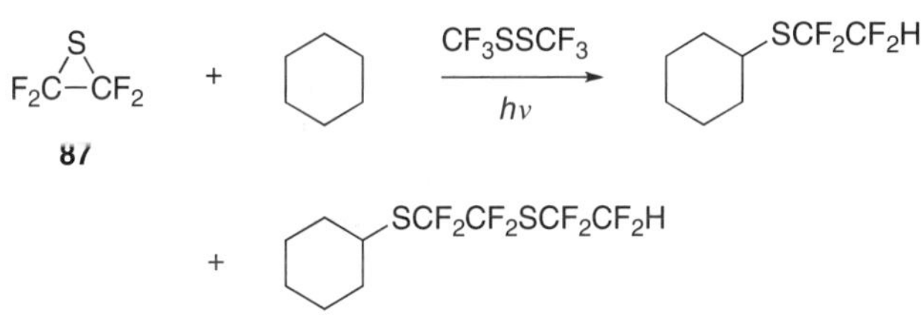

87

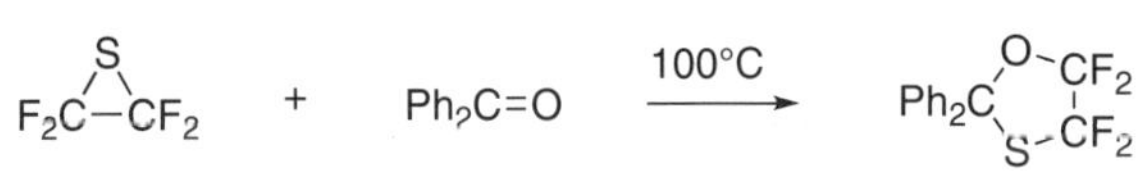

87 **100**

30 **101**

Fluorothiiranes are readily desulfurized by a variety of reagents. Triphenylphosphine, for example, reacted at RT with aminosulfide **102** to yield Dewar pyrrole **103**.[100] For the desulfurization of azo compound **104**, a nonbasic thiophile was required because of the extremely facile azo-to-hydrazone isomerization. An imidazolethione accomplished this catalytically and cleanly to give **105** plus elemental sulfur, presumably via zwitterion **106**.[101] Thiirane **30** was transformed into perfluoroisobutylene (**107**) by both amines and antimony pentafluoride.[99]

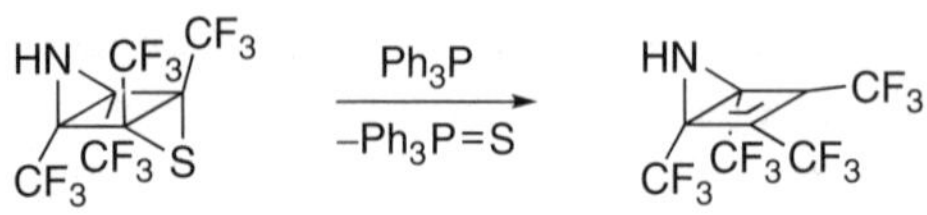

102 103

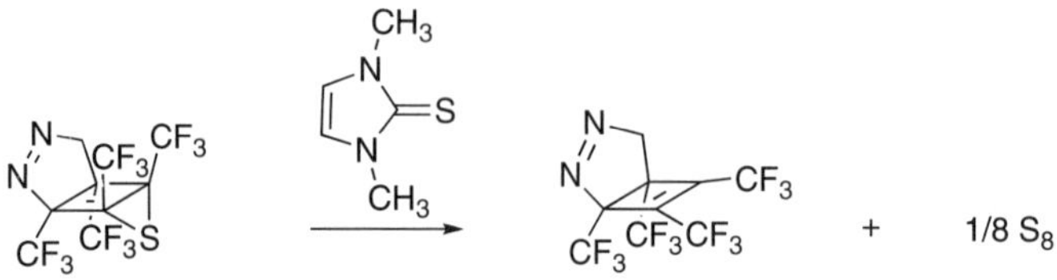

104 105

Via

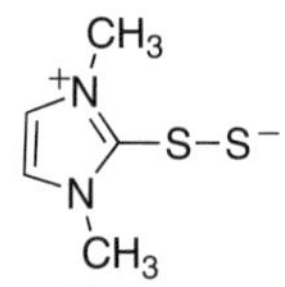

106

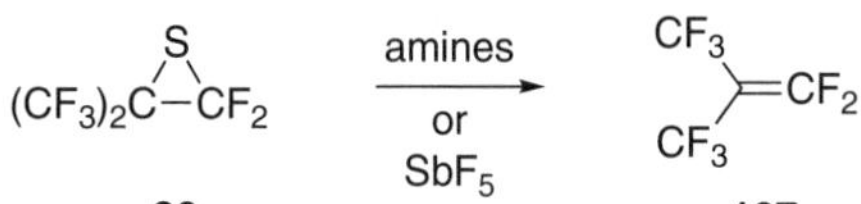

30 107

Cycloaddition of cyclopentadiene at 50°C to Dewar thiophene **108** yields a Diels–Alder adduct (**109**) and a [2 + 2] adduct (**110**) in the ratio 2:1.[102] Both have lost the sulfur atom, despite the fact that **108** retains its sulfur at 200°C. The much greater thermal stability of the thiirane ring in the starting material than in the product reflects the fact that desulfurization of the former would yield an antiaromatic cyclobutadiene. Desulfurization also occurs when Dewar thiophene **108** is treated with diiron nonacarbonyl, and a cyclobutadienyliron complex (**111**) is formed.[102]

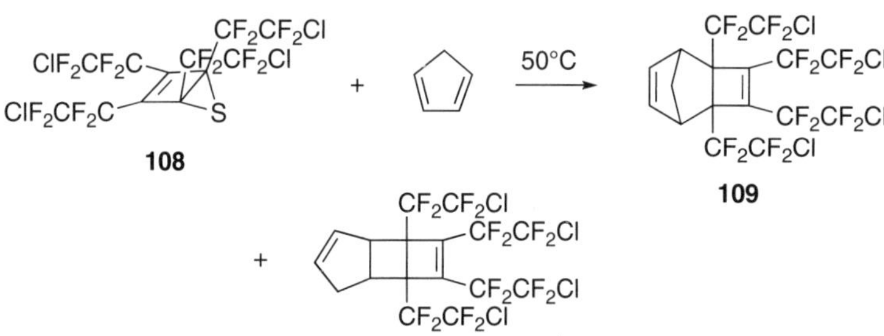

108 109 110

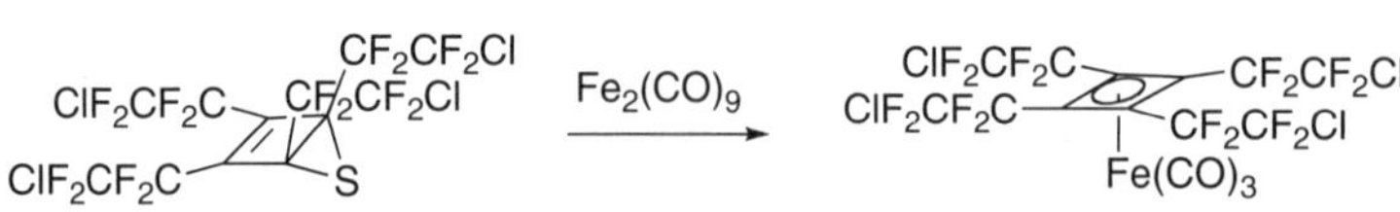

108 111

Instead of undergoing desulfurization with triphenylphosphine and other trivalent phosphorus derivatives, which would yield a cyclobutadiene, Dewar thiophene **99** is aromatized to thiophene **98**.[103] Thermal aromatization of **99**, an orbital topology-forbidden process, has a half-life of 5.1 h at 160°C, but triphenylphosphine brings about this change quite rapidly at RT.

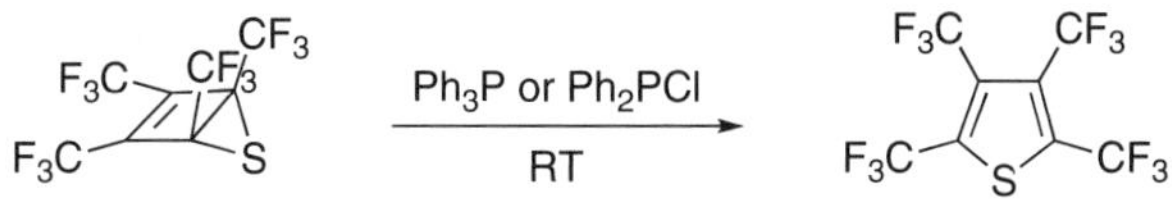

99 **98**

A variable temperature NMR study revealed that **99** undergoes a "walk" rearrangement with $\Delta H^{\ddagger} = 18.8 \pm 0.3$ kcal/mol and $\Delta S^{\ddagger} = -7.7 \pm 0.8$ cal/(mol K).[104] Peroxytrifluoroacetic acid oxidizes **99** to *exo*-sulfoxide **112**,[105] which undergoes degenerate rearrangement similarly with retention of configuration at sulfur, but with $\Delta H^{\ddagger} = 6.6 \pm 0.2$ kcal/mol and $\Delta S^{\ddagger} = -0.5 \pm 0.6$ kcal/(mol K). At 25°C, the rearrangement of **112** is about 3×10^{10} times faster than that of **99**, signifying a remarkable substituent effect by the oxygen atom.[104] Recent calculations point to the conclusion that the rearrangement of **112** is a pericyclic process.[106]

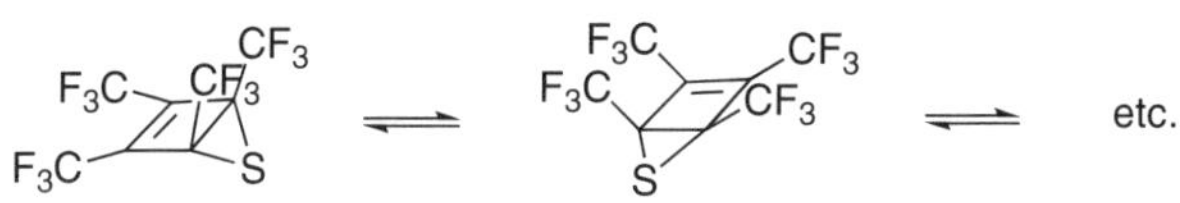

99 **99**

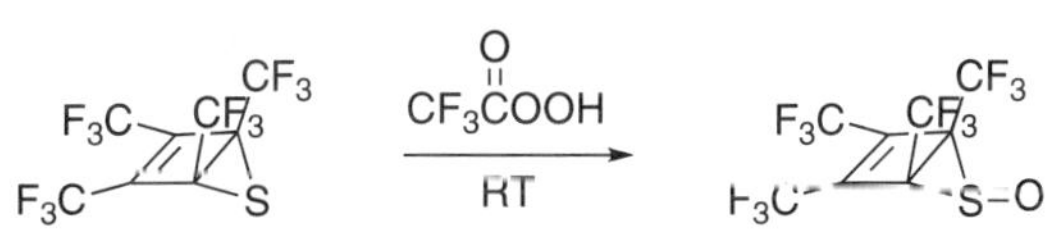

99 **112**

UV irradiation of 2,3-bis(trifluoromethyl)-thiophene (**113**) yielded Dewar isomers **114** and **115** in the ratio 8:1, together with all four bis(trifluoromethyl)thiophenes.[107] Evidence that **114** and **115** were thermally equilibrating with one another was provided by reaction of the mixture with 2,5-dimethylfuran. Only **115** underwent Diels–Alder reaction with the diene, giving adduct **116**, but **114** slowly disappeared with concomitant formation of more **116**. "Walk" rearrangement can account for formation of two of the three new thiophenes, but not the 2,4-isomer. Presumably either **117** or **118** is an intermediate *en route* from the 2,5-isomer. Thermolysis of triazoline **119** affords diazothiirane **120**, a *retro*-1,3-dipolar cycloaddition.[108] On continued heating, **120** undergoes ring expansion with loss of nitrogen to give thietene **121**. In a related transformation, pyrazoline **122** fragments at higher temperatures to yield acyclic thione **123**.[109] In this case, the presumed intermediates diazothiirane **124** and thietene **125** are not observed.

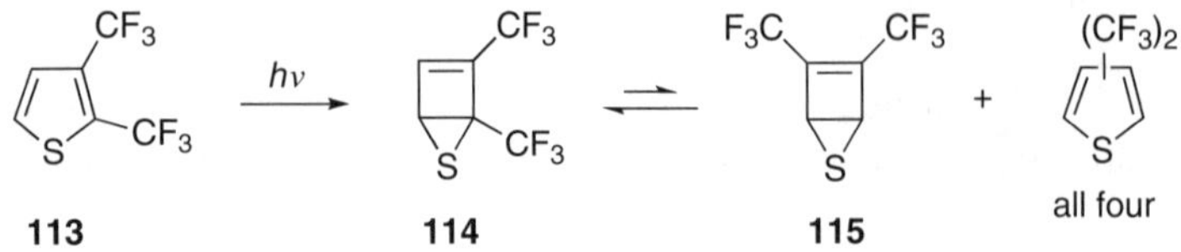

113 **114** **115**

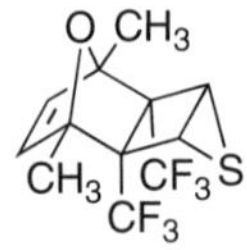

116

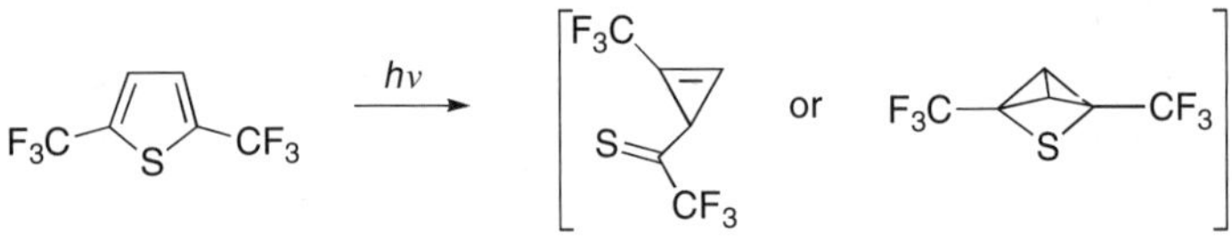

117 **118**

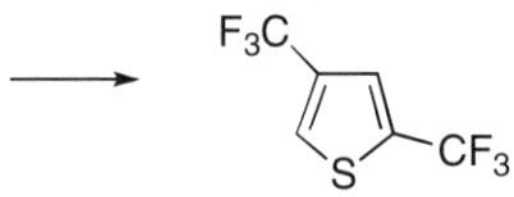

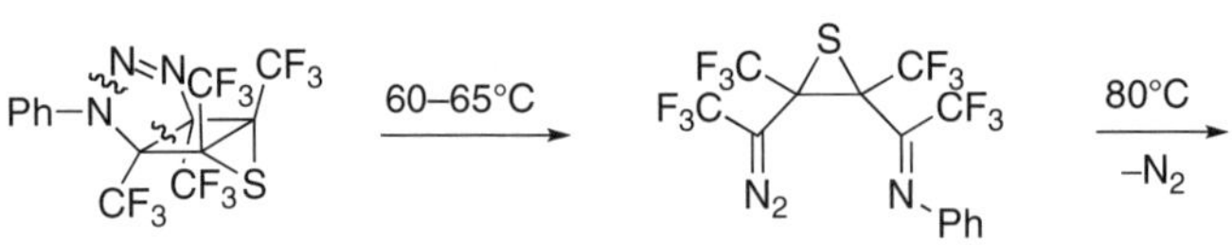

119 **120**

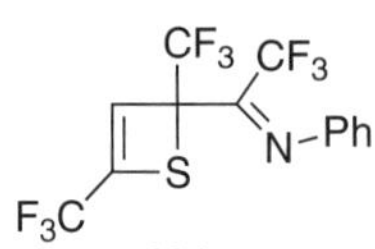

121

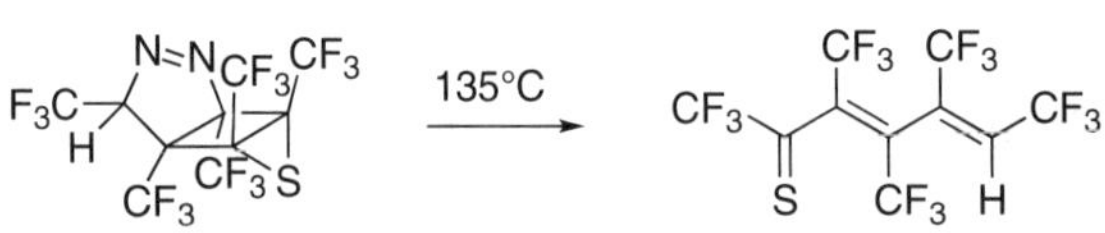

122 **123**

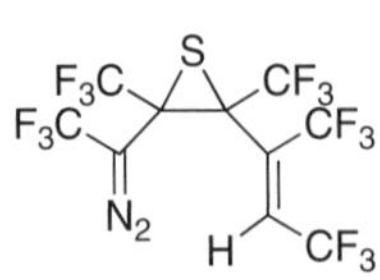

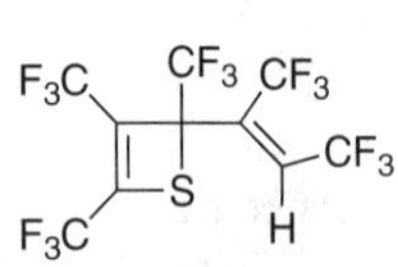

124 **125**

1.4 FLUOROAZIRIDINES

1.4.1 Synthesis

Aziridines are commonly prepared from imine precursors. Carbene addition to the C=N bond is illustrated by the formation of aziridine **127** from imine **126**.[110] Difluorocarbene, generated from HFPO (**1**), also adds to imines such as **128** to give the highly fluorinated aziridine **129**.[39] In the presence of Lewis acids, diazo compounds react with imines to produce aziridines. Ethyl diazoacetate and imine **130** gave aziridine **131** in 93% yield, with a *cis/trans* ratio of 95:5.[111] Chiral diazo compound **132** reacted with the aldimine precursor **133** to afford aziridine **134** in 81% yield.[112] The reaction displayed both high *cis* selectivity (>95:5) and excellent diasteroselectivity (94% de). Reductive removal of the chiral auxiliary gave the optically active hydroxymethylaziridine **135**.

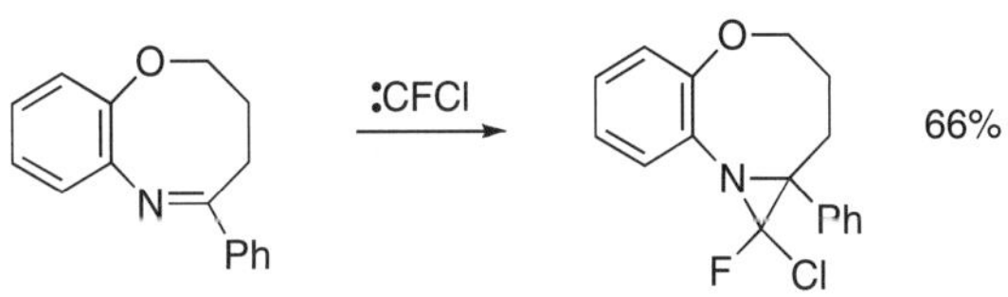

126 **127**

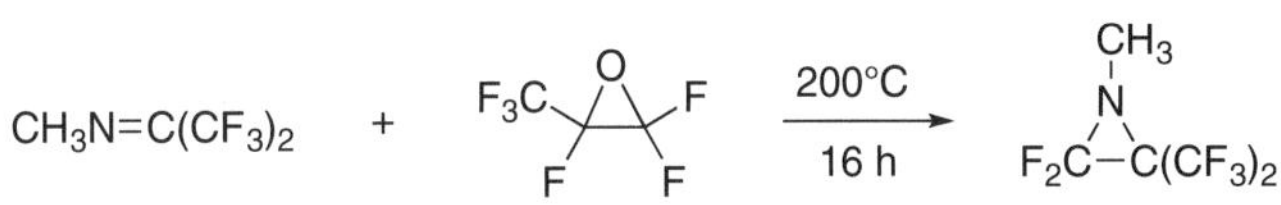

128 **1** **129**

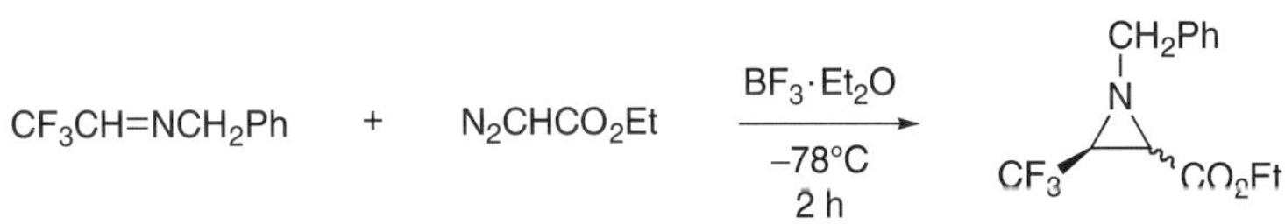

130 **131**

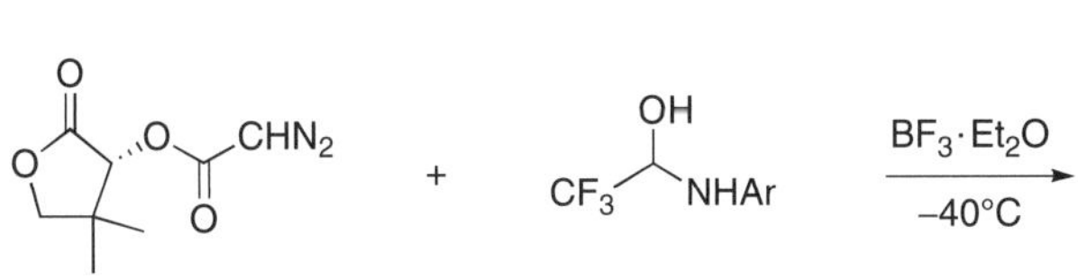

132 **133**

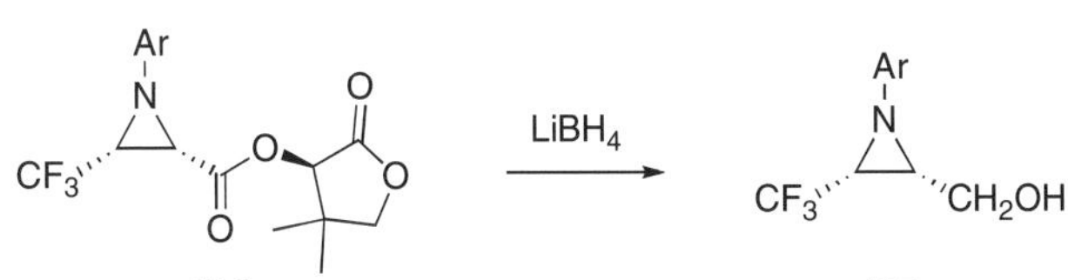

134 **135**

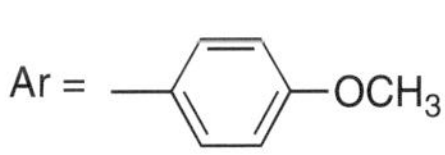

Reaction of an azide with an alkene provides another route to aziridines. Aziridine **31**, mentioned earlier as the product of difluorocarbene addition to an imine, has also been prepared from phenyl azide and perfluoroisobutylene via a triazoline intermediate.[113] Loss of nitrogen from a triazoline can also be accomplished photochemically, as illustrated with the preparation of novel aziridine **136**.[114] In an unusual transformation, an aziridine ring was introduced into the fullerene $C_{60}F_{18}$ with (methoxyethoxy)methyl azide in refluxing toluene.[115] The new C–N bonds were formed at the expense of two C–F bonds.

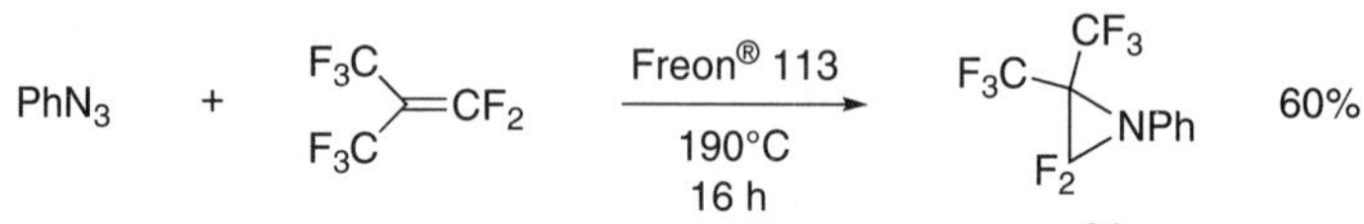

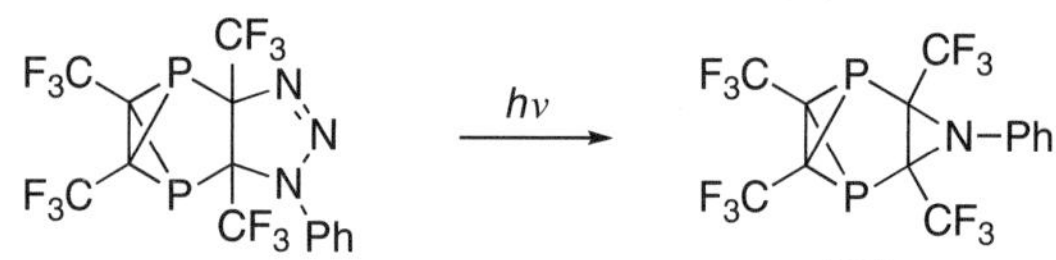

Aziridination of α-trifluoromethyl-substituted acrylates such as **137** can be accomplished with ethyl nosyloxycarbamate. With a weak base such as calcium oxide, a Michael adduct is obtained (**138**), but with sodium hydride Michael addition is followed by cyclization with extrusion of nosylate ion to give **139**.[116]

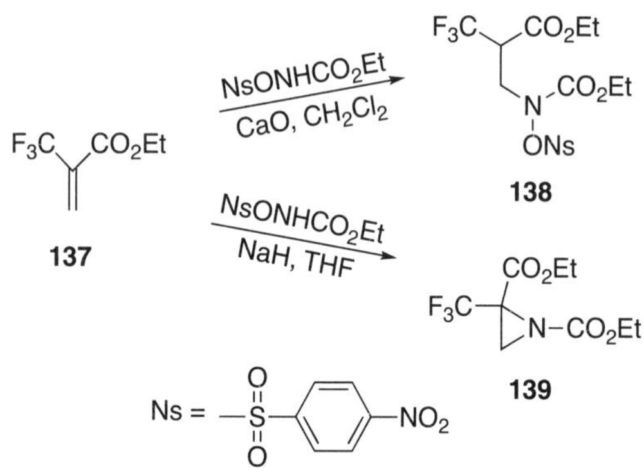

The aza-Corey–Chaykovsky reaction yields aziridines via the interaction of oxosulfonium or sulfonium ylides with imines. The reaction mechanism has not been established, and no intermediate has been observed or isolated. The fact that stable thiazetidine **140** fragments thermally into aziridine **141** and sulfinate ester **142** supports the hypothesis that the aza-Corey–Chaykovsky reaction proceeds via such a four-membered ring intermediate formed by [2 + 2] cycloaddition.[117]

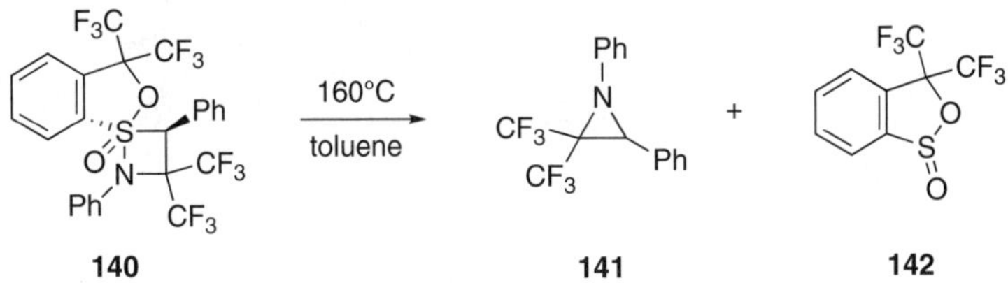

An interesting variation of the aza-Corey–Chaykovsky reaction provides a pathway to difluoromethyl-substituted aziridines. Reduction of trifluoromethylimine **143** with magnesium in the presence of chlorotrimethylsilane gave the enamine **144**.[118] Treated with dimethyloxosulfonium methylide in DMSO, **144** first underwent desilylation and tautomerization to produce imine **145**, which then afforded aziridine **146** via the aza-Corey–Chaykovsky reaction in 92% yield.

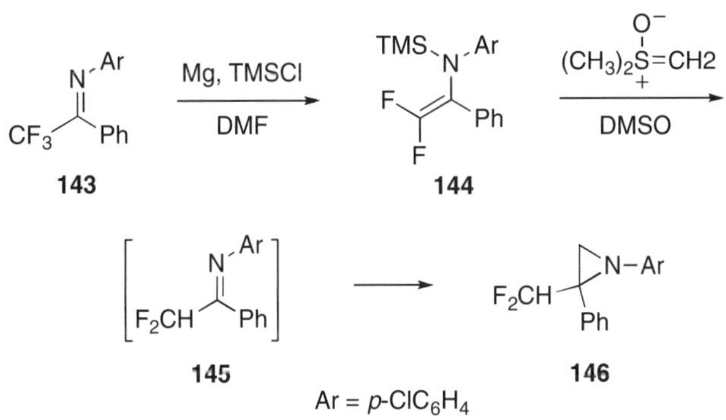

Chiral aziridines have become available starting from oxirane **34** with inversion at the chiral center, as illustrated with the synthesis of **147**.[119] As described below, these aziridines can be elaborated with preservation of stereochemistry to afford a wide variety of derivatives.

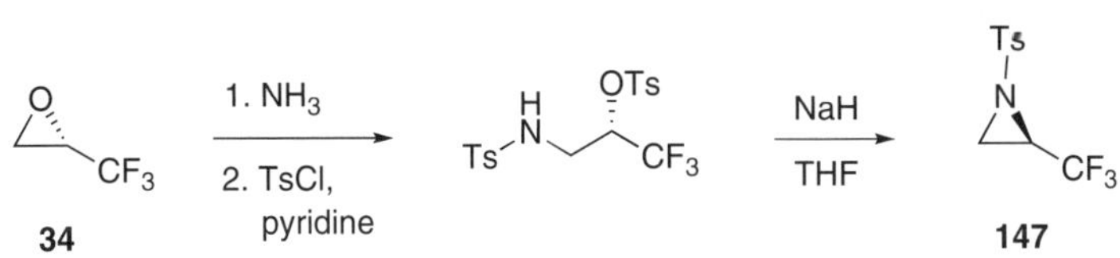

Thermal rearrangement of isoxazolines provides yet another route to aziridines. Prepared by 1,3-dipolar cycloaddition of *N*-methyl-*C*-trifluoromethylnitrone (**148**) to phenylacetylene, isoxazoline **149** was transformed in refluxing toluene into *cis*-aziridine **150** (81%).[120] The clean *cis* stereochemistry is consistent with interpretation of the mechanism as an allowed 1,3-sigmatropic rearrangement, a four-electron process with a Möbius transition state.[121]

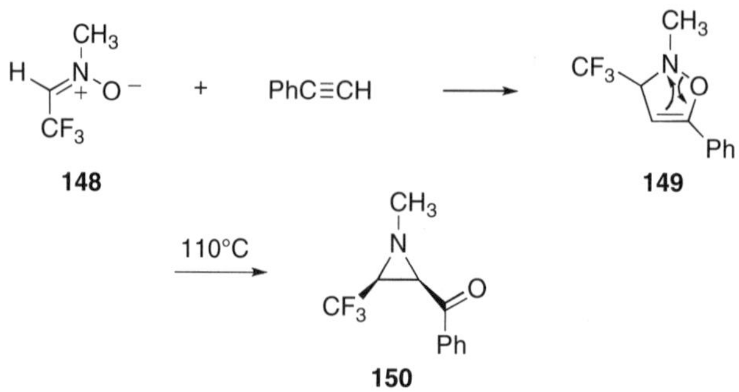

Aziridines can be formed by valence isomerization. UV irradiation in Pyrex of Dewar pyridine **151** produced prismane **152** (86%), which incorporates a highly strained aziridine ring.[122]

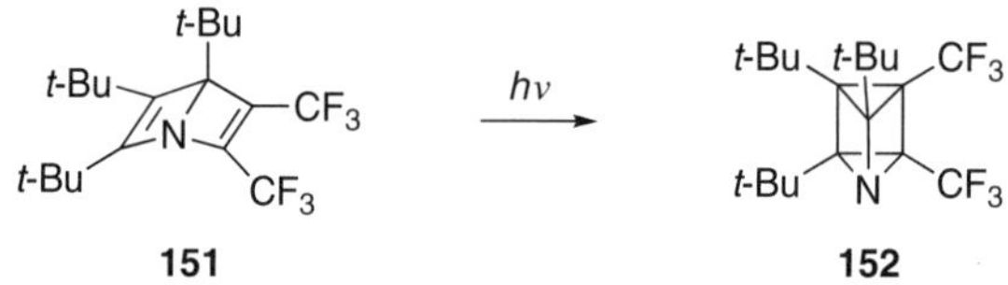

Azirines serve as precursors for aziridines. Treatment of tetrafluoroethylene pentamer **153** with azide ion at RT gave azirine **154** in 71% yield.[123] Reaction of the azirine with methylmagnesium iodide produced aziridine **156**. The first addition of the Grignard reagent resulted in elimination of the perfluoro-3-methylpentyl group as a rather stable anion, giving the new azirine **155**, to which a second Grignard addition occurred.

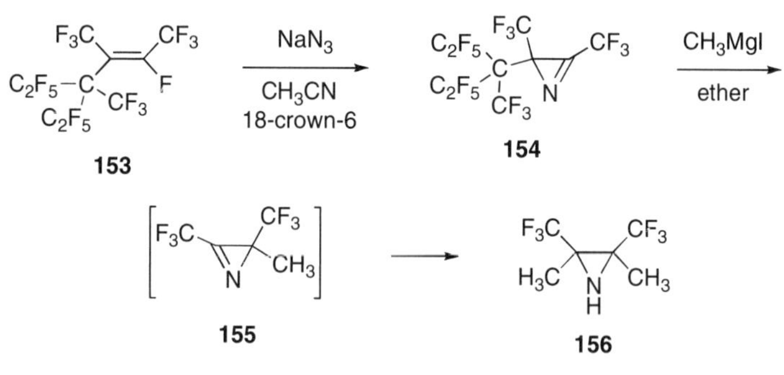

Aziridines can also be synthesized via azirine intermediates generated from oxime derivatives. Treated with triethylamine in methanol, oxime tosylate **157** cyclizes to azirine **158**, which adds methoxide on the less hindered face to give aziridine **159** in 70% yield.[124]

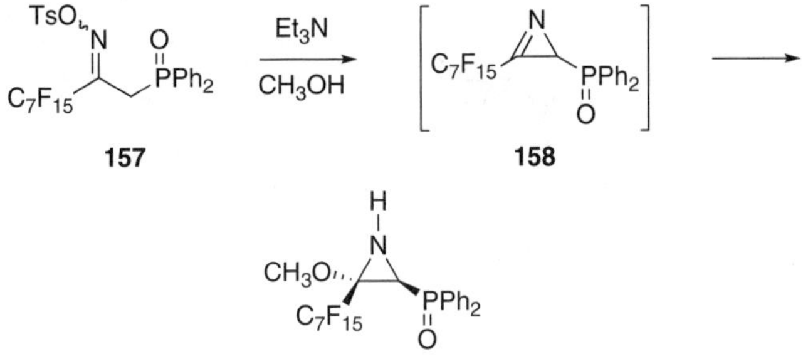

157 158 159

Fluoroaziridinones have also been synthesized. The highly electrophilic example **161** was formed by the reaction of oxaziridine **160** with bis(trifluoromethyl)ketene.[125] A likely pathway for this transformation involves spontaneous ring opening of the oxaziridine to a 1,3-dipolar species followed by attack of the nitrogen on the ketene carbonyl carbon. The resulting zwitterion **162** then fragments, forming the new three-membered ring with extrusion of benzophenone.

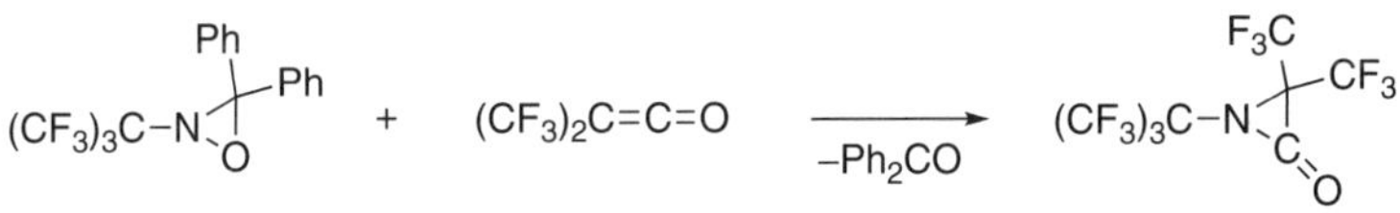

160 161

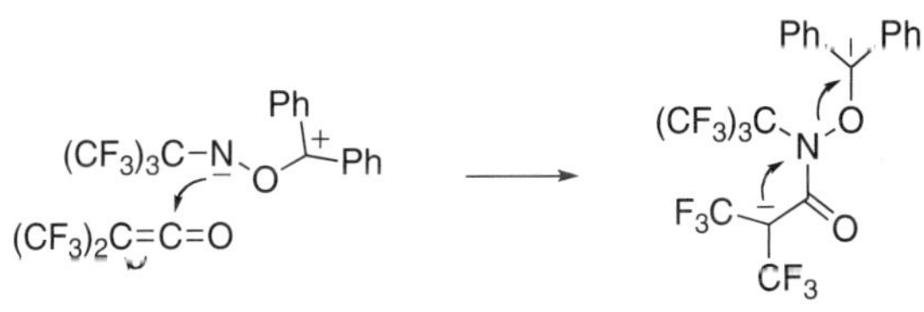

162

1.4.2 Reactions

Nucleophilic ring opening of fluoroaziridines requires much more vigorous conditions than that of fluoroepoxides unless activation is provided by protonation or the presence of an electron-withdrawing group on the nitrogen. Treatment of aziridine **31** with cesium fluoride at 100–110°C for 50 h resulted in rearrangement to imine **164** via anion **163**.[39] Here, as in nucleophilic attack on the corresponding oxirane (**1**), fluoride ion reacted at the more hindered carbon. As an example of nucleophilic ring opening with acid catalysis, *cis*-aziridine **131** reacts with benzyl mercaptan in the presence of triflic acid to give aminoester **165** in 98% yield.[111] Nucleophilic attack can occur transannularly, as in the transformation of aziridine **166** into **167** with zinc chloride catalysis via intermediates **168–170**.[110]

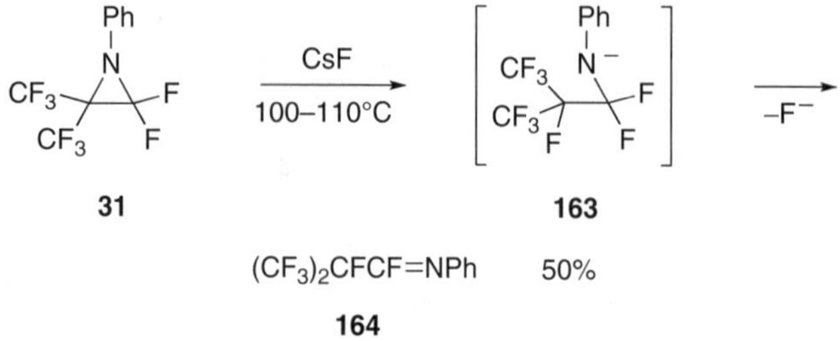

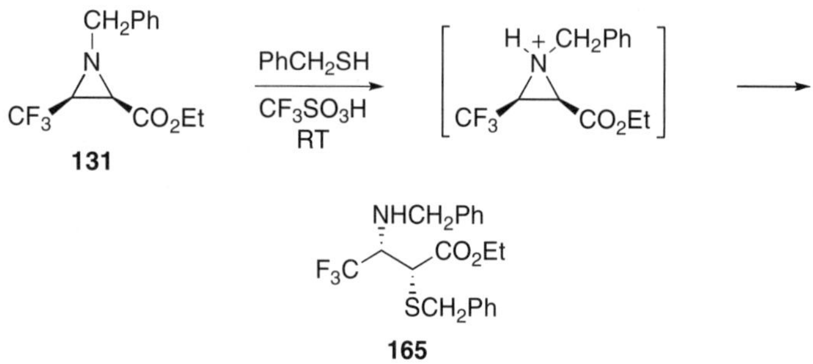

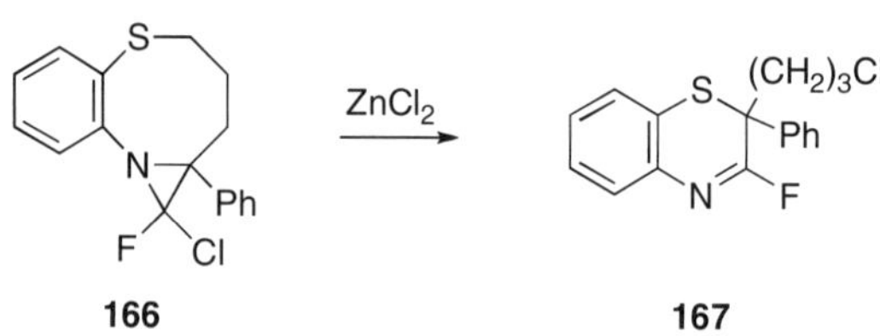

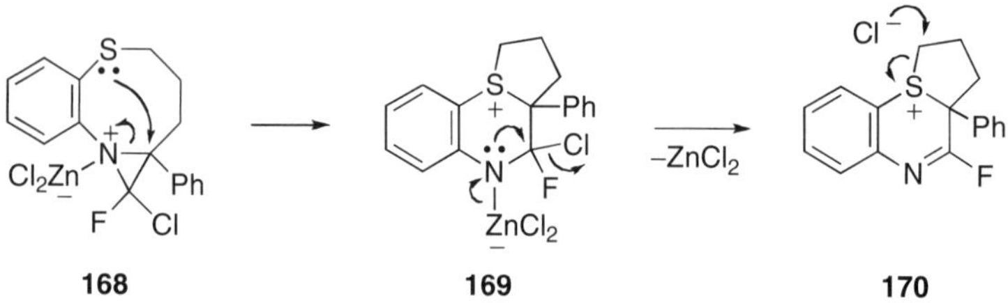

The activated aziridine derivative that suffers nucleophilic attack can also be formed *in situ* by ring closure. Chiral aminoalcohol **171** was converted via aziridinium chloride **172** to chloramine **173** with inversion of configuration in 91% yield.[126]

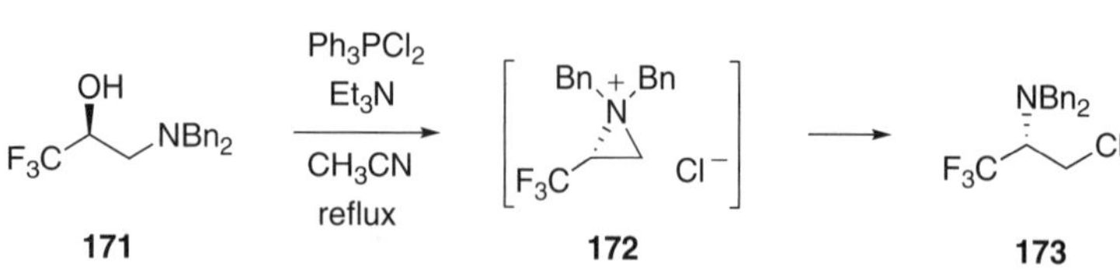

Electrophilic ring opening of highly fluorinated aziridines requires vigorous conditions. Treated with anhydrous HF in the presence of BF_3 as catalyst, aziridine **174** affords acyclic amine **175** in 95% yield.[39] A modest change in the substituent on nitrogen alters the reaction course completely. With a *p*-fluorophenyl substituent (**176**), the aziridine is transformed in 88% yield under the same reaction conditions into indolinone **177**, after hydrolytic workup. The increased basicity of the nitrogen may suffice to allow loss of HF, giving highly strained aziridinium ion **178**, which opens to the conjugated cation **179**. Alternatively, **179** may form in fully concerted fashion. Recyclization of the cation (effectively a pentadienyl cation electrocyclization[48]) and proton loss would produce imidoyl fluoride **180**. Hydrolysis would then afford the indolinone.

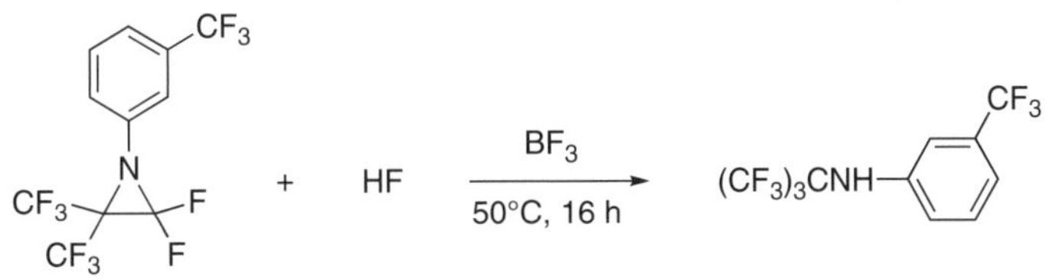

174 175

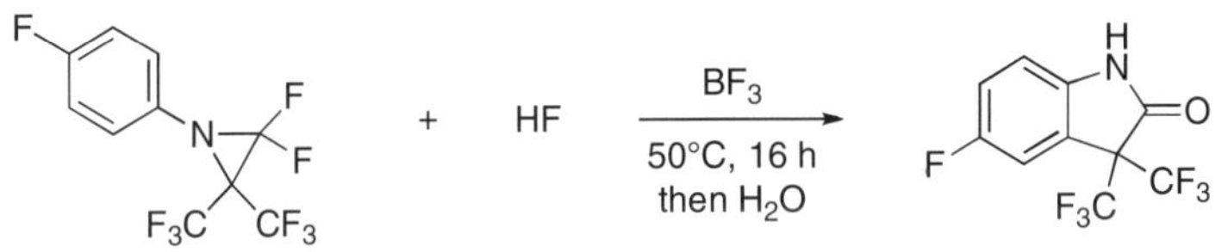

176 177

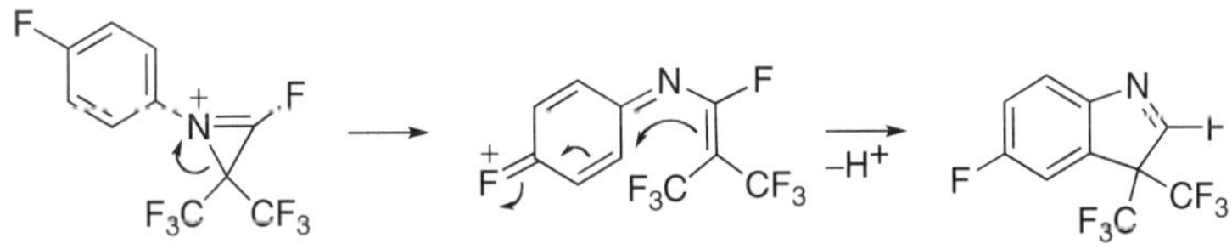

178 179 180

Opening of fluoroaziridines has also been accomplished with catalytic hydrogenation. Aziridine **181** was reduced to the acyclic amine **182** over palladium-on-carbon with ammonium formate as the hydrogen source.[124]

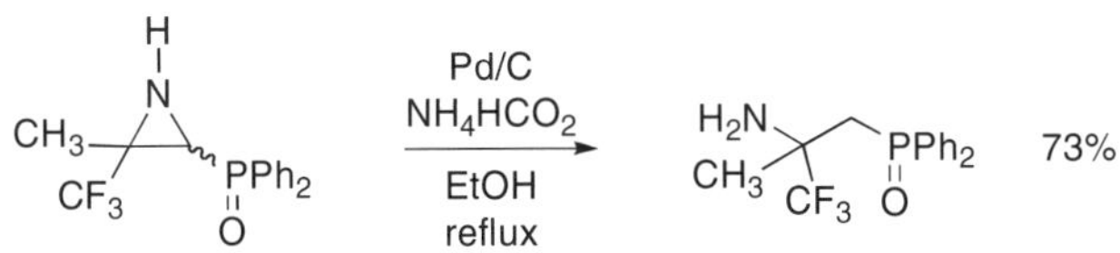

181 182

Thermolysis of fluoroaziridines can follow a variety of pathways. *N*-Benzoylaziridine **183** rearranged slowly but quantitatively at 152°C to give a 3:1 mixture of stereoisomers **184** and **185**.[127] It is not clear whether the reaction proceeds via a diradical or a zwitterion. Dewar pyrrole **186** rearranges spontaneously at RT via **187** to the indoline **188**, an example of an aza-Cope rearrangement facilitated by the strain in the three-membered ring.[100b] Azaprismane **152** aromatizes highly selectively upon heating, affording pyridine **189**.[122] It is likely that the reaction proceeds via scission of the indicated bond to give diradical **190**, as that bond is especially strained by nonbonded repulsion between the attached *t*-butyl groups.

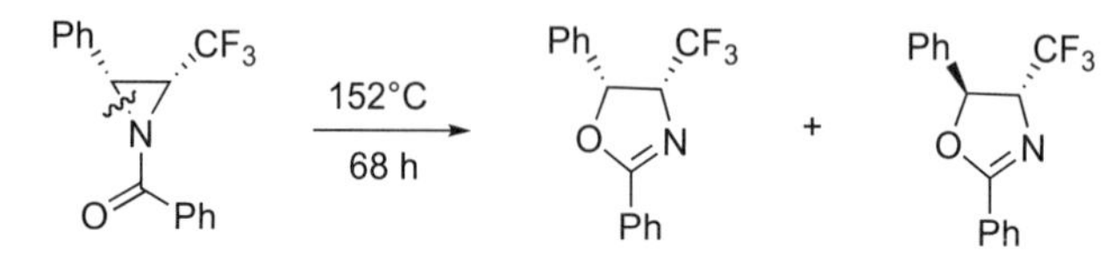

183 **184** **185**

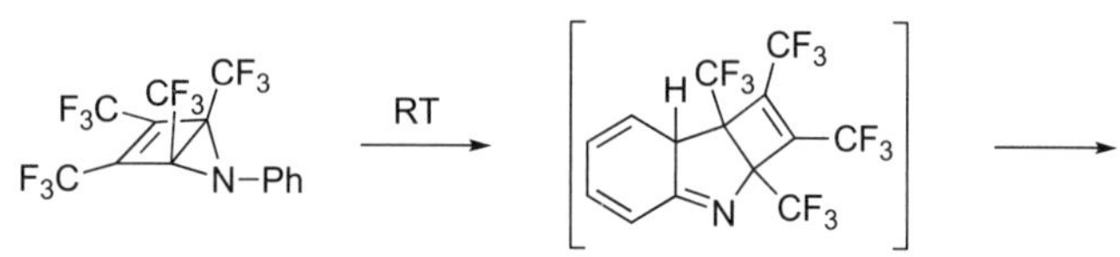

186 **187**

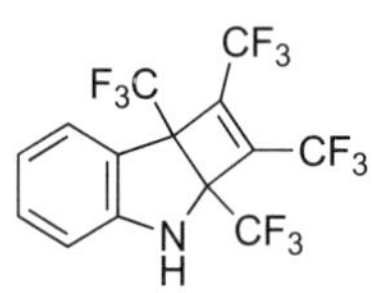

188

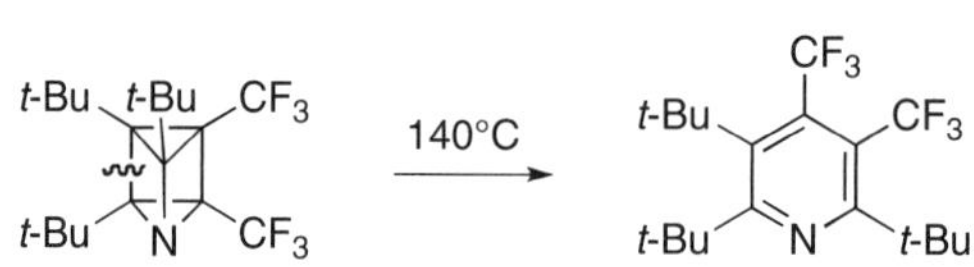

152 **189**

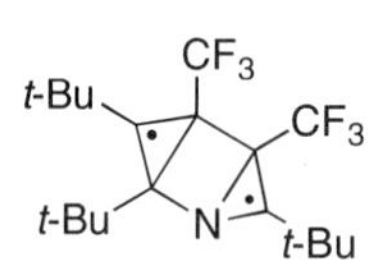

190

Upon heating, aziridine **191** opened in the conrotatory manner to give azomethine ylides **192** and/or **193**, which underwent 1,3-dipolar cycloaddition reactions with alkenes and acetylenes.[128] With styrene, for example, pyrrolidine **194** was formed exclusively in 81% yield, and the regiochemistry of the cycloaddition was ascribed to control by the LUMO of the electron-deficient azomethine ylide. The *cis* relationship of the phenyl and benzoyl groups was attributed to secondary orbital interactions between them in the transition state.

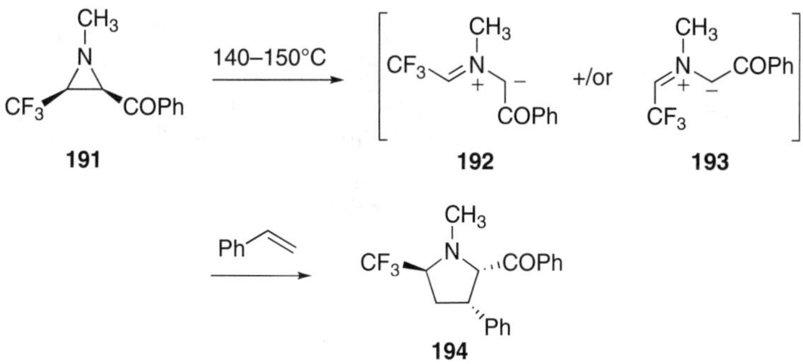

Though very stable thermally, aziridine **31** slowly decomposes at 220°C, producing imidoyl fluoride **195** and amine **196** in a 1:1 ratio in 95% yield.[39] A likely course of events entails reversible ring opening to azomethine ylide **197**, which abstracts a fluoride ion from the starting aziridine to form anion **198** and cation **199** (akin to **179** above). Ring closure of **199** yields **200**, and then transfer of a proton from **200** to **198** gives the final products.

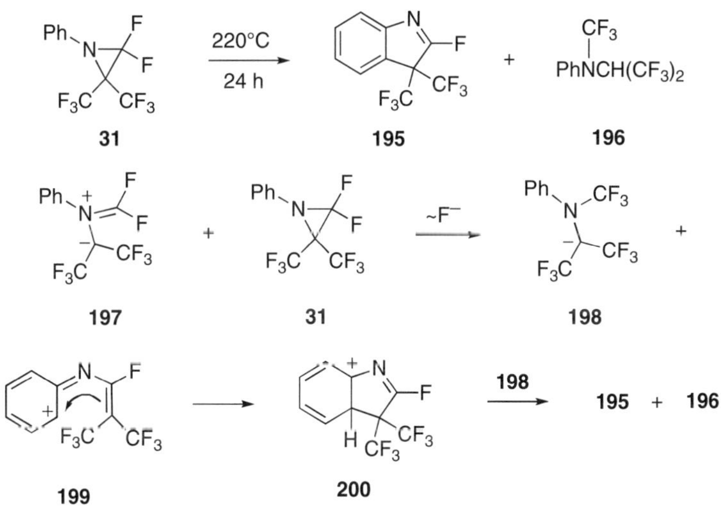

All of the reactions of fluoroaziridines described above involve ring opening, but introduction of substituents onto an intact ring is an important addition to the reaction repertoire. Paralleling oxirane chemistry (e.g., **34–36**), chiral aziridine **147** was deprotonated with butyllithium at −102°C, and the resulting anion **201** was allowed to react with a variety of electrophiles.[42,119] They included aldehydes, ketones, halides, and a disulfide. Reaction with benzaldehyde, for example, afforded aziridinyl alcohol **202** with preservation of configuration in 82% yield.

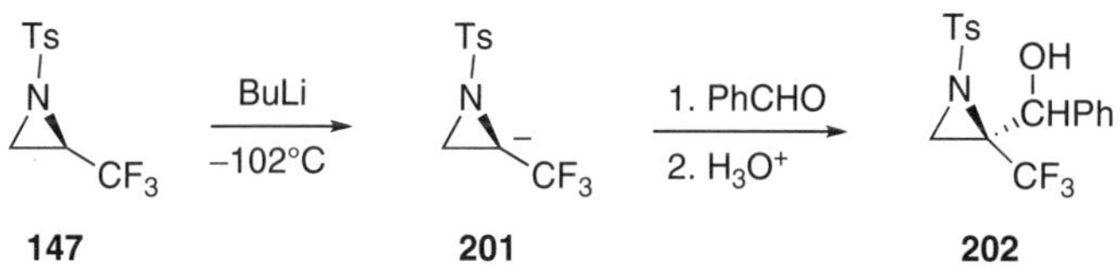

Aziridinone **161** undergoes rapid ring-opening rearrangement to acyl fluoride **203** in the presence of tributylamine.[129] Presumably, intermediate **204** expels fluoride ion to give **205**, and then fluoride attacks at the carbonyl carbon, eliminating tributylamine. Treatment of aziridinone **161** with pyridine *N*-oxide resulted in ring expansion to give oxazetidinone **206**, again with cleavage of the ring C–C bond.[130] However, reaction of **161** with acetone produced oxazolidinone **207** with opening of the ring N–CO bond.[131]

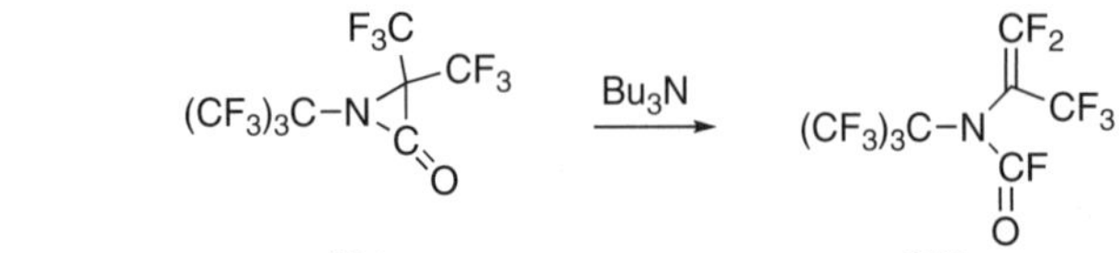

161 **203**

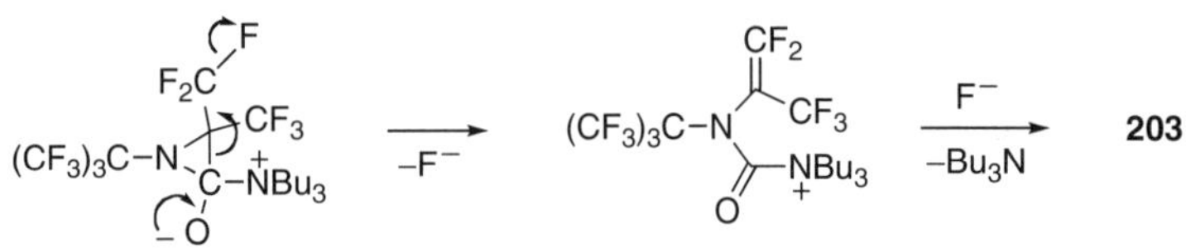

204 **205**

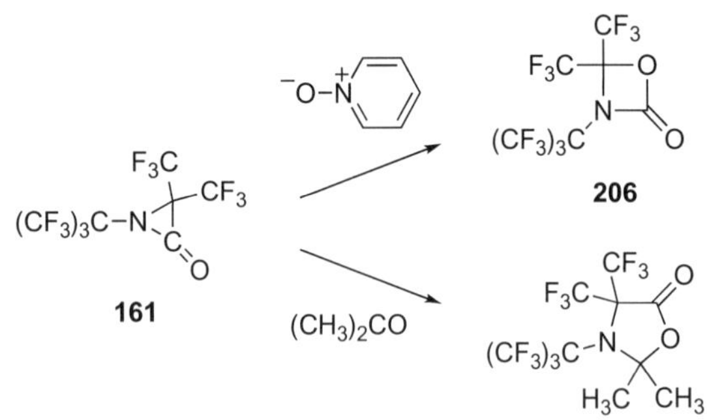

161 **206** **207**

1.5 FLUOROAZIRINES

1.5.1 Synthesis

Fluoroazirines are generally prepared from vinyl azides, whether they be stable starting materials or reaction intermediates. *trans*-1-Azidopentafluoropropene (**208**) was obtained in 21% yield from sodium azide and hexafluoropropene in an addition–elimination reaction.[132,133] This yellowish-green liquid is very unstable, decomposing smoothly at 20°C to give 2,3-difluoro-2-trifluoromethyl-2*H*-azirine (**209**), the first fluoroazirine. It appears that formation of **209** is a concerted process in which the double bond assists the loss of nitrogen. 2,2-Bis(trifluoromethyl)-3-methoxy-2*H*-azirine

(**211**) was obtained similarly by the reaction of vinyl ether **210** with sodium azide.[134] In this case, the intermediate vinyl azide began losing nitrogen even at 0°C. The vinyl azide formed *en route* to azirine **154** (see Section 1.4) also decomposed during the reaction, which was run at RT.

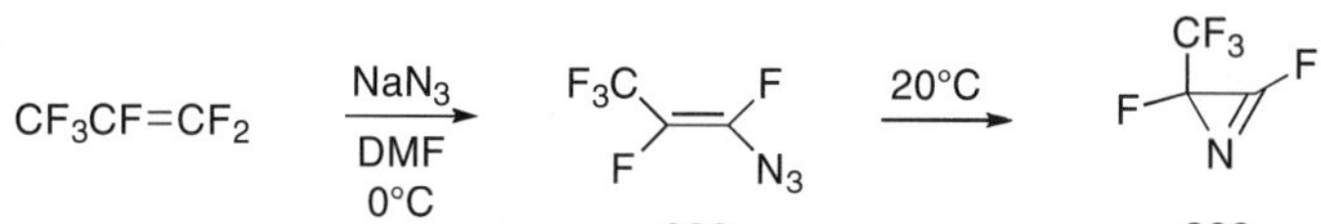

208 **209**

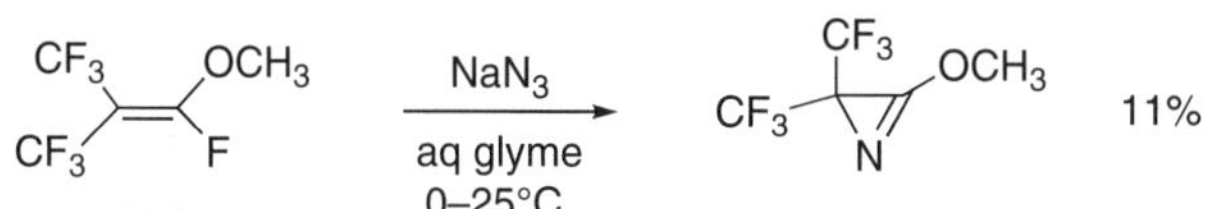

210 **211**

Fluorinated vinyl azides have also been prepared by Michael addition to alkynyl esters. Azide ion addition to ester **212** at RT afforded vinyl azide **213** as a stereoisomeric mixture, $Z/E = 4{:}1$.[135,136] The azide decomposed in refluxing toluene to give azirinyl ester **214** in high yield. Dehydrohalogenation has provided another route to vinyl azides. β-Iodoazide **215** of carbohydrate origin reacted with DBU to afford **216** as a 1:3 *E*/*Z* mixture.[137] Thermolysis of the *E* isomer occurred readily at 80°C and that of the *Z* isomer at 120°C, giving the rather unstable fluoroazirine **217** in each case as a 7:3 mixture of diastereomers.

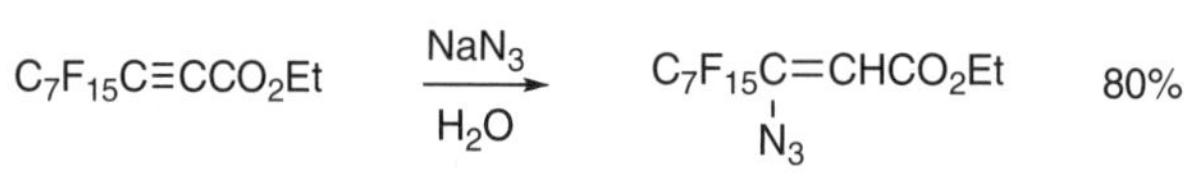

212 **213**

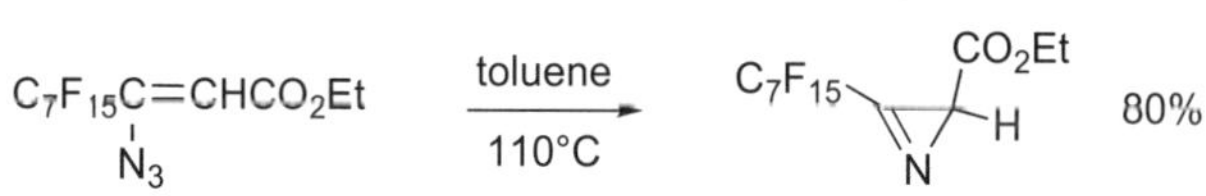

213 **214**

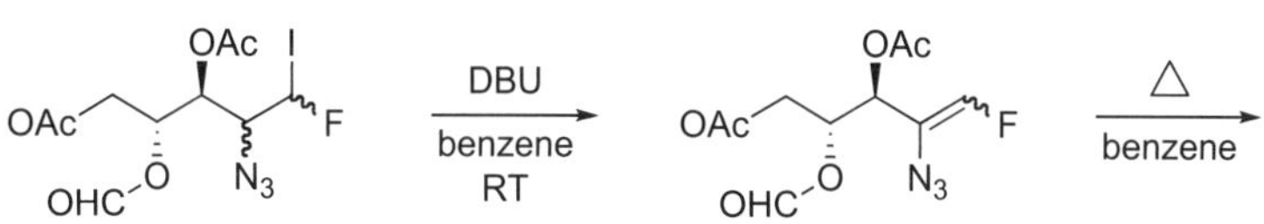

215 **216**

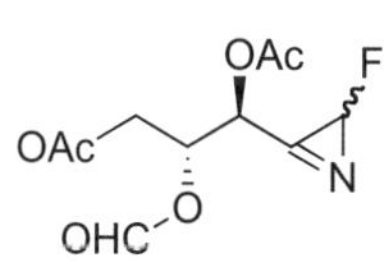

217

Azirines are sometimes unobserved reaction intermediates, as in the transformation of oxime tosylate **157** via azirine **158** into aziridine **159**. A related example is the reaction of oxime **218** with ethylmagnesium bromide.[138] The intermediate azirine **219** undergoes reduction to aziridine **220** by the Grignard reagent instead of the expected introduction of an ethyl substituent. Hydride addition occurs on the less hindered face of the azirine.

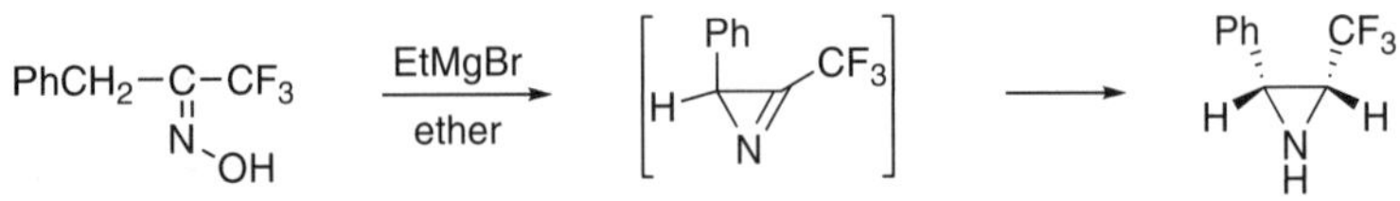

Photolysis of phenyl azide and its fluorinated derivatives results in a series of transformations in which transient azirines play an important role. *o*-Fluorophenyl azide (**221**) has been studied in detail by laser flash photolysis.[139] Loss of nitrogen generates singlet nitrene **222**, which suffers either of two fates, reversible cyclization to form azirine **224** or intersystem crossing to the triplet state (**223**). The azirine opens reversibly to the ketenimine **225**, which polymerizes. Azirine generation is faster than intersystem crossing, but because both azirine and ketenimine formation are reversible processes, under the flash photolysis conditions most of the ultimate product is *o*-fluoroazobenzene (**226**, 76%), the result of dimerization of the triplet nitrene. Cyclization of the singlet nitrene toward the *o*-fluorine substituent has a substantially higher barrier than cyclization to give **224**, primarily because of steric repulsion in the transition state.[140] If both *ortho* positions of the nitrene are occupied by fluorine, azirine formation is strongly inhibited. The resulting increase in lifetime of the nitrene is an important advantage for photoaffinity labeling of biological molecules, as the nitrene has more opportunity to attack the desired target before decaying.

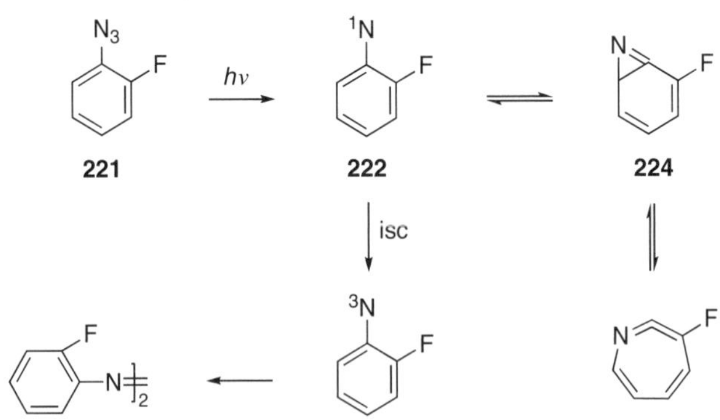

1.5.2 Reactions

The strain energy of the parent 2*H*-azirine has been reported as 44.6 and 46.7 kcal/mol based on calculations at the MP2/6-31G* and B3LYP/6-31G* levels of theory, respectively.[141] As a consequence of their great ring strain, fluoroazirines are very reactive molecules. Nucleophilic attack resulting in addition across the C=N bond is particularly facile. Examples of this reaction type have appeared above: **154** → **156**, **158** → **159**, and **219** → **220**. At RT, azirine **214** adds water readily at pH 7.5, affording hydroxyaziridine **227** in 60% yield.[135,136] The fact that this hemiaminal is stable against opening despite the strain remaining in the three-membered ring reflects the ability of the perfluoroalkyl substituent, by virtue of its electron-withdrawing effect, to prevent formation of a carbonyl group. Treated with aqueous HCl, aziridine **227** does ring open and the resulting α-aminoketone **228** dimerizes to give piperazine **229**.

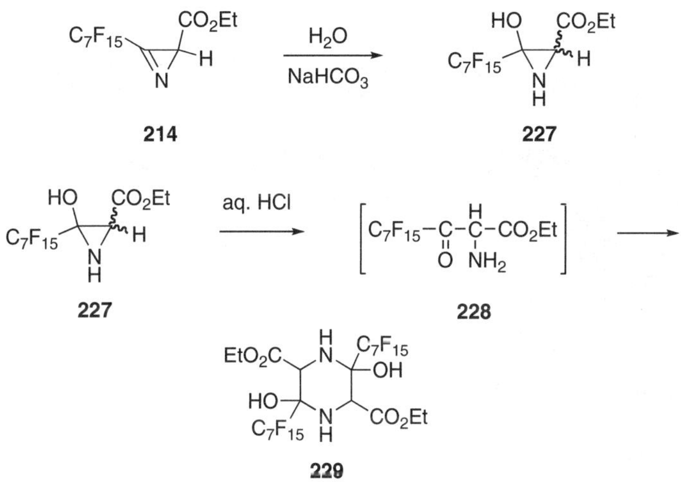

Hydrolysis of azirine **211** presents an interesting contrast, as the intermediate hydroxyaziridine **230** undergoes cleavage of the ring C–C bond, yielding urethane **231**, instead of the C–N bond as occurs with **227**.[134] This result reflects again the ability of CF_3 groups to stabilize negative charge.

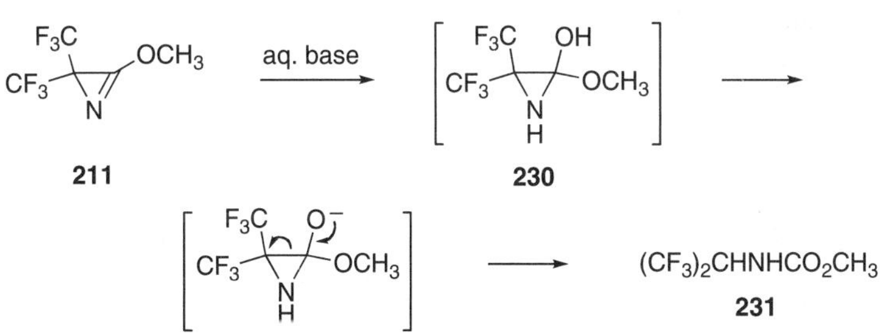

Fluoroazirines can undergo isomerization as well as addition reactions. The original fluoroazirine **209** was transformed in the vapor phase into the more stable perfluoro-3-methyl-2*H*-azirine (**233**) by a catalytic amount of anhydrous hydrogen fluoride.[132,133] The intermediate HF adduct **232** was obtained in 87% yield when excess reagent was present. Azirine **211** suffered a different kind of isomerization upon heating.[134] The product was isocyanate **234**, possibly formed via reversible ring scission to give the resonance-stabilized ylide **235**. The geometry of **235** is obviously ill suited for intramolecular transfer of the methyl group to the anionic carbon to form **234**, but the transfer might be accomplished intermolecularly.

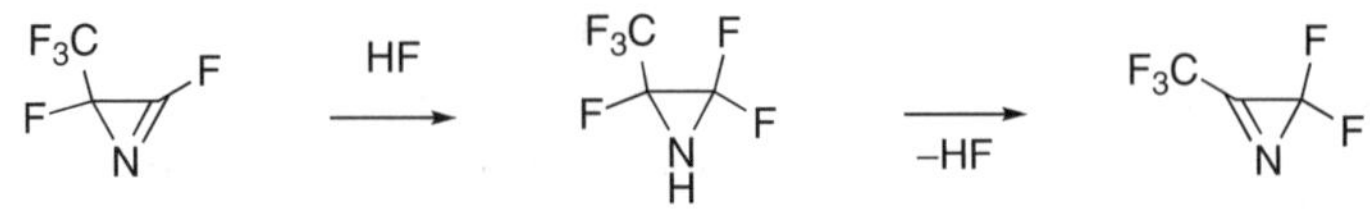

209 **232** **233**

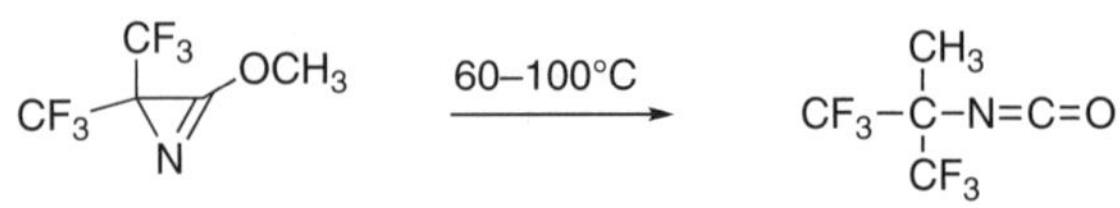

211 **234**

235

1.6 FLUORODIAZIRIDINES

1.6.1 Synthesis

These compounds are most often prepared as precursors for diazirines, which serve as a source of carbenes. Fluorodiaziridines made for this purpose are generally synthesized from oxime tosylates, as illustrated with that derived from α,α,α-trifluoro-3-methoxyacetophenone (**236**), which gives diaziridine **237** in 74% yield.[142] Similarly, the *N*-substituted diaziridine **239** was prepared from hexafluoroacetone oxime tosylate (**238**) and obtained 85.5% optically pure by resolution with D-camphor-3-carboxylic acid.[143] Racemization kinetics, measured polarimetrically in benzene, revealed that the barrier to double nitrogen inversion in **239** is $\Delta G^{\ddagger} = 23.13$ kcal/mol (25.5°C).

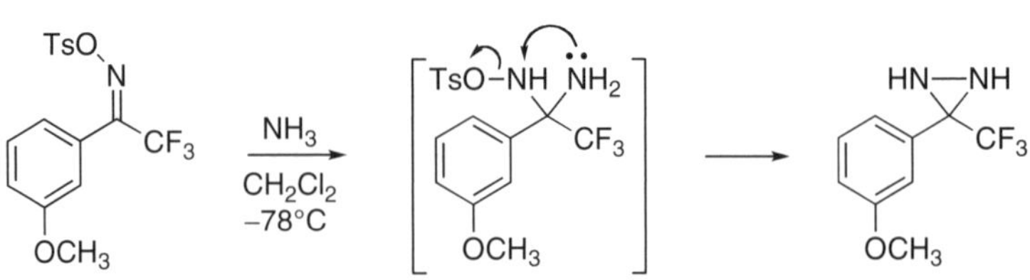

236 **237**

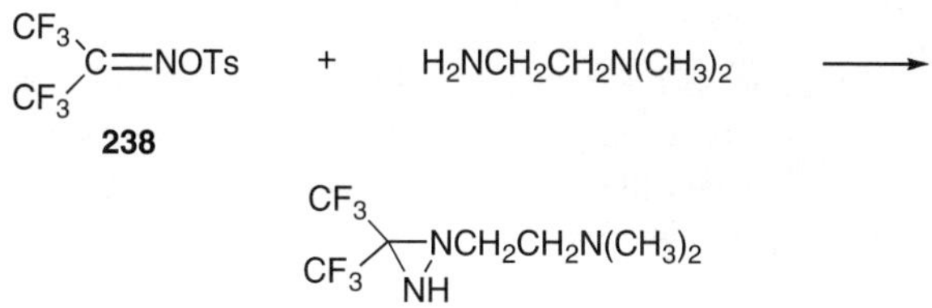

238 239

Fluorodiaziridines can also be synthesized by fluoride ion-catalyzed isomerization of fluorinated amidines, the first example of which was formation of tetrafluorodiaziridine (**241**) from tetrafluoroformamidine (**240**).[144] Treatment of trifluoromethanimine (**242**) with KF results in dimerization to perfluoro-*N*-methylformamidine (**243**, 80–90% yield).[145,146] The more reactive fluoride CsF effects a second addition of fluoride ion, thereby isomerizing **243** to the diaziridine **244**. In analogous fashion, imine **242** reacts with imines **245** to afford diaziridines **246**.[147,148] *N*-Fluoroalkanimines of the type $R_fCF{=}NF$ also dimerize and then cyclize to diaziridines in a polar solvent under the influence of CsF, for example, **247** → **248**.[149] Again, the ring closure step entails elimination of fluoride ion from nitrogen.

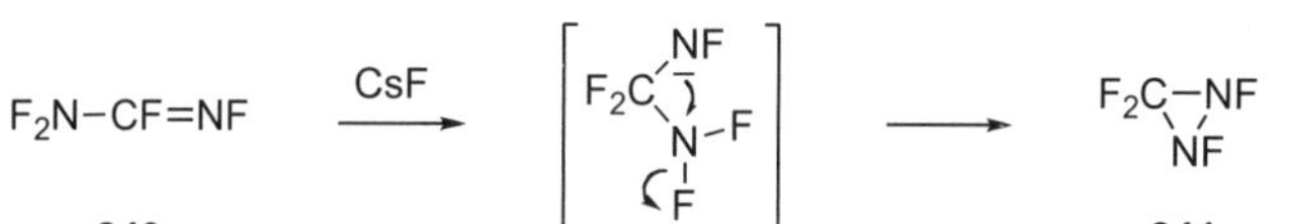

240 241

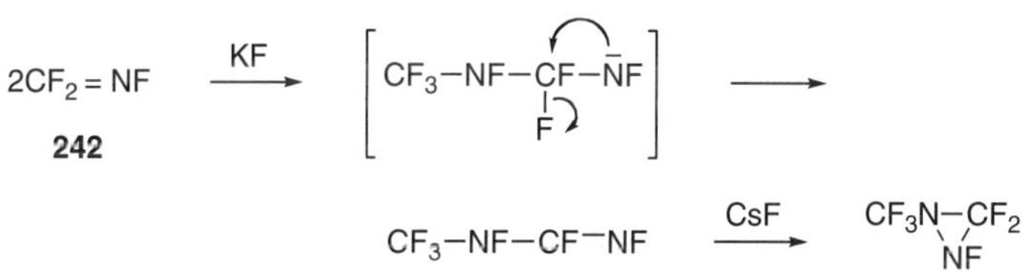

243 244

242 245 246

X = Br, Cl, CF_3

Via $CF_3{=}\bar{N}F$

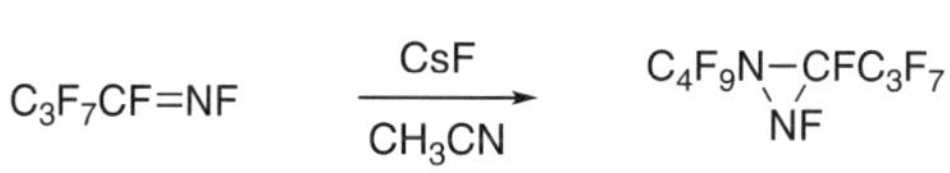

86%

247 248

1.6.2 Reactions

By far the most common reaction of fluorodiaziridines is oxidation to diazirines. The reaction has been carried out with many different reagents, particularly with iodine/triethylamine in recent literature.[150] Oxidation of **237** to diazirine **249** with silver oxide is a representative example.[142]

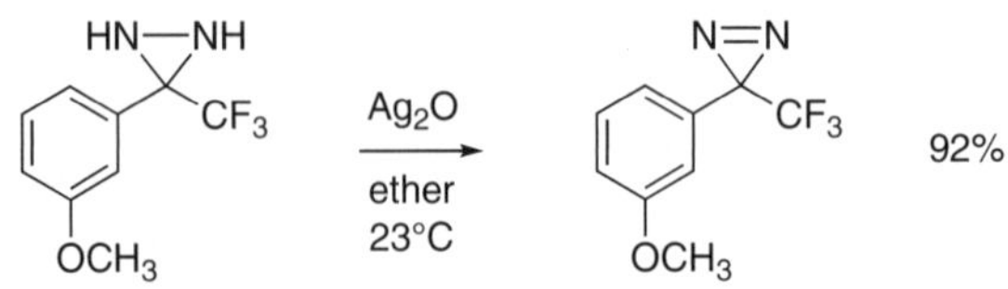

237 **249**

Antimony pentafluoride catalyzes a highly exothermic rearrangement of diaziridine **248** to azo compound **252**, probably by way of intermediates **250** and **251**.[149]

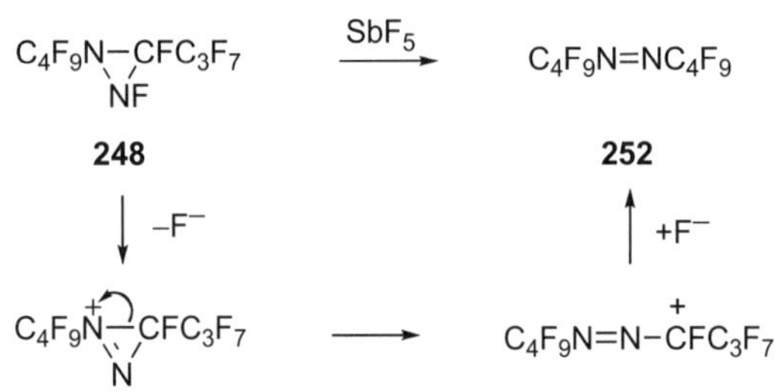

250 **251**

Selective mechanism-based inactivators can be valuable for exploring an enzyme's structure and function and for modulating its activity in a therapeutic situation. Several trifluoromethyl-substituted diaziridines, including **253**, have been found to irreversibly inhibit a human cytochrome P450 enzyme.[151] The proposed mechanism of inactivation entails one-electron oxidation by the enzyme to radical cation **254** followed by ring opening and proton loss to give **255**, and a second oxidation step leading to loss of another proton and nitrogen. The resulting carbene **256** then undergoes insertion at a nucleophilic site on the enzyme.

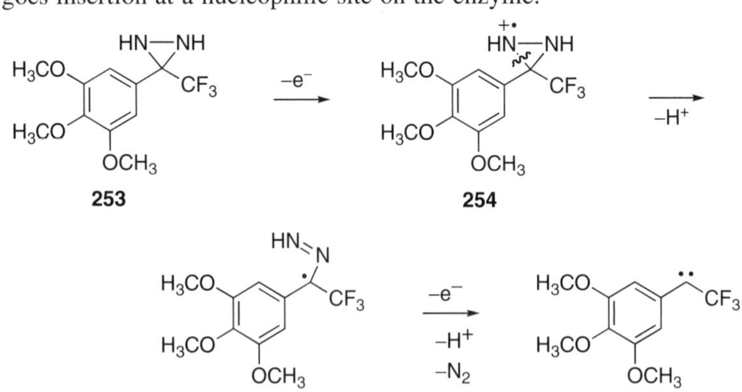

255 **256**

1.7 FLUORODIAZIRINES

1.7.1 Synthesis

Oxidation of diaziridines, a widely used method for preparing diazirines, has been described above. In an important alternative approach, the three-membered ring is generated by oxidation of an amidine in a hypochlorite or hypobromite solution containing the respective halide ion (Graham reaction).[152] Chlorotrifluoromethyl-3*H*-diazirine (**258**) was synthesized in this way from trifluoroacetamidine (**257**).[153] The reaction is believed to proceed via intermediates **259–261**; chloride ion attack on **261** in S_N2' fashion completes the transformation to diazirine **258**.

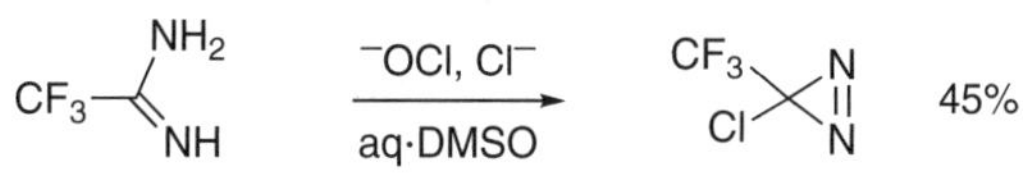

257 **258**

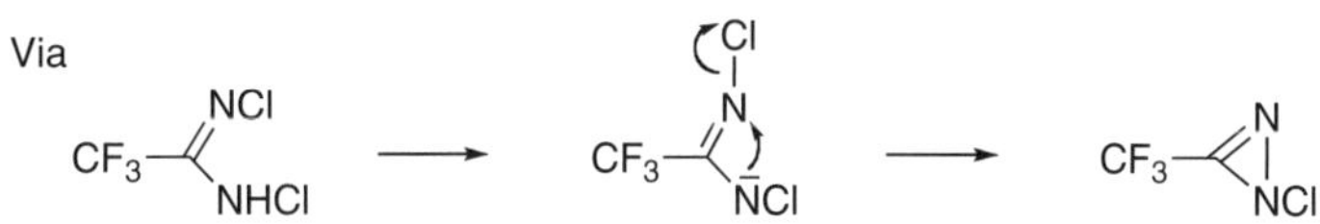

259 **260** **261**

Prepared by lead tetraacetate oxidation of the corresponding diaziridine,[154] bis(trifluoromethyl)diazirine (**263**) has also been made by hypochlorite oxidation of hexafluoroacetone aminal **262**.[155] Instead of by oxidation, ring formation in the case of difluorodiazirine (**265**) has been achieved by mild reduction of bis(difluoroamino)difluoromethane (**264**).[156,157] Alternatively, **265** can be prepared by fluoride-catalyzed rearrangement of difluorocyanamide (**266**).[158]

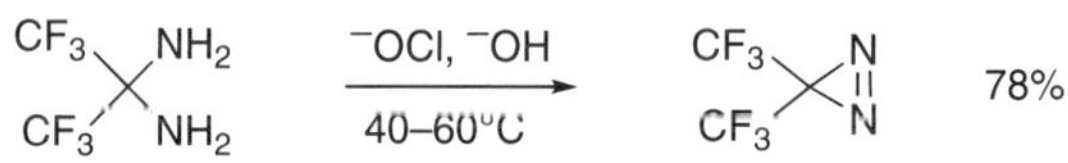

262 **263**

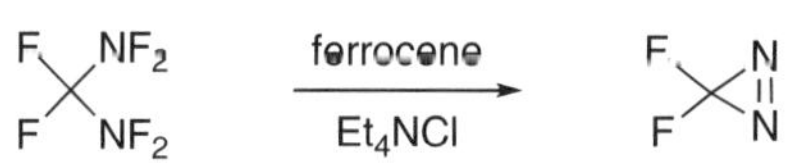

264 **265**

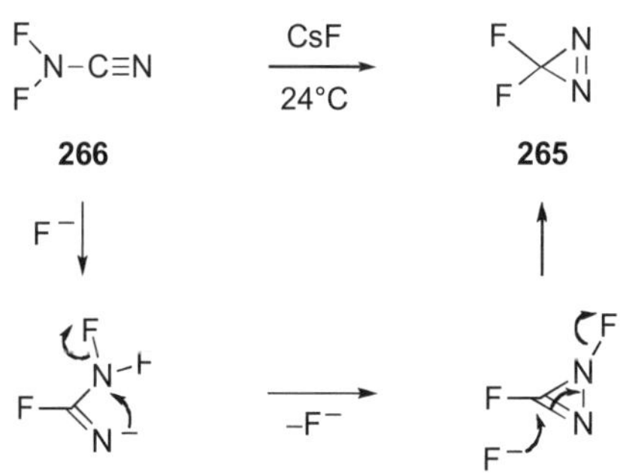

266 **265**

A fluoro substituent can be introduced into a chloro- or bromodiazirine by an exchange reaction.[159–162] For example, bromo-*p*-tolyldiazirine (**267**) was transformed into fluoro-*p*-tolyldiazirine (**268**) by treatment with tetrabutylammonium fluoride at RT.[160] It was proposed that the reaction proceeds via formation of a diazirinium ion pair followed by attack by fluoride, but the finding that electron-deficient diazirine **269** undergoes exchange with fluoride ion at RT to give fluorotrifluoromethyldiazirine (**271**) militated against an ionization mechanism and suggested that the reaction proceeds via sequential S_N2' reactions with *N*-fluorodiazirine **270** as an intermediate.[162] Subsequent elegant labeling studies of the reaction of azide ion with bromophenyldiazirine, together with quantum chemical calculations, provided powerful support for this pathway for exchange reactions of halodiazirines.[163,164] In cases where the halide exchange reaction with fluoride ion has failed, it has been possible to obtain the desired fluorodiazirine in low yield by a "modfied" Graham reaction in which high concentrations of fluoride are present.[165]

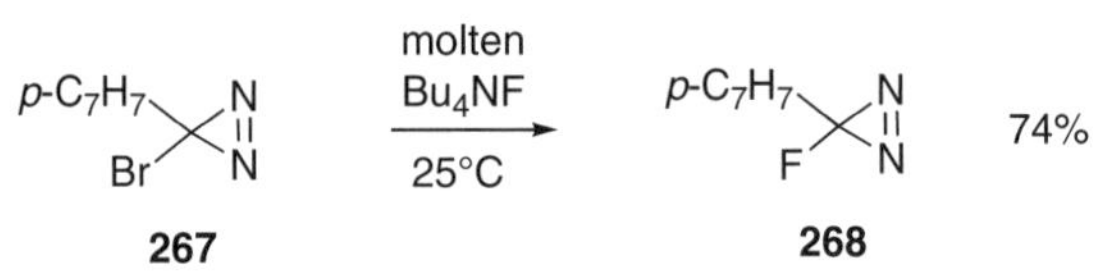

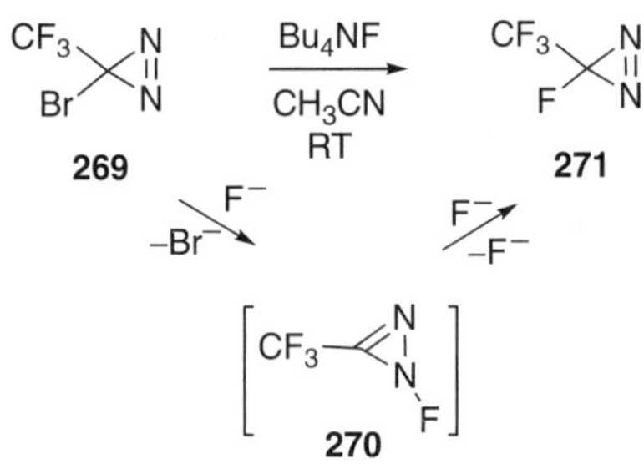

Diazirines can be formed by photoisomerization of diazo compounds. Irradiation of perfluoro-3-diazo-2-butanone (**272**) at wavelengths >320 nm either at RT in carbon tetrachloride solution or at 77K in an argon matrix yielded perfluoroacetylmethyldiazirine (**273**), together with products resulting from decomposition of the diazo compound to a carbene.[166]

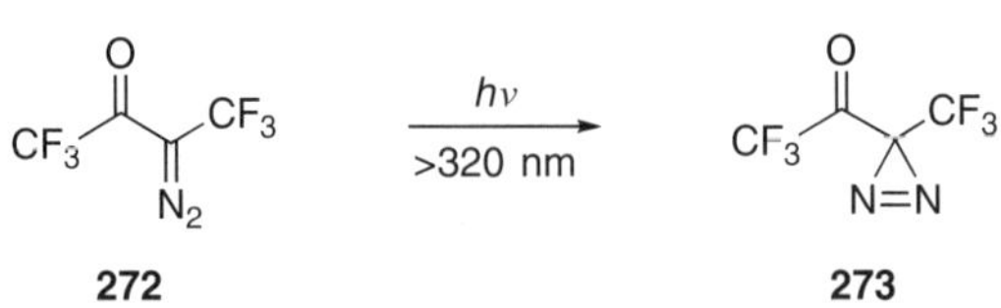

1.7.2 Reactions

By far the most important reaction of fluorodiazirines is their thermal or photochemical fragmentation with loss of nitrogen to generate carbenes. The literature abounds with papers describing the synthesis of photoaffinity labels that incorporate a fluorinated diazirine[142,150a,167] and with their use in tagging a great variety of biological molecules.[168] The photogenerated species are generally aryltrifluoromethylcarbenes, as the trifluoromethyl group offers important advantages. It enhances the carbene's electrophilicity, and thus its voracity in attacking and binding to an electron-rich site on the biomolecule. Because of fluorine's reluctance to migrate,[169] the carbene has a longer lifetime to find its target than a carbene with α-hydrogen atoms. As an example of a photoaffinity probe based on a fluorodiazirine, the potent intravenous anesthetic agent etomidate (**274**) has been derivatized as **275**.[168a] Shown to be an effective and selective photolabel for the *Torpedo* nicotinic acetylcholine receptor, **275** holds promise for identifying the site of etomidate-induced anesthesia.

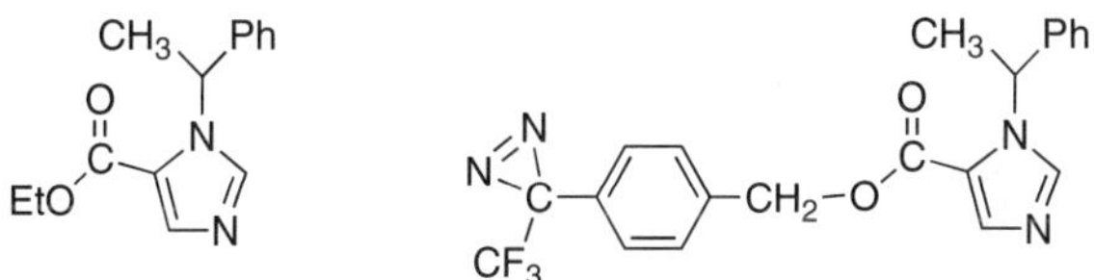

274 **275**

Fluorodiazirines have also been used extensively as carbene photoprecursors for cryogenic matrix isolation[169–171] and laser flash photolysis studies.[165,170,172] Photolysis of fluorophenoxydiazirine has made observation of the corresponding carbene directly observable in yet another fashion: encapsulated in a hemicarcerand.[173] Diazirines have been used for studies in solution of carbene philicities as well as for synthesis of cyclopropanes and cyclopropenes.[174] Of course, diazo compounds can also serve as carbene photoprecursors, but for carbenes such as difluorocarbene, the diazo precursor is either nonexistent or extremely unstable.

Azine formation is common when fluorodiazirines are decomposed, whether thermally or photochemically. Gas-phase pyrolysis of bis(trifluoromethyl)diazirine (**263**) gave hexafluoropropene, from rearrangement of bis(trifluoromethyl)carbene (**276**), and azine **277**, from attack of the carbene on starting material.[175]

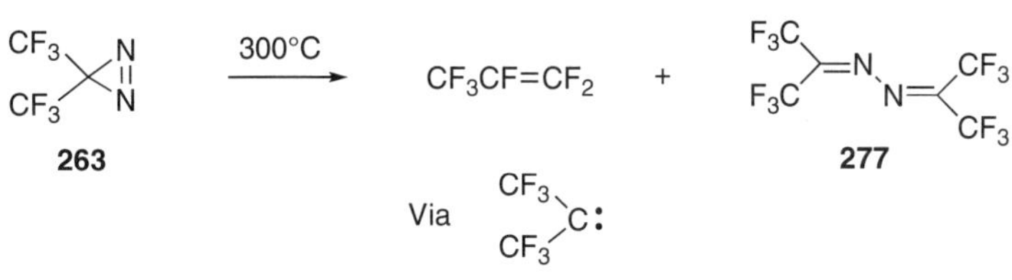

263 **277** **276**

Fluorodiazirines may photoisomerize to diazo compounds, as illustrated with diazirine **278**. Photolysis of **278** at 334 nm in a nitrogen matrix at 8K primarily produced diazo compound **279**, which was transformed by visible light into cyclobutylcarbene **280** and cyclopentene **281**.[171a] This chemistry is of special interest because **280** was found to rearrange to **281** at 8K in the absence of light by way of heavy-atom tunneling. Because diazo compounds absorb at longer wavelengths than the corresponding diazirines, choice of wavelength can dictate the direction of their interconversion, as is the case for **272** and **273**.

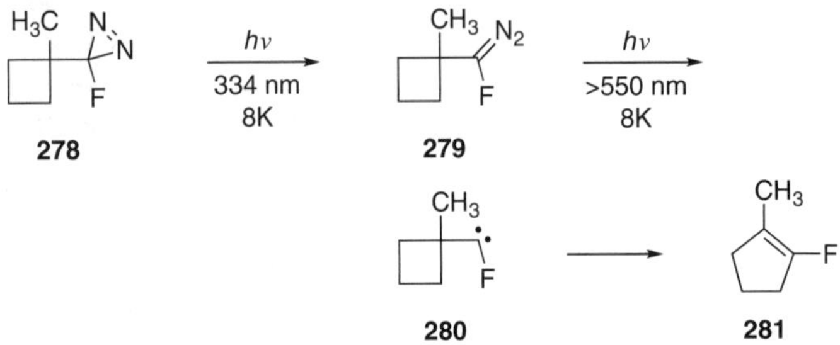

278 **279** **280** **281**

A method for transforming alcohols into amines and carboxylic acids into noramines was developed based on fluorodiazirines.[176,177] The conversion of cyclohexanecarboxylic acid into cyclohexylamine is illustrative. Photolysis of thiopyridone derivative **282**, prepared from the acid, generated the cyclohexyl radical, which added to the double bond of (trifluoromethyl)phenyldiazirine **283** to give diazirinyl radical **284**. Dimerization of **284** to tetraazo compound **285** was followed by fragmentation with loss of nitrogen, forming two molecules of imine **286** (NMR yield, 89%). Hydrolysis of **286** then yielded cyclohexylamine. As evidence for this mechanism, in one reaction series the tetraazo compound was isolated and its crystal structure was obtained.

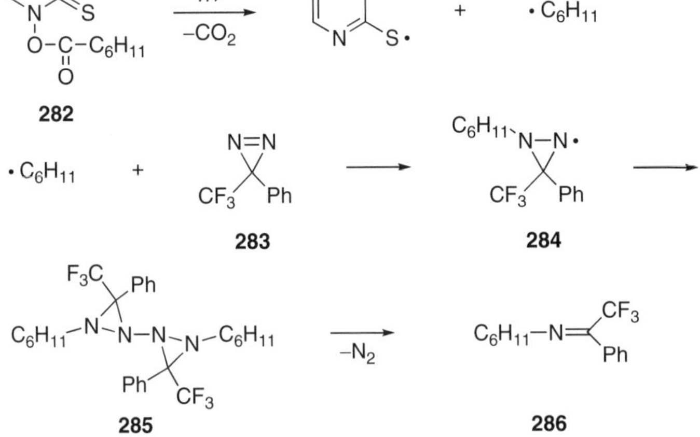

282 **283** **284** **285** **286**

1.8 FLUOROOXAZIRIDINES[178]

1.8.1 Synthesis

The earliest method for synthesizing perfluorinated oxaziridines entailed addition of trifluoromethyl hydroperoxide (**287**) to a perfluoroimine, followed by elimination of HF mediated by a metal fluoride.[179] The fluoride of choice was the mild reagent KHF_2 because it was basic enough to effect the elimination but too weak to ring open the oxaziridine product.[180] Oxidation of imine **288** by this method gave perfluoro-*N*-methyloxaziridine (**289**), the first perfluorooxaziridine. In the case of hindered imine **290**, no adduct was obtained with hydroperoxide **287**, but oxidation to oxaziridine **291** was successful nonetheless with the hydroperoxide in the presence of KF.[181] Presumably the imine was attacked by the hydroperoxide anion, forming the intermediate anion **292**.

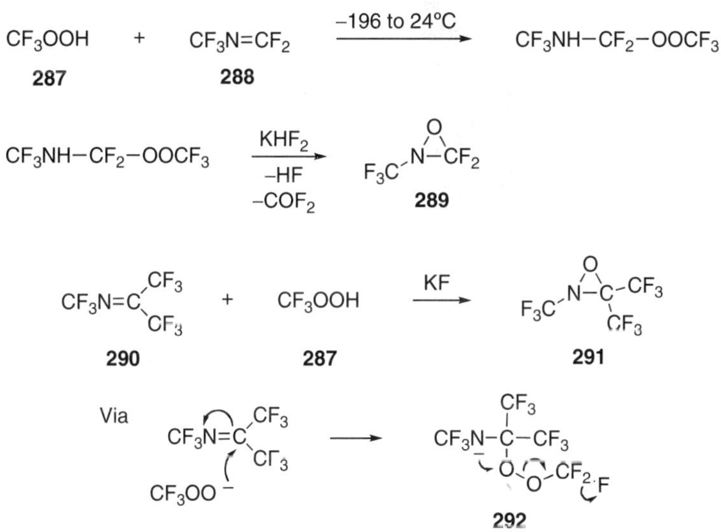

Other reagents found capable of oxidizing highly fluorinated imines were chlorine gas in the presence of a metal carbonate,[182] 50% hydrogen peroxide,[183] $CF_3OOC(O)F$, and $CF_2(OF)_2$.[184] All, including CF_3OOH, are highly reactive oxidizers and most pose an explosion hazard and are not readily available. Thus, it was a welcome finding that the relatively safe and convenient reagent *m*-chloroperoxybenzoic acid (MCPBA) is also effective if concentrated (>80%) and well dried, with acetonitrile as solvent. Imine **293**, for example, was oxidized to oxaziridine **294** in 77% yield by this method,[185] and highly fluorinated *N*-aryloxaziridines have been obtained similarly with the solvent sulfolane.[186] Fluorinated *N*-sulfonyloxaziridines have been synthesized from sulfonamides with MCPBA in either acetonitrile or sulfolane.[187] The sulfonamide apparently existed in equilibrium with a small amount of the corresponding imine plus HF, and oxidation of the imine drove the equilibrium. In this way, oxaziridine **296** was obtained in 73% yield from sulfonamide **295** via imine **297**.

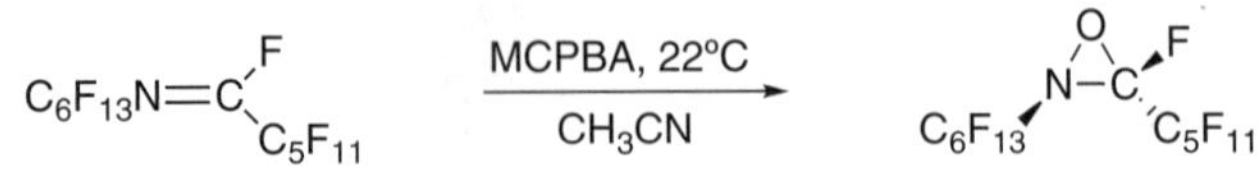

293 294

295 296

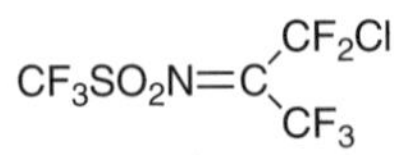

297

Biphasic oxidation of imine **298** to oxaziridine **299** in 90% yield was accomplished with 50% MCPBA and saturated potassium carbonate.[188] Here the electron-deficient imine was presumably attacked in nucleophilic fashion by the conjugate base of MCPBA, and reaction occurred on the less hindered face of the double bond.

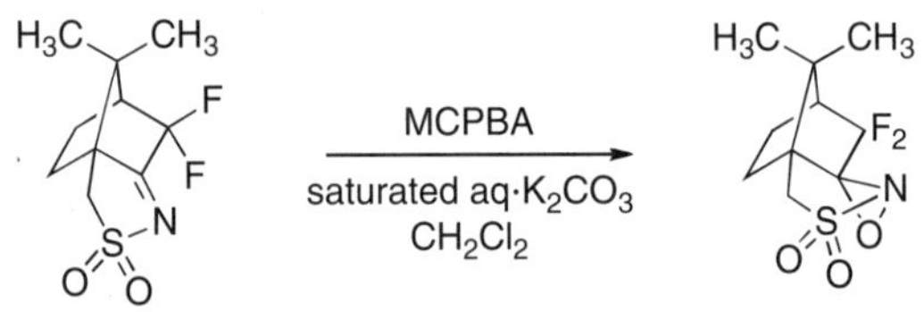

298 299

1.8.2 Oxygen Atom Transfer Reactions

Fluorooxaziridines are versatile reagents, the most important transformations of which entail donation of their oxygen atom. Considering first oxygenation at carbon, olefins undergo epoxidation under reaction conditions that vary greatly with their nucleophilicity. (*R*)-Carvone (**300**) reacted quickly and selectively at the isolated double bond with oxaziridine **301** at −40°C in dichloropentafluoropropane, giving epoxide **302** as a 1:1 mixture of diastereomers. In contrast, chlorotrifluoroethylene required 16 h at 100°C with the same reagent to afford the corresponding epoxide (60% yield).[189]

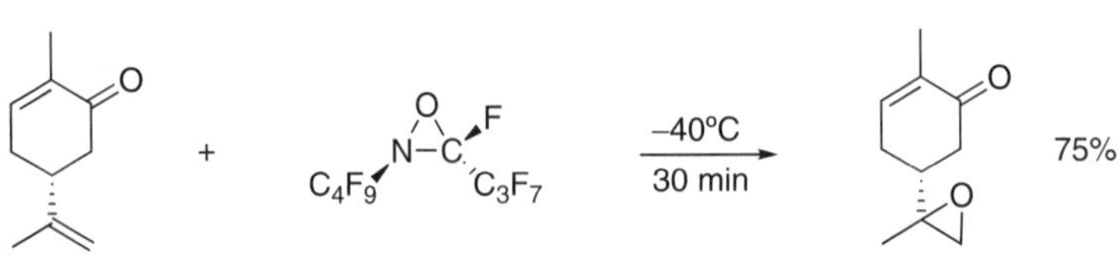

300 301 302

Enolates are hydroxylated by fluorooxaziridines rapidly under very mild conditions, as exemplified by the reaction of the *Z*-enolate **303** of propiophenone with the chiral oxaziridine **299**. Hydroxyketone **304** was obtained in 94% ee and 71% yield.[188]

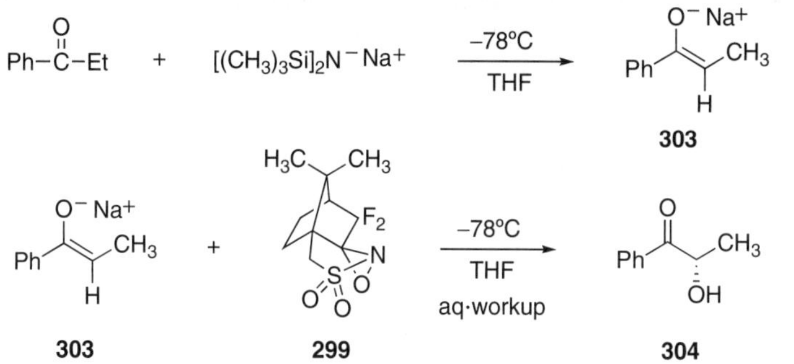

Fluorooxaziridines are also capable of inserting an oxygen atom into an unactivated C–H bond, a transformation that reveals both their considerable power and striking selectivity. Both characteristics are apparent in the oxidation at RT of steroid **305** by oxaziridine **301**, giving the 5β-hydroxy derivative **306** in 75% yield with little by-product.[190] While selectivity for tertiary C-H bonds is not surprising, such selectivity for one of the five 3° C–H bonds is remarkable. Hydroxylation of (*S*)-halocarbon **307** with the same oxaziridine, yielding (*R*)-alcohol **308** with 98% retention of configuration, also attests to the reagent's selectivity.[191] As further evidence of its power, mass spectral data indicate that at 80°C it has introduced up to 16 hydroxyl groups onto the 18 available 3° carbons of a dodecahedrane diester.[192] Fluorooxaziridines also oxygenate the α-carbon of secondary alcohols[193] and ethers,[194] affording ketones and α-hydroxy ethers, respectively.

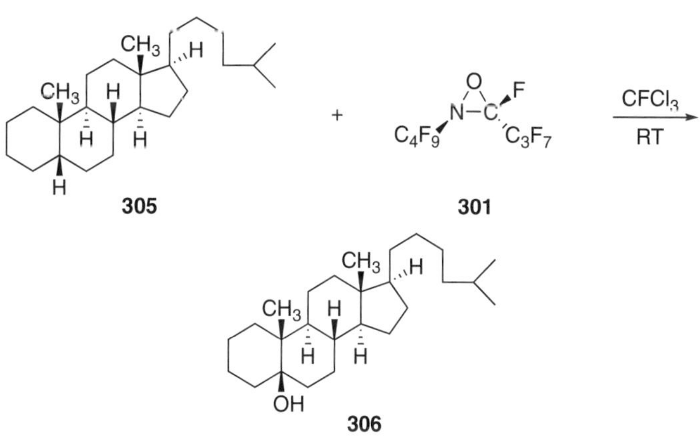

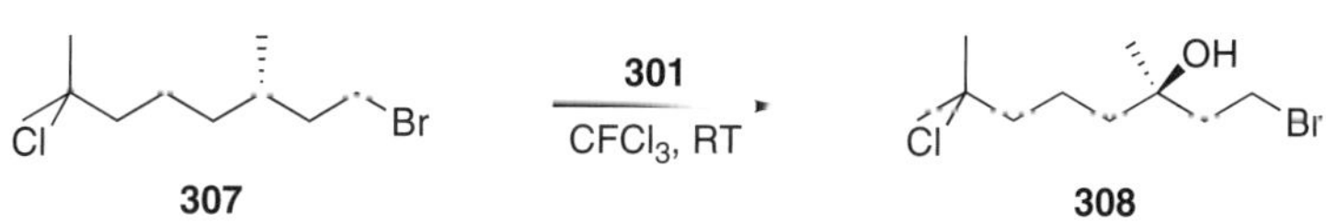

Regarding oxygenation at atoms other than carbon, silanes are transformed into silanols,[195] and sulfur is a common target for attack. Sulfides are oxidized in excellent yields to sulfoxides or sulfones, as illustrated with the transformation under mild conditions of allyl methyl sulfide (**309**) into **310** and **311**.[196] Similarly, promazine (**312**) was converted into its sulfoxide (**313**) in >90% yield at −40°C. Thus, other oxidizable sites in the molecule were left untouched.

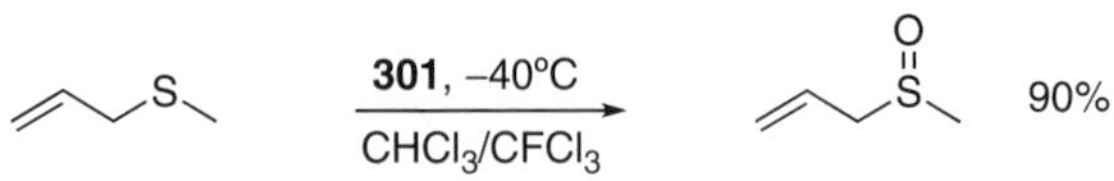

309 **310**

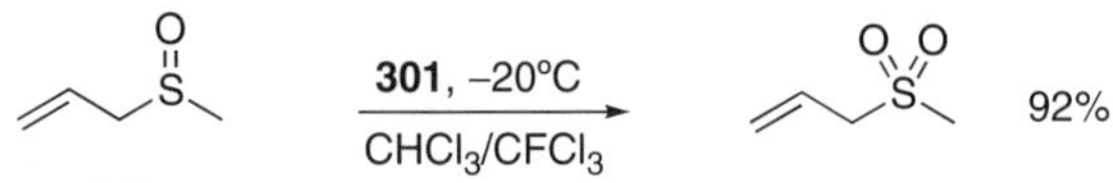

310 **311**

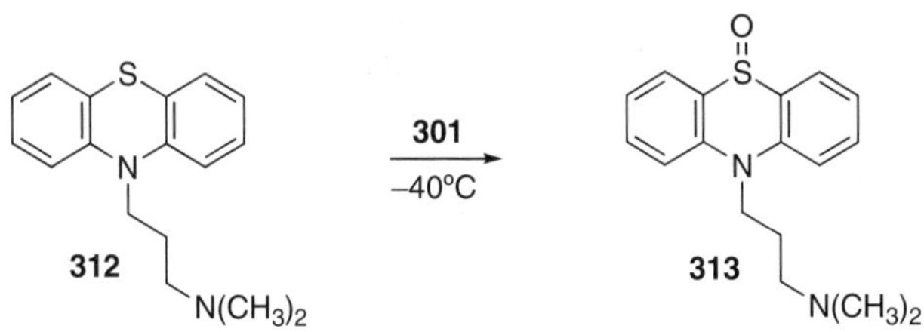

312 **313**

Variously substituted phosphorus(V) sulfides and selenides were converted to the corresponding oxides by fluorooxaziridines. For example, reaction of triisobutylphosphine sulfide (**314**) with oxaziridine **294** in dichloropentafluoropropane gave the oxide **315** in 92% yield, together with elemental sulfur.[197] Surprisingly, it was found with a chiral sulfide that the reaction occurred with clean inversion of configuration, whereas retention was observed when the oxidant was MCPBA.

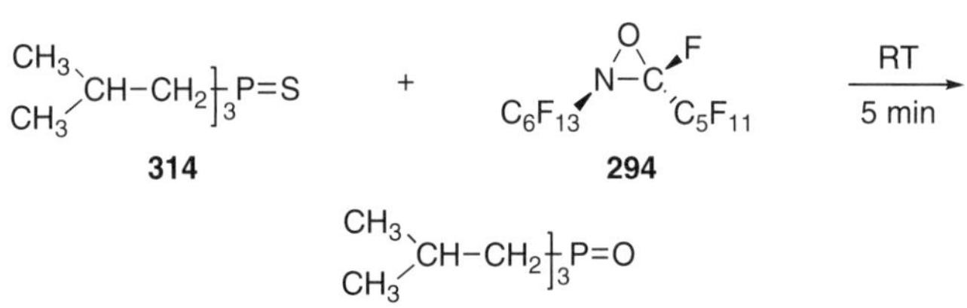

314 **294**

315

Nitrogen also can be a site of attack for oxygen atom transfer from a fluorooxaziridine. 2-Methylpyridine (**316**) was converted to its *N*-oxide (**317**) in 70% yield by oxaziridine **301** under very gentle conditions.[198]

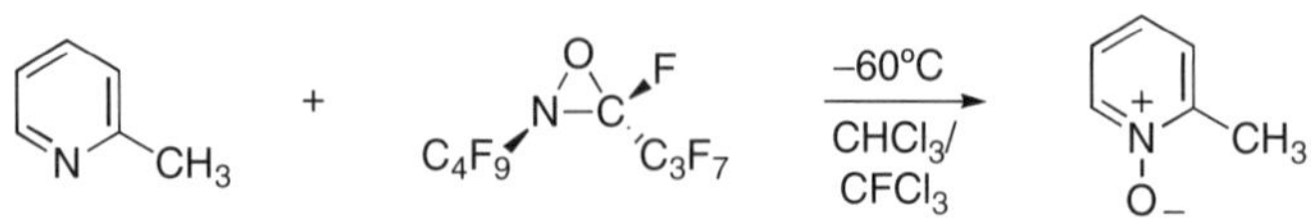

316 **301** **317**

1.8.3 Other Reactions

Fluorooxaziridines also function as aminating agents. In the case of **316** and other 2-substituted pyridines, oxidation is the exclusive reaction because of steric hindrance, but with pyridine itself and with 3- and 4-substituted pyridines as well as other aromatic nitrogen heterocycles, amination at nitrogen is a competing process. With quinoxaline (**318**), for example, the product of amination (**319**) is the major one (45%) and that of oxidation (**320**) the minor (30%).[198] The pathway proposed for formation of the aminides, shown for pyridine, involves nucleophilic attack at the oxaziridine nitrogen to produce zwitterion **321**, followed by fragmentation into acyl fluoride **322** and aminide **323**. Fluoride elimination gives pyridinium fluoride **324**, which suffers hydrolysis to **325** as a result of either deliberate or adventitious contact with water.

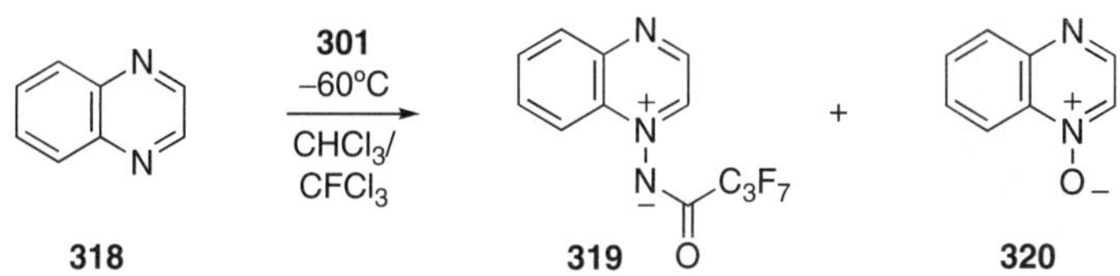

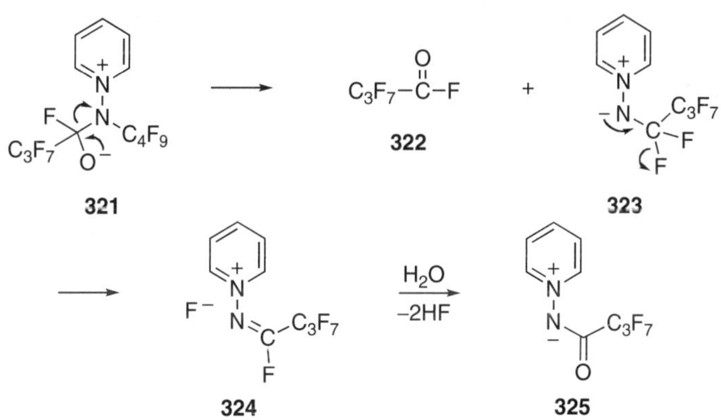

Ring opening of perfluoro-*N*-methyloxaziridine (**289**) by attack at nitrogen has been shown to occur with many kinds of nucleophiles, including amines, alcohols, carboxylic acids, mercaptans, KCN, KF, and $CsOCF_3$.[199,200] The facility of this process is apparent from the fact that *t*-butanol reacts with it in a 1:1 ratio at ≤22°C to give acyl fluoride **326** in 93% yield. This mode of ring opening is reminiscent of nucleophilic attack on hexafluoropropylene oxide (**1**), which also occurs at the CF_3-bearing atom. In both heterocycles, stabilization of the developing oxide ion by negative hyperconjugation in the transition state probably accounts, at least in part, for the direction of ring opening.[15]

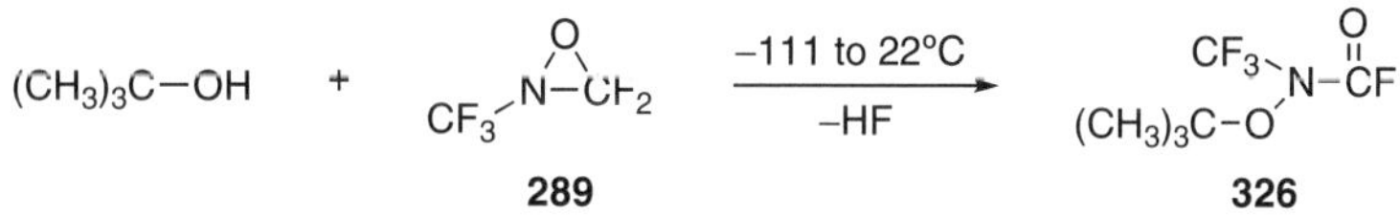

Oxaziridine **289** reacts with an array of *gem*-difluoroalkenes to afford 1,3-oxazolidines, instead of transferring an oxygen atom to form oxiranes.[200–202] With bromotrifluoroethylene, for example, cycloaddition to give **327** occurs in 85% yield. The reaction probably proceeds via ring opening to biradical **328**, and then attack by the oxygen at the CF_2 carbon to form the most stable biradical (**329**).

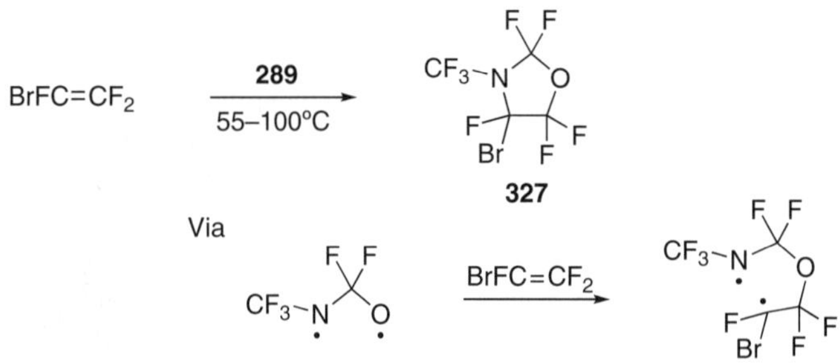

327 **328** **329**

The same oxaziridine (**289**) also cycloadds to ketones, giving 1,3,4-dioxazolidines. The adduct with acetone (**331**) was obtained in 75% yield at RT. Given the ease with which nucleophiles attack **289**, it is likely that the reaction occurred by attack of the carbonyl oxygen at the nitrogen of **289**, generating zwitterionic intermediate **330**.

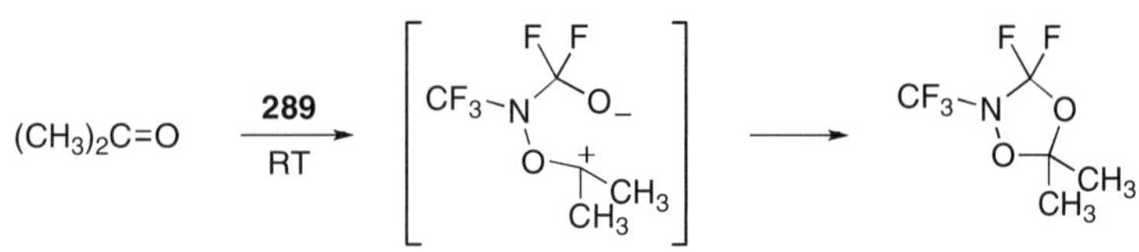

330 **331**

Tris(perfluoroalkyl)oxaziridines such as **291** undergo thermal rearrangement to amides. Scission of the N–O bond to give biradical **332** followed by migration to nitrogen of a CF_3 yields amide **333** quantitatively.[181,186] If one of the substituents at C3 is a fluorine, however, migration of the perfluoroalkyl group in the biradical occurs in the opposite direction.[203] Oxaziridine **334**, for example, rearranges to alkoxyimine **335**.

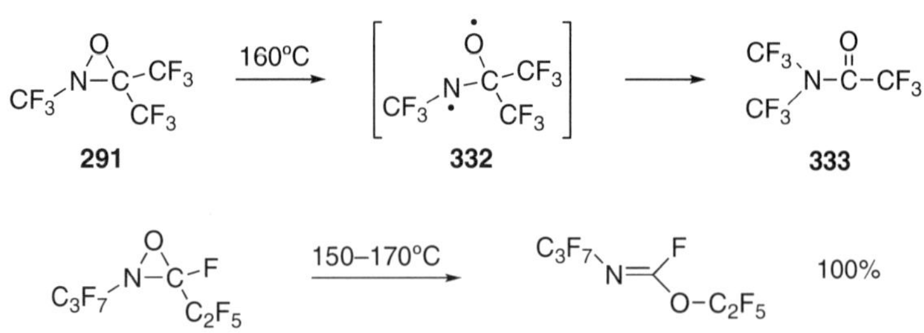

291 **332** **333**

334 **335**

At lower temperatures in the presence of antimony pentafluoride, oxaziridine **334** rearranges to a different alkoxyimine, **337**. A likely pathway for this transformation involves abstraction of a fluoride ion to form immonium ion **335** followed by or concerted with ring opening to carbocation **336**, and then back donation of the fluoride.

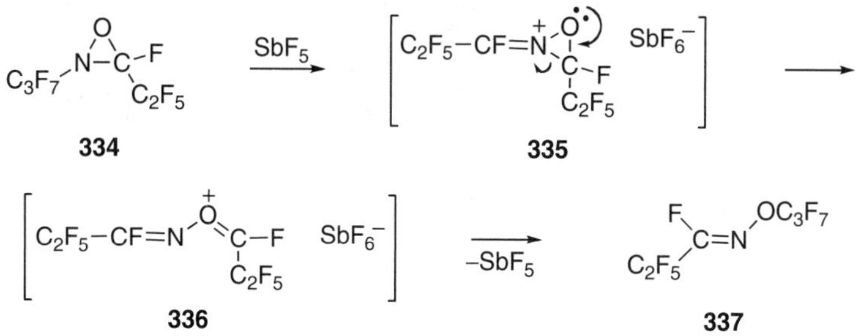

Treatment of oxaziridine **301** with dry cesium fluoride at 100°C gave in high yield a 1:1 mixture of imine **338** and acyl fluoride **339**.[203] Presumably ring opening at the nitrogen by fluoride to generate anion **340** was followed by loss of fluoride to produce amide **341**. The reversibility of the latter step made possible an alternative pathway, cleavage of the bond to nitrogen. Elimination of fluoride from the resulting anion **342** completed the reaction sequence.

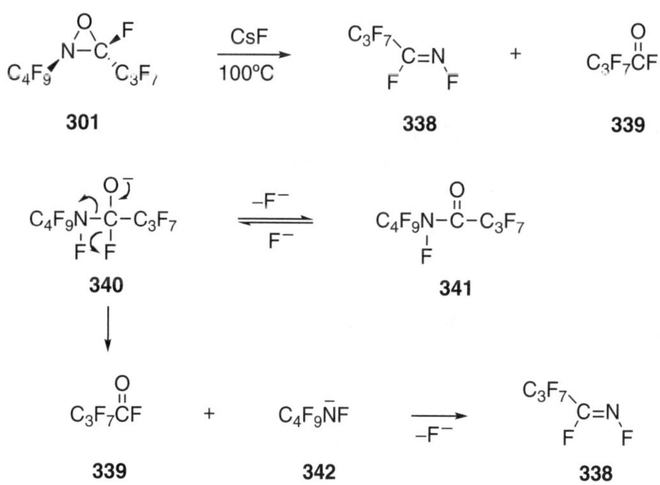

The same reactants, oxaziridine **301** and CsF, in acetonitrile at 22°C yield a single product, alkoxyimine **343**. This surprising result can be explained in terms of a nucleophilic chain process, initiation of which entails formation of imine **338** and acyl fluoride **339** followed by fluoride addition to the latter to give alkoxide **344**. Propagation involves attack by anion **344** on the oxaziridine to form **345**, scission of the C–N bond to produce anion **346** plus **339**, elimination of fluoride ion from **346**, and regeneration of **344** from **339**. Thus, alkoxide ion **344** carries the chain. The contrast in outcome with versus without acetonitrile is understandable because reaction of CsF with acyl fluorides is facilitated by dipolar aprotic solvents.

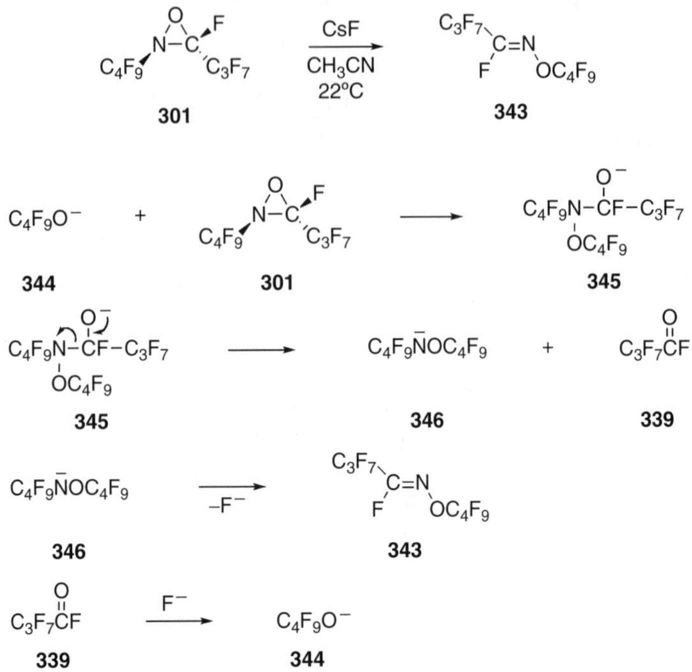

This interpretation is consistent with the finding that hexafluoroacetone (**20**) reacts with oxaziridine **301** and CsF under the same conditions to give alkoxyimine **347** (95% yield) and acyl fluoride **339**.[203] In this case, no chain is established because the alkoxide ion **348** is less reactive than **344**.

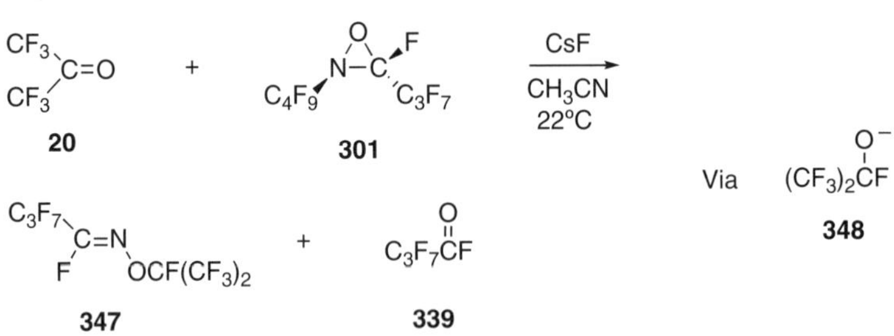

1.9 AFTERWORD

The combination of ring strain and electron deficiency imbues fluorinated three-membered ring heterocycles with unusual reactivity. As a result, the chemistry of this class of compounds is especially rich. Together with a description of their origins, this chapter has presented a sampling from the cornucopia of transformations that these molecules undergo. Their reactivity extends our knowledge and understanding of organofluorine chemistry, and important aspects of it affect other areas of organic as well as biochemistry, where fluorodioxiranes, -diazirines, and oxaziridines serve as valuable reagents. Surely a wealth of new chemistry of these smallest fluorinated heterocycles awaits discovery.

ACKNOWLEDGMENT

The authors thank the National Science Foundation for financial support provided by grant no. CHE-0653935.

REFERENCES

1. Chambers, R.D. *Fluorine in Organic Chemistry*; Blackwell: Oxford, 2004; pp 259–264.
2. Paleta, O. Oxidation. In *Chemistry of Organic Fluorine Compounds II*; Hudlicky, M.; Pavlath, A., Eds.; American Chemical Society: Washington, DC, 1995; pp 322–327.
3. Millauer, H.; Schwertfeger, W.; Siegemund, G. Hexafluoropropene oxide: a key compound in organofluorine chemistry. *Angew. Chem., Int. Ed. Engl.* **1985**, *24*, 161.
4. Tarrant, P.; Allison, C.G.; Barthold, K.P.; Stump, E.C. Fluorine-containing epoxides. *Fluorine Chem. Rev.* **1971**, *5*, 77.
5. Eleuterio, H.S. Polymerization of perfluoro epoxides. *J. Macromol. Sci. Chem.* **1972**, *6*, 1027.
6. Makarov, K.N.; Pletnev, S.I.; Gervits, L.L.; Natarov, V.P.; Prokudin, I.P.; Knunyants, I.L. Functional derivatives of perfluorodecalin. *Izv. Akad. Nauk SSSR, Ser. Khim.* **1986**, 1313; *Chem. Abstr.* **1987**, *107*, 115268.
7. Petrov, V.A.; Marshall, W.J.; Krespan, C.G.; Cherstkov, V.F.; Avetisian, E.A. New partially fluorinated epoxides by oxidation of olefins with sodium hypohalites under phase transfer catalysis. *J. Fluorine Chem.* **2004**, *125*, 99.
8. Ono, T.; Henderson, P. Novel epoxidation reaction of perfluoroalkenes with trimethylamine *N*-oxide and iodosylbenzene. *Tetrahedron Lett.* **2002**, *43*, 7961.
9. Kolenko, I.P.; Filyakova, T.I.; Zapevalov, A.Y.; Mochalina, E.P.; German, L.S.; Polishchuk, V.R. Synthesis and rearrangement of [(fluorosulfonyl)oxy]pentafluoroacetone. *Izv. Akad. Nauk SSSR, Ser. Khim.* **1979**, 667; *Chem. Abstr.* **1979**, *91*, 38866.
10. Liu, Y. Part I. Tetraaminoethylenes and diaminocarbenes. Part II. The attempted synthesis of perfluorobicyclo[1.1.0]butane and perfluorobicyclo[2.1.0]pentane. Ph.D. Dissertation, Dartmouth College, **2001**.
11. Kadyrov, A.A.; Rokhlin, E.M.; Knunyants, I.L. Epoxidation of highly electrophilic fluorine-containing unsaturated compounds with peracids. *Izv. Akad. Nauk SSSR, Ser. Khim.* **1982**, *10*, 2344; *Chem. Abstr.* **1983**, *98*, 125776.
12. Vilenchik, Y.M.; Mitrofanova, L.N.; Senichev, Y.N. Photochemical synthesis of tetrafluoroethylene oxide. *Zh. Org. Khim.* **1978**, *14*, 1587; *Chem. Abstr.* **1978**, *89*, 197230.
13. Hung, M.-H.; Resnick, P.R. The chemistry of fluorinated dioxoles and dioxolanes. Part 1. Thermal rearrangement of fluorinated dioxoles. *J. Am. Chem. Soc.* **1990**, *112*, 9671.

14. Sianesi, D.; Pasetti, A.; Tarli, F. The chemistry of hexafluoropropene epoxide. *J. Org. Chem.* **1966**, *31*, 2312.

15. Farnham, W.B.; Smart, B.E.; Middleton, W.J.; Calabrese, J.C.; Dixon, D.A. Crystal and molecular structure of tris(dimethylamino)sulfonium trifluoromethoxide. Evidence for negative fluorine hyperconjugation. *J. Am. Chem. Soc.* **1985**, *107*, 4565.

16. Daikin Kogyo, Jpn. Kokai Tokkyo Koho. Pentafluoropropionyl fluoride. JP 5838231, **1983**; *Chem. Abstr.* **1983**, *99*, 5225f.

17. Tyul'ga, G.M.; Gubanov, V.A.; Rondarev, D.S.; Ryanazova, R.M.; Dolgopol'skii, I.M. Reactivity of α-oxides of fluorinated olefins with respect to cesium fluoride in tetrahydrofuran. *Zh. Org. Khim.* **1980**, *16*, 300.

18. Du Boisson, R.A. The fluorochemical industry in the United States. In *Organofluorine Chemistry, Principles and Commercial Applications*; Banks, R.E.; Smart, B.E.; Tatlow, J.C., Eds.; Plenum: New York, 1994; p 589.

19. Kvícalá, J.; Hovorková, P.; Paleta, O. Reactions of 3-chloropentafluoropropene-1,2-oxide with bifunctional nucleophiles: a new route to chlorodifluoromethyl-substituted heterocycles. *J. Fluorine Chem.* **2001**, *108*, 1.

20. Knunyants, I.L.; Shokina, V.V.; Tyuleneva, U.V.; Cheburkov, Y.A.; Aronov, Y.E. Derivatives of perfluoroisobutyric acid. *Izv. Akad. Nauk SSSR, Ser. Khim.* **1966**, 1831.

21. Bekker, R.A.; Astratyan, G.V.; Dyatkin, B.L. Polyhalogenated α-oxides. IV. Role of steric factors in reactions of polyfluorinated α-oxides with amines. *Zh. Org. Khim.* **1973**, *9*, 1644.

22. Bekker, R.A.; Astratyan, G.V.; Dyatkin, B.L.; Knunyants, I.L. Polyhalogenated α-oxides. V. Role of steric factors in reactivity of polyfluorinated α-oxides. *Tetrahedron* **1974**, *30*, 3539.

23. Barten, J.A.; Kadyrov, A.A.; Lork, E.; Röschenthaler, G.-V. Novel and facile synthesis of fluorinated enamino ketones and amino alcohols. *J. Fluorine Chem.* **2002**, *116*, 87.

24. Petrov, V.A. Synthesis of polyfluorinated tertiary alcohols using ring opening reactions of 2,2-bis(trifluoromethyl)oxirane. *Synthesis* **2002**, 2225.

25. Petrov, V.A. Reactions of 2,2-bis(trifluoromethyl)oxirane with alcohols under phase transfer catalysis. *J. Fluorine Chem.* **2004**, *125*, 531.

26. Feiring, A.E.; Crawford, M.K.; Farnham, W.B.; French, R.H.; Leffew, K.W.; Petrov, V.A.; Schadt, F.L., III; Tran, H.V.; Zumsteg, F.C. Bis(fluoroalcohol) monomers and polymers: improved transparency fluoropolymer photoresists for semiconductor lithography at 157 nm. *Macromolecules* **2006**, *39*, 1443; See also Feiring, A.E.; Crawford, M.K.; Farnham, W.B.; Feldman, J.; French, R.H.; Junk, C.P.; Leffew, K.W.; Petrov, V.A.; Qiu, W.; Schadt, F.L., III; Tran, H.V.; Zumsteg, F.C. New amorphous fluoropolymers of tetrafluoroethylene with fluorinated and non-fluorinated tricyclononenes. Semiconductor photoresists for imaging at 157 and 193 nm. *Macromolecules* **2006**, *39*, 3252.

27. Pavlik, F. Decyclization of fluorinated cyclic ethers to perfluorinated tertiary alcohols. *US Patent* 3,385,904, **1968**; *Chem. Abstr.* **1968**, *69*, 26753.

28. Moore, E.D.; Milian, A.S. Fluorinated carbonyl compounds. *US Patent* 3,321,515, **1967**; *Chem. Abstr.* **1967**, *66*, 116587.

29. Morin, D.E. Decyclization of fluorinated cyclic ethers. *US Patent* 3,213,134, **1965**; *Chem. Abstr.* **1965**, *63*, 18029.

30. Squire, E.N. Isomerization of hexafluoropropylene oxide to hexafluoroacetone. *US Patent* 4,302,608, **1981**; *Chem. Abstr.* **1982**, *96*, 68343.

31. Petrov, V.A.; Smart, B.E. Process and catalyst for the preparation of fluoroketones from fluoroepoxides. *US Patent* 5,457,238, **1995**.
32. Krusic, P.J.; Roe, D.C.; Smart, B.E. Kinetics of hexafluoropropylene oxide pyrolysis studied by gas-phase NMR. Kinetic measurements made easy. *Isr. J. Chem.* **1999**, *39*, 117.
33. Sargeant, P.B. Fluorocyclopropanes. I. Preparation and nuclear magnetic resonance spectra. *J. Org. Chem.* **1970**, *35*, 678.
34. Mahler, W.; Resnick, P.R. Reversible difluorocarbene elimination from hexafluoropropylene epoxide. *J. Fluorine Chem.* **1973**, *3*, 451.
35. Dolbier, W.R., Jr.; Sellers, S.F.; Al-Sader, B.H.; Fielder, T.H., Jr.; Elsheimer, S.; Smart, B.E. Thermal isomerizations of fluorinated spiropentanes. *Isr. J. Chem.* **1981**, *21*, 176.
36. Dailey, W.P.; Ralli, P.; Wasserman, D.; Lemal, D.M. 1,2-Cycloadditions to *cis*-5,6-dichlorohexafluorocyclohexa-1,3-dienes. *J. Org. Chem.* **1989**, *54*, 5516.
37. De Pasquale, R.J.; Baucom, K.B.; Patton, J.R.; Brey, W.S., Jr. Perfluoro [cyclobutanespiro-2′-(epoxyethane)]. *Tetrahedron Lett.* **1974**, *15*, 1111.
38. Krespan, C.G.; Brasen, W.R.; Cripps, H.N. Fluorinated sulfide polymers. *Adv. Chem. Ser.* **1972**, *110*, 179.
39. Petrov, V.A. Chemistry of *N*-substituted 2,2-difluoro-3,3-bis(trifluoromethyl)aziridines. *J. Fluorine Chem.* **2000**, *106*, 25.
40. Yang, Z.-Y. Environmentally benign processes for making useful fluorocarbons: nickel- or copper(I) iodide-catalyzed reaction of highly fluorinated epoxides with halogens in the absence of solvent and thermal addition of CF_2I_2 to olefins. *J. Org. Chem.* **2004**, *69*, 2394.
41. Yamauchi, Y.; Katagiri, T.; Uneyama, K. The first generation and stereospecific alkylation of alpha-trifluoromethyl oxiranyl anion. *Org. Lett.* **2002**, *4*, 173.
42. Yamauchi, Y.; Kawate, T.; Katagiri, T.; Uneyama, K. Trifluoromethyl-stabilized optically active oxiranyl and aziridinyl anions for stereospecific syntheses of trifluoromethylated compounds. *Tetrahedron* **2003**, *59*, 9839.
43. Katagiri, T.; Uneyama, K. A chemistry of 2,3-epoxy-1,1,1-trifluoropropane. *J. Fluorine Chem.* **2000**, *105*, 285.
44. Takenaka, N.E.; Hamlin, R.E.; Lemal, D.M. Hexafluorobenzene oxide and hexafluorooxepin. *J. Am. Chem. Soc.* **1990**, *112*, 6715.
45. Takenaka, N.E.; Lemal, D.M. The perfluorobenzene oxide–perfluorooxepin system. In *Synthetic Fluorine Chemistry*; Olah, G.A.; Chambers, R.D.; Prakash, G.K.S., Eds.; Wiley: New York, 1992; p 287.
46. Günther, H. P.M.R. Spectroscopy of unsaturated ring systems. II. Detection and kinetics of benzene oxide–oxepine equilibrium. *Tetrahedron Lett.* **1965**, *6*, 4085.
47. Wirth, D.; Lemal, D.M. A cyclobutadiene oxide (Dewar furan). *J. Am. Chem. Soc.* **1982**, *104*, 847.
48. Woodward, R.B.; Hoffmann, R. The conservation of orbital symmetry. *Angew. Chem., Int. Ed. Engl.* **1969**, *8*, 781.
49. Adam, W.; Liu, J.C.; Rodriguez, O. Bis(trifluoromethyl)acetolactone [synthesis]. *Org. Photochem. Synth.* **1976**, *2*, 13.
50. Adam, W.; Liu, J.C.; Rodriguez, O. Cyclic peroxides. XXIII. Bis-(trifluoromethyl) acetolactone, an unusually stable α-lactone. *J. Org. Chem.* **1973**, *38*, 2269.
51. Lemal, D.M. Perspective on fluorocarbon chemistry. *J. Org. Chem.* **2004**, *69*, 1.

52. Rahman, M.; McKee, M.L.; Shevlin, P.B.; Sztyrbicka, R. Oxygen donation by an intermediate in the reaction of $^3CF_2$ with oxygen. *J. Am. Chem. Soc.* **1988**, *110*, 4002; See also Dils, B.; Carl, S.A.; Peeters, J. Absolute rate coefficient of the reaction of $CF_2(\tilde{a}^3B_1)$ with O_2 between 288 and 600K. *Phys. Chem. Chem. Phys.* **2003**, *5*, 2376.
53. Russo, A.; DesMarteau, D.D. Difluorodioxirane. *Angew. Chem., Int. Ed. Engl.* **1993**, *32*, 905.
54. Huang, Q.; DesMarteau, D.D. The reaction of difluorodioxirane with cesium trifluoromethoxide. *Chem. Commun.* **1999**, 1671.
55. Bierbrauer, K.L.; Codnia, J.; Azcárate, M.L.; Argüello, G.A. New route to the synthesis of difluorodioxirane from FC(O)OF. *Inorg. Chem.* **2001**, *40*, 2922.
56. Sung, K. Substituent effects on stability of oxiranes, oxirenes, and dioxiranes. *Can. J. Chem.* **2000**, *78*, 562.
57. Kraka, E.; Konkoli, Z.; Cremer, D.; Fowler, J.; Schaefer, H.F., III; Difluorodioxirane: an unusual cyclic peroxide. *J. Am. Chem. Soc.* **1996**, *118*, 10595.
58. Casper, B.; Christen, D.; Mack, H.-G.; Oberhammer, H.; Argüello, G.A.; Jülicher, B.; Kronberg, M.; Willner, H. Electron diffraction and vibrational spectra of difluorodioxirane, CF_2O_2. *J. Phys. Chem.* **1996**, *100*, 3983.
59. (a) Curci, R.; Dinoi, A.; Rubino, M.F. Dioxirane oxidations: taming the reactivity–selectivity principle. *Pure Appl. Chem.* **1995**, *67*, 811. (b) Murray, R.W. Dioxiranes. *Chem. Rev.* **1989**, *89*, 1187.
60. Mello, R.; Fiorentino, M.; Sciacovelli, O.; Curci, R. On the isolation and characterization of methyl(trifluoromethyl)dioxirane. *J. Org. Chem.* **1988**, *53*, 3890.
61. Schulz, M.; Liebsch, S.; Kluge, R.; Adam, W. Organo sulfonic peracids. 4. The reaction of arenesulfonylimidazoles with H_2O_2 in the presence of ketones—a new entry to dioxiranes. *J. Org. Chem.* **1997**, *62*, 188.
62. Adam, W.; Curci, R.; González-Nunez, M.E.; Mello, R. Thermally and photochemically initiated radical chain decomposition of ketone-free methyl(trifluoromethyl)dioxirane. *J. Am. Chem. Soc.* **1991**, *113*, 7654.
63. Bach, R.D.; Dmitrenko, O.; Adam, W.; Schambony, S.; *J. Am. Chem. Soc.* **2003**, *125*, 924.
64. D'Accolti, L.; Fusco, C.; Rella, M.R.; Curci, R. Concerning synthesis of ring-A fluorinated anthracyclines. The dioxirane shunt. *Synth. Commun.* **2003**, *33*, 3009.
65. Mello, R.; Ciminale, F.; Fiorentino, M.; Fusco, C.; Prencipe, T.; Curci, R. Oxidations by methyl(trifluoromethyl)dioxirane. 4. Oxyfunctionalization of aromatic hydrocarbons. *Tetrahedron Lett.* **1990**, *31*, 6097.
66. Fusco, C.; Seraglia, R.; Curci, R. Oxyfunctionalization of non-natural targets by dioxiranes. 3. Efficient oxidation of Buckminsterfullerene C_{60} with methyl (trifluoromethyl)dioxirane. *J. Org. Chem.* **1999**, *64*, 8363.
67. D'Accolti, L.; Fiorentino, M.; Fusco, C.; Crupi, P.; Curci, R. Selective oxidation of acetylenic 1,4-diols with dioxiranes in comparison with the methyltrioxorhenium–hydrogen peroxide oxidant. *Tetrahedron Lett.* **2004**, *45*, 8575.
68. Mello, R.; Fiorentino, M.; Fusco, C.; Curci, R. Oxidations by methyl(trifluoromethyl) dioxirane. 2. Oxyfunctionalization of saturated hydrocarbons. *J. Am. Chem. Soc.* **1989**, *111*, 6749.
69. Grabovskiy, S.A.; Antipin, A.V.; Ivanova, E.V.; Dokichev, V.A.; Tomilov, Y.V.; Kabal'nova, N.N. Oxidation of some cage hydrocarbons by dioxiranes. Nature of the transition structure for the reaction of C–H bonds with dimethyldioxirane: a

comparison of B3PW91 density functional theory with experiment. *Org. Biomol. Chem.* **2007**, *5*, 2302.

70. Curci, R.; D'Accolti, L.; Fusco, C. On the hydroxylation of bicyclo[2.1.0]pentane using dioxiranes. *Tetrahedron Lett.* **2001**, *42*, 7087.
71. D'Accolti, L.; Fiorentino, M.; Fusco, C.; Capitelli, F.; Curci, R. Stereoselective dioxirane hydroxylations and the synthesis of tripod boronic acid esters. *Tetrahedron Lett.* **2007**, *48*, 3575.
72. Mello, R.; Cassidei, L.; Fiorentino, M.; Fusco, C.; Curci, R. Oxidations by methyl (trifluoromethyl)dioxirane. 3. Selective polyoxyfunctionalization of adamantane. *Tetrahedron Lett.* **1990**, *31*, 3067.
73. Kazakov, D.V.; Barzilova, A.B.; Kazakov, V.P. A novel chemiluminescence from the reaction of dioxiranes with alkanes. Proposed mechanism of oxygen-transfer chemiluminescence. *Chem. Commun.* **2001**, 191.
74. D'Accolti, L.; Fusco, C.; Annese, C.; Rella, M.R.; Turtletaub, J.S.; Willard, P.G.; Curci, R. Concerning the efficient conversion of epoxy alcohols into epoxy ketones using dioxiranes. *J. Org. Chem.* **2004**, *69*, 8510.
75. Mello, R.; Cassedei, L.; Fiorentino, M.; Fusco, C.; Hümmer, W.; Jäger, V.; Curci, R. Oxidations by methyl(trifluoromethyl)dioxirane. 5. Conversion of alcohols into carbonyl compounds. *J. Am. Chem. Soc.* **1991**, *113*, 2205.
76. Altamura, A.; Curci, R.; Edwards, J.O. Facile regeneration of carbonyl compounds by oxidative cleavage of hydrazones using dioxiranes. *J. Org. Chem.* **1993**, *58*, 7289.
77. González-Nunez, M.E.; Mello, R.; Royo, J.; Asensio, G.; Monzó, I.; Tomás, F.; López, J.G.; Ortiz, F.L. Mechanism of the oxidation of sulfides by dioxiranes: conformational mobility and transannular interaction in the oxidation of thianthrene 5-oxide. *J. Org. Chem.* **2004**, *69*, 9090.
78. (a) González-Nunez, M.E.; Mello, R.; Royo, J.; Ríos, J.V.; Asensio, G. Mechanism of the oxidation of sulfides by dioxiranes: 1. Intermediacy of a 10-*S*-4 hypervalent sulfur adduct. *J. Am. Chem. Soc.* **2002**, *124*, 9154. (b) Asensio, G.; Mello, R.; González-Nunez, M.E. Evidence for the involvement of a sulfurane intermediate in the oxidation of simple sulfides by methyl(trifluoromethyl)dioxirane. *Tetrahedron Lett.* **1996**, *37*, 2299.
79. Adam, W.; Mello, R.; Curci, R. Oxygen atom insertion into silicon–hydrogen bonds with assistance by dioxiranes. A stereospecific and direct rearrangement of silanes to silanols. *Angew. Chem.* **1990**, *102*, 916.
80. Detomaso, A.; Curci, R. Oxidation of natural targets by dioxiranes. Part 4. A novel approach to the synthesis of *N*-hydroxyamino acids using dioxiranes. *Tetrahedron Lett.* **2001**, *42*, 755.
81. Paquette, L.A.; Kreilein, M.M.; Bedore, M.W.; Friedrich, D. Oxidative cleavage of *p*-methoxybenzyl ethers with methyl(trifluoromethyl)dioxirane. *Org. Lett.* **2005**, *7*, 4665.
82. Ferrer, M.; Sánchez-Baeza, F.; Messeguer, A.; Adam, W.; Golsch, D.; Görth, F.; Kiefer, W.; Nagel, V. The release of singlet oxygen in the reaction of dioxiranes with amine *N*-oxides. *Eur. J. Org. Chem.* **1998**, 2527.
83. Adam, W.; Kazakov, D.V.; Kazakov, V.P.; Kiefer, W.; Latypova, R.R.; Schlücker, S. Singlet-oxygen generation in the catalytic reaction of dioxiranes with nucleophilic anions. *Photochem. Photobiol. Sci.* **2004**, *3*, 182.
84. Sander, W. Effects of electron-withdrawing groups on carbonyl *O*-oxides. Trifluoroacetophenone *O*-oxide and hexafluoroacetone *O*-oxide. *J. Org. Chem.* **1988**, *53*, 121.

85. Talbott, R.I.; Thompson, P.G. Fluorinated organic cyclic peroxides. *US Patent* 3,632,606, **1972**.

86. Middleton, W.J. Bentley fluoroalkylthiiranes. *US Patent* 3,136,781, **1964**.

87. Mloston, G.; Gendek, T.; Heimgartner, H. Syntheses of trifluoromethyl-substituted sulfur heterocycles using 3,3,3-trifluoropyruvic acid derivatives. *Helv. Chim. Acta* **1996**, *79*, 1537.

88. Middleton, W.J. Reactions of bis(trifluoromethyl)diazomethane with perfluorothiocarbonyl compounds. *J. Org. Chem.* **1969**, *34*, 3201.

89. Brasen, W.R.; Cripps, H.N.; Bottomley, C.G.; Farlow, M.W.; Krespan, C.G. Tetrafluorothiirane. *J. Org. Chem.* **1965**, *30*, 4188.

90. Rall, K.; Sundermeyer, W. Synthesis of perhalogenated 1,3-dithietane *S*-oxides and thiiranes. *J. Fluorine Chem.* **1990**, *47*, 121.

91. Bekker, R.A.; Popkova, V.Ya.; Knunyants, I.L. Perfluorothiiranes. *Izv. Akad. Nauk SSSR, Ser. Khim.* **1980**, 1692.

92. Popkova, V.Ya.; Galakhov, M.V.; Knunyants, I.L. Reaction of polyhalo disulfides with hexaethylphosphorus triamide. *Izv. Akad. Nauk SSSR, Ser. Khim.* **1989**, 116.

93. Popkova, V.Ya.; German, L.S. Preparation of a new precursor of perfluorinated surfactants: (perfluorohexyl)thiirane. *J. Fluorine Chem.* **1990**, *49*, 159.

94. Wiebe, H.A.; Braslavsky, S.; Heicklen, J. Photolysis of tetra(trifluoromethyl)thiophene vapor. *Can. J. Chem.* **1972**, *50*, 2721.

95. Kobayashi, Y.; Kumadaki, I.; Ohsawa, A.; Sekine, Y.; Ando, A. Reactions of tetrakis (trifluoromethyl)Dewarthiophene. *Heterocycles* **1977**, *6*, 1587.

96. Kobayashi, Y.; Kumadaki, I.; Ohsawa, A.; Sekine, Y.; Mochizuki, H. Organic fluorine compounds. XIX. Synthesis and reactions of Dewar thiophene. *Chem. Pharm. Bull.* **1975**, *23*, 2773.

97. Wiebe, H.A.; Knight, A.R.; Strausz, O.P.; Gunning, H.E. The reactions of sulfur atoms. V. Further studies on the reactions with olefins. *J. Am. Chem. Soc.* **1965**, *87*, 1443.

98. Verkoczy, B.; Sherwood, A.G.; Safarik, I.; Strausz, O.P. Thiirene formation in the reactions of sulfur atoms with alkynes. *Can. J. Chem.* **1983**, *61*, 2268.

99. Bekker, R.A.; Rozov, L.A.; Popkova, V.Ya. Dehydrohalogenation of fluorine-containing sulfenyl chlorides; synthesis and properties of perfluoroisobutylene sulfide. *Izv. Akad. Nauk SSSR, Ser. Khim.* **1983**, 2575.

100. (a) Kobayashi, Y.; Ando, A.; Kumadaki, I. Synthesis of the *N*-unsubstituted Dewar pyrrole 1,2,3,4-tetrakistrifluoromethyl-5-azabicyclo[2.1.0]pent-2-ene. *J. Chem. Soc., Chem. Commun.* **1978**, 509. (b) Kobayashi, Y.; Ando, A.; Kawada, K.; Kumadaki, I. Organic fluorine compounds. 33. Derivation of tetrakis(trifluoromethyl)(Dewar pyrroles) from tetrakis(trifluoromethyl)(Dewar thiophene). *J. Org. Chem.* **1980**, *45*, 2966. See also Laganis, E.D.; Lemal, D.M. 5*H*-(Perfluoropentamethyl)cyclopentadiene, an extraordinary carbon acid. *J. Am. Chem. Soc.* **1980**, *102*, 6633 and Ref. 47.

101. Janulis, E.P., Jr.; Arduengo, A.J., III. Diazotetrakis(trifluoromethyl)cyclopentadiene and ylides of electronegative elements. *J. Am. Chem. Soc.* **1983**, *105*, 3563.

102. Smith, J.R.; Lemal, D.M. Synthetic approaches to a fluorinated tetrahedrane. *J. Fluorine Chem.* **2000**, *102*, 323.

103. Kobayashi, Y.; Kumadaki, I.; Ohsawa, H.; Sekine, Y. Isomerization of Dewar thiophene catalyzed by phosphorus compounds. *Tetrahedron Lett.* **1975**, *16*, 1639.

104. Bushweller, C.H.; Ross, J.A.; Lemal, D.M. Automerization of a Dewar thiophene and its *exo-S*-oxide. *J. Am. Chem. Soc.* **1977**, *99*, 629.

105. Ross, J.A.; Seiders, R.P.; Lemal, D.M. An extraordinarily facile sulfoxide rearrangement. *J. Am. Chem. Soc.* **1976**, *98*, 4325.

106. Rodríguez-Otero, J.; Cabaleiro-Lago, E.M.; Peña-Gallego, A. Theoretical study of the walk rearrangement in perfluorotetramethyl (Dewar thiophene) *exo-S*-oxide. *Tetrahedron* **2007**, *63*, 2191.

107. Kobayashi, Y.; Kawada, K.; Ando, A.; Kumadaki, I. Photoisomerization of bis (trifluoromethyl)thiophenes. *Tetrahedron Lett.* **1984**, *25*, 1917.

108. Kobayashi, Y.; Ando, A.; Kawada, K.; Ohsawa, A.; Kumadaki, I. Organic fluorine compounds. Part 32. A 1,3-dipolar cycloaddition reaction of tetrakis(trifluoromethyl) (Dewar thiophene) and some reactions of the cycloadducts. *J. Org. Chem.* **1980**, *45*, 2962.

109. Laganis, E.D.; Lemal, D.M. Stereospecific 1,3-dipolar cycloelimination in strained pyrazolines. *J. Am. Chem. Soc.* **1980**, *102*, 6634.

110. Khlebnikov, A.F.; Novikov, M.S.; Shinkevich, E.Yu.; Vidovic, D. Selective transannular ring transformations in azirino-fused eight-membered *O,N*- or *S,N*-heterocycles. *Org. Biomol. Chem.* **2005**, *3*, 4040.

111. Crousse, B.; Narizuka, S.; Bonnet-Delpon, D.; Bégué, J.-P. First stereoselective synthesis of *cis* 3-CF_3-aziridine-2-carboxylates. A route to new (trifluoromethyl) α-functionalized β-amino acids. *Synlett* **2001**, 679.

112. Akiyama, T.; Ogi, S.; Fuchibe, K. Stereoselective synthesis of CF_3-substituted aziridines by Lewis acid-mediated aziridination of aldimines with diazoacetates. *Tetrahedron Lett.* **2003**, *44*, 4011.

113. Zeifman, Yu.V.; Lantseva, L.T. Synthesis of 1-phenyl-2,2-difluoro-3,3-bis(trifluoromethyl)aziridine. *Izv. Akad. Nauk SSSR, Ser. Khim.* **1986**, 248.

114. Kobayashi, Y.; Fujino, S.; Hamana, H.; Hanzawa, Y.; Morita, S.; Kumadaki, I. Studies on organic fluorine compounds. 36. Synthesis and reactions of trifluoromethylated diphosphabenzvalene. *J. Org. Chem.* **1980**, *45*, 4683.

115. Burley, G.A.; Darwish, A.D.; Street, J.M.; Taylor, R. Novel formation of a fluorinated aziridino[60]fullerene. *Tetrahedron Lett.* **2004**, *45*, 3617.

116. Fioravanti, S.; Colantoni, D.; Pellacani, L.; Tardella, P.A. Aziridines versus vinyl carbamates from the direct amination of electron-withdrawing group-substituted trifluoromethyl enoates. *J. Org. Chem.* **2005**, *70*, 3296.

117. Kano, N.; Daicho, Y.; Kawashima, T. A potential intermediate for the aza-Corey–Chaykovsky reaction: synthesis, structure, and thermolysis of a pentacoordinate 1,2-thiazetidine 1-oxide. *Org. Lett.* **2006**, *8*, 4625.

118. Mae, M.; Matsuura, M.; Amii, H.; Uneyama, K. A facile synthesis of difluoromethylaziridines. *Tetrahedron Lett.* **2002**, *43*, 2069.

119. Yamauchi, Y.; Kawate, T.; Itahashi, H.; Katagiri, T.; Uneyama, K. Generation and reactions of α-trifluoromethyl stabilized aziridinyl anion, a general synthetic precursor for stereospecific construction of α-amino-α-trifluoromethylated quaternary carbon. *Tetrahedron Lett.* **2003**, *44*, 6319.

120. Tanaka, K.; Ohsuga, M.; Sugimoto, Y.; Okafuji, Y.; Mitsuhashi, K. Applications of the fluorinated 1,3-dipolar compounds as the building blocks of the heterocycles with fluorine groups. Part XII. Synthesis of trifluoromethylisoxazolines and their rearrangement into trifluoromethylaziridines. *J. Fluorine Chem.* **1988**, *39*, 39.

121. Zimmerman, H.E. Moebius–Hueckel concept in organic chemistry. Application of organic molecules and reactions. *Acc. Chem. Res.* **1971**, *4*, 272.

122. Hees, U.; Vogelbacher, U.-J.; Michels, G.; Regitz, M. Antiaromatic compounds. 24. Steric effects on valence isomerizations in the Dewar pyridine/azaprismane/pyridine system. *Tetrahedron* **1989**, *45*, 3115.

123. Coe, P.L.; Cook, M.I. Reactions of tetrafluoroethene oligomers. Part 16. The synthesis and reactions of perfluoro[2,3-dimethyl-2-(3-methylpentan-3-yl)]-2*H*-azirine: a source of 2,3-bis(trifluoromethyl)aziridine derivatives. *J. Chem. Soc., Perkin Trans. 1* **2000**, 1537.

124. Palacios, F.; Ochoa de Retana, A.M.; Alonso, J.M. Regioselective synthesis of fluoroalkylated β-aminophosphorus derivatives and aziridines from phosphorylated oximes and nucleophilic reagents. *J. Org. Chem.* **2006**, *71*, 6141.

125. Del'tsova, D.P.; Gambaryan, N.P.; Lur'e, E.P. Fluorine-containing heterocumulenes. 19. Synthesis and reactions of 2-perfluoro-*tert*-butyl-3,3-diphenyloxaziridine. *Izv. Akad. Nauk SSSR, Ser. Khim.* **1979**, 1788.

126. Katagiri, T.; Takahashi, M.; Fujiwara, Y.; Ihara, H.; Uneyama, K. General syntheses of optically active α-trifluoromethylated amines via ring-opening reactions of *N*-benzyl-2-trifluoromethylaziridine. *J. Org. Chem.* **1999**, *64*, 7323.

127. Quinze, K.; Laurent, A.; Mison, P. Synthesis of five-membered nitrogen heterocycles bearing a trifluoromethyl substituent. *J. Fluorine Chem.* **1989**, *44*, 233.

128. Tanaka, K.; Nagatani, S.; Ohsuga, M.; Mitsuhashi, K. Regio- and stereoselectivity of cycloadditions of trifluoromethylated azomethine ylide. *Bull. Chem. Soc. Jpn.* **1994**, *67*, 589.

129. Del'tsova, D.P.; Gambaryan, N.P. Isomerization of a perfluoro-α-lactam to a substituted carbamoyl fluoride. *Izv. Akad. Nauk SSSR, Ser. Khim.* **1984**, 1367.

130. Del'tsova, D.P.; Gambaryan, N.P.; Mysov, E.I.; Knunyants, I.L. Synthesis of four-membered heterocycles from a perfluoro-α-lactam. *Dokl. Akad. Nauk SSSR* **1980**, *253*, 886.

131. Del'tsova, D.P.; Gambaryan, N.P.; Mysov, E.I. Synthesis of five-membered heterocycles from perfluoro-α-lactam. *Izv. Akad. Nauk SSSR, Ser. Khim.* **1981**, 2343.

132. Banks, R.E.; Moore, G.J. Azide chemistry. I. Synthesis of perfluoropropenyl azide and its conversion into perfluoro(2- and 3-methyl-2*H*-azirine). *J. Chem. Soc. (C)* **1966**, 2304.

133. Cleaver, C.S.; Krespan, C.G. Fluoroazirines. Synthesis and polymerisation. *J. Am. Chem. Soc.* **1965**, *87*, 3716.

134. Krespan, C.G. Fluorinated cyanates and isocyanates. New type of rearrangement. *J. Org. Chem.* **1969**, *34*, 1278.

135. Haddach, M.; Pastor, R.; Riess, J.G. Abnormal stability and reactivity of perfluoroalkyl azirines and aziridines. *Tetrahedron Lett.* **1990**, *31*, 1989.

136. Haddach, M.; Pastor, R.; Riess, J.G. Synthesis and reactivity of novel synthons of perfluoroalkyls. A typical behavior of *F*-alkyl azirines and aziridines. *Tetrahedron* **1993**, *49*, 4627.

137. Alonso-Cruz, C.R.; Kennedy, A.R.; Rodríguez, M.S.; Suárez, E. Alkoxyl radical fragmentation of 3-azido-2,3-dideoxy-2-halo-hexopyranoses: a new entry to chiral poly-hydroxylated 2-azido-1-halo-1-alkenes. *Tetrahedron Lett.* **2007**, *48*, 7207.

138. Quinze, K.; Laurent, A.; Mison, P. Synthesis of secondary aziridines substituted with a trifluoromethyl group. *J. Fluorine Chem.* **1989**, *44*, 211.

139. Gritsan, N.P.; Gudmundsdottir, A.D.; Tigelaar, D.; Zhu, Z.; Karney, W.L.; Hadad, C.M.; Platz, M.S. A laser flash photolysis and quantum chemical study of the fluorinated derivatives of singlet phenylnitrene. *J. Am. Chem. Soc.* **2001**, *123*, 1951.

140. Karney, W.L.; Borden, W.T. Why does *o*-fluorine substitution raise the barrier to ring expansion of phenylnitrene? *J. Am. Chem. Soc.* **1997**, *119*, 3347.

141. Calvo-Losada, S.; Quirante, J.J.; Suárez, D.; Sordo, T.L. Rearrangement of azirine intermediates to nitriles: theoretical study of cleavage of 3,4-dihydro-1α*H*-azirine[2,3-*c*]pyrrol-2-one to cyanoketene–formaldimine complex. *J. Comput. Chem.* **1998**, *19*, 912.

142. Mayer, T.; Maier, M.E. Design and synthesis of a tag-free chemical probe for photoaffinity labeling. *Eur. J. Org. Chem.* **2007**, 4711.

143. Kostyanovsky, R.G.; Shustov, G.V.; Zaichenko, N.L. Asymmetrical nonbridgehead nitrogen. XXIII. Chiral 3,3-bis(trifluoromethyl)diaziridines. *Tetrahedron* **1982**, *38*, 949.

144. Firth, W.C., Jr. Reactions of pentafluoroguanidine and tetrafluoroformamidine with alkali metal fluorides. Fluorine-containing diaziridines. *J. Org. Chem.* **1968**, *33*, 3489.

145. Chang, S.-C.; DesMarteau, D.D. Fluoride-catalyzed reactions of perfluoromethanimine. Novel chemistry of the perfluoromethanamine ion. *J. Org. Chem.* **1983**, *48*, 771.

146. Chang, S.-C.; DesMarteau, D.D. Perfluoromethanamine ion. *Polyhedron* **1982**, *1*, 129.

147. Bauknight, C.W., Jr.; DesMarteau, D.D. Reactions of *N*-bromodifluoromethanimine. *J. Org. Chem.* **1988**, *53*, 4443.

148. Zheng, Y.Y.; Baukknight, C.W., Jr.; DesMarteau, D.D. Some novel reactions of *N*-chlorodifluoromethanimine. *J. Org. Chem.* **1984**, *49*, 3590.

149. Petrov, V.A.; DesMarteau, D.D. New reactions of perfluorinated *N*-fluoroalkanimines. *J. Fluorine Chem.* **1998**, *89*, 111.

150. (a) Bentz, E.L.; Gibson, H.; Hudson, C.; Moloney, M.G.; Seldon, D.A.; Wearmouth, E.S. Aryldiazirine-modified pyroglutamates: photoaffinity labels for glutamate. *Synlett* **2006**, 247. (b) Biasotti, B.; Dallavalle, S.; Merlini, L.; Farina, C.; Gagliardi, S.; Parini, C.; Belfiore, P. Synthesis of photoactivable inhibitors of osteoclast vacuolar ATPase. *Bioorg. Med. Chem.* **2003**, *11*, 2247. (c) Sammelson, R.E.; Casida, J.E. Synthesis of a tritium-labeled, fipronil based, highly potent, photoaffinity probe for the GABA receptor. *J. Org. Chem.* **2003**, *68*, 8075. (d) Lu, X.; Cseh, S.; Byun, H.-S.; Tigyi, G.; Bittman, R. Total synthesis of two photo-activatable analogues of the growth-factor-like mediator sphingosine 1-phosphate: differential interaction with protein targets. *J. Org. Chem.* **2003**, *68*, 7046.

151. Sridar, C.; Kobayashi, Y.; Brevig, H.; Kent, U.M.; Puppali, S.G.; Rimoldi, J.M.; Hollenberg, P.F. Synthesis of substituted phenyl diaziridines and characterization as mechanism-based inactivators of human cytochrome P450 2B6. *Drug Metab. Dispos.* **2006**, *34*, 1849.

152. Graham, W.H. Halogenation of amidines. I. Synthesis of 3-halo- and other negatively substituted diazirines. *J. Am. Chem. Soc.* **1965**, *87*, 4396.

153. Grayston, M.W.; Lemal, D.M. Perfluorohexamethylbicyclopropenyl. *J. Am. Chem. Soc.* **1976**, *98*, 1278.

154. Gale, D.M.; Middleton, W.J.; Krespan, C.G. Perfluorodiazo compounds. *J. Am. Chem. Soc.* **1966**, *88*, 3617.

155. Minasyan, R.B.; Rokhlin, E.M.; Gambaryan, N.P.; Zeifman, Y.V.; Knunyants, I.L. Bis(trifluoromethyl)cyclodiazomethane. *Izv. Akad. Nauk SSSR, Ser. Khim.* **1965**, 761.

156. Mitsch, R.A. Difluorodiazirine. VI. Synthesis of difluorodiazirine. *J. Heterocycl. Chem.* **1966**, *3*, 245.

157. Mitsch, R.A. Organic fluoronitrogens. X. Reductive defluorination–cyclization. *J. Org. Chem.* **1968**, *33*, 1847.

158. Meyers, M.D.; Frank, S. Difluorocyanamide. *Inorg. Chem.* **1966**, *5*, 1455.

159. Cox, D.P.; Moss, R.A.; Terpinski, J. Exchange reactions of halodiazirines. Synthesis of fluorodiazirines. *J. Am. Chem. Soc.* **1983**, *105*, 6513.

160. Moss, R.A.; Terpinski, J.; Cox, D.P.; Denney, D.Z.; Krogh-Jespersen, K. Azide and fluoride exchange reactions of halodiazirines. *J. Am. Chem. Soc.* **1985**, *107*, 2743.

161. Moss, R.A.; Fedorynski, M.; Kmiecik-Lawrynowicz, G.; Terpinski, J. Exchange reactions of diazirines: 3-X-3-methyldiazirines and derived carbenes. *Tetrahedron Lett.* **1986**, *27*, 2707.

162. Dailey, W.P. 3-Fluoro-3-(trifluoromethyl)diazirine. *Tetrahedron Lett.* **1987**, *28*, 5801.

163. Bainbridge, K.E.; Dailey, W.P. The mechanism of the diazirine exchange reaction with azide anion. *Tetrahedron Lett.* **1989**, *30*, 4901.

164. Creary, X.; Sky, A.F. Reaction of arylbromodiazirines with azide ion. Evidence for *N*-azidodiazirine intermediates. *J. Am. Chem. Soc.* **1990**, *112*, 368.

165. (a) Moss, R.A.; Xue, S.; Sauers, R.R. A remarkably specific π-mediated rearrangement of benzocyclobutenylfluorocarbene. *J. Am. Chem. Soc.* **1996**, *118*, 10307. (b) Moss, R.A.; Maksimocic, L.; Merrer, D.C. Benzylfluorocarbene: reactions and kinetics. *Tetrahedron Lett.* **1997**, *40*, 7049.

166. Laganis, E.D.; Janik, D.S.; Curphey, T.J.; Lemal, D.M. Photochemistry of perfluoro-3-diazo-2-butanone. *J. Am. Chem. Soc.* **1983**, *105*, 7457.

167. Recent examples: (a) Sato, S.; Yamada, M.; Wakahara, T.; Tsuchiya, T.; Ishitsuka, M.O.; Akasaka, T.; Liu, M.T.H. Photo-labeling of C_{60} with 3-trifluoromethyl-3-phenyldiazirine. *Tetrahedron Lett.* **2007**, *48*, 6290. (b) Zumbuehl, A.; Stano, P.; Sohrmann, M.; Peter, M.; Walde, P.; Carreira, E.M. A novel strategy for bioconjugation: synthesis and preliminary evaluation with amphotericin B. *Org. Biomol. Chem.* **2007**, *5*, 1339. (c) Lee, J.-S.; Han, S.Y.; Kim, M.S.; Yu, C.-M.; Kim, M.H.; Kim, S.H.; Min, Y.K.; Kim, B.T. Synthesis of novel chemical probes for the study of tanshinone binding proteins. *Bioorg. Med. Chem. Lett.* **2006**, *16*, 4733. (d) Hashimoto, M.; Kato, Y.; Hatanaka, Y. Simple method for the introduction of iodo-label on (3-trifluoromethyl)phenyldiazirine for photoaffinity labeling. *Tetrahedron Lett.* **2006**, *47*, 3391. (e) Sanderson, J.M.; Findlay, J.B.C.; Fishwick, C.W.G. Preparation of novel 3*H*-trifluoromethyldiazirine-based photoactivatable potassium channel antagonists. *Tetrahedron* **2005**, *61*, 11244. (f) Hashimoto, M.; Komano, T.; Nabeta, K.; Hatanaka, Y. Stereoselective synthesis of (*E*)- and (*Z*)-β-bromostyrene containing trifluoromethyldiazirine for photoaffinity labeling. *Chem. Pharm. Bull.* **2005**, *53*, 140.

168. Recent examples: (a) Husain, S.S.; Nirthanan, S.; Ruesch, D.; Solt, K.; Cheng, Q.; Li, G.-D.; Arevalo, E.; Olsen, R.W.; Raines, D.E.; Forman, S.A.; Cohen, J.B.; Miller, K. W. Synthesis of trifluoromethylaryl diazirine and benzophenone derivatives of etomidate that are potent general anesthetics and effective photolabels for probing sites on ligand-gated ion channels. *J. Med. Chem.* **2006**, *49*, 4818. (b) Bender, T.; Huss, M.; Wieczorek, H.; Groond, S.; von Zezschwitz, P. Convenient synthesis of a [1-^{14}C]diazirinylbenzoic acid as a photoaffinity label for binding studies of V-ATPase inhibitors. *Eur. J. Org. Chem.* **2007**, 3870. (c) Kuroda, T.; Suenaga, K.; Sakakura, A.; Handa, T.; Okamoto, K.; Kigoshi, H. Study of the interaction between actin and antitumor substance aplyronine A with a

novel fluorescent photoaffinity probe. *Bioconj. Chem.* **2006**, *17*, 524. (d) Oberfeld, B.; Brunner, J.; Dimroth, P. Phospholipids occupy the internal lumen of the c ring of the ATP synthase of *Escherichia coli*. *Biochemistry* **2006**, *45*, 1841. (e) Qiu, W.-W.; Xu, J.; Liu, D.-Z.; Li, J.-Y.; Ye, Y.; Zhu, X.-Z.; Li, J.; Nan, F.-J. Design and synthesis of a biotin-tagged photoaffinity probe of paeoniflorin. *Bioorg. Med. Chem. Lett.* **2006**, *16*, 3306. (f) Xi, J.; Liu, R.; Rossi, M.J.; Yang, J.; Loll, P.J.; Dailey, W.P.; Eckenhoff, R.G. Photoactive analogues of the haloether anesthetics provide high-resolution features from low-affinity interactions. *ACS Chem. Biol.* **2006**, *1*, 377. (g) Fillion, D.; Deraët, M.; Holleran, B.J.; Escher, E. Stereospecific synthesis of a carbene-generating angiotensin II analogue for comparative photoaffinity labeling: improved incorporation and absence of methionine selectivity. *J. Med. Chem.* **2006**, *49*, 2200. (h) Riber, D.; Venkataramana, M.; Sanyal, S.; Duvold, T. Synthesis and biological evaluation of photoaffinity labeled fusidic acid analogues. *J. Med. Chem.* **2006**, *49*, 1503. (i) Nakashima, H.; Hashimoto, M.; Sadakane, Y.; Tomohiro, T.; Hatanaka, Y. Simple and versatile method for tagging phenyldiazirine photophores. *J. Am. Chem. Soc.* **2006**, *128*, 15092. (j) Hashimoto, M.; Hatanaka, Y. Positively coded photoaffinity label for altering isoelectric points of proteins. *Bioorg. Med. Chem. Lett.* **2006**, *16*, 5998. (k) Park, J.-J.; Dadakane, Y.; Masuda, K.; Tomohiro, T.; Nakano, T.; Hatanaka, Y. Synthesis of diazirinyl photoprobe carrying a novel cleavable biotin. *ChemBioChem* **2005**, *6*, 814. (l) Hashimoto, M.; Hatanaka, Y. Post-biotinylation of photocrosslinking by Staudinger–Bertozzi ligation of preinstalled alkylazide tag. *Chem. Pharm. Bull.* **2005**, *53*, 1510. (m) Fujii, T.; Manabe, Y.; Sugimoto, T.; Ueda, M. Detection of 210 kDa receptor protein for a leaf-movement factor by using novel photoaffinity probes. *Tetrahedron* **2005**, *61*, 7874. (n) Chattopadhaya, S.; Chan, E.W.S.; Yao, S.Q. An affinity-based probe for the proteomic profiling of aspartic proteases. *Tetrahedron Lett.* **2005**, *46*, 4053. (o) Hosoya, T.; Hiramatsu, T.; Ikemoto, T.; Aoyama, H.; Ohmae, T.; Endo, M.; Suzuki, M. Design of dantrolene-derived probes for radioisotope-free photoaffinity labeling of proteins involved in the physiological Ca^{2+} release from sarcoplasmic reticulum of skeletal muscle. *Bioorg. Med. Chem. Lett.* **2005**, *15*, 1289.

169. O'Gara, J.E.; Dailey, W.P. Direct observation, reactions under matrix-isolation conditions, and *ab initio* calculations for halo(trifluoromethyl)carbenes. Evidence for the photochemical addition of a carbene to dinitrogen. *J. Am. Chem. Soc.* **1992**, *114*, 3581.
170. Brahms, D.L.S.; Dailey, W.P. Fluorinated carbenes. *Chem. Rev.* **1996**, *96*, 1585.
171. (a) Zuev, P.S.; Sheridan, R.S.; Albu, T.V.; Truhlar, D.G.; Hrovat, D.A.; Borden, W.T. Carbon tunneling from a single quantum state. *Science* **2003**, *299*, 867. (b) Wang, J.; Sheridan, R.S. A singlet aryl-CF_3 carbene: 2-benzothienyl(trifluoromethyl)carbene and interconversion with a strained cyclic allene. *Org. Lett.* **2007**, *9*, 3177. (c) Enyo, T.; Arai, N.; Nakane, N.; Nicolaides, A.; Tomioka, H. *o*-Phenylene halocarbenonitrenes and *o*-phenylene chlorocarbenocarbene: a combined experimental and computational approach. *J. Org. Chem.* **2005**, *70*, 7744. (d) Enyo, T.; Nicolaides, A.; Tomioka, H. Halogen derivatives of *m*-phenylene(carbeno)nitrene: a switch in ground-state multiplicity. *J. Org. Chem.* **2002**, *67*, 5578. (e) Nicolaides, A.; Enyo, T.; Miura, D.; Tomioka, H. *p*-Phenylenecarbenonitrene and its halogen derivatives: how does resonance interaction between a nitrene and a carbene center affect the overall electronic configuration? *J. Am. Chem. Soc.* **2001**, *123*, 2628. (f) Zuev, P.; Sheridan, R.S. Substituent switching of biscarbene electronic configurations: *p*-phenylenebis (fluoromethylene). *J. Am. Chem. Soc.* **1994**, *116*, 9381. (g) McMahon, R.J.; Abelt, C. J.; Chapman, O.L.; Johnson, J.W.; Kreil, C.L.; LeRoux, J.-P.; Mooring, A.M.; West, P.R.

1,2,4,6-Cycloheptatetraene: the key intermediate in arylcarbene interconversions and related C_7H_6 rearrangements. *J. Am. Chem. Soc.* **1987**, *109*, 2456.

172. (a) Moss, R.A.; Wang, L.; Sauers, R.R. Presumptive evidence for an intermediate oxirane in the reaction of phenylfluorocarbene with cyclohexenone. *Tetrahedron Lett.* **2007**, *48*, 5551. (b) Moss, R.A.; Tian, J.; Sauers, R.R.; Skalit, C.; Krogh-Jespersen, K. Chlorofluorocarbene: first UV observation of a dihalocarbene in solution. *Org. Lett.* **2007**, *9*, 4053. (c) Moss, R.A.; Tian, J. Modulation of selectivity in a fluorocarbene cyclopropanation reaction: a catalytic role for bromide ion. *Tetrahedron Lett.* **2006**, *47*, 3419. (d) Makihara, T.; Nojima, T.; Ishiguro, K.; Sawaki, Y. Spin selectivity in the oxygenation of singlet phenylhalocarbenes with oxygen. *Tetrahedron Lett.* **2003**, *44*, 865. (e) Ge, C.-S.; Jang, E.G.; Jefferson, E.A.; Liu, W.; Moss, R.A.; Wlostowska, J.; Xue, S. The kinetic range of carbene–pyridine ylide forming reactions. *J. Chem. Soc., Chem. Commun.* **1994**, 1479. (f) Chidester, W.; Modarelli, D.A.; White, W.R., III; Whitt, D.E.; Platz, M.S. Exploratory photochemistry of alkylbromo- and alkylfluorodiazirines. Excited-state hydrogen migration and carbene formation. *J. Phys. Org. Chem.* **1994**, *7*, 24. (g) Moss, R.A.; Ho, G.-J.; Liu, W.; Sierakowski, C. An extraordinary isotope effect in a carbene rearrangement. *Tetrahedron Lett.* **1992**, *33*, 4287. (h) Moss, R.A.; Ho, G.-J.; Sierakowski, C. Direct observation of the formation and rearrangement of carbene/allyl sulfide ylides. *J. Am. Chem. Soc.* **1992**, *114*, 3128. (i) Moss, R.A.; Jang, E.C.; Kim, H.-R.; Ho, G.-J.; Baird, M.S. Pyridylhalocarbenes and pyridinium halocarbenes. *Tetrahedron Lett.* **1992**, *33*, 1427. (j) Moss, R.A.; Fan, H.; Gurumurthy, R.; Ho, G.-J. Reactions of phenylfluorocarbene with lithium salts. Absolute kinetics of carbenoid formation. *J. Am. Chem. Soc.* **1991**, *113*, 1435. (k) Moss, R.A. Carbenic reactivity revisited. *Acc. Chem. Res.* **1989**, *22*, 15. (l) Turro, N.J.; Okamoto, M.; Gould, I.R.; Moss, R.A.; Lawrynowicz, W.; Hadel, L.M. Volumes of activation for the cycloaddition reactions of phenylhalocarbenes to alkenes. *J. Am. Chem. Soc.* **1987**, *109*, 4973. (m) Moss, R.A.; Lawrynowicz, W.; Turro, N. J.; Gould, I.R.; Cha, Y. Activation parameters for the additions of arylhalocarbenes to alkenes. *J. Am. Chem. Soc.* **1986**, *108*, 7028. (n) Cox, D.P.; Gould, I.R.; Hacker, N.P.; Moss, R.A.; Turro, N.J. Absolute rate constants for the additions of halophenylcarbenes to alkenes; a reactivity–selectivity relation. *Tetrahedron Lett.* **1983**, *24*, 5313.

173. Liu, X.; Chu, G.; Moss, R.A.; Sauers, R.R.; Warmuth, R. Fluorophenoxycarbene inside a hemicarcerand: a bottled singlet carbene. *Angew. Chem., Int. Ed.* **2005**, *44*, 1994.

174. Merrer, D.C.; Moss, R.A. In *Advances in Carbene Chemistry*; Brinker, U.H., Ed.; Elsevier: Amsterdam, 2001; Vol. 3, p 53f.

175. Gale, D.M.; Middleton, W.J.; Krespan, C.G.; *J. Am. Chem. Soc.* **1966**, *88*, 3617.

176. Barton, D.H.R.; Jaszberenyi, J.Cs.; Theodorakis, E.A.; Reibenspies, J.H. The invention of radical reactions. 30. Diazirines as carbon radical traps. Mechanistic aspects and synthetic applications of a novel and efficient amination process. *J. Am. Chem. Soc.* **1993**, *115*, 8050.

177. Barton, D.H.R.; Jaszberenyi, J.Cs.; Theodorakis, E.A. Nitrogen transfer to carbon radicals. *J. Am. Chem. Soc.* **1992**, *114*, 5904.

178. Petrov, V.A.; Resnati, G. Polyfluorinated oxaziridines: synthesis and reactivity. *Chem. Rev.* **1996**, *96*, 1809.

179. Falardeau, E.R.; DesMarteau, D.D. Direct synthesis of fluorinated peroxides. 6. The addition of fluorinated hydroperoxides to perfluoro-2-azapropene and the preparation of the first perfluorooxazirine. *J. Am. Chem. Soc.* 1976, *98*, 3529.

180. Sekiya, A.; DesMarteau, D.D. Reaction of metal fluorides with $CF_3OOCF_2N(H)CF_3$. *Inorg. Chem.* **1979**, *18*, 919.

181. Bragante, L.; DesMarteau, D.D. The chemistry of 3-(trifluoromethyl)perfluoroaza-2-butene and the synthesis of a new oxaziridine: 3,3-bis(trifluoromethyl)-2-(trifluoromethyl)oxaziridine. *J. Fluorine Chem.* **1991**, *53*, 181.

182. Ratcliffe, C.T. Epoxides. *US Patent* 4,287,128, **1981**; *Chem. Abstr.* **1982**, *96*, 6550h.

183. Navarrini, W.; DesMarteau, D.D. Preparation of perfluoroalkylaminooxaziridines as monomers and intermediates for nitrons and photopolymerization initiators. *US Patent* 4,874,875, **1989**; *Chem. Abstr.* **1990**, *112*, 159140y.

184. (a) Petrov, V.A.; DesMarteau, D.D. New methods for the synthesis of perfluorooxaziridines. Preparation of *cis*-perfluoro-2,3-dialkyloxaziridines. *Mendeleev Commun.* **1993**, 87. (b) Petrov, V.A.; DesMarteau, D.D.; Bragante, L. Preparation of perfluorooxaziridines. *Eur. Pat. Appl.* EP 496414, **1992**; *Chem. Abstr.* **1992**, *117*, 191830n.

185. (a) Petrov, V.A.; DesMarteau, D.D. A new method for the synthesis of perfluorooxaziridines. Preparation of perfluoro-*cis*-2,3-dialkyloxaziridines. *J. Org. Chem.* **1993**, *58*, 4754. (b) Petrov, V.A.; DesMarteau, D.D.; Navarrini, W. Preparation of perfluorooxaziridines. *Eur. Pat. Appl.* EP 496413, **1992**; *Chem. Abstr.* **1992**, *117*, 191831.

186. Petrov, V.A.; DesMarteau, D.D. Polyfluoro *N*-aryloxaziridines. Synthesis and thermal rearrangement. *J. Fluorine Chem.* **1996**, *77*, 175.

187. Petrov, V.A.; Mlsna, T.E.; DesMarteau, D.D. Synthesis of polyfluoro-2-alkanesulfonyl-3,3-dialkyloxaziridines. *J. Fluorine Chem.* **1994**, *68*, 277.

188. Davis, F.A.; Weismiller, M.C.; Murphy, C.K.; Reddy, R.T.; Chen, B.-C. Chemistry of oxaziridines. 18. Synthesis and enantioselective oxidations of the [(8,8-dihalocamphoryl) sulfonyl]oxaziridines. *J. Org. Chem.* **1992**, *57*, 7274. See also Davis, F.A.; Reddy, R.T.; Han, W.; Carroll, P.J. Chemistry of oxaziridines. 17. *N*-(Phenylsulfonyl)(3,3-dichlorocamphoryl)oxaziridine: a highly efficient reagent for the asymmetric oxidation of sulfides to sulfoxides. *J. Am. Chem. Soc.* **1992**, *114*, 1428.

189. Arnone, A.; DesMarteau, D.D.; Novo, B.; Petrov, V.A.; Pregnolato, M.; Resnati, G. Selective epoxidation of olefins by perfluoro-*cis*-2,3-dialkyloxaziridines. *J. Org. Chem.* **1996**, *61*, 8805.

190. Arnone, A.; Cavicchioli, M.; Montanari, V.; Resnati, G. Direct oxyfunctionalization at unactivated sites. Synthesis of 5β-hydroxysteroids by perfluorodialkyloxaziridines. *J. Org. Chem.* **1994**, *59*, 5511.

191. Arnone, A.; Foletto, S.; Metrangelo, P.; Pregnolato, M.; Resnati, G. Highly enantiospecific oxyfunctionalization of nonactivated hydrocarbon sites by perfluoro-*cis*-2-*n*-butyl-3-*n*-propyloxaziridine. *Org. Lett.* **1999**, *1*, 281.

192. Wahl, F.; Weiler, A.; Landenberger, P.; Sackers, E.; Voss, T.; Haas, A.; Lieb, M.; Hunkler, D.; Wörth, J.; Knothe, L.; Prinzbach, H. Towards perfunctionalized dodecahedranes: *en route* to C_{20} fullerene. *Chem. Eur. J.* **2006**, *12*, 6255.

193. DesMarteau, D.D.; Petrov, V.A.; Montanari, V.; Pregnolato, M.; Resnati, G. Oxidation of alcohols by perfluoro-*cis*-2,3-dialkyloxaziridines. *Tetrahedron Lett.* **1992**, *33*, 7245.

194. Arnone, A.; Bernardi, R.; Cavicchioli, M.; Resnati, G. Perfluoro-*cis*-2,3-dialkyloxaziridines: effective reagents for the selective oxidation of ethers to carbonyl compounds. *J. Org. Chem.* **1995**, *60*, 2314.

195. Cavicchioli, M.; Montanari, V.; Resnati, G. Oxyfunctionalization reactions by perfluoro *cis*-2,3-dialkyloxaziridines. Enantioselective conversion of silanes into silanols. *Tetrahedron Lett.* **1994**, *35*, 6329.

196. DesMarteau, D.D.; Petrov, V.A.; Montanari, V.; Pregnolato, M.; Resnati, G. Mild and selective oxygenation of sulfides to sulfoxides and sulfones by perfluoro-*cis*-2,3-dialkyloxaziridines. *J. Org. Chem.* **1994**, *59*, 2762.

197. Arnone, A.; Novo, B.; Pregnolato, M.; Resnati, G.; Terreni, M. Conversion of thio- and selenophosphoryl into phosphoryl group by perfluoro *cis*-2,3-dialkyloxaziridines. *J. Org. Chem.* **1997**, *62*, 6401.

198. Bernardi, R.; Novo, B.; Resnati, G. Reactivity of perfluoro-(*cis*-2,3-dialkyloxaziridines) with heteroaromatic nitrogen compounds. *J. Chem. Soc., Perkin Trans. 1* **1996**, 2517.

199. (a) Sekiya, A.; DesMarteau, D.D. The reaction of 2-(trifluoromethyl)-3,3-difluorooxaziridine with nucleophiles. *J. Fluorine Chem.* **1979**, *14*, 289. (b) Sekiya, A.; DesMarteau, D.D. Reaction of 2-trifluoromethyl-3,3-difluorooxaziridine with some fluorinated nucleophiles. *J. Org. Chem.* **1979**, *44*, 1131.

200. O'Brien, B.A.; Lam, W.Y.; DesMarteau, D.D. Cycloaddition and oxygen-transfer reactions of 2-(trifluoromethyl)-3,3-difluorooxaziridine. *J. Org. Chem.* **1986**, *51*, 4466.

201. Lam, W.Y.; DesMarteau, D.D. Unusual cycloaddition reactions with 2-(trifluoromethyl)-3,3-difluorooxaziridine. *J. Am. Chem. Soc.* **1982**, *104*, 4034.

202. Misna, T.E.; Young, J.A.; DesMarteau, D.D. Synthesis and chemistry of novel perhalogenated imines, oxaziridines, and oxazolidines. *Z. Anorg. Allgem. Chem.* **2002**, *628*, 1789.

203. Petrov, V.A.; DesMarteau, D.D. Some novel reactions of perfluoro-2,3-dialkyloxaziridines. *J. Chem. Soc., Perkin Trans. 1* **1993**, 505.

2

FLUORINATED FOUR-MEMBERED HETEROCYCLES

ROMAN B. LARICHEV AND VIACHESLAV A. PETROV

DuPont Central Research and Development, Experimental Station, P.O. Box 80500, Wilmington, DE 19880-0500, USA

2.1 FLUORINATED OXETANES

Preparation of *F*-oxetane (**1**), disclosed in the patent by Kauck and Simons in 1952,[1] was the first report on synthesis of fluorinated oxetane. It was obtained from parent hydrocarbon by electrochemical fluorination in anhydrous HF and isolated as nonflammable, water-insoluble, colorless gas with boiling point of −38°C. Much later, a direct fluorination by fluorine gas was employed to convert 2,2,3,3-tetrafluorooxetane (**2**) into *F*-oxetane (**1**) or 2,2,3,3,4-pentafluorooxetane (Scheme 2.1).[2] Varying the conditions of the fluorination, each of the two can be obtained as the major product.

Formation of the four-membered ring of fluorinated oxetanes as for their hydrocarbon congeners is normally accomplished by one of the two general approaches: the cyclization of γ-substituted alcohols or [2 + 2] cycloaddition reaction.

Alcohols with a good leaving group in the γ-position can be cyclized into oxetanes. This approach was successfully applied to the synthesis of fluorinated oxetanes either by electrophilic cyclization of 2-perfluoroalkyl 1,3-diols in concentrated sulfuric acid[3] or by nucleophilic ring closure of fluorinated γ-chloro[4] or γ-fluoro[5] alcohols under action of a strong base (see Scheme 2.2). The ease of intramolecular cyclization is influenced by the nucleophilicity of alkoxy group and the best yields are obtained for tertiary alcohols.

Fluorinated Heterocyclic Compounds: Synthesis, Chemistry, and Applications, Edited by Viacheslav A. Petrov

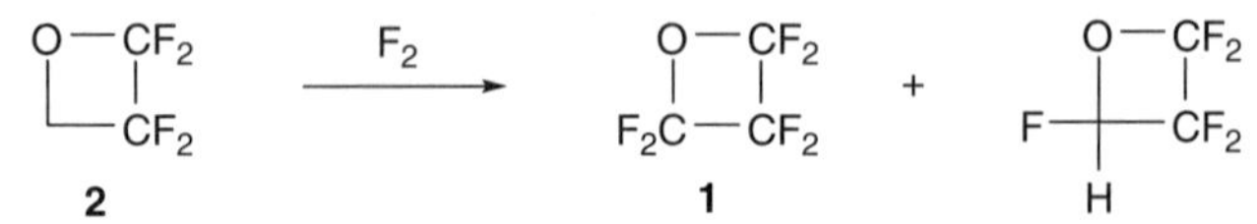

SCHEME 2.1 Fluorination of tetrafluorooxetane by fluorine gas.

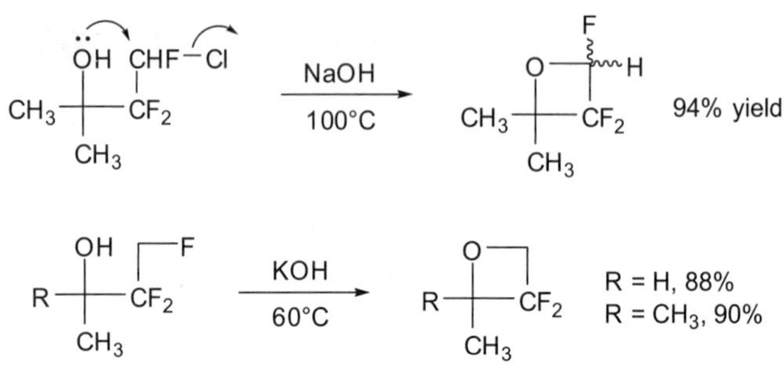

SCHEME 2.2 Nucleophilic cyclization of fluorinated alcohols.

2.1.1 Photochemical Cycloaddition

By far the most important method of preparation of fluorinated oxetanes is [2 + 2] cycloaddition that can either be photoinitiated (Paterno–Buchi reaction) or result from electrophilic condensation.

A successful photochemical cycloaddition of various fluoroalkyl ketones, aldehydes, and acyl fluorides to fluoroolefins was reported by Harris and Coffman.[6] Originally it was reported that among four possible stereoisomers, only the regioisomers of type A with CF_2 unit next to oxygen were formed (Scheme 2.3). It was rationalized that the reaction proceeds through the most stable diradical intermediate. The yields of oxetanes in this reaction varied mostly from 30% to 70% when $X = CF_3$ and were about 15% for $X = Cl$. Presumably the mechanism of this reaction involves

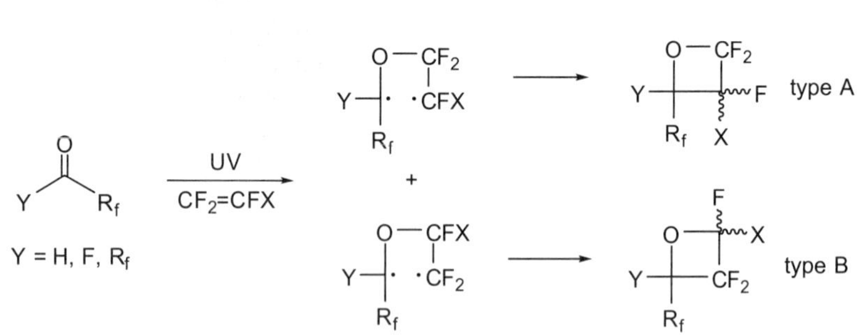

SCHEME 2.3 Photochemical cycloaddition of fluorinated carbonyl compounds to fluoroolefins.

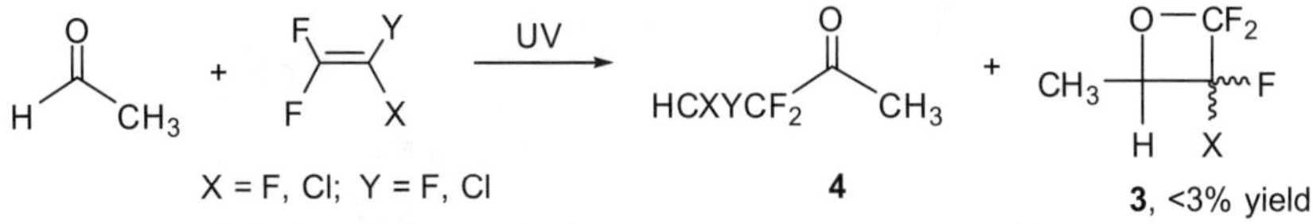

SCHEME 2.4 UV-initiated reaction of acetaldehyde and fluoroethylenes.

the formation of diradical from the carbonyl compound followed by radical addition of the olefin and cyclization.

Very recently, however, it has been shown by Lemal and Raghavanpillai that in case of the photochemical reaction of hexafluoroacetone (HFA) or trifluoroacetaldehyde with various polyfluorinated ethylenes and propylenes, *both* regioisomers (types A and B) are formed with the combined yields ranging from 50% to 95% (Scheme 2.3).[7,8]

Bissell and Fields reported that very low yields (<3%) of fluorinated oxetanes of type **3** from radical addition of acetaldehyde to perhalogenated ethylenes can be achieved, while the major (but still low yield) products were the ketones of type **4** (Scheme 2.4).[9]

Hydrocarbon and partially fluorinated olefins also undergo cycloaddition with fluorinated ketones. Thus, UV-induced addition of HFA to ethylene, vinyl fluoride, vinylidene fluoride,[10] 1,2-difluoroethylene,[11] trifluoroethylene,[7] and 3,3,3-trifluoropropylene[12] was reported to afford corresponding oxetanes in the yield >70%.[7] In all cases where it is possible, the formation of both regioisomers was observed. For example, irradiation of the equimolar mixture of HFA and $CF_2{=}CH_2$ resulted in high-yield formation of regioisomers **5** and **6** (Scheme 2.5).

Noteworthy, very few reports can be found on UV light-initiated cycloaddition of carbonyl compounds to tetrafluoroethylene (TFE).[13,14] The yields of oxetanes in these processes are generally low due to facile polymerization of TFE under the reaction conditions.[15]

2.1.2 Electrophilic Cycloaddition

Fluorinated oxetanes can be obtained in decent yield through an electrophilic [2 + 2] cycloaddition of carbonyl compounds and fluoroolefins. The formation of oxetane **2** was first reported by Weinmayr in 1963.[16] The reaction of TFE with paraformaldehyde

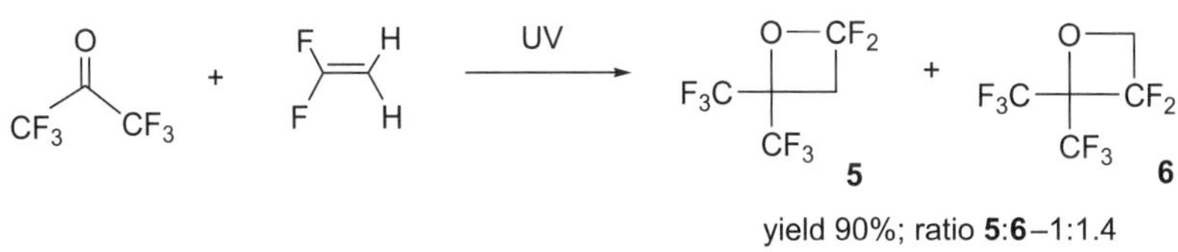

SCHEME 2.5 The formation of regioisomers in photochemical [2 + 2] reaction of hexafluoroacetone and $CH_2{=}CF_2$.

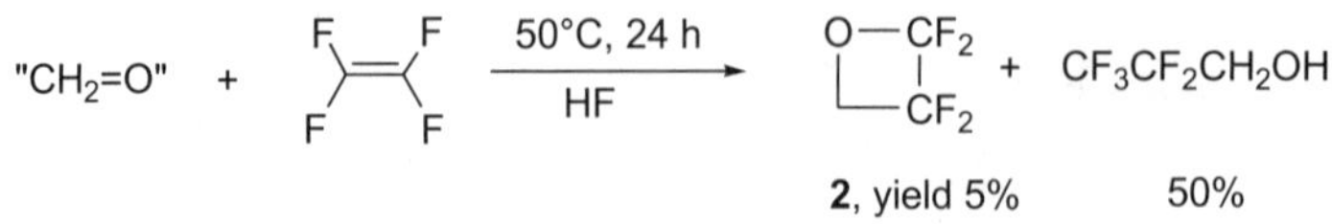

SCHEME 2.6 Reaction of formaldehyde with TFE in anhydrous HF.

in hydrogen fluoride solvent at 60°C resulted in low-yield formation of **2**, along with the major product of this reaction 2,2,3,3,3-pentafluoropropanol isolated in 50% yield (Scheme 2.6).

The yield of **2** in this reaction can be improved up to 40% by adjusting reaction conditions and using some additives.[17] An alternative route is based on gas-phase oxidation of $CH_3CF_2CF_2H$ by O_2 at 500°C, leading to oxetanes **2** in a good yield.[17] Perfluoroalkoxy-substituted oxetanes can be prepared in low yield (<20%) by electrophilic cycloaddition of formaldehyde to perfluorovinyl ethers.[18] Both regioisomers are formed in this process.

It should be pointed out that the electrophilic [2 + 2] cycloaddition of $CH_2{=}O$ and fluoroolefins carried out in HF is rather limited in the scope. For example, reaction of formaldehyde with hexafluoropropylene exclusively affords $CF_3CF(CF_3)CH_2OH$, and in reaction with vinylidene fluoride, the ether $(CF_3CH_2CH_2)_2O$ is formed as the major product.[16] More on the electrophilic reactions of fluoroolefins in anhydrous HF can be found in detailed review.[19]

The electrophilicity of a carbonyl compound can be enhanced by addition of strong Lewis acid. The first example of the formation of polyfluorinated oxetanes in reaction of 1,1,3-trichlorotrifluoroacetone with TFE or trifluoroethylene catalyzed by SbF_5 was reported in 1978.[20] The process leads to a mixture of two major products due to the competition between the insertion of fluoroolefin into C–F bond of the CCl_2F group of ketone, leading in the case of TFE to product **7**, and the cycloaddition process, resulting in the formation of **8**.

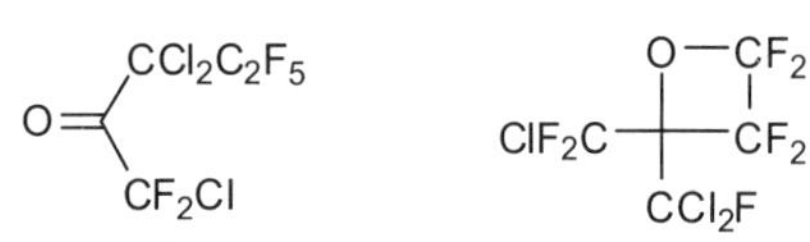

7, 16% **8**, 29%

Antimony pentafluoride-catalyzed reaction is limited to 1,1,3-trichlorotrifluoroacetone, since both hexafluoro- and 1,3-dichlorotetrafluoroacetones were found to be inactive in reaction with TFE under similar conditions.[20]

An interesting transformation leading to fluorinated oxetanes **9** and **10** was reported by Zeifman et al. in 1996.[21] The treatment of $(CF_3)_2C(OH)CCl_2CF_3$ by antimony pentafluoride at low temperature (−20°C) leads to the formation of oxetanes **9** in 26% yield, while the reaction carried out at elevated temperature (80°C) results in the oxetane **10** formed from **9** as the result of the chlorine–fluorine exchange under action of SbF_5.

9, 26% 10, 58%

Significantly stronger Lewis acid aluminum chlorofluoride (ACF, $AlCl_xF_y$)[22] was shown to be advantageous in promoting electrophilic [2 + 2] cycloadditions of hexafluoroacetone. ACF, known to be an effective catalyst for the condensation of halomethanes[23] and fluoroolefins[24] with fluoroethylenes, was also found to catalyze the reaction of HFA with fluoroethylenes. For example, the reaction of HFA and TFE gives the corresponding *F*-(2,2-dimethyloxetane) **11** in moderate yield along with some polytetrafluoroethylene (Scheme 2.7).[25]

In an analogous way, HFA undergoes cycloaddition with halogenated ethylenes $CF_2{=}CFX$ (X = Cl, Br, H) in the presence of ACF catalyst to afford the corresponding oxetanes. The selective formation of only one regioisomer **5** (along with acyclic alcohols as minor by-products) in the reaction of $CH_2{=}CF_2$ is consistent with electrophilic mechanism of the process, presented by Scheme 2.8.[25]

Some electron-rich hydrocarbon compounds can undergo the cycloaddition with electron-deficient fluorinated carbonyls spontaneously. For example, recently it was

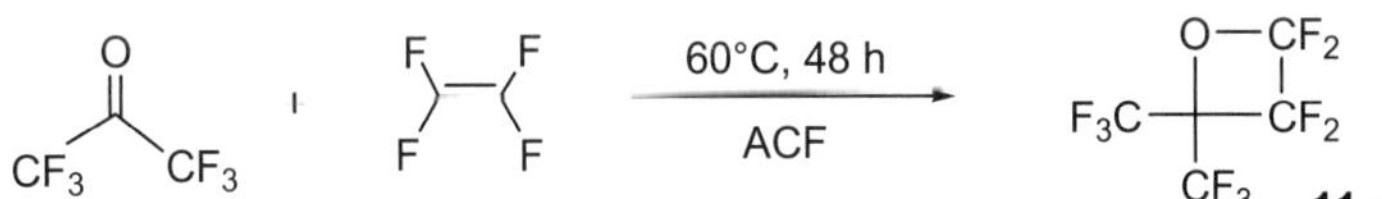

11, 66%

SCHEME 2.7 Electrophilic [2 + 2] cycloaddition of hexafluoroacetone and TFE catalyzed by ACF.

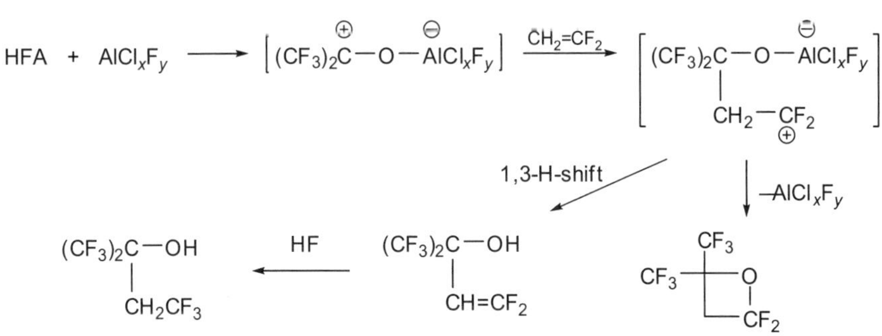

5

SCHEME 2.8 Mechanism of the reaction of hexafluoroacetone and $CH_2{=}CF_2$ catalyzed by ACF.

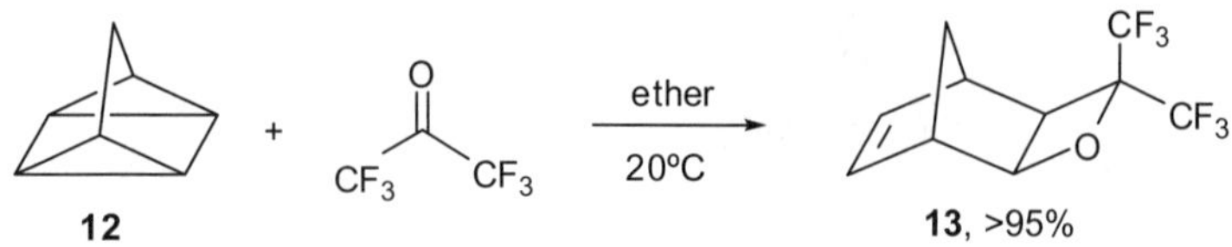

SCHEME 2.9 Reaction of quadricyclane and hexafluoroacetone.

shown that quadricyclane (**12**) reacts with HFA at subambient temperature to afford *exo*-oxetane **13** (Scheme 2.9).[26]

The reaction is not limited to HFA and a number of norbornenoxetanes, similar to **13**, were synthesized using this reaction. Reported examples include cycloadducts of quadricyclane and various fluorinated ketones, $CF_3C(O)X$ (X = F, Cl), $FSO_2C(O)F$, $C(O)F_2$, $CF_3C(O)H$, and $(CF_3)_2C{=}C{=}O$.[26,27]

The highly electrophilic polyfluorinated ketones, such as HFA, nitropentafluoroacetone, or chloro(fluoro)acetones also react with electron-rich vinyl ethers producing oxetanes, without a catalyst or irradiation. For example, HFA forms the corresponding oxetanes **14** in reaction with ethyl or *n*-butyl vinyl ethers at room temperature (Scheme 2.10).[28]

Other fluorinated ketones such as nitropentafluoro-[28] and 1,3-dichlorotetrafluoroacetones[29] undergo cycloaddition with various vinyl ethers to form oxetanes similar to **14**. Alkoxy oxetanes **14** are highly reactive materials. For example, they undergo rapid isomerization into vinyl ethers **15** either upon heating or under action of catalytic amount of H_2SO_4.[28]

Ethoxyacetylene reacts with HFA even at low temperature to give unstable oxetene **16**, which at room temperature rearranges into the ester **17** (Scheme 2.11).[30]

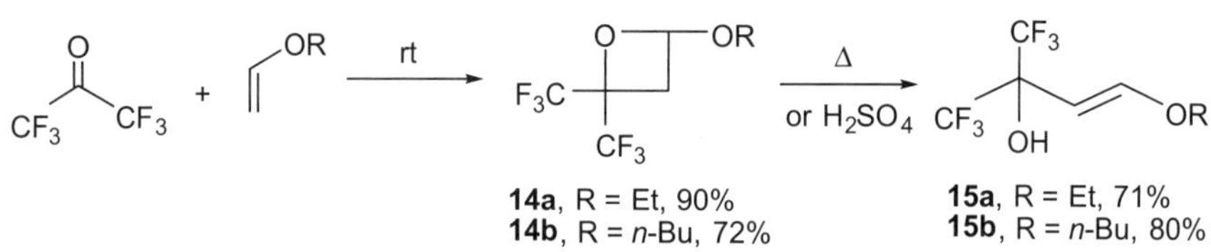

SCHEME 2.10 [2 + 2] Cycloaddition of hexafluoroacetone and vinyl ethers.

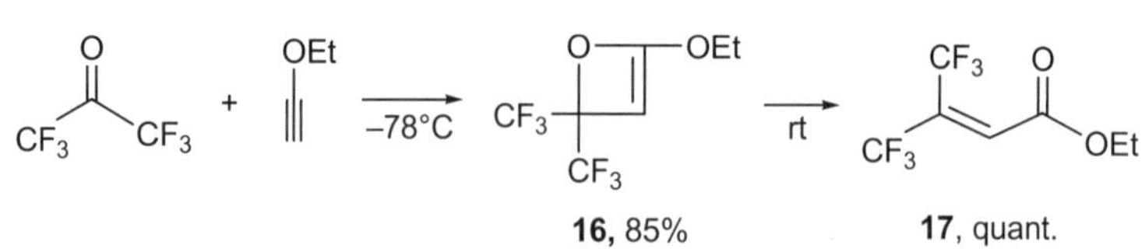

SCHEME 2.11 Cycloaddition of hexafluoroacetone to ethoxyacetylene and isomerization of oxetene **16**.

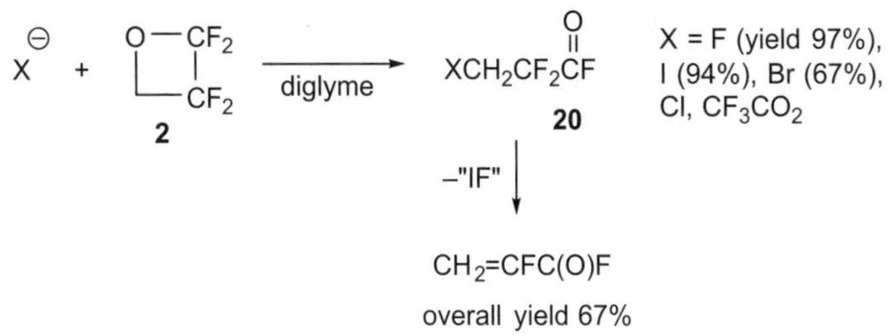

SCHEME 2.12 Nucleophilic ring-opening reactions of tetrafluorooxetane **2**.

The other methods used to prepare oxetenes involve either dehydrohalogenation[31] or dehalogenation[32] of appropriate precursors. For example, oxetene **18** was prepared in high yield by dechlorination of 3,4-dichlorooxetane **19** with treatment with zinc.[32]

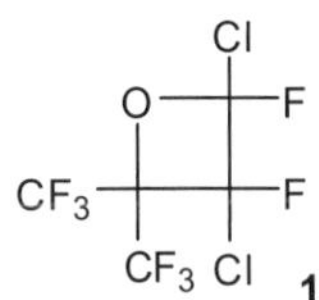

19

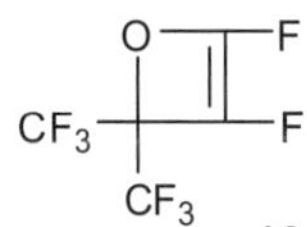

18, 92%

Due to the substantial ring strain, polyfluorinated oxetanes are able to undergo ring-opening reactions under the action of nucleophilic or electrophilic reagents. The compound drawing most of the attention in this respect is tetrafluorooxetane **2** owing to its high and unique reactivity. Ring-opening polymerization of **2** is used as the initial step in commercial synthesis of linear perfluoropolyether fluids Demnum®[33] (see also Chapter 13). With the stoichiometric amount of nucleophile, the corresponding acyl fluorides **20** are formed (Scheme 2.12).[34] Dehalogenation of **20** is a convenient route to a useful monomer precursor—$CH_2{=}CFC(O)F$.[35]

At the same time, even highly fluorinated oxetanes are able to react with electrophiles. For example, the reaction of *F*-(2,2-dimethyloxetane) **11** with HF/SbF_5 leads to high-yield formation of $C_2F_5C(CF_3)_2OH$.[14]

2.2 FLUORINATED β-SULTONES

Generally, sulfur trioxide reacts with hydrocarbon olefins quite vigorously and has to be either diluted or used in complexes with pyridine or dioxane to moderate its reactivity. The corresponding cycloadducts, β-sultones, easily undergo ring opening and have not yet been isolated. In sharp contrast, fluorinated β-sultones are relatively stable materials and over 50 compounds of this type have been prepared and isolated. Synthesis and chemistry of this class of heterocycles is well documented and has been previously reviewed.[36,37]

The original reports on synthesis of fluorinated β-sultones appeared from three different groups at approximately the same time in late 1950s.[38–40] Addition of freshly distilled sulfur trioxide to TFE, chlorotrifluoroethylene, hexafluoropropylene, and

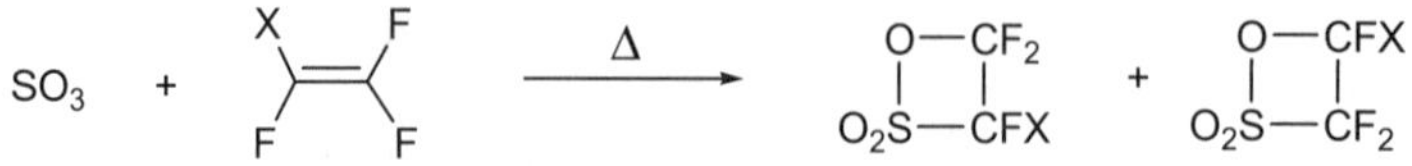

X = F (**21**), Cl (**22a**), CF_3 (**23**)

X = Cl (**22b**)

SCHEME 2.13 Formation of β-sultones in reaction of SO_3 and fluoroolefins.

some other fluorinated olefins affords the corresponding β-sultones **21–23** in almost quantitative yields (Scheme 2.13).[41] In some cases, the cycloaddition is regioselective, for example, reaction of hexafluoropropylene produces a single isomer **23**. However, chlorotrifluoroethylene gives approximately equal amounts of the two possible isomeric sultones **22a** and **22b**.

In the series of fluoroethylenes with successive replacement of fluorine by hydrogen, a decreasing tendency to β-sultone formation was observed.[41] While TFE gave almost quantitative yield of sultone **21**, trifluoroethylene gave only 60% yield of the corresponding product **24** along with high-boiling compound, identified as the cyclic sulfonate–sulfate anhydride **25**. Reactions of vinylidene fluoride and vinyl fluoride with sulfur trioxide produced only viscous liquids and tars, presumably containing compounds analogous to **25** and higher oligomers. It is very important to use freshly distilled sulfur trioxide in synthesis of sultones, because even trace amount of water promotes polymerization of sulfur trioxide inhibiting its reaction with fluorinated olefin lowering the yield of β-sultones and consequently increasing the amount of by-products, such as **26–28**, in case of TFE/SO_3 reaction.[36,41]

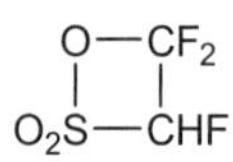

24

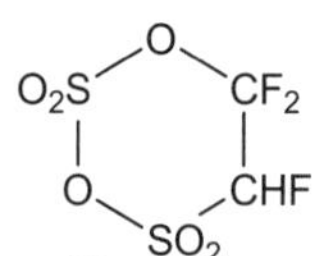

25

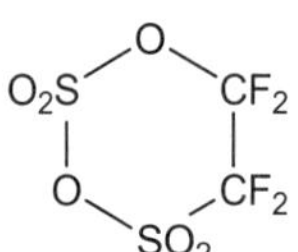

26

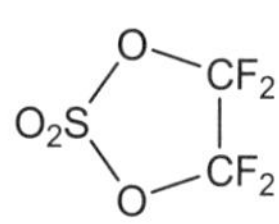

27

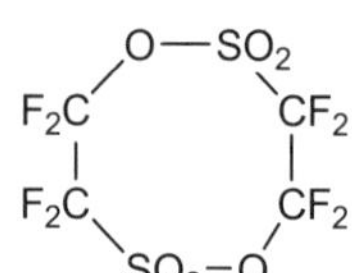

28

An electrophilic cycloaddition process was proposed as the mechanism of β-sultone formation.[36] The initial step is an electrophilic attack of sulfur trioxide on the double bond of fluoroolefin to afford a dipole intermediate (**29**), which then undergoes intramolecular cyclization (Scheme 2.14).

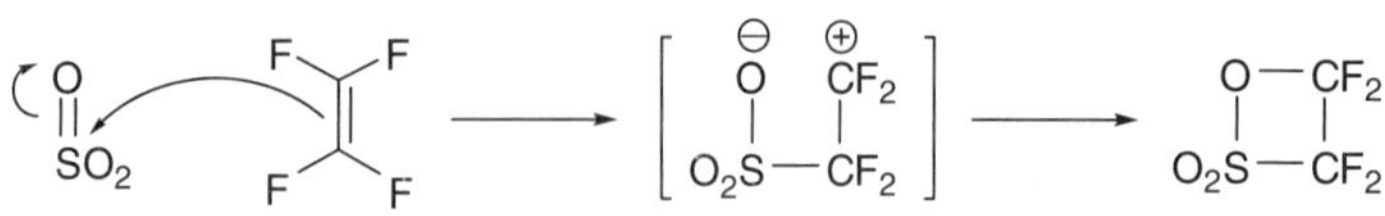

29 **21**

SCHEME 2.14 Mechanism of electrophilic [2 + 2] cycloaddition of SO_3 to fluoroolefins.

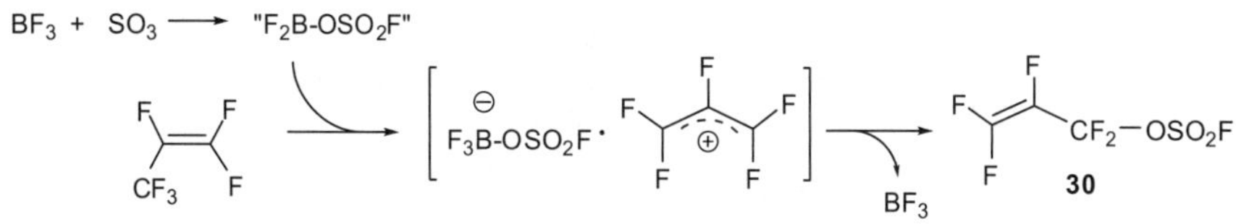

SCHEME 2.15 Stepwise mechanism of electrophilic insertion of SO_3 into allylic C–F bond of hexafluoropropene.

The addition of catalytic amount of a Lewis acid, such as BF_3, completely changes the course of the reaction of hexafluoropropene and sulfur trioxide. Allyl fluorosulfate **30** (rather than sultone **23**) is the major product of the reaction in this case.[42] The proposed mechanism of this reaction involves the formation of fluorosulfated boron fluoride intermediate by abstracting an allylic fluorine from hexafluoropropene to form the perfluoroallyl cation, which is further stabilized by addition of -OSO_2F group affording **30** (Scheme 2.15).[42] It should be pointed out, however, that the concerted mechanism of SO_3 insertion into allylic C–F bond of CF_3 group also cannot be ruled out.

2.2.1 Reactions of β-Sultones

In the presence of catalytic amount of base, β-sultones undergo facile rearrangement.[41] Scheme 2.16 illustrates base-catalyzed rearrangement of compound **21**. In the first step, a base attacks at the sulfur atom to induce the ring opening by the cleavage of S–O bond to give the linear perfluoroalkoxy anion intermediate, which further loses fluoride anion to afford the corresponding acyl fluoride. The fluoride anion displaces the base at the sulfonyl group to give the final product **31** or attacks the sultone **21** to initiate the next ring opening.

Reactions of **21** with nucleophilic reagents such as alcohols, thiols, and primary and secondary amines lead to the corresponding derivatives of fluorosulfonyldifluoroacetic acid.[41] The acid itself can be prepared by controlled hydrolysis of **21**. All these products are apparently formed by initial rearrangement of **21** into **31** followed by subsequent nucleophilic displacement of fluoride in –C(O)F group, since this group is much more reactive compared to –SO_2F. The chemistry of β-sultones with the side chain, such as compound **23**, is analogous to the chemistry of **21** with one exception:

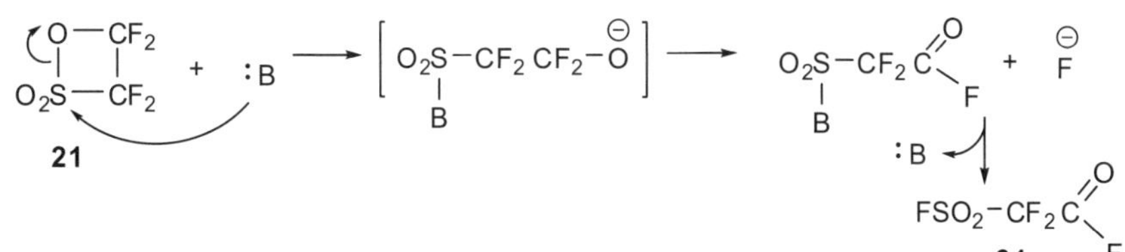

SCHEME 2.16 Base-catalyzed ring opening of sultone **21**.

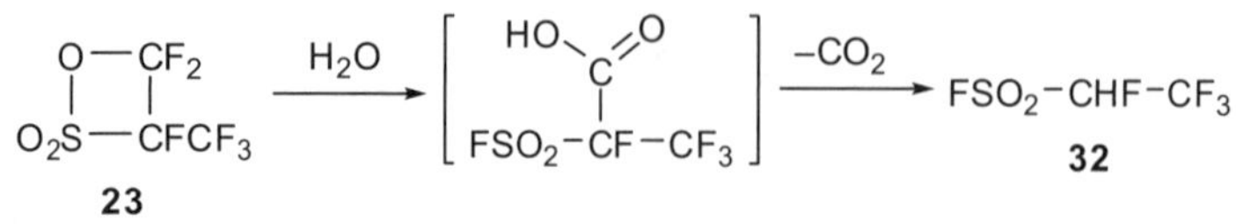

SCHEME 2.17 Ring-opening reaction of sultone **23**.

on the treatment of such sultones with water, the corresponding secondary carboxylic acid readily undergoes decarboxylation giving the linear sulfonyl fluoride **32** (Scheme 2.17).

Among all β-sultones, compound **21** has drawn the greatest attention. Its rearranged isomer **31** is the precursor for the synthesis of perfluorinated vinyl ether **33**, which is a key monomer in the synthesis of DuPont's Nafion® polymer used for preparation of commercial ion-exchange membranes[43,44] (see also Chapter 13).

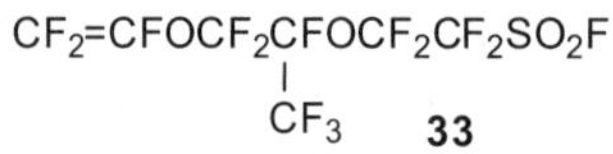

33

2.3 FLUORINATED DITHIETANES AND THIETANES

The most easily accessible compounds of this class are 1,3-dithietanes. They are formed by a spontaneous dimerization of fluorinated thioketones or acyl halides. Thus, hexafluorothioacetone **34**, generated in the reaction of bis(perfluoroisopropyl) mercury with molten sulfur at about 450°C (Scheme 2.18),[45] undergoes at ambient temperature slow dimerization to form dithietane **35**. This dimerization is significantly accelerated by catalytic amount of base, such as pyridine. At elevated temperature, the dimer **35** can be converted back into **34**.

Compound **35** also can be produced from hexafluoropropene by heating it with sulfur in the presence of either activated carbon[46] or potassium fluoride.[47] Another route to **35** and other perfluorinated thioketones involves sulfurization of secondary perfluoroalkyl iodides with refluxing phosphorus pentasulfide.[48] However, the most practical synthesis of **35** reported by England is based on the KF-catalyzed reaction of commercially available hexafluoropropene with sulfur in DMF solvent at atmospheric pressure.[49] The reaction produces **35** in over 80% yield and is simple and easy to scale up.

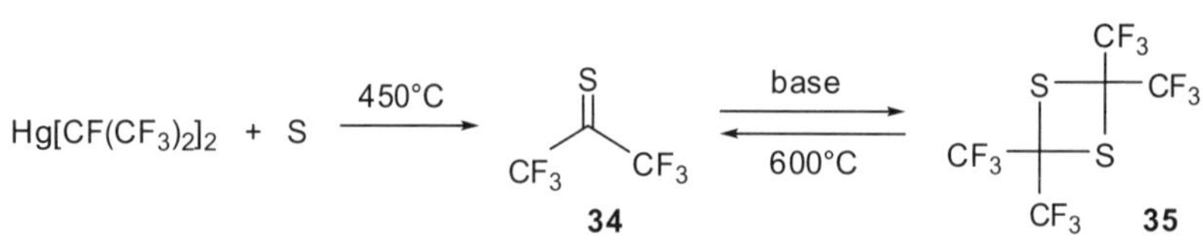

SCHEME 2.18 Preparation of 1,3-dithietane **35**.

Facile dimerization to 1,3-dithietanes is also characteristic of other fluorinated thioketones, such as 4-hydro- and 4-chloroperfluorobutan-2-thiones. Trifluorothioacetyl fluoride does not dimerize spontaneously; however, when it is irradiated by UV light, it is converted to a mixture of *cis* and *trans* isomers of compound **36a**. Dithietanes **36b**, **37a**, and **37b** were also obtained by photochemical dimerization of the corresponding thiocarbonyls.[48]

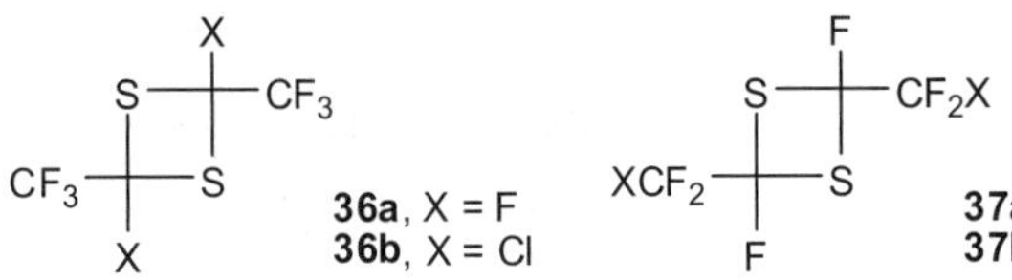

36a, X = F
36b, X = Cl

37a, X = Cl
37b, X = Br

Hexafluorothioacetone is extremely active dienophile in the Diels–Alder reaction,[50] but it also readily undergoes [2 + 2] cycloaddition with variety of electron-rich olefins that lack allylic hydrogen. For example, methyl vinyl ether and methyl vinyl sulfide react rapidly with **34** to form the corresponding thietanes (Scheme 2.19).[51]

When **34** reacts with olefins possessing allylic hydrogens, the products are the corresponding allyl sulfides, forming as the result of insertion of **34** into allylic C–H bond (Scheme 2.19). It should be pointed out that similar process known for hexafluoroacetone leads to the corresponding allylhexafluoro-*i*-propanol. Thus, it appears that the sulfur atom in thioketone **34** is at the positive end of the C–S dipole.

Since the simplest *F*-1,3-dithietane (**38**) cannot be prepared by dimerization of thiocarbonyl fluoride, probably due to propensity of the latter to polymerize, a different synthetic approach is employed for its synthesis. Two-step process includes photocyclodimerization of $Cl_2C{=}S$ and Swarts-type fluorination of the dimer by antimony trifluoride to give **38** (Scheme 2.20). Compound **38** can be a source of very pure thiocarbonyl fluoride, when pyrolyzed at 500°C.[45]

Fluorinated 1,2-dithietenes **39a–c** are the only known representatives of fluorinated four-membered ring systems containing two sulfur atoms in 1,2-position, while fluorinated 1,2-dithietanes have not been prepared so far.

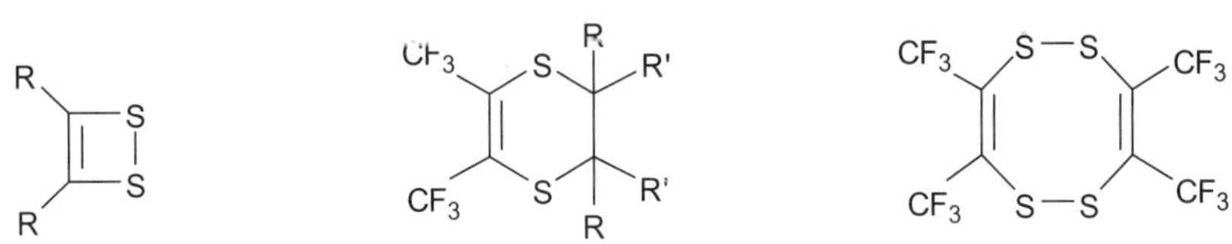

39a, R = CF_3, 80%
39b, R = CF_2Cl, 82%
39c, R = $(CF_2)_5H$, 40%

40a, R = R' = H, 24%
40b, R = R' = CH_3, 47%
40c, R = H, R' = R" = C_6H_5, 36%
40d, R = R' = H, R" = OC_2H_5, 50%
40e, R = R' = H, R" = SC_4H_9, 31%

41, 87%

Compounds **39a–c** were synthesized by passing the vapor of corresponding fluoroalkylacetylene through refluxing sulfur.[52] Compound **39a** is the most studied representative of this group. It is reported to undergo unusual [4 + 2] cycloaddition

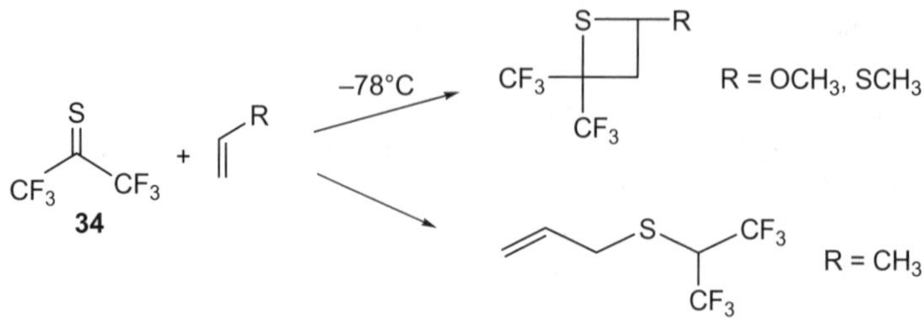

SCHEME 2.19 Typical reactions of hexafluorothioacetone.

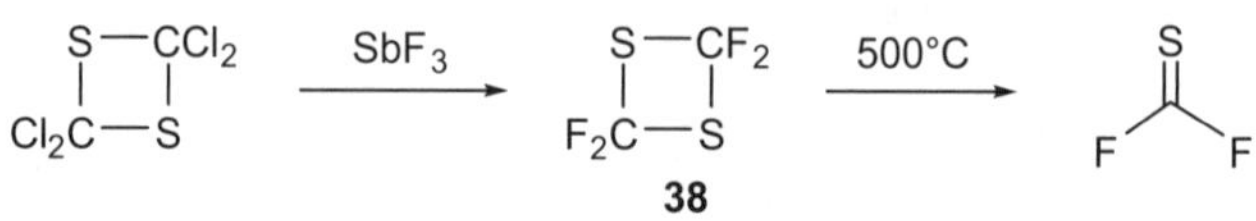

SCHEME 2.20 Preparation of cyclic dimer thiocarbonyl fluoride.

reaction with electron-rich olefins with the formation of the corresponding dihydro-1,4-dithiines **40a–e**. Under basic conditions, **39a** rapidly forms cyclic dimer **41**.[53]

Another quite particular 1,3-dithietane **42** can be synthesized by several methods.[54–56] One of them involves fluorination of the adduct of diethyl malonate and thiophosgene by sulfur tetrafluoride in anhydrous HF (Scheme 2.21). On pyrolysis at 600–750°C, compound **42** breaks down affording bis(trifluoromethyl)thioketene **43**, a reddish-orange liquid, stable enough to be handled and stored without special precautions at ambient temperature. Rapid dimerization of **43** back to **42** is initiated by catalytic amounts of tertiary amines. The chemistry of this thioketene in many respects is very similar to the chemistry of hexafluorothioacetone (**34**). For example, **43** undergoes cyclodimerization to form **42** and reacts with a variety of other thiocarbonyl compounds to form dithietanes **44** in good yields (Scheme 2.21).[56]

Although, it does not react with carbonyl compounds, **43** undergoes the cycloaddition with vinyl ethers, styrenes, and ketenes to form variety of unsaturated thietanes **45** (Scheme 2.21).

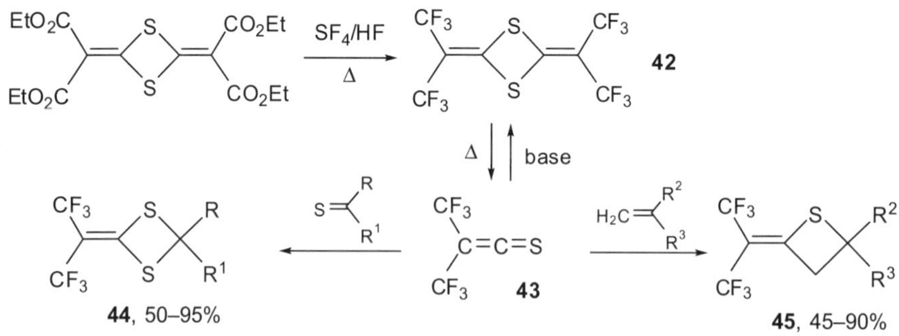

SCHEME 2.21 Preparation and reactions of bis(trifluoromethyl)thioketene (**43**).

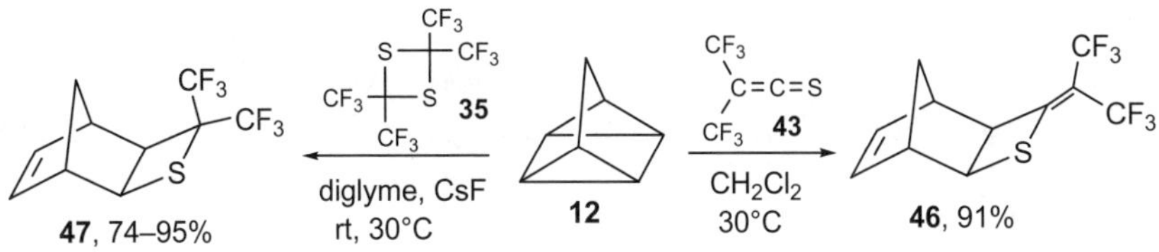

SCHEME 2.22 Cycloaddition reaction of quadricyclane (**12**) with hexafluorothioacetone and bis(trifluoromethyl)thioketene.

The uncatalyzed reaction of **43** with quadricyclane (**12**) proceeds readily to give the *exo*-adduct **46** (Scheme 2.22).[56] Analogous cycloaddition with formation of adduct **47** was also observed for hexafluorothioacetone generated *in situ* from dithietane **35** (Scheme 2.22).[57]

F-(2,2,4,4-Tetrakis(trifluoromethyl)thietane is the only known example of perfluorinated thietanes. It can be prepared by high-yield reaction of $(CF_3)_2C{=}CF_2$ with sulfur, catalyzed by CsF.[58] First perfluorinated thiete was prepared by the reaction of hexafluorpropene trimer with $(CH_3)_3CSH$[58a] and later by dehydrofluorination of unsaturated thiol $[(CF_3)_2CF]_2C{=}C(SH)CF_3$ (X=H).[58b] The hydrolysis of sulfenul chloride $[(CF_3)_2CF]_2C{=}C(SCl)CF_3$ afforded perfluorinated thiete oxide.[58b]

Synthetic applications of dithietane **35** are not limited to its use as a source of hexafluorothioacetone. This compound is also a valuable precursor for the synthesis of a variety of fluorinated sulfur-containing derivatives. For example, the pyrolysis of **35** at 325°C results in the high-yield formation of $(CF_3)_2C{=}C(CF_3)_2$.[59] The reaction of **35** with Ph_3P was found to be a convenient route to $(Ph)_3P{=}C(CF_3)_2$, which was used for the preparation of 1,1-bis(trifluoromethyl) alkenes[60,61] and $CF_2{=}C(CF_3)P(O)(OR)_2$.[62]

Recently reported transformation of fluorinated thietanes involves reductive ring expansion to substituted dihydrothiophenes. For example, the treatment of compound **47** with aluminum powder in the presence of a catalytic amount of $PbCl_2$ resulted in an interesting ring expansion process leading to the formation of compound **48** in excellent yield (Scheme 2.23).[63] Readily available fluorinated cycloadducts **49**

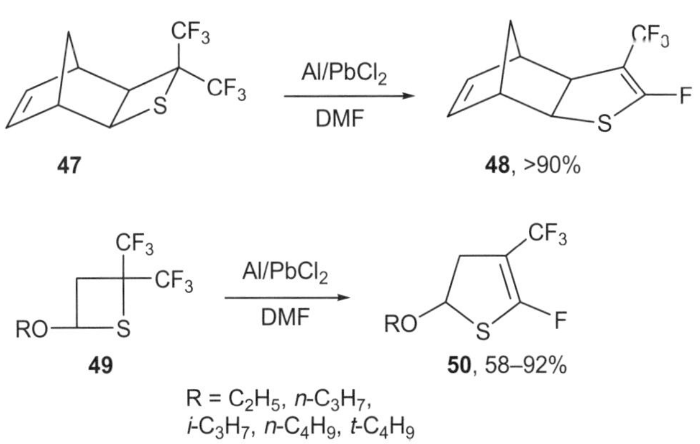

SCHEME 2.23 Reductive ring expansion of fluorinated thietanes.

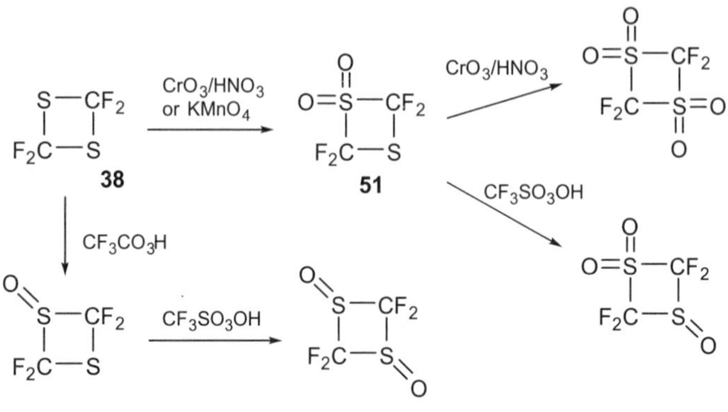

SCHEME 2.24 Oxidation of dithietane **38**.

derived from hexafluorothioacetone and vinyl ethers[64] were converted into dihydrothiophenes **50** in high yield under similar conditions (Scheme 2.23).[63a]

A method for conversion of **38** into monosulfone **51** by oxidation with chromium trioxide in nitric acid was first claimed by Carboni and Kauer.[65] Controlled oxidation of **38** affords a range of 1,3-dithietane *S*-oxides and sulfones (Scheme 2.24).[66] The perfluorinated dithietane *S*-oxides are extremely thermally stable compounds. They completely decompose into smaller fragments only at about 500°C.

The cycloadduct **35** can also be selectively oxidized to tetrakis(trifluoromethyl)-1,4-dithietane *S*-oxides **52a–c**.[67] The oxidation of cycloadduct **47** depending on the reaction conditions may result either in the selective formation of the sulfoxide **53** or in the complete oxidation leading to the corresponding epoxysulfone **54**.[68]

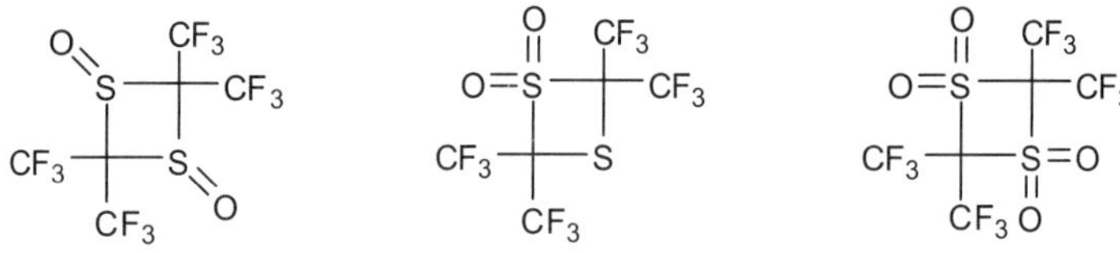

52a, 30% **52b**, 55% **52c**, 56%

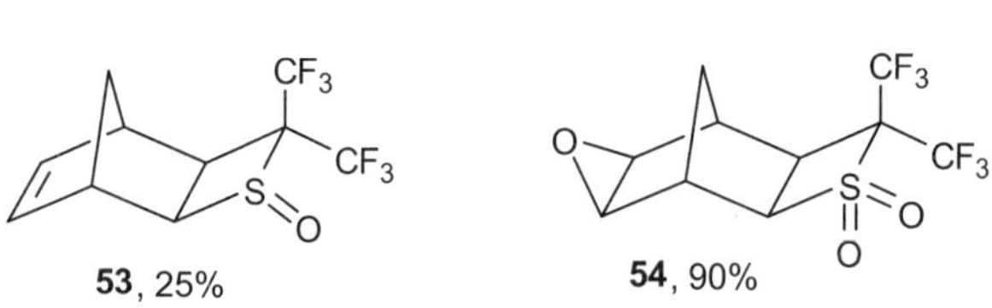

53, 25% **54**, 90%

It should be pointed out that the oxidation of quadricyclane–bis(trifluoromethyl)-thioketene cycloadduct **46** by MCPBA (25°C, 2 days) was reported to proceed without oxidation of either hydrocarbon or fluorinated double bonds, leading to selective formation of the corresponding sulfone, isolated in 92% yield.[69]

The 1,3-dioxide **52a** can be quantitatively cleaved into bis(trifluoromethyl)sulfine (**55**) by pyrolysis at 450°C (Scheme 2.25).[70] Alternative synthesis of sulfine **55** involves oxidation of Diels–Alder adduct of hexafluorothioacetone and anthracene to form oxide

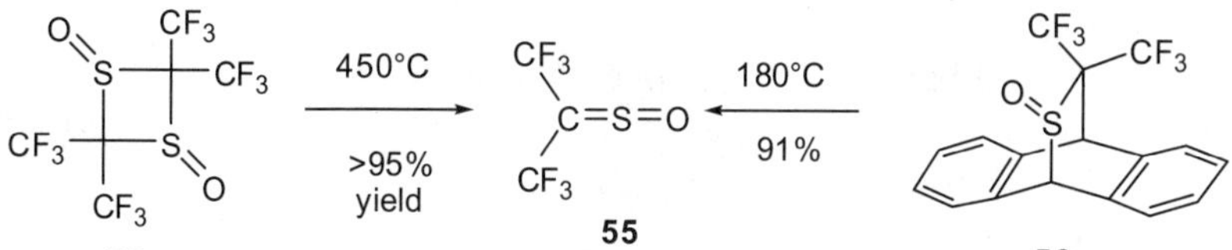

52a **55** **56**

SCHEME 2.25 Preparation of bis(trifluoromethyl)sulfine.

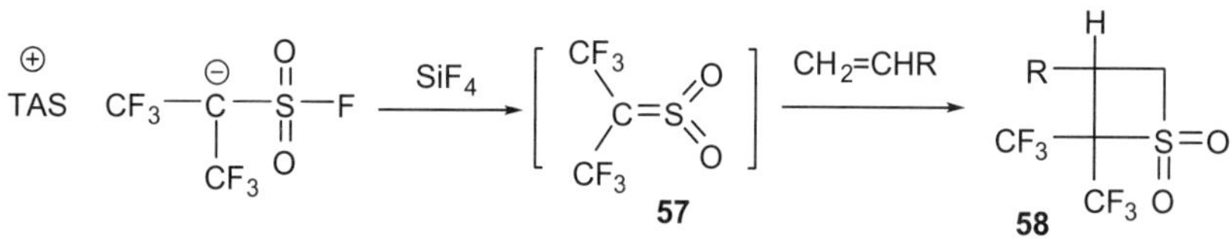

57 **58**

TAS = tris(dimethylamino)sulfonium

R = OCH_3 (33%), 4-OCH_3–C_6H_4(90%)

SCHEME 2.26 Generation and reactions of bis(trifluoromethyl)sulfene.

56, followed its thermal retro Diels–Alder reaction.[71] Compound **55** is a very reactive dienophile and undergoes various cycloadditions with unsaturated hydrocarbons.[71]

Fluorinated thietane *S*-dioxides **58** were synthesized by intercepting bis(trifluoromethyl)sulfene (**57**) with hydrocarbon olefins (Scheme 2.26).[72] Highly reactive **57** has not been isolated so far. When its precursor, anion $FSO_2C(CF_3)_2^-$, was treated with BF_3 or SiF_4 at 0°C, isomeric $CF_2{=}C(CF_3)SO_2F$ was isolated as the principal product. However, when electron-rich olefin is present in the reaction mixture, it efficiently traps **57** generated *in situ*, affording [2 + 2] cycloadducts **58** in moderate to high yield (Scheme 2.26).[72] Structurally similar to **58** *mono*-S-oxides were prepared recently by highly selective oxidation of thietanes **49** with MCPBA.[63a]

2.4 FLUORINATED β-LACTONES

Synthetic approaches to fluorinated β-lactones (oxetane-2-ones) are very similar to oxetane synthesis. Usually β-lactones are prepared either by cycloaddition of ketenes to aldehydes or ketones or by cyclodehydration of β-hydroxycarboxylic acids.

The first β-lactone with fluorinated substituents reported in 1960 by Cheburkov and Knunyants was prepared by cycloaddition of the ketene with hexafluoroacetone.[73] The reaction takes place at −78°C in the absence of catalyst to give β-lactone **59** in quantitative yield (Scheme 2.27). Other methods of synthesis of lactones **59** include dehydration of the appropriate hydroxybutyric acid with phosphorus pentoxide or dehydrobromination of the corresponding butyryl bromide.

Pyrolysis of **59** typically results in fluorinated β-lactones extrusion of CO_2 and high-yield formation of $(CF_3)_2C{=}CH_2$. The reaction is used for the synthesis of this olefin on commercial scale.

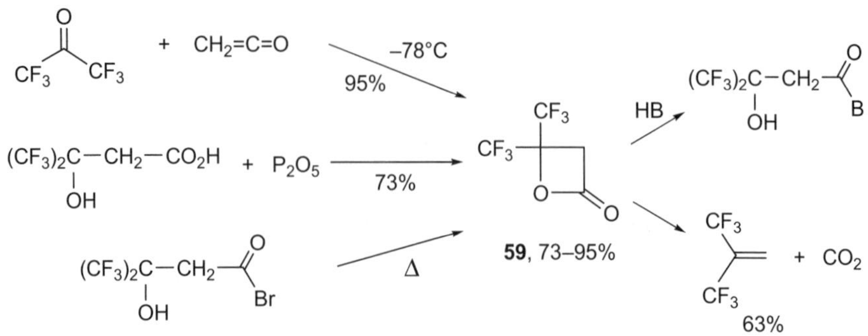

SCHEME 2.27 Preparation and some reactions of lactone **59**.

The isomeric β-lactone **60** bearing two trifluoromethyl groups at α-carbon was obtained by cycloaddition of bis(trifluoromethyl)ketene (**61**) to formaldehyde generated from trioxane (Scheme 2.28).[74]

Ketene **61** upon treatment with catalytic amount of fluoride ion in a condensed phase forms cyclic dimer **62** (Scheme 2.28).[75] The pyrolysis of **62** proceeds with the loss of CO_2 affording tetrakis(trifluoromethyl)allene (**63**).

The first example of β-lactone bearing fluorine in the ring was prepared by trapping difluoroketene with acetone, to form α,α-difluoro-β,β-dimethyl β-lactone **64** (Scheme 2.29).[76]

Later, the route that employs cyclization of β-hydroxy carboxylic acids under action of benzenesulfonyl chloride and pyridine was used to prepare a series of α,α-difluoro-β-lactones **65** (Scheme 2.30).[77] The compounds **65** were shown to be convenient precursors to the corresponding 1,1-difluoroalkenes **66** afforded in excellent yields by extrusion of CO_2 in solution at 150–180°C.

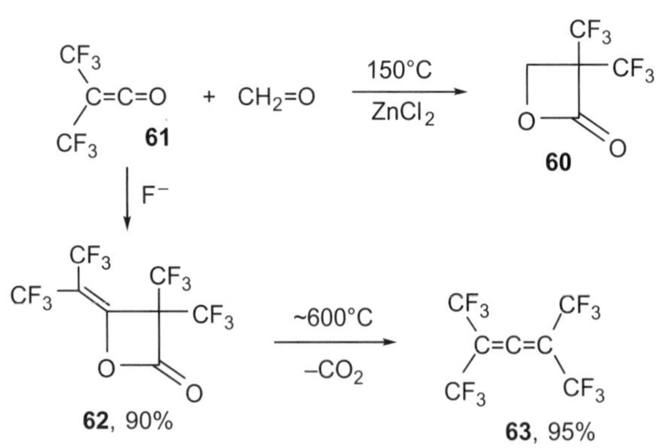

SCHEME 2.28 Dimerization of bis(trifluoromethyl)ketene (**61**) and its cycloaddition to formaldehyde.

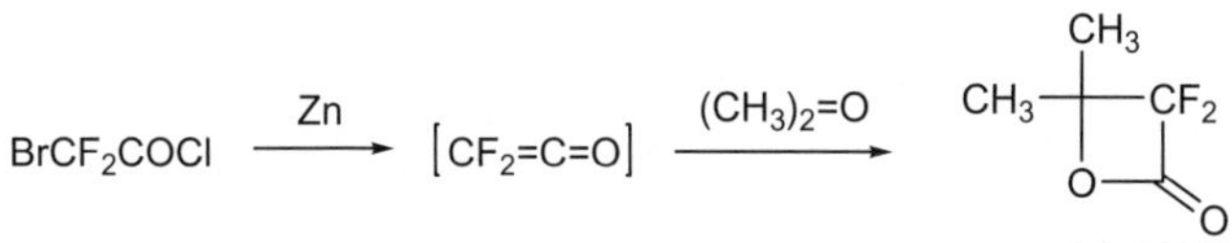

64, 50%

SCHEME 2.29 Synthesis of α,α-difluoro-β,β-dimethyl β-lactone.

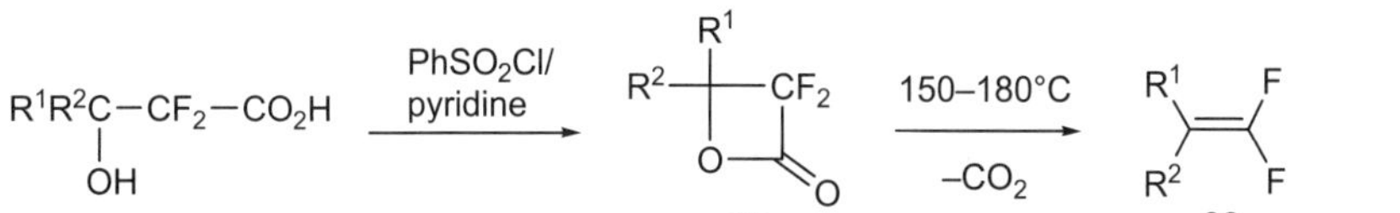

65 $R^1 = R^2$ = Me (89%), Et (95%), Bn (100%)

66 $R^1 = R^2$ = Me (87%), Et (98%), Bn (95%)

SCHEME 2.30 Synthesis of substituted α,α-difluoro-β-lactones **65** and their pyrolysis.

2.5 FLUORINATED β-LACTAMS, AZETIDINES, AND OXAZETIDINES

2.5.1 β-Lactams

Among the fluorinated nitrogen-containing four-membered heterocycles, the most studied are β-lactams. This class of compounds attracted considerable attention since they can serve as precursors for β-amino acids, valuable building blocks for the synthesis of new β-peptides and potential therapeutic drugs. The development of new synthetic procedures for preparation of fluorinated β-amino acids and investigation of their biological implications is of particular interest.[78–80] The synthesis of fluorinated β-lactams is analogous to the synthesis of β-lactones described earlier in this chapter. Usually the β-lactams are prepared either by [2 + 2] cycloaddition of fluorinated imines to ketenes or through the intramolecular cyclization of fluorinated amides.

Reactive ketenes (usually generated *in situ* by treating an appropriate acyl chloride with a base) undergo the [2 + 2] cycloaddition with fluorinated imine to afford the mixture of stereoisomers of β-lactam **67** (Scheme 2.31).[81,82] Then the kinetic optical resolution is performed on the product to obtain pure enantiomers.

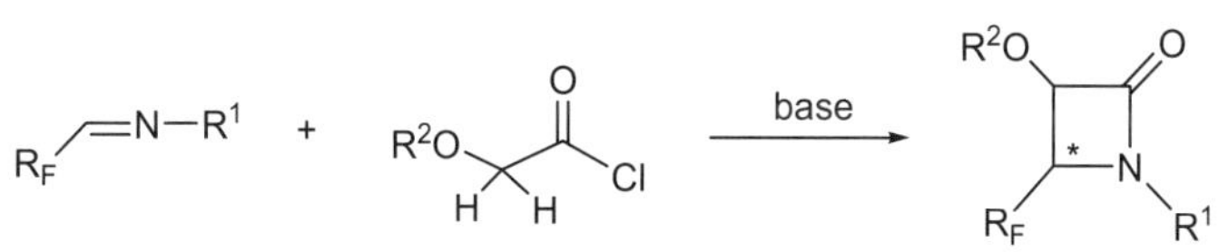

67

SCHEME 2.31 Preparation of 4-fluoroalkyl-β-lactams.

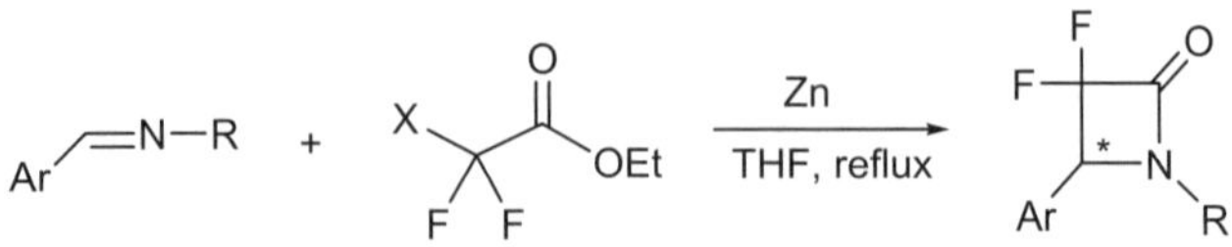

SCHEME 2.32 Synthesis of 3,3-difluoro-β-lactams.

The 3,3-difluoro-β-lactams **68** are made by a similar reaction, where difluoroketene is generated from either bromo- or iododifluoroacetate and then intercepted by imine to afford the corresponding lactams (Scheme 2.32). It was noted that the imines derived from aromatic aldehydes usually give better yields.[83] Later, this cycloaddition was accomplished under milder conditions using diethyl zinc and rhodium catalyst.[84]

Very recently, β-lactams **69** were obtained with high diastereomeric excess (85–98%) from 1,3-oxazolidines **70**, existing in equilibrium with the corresponding imino alcohols (Scheme 2.33).[85] The high asymmetric induction affected by the chiral auxiliary on the substrate can be explained by the strongly chelated intermediate proposed by Pridgen.[86]

The preparation of 3,3-difluoro-β-lactams using the cyclization of β-hydroxy or β-bromo amides is briefly reviewed in Ref. 85.

Very recently, a different synthetic approach to fluorinated β-lactams involving catalytic hydrogenolysis of the N–O bond of isoxazolidines appeared in the literature. The method involves the synthesis of isoxazolidines (**71**) by 1,3-cycloaddition of aromatic nitrones to fluorinated alkenes,[87,88] followed by N–O bond cleavage by hydrogenolysis using palladium catalyst (Scheme 2.34).[88]

2.5.2 Azetidines

Similarly to fluorinated ketones, electrophilic polyfluorinated azines[89] and imidoyl fluorides[90] undergo [2 + 2 + 2] cycloaddition reaction with quadricyclane (**12**) to afford fluorinated norbornenazetidines **73–74** (Scheme 2.35).

Recently, it was also found that the perfluoroalkyl nitriles react with quadricyclane at elevated temperature to give the corresponding cyclic adducts **75** (Scheme 2.36).[90]

Cyclic azadienes **75** containing electrophilic C=N double bond rapidly react with $CF_3Si(CH_3)_3$ in the presence of CsF catalyst forming stable silanes **76** (Scheme 2.36). Less electrophilic C=C bond remains unchanged in this process. It should be pointed

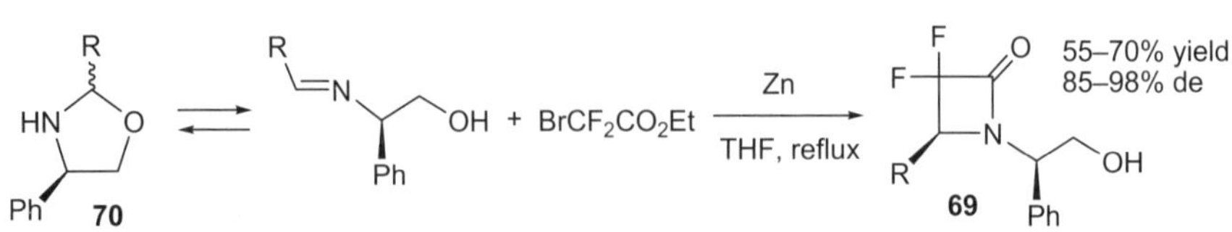

SCHEME 2.33

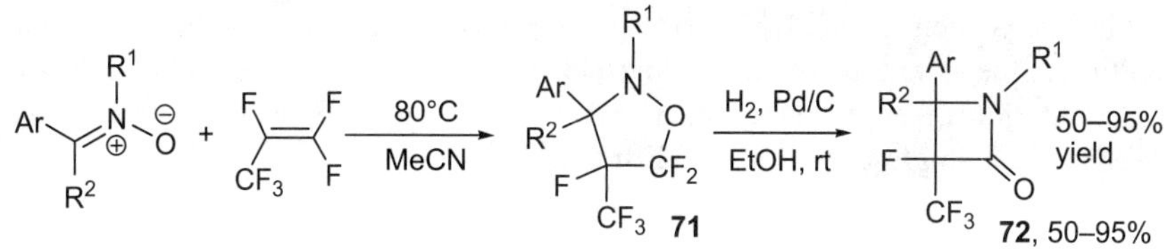

SCHEME 2.34 Synthesis of *β*-lactams **72**.

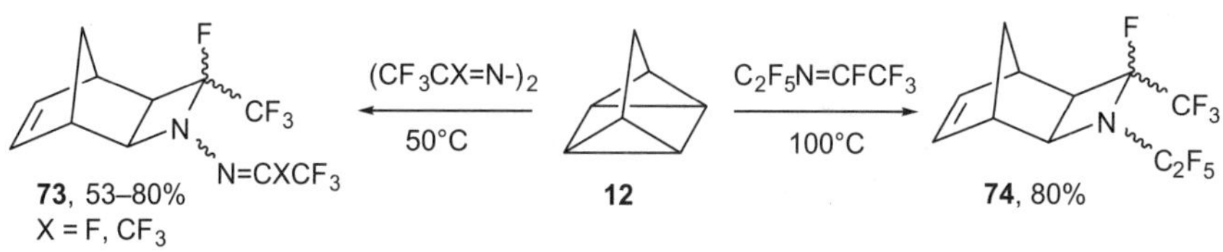

SCHEME 2.35 Synthesis of fluorinated norbornenazetidines.

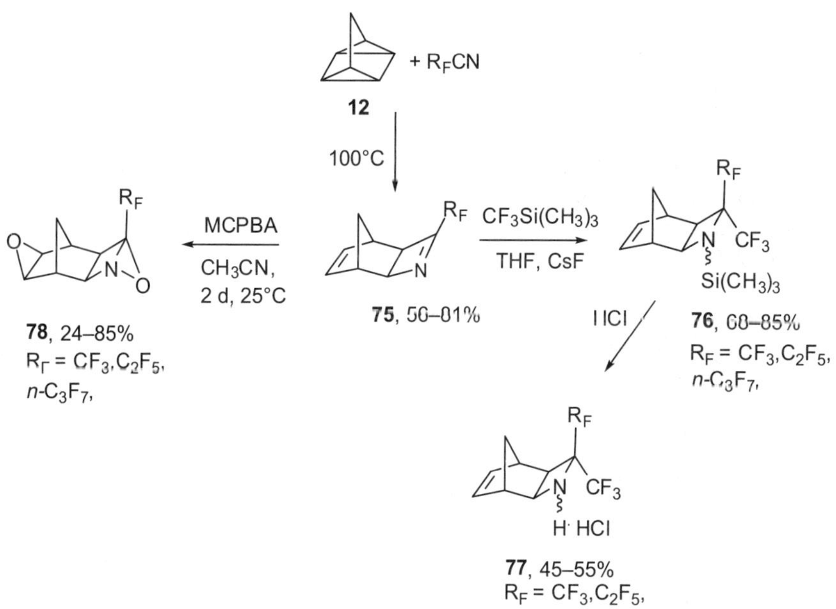

SCHEME 2.36 Preparation and chemical transformations of cycloadducts **75**.

out that the addition of silane is highly stereoselective, proceeding exclusively from less hindered *anti*-face of azabutane ring and resulting in the formation of single isomer.[90] Hydrolytically stable silanes **76** can be converted into the corresponding hydrochlorides **77** upon treatment with hydrogen chloride.[90] The reaction of **75** with *m*-chloroperoxibenzoic acid (MCPBA) proceeds with oxidation of both double bonds resulting in the formation of oxaziridines **78**. The structure of **78** ($R_F = C_2F_5$) having *exo*-orientation of epoxide fragment and *anti*-orientation of oxaziridine ring was established by single-crystal X-ray diffraction.[90]

2.5.3 Oxazetidines

The first report on the synthesis of fluorinated oxazetidine was published in 1955 by Barr and Haszeldine.[91] They isolated perfluoro-2-methyloxazetidine (**79**) from reaction of tetrafluoroethylene with trifluoronitrosomethane, along with its polymer. This reaction proceeds even at room temperature, but higher temperature (about 100°C) favors formation of **79**. The compound is inert to aqueous base or acid and UV light, but its pyrolysis at 550°C yields equimolar amounts of carbonyl fluoride and perfluoro-2-azapropene (Scheme 2.37).

Later, the reactions of other fluorinated ethylenes with different perfluoroalkylnitroso compounds to give analogous oxazetidines were reported,[92,93] followed by a comprehensive review on synthesis and properties of fluorinated nitroso polymers.[94] The reactions of oxazetidines are limited. Similarly to **79**, these compounds undergo retrocycloaddition reaction at elevated temperature with extrusion of $CF_2{=}O$ and clean formation of the corresponding imidoyl fluorides $CF_2{=}N\text{–}R_F$. Functional oxazetidines can be prepared by the reaction of highly active $ONOSO_2F$[95] and $ONOSO_2CF_3$[96] with two equivalents of fluoroethylenes. For example, when TFE reacts with $ONOSO_2F$, the first molecule of the olefin inserts into N–O bond to give nitroso derivative **80** followed by cycloaddition of the second molecule to the nitroso group to produce **81**. It should be pointed out that substantial amount of polymeric materials is always formed in these reactions (Scheme 2.38).[95]

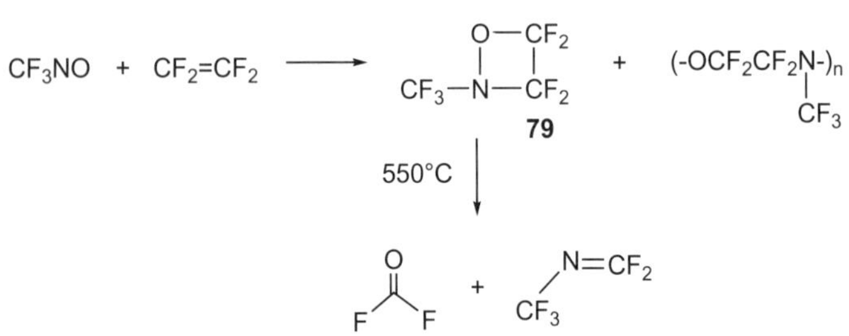

SCHEME 2.37 Synthesis of azetidine **79** and its pyrolysis.

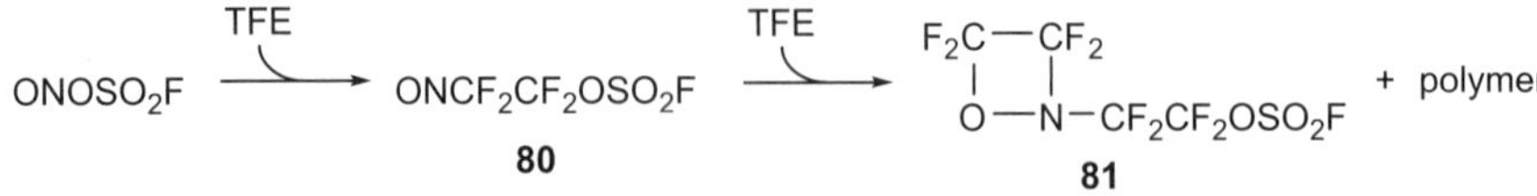

SCHEME 2.38 Reaction of $NOSO_3F$ with tetrafluoroethylene.

However, the hexafluoropropene does not undergo cycloaddition with $ONOSO_2F$ even at elevated temperature and the cycloaddition of $CFCl{=}CF_2$ proceeds regioselectively, leading to the isomer with CFCl group in α-position to nitrogen.[95]

REFERENCES

1. Kauck, E.A.; Simons, J.H. Cyclic perfluoroalkylene oxides. *US Patent* 2,594,271, **1952**.
2. Otsuka, T.; Aoyama, H. Preparation of 2,2,3,3,4-pentafluorooxetane and/or 2,2,3,3,4,4-hexafluorooxetane from 2,2,3,3-tetrafluorooxetane. *US Patent* 5,965,779, **1998**.
3. Case, L.C.; Todd, C.C. Poly(perfluoroalkyloxetanes): a new class of thermally stable polymers. *J. Polym. Sci.* **1962**, *58*, 633–638.
4. Liska, F.; Dedek, V.; Holik, M. Organic fluorine compounds. VIII. Cyclization of fluorochloroalkanols to fluorinated derivatives of oxetane and tetrahydropyran. *Collect. Czech. Chem. Commun.* **1970**, *35*(4), 1208–1215.
5. Chambers, R.D.; Gilani, A.H.S.; Gilbert, A.F.; Hutchinson, J.; Powell, R.L. Free-radical chemistry. Part XII. Radical reactions of trifluoroethene. *J. Fluorine Chem.* **2000**, *106*(1), 53–67.
6. Harris, J.F., Jr.; Coffman, D.D. Synthesis of polyfluorooxetanes by photoinitiated addition of fluorocarbonyl compounds to fluoro olefins. *J. Am. Chem. Soc.* **1962**, *84*, 1553–1561.
7. Eger, E.I., II; Lemal, D.; Laster, M.J.; Liao, M.; Jankowska, K.; Raghavanpillai, A.; Popov, A.V.; Gan, Y.; Lou, Y. Anesthetic properties of some fluorinated oxolanes and oxetanes. *Anesth. Analg.* **2007**, *104*(5), 1090–1097.
8. Lemal, D.M.; Raghavanpillai, A.Unpublished results.
9. Bissell, E.R.; Fields, D.B. Addition of acetaldehydes to fluoroethylenes. *J. Org. Chem.* **1964**, *29*(1), 249–252.
10. Cook, E.W.; Landrum, B.F. Synthesis of partially fluorinated oxetanes. *J. Heterocycl. Chem.* **1965**, *2*(3), 327–328.
11. Barlow, M.G.; Coles, B.; Haszeldine, R.N. Heterocyclic polyfluoro-compounds. Part 33. Photochemical oxetane formation from fluoro ketones and perfluoro aldehydes and 1,2-difluoroethylene. *J. Chem. Soc., Perkin Trans. 1* **1980**, (10), 2258–2267.
12. Tarrant, P.; Bull, R.N. The reaction of some 3- and 4-fluorooxetanes with acids. *J. Fluorine Chem.* **1988**, *40*(2–3), 201–215.
13. Schlechter, M.M.; Woolf, C. 2,2-Bis(trifluoromethyl)oxetanes. *FR Patent* 1,391,493, **1965**.
14. Kotun, S.P.; DesMarteau, D.D. Superacid-induced ring-opening reactions of fluorinated heterocycles. *Can. J. Chem.* **1989**, *67*(11), 1724–1728.
15. Lemal, D.M.; Raghavanpillai, A.Unpublished results.
16. Weinmayr, V. Hydrogen fluoride as a condensing agent. VI. Reactions of fluoro olefins with formaldehyde in hydrogen fluoride. *J. Org. Chem.* **1963**, *28*, 492–494.
17. Ikeda, S.; Sonoi, T. Process for producing 2,2,3,3-tetrafluorooxetane. *WO Patent* 2005080365, **2005**.
18. Ohsaka, Y.; Takaki, S. Substituted trifluorooxetanes. *US Patent* 4,709,060, **1987**.
19. German, L.S.; Knunyants, I.L. Conjugated additions in hydrogen fluoride. *Angew. Chem., Int. Ed. Engl.* **1969**, *8*(5), 349–356.

20. Belen'kii, G.G.; Savicheva, G.I.; Lur'e, E.P.; German, L.S. Condensation of trifluorotrichloroacetone with fluoroolefins in the presence of antimony pentafluoride. *Izv. Akad. Nauk SSSR, Ser. Khim.* **1978**, (6), 1430–1432.
21. Postovoi, S.A.; Kagramanova, E.M.; Mysov, E.I.; Zeifman, Y.V. Reductive addition of polychloro(fluoro)alkanes to fluoro carbonyl compounds. *Izv. Akad. Nauk SSSR, Ser. Khim.* **1996**, (6), 1496–1498.
22. Krespan, C.G.; Petrov, V.A. The chemistry of highly fluorinated carbocations. *Chem. Rev.* **1996**, *96*(8), 3269–3301.
23. Krespan, C.G.; Sievert, A.C.; Weigert, F.J. Process for chlorofluoropropanes. *WO Patent* 9105753, **1991**.
24. Krespan, C.G.; Dixon, D.A. Fluoroolefin condensation catalyzed by aluminum chlorofluoride. *J. Fluorine Chem.* **1996**, *77*(2), 117–126.
25. Petrov, V.A.; Davidson, F.; Smart, B.E. A new synthesis of fluorinated oxetanes. *J. Org. Chem.* **1995**, *60*(11), 3419–3422.
26. Petrov, V.A.; Davidson, F.; Smart, B.E. Quadricyclane—thermal cycloaddition to polyfluorinated carbonyl compounds. *J. Fluorine Chem.* **2004**, *125*(10), 1543–1552.
27. Petrov, V.A. Stereoselective synthesis of polyfluorinated *exo*-tricyclononenes and norbornenes. *ACS Symp. Ser.* **2007**, *949* (Current Fluoroorganic Chemistry), 113–140.
28. Gambaryan, N.P.; Simonyan, L.A.; Petrovskii, P.V. Reaction of fluoro ketones with vinyl esters. *Izv. Akad. Nauk SSSR, Ser. Khim.* **1967**, (4), 918–921.
29. Davis, H.R.Fluorinated oxetanes. *US Patent* 3,164,610, **1965**.
30. Middleton, W.J. Isolation of a cyclic intermediate in the ketone–alkoxyacetylene reaction. *J. Org. Chem.* **1965**, *30*(4), 1307.
31. Barlow, M.G.; Coles, B.; Haszeldine, R.N. Heterocyclic polyfluoro compounds. Part 34. Two-way photochemical addition of hexafluoroacetone to 1,2-dichlorofluoroethylene and the preparation of an oxete. *J. Fluorine Chem.* **1980**, *15*(5), 381–386.
32. Hollander, J.; Woolf, C. Polyfluoro-oxetanes and derivatives thereof. *US Patent* 3,362,963, **1968**.
33. Ohsaka, Y. Perfluoropolyether fluids (Demnum) based on oxetanes. In Organofluorine Chemistry. Principles and Commercial Applications; Banks, R.E., Ed.; Plenum Press: New York, **1994**; pp 463–467.
34. Ohsaka, Y.; Tohzuka, T.; Takaki, S.; Negishi, Y.; Kohnos, S.2,2-Difluoropropionic acid derivatives. *US Patent* 4,719,052, **1988**.
35. Ohmori, A.; Takaki, S.; Kitahara, T.α-Fluoroacrylic acid esters and acid fluoride. *US Patent* 4,604,482, **1986**.
36. Knunyants, I.L.; Sokolskii, G.A. Fluorinated β-sultones. *Angew. Chem., Int. Ed. Engl.* **1972**, *11*(7), 583–595.
37. Mohtasham, J.; Gard, G.L. β-Fluorosultones: synthesis, reactivity, structure and uses. *Coord. Chem. Rev.* **1992**, *112*, 47–79.
38. Jiang, S.H.-K. Addition reactions of sulfur trioxide to carbon–carbon double bonds. I. Addition products of sulfur trioxide and some perfluoro olefins. *Huaxue Xuebao* **1957**, *23*, 330–339.
39. England, D.C. α-Sulfopolyfluorocarboxylic acids and derivatives. *US Patent* 2,852,554, **1958**.

40. Dmitriev, M.A.; Sokol'skii, G.A.; Knunyants, I.L. Addition of sulfur trioxide to fluoro olefins. *Khim. Nauka i Prom.* **1958**, *3*, 826–828.

41. England, D.C.; Dietrich, M.A.; Lindsey, R.V., Jr. Reactions of fluoroolefins with sulfur trioxide. *J. Am. Chem. Soc.* **1960**, *82*, 6181–6188.

42. Krespan, C.G.; England, D.V. Perfluoroallyl fluorosulfate, a reactive new perfluoroallylating agent. *J. Am. Chem. Soc.* **1981**, *103*(18), 5598–5599.

43. Yamabe, M.; Miyake, H. Fluorinated membranes. In Organofluorine Chemistry: Principles and Commercial Applications; Banks, R.E., Ed.; Plenum: New York, **1994**; pp 403–411.

44. Sondheimer, S.J.; Bunce, N.J.; Fyfe, C.A. Structure and chemistry of Nafion-H: a fluorinated sulfonic acid polymer. *J. Macromol. Sci. Rev.* **1986**, *C26*(3), 351–411.

45. Middleton, W.J.; Howard, E.G.; Sharkey, W.H. Perfluorothiocarbonyl compounds. *J. Am. Chem. Soc.* **1961**, *83*, 2589–2590.

46. Martin, K.V. Reaction of fluoro olefins with sulfur. *J. Chem. Soc.* **1964**, (September), 2944–2947.

47. Dyatkin, B.L.; Sterlin, S.R.; Zhuravkova, L.G.; Martynov, B.I.; Mysov, E.I.; Knunyants, I.L. Reactions of perfluoroalkyl carbanions with sulfur. *Tetrahedron* **1973**, *29*(18), 2759–2767.

48. Middleton, W.J.; Howard, E.G.; Sharkey, W.H. Fluorothiocarbonyl compounds. I. Preparation of thio ketones, thioacyl halides, and thio esters. *J. Org. Chem.* **1965**, *30*(5), 1375–1384.

49. England, D.C. Fluoroketenes. 11. Synthesis and chemistry of a perfluoroacylketene and related compounds containing a perfluoroisopropyl sulfide group. *J. Org. Chem.* **1981**, *46*(1),153–157.

50. Middleton, W.J. Fluorothiocarbonyl compounds. III. Diels–Alder reactions. *J. Org. Chem.* **1965**, 1390–1394.

51. Middleton, W.J. Fluorothiocarbonyl compounds. IV. Hexafluorothioacetone–olefin adducts. *J. Org. Chem.* **1965**, 1395–1398.

52. Krespan, C.G.; McKusick, B.C.; Cairns, T.L. Dithietene and bicyclooctatriene ring systems from bis(fluoroalkyl)acetylenes. *J. Am. Chem. Soc.* **1960**, *82*, 1515–1516.

53. Krespan, C.G. Bis(polyfluoroalkyl) acetylenes. IV. Fluorinated dithietenes and related heterocyclic compounds from bis(polyfluoroalkyl) acetylenes and sulfur. *J. Am. Chem. Soc.* **1961**, *83*, 3434–3437.

54. England, D.C. 2,4-Bis(hexafluoroisopropylidene)-1,3-dithietane and fluorinated polysulfides useful as solvents for fluorinated polymers. *US Patent* 3,544,591, **1970**.

55. England, D.C.; Raasch, M.S.; Sheppard, W.A.Preparation of 2,4-bis(hexafluoroisopropylidene)-1,3-dithietane by reaction of perfluoroisobutene with alkali metal salts. *US Patent* 3,694,460, **1972**.

56. Raasch, M.S. Bis(trifluoromethyl)thioketene. I. Synthesis and cycloaddition reactions. *J. Org. Chem.* **1970**, *35*(10), 3470–3483.

57. Petrov, V.A.; Krespan, C.G.; Marshall, W. Reaction of quadricyclane with fluorinated sulfur-containing compounds. *J. Fluorine Chem.* **2005**, *126*(9–10), 1332–1341.

58. Postovoi, S.A.; Vol'pin, I.M.; Delyagina, N.I.; Galakhov, M.V.; Zeifman, Y.V.; German, L. S. Sulfur anions from 2,2,4,4-tetrakis(trifluoromethyl)-1,3-dithietane and their reactions. *Izv. Akad. Nauk SSSR, Ser. Khim.* **1989**, (6), 1380–1383. (a) Nikolaeva, E. E.;

Snegirev, V. F.; Makarov, K. N. New stable fluorine-containing thiete. *Izv. Akad. Nauk SSSR, Ser. Khim.* **1987**, (4), 957–958. (b) Popkova, V. Ya.; Snegirev, V. F.; Antipin, M. Yu.; Struchkov, Yu. T. Perfluorinated aliphatic enethiols and a, b -unsaturated sulfenyl chlorides. *Izv. Akad. Nauk SSSR, Ser. Khim.* **1989**, (5), 1123–1129.

59. Bell, A.N.; Fields, R.; Haszeldine, R.N.; Moran, D. Fluoro olefin chemistry. Part 13. A further route to perfluoro-2,3-dimethylbut-2-ene and the photochemical rearrangement of some perfluoroalkyl olefins. *J. Chem. Soc., Perkin Trans. 1* **1980**, (2), 487–489.
60. Burton, D.J.; Inouye, Y. The *in situ* generation of 2-(triphenylphosphoranylidene)hexafluoropropane. A facile one pot conversion of aldehydes to bistrifluoromethyl olefins. *Tetrahedron Lett.* **1979**, (36), 3397–3400.
61. Kadyrov, A.A.; Rokhlin, E.M. (Fluoroalkenyl)phosphonates. *Usp. Khim.* **1988**, *57*(9), 1488–1509.
62. Kadyrov, A.A.; Rokhlin, E.M. Perfluoroisopropylidene and perfluoroisopropenyl derivatives of phosphorus. 1. Synthesis of perfluoroisopropenyl phosphonates. *Izv. Akad. Nauk SSSR, Ser. Khim.* **1981**, (11), 2583–2589.
63. Petrov, V.A. Abstracts of 15th European Symposium on Fluorine Chemistry, Prague, Czech Republic, July 15–20, **2007**, p 161. (a) Petrov, V.A.; Marshall, W.J. Reactions of Polyfluorinated Thietanes. Selective Synthesis of Polyfluorinated Thietanes S-Oxides and 2-Substituted 5-fluoro-4-(trifluoromethyl)-2,3-dihydrothiophenes, *J. Fluorine Chem.* **2009**, doi:10.1016/j.jfluchem.2009.06.011.
64. Kitazume, T.; Otaka, T.; Takei, R.; Ishikawa, N. Preparation and reactions of 4-alkoxy-2,2-bis(trifluoromethyl)thietanes and 5-alkoxy-3,3-bis(trifluoromethyl)dithiolanes. *Bull. Chem. Soc. Jpn.* **1976**, *49*(9), 2491–2494.
65. Carboni, R.A.; Kauer, J.C. Polyfluorinated cyclic sulfones. *US Patent* 3,058,993, **1962**.
66. Sundermeyer, W. Perhalogenated sulfines (thione oxides) and sulfenes (thione dioxides): their synthesis and synthetic potential. *Synthesis* **1988**, (5), 349–359.
67. Elsaesser, A.; Sundermeyer, W.; Stephenson, D.S. Synthesis and properties of tetrakis (trifluoromethyl)-1,3-dithietane *S*-oxides and bis(trifluoromethyl)sulfine. *Chem. Ber.* **1985**, *118*(1), 116–123.
68. Petrov, V.A.; Lustig, S.; Marshall, W. Reactivity of fluorinated sulfur-containing heterocycles towards nucleophilic and oxidizing reagents. *J. Fluorine Chem.* **2007**, *128*(10), 1227–1234.
69. Raasch, M.S. Heteroatom participation during addition–rearrangement reactions of 2-thia- and 2-azanorbornenes. *J. Org. Chem.* **1975**, *40*(2), 161–172.
70. Elsaesser, A.; Sundermeyer, W. Bis(trifluoromethyl)sulfine, $(CF_3)_2C{=}SO$. *Tetrahedron Lett.* **1983**, *24*(21), 2141–2142.
71. Elsaesser, A.; Sundermeyer, W. Bis(trifluoromethyl)sulfine, $(CF_3)_2C{=}SO$. New syntheses and some reactions. *Chem. Ber.* **1985**, *118*(11), 4553–4560.
72. Smart, B.E.; Middleton, W.J. Bis(trifluoromethyl)sulfene: generation and cycloaddition reactions. *J. Am. Chem. Soc.* **1987**, *109*(16), 4982–4992.
73. Knunyants, I.L.; Cheburkov, Y.A. Fluorine containing β-lactones. I. β,β-Bis(trifluoromethyl)-β-propiolactone and its properties. *Izv. Akad. Nauk SSSR, Ser. Khim.* **1960**, 678–685.
74. England, D.C.; Krespan, C.G. Bis(trifluoromethyl)ketene. *J. Am. Chem. Soc.* **1965**, *87*(17), 4019–4020.
75. England, D.C.; Krespan, C.G. Fluoroketenes. I. Bis(trifluoromethyl)ketene and its reactions with fluoride ion. *J. Am. Chem. Soc.* **1966**, *88*(23), 5582–5587.

76. England, D.C.; Krespan, C.G. Fluoroketenes. II. Difluoroketene. *J. Org. Chem.* **1968**, *33*(2),816–819.
77. Dolbier, W.R., Jr.; Ocampo, R.; Paredes, R. α,α-Difluoro β-lactones and their thermal decarboxylation to 1,1-difluoroalkenes. *J. Org. Chem.* **1995**, *60*(17), 5378–5379.
78. Ojima, I.; McCarthy, J.R.; Welch, J.T. (Eds.) Biomedical frontiers of fluorine chemistry. ACS Symp. Ser. **1996**, *639*, 356.
79. Ojima, I.; Inoue, T.; Slater, J.C.; Lin, S.; Kuduk, S.D.; Chakravarty, S.; Walsh, J.J.; Cresteil, T.; Monsarrat, B.; Pera, P.; Bernacki, R.J. Synthesis of enantiopure fluorine-containing taxoids and their use as anticancer agents as well as probes for biomedical problems. *ACS Symp. Ser.* **2000**, *746*, 158–181.
80. Ojima, I. Use of fluorine in the medicinal chemistry and chemical biology of bioactive compounds: a case study on fluorinated taxane anticancer agents. *ChemBioChem* **2004**, *5*(5), 628–635.
81. Ojima, I.; Kuznetsova, L.; Ungureanu, I.M.; Pepe, A.; Zanardi, I.; Chen, J. Fluoro-β-lactams as useful building blocks for the synthesis of fluorinated amino acids, dipeptides, and taxoids. *ACS Symp. Ser.* **2005**, *911*, 544–561.
82. Ojima, I.; Kuznetsova, L.V.; Sun, L. Organofluorine chemistry at the biomedical interface: a case study on fluoro-taxoid anticancer agents. *ACS Symp. Ser.* **2007**, *949*, 288–304.
83. Taguchi, T.; Kitagawa, O.; Suda, Y.; Ohkawa, S.; Hashimoto, A.; Iitaka, Y.; Kobayashi, Y. Synthesis of 3,3-difluoro-2-azetidinones and 2,3-dideoxy-2,2-difluoro-3-aminosugars through the Reformatskii reaction of difluoroacetate with imine. *Tetrahedron Lett.* **1988**, *29*(41), 5291–5294.
84. Sato, K.; Tarui, A.; Matsuda, S.; Omote, M.; Ando, A.; Kumadaki, I. The simple and selective synthesis of 3-amino 2,2-difluoro carboxylic esters and difluoro β-lactams using ethyl bromodifluoroacetate in the presence of rhodium catalyst. *Tetrahedron Lett.* **2005**, *46*(45), 7679–7681.
85. Boyer, N.; Gloanec, P.; De Nanteuil, G.; Jubault, P.; Quirion, J.-C. Chemoselective and stereoselective synthesis of *gem*-difluoro-β-aminoesters or *gem*-difluoro-β-lactams from ethyl bromodifluoroacetate and imines during Reformatsky reaction. *Tetrahedron* **2007**, *63*(50), 12352–12366.
86. Wu, M.J.; Pridgen, L.N. Synthesis of chiral α-alkyl phenethylamines via organometallic addition to chiral 2-aryl-1,3-oxazolidines. *J. Org. Chem.* **1991**, *56*(3), 1340–1344.
87. Bigdeli, M.A.; Tipping, A.E. Reaction of nitrones with internal fluoroalkenes and with tetrafluorobenzyne. *J. Fluorine Chem.* **1992**, *58*(1), 101–110.
88. Jakowiecki, J.; Loska, R.; Makosza, M. Synthesis of α-trifluoromethyl-β-lactams and esters of β-amino acids via 1,3-dipolar cycloaddition of nitrones to fluoroalkenes. *J. Org. Chem.* **2008**, *73*(14), 5436–5441.
89. Barlow, M.G.; Suliman, N.N.E.; Tipping, A.E. Unsaturated nitrogen compounds containing fluorine. Part 18. Novel cycloaddition products from the reaction of quadricyclane with fluorinated azines and the azomethine imide 2-hexafluoroisopropylidene-3-methyl-3-phenyl-5,5-bis(trifluoromethyl)tetrahydropyrazol-2-ium-1-ide. *J. Fluorine Chem.* **1995**, *72*(1), 147–156.
90. Petrov, V.A.; Davidson, F.; Marshall, W. Reactions of quadricyclane with fluorinated nitrogen-containing compounds. Synthesis of 3-aza-4-perfluoroalkyl-tricyclo[4.2.1.$0^{2,5}$] non-3,7-dienes. *J. Fluorine Chem.* **2004**, *125*(11), 1621–1628.

91. Barr, D.A.; Haszeldine, R.N. Perfluoroalkyl derivatives of nitrogen. I. Perfluoro-2-methyl-1,2-oxazetidine and perfluoro (alkylene alkylamines). *J. Chem. Soc.* **1955**, 1881–1889.
92. Barr, D.A.; Haszeldine, R.N. Perfluoroalkyl derivatives of nitrogen. III. Heptafluoronitrosopropane, perfluoro-2-*n*-propyl-1,2-oxazetidine, perfluoro(methylene-*n*-propylamine), and related compounds. *J. Chem. Soc.* **1956**, 3416–3428.
93. Barr, D.A.; Haszeldine, R.N. Perfluoroalkyl derivatives of nitrogen. VII. Reactions of nitrosyl halides with tetrafluoroethylene. *J. Chem. Soc.* **1960**, 1151–1155.
94. Henry, M.C.; Griffis, C.B.; Stump, E.C. Synthesis, compounding, and properties of nitroso rubbers. *Fluorine Chem. Rev.* **1967**, *1*(1), 1–75.
95. Fokin, A.V.; Rapkin, A.I.; Chilikin, V.G.; Verenikin, O.V.; Studnev, Y.N. Synthesis and properties of 2-nitrosopolyfluoroalkyl fluorosulfates. *Izv. Akad. Nauk SSSR, Ser. Khim.* **1986**, (4), 891–898.
96. Savostin, V.S.; Krylov, I.I.; Kutepov, A.P.; Rapkin, A.I.; Fokin, A.V. Interaction of nitrosonium trifluoromethanesulfonate with fluoro olefins and properties of the products. *Izv. Akad. Nauk SSSR, Ser. Khim.* **1990**, (8), 1851–1855.

3

FLUORINATED FIVE-MEMBERED NITROGEN-CONTAINING HETEROCYCLES

KENNETH L. KIRK
Laboratory of Bioorganic Chemistry, National Institute of Diabetes and Digestive and Kidney Diseases, National Institutes of Health, DHHS, Bethesda, MD 20892, USA

3.1 INTRODUCTION

There is much published material on the chemistry and biological properties of fluorinated analogues of five-membered heterocycles containing nitrogen. For this review, the decision was made to organize the material according to compound class, proceeding logically from heterocycles containing one nitrogen (pyrroles and relatives) to tetrazoles followed by a short discussion of selected perfluorinated derivatives. An emphasis on synthetic methodology will be apparent to the exclusion of some important applications.

3.2 FLUORINATED HETEROCYCLES CONTAINING ONE NITROGEN

3.2.1 Fluorinated Pyrrolidines

The pyrrolidine ring is found abundantly in both naturally occurring and synthetic biologically active compounds. The alkaloid nicotine, an agonist for the nicotinic acetylcholine receptor, is a notorious example of the former. Pyrrolidine itself is found in carrot and tobacco leaves. The prevalence of pyrrolidines in bioactive compounds has led to much interest in the development of flouropyrrolidine building blocks. The

Fluorinated Heterocyclic Compounds: Synthesis, Chemistry, and Applications, Edited by Viacheslav A. Petrov

chemistry of fluorinated analogues of proline, the pyrrolidine-containing amino acid, will be discussed in a subsequent section.

In a majority of cases, fluorinated pyrrolidines have been made by nucleophilic fluorination of precursor alcohols or ketones. Giardina and coworkers synthesized (*R*, *S*)-, (*R*)-, and (*S*)-3-fluoropyrrolidines and 3,3-difluoropyrrolidine to study the effects of fluorine substitution on a series of kappa opioid agonist analgesics (**1**).[1]

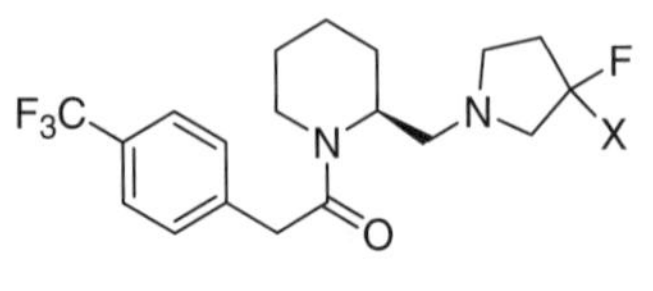

X = F, H

1

For the purpose of preparing such analogues, racemic 3-fluoropyrrolidine (*R*,*S*)-**2** was prepared by nucleophilic displacement of *N*-benzyl-3-tosyloxypyrrolidine with KF. (2*S*,4*R*)-3-Hydroxyproline was the starting point for both (3*R*)- and (3*S*)-3-fluoropyrrolidine. Decarboxylation and protection with the carbobenzloxy (Cbz) group was followed by tosylation and displacement with fluoride and deprotection. A Mitsunobo inversion of the intermediate 3-hydroxypyrrolidine provided access to the other enantiomer by the same route (Fig. 3.1).[1]

3,3-Difluoropyrrolidine (**3**) was prepared by oxidation of the intermediate 3-hydroxypyrrolidine and subsequent fluorodeoxygenation with diethylaminosulfur trifluoride(DAST) (Fig. 3.2).[1]

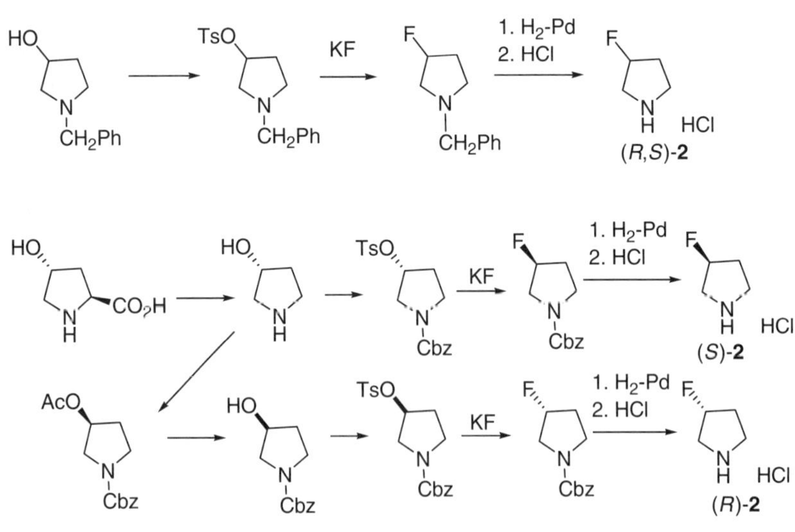

FIGURE 3.1 Synthesis of 3-fluoropyrrolidine.

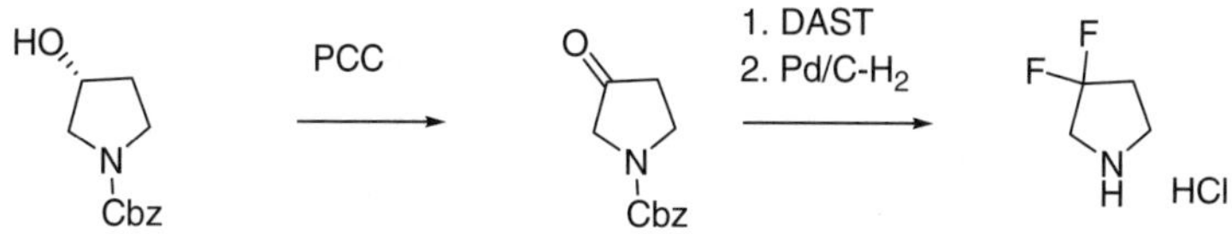

FIGURE 3.2 Synthesis of 3,3-difluoropyrrolidine.

A recent important application of fluorinated pyrrolidines in drug design has sustained interest in their synthesis. This relates to the identification of a series of fluoropyrrolidine amides as important leads for the development of peptidyl peptidase IV (DPP-IV) inhibitors.[2]Glucagon-like peptide 1 (GLP-1) regulates glucose homeostasis by stimulating biosynthesis and release of insulin and inhibiting glucagon release. Since it is rapidly deactivated by DPP-IV, inhibitors of this protease have been developed as a strategy to treat type 2 diabetes. Examples of fluoropyrrolidine-based inhibitors include the α-amino amide **4**,[3] the oxadiazole **5**,[4] the fluoropyrroldine amide **6** developed by Merck & Co.[5] and the amides **7** and **8** developed by Pfizer (Fig. 3.3).[6] The synthesis of these and other fluoropyrrolidine-based leads required preparation of the various fluoropyrrolidine building blocks.

In their work on DPP-IV inhibitors, Caldwell et al.[5] prepared the requisite monofluorinated pyrrolidines (*R*)- and (*S*)-**2** by a variation of the procedure used

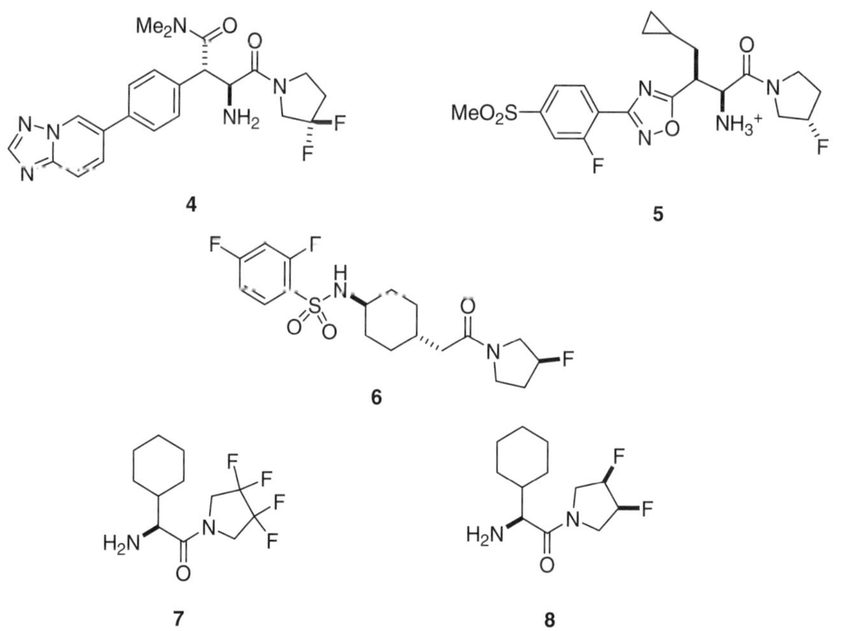

FIGURE 3.3 Fluoropyrrolildine amides in the development of peptidyl peptidase IV inhibitors.

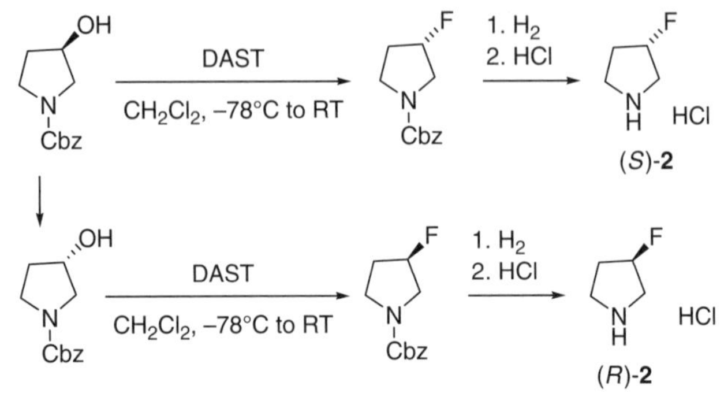

FIGURE 3.4 Synthesis of the enantiomers of 3-fluoropyrrolidine.

by Giardina et al.[1] described above. In their synthesis, Caldwell and coworkers used fluorodeoxygenation with DAST in the fluorination step (Fig. 3.4).

The two enantiomers of *trans*-3,4-difluoropyrrolidine, (3*R*,4*R*)-**9** and (3*S*,4*S*)-**9**, were prepared from D- and L-tartaric acids. Figure 3.5 illustrates the sequence starting with L-tartaric acid[5] to produce (3*R*,4*R*)-**9**. This was conceptually similar to the first enantioselective syntheses of vicinal *N*-substituted difluoropyrrolidines carried out previously.[7] In this previous work, the requisite *trans*-2,3-dihydroxypyrrolidine derivatives were made by reduction of diacetoxy succinimides.

Hulin and coworkers also prepared dipeptidyl peptidase inhibitors incorporating fluorinated pyrrolidines.[6] These fluoropyrrolidines included tetrafluoropyrrolidine (**10**) as well as *trans*- (**9**) and *cis*-3,4-difluoropyrrolidine (**11**). Tetrafluoropyrrolidine (**10**) was prepared by cyclization of tetrafluorobutane diol, as shown in Fig. 3.6. Compound **10** had previously been prepared by reduction of tetrafluorosuccinimide with $LiAlH_4$[8] or borane.[9]

Hulin and coworkers[6] prepared the enantiomers of *trans*-3,4-difluoropyrrolidines **9** by a sequence similar to that used by Caldwell et al. The previously unreported *meso*-isomer, *cis*-3,4-difluoropyrrolidine **11**, was prepared from the epoxide derived

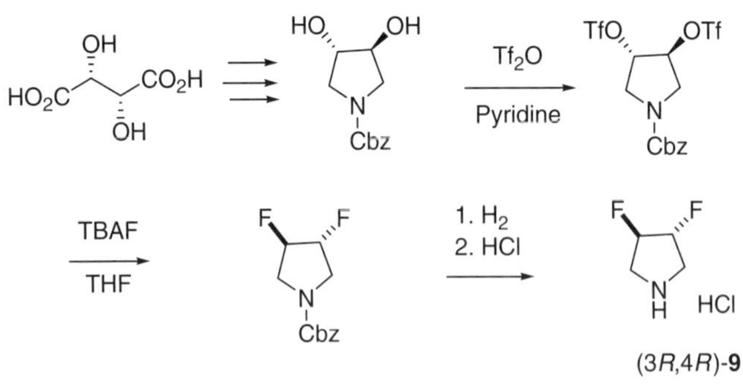

FIGURE 3.5 Synthesis of difluoropyrrolidines.

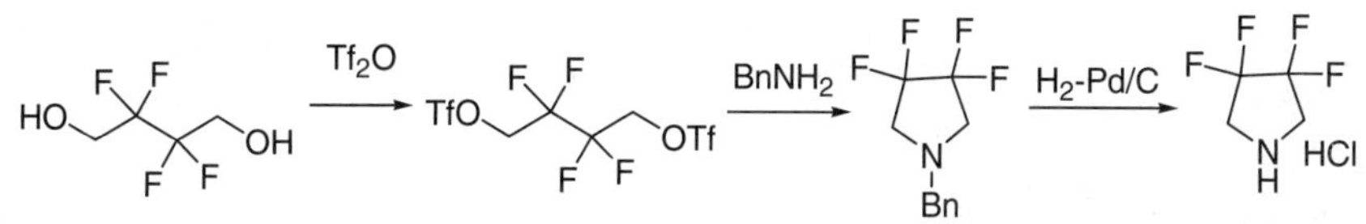

10

FIGURE 3.6 Synthesis of tetrafluoropyrrolidine **10**.

from the olefin prepared by Grubbs cyclization of *t*-butyloxycarbonyl (Boc)-protected diallylamine. The epoxide was opened by nucleophilic fluoride displacement and the resulting fluorohydrin was treated with DAST (Fig. 3.7).

In the various projects already described, a fairly complete inventory of simple fluorinated pyrrolidines has been assembled. Trifluoropyrrolidines have apparently not been made. Many fluoropyrrolidines possessing additional C-functionality have been prepared, including those derived from hydroxyproline through fluoroproline intermediates, as discussed below.

The syntheses of simple fluoropyrrolidines described above were based on the installation of the C–F bond through nucleophilic fluorination. An example of preparation of fluoropyrrolidine derivatives through electrophilic fluorination is found in the synthesis of a new fluoroquinolone antibiotic **12**.[10] The intermediate lactam **13** was fluorinated with *N*-fluorobenzenesulfonimide (NFSi) to give a single diastereomer, the monofluoropyrrolidinone **14**. Treatment of **14** with base followed by quenching with 2,6-di-*tert*-butylphenol produced the other diastereomer **15**. Further fluorination of **14** gave the difluoropyrrolidinone **16**. Following deoxygenation, the resulting fluoropyrrolidines **17** were used for the preparation of **12** (Fig. 3.8).

3.2.2 Fluorinated Prolines

(*S*)-Proline (**18**) is a nonessential proteogenic amino acid that plays important and unique roles in protein structure because of its conformational rigidity. In proteins, it is commonly found at the beginning of an α-helix, in turns, and it disrupts secondary

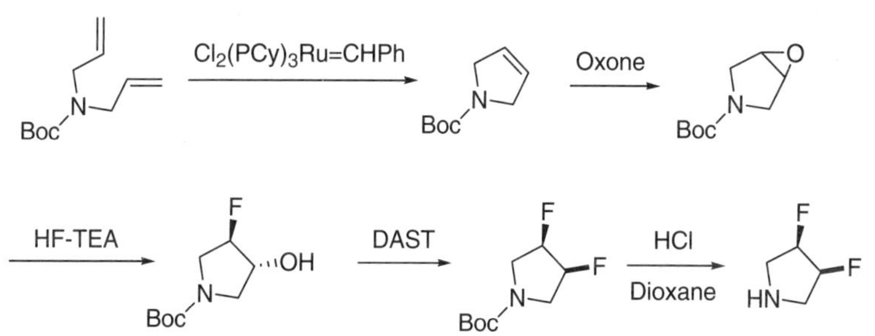

11

FIGURE 3.7 Synthesis of *meso*-3,4-difluoropyrrolidine.

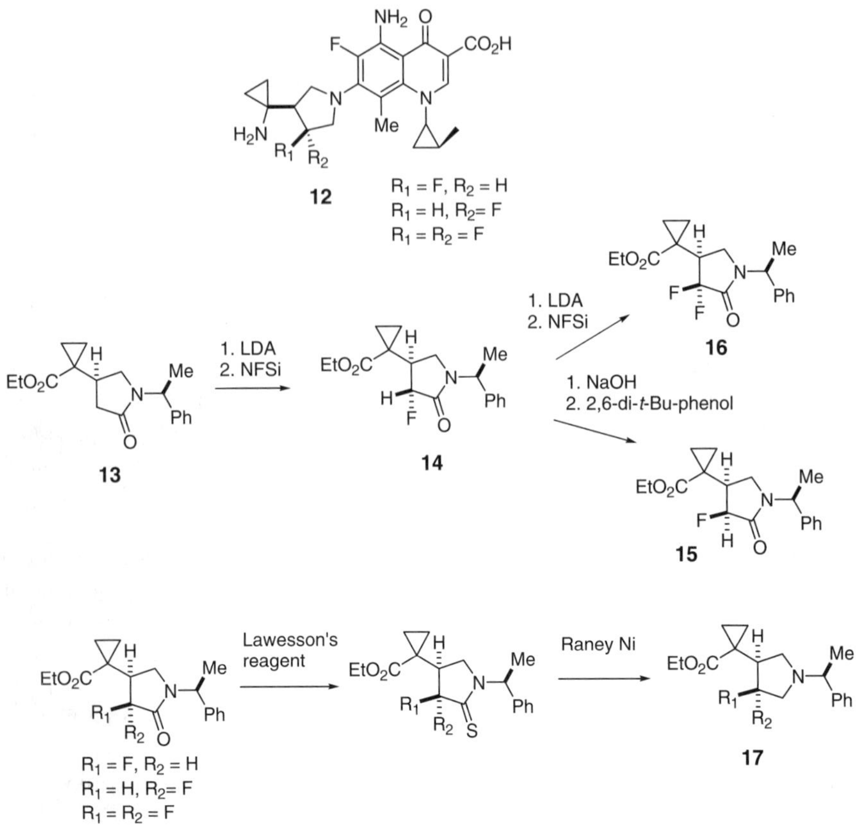

FIGURE 3.8 Preparation of an intermediate for synthesis of the new fluoroquinolone antibiotics **12**.

structural elements such as α-helices and β-sheets. Proline and (2*S*,4*R*)-hydroxyproline (**19**) are critical components of collagen. Accordingly, fluorinated analogues of proline and hydroxyproline have received much attention as tools to study the special properties of proline-containing peptides and proteins. In addition, owing to the convenient presence of ring functionality (the carboxyl group), fluorinated prolines have been used as key intermediates for other fluoropyrrolidine building blocks.

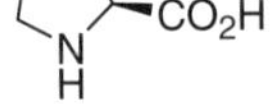

18

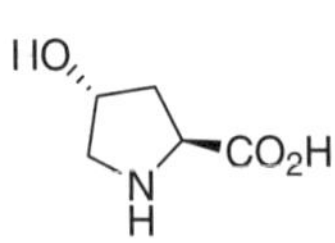

19

3.2.2.1 Synthesis of Fluoroprolines The first syntheses of (2*S*,4*R*)-fluoroproline (4*R*-FPro) (**20**) and (2*S*,4*S*)-fluoroproline (4*S*-FPro) (**21**) were reported in 1965[11] in

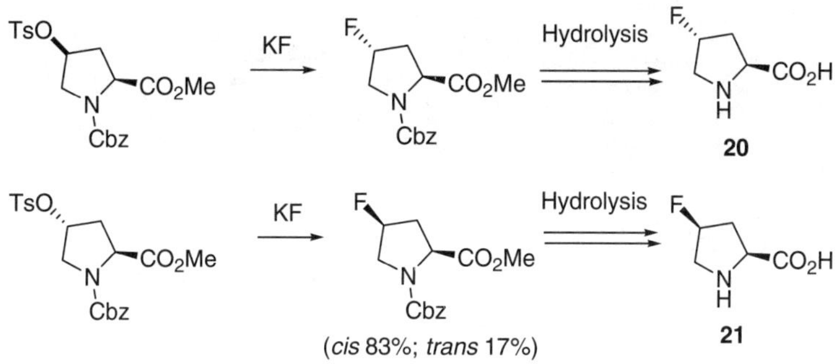

FIGURE 3.9 Synthesis of 4-fluoroprolines.

research on the incorporation of these analogues into protein. The fluorinated prolines were prepared by fluoride displacement of sulfonic esters derived from the two diastereomers of *N*-protected hydroxyproline (Fig. 3.9). Erosion of stereoselectivity of the *trans*-sulfonate was ascribed to participation of the ester carbonyl oxygen.

(2*S*)-4,4-Difluoroproline (**22**) was first reported in 1977, synthesized as a potential inhibitor of collagen biosynthesis.[12] The chiral tricyclic diketopiperazine was prepared from (2*S*,4*R*)-4-hydroxyproline methyl ester and oxidized to the diketone. Fluorination with sulfur tetrafluoride and hydrolysis produced **22** (Fig. 3.10).

Burger and coworkers reported syntheses of (2*S*)-4,4-difluoroproline **22** and (2*S*,4*R*)-fluoroproline **20** from (*S*)-aspartic acid.[13] This amino acid was converted to the diazoketone **23** in three steps, an intermediate that contains the requisite carbon skeleton for the proline derivatives. Rhodium-catalyzed cyclization produced the diketoproline derivative **24**, and this was fluorinated with DAST to give **25**, hydrolysis of which produced **22**. Stereoselective reduction of **24** was directed by the concave face to give **26**. Fluorination with inversion and hydrolysis produced **20** (Fig. 3.11).

The syntheses of Boc- and Fmoc (9*H*-fluoren-9-ylmethoxycarbonyl)-protected (2*S*,4*R*)- and (2*S*,4*S*)-fluoroprolines from (2*S*,3*R*)-hydroxyproline have been reported.[14] This synthesis featured a Mitsunobo inversion of the hydroxyproline and fluorination with DAST. A similar synthesis was reported that used morpholinosulfur trifluoride as the fluorinating agent.[15]

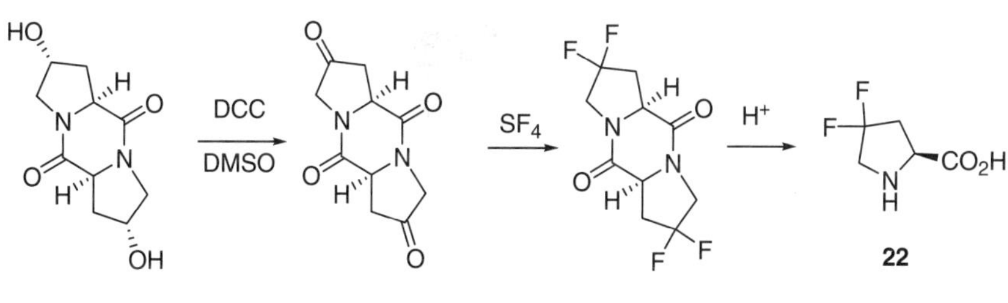

FIGURE 3.10 Synthesis of (2*S*)-4,4-difluoroproline.

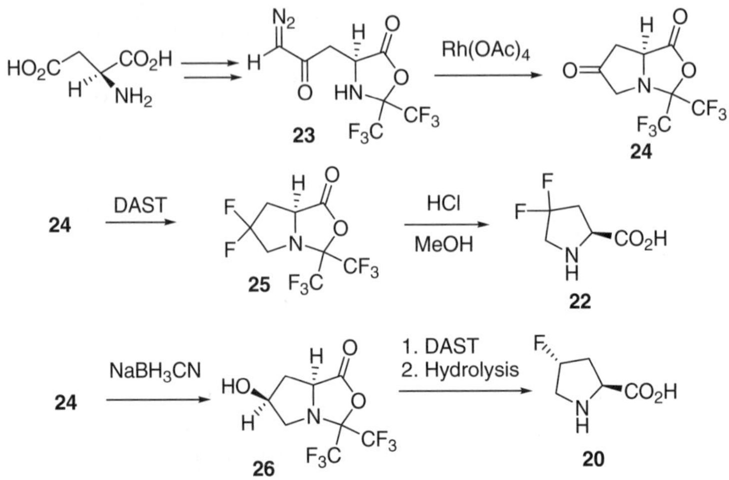

FIGURE 3.11 Synthesis of fluoroprolines from aspartic acid.

β,β-difluoro-α-amino acids have been prepared by transformations of δ,ε-unsaturated-β,β-difluoro-α-keto esters. Among the fluorinated amino acids synthesized was 3,3-difluoroproline **27** (Fig. 3.12).[16]

For their research on the effects of fluorine substitution on peptide conformations (see below), Zondlo and coworkers developed a divergent synthesis of peptides (Thr-Tyr-X-Asp) (TYXN) containing an internal substituted Tyr-X (X = Pro or substituted Pro) sequence.[17] Polymer-supported peptides containing hydroxyproline were prepared, modified, and released by standard acidic cleavage and deprotection. This is summarized in Fig. 3.13.

1,3-Dipolar addition of azomethine ylides to electron-deficient olefins is a versatile route to nitrogen-containing heterocycles. Applying this approach to (*E*)-ethyl 3-fluoroacrylates utilizing L-menthol as a chiral auxiliary provides a stereoselective and regioselective synthesis of enantiopure-fluorinated prolines **28** and **29**.[18] Careful

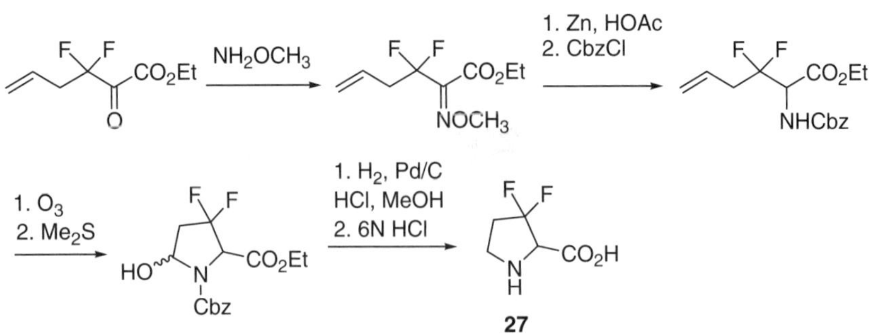

FIGURE 3.12 Synthesis of 3,3-difluoroproline.

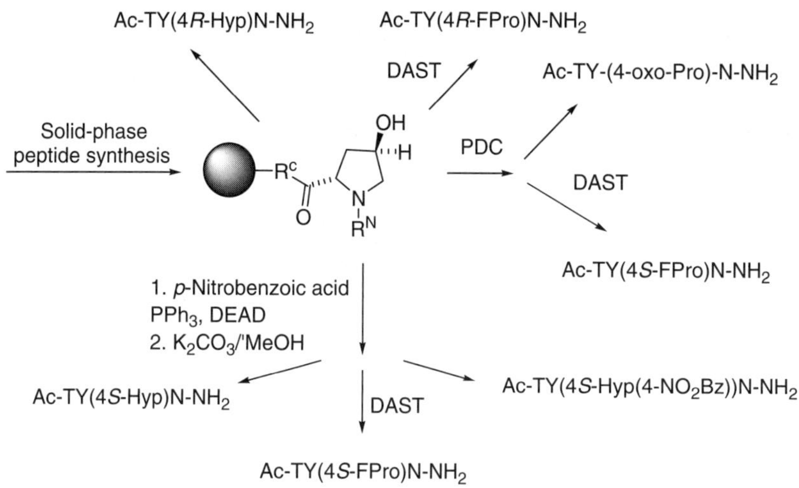

FIGURE 3.13 Polymer-supported synthesis of a series of peptides containing proline analogues.

removal of the chiral auxiliary and methyl ester formation gave the final fluorinated proline derivatives (Fig. 3.14).

Ichikawa and coworkers synthesized a series of trifluoromethyl-, difluoromethyl-, and difluoromethylene-substituted prolines based on a 5-*endo-trig*-S_N2' cyclization reaction. The terminal 2-trifluoromethyl-substituted olefin containing an *N*-tosylamide in the 4-position underwent intramolecular addition in the normally disfavored 5-*endo–trig* fashion to give trifluoromethyl- or difluoromethylene-substituted pyrrolidines, depending on conditions. α- Or β-face hydrogenation of the latter produced the difluoromethyl derivatives. Oxidation of the terminal aryl-substituted ethers produced the proline analogues **30**, **31**, **32**, and **33**.[19] Final products in the key cyclization step depended on reaction conditions (Fig. 3.15). Details of the reactions can be found in the original literature.

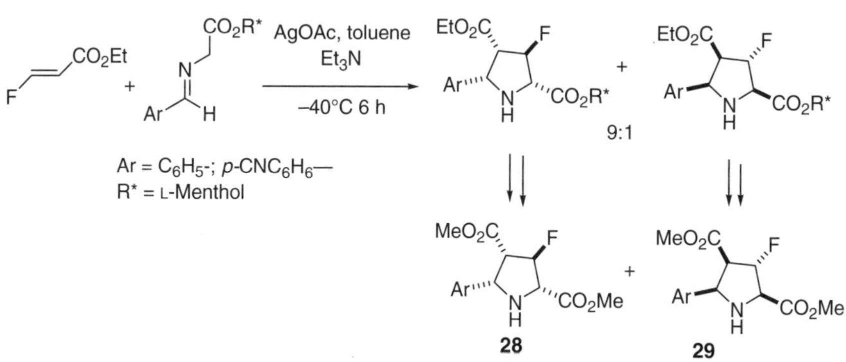

FIGURE 3.14 An enantioselective synthesis of fluorinated proline derivatives.

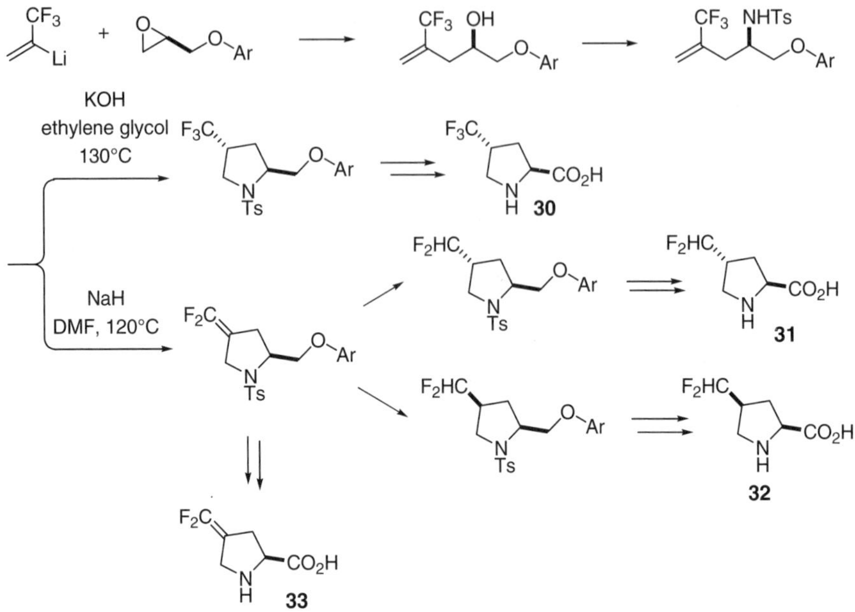

FIGURE 3.15 Asymmetric synthesis of side chain-fluorinated prolines.

3.2.2.2 Fluoroproline in Peptides and Proteins A major proportion of fluoroproline-containing biomolecules understandably consists of those derived from peptide bond formation. As noted above, the original synthesis of fluoroprolines[11] was prompted by the potential of these analogues as tools to study collagen biochemistry. Fluoroprolines have proven to be very effective probes for delineating important aspects of protein structures, particularly proline-rich proteins such as collagen, and protein–protein interactions. A very brief summary of this work will be presented in this section.

Incorporation of fluoroproline can dramatically affect the stability of collagen and related peptides and proteins. Collagen is a family of fibrous proteins that are the most abundant proteins in humans. Collagen has many important functions, many of which are related to its stability and tensile strength. For example, collagen fibers are important in contributing to the external structure of cells and help maintain the integrity of many tissues (for further discussions, see Ref. 20). In each of the approximately 19 different types of collagen, the polypeptides contain about 300 repeats of the sequence Gly-X-Y, where X is often Pro and Y is often (2*S*,4*R*)-hydroxyproline (Hyp). Hyp residues are incorporated posttranscriptionally by prolyl hydroxylase before the chains form the very stable triple helices.[21] In 1973, Prockop and coworkers showed that the hydroxyl groups of Hyp caused a dramatic increase in thermal stability to triple helical collagen.[22] Since models of triple helical collagen and calculations indicated that the Hyp residue hydroxyl groups could not form hydrogen bonds with the main chain of the helix, other models were proposed. For example, one or more water molecules were proposed to bridge the Hyp hydroxyl group and a main-chain oxygen. High-resolution structures determined by X-ray diffraction in 1994 confirmed the presence of these water molecules.[23]

Citing contraindicating experimental and theoretical arguments, including entropic issues, Raines and coworkers initiated studies to examine other explanations for the role of Hyp in collagen stability.[24] To distinguish between hydrogen bonding effects and inductive effects, peptides were prepared that contained (2*S*,4*R*)-fluoroproline (4*R*-FPro) instead of Pro or Hyp and a comparison of triple helices formed from three peptides was made. The presence of 4*R*-FPro dramatically increased the stability in the series: $(\text{Pro-Pro-Gly})_{10} < (\text{Pro-Hyp-Gly})_{10} < (\text{Pro-4}R\text{-FPro-Gly})_{10}$. This dramatic increase in stability is not consistent with hydrogen bonding associated with Hyp but indicates the importance of inductive effects. The authors ascribed this behavior to a combination of a *gauche* effect, an increased preference for C^{γ}-*exo* ring pucker, and an affect on the *trans–cis* ratio caused by the 4*R*-fluoro substituent.[21,24]

In subsequent work, the effects of isomeric fluoroprolines, either in the X- or Y-position of the X-Y-Gly motif, have been extensively studied by several groups. For example, Raines and coworkers determined that in contrast to 4*R*-FPro in the Y-position that enhances helix stability, 4*S*-FPro in the Y-position precludes triple helix formation.[25] Furthermore, replacing Pro with 4*R*-FPro in the X-position decreased helical stability of $(4R\text{-FPro-Pro-Gly})_7$[26] and $(4R\text{-FPro-Pro-Gly}])_{10}$,[27] whereas 4*S*-FPro in the corresponding peptides increased stability. The hyperstability of the 4*S*-FPro and 4*R*-FPro-containing peptides was ascribed to stereoelectronic effects that preorganize main-chain dihedral angles to correspond to the conformation found in the triple helix.[28,29] The effects were not additive because peptides having both 4*S*-FPro in the X-position and 4*R*-FPro in the Y-position did not form helices.[29]

The probing of triple helical stability with fluorinated prolines was extended to (2*S*,3*R*)- and (2*S*,3*S*)-fluoroprolines. It was found that peptide having 3*S*-FPro in the X-position and 4*R*-FPro in the Y-position do form a triple helix, albeit less stable than the helix having Pro in the X-position.[29]

These studies have not been confined to collagen. Conticello and coworkers studied the effects of fluoroprolines on the conformational properties of peptides related to elastin, a relative of collagen.[30]

Using a divergent synthetic strategy (see above, Fig. 3.13), Zondlo and coworkers prepared a series of peptides (Thr-Tyr-X-Asp) by postsynthetic modifications wherein X consists of hydroxyl, oxo, monofluoro, and difluoroprolines.[17] The stabilities were compared to TYProN and significant differences were observed. The term "proline editing" was introduced to refer to the strategy of altering backbone and side chain conformations in peptides based on selective incorporation of substituents, especially fluorine.[17]

3.2.2.3 Fluoroprolines as Synthetic Intermediates There are several examples of fluoropyrrolidine-containing biologically active compounds that have been synthesized. Included are the benzodiazepines **34**[31] that show anticancer activity and the DNA-interactive benzodiazepine dimers **35**.[32] The fluoropyrrolidine moicty was prepared from fluoroprolines derived from hydroxyproline. Likewise, the dimeric

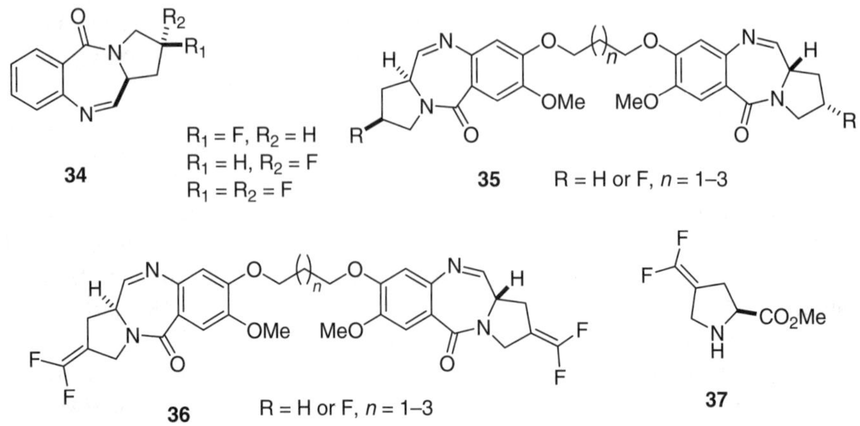

FIGURE 3.16 Fluoropyrrolidine-containing biologically active compounds.

benzodiazepine **36** featured difluoromethylenation of a 4-oxo-proline derivative to produce a difluoromethylene-containing proline ester **37** that was used to construct the tricyclic ring system[33] (Fig. 3.16).

3.2.2.4 [¹⁸F]Fluoroproline in Positron Emission Tomography There have been several reports of the synthesis of [^{18}F]4-fluoro-L-proline. The first synthesis reported in 1983 involved reaction of tetraethylammonium [^{18}F]fluoride with 4-*N*-tosly-4-trifluoromethanesulfonyloxy-L-proline methyl ester[34] (Fig. 3.17). Hamacher and coworkers reported the syntheses of no-carrier-added (NCA) *cis*- and *trans*-[^{18}F]4-fluoro-L-prolines by Kryptofix 222-mediated nucleophilic [^{18}F] fluorination of *cis*- and *trans*- *N*-*t*-Boc-4-tosyloxy-L-prolines followed by

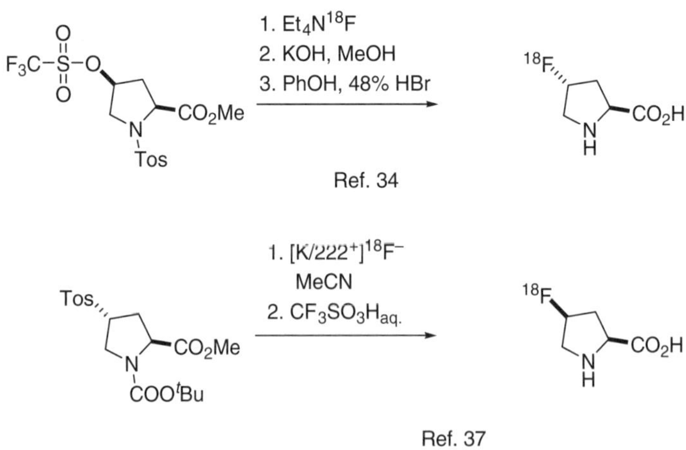

FIGURE 3.17 Syntheses of [^{18}F]4-fluoro-L-prolines.

deprotection (illustrated in Fig. 3.17 for the *cis*-isomer).[35] Using similar chemistry, Mazza reported a semiautomated synthesis of NCA *cis*- and *trans*-4-[^{18}F]fluoro-L-prolines.[36]

The *cis*-isomer of 4-[^{18}F]proline, but not the *trans*-isomer, is incorporated into murine protein, and only the *cis*-isomer is taken up into tumors.[37] This isomer has been studied as a tumor imaging agent, with collagen-rich tumor structural proteins being targeted. Initial results with urologic tumors were not promising.[38] Uptake of the L-isomer of *cis*-4-[^{18}F]proline into brain tumors appeared to be limited to areas where the blood–brain barrier was disrupted.[39] Later studies showed preferential uptake of the D-isomer of *cis*-4-[^{18}F]proline into the brains of human subjects and apparently this isomerizes in the brain to the L-isomer.[40] This is consistent with previous observations with the enantiomers of proline.

3.2.3 Fluorinated Derivatives of Pyroglutamic Acid

Pyroglutamic acid (PyroGlu), the five-membered nitrogen-containing heterocycle that is the cyclic form of glutamic acid, has important functions in its own right, being present, for example, in thyrotropin-releasing hormone (TRH) (PyroGlu-His-ProNH$_2$). As such, fluorinated analogues of pyroglutamic acid have received considerable attention, often as precursors of fluorinated glutamic acids, the latter of particular interest for preparation of tetrahydrofolate analogues as potential anticancer agents. Fluorinated prolines also are intermediates in certain of these procedures, so there is some overlap of this section with the above discussion of syntheses of fluorinated prolines. For example, in an early example, Hudlický reported a stereospecific synthesis of isomers of 4-fluoroglutamic acids based on initial preparation of 4-fluoroprolines (Fig. 3.18).[41] The procedure is illustrated with the synthesis of the 2*S*,4*R*-isomer **38**.

Coward and coworkers used a similar approach in an improved synthesis of DL-3,3-Difluoroglutamic acid from 3-oxoprolinol and difluoropyroglutamic acid derivative (**39**) intermediates (Fig. 3.19).[42] DL-3,3-difluoroglutamic acid had been previously studied as a substrate for folylpoly-γ-glutamate synthase.[43]

In the later work, Coward and coworkers carried out a study of electrophilic fluorination of pyroglutamic acid derivatives. The results provided an alternative route to fluorinated pyroglutamic acids and derived fluoroglutamic acids.[44] Conditions were found that led to diastereoselective monofluorination to produce **40**, but attempts to carry out difluorination of the substrate were unsuccessful. The bicyclic

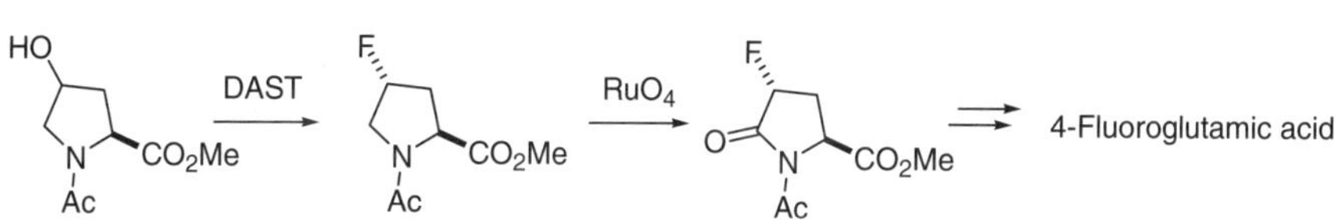

38

FIGURE 3.18 Fluorinated pyroglutamic acid as a precursor of fluoroglutamic acid.

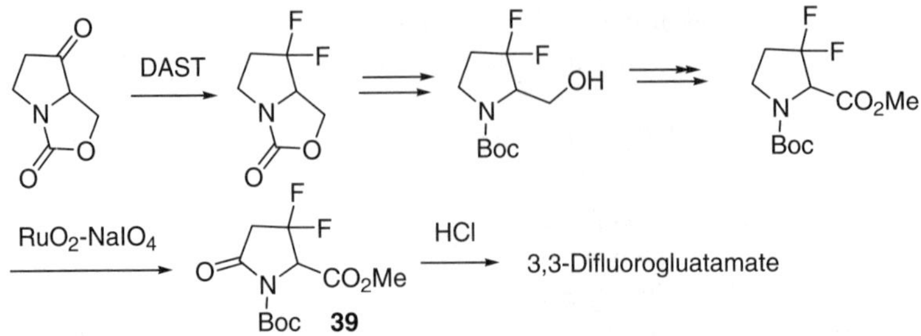

FIGURE 3.19 Difluoropyroglutamic acid as a precursor of difluoroglutamic acid.

substrate, on the other hand, was converted to the difluoroglutamate precursor **41** with excess fluorinating agent (Fig. 3.20).

Trifluoromethyl and difluoromethylpyroglutamic acids (**42** and **43**) were prepared from the corresponding fluorinated proline derivatives by ruthenium-mediated oxidation. The process was made difficult by the tendency of the substrates to form pyrrole derivatives under the conditions of the oxidation (Fig. 3.21).[45]

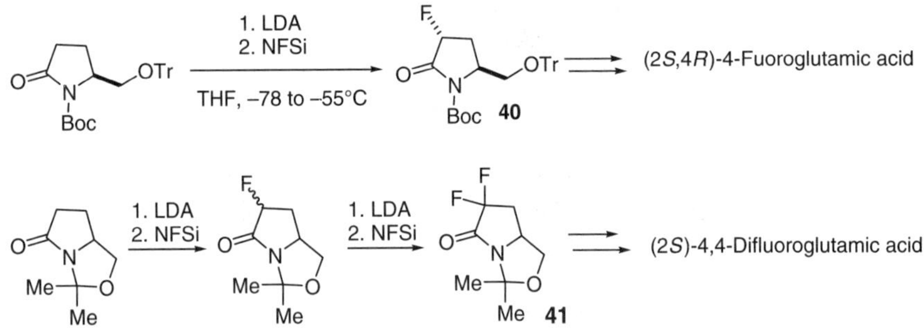

FIGURE 3.20 Additional approaches to fluorinated glutamic acids through proline derivatives.

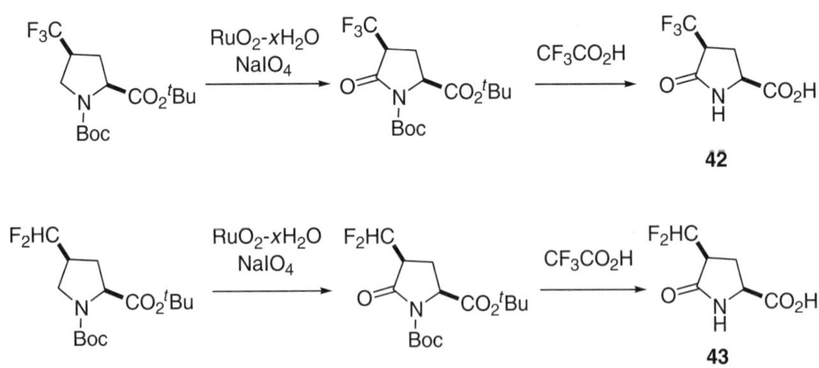

FIGURE 3.21 Trifluoromethyl- and difluoromethyl-substituted pyroglutamic acid.

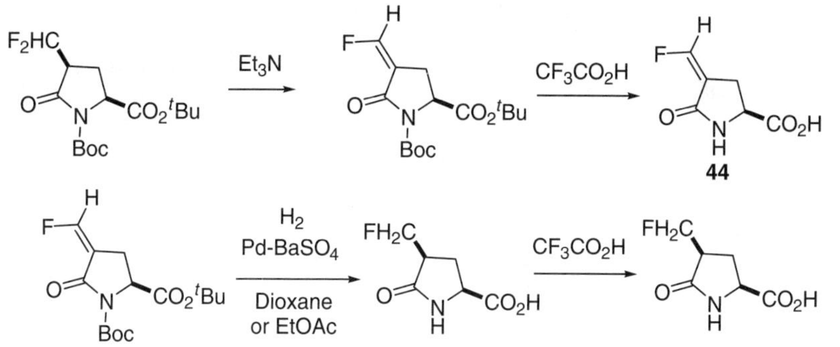

FIGURE 3.22 Additional fluorinated analogues of pyroglutamic acid.

The initial difluoromethyl derivative was found to undergo dehydrofluorination under basic conditions to produce the fluoromethylene analogue **44**. This, in turn, could be hydrogenated to produce the monofluoromethyl derivative **45** (Fig. 3.22).[46] Use of dioxane or ethyl acetate mainly produced *cis*-reduction product.

3.2.4 Fluorinated Pyrroles

Pyrrole (**46**) is a relatively nonbasic aromatic heterocycle, the low pK_a reflecting delocalization of the nitrogen lone pair into the aromatic system. (In this discussion, unless otherwise noted, we refer to 1*H*-pyrroles.) This heterocycle often occurs in nature, in particular in complex macrocycles including the porphyrins of heme, the chlorins, and the bacteriochlorins of chlorophyll and porphyrinogens. Porphobilinogen (**47**) is a trisubstituted pyrrole that is the biosynthetic precursor of many natural products. Bicyclic pyrroles, including indolizines (**48**), indoles (**49**), and isoindoles (**50**) (Fig. 3.23) are important in many biological processes. In addition, pyrrolo[2,3-*d*]pyrimidine nucleosides such as tubercidin (**51**) and analogues have received much attention, reflecting their wide range of antimetabolic properties, including antitumor and antiviral activities.

In reviewing fluorinated pyrroles, discussion will be confined to the compounds having fluorine or perfluoroalkyl groups directly bonded to the heteroaromatic ring. The considerable literature dealing with compounds having fluorine substituted on the benzenoid portion of bicyclic pyrroles (e.g., fluorinated tryptophans) will not be covered.

3.2.4.1 Synthesis of Fluorinated Pyrroles As described in the previous Section 3.2.1, the availability of hydroxypyrrolidines provided convenient functionality for introduction of fluorine, and many syntheses were based on either nucleophilic displacement of oxygen functionality by fluoride or deoxyfluorination with DAST or other such reagents. In contrast, many fundamentally different methods have been used to prepare fluorinated pyrroles.

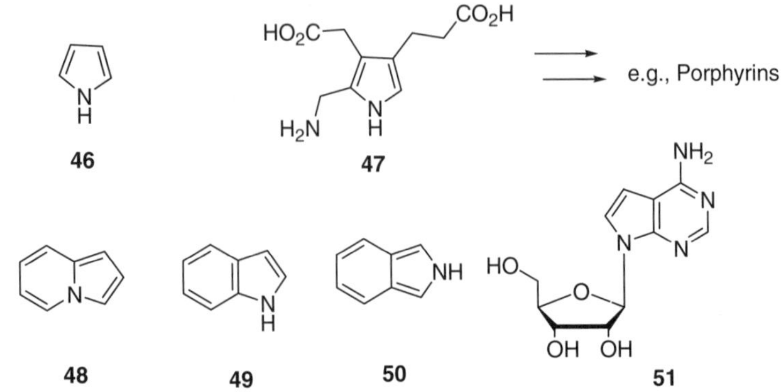

FIGURE 3.23 The pyrrole ring in nature.

Porphyrins can be conveniently assembled by condensation of simple pyrroles with aldehydes under Lewis acid catalysis (Lindsey conditions). Accordingly, 3-fluoro- and 3,4-difluoropyrroles have received much attention as precursors of fluorinated porphyrin derivatives. For example, in work on the synthesis of partially β-fluorinated 5,10,15,20-tetraphenylporphyrins and derivatives, previously unreported 3-fluoro-1*H*-pyrrole (**52**) was required (see Fig. 3.24).[47]

Barnes and coworkers had prepared *N*-triisopropylsilyl-3-fluoropyrrole **53** from the corresponding bromo derivative but did not report the isolation of the *N*-unsubstituted parent **52**.[48] However, **53** could be desilylated to produce **52**, which was used *in situ* to produce partially β-fluorinated porphyrins and derivatives (Fig. 3.24).[49] Attempts to isolate and purify **52** either from the desilylated crude product or through deprotection of the derived Boc-protected intermediate **54** met with limited success, although sufficient material for characterization was produced

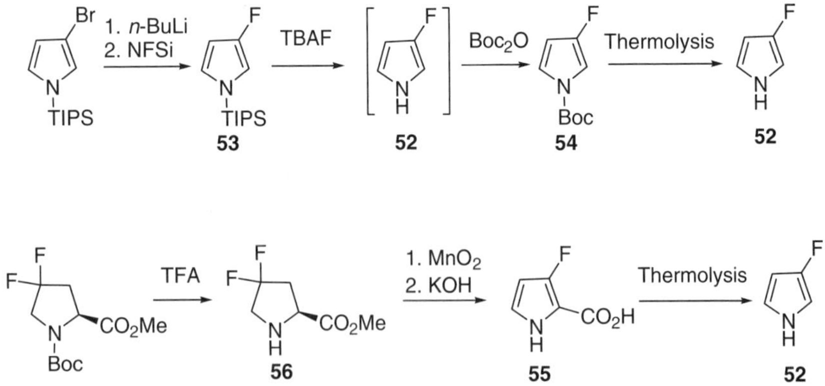

FIGURE 3.24 Synthetic approaches to 3-fluoropyrrole.

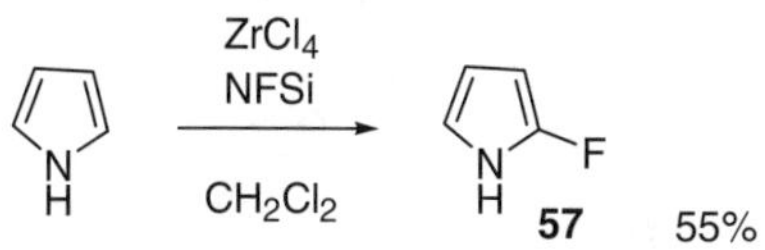

FIGURE 3.25 Synthesis of 2-fluoropyrrole.

(Fig. 3.24).[47] An alternative approach through decarboxylation of 3-fluoropyrrole 2-carboxylic acid (**55**), prepared by aromatization of methyl 3,3-difluoropyrrole-2-carboxylate (**56**), also proved to be problematic (Fig. 3.24). Technical difficulties were ascribed to volatility and possible instability of the final product.[47]

A recently reported procedure for a Lewis acid-catalyzed selective halogenation of aromatic compounds was applied to the synthesis of 2-fluoropyrrole (**57**).[50] Thus, reaction of pyrrole with NFSi in the presence of a catalytic amount of $ZrCl_4$ in CH_2Cl_2 for 12 h gave a 55% yield of 2-fluoropyrrole (Fig. 3.25).

3,4-Difluoropyrrole (**58**) has been extensively used in the syntheses of octafluoroporphyrins and other calyx(*n*)pyrroles. This was first accessed by Leroy and Wakselman by barium-promoted copper chromite decarboxylation of 3,4-difluoropyrrole-2-carboxylic acid in quinoline at 200°C.[51] The acid was prepared in four steps beginning with a cycloaddition reaction of the protected aziridine **59** and chlorotrifluoroethylene (Fig. 3.26).

DiMagno and coworkers reported an efficient route to **58** that also employs a double elimination step, but this is performed on the readily available tetrafluoropyrrolidine **10** (see Section 3.2.1 above) (Fig. 3.27).[52] Optimization of conditions provided **58** in good yield in a reaction that is readily scaled up, thus providing ready access to this fluorinated building block. This has been particularly valuable for the synthesis of perfluorinated porphyrins (see the discussion below).

Several routes to functionalized 3-fluoropyrroles have been reported, examples of which will be summarized in this section (Fig. 3.28). The first example involved a

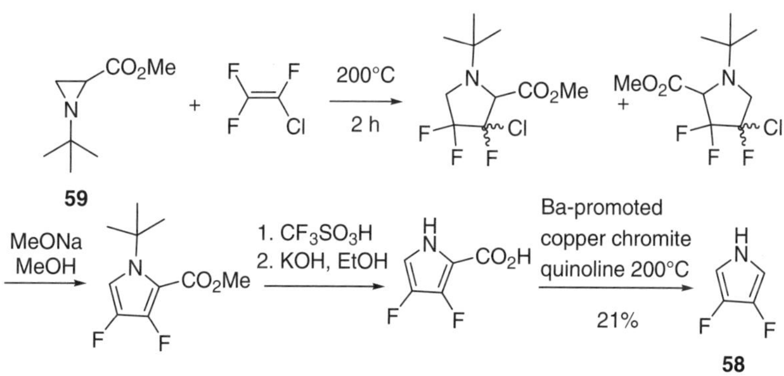

FIGURE 3.26 Synthesis of 3,4-difluoropyrrole.

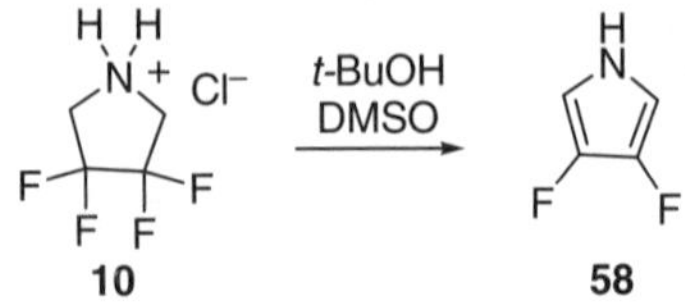

FIGURE 3.27 A convenient synthesis of 3,4-difluoropyrrole.

thermal or photochemically induced ring expansion of 2-azido-3,3-difluorocyclobutenes **60** in the presence of such nucleophilic solvents as benzene or furan to produce substituted pyrroles **62** by addition to intermediate **61**.[53] A photochemical Schiemann reaction was used to prepare 3-fluoropyrroles **63** from 3-aminopyrroles **64**, available from the nitro compounds.[54] Burton and coworkers had prepared a series of α,α-difluoro-γ-EWG pyrroles **65** by photochemical addition of iododifluoromethyl ketones to electron-deficient olefins. Treatment of these products with aqueous ammonia gave excellent yields of 3-fluoropyrrole derivatives **66**.[55]

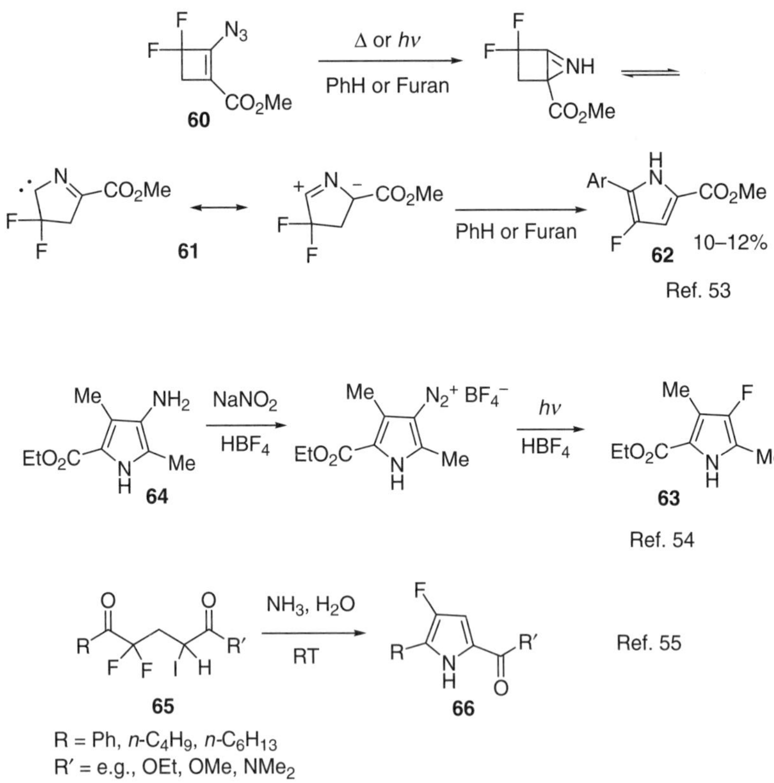

FIGURE 3.28 Synthetic routes to functionalized fluoropyrroles.

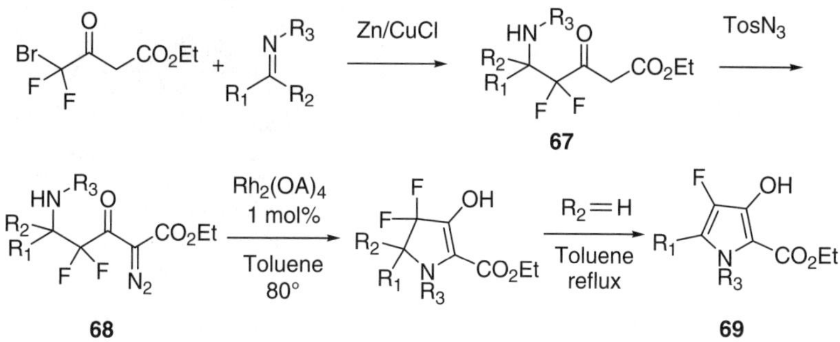

FIGURE 3.29 Rhodium(II)-catalyzed cyclization in pyrrole ring construction.

More recent approaches include a rhodium(II)-catalyzed intramolecular insertion reaction to form the five-membered heterocycle.[56] Reformatsky-imine addition of 4-bromo-4,4-difluoroacetoacetate with aldimines gave δ-amino-γ,γ-difluoroacetoacetates **67** that were readily converted to the key diazo intermediates **68** through the action of tosyl azide and molecular sieve. Rhodium(II)-catalyzed intramolecular insertion followed by aromatization through loss of HF gave the functionalized pyrroles **69** (Fig. 3.29).

Direct fluorination of the pyrrole ring with xenon difluoride and fluorodecarboxylation of pyrrole carboxylic acids essentially remained the sole routes to 2-fluoropyrroles until the late 1990s.[57–59] Using the first approach, Wang and Scott studied fluorination of variously substituted pyrroles with a variety of fluorinating agents and solvents and found that xenon difluoride allowed fluorination in the 2-position of pyrroles if electron-withdrawing groups (EWGs) were present as in **70** to block oxidation. A key result of this work was the successful synthesis of 3-(4-carboxymethyl-2-fluoro-5-hydroxymethyl-1*H*-pyrrol-3-yl)-propionic acid (**71**), a fluorinated analogue of porphobilinogen **47**, by fluorination of the aldehyde **70** followed by hydrolysis and reduction to give **71** (Fig. 3.30).[59] This functioned as a suicide

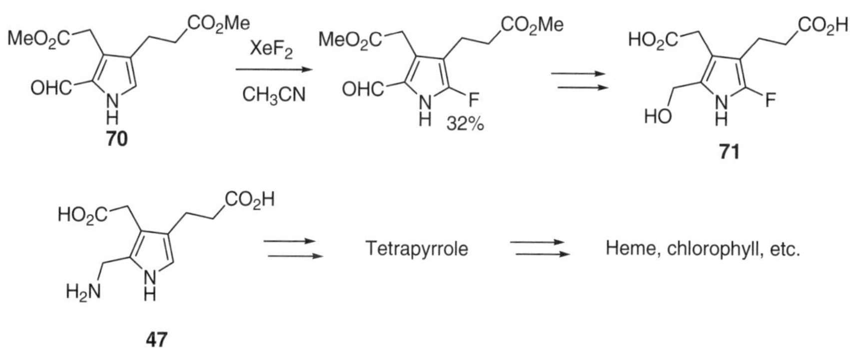

FIGURE 3.30 Synthesis of a fluorinated analogue of porphobilinogen.

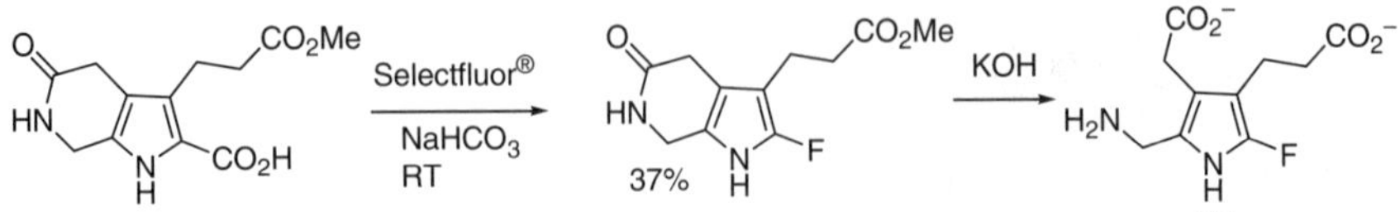
72

FIGURE 3.31 A synthesis of 2-fluoroporphobilinogen.

inhibitor of porphobilinogen deaminase. This enzyme catalyzes the stepwise polymerization of four molecules of porphobilinogen to a linear tetrapyrrole that is a key biosynthetic intermediate of heme, chlorophylls, vitamin B_{12}, and related macrocycles.

The same group applied fluorodecarboxylation to this synthetic goal. Reaction of a range of pyrrole-2-carboxylic acids, substituted by either electron-withdrawing or electron-donating groups, with Selectfluor® gave 2-fluoropyrroles in 32–47% yields.[60] A new synthesis of the fluoroporphobilinogen **72** illustrates this procedure (Fig. 3.31).

A conceptually different approach was taken by Novikov and coworkers using difluorocarbene as a building block. 1,3-Dipolar addition of dipolarophiles to azomethine ylides derived by addition of difluorocarbene to imines produces substituted 2-fluoropyrroles.[61,62] The process is illustrated below with the imine derived from benzaldehyde and aniline. Using dimethyl acetylenedicarboxylate (DMAD) as the dipolarophile produces the 2-fluoropyrrole derivative **73** (Fig. 3.32). Several substrates were studied and such issues as regiochemistry were addressed.

Using the same approach employing olefinic rather than acetylenic dipolarophiles produces 2-fluoro-2-pyrrolines.[63]

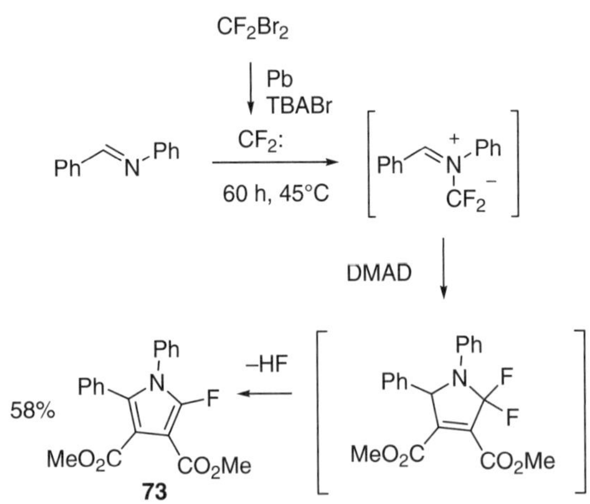
73

FIGURE 3.32 A 1,3-dipolar addition approach to 2-fluoropyrroles.

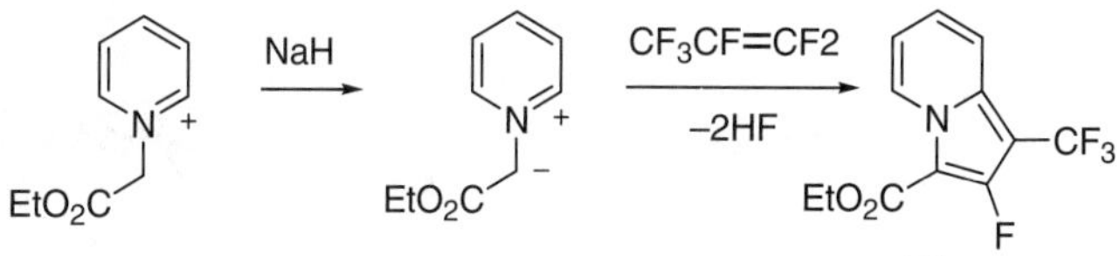

74

FIGURE 3.33 Synthesis of a fluorinated indolizine.

3.2.5 Fluorinated Indolizines, Indolizidines, Indoles, and Related Compounds

Indolizines, indolizidines, indoles, and other bicyclic or polycyclic compounds containing a fused pyrrole or pyrrolidine moiety often possess potent and useful biological activities. There have been several approaches to the preparation of fluorinated analogues of these heterocycles, mainly based on 1,3-dipolar addition reactions using fluorinated synthons either as the dipole or as the dipolarophile. Several examples of this strategy are given below.

Banks and coworkers published a series of papers describing the reaction of pyridinium methylides with prefluoroalkenes, perfluoroazaolefins, and trifloroacetonitrile to produce fluorinated heterocycles, exemplified by the synthesis of indolizine **74** (Fig. 3.33).[64] Modest yields isolated from complex mixtures were reported.

Using DMF as a solvent to facilitate solubility of fluorocarbons and thus ease conditions of the reaction, Wu and Chen prepared indolizines and 4*H*-pyrrolo[1,2-*a*] benzimidazoles in good yields by 1,3-dipolar cycloaddition of fluoroalkenes to *N*-ylides, exemplified below by the synthesis of indolizine **75** (Fig. 3.34).[65]

In a similar approach, Huang and coworkers published a series of papers describing the cycloaddition reactions of fluorinated olefins, generated *in situ* from fluorinated alkanoates, with pyridinium and other nitrogen-based ylides.[66–70] The synthesis of the fluorinated indolizine **76** that proceeds by cycloaddition followed by a dehydrochlorination step provides an example of this chemistry (Fig. 3.35).[67]

Pyridinium ylides were used to prepare a series of 1-trifluoroacetyl indolizines by cycloaddition with 4-ethoxy-1,1,1-trifluorobut-3-one.[71] After cycloaddition, loss of ethanol and spontaneous aromatization were proposed for the formation of the indolizine. An example of this chemistry is shown in the synthesis of indolizines **77** (Fig. 3.36). Good yields were reported.

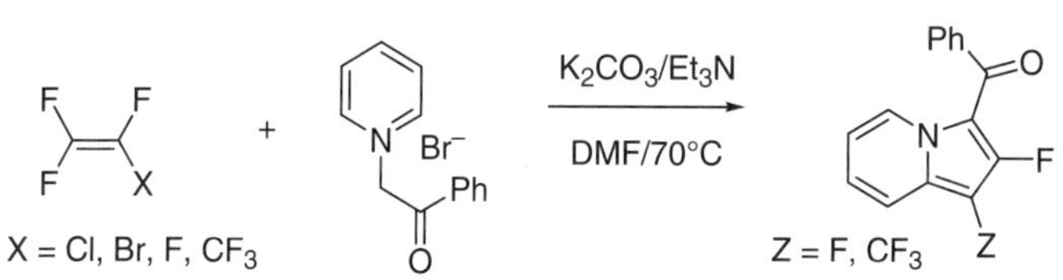

75

FIGURE 3.34 Synthesis of fluorinated indolizines.

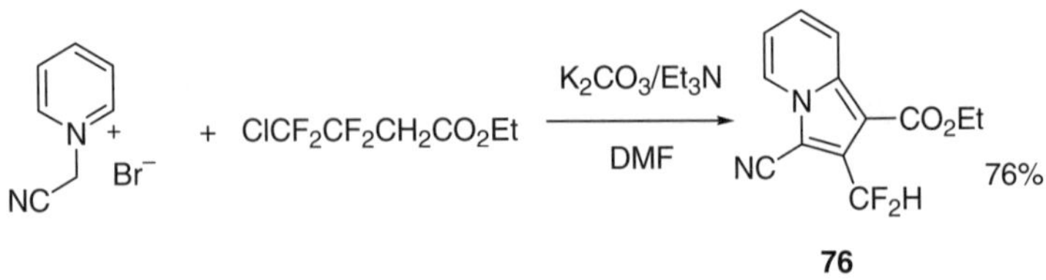

FIGURE 3.35 Nitrogen-based ylides in the preparation of fluorinated indolizines.

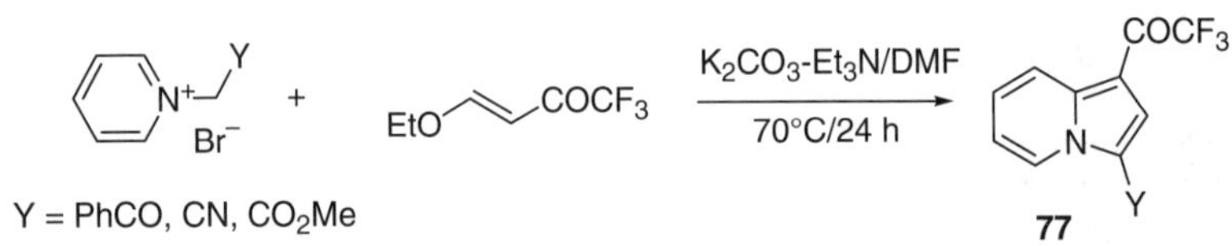

FIGURE 3.36 Pyridinium ylides as precursors of fluorinated indolizines.

1,3-Dipolar addition reactions of *N*-ylides with 2,2-difluorovinyl tosylate provides a convenient route to moderate yields of monofluorinated indolizines and related structures. *N*-Ylides derived from pyridinium, isoquinolinium, and bezimidazolium salts were generated *in situ* from halide salts (Fig. 3.37).[72]

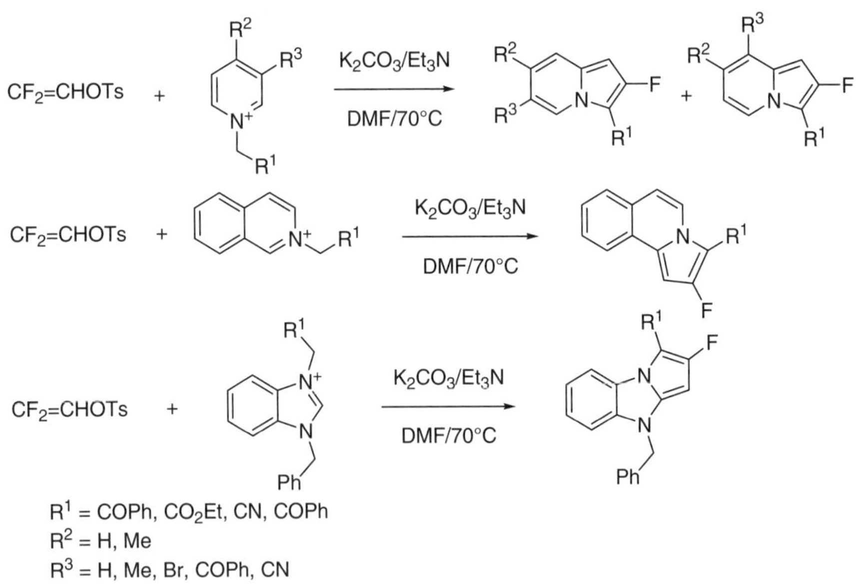

FIGURE 3.37 Further examples of nitrogen-based ylides in the synthesis of fluorinated indolizines and related structures.

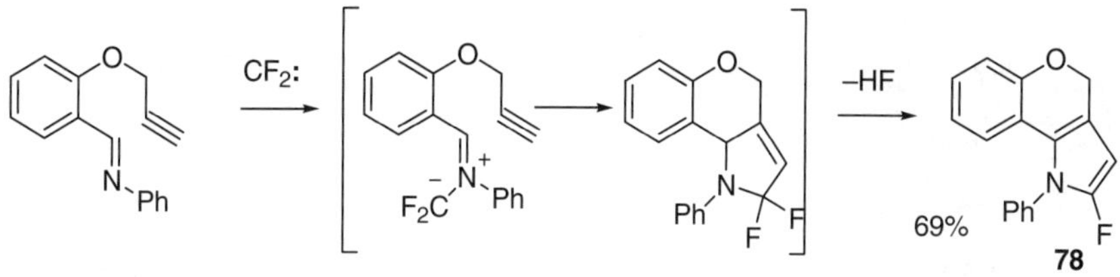

FIGURE 3.38 A difluorocarbene approach to chromeno[3.4-*b*]pyrroles.

Novikov and coworkers described an intramolecular variant of their difluorocarbene/Schiff base chemistry discussed above. Iminodifluoromethanides derived from the reaction of difluorocarbene with Schiff bases react with tethered acetylenes to produce fluorinated chromeno[4,3-*b*]pyrroles, for example, **78** (Fig. 3.38).[73]

The above discussion highlights the use of fluorinated synthons in the synthesis of fused ring systems containing a fluorinated pyrrole moiety. Direct fluorination has also been used to fluorinate similar fused systems. For example, a recent report describes the synthesis of 8-fluorotubercidine **79**, a fluorinated analogue of the naturally occurring pyrrolo[2,3-*d*]pyrimidine nucleoside tubercidine **51**. 4-Chloro [2,3-*d*]pyrimidine **80** was treated with Selectfluor® in wet acetonitrile to give the fluorohydrin **81**, which could be converted *in situ* to the 5-fluoro derivative **82**. Coupling to the protected ribose and treatment of the intermediate nucleoside **83** with ammonia in dioxane completed the synthesis (Fig. 3.39).[74]

An example of nucleophilic fluorination to prepare an indolizidine derivative is found in the synthesis of a fluorinated analogue of (+)-castanospermine, a naturally occurring glucosidase inhibitor. The epoxylactam **84** was treated with HF-Et_3N. Stereospecific attack at position 6 gave, after acylation, the acetyl-protected

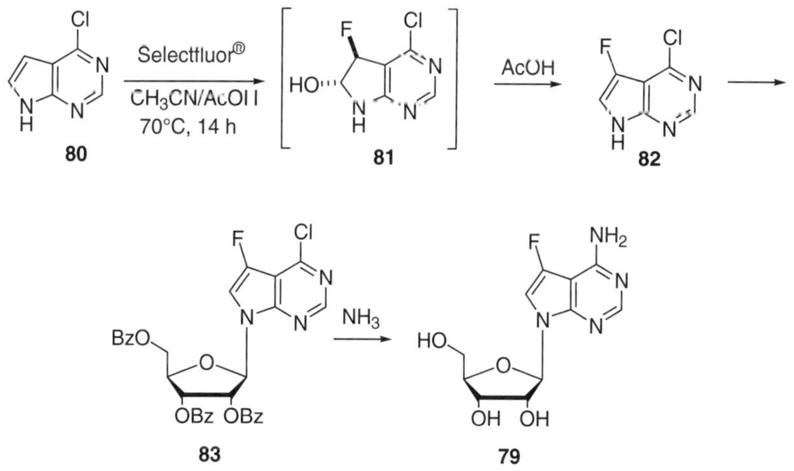

FIGURE 3.39 Synthesis of 5-fluorotubercidin.

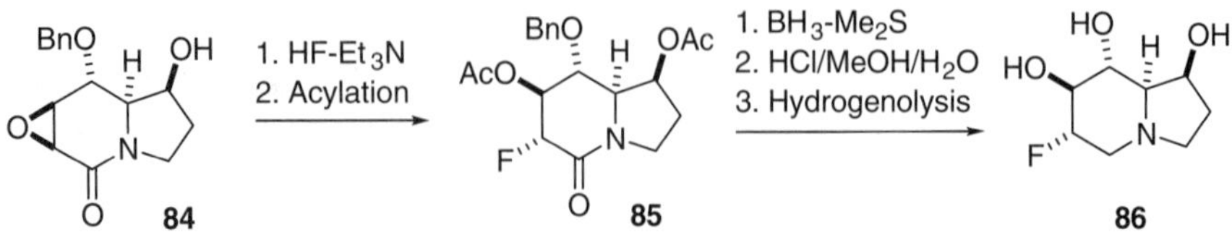

FIGURE 3.40 Synthesis of a fluorinated analogue of (+)-castanospermine.

derivative **85**. This was converted to (+)-6-deoxy-6-fluorocastanospermine **86** by reduction and deprotection (Fig. 3.40).[75]

3.2.5.1 Fluorinated Porphyrins and Related Macrocycles Porphyrins, essential to living organisms, are heterocyclic macrocycles derived from four pyrrole-like subunits. Derivatives of porphine, **87**, the simplest porphyrin, include the heme of hemoglobin, the ring-contracted chlorin found in chlorophyll, and the corrin structure of vitamin B_{12}. Many synthetic variants are known, such as ring-expanded calixpyrroles. The metal complexes of porphyrins are very stable and form the bases of many of the biological properties, including redox activity and ligand binding. The importance of these macrocycles has prompted much research interest in synthetic analogues to study and modulate physicochemical and biological properties. The study of fluorinated analogues has been particularly useful for isolating electronic and steric effects of substituents on such parameters as redox potentials and spectral properties. In this section, a brief summary of some of the notable achievements in this research will be given. No attempt will be made to be comprehensive in the treatment of this expanding area of research.

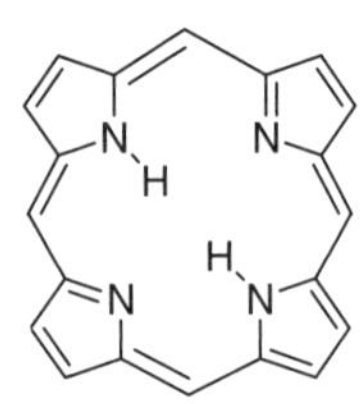

87

An early report described the preparation in low yield of two isomeric *meso*-monofluorinated derivatives of porphyrins using the Schiemann reaction.[76] Electrophilic fluorinating agents have been a more effective way to carry out direct fluorination of the porphyrin system. For example, reaction of octaethylporphyrin with *N*-fluoropyridinium salts such as *N*-fluoro-2,3,4,5,6-pentachloropyridinium triflate gave a mixture of mono, di, tri, and tetrafluoro derivatives **88** (Fig. 3.41).[77] The perfluorinated derivative was formed in 20% yield. *meso*-Fluorination of octaethylporphyrin had minimal effects on oxidation potential and spectral properties.

DiMagno and Williams developed a general route to highly electron-deficient *meso*-substituted porphyrins based on condensation of perfluoroaldehyde hydrates

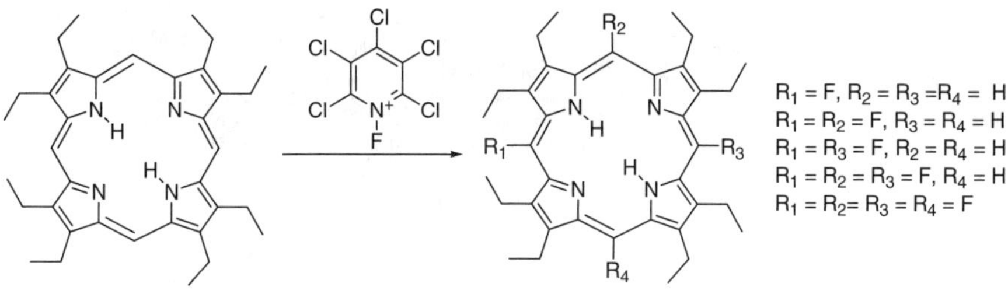

88

FIGURE 3.41 Electrophilic fluorination of octaethylporphyrin.

(e.g., **89**) with pyrrole. The success of the reaction depends on the efficient removal of the water formed in the reversible condensation step.[78] The tetrakis(heptafluoropropyl) substitution present in porphyrin **90** imparts interesting and useful properties to the porphyrin system, including enhanced solubility in a variety of solvents and a large increasc in redox potential. This new class of porphyrin is both extremely electron deficient and nonplanar. The absence of π-conjugative interaction with the ring by the perfluoroalkyl substituents permits the study of ring-distortion effects in the absence of conjugative effects in metalloporphyrins (Fig. 3.42).[79]

There have been several reports of the synthesis of β-fluorinated porphyrins by construction of the macrocyclic ring using fluorinated pyrrole precursors. For

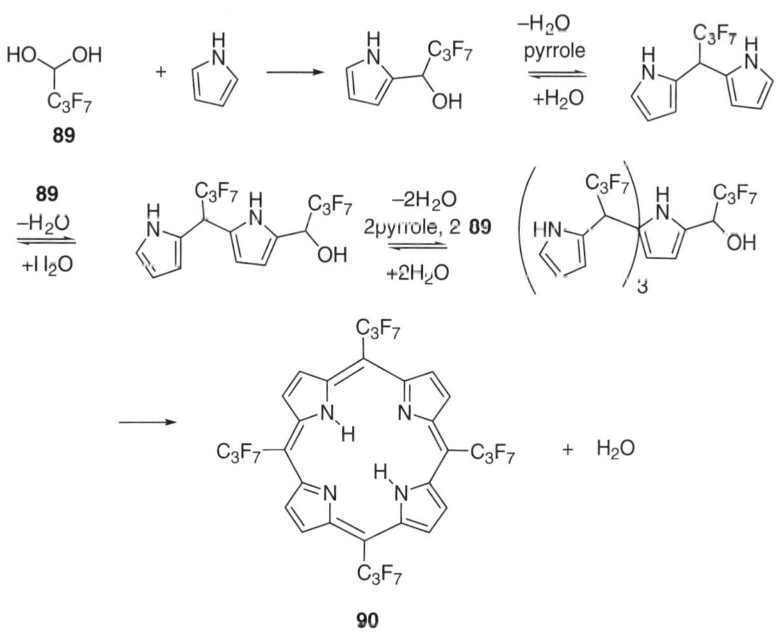

90

FIGURE 3.42 Synthesis of perfluoroalkylporhyrins.

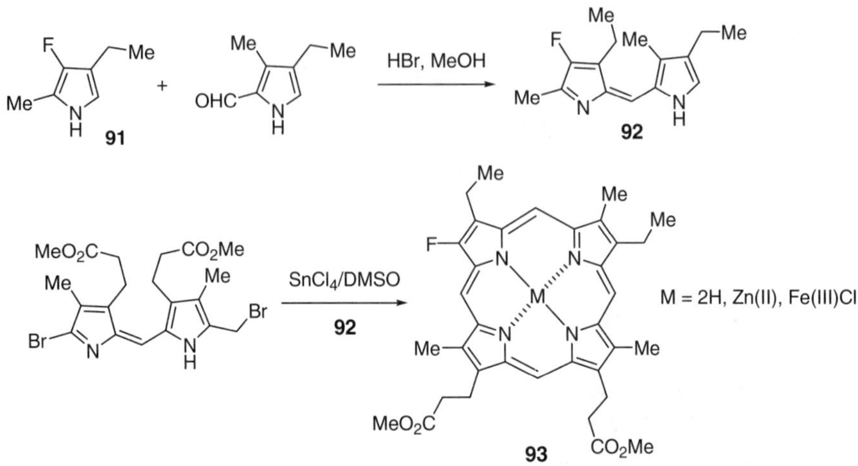

FIGURE 3.43 Synthesis of a mono-fluoroprophyrin from a fluorinated pyrrole.

example, 4-ethyl-3-fluoro-2-methylpyrrole **91** was converted through the intermediate dipyrrole **92** to a monofluorinated derivative of 1-demethylmesoporphyrin-IX (**93**) (Fig. 3.43).[80]

Fluorinated analogues of these and related porphyrins were later used to reconstitute myoglobins that contain fluorinated hemes. These were used for NMR studies of the effects of distortion of the porphyrin π-system by the chemical properties of the peripheral side chains.[81] A similar strategy involving the oxidative coupling of similar dipyrromethenes and monofluorodipyrromethenes produced 6-monofluorinated porphyrins. Reconstituted myoglobins and iron complexes prepared from these were studied.[82]

There has been much interest in the development of electron-deficient porphyrins as tools to study electronic effects on metal-centered chemistry, redox properties, and structure. Both β-fluorinated and perflouroalkylated derivatives have been made. Ogoshi and coworkers used a modified Knorr condensation to prepare the trifluoromethyl analogue of etioporphyrin along with its metal complexes **94** (Fig. 3.44).[83]

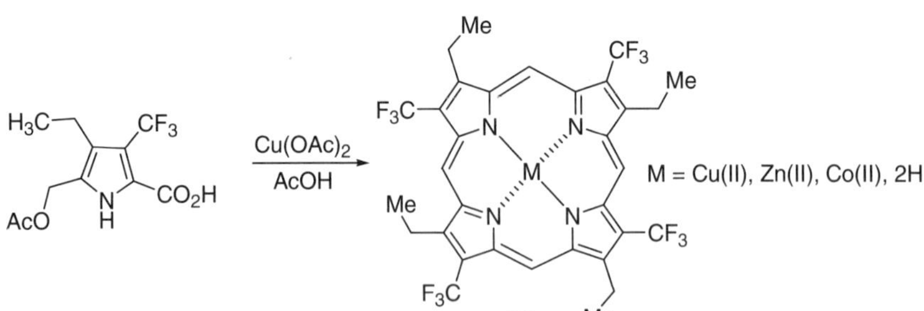

FIGURE 3.44 Synthesis of metal salt complexes of trifluoromethyl-substituted porphyrins.

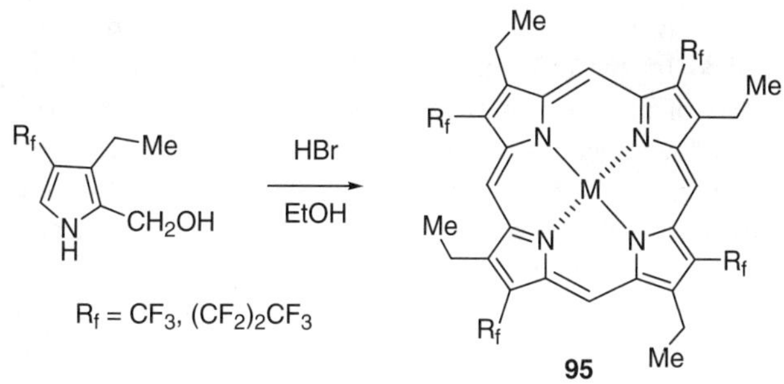

FIGURE 3.45 Synthesis of perfluoroalkyl-substituted porphyrins.

The same group adopted a slightly different strategy to prepare a series of perfluoroalkyl porphyrins **95** (Fig. 3.45).[84]

Condensation of pyrroles with carbonyl compounds under Lewis acid catalysis followed by oxidation provides an efficient route to the porphyrin ring system. This has been effectively used to prepare polyfluorinated porphyrins from 3-fluoro and 3,4-difluoropyrroles, as well as pyrroles substituted with fluorinated side chains in these positions. For example, DiMagno and coworkers extended their studies of highly electron-deficient porphyrins to a series of 2,3,7,8,12,13,17,18-octafluoro-5,10,15,20-tetraarylporphyrins **96**, including the perfluorinated compound, by BF_3-catalyzed condensation of 3,4-difluoropyrrole with benzaldehydes followed by dichloro dicyano quinone (DDQ) oxidation (Fig. 3.46).[85] A similar approach was used by Leroy and coworkers.[86]

These electron-deficient porphyrins and their metal complexes have proven to be valuable tools to study many aspects of porphyrin structure and function. An in-depth discussion of this area is beyond the scope of this review but examples will be cited. Thus, large substituent-dependent variations in electron-transfer properties of cobalt complexes were ascribed to widely varying inner sphere reorganization energies related to core expansion and contraction.[87] In another example, a ruthenium complex functioned as a potential methane functionalization catalyst by virtue of the ability of

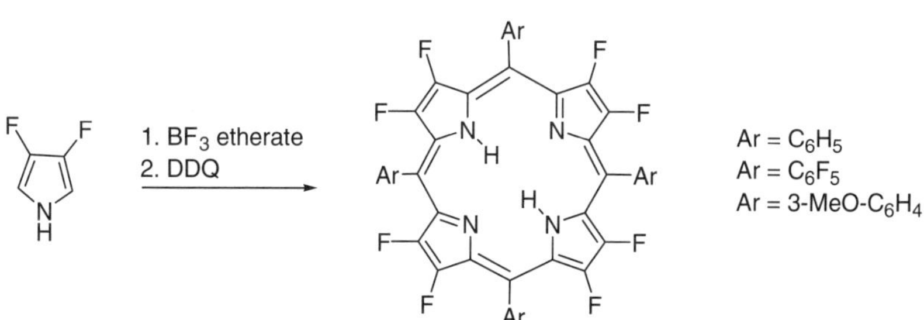

FIGURE 3.46 Synthesis of perfluoroporphyrins from difluoropyrrole.

the $[RH(F_{28}TPP)]^-$ to serve as a leaving group. This represents umpolung of the metal–carbon bond by the electron demand of the perfluoroporphyrin nucleus.[88] Additional studies of the use of ruthenium complexes of fluorinated porphyrins as oxidation catalysis have been reported.[89]

Partially fluorinated porphyrins were made by a similar Lindsey cyclization. Condensation of benzaldehyde with mixtures of 3,4-difluoropyrrole and pyrrole gave mixtures of di-, tetra-, and hexafluoro *meso*-tetraphenylporphyrin with fluorine atoms in pairs per pyrrole. On the other hand, condensation of 3-fluoropyrrole with benzaldehyde gave a mixture of tetrafluoro-*meso*-tetraphenylporphyrin with one fluorine atom per pyrrole.[47,90] These partially β-fluorinated porphyrins were used to study the differential effects of fluorine substituents on spectral properties and core deformations of the porphyrin system.

The above discussion has concentrated mainly on porphyrins having fluorine directly substituted on the macrocyclic ring. Perfluoroalky- and fluoroalkenyl-substituted porphyrins have also received much attention, but will be covered only briefly here. For example, Kumadaki and coworkers synthesized a series of natural porphyrins substituted variously with fluorovinyl, trihalovinyl, and trifluoroethoxy groups to be used as sensitizers in cancer phtotodynamic therapy (reviewed in Ref. 91).

Sessler and coworkers have prepared fluorinated calix[4]pyrroles (**97**; $n = 1$) and expanded calixpyrroles (**98**; $n = 2$–5) by the condensation of 3,4-difluoropyrrole with acetone under acid catalysis (Fig. 3.47).[92,93] The fluorinated dipyrrolylquinoxaline **99** was also synthesized from 3,4-difluoropyrrole.[92] These were studied as neutral anion receptors that had higher affinity and greater selectivity relative to their nonfluorinated congeners.

Fluorinated analogues of other pyrrolic macrocyclic systems are also receiving current attention. The recent development of a short and efficient synthesis of substituted corroles has facilitated the syntheses of several fluorinated and perfluorinated analogues of this tetrapyrrolic macrocycle.[94,95] Gross and coworkers found that

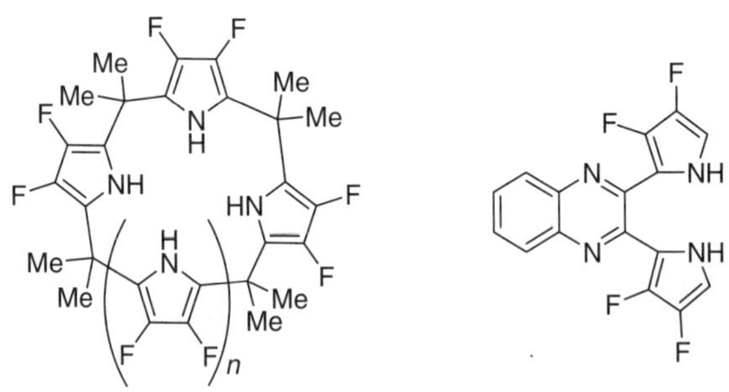

97, $n = 1$
98, $n = 2$–5

99

FIGURE 3.47 Fluorinated calix[4]pyrroles and expanded calixpyrroles and a fluorinated dipyrrolylquinoxaline.

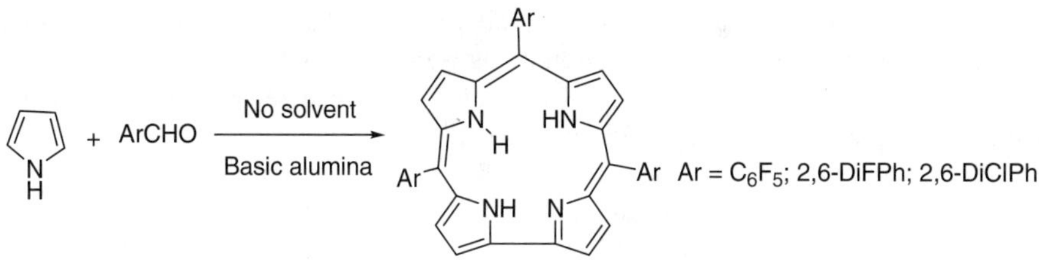

100

FIGURE 3.48 Synthesis of fluorinated corroles.

heating electron-withdrawing benzaldehydes with pyrrole in the absence of solvent gave triarylcorroles **100** in acceptable yields (Fig. 3.48).

This was initially described as being confined to these electron-deficient benzaldehydes. However, Ghosh and coworkers expanded this to a wider range of benzaldehydes and applied the reaction to 3,5-difluoropyrrole to access a series of octafluorocorroles **101** (Fig. 3.49).[96]

Chang and coworkers condensed 3,4-difluoropyrrole with pentafluorobenzaldehyde under similar but somewhat altered conditions to give perfluorinated corrole (**101**; Ar=C_6F_5) (Fig. 3.50).[97] An intermediate linear bilene **102** was cyclized to the

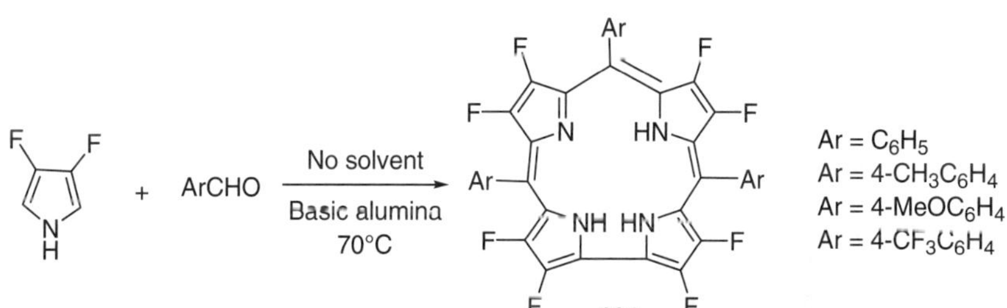

101

FIGURE 3.49 Synthesis of octafluorocorroles.

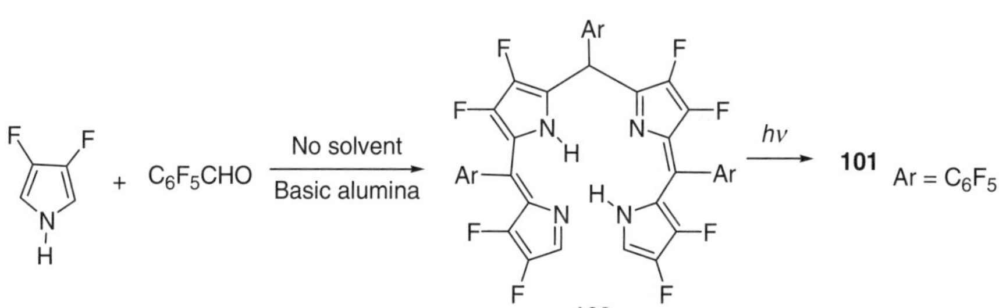

102

FIGURE 3.50 Synthesis of perfluorinated corrole.

corrole with floodlight irradiation. The corrole was formed directly in 5% yield under the original Gross conditions if reaction times were increased. The Mn=O complex had a 28-fold increased rate of oxidation of cyclooctene relative to the Mn=O complex of the non-β-fluorinated F-15 corrole.

As can be seen from the brief overview in this section, fluorinated pyrroles have played a major role in the development of new tools for the study of porphyrin chemistry. The high electronegativity of fluorine, especially in polyfluorinated analogues, has been particularly useful to alter properties and also to study the electronic effects in porphyrin structure and function.

3.3 FLUORINATED HETEROCYCLES CONTAINING TWO NITROGENS

3.3.1 Fluorinated Imidazoles

3.3.1.1 Synthesis of Ring-Fluorinated Imidazoles Synthetic routes to ring-fluorinated imidazoles are somewhat limited. The only general procedure to date is the photochemical Schiemann reaction first reported in 1971,[98] developed after a range of other fluorination procedures, including the thermal Schiemann reaction, had proven totally unfruitful. The sequence involved *in situ* formation and irradiation of an imidazole diazonium salt in aqueous fluoroboric acid. This was initially applied to the stable 2-aminoimidazoles and 4-aminoimidazoles that were stabilized by an electron-withdrawing group (carboxylic ester) in the 5-position.[98–100] The procedure proved to be applicable to a wide range of imidazole derivatives as well as to other aromatic and heteroaromatic systems.

Included in biologically relevant members of this new class of fluorinated heterocycles were 2- and 4-fluorohistidines, 2- and 4-fluorohistamines, 2- and 4-fluorourocanic acids, and 4-fluoroimidazole-5-carboxamide riboside (Fig. 3.51).[99–101]

Functionalized 2-aminoimidazoles required for direct synthesis of more complex derivatives such as 2-fluorohisitidine and 2-fluorohistamine were prepared by catalytic reduction of 2-arylazoimidazoles, the products of coupling of the imidazole ring with aryl diazonium salts.[102] 4-Aminoimidazoles are unstable unless an electron-withdrawing group is also present on the imidazole ring. 4-Fluoroimidazole was initially prepared by *in situ* deprotection of the Boc-protected amine in cold fluoroboric acid followed by immediate diazotization and irradiation.[98,99] A more direct route was developed based on reduction of 4-nitroimidazoles with zinc dust in cold fluoroboric acid followed by *in situ* diazotization and irradiation (Fig. 3.52).[103] Although the overall yields over the three transformations are modest, the directness of the process makes this an effective route to 4-fluoroimidazole derivatives. In more recent work to be published, the *N*-acetyl group was replaced with the *N*-trifluoroacetyl group to provide better organic solvent solubility and easier deprotection.

Both 4,5-difluoro- and 2,4-difluoroimidazoles have been prepared. The former was synthesized from ethyl 4-fluoroimidazole-5-carboxylate through a Curtius rearrangement that provided the Boc-protected 4-fluoro-5-aminoimidazole. *In situ* depro-

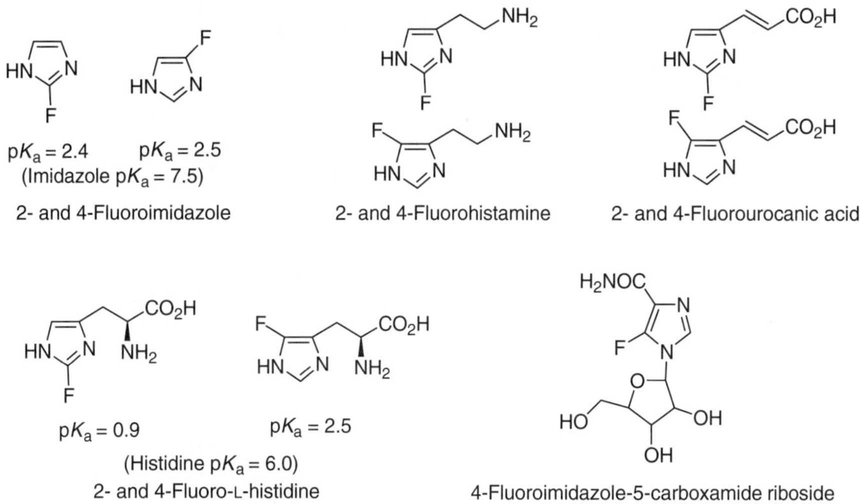

FIGURE 3.51 Structures of ring-fluorinated imidazoles.

tection, diazotization, and photolysis produced 4,5-difluoroimidazole (**103**) (Fig. 3.53).[104]

Attempts to prepare 2,4-difluoroimidazoles met with difficulty in large part due to the lability of the fluorine in the 2-position. Ethyl 2,4-difluoroimidazole carboxylate **104** was prepared by a sequence that involved arylazo coupling of ethyl 4-fluoroimidazole-5-carboxylate (**105**) to give the arylazo intermediate **106**, reduction to the 2-amino derivative **107** using formamidinesulfinic acid (FASA), and the photochemical Schiemann procedure (Fig. 3.54).[105] A similar sequence using 4-fluoroimidazole carboxamide gave 2,4-difluoroimidazole-5-carboxamide. Attempts to elaborate the amino acid side chain of **104** as a route to 2,4-difluorohistidine were thwarted by loss of the 2-fluoro substituent during reduction of the ester to the hydroxymethyl intermediate.

Other routes to fluoroimidazoles have been reported, some of which are confined to specific structural motifs. For example, methyl and fluorine substituents were used in a study of the phototranspositions of the 1-methylpyrazole ring. The "nitrogen walk" mechanism produced ring-fluorinated imidazoles from the ring-fluorinated pyrazoles (Fig. 3.55).[106]

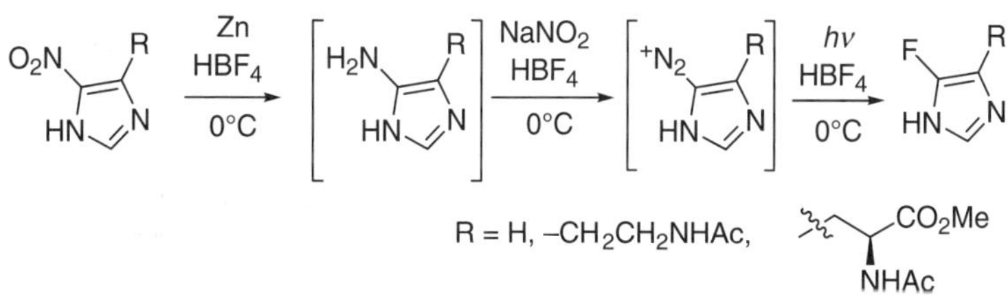

FIGURE 3.52 Synthesis of 4-fluoroimidazoles from 4-nitroimidiazoles.

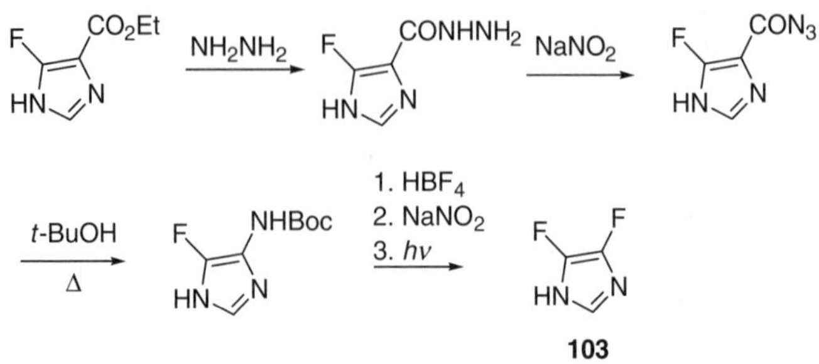

FIGURE 3.53 Synthesis of 4,5-difluoroimidazole.

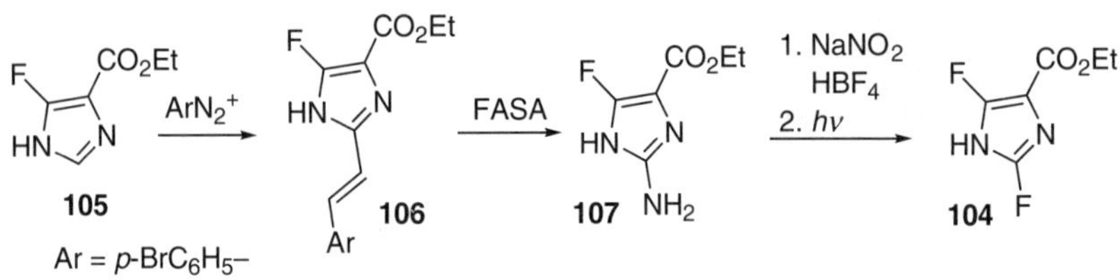

FIGURE 3.54 Synthesis of 2,4-difluoroimidazoles.

In initial attempts to prepare ring-fluorinated imidazoles, nucleophilic displacement of activated halogens with fluoride was unsuccessful.[98] However, if sufficiently activated, halide exchange can occur. Thus, 1-methyl-2-fluoro-4,5-dicyanoimidazole was prepared from 1-methyl-2-bromo-4,5-dicyanoimidazole by reaction with spray-dried KF in the presence of 18-crown-6 ether as catalyst.[107]

The recent development of anhydrous highly nucleophilic fluoride salts has made such nucleophilic substitutions a facile transformation.[108] 1-Methyl-5-chloro-4-nitroimidazole **108**, one of the several substrates that underwent rapid nucleophilic

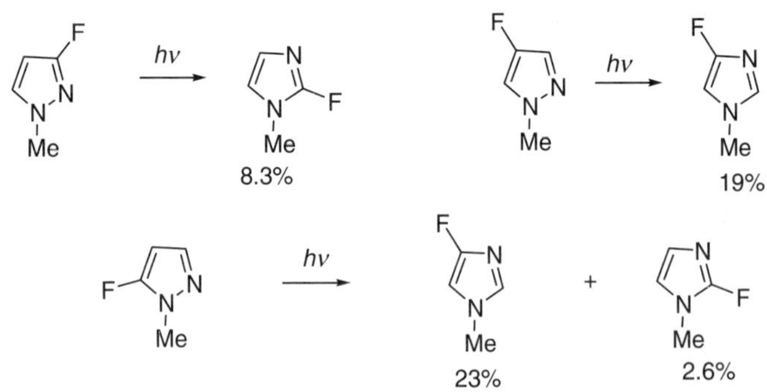

FIGURE 3.55 A photochemically induced “nitrogen walk” to access fluoroimidazoles.

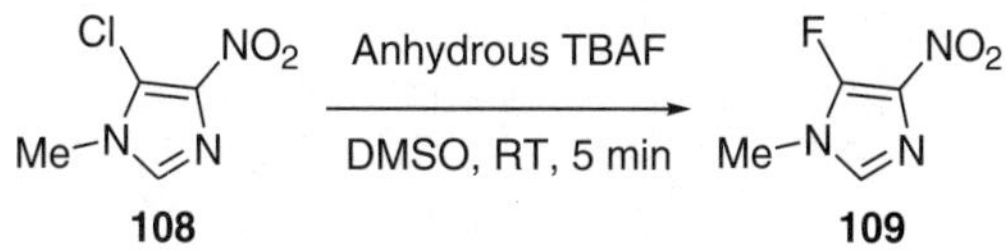

FIGURE 3.56 Use of anhydrous TBAF to prepare fluorinated imidazoles.

fluorination, was converted to the fluorinated imidazole **109**. This new procedure may represent a second general route to ring-fluorinated imidazoles (Fig. 3.56).

Fluorinated synthons have been used to access fluorinated imidazoles. Reduction of Schiff bases derived from amidines and hexafluoroacetone with $SnCl_2$ gives rise to 5-fluoro-4-trifluoromethylimidazoles **110** (Fig. 3.57).[109,110] Because the 5-fluoro substituent is readily displaced by nucleophiles such as alkoxide, cyanide, and others, this procedure provides access to a range of trifluoromethyl-substituted imidazoles (also see below).[110]

There have also been reports of preparation of fluoroimidazoles by electrophilic fluorination of metalated intermediates, although results have been modest. There exists an unpublished report that a more than 50% yield of 2-fluoro-1-methylimidazole was produced by reaction of 2-lithio-1-methylimidazole with perchloryl fluoride.[111] In an another attempt to prepare fluoroimidazoles by electrophilic substitution, reaction of 1-methyl-2-trimethylstannylimidazole, 1-methyl-4-trimethylstannylimidazole, or 1,2-dimethyl-5-trimethylstannylimidazole with dilute fluorine gas at −78°C gave crude product mixtures that contained the corresponding fluorinated imidazole as detected by NMR spectroscopy. The products were not isolated.[112]

3.3.1.2 Synthesis of Trifluoromethylimidazoles Several approaches to imidazoles substituted on the ring with a trifluoromethyl group are available. These include direct introduction of the trifluoromethyl group on the ring as well as construction of the imidazole ring from a trifluoromethylated building block. The work of Burger cited above is an example of the latter strategy. Condensation of aldehydes with 3,3-dibromo-1,1,1-trifluoroacetone in the presence of ammonia provides a facile synthesis of 2-substituted 4-trifluoromethylimidazoles (Fig. 3.58).[113] Hydrolysis of the CF_3 group provides a route to the corresponding imidazole carboxylic acids.

Reaction of imidazole and substituted imidazoles with benzoyl chloride produces dibenzamidoethene derivatives (the Bamberger cleavage). Reaction of this

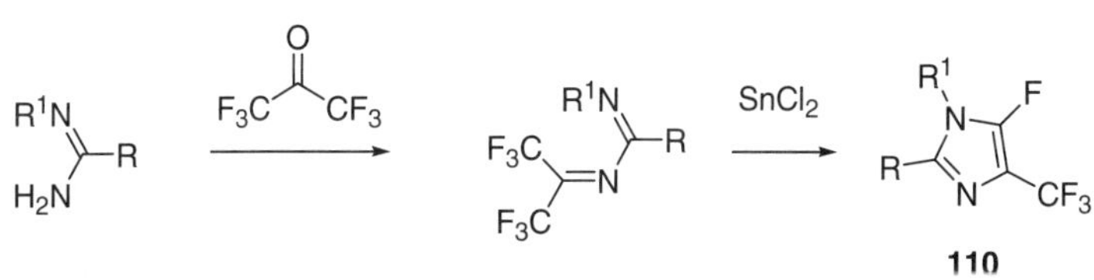

FIGURE 3.57 A fluorinated building block approach to fluorinated imidazoles.

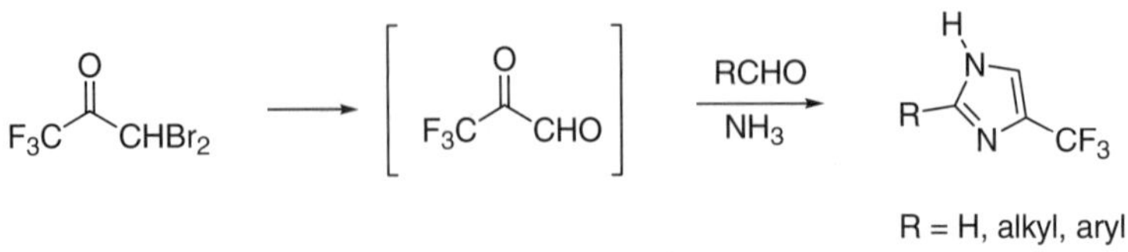

FIGURE 3.58 Synthesis of 2-substituted-4-trifluoromethylimidazoles.

intermediate with trifluoroacetic anhydride (Tf_2O) provides a route to 2-trifluoromethylimidazoles (Fig. 3.59).[114]

Direct introduction of the trifluoromethyl group provides another general route. Irradiation of a methanolic solution of trifluoromethyl iodide and imidazole produces a mixture of 2- and 4-trifluoromethylimidazoles and 2,4-bistrifluoromethylimidazole. This can be applied to simple imidazoles and substituted imidazoles. The photochemical procedure provides a convenient synthesis of trifluoromethyl derivatives of histidine and histamine.[115,116]

3.3.1.3 Fluorinated Purines and Benzimidazoles The importance of fluorinated purine and pyrimidine derivatives as antiviral and anticancer agents has increased the need for efficient methods for introduction of fluorine into these heterocyclic systems. This discussion will be confined to examples wherein the fluorine functionality is on the imidazole ring of purines and related compounds. A review of the substantial literature dealing with 2-fluoro-substituted purines, for example, will not be included.

Routes to 8-fluoropurines had been limited until recently. However, an efficient one-step synthesis was reported in 1996 that has made available these compounds for a variety of studies. Thus, reaction of a series of protected oxopurines and protected adenosine with 1% F_2 in He gave the corresponding 8-fluoro derivatives in 25–30% yields (Fig. 3.60).[117] 2′,3′,5′-Tri-*O*-acetyl-8-fluoroadenosine so prepared was found to undergo defluorination under normal conditions of deprotection. A thermal labile hydrolase was subsequently used to effect this deprotection to produce, for the first time, free 8-fluoroadenosine.[118] More detailed studies on the mechanism of defluorination of 8-fluoropurines indicated that acidic conditions are incompatible with 8-fluoroguanine and 8-fluoroadenosine while basic conditions are also incompatible with 8-fluoroadenosine.[119]

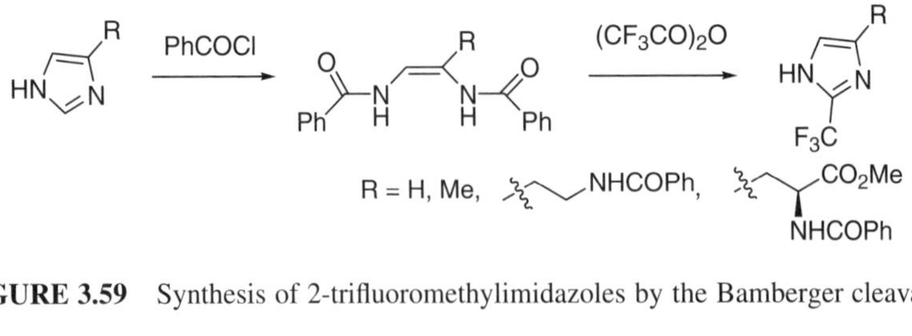

FIGURE 3.59 Synthesis of 2-trifluoromethylimidazoles by the Bamberger cleavage.

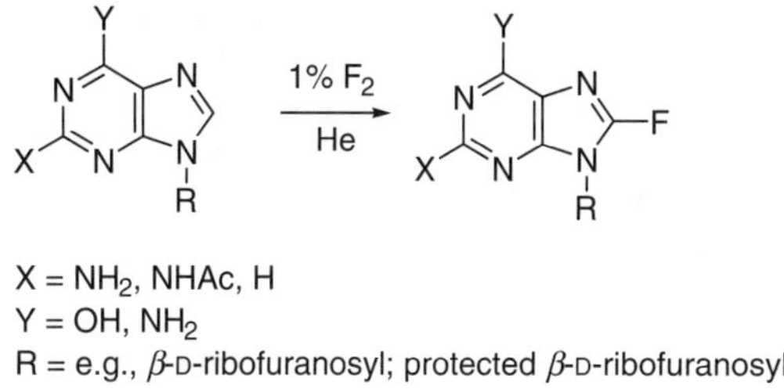

FIGURE 3.60 Synthesis of 8-fluoropurines by direct fluorination.

Direct fluorination with dilute elemental $[^{18}F]F_2$ was used to prepare 8-$[^{18}F]$ fluoroguanine, a positron emission tomography (PET) scanning probe that was used to image gene expression.[120]

The elusive 8-fluoroadenosine has recently been prepared by halogen exchange of the bromopurine **111** followed by nonenzymatic removal of protecting groups from the intermediate **112** to give 8-fluoroadenosine **113** (Fig. 3.61).[121] Among the obstacles to be overcome were the intramolecular displacement of an intermediate fluorinated product by the 5′-hydroxyl group, displacement of fluoride by the exocyclic purine nitrogen, and acid lability. Careful choice of protecting groups and reaction conditions resulted in an efficient route to 8-fluoroadenosine.

There are surprisingly few reports on the synthesis of the related 2-fluorobenzimidazoles. Reaction of 1-(4-fluorobenzyl)-2-chlorobenzimidazole **114** with CsF in the presence of 18-crown 6 for 4 h gave a good yield of the 2-fluoro derivative **115**. Since the fluorine was rapidly displaced with amine nucleophiles, a sequence was developed wherein the chlorobenzimidazole was converted directly to the antihistaminic norastemizole **116**. Fluoride functions as a catalyst by generating the highly reactive **115**, and is reformed by displacement by the amine (Fig. 3.62).[122]

Sun and DiMagno also used 1-(4-fluorobenzyl)-2-chlorobenzimidazole as a substrate for studying their recently developed anhydrous TBAF and reported a >90% conversion to the 2-fluoro derivative in 30 min at room temperature.[123] The milder conditions reveal again the reactivity of anhydrous TBAF.

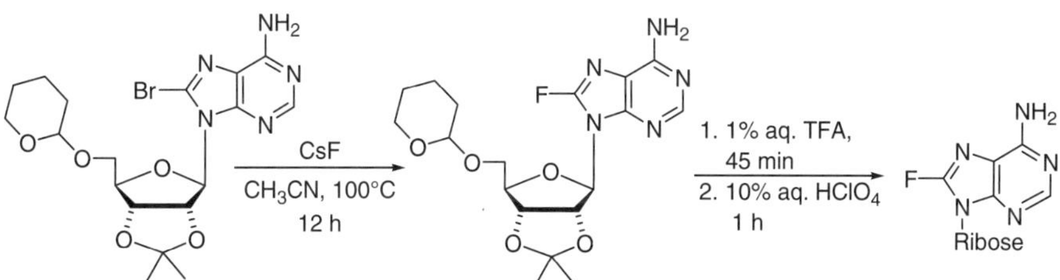

111 **112** **113**

FIGURE 3.61 Synthesis of 8-fluoroadenosine by halogen exchange.

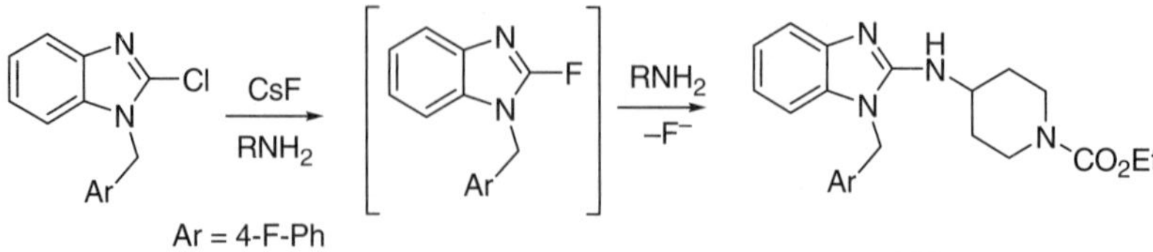

FIGURE 3.62 Fluoride catalysis in the preparation of an antihistaminic 2-aminobenzimidazole.

3.3.1.4 Chemical Reactivity A major consequence of fluorination of the imidazole ring is a dramatic drop in pK_a by several pH units. For example, the pK_a's of 2-fluoro- and 4-flurohistidines were determined by NMR chemical shift titrations to be 0.9 and 2.5, respectively.[124] A value of 6.0 is reported for histidine.

4-Fluoroimidazoles are quite stable under a variety of conditions unless activated by electron-withdrawing groups, such as CF_3 (see above). In contrast, fluorine in the 2-position is subject to nucleophilic displacement even in the absence of activation by electron-withdrawing groups.[100] Of particular interest was the observation that solid samples of 2-fluoroimidazoles trimerize when stored at ambient temperatures, more rapidly when heated, to give "trimidazoles" (Fig. 3.63).[125] Protonation and activation of the imidazole ring by the HF that is generated presumably results in an autocatalytic process.

Advantage was taken of the lability of fluorine in 2-fluoroimidazoles to synthesize a series of 7-imidazoylaminocephalosporin analogues **117**. A solution of 2-fluoroimidazole was first treated with an appropriate *C*-7-aminocephalosporin ester **118** to effect displacement of fluoride (Fig. 3.64). A broad range of substituents on the 4-position of the imidazole ring were used to provide a series of compounds that were studied with respect to β-lactamase stability and for activity against gram-positive organisms.[126]

3.3.1.5 Biological Properties of Fluorinated Imidazoles The importance of the imidazole ring in biological structure and function combined with the profound effects on physicochemical properties that result from fluorine substitution combine to make fluorinated imidazoles a class of compounds with a wide range of biological

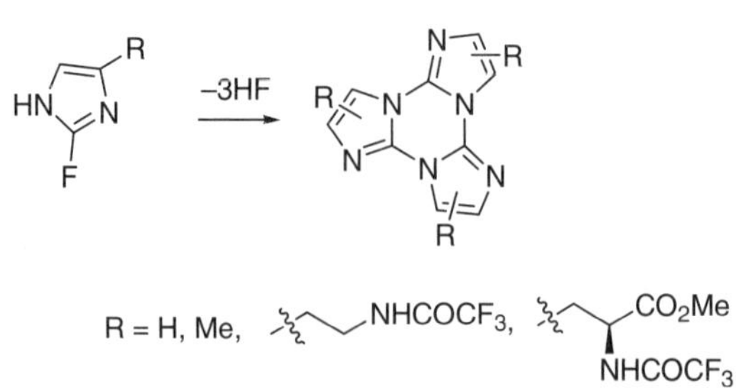

FIGURE 3.63 Spontaneous trimerization of 2-fluoroimidazoles.

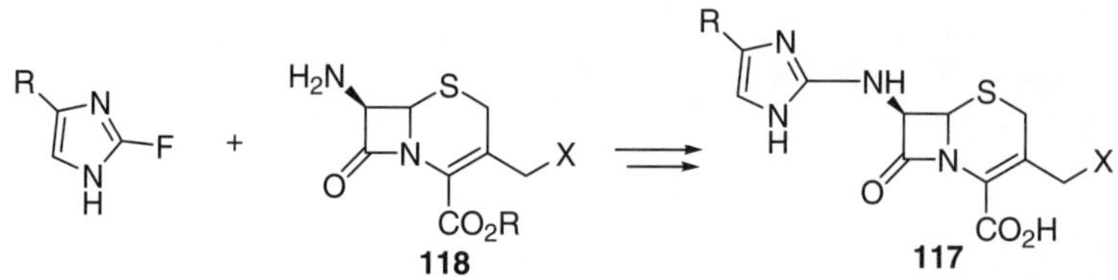

FIGURE 3.64 Synthesis of cephalosporin analogues by displacement of fluorine.

activities. These have been studied extensively, but here only the properties of fluorohistidine will be briefly reviewed.

2- Fluorohistidine, but not the isomeric 4-fluorohistidine, is a substrate for protein biosynthesis, and is incorporated into bacterial, viral, and mammalian proteins *in vitro* and *in vivo*. This difference in behavior may explain in part the fact that the 2-fluoro isomer displays antibacterial, antiviral, and antimalarial activity, while the 4-fluoro isomer is essentially inactive.[101] 2-Fluorohistidine also inhibits enzyme induction, presumably as a result of incorporation into protein.[127] Similarly, the dramatic drop in murine leukocyte levels[128] may be related to incorporation of 2-fluorohistidine into murine proteins *in vivo*.[129] Recent research has exploited the special properties of 2-fluorohistidine to examine the role of histidine residues in the mechanism of anthrax intoxication.[130]

3.3.2 Pyrazoles and Pyrazolines

Introduction of a pyrazole ring is a popular choice for molecular modification in designing medicinal and agrochemical agents, and the use of fluorinated pyrazoles in particular has received much recent attention. The development of celecoxib provides a notable example.[131] Many procedures have been developed for the preparation of fluorinated pyrazoles, but a majority of early work was concentrated on perfluoroalkyl groups substituted on the ring. However, in the past several years, new routes to 3-, 4-, and 5-fluoropyrazoles have been reported.

3.3.2.1 Synthesis of Ring-Fluorinated Pyrazoles Early procedures for accessing fluoropyrazoles included the photochemical Schiemann reaction that had been developed earlier for preparation of ring-fluorinated imidazoles, as discussed above. 3-Fluoro-, 4-fluoro-, and 5-fluoro-1-methylpyrazoles were obtained in low yields (Fig. 3.65).[132] This procedure was also used to prepare 3,4-, 4,5-, and 3,5-difluoropyrazoles.[133] An electrochemical procedure was reported that lacked generality.[134]

The fluorinated building blocks have been more often used in more general approaches to ring-fluorinated pyrazoles. For example, reaction of hydrazines with 2-fluoro-1,3-diketones leads to efficient formation of 4-fluoropyrazoles (Fig. 3.66).[135] Observed regiochemistry of products formed from unsymmetrical diketones was ascribed to initial attack of the more nucleophilic β-nitrogen of the hydrazine at the more electrophilic carbonyl group.

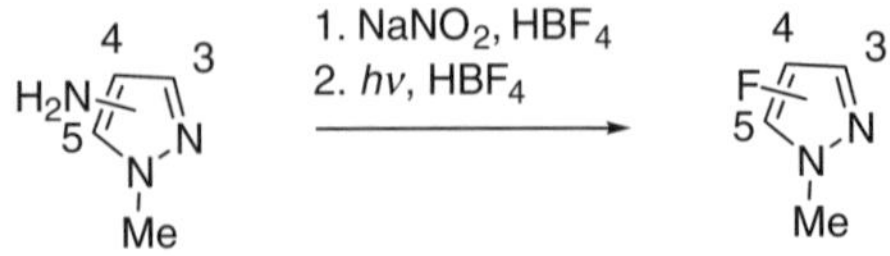

FIGURE 3.65 Preparation of fluorinated pyrazoles by the photochemical Schiemann reaction.

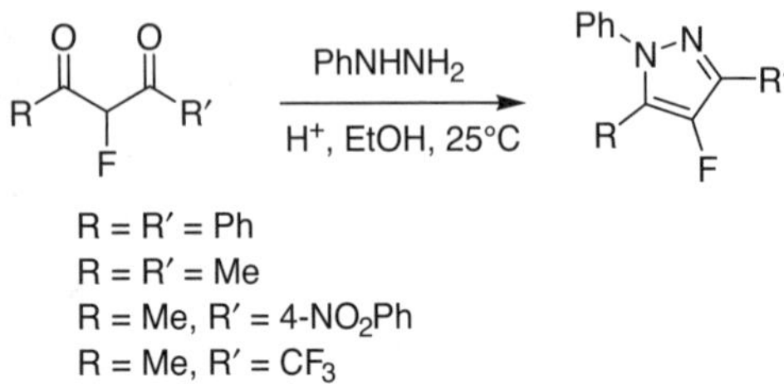

FIGURE 3.66 Synthesis of substituted 4-fluoropyrazoles from fluorinated 1,3-diketones.

If a protected fluoromalonaldehyde is used as the dicarbonyl component, direct access to unsubstituted 4-fluoropyrazole is realized. This approach was made more efficient by the development of a new synthesis of protected fluoromalonaldehyde using chlorofluorocarbene chemistry. Reaction of the carbene, generated from dichlorofluoromethane by phase-transfer catalysis (PTC), with dihydropyran followed by acid-catalyzed ethanolysis gave a mixture of acetals that could be directly used for condensation with hydrazine to give 4-fluoropyrazole **119** in good yield (Fig. 3.67).[136]

4-Fluoro-5-(perfluoroalkyl)pyrazoles have been prepared from organo(per)fluorosilicon building blocks. The perfluorinated enone **120** undergoes reactions with methylhydrazine with loss of water and HF to produce 4-fluoro-5-perfluoroalkylpyrazoles **121**. The enone could be generated *in situ* by the action of hydrazine on a silylated alcohol **122**, itself generated from an acyl silane **123** through a Brooke rearrangement or from the silyl enol ether **124**. The reactions were regiospecific to

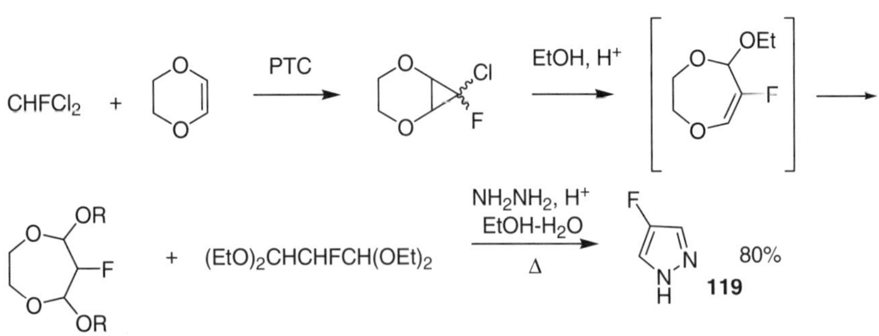

FIGURE 3.67 Synthesis of 4-fluoropyrazole from a fluoromalonaldehyde equivalent.

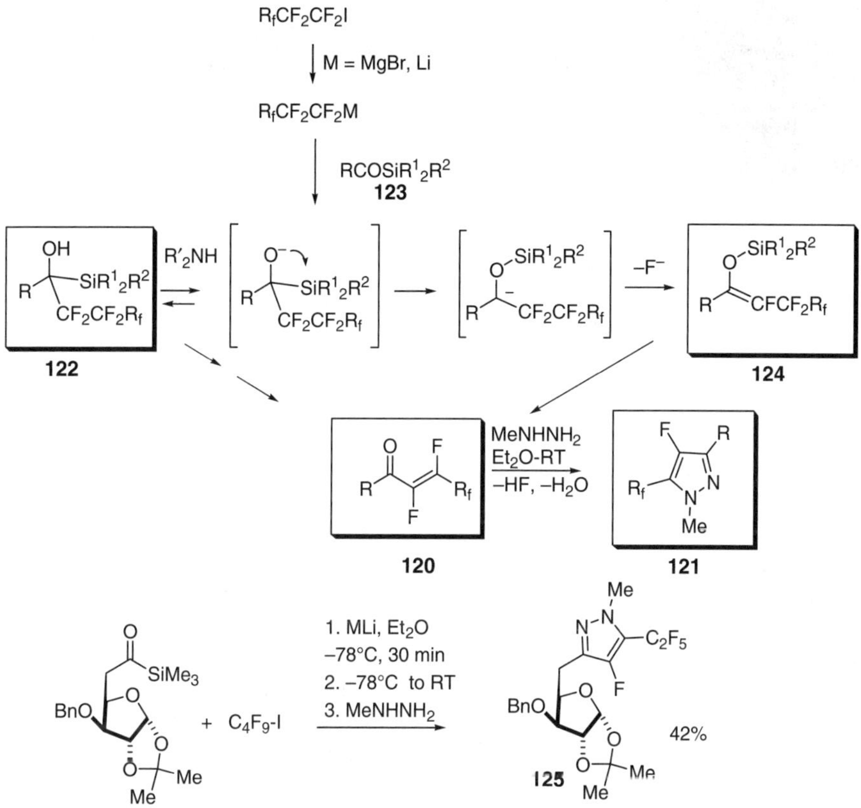

FIGURE 3.68 Synthesis of 4-fluoropyrazoles using organo(per)fluorosilicon building blocks.

give the 4-fluoropyrazoles **121** (Fig. 3.68). A one-pot synthesis from the acyl silane was also demonstrated, for example, through the synthesis of fluoropyrazoles attached to sugar moieties, as shown in Fig. 3.68 for the synthesis of **125**.[137]

Syntheses of 3- and 5-fluoropyrazoles have also been accomplished using fluorinated synthons. 5-Fluoropyrazoles are formed by reaction of 2,2-difluorovinyl ketones with substituted hydrazines. The regiochemistry was ascribed to initial 1,4-addition of the *N*-1 (substituted) nitrogen of the hydrazine followed by cyclization and dehydrofluorination. Yields were improved with excess hydrazine to neutralize released HF. The lithium salts of arylhydrazines were used in THF to effect the transformation with these, otherwise, less nucleophilic reactants (Fig. 3.69).[138]

Reaction of 2,2-difluorovinyl ketones with unsubstituted hydrazine in the presence of trifluoroacetic acid produces 3(5)-fluoropyrazoles in good yield, existing as the 3-fluoro tautomer as evidenced by C^{13}-NMR. Deprotonation gives the ambident anion that almost exclusively gives 3-fluoropyrazoles on alkylation. Repulsive interactions between the C–F dipole and the negative charge were invoked to explain the regiochemistry of alkylation (Fig. 3.70).[138]

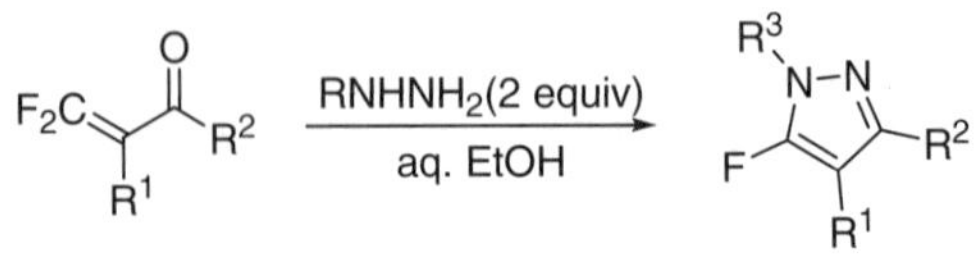

FIGURE 3.69 Synthesis of 1-substituted 5-fluoropyrazoles from 2,2-difluorovinyl ketones.

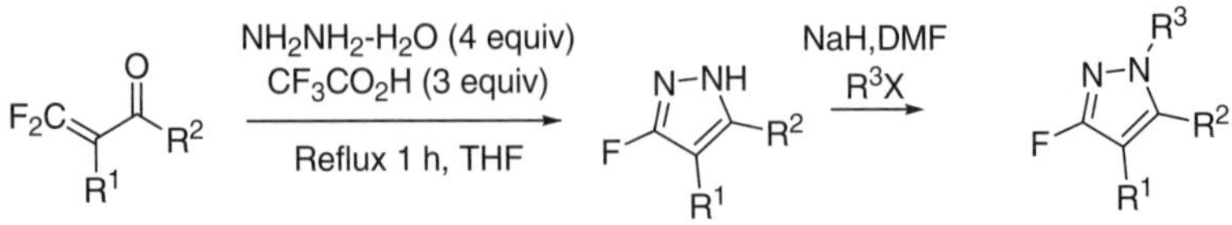

FIGURE 3.70 Synthesis and alkylation of 3(5)-fluoropyrazoles.

Reactions of fluorinated vinamidinium salts with hydrazines provide another route to 4-fluoropyrazoles. The salt is prepared by reaction of *N*-(2,3,3-trifluoro-1-propenyl)trimethylammonium iodide with diethyl amine and treated *in situ* with the hydrazine (Fig. 3.71).[139]

Schlosser developed a "fluorine sacrificial" route to 5-fluoropyrazoles.[140] Treatment of methyl 3-methoxy-2-trifluoromethyl-2-propenoate with aryl or heteroaryl hydrazines resulted in nucleophilic displacement of the methoxy group. Under weakly basic conditions, cyclization occurred with fluorine elimination to give 1-(het)aryl-4-carboxymethyl-5-fluoropyrazoles **126** in moderate yields. The reaction with alkyl hydrazines resulted in no heterocyclization (Fig. 3.72).

1,3-Dipolar addition reactions have been extensively used for the synthesis of five-membered heterocyclic compounds. A recent application of this approach to fluorinated pyrazole syntheses has been reported. Fluoro(tributylstannyl)acetylene (**127**), prepared *in situ* from 1,1-difluoroethylene, undergoes cycloaddition with diazomethane to give 5-tributylstannyl-4-fluoropyrazole (**128**).[141] Palladium-catalyzed cross-coupling reactions with aryl iodides provided high yields of the corresponding 5-aryl-4-fluoro-1*H*-pyrazoles **129**.[142] If the cross-coupling reaction is carried out under an atmosphere of CO, insertion takes place to give the corresponding acyl derivatives **130** (Fig. 3.73).

A series of fused pyrazoles were investigated in the search for agonists selective for the estrogen receptor-β (ERβ) versus estrogen receptor-α (ERα) for hormone replacement therapy. As part of this study, direct fluorination of the fused pyrazole

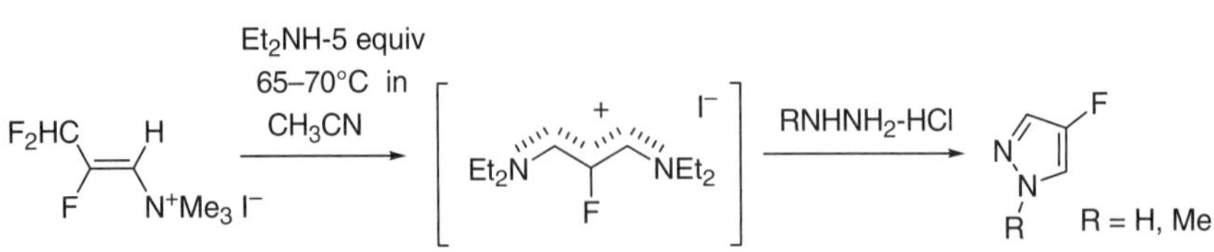

FIGURE 3.71 Synthesis of 4-fluoropyrazoles from vinamidinium salts.

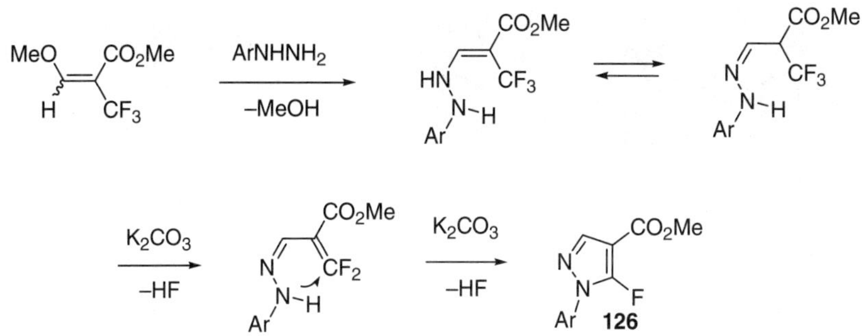

FIGURE 3.72 A "fluorine sacrificial" route to 5-fluoropyrazoles.

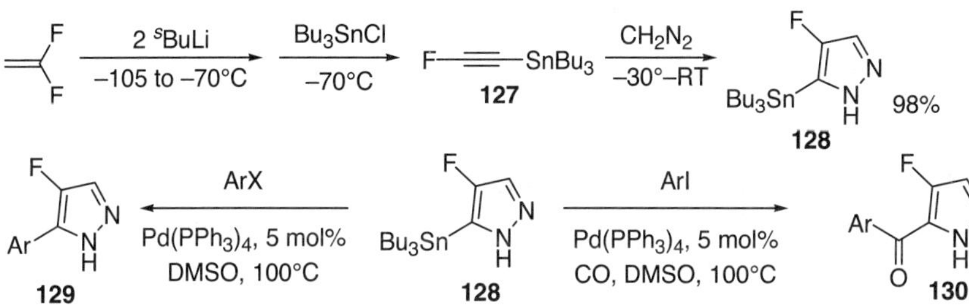

FIGURE 3.73 Synthesis and reactions of 5-tributylstannyl-4-fluoropyrazole.

moiety produced low but useful yields of fluorinated analogues. Some of these, for example **131**, showed increased affinity and selectivity for ERβ (Fig. 3.74).[143]

3.3.2.2 Ring-Fluorinated Pyrazolines Fluorinated olefins undergo 1,3-dipolar cycloaddition reactions with diazomethane to produce fluorinated pyrazolines. In some cases, these are not isolated but are directly converted to fluorinated cyclopropanes by extrusion of nitrogen, either thermally or photochemically. For example, reaction of a series of 2-aryl-3-fluoroacrylates with diazomethane produced intermediate fluorinated pyrazolines that were converted to a series of 1-aryl-1-cyclopropane carboxylates by irradiation at 3500 Å (Fig. 3.75). The esters in turn were

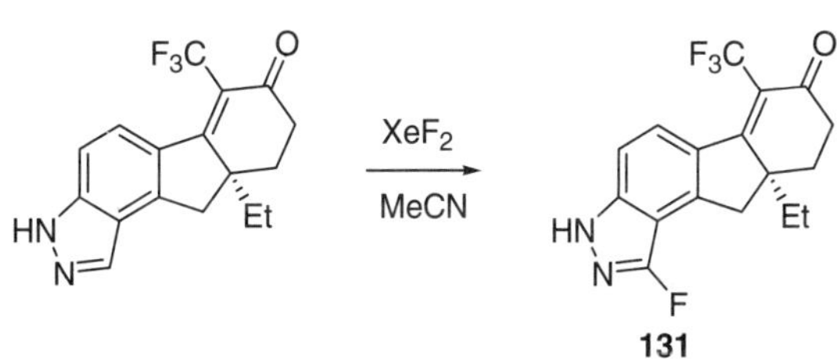

FIGURE 3.74 Synthesis of a fluorinated pyrazole as an estrogen receptor agonist.

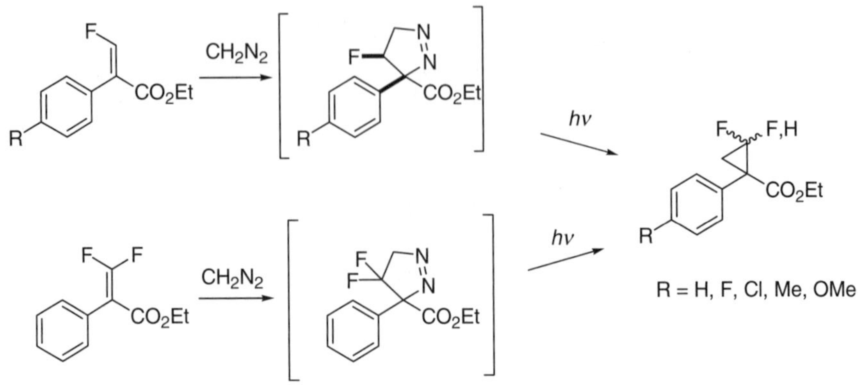

FIGURE 3.75 Ring-fluorinated pyrazolines as synthetic intermediates.

converted to cyclopropyl amines and evaluated as inhibitors of monoamine oxidases.[144]

The structures of the intermediate pyrazolines were not determined. Although mainly *E*-2-fluorocyclopropyl esters were obtained from the monofluoroolefin, some loss of stereochemistry was observed during the photochemical step and *E*-isomers also were obtained.

In another example, an α-(difluoromethylene)-γ-lactone reacted slowly with diazomethane to give a mixture of difluorocyclopropane and difluoropyrazoline. The latter was readily transformed to the difluorocyclopropane by irradiation with a high pressure mercury lamp (Fig. 3.76).[145]

In an example of a fluorine sacrificial procedure, 5-fluoropyrazolin-3-ones have been prepared from α-trifluoromethylated α-arylacetates. The conversions proceed regiospecifically and in high yield, with choice of solvent and temperature being critical. The proposed mechanism is shown in the representative synthetic example (Fig. 3.77).[146]

3.3.2.3 Trifluoromethyl- and Perfluoroalkyl-Substituted Pyrazoles Trifluoromethyl-substituted pyrazoles are readily available through reactions of hydrazines with α,β-unsaturated trifluoromethyl ketones and trifluoromethylated β-dicarbonyl compounds. In addition, 1,3-dipolar addition reactions can be applied. Such transformations have been

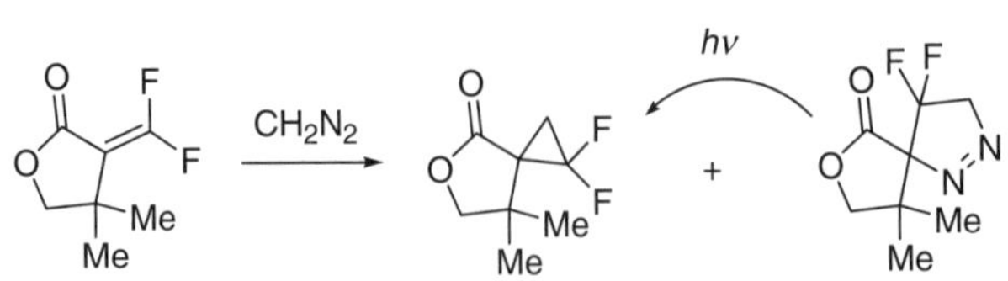

FIGURE 3.76 Diazomethane addition to give fluorinated pyrazolines and fluorinated cyclopropane.

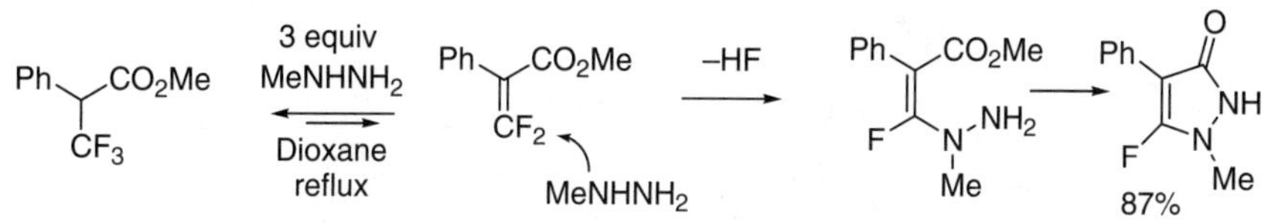

FIGURE 3.77 A "fluorine sacrificial" route to 5-fluoropyrazolin-3-ones.

particularly useful for installing the trifluoromethyl pyrazole moiety into biologically active compounds for drug optimization.

A noted example is found in the development of the selective cyclooxygenase-2 (COX-2) inhibitor celecoxib (**132**) for the treatment of rheumatoid arthritis.[131] Synthetic routes to celecoxib demonstrate two important strategies for constructing the trifluoromethyl pyrazole moiety. In the initial synthesis, a key step was the reaction of an arylhydrazine with the enol ether **133** prepared by condensation of ethyltrifluoroacetate with the substituted acetophenone. Two regioisomers are formed, but the desired 1,5-pyrazole predominates when the condensation is performed with the arylhydrazine hydrochloride in refluxing ethanol (Fig. 3.78).[131]

An alternative approach to celecoxib using a 1,3-dipolar addition between a trifluoromethyl-containing nitrile imine and an eneamine derived from *p*-methylacetophenone avoids the issue of regiochemistry. The nitrile imine was readily prepared by reaction of trifluoroacetylated sulfonamidophenylhydrazine with benzoyl chloride to give the benzenesulfonate **134**. Treatment with triethylamine causes elimination of sulfonate to generate the 1,3-dipole for the cycloaddition reaction. The reaction was regiospecific for the formation of celecoxib (Fig. 3.79).[147]

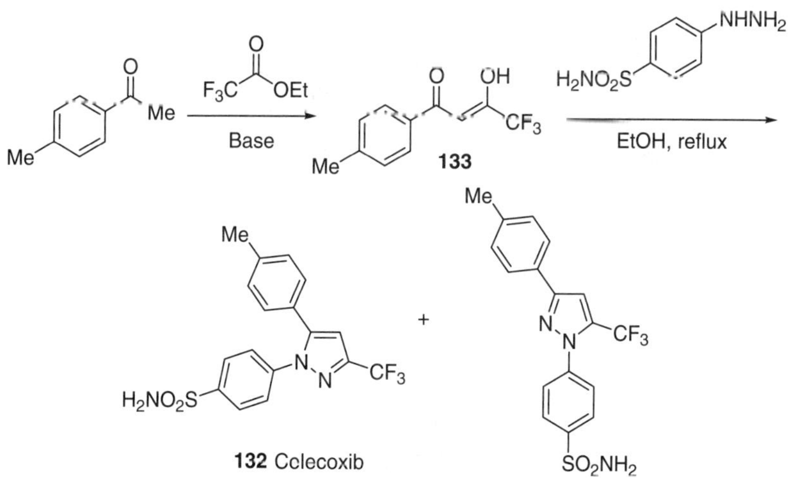

FIGURE 3.78 Synthesis of celecoxib.

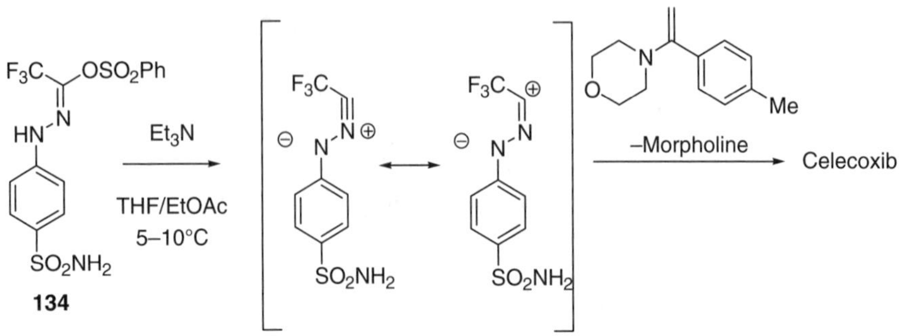

FIGURE 3.79 An alternative synthesis of celecoxib.

Fluorinated 1,3-dicarbonyl reagents or their equivalents can be used as building blocks for the preparation of trifluoromethyl-substituted pyrazoles in a strategy similar to that used for the synthesis of ring-fluorinated pyrazoles. Trifluoroacetyl dihydrofuran and pyran (**135** or **136**), as well as trifluoro-4-ethoxy-3-butene-2-one (**137**), readily react with hydrazines to give 5-trifluoromethylpyrazole derivatives (Fig. 3.80).[148] With certain fluorinated hydrazines, a separate dehydration step became necessary to produce the pyrazoles.

A subsequent report describes similar reactions with unmasked trifluoromethyl-1,3-diketones. Following dehydration of the intermediate product, the 5-trifluoromethylpyrazole was produced regioselectively (Fig. 3.81).[149]

The pentafluorosulfanyl group has received much attention for a variety of applications. Shreeve and coworkers recently prepared 4-pentasulfanyl pyrazole (**138**) by reaction of 1-pentafluorosulfanyl-2-triisopropylsilylacetylene with diazomethane followed by fluoride–mediated desilylation (Fig. 3.82).[150] This was

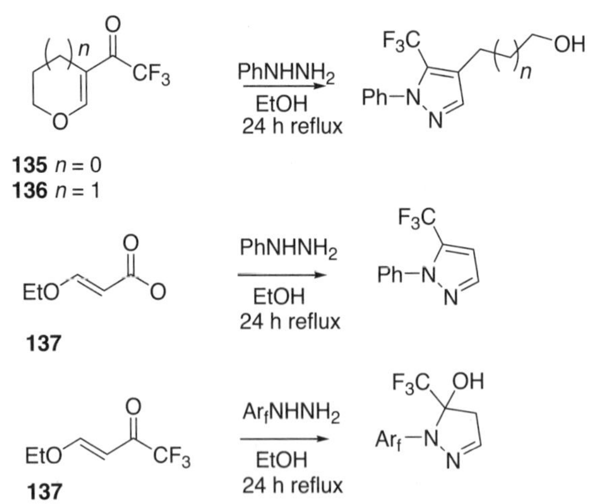

FIGURE 3.80 β-Alkoxyvinyl trifluoroketones as fluorinated pyrazole precursors.

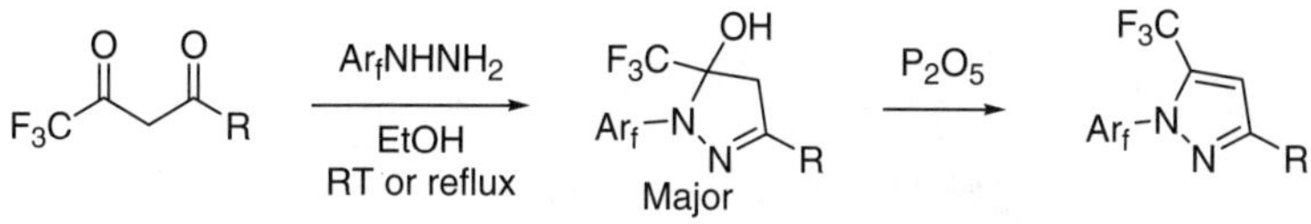

FIGURE 3.81 Pyrazoles from trifluoromethyl-1,3-diketones.

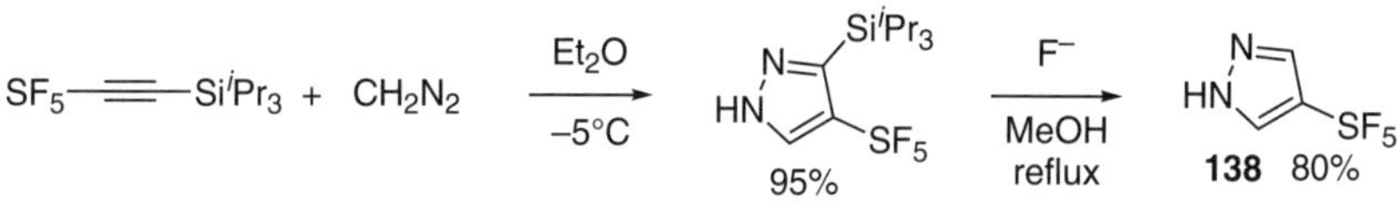

138 80%

FIGURE 3.82 Synthesis of pentafluorosulfanyl pyrazole.

extended to derivatives with high-nitrogen content to be studied as energetic materials. Pentafluorosulfanyl triazole derivatives were also prepared (see below).

3.4 FLUORINATED TRIAZOLES

3.4.1 Fluorinated 1,2,4-Triazoles

The first synthesis of ring-fluorinated 1,2,4-triazoles was reported in 1973.[151] Reaction of 3(5)-nitrotriazoles with liquid HF at 150°C produces the corresponding 3(5)-fluoro-1,2,4-triazole in good yield (Fig. 3.83). Although an efficient process, the reaction conditions make routine application problematic.

3(5)-Fluoro-1,2,4-traizole was also prepared by the photochemical Schiemann reaction.[152] Yields were improved over the original method[98] by increasing the concentration of fluoroborate ion in the reaction mixture.

Advantage was taken of the greater reactivity of the 5-position compared to the 3-position of 1-alkyl-3,5-dibromo-1,2,4-triazoles (e.g., **139**) toward nucleophilic substitution to prepare 1-alkyl-3-fluoro-1,2,4 triazoles **140** by halogen exchange.[153] The initial fluorination product was deprotected to give the bromo–fluoro derivative **141** and this was used for subsequent synthetic manipulations. Included was the preparation of a 3,5-difluoro derivative **142** (Fig. 3.84).

5-Perfluoroalkyl-1,2,4-oxadizoles are readily prepared by the reaction of an amidoxime with a fluorinated acylating agent. Reaction of the oxadiazole with

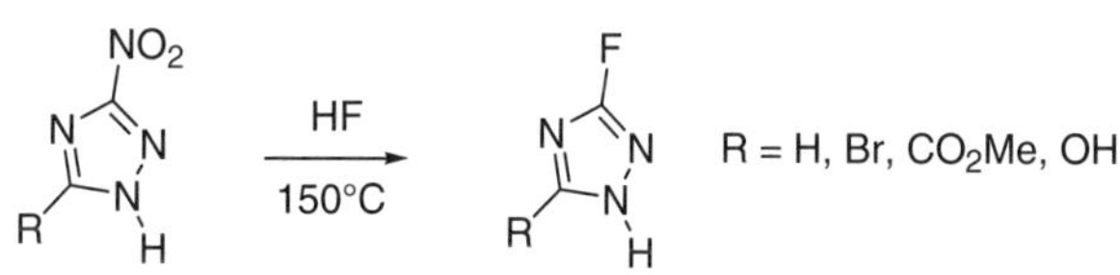

FIGURE 3.83 Synthesis of fluorinated 1,2,4-triazole.

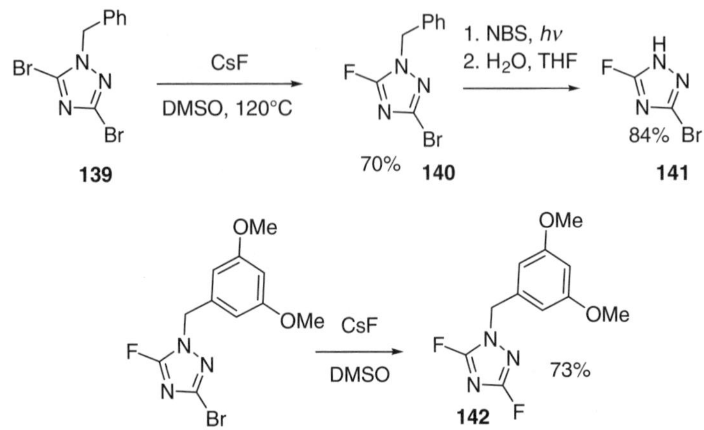

FIGURE 3.84 Synthesis of fluorinated 1,2,4-triazoles by halogen exchange.

hydrazine in turn provides a convenient synthesis of 5-perfluoroalkyl-1,2,4-triazoles (Fig. 3.85).[154] Reaction with methylhydrazine produced the 1-methyltriazole regioisomer. However, with the perfluoropropyl derivative, a significant amount of demethylated product was formed. Triazole formation is explained by the initial nucleophilic attack on the perfluoroalkyl-substituted carbon of the oxazole ring, ring opening, and recyclization with loss of hydroxylamine.

Using similar chemistry, the construction of perfluoroalkyl-substituted 1,2,4-triazoles can be readily accessed through a three-component condensation reaction of a perfluoroalkyl ester, a hydrazine, and an amidine.[155] For example, Shreeve and coworkers prepared a series of alkylated 3-perfluoroalkyl-4,5-dimethyl-1,2,4-triazolium salts

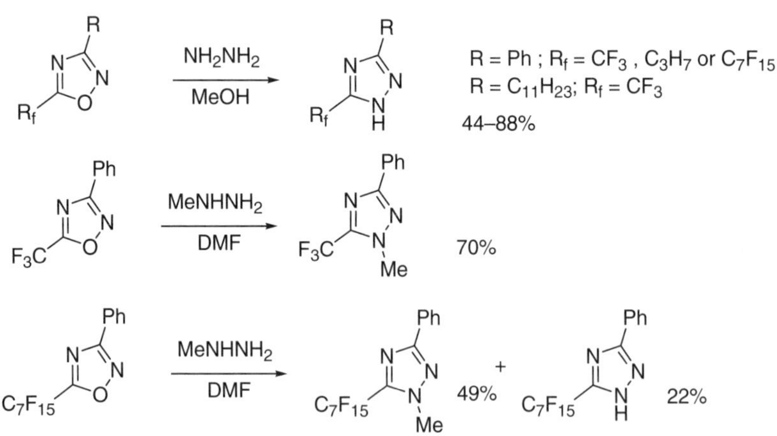

FIGURE 3.85 Synthesis of perfluoroalkyl-substituted 1,2,4-triazoles from oxazoles.

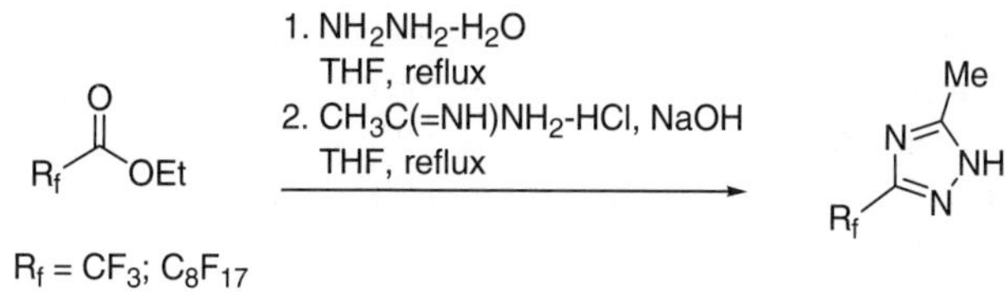

FIGURE 3.86 A three-component condensation reaction to prepare perfluoroalkylated 1,2,4-triazoles.

in their research on the development of ionic liquids.[156] The substrates for alkylation were prepared by a three-component condensation reaction (Fig. 3.86).

An example of a photochemical conversion of one fluorinated heterocyclic system (fluoropyrazole) to another (fluoroimidazole) was described above. In an another example of photochemical isomerization to access fluorinated heterocyclic systems, irradiation of perfluoroalkylated 3-*N*-methylamino-oxadiazoles **143** in the presence of an excess of methylamine produces perfluoroalkylated 1,2,4-triazoles **144** along with the isomeric 1,3,4-oxadiazole **145**.[157] The formation of an intermediate exocyclic diazerene **146** or the open-chain carbodiimide **147** was proposed (Fig. 3.87). If the irradiation is carried out in the presence of methanol and triethylamine,

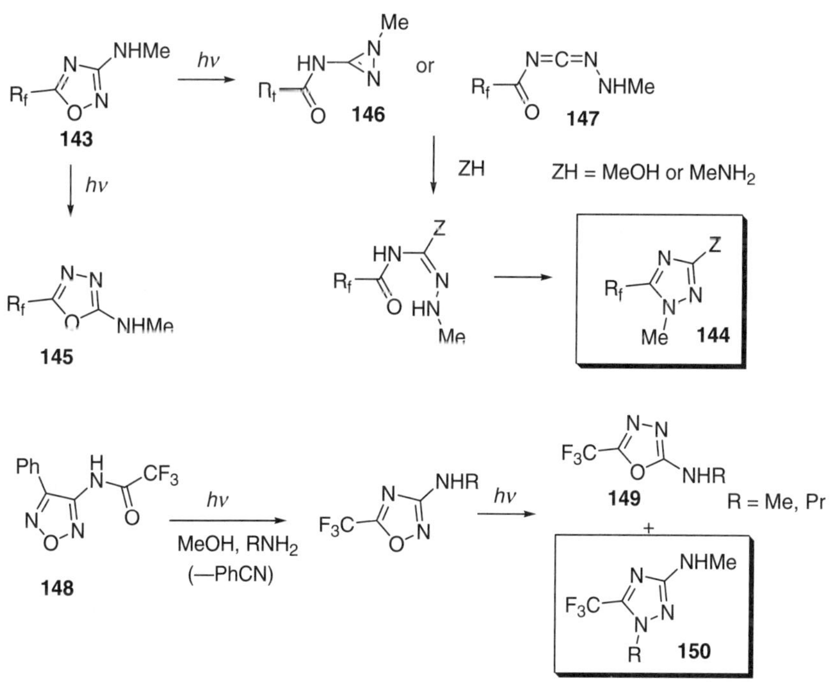

FIGURE 3.87 Photochemical isomerization of fluorinated heterocyclic systems leading to perfluoroalkyl-substituted 1,2,4-triazoles.

photo-rearranged 1,3,4-oxadiazole is formed along with 3-methoxytriazole, consistent with such intermediates. Based on the previous photochemistry developed, a one-pot photoisomerization of 3-perfluoroalkanoylaminofurzins **148** to 1,2,4-oxadiazoles **149** and 1,2,4-triazoles **150** was demonstrated (Fig. 3.87).

Brief mention will be made here of the use of fluorinated 1,2,4-triazolones as agrochemical agents. Novel 4-difluoromethyl-1-aryl-1,2,4-triazolin-5-ones function as important herbicidal compounds for broad-spectrum weed control for use on such crops as corn, soybean, and wheat. The mode of action has been identified as inhibition of protoporphyrinogen oxidase. Two examples are sulfentrazone **151** and carfentrazone-ethyl **152**.[158] A more detailed and comprehensive discussion of agrochemical applications of fluorinated heterocyclic compounds can be found in another chapter of this book.[159]

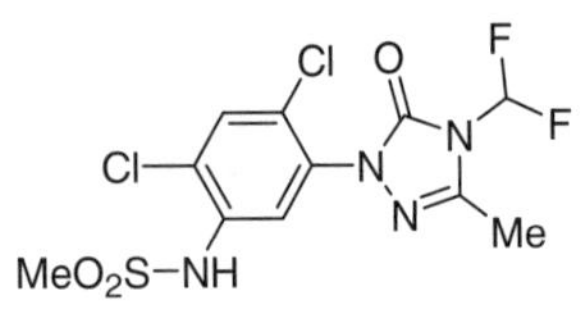

Sulfentrazone
151

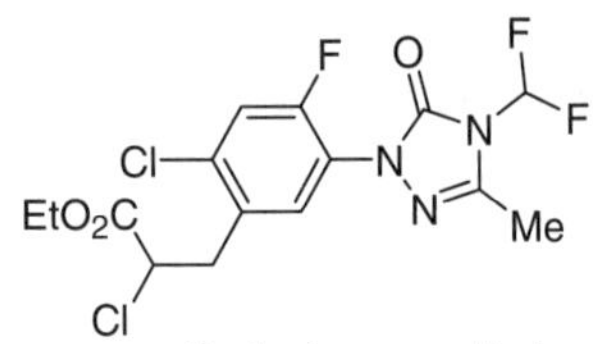

Carfentrazone-ethyl
152

3.4.2 1,2,3-Triazoles

A search of the literature revealed no reports of unsubstituted 4(5)-fluoro-1,2,3-triazole. 4(5)-Trifluoromethyl-1,2,3-triazole (**153**) was prepared by the reaction of $TMSCHN_2$ with trifluoroacetonitrile (Fig. 3.88). The formation of the *N*-silylated intermediate was ascribed to TMS-migration.[160]

Shreeve and coworkers recently reported that reaction of trimethylsilylcyanide with trifluoromethylacetylene gives 4-trifluoromethyl-1,2,3-triazole in good yield.[150]

The SF_5 group has become a focus of much recent attention in organofluorine chemistry. Shreeve and coworkers have also used "click chemistry" to prepare a series of SF_5-substituted 1,2,3-triazoles by reaction of pentafluorosulfanyl acetylene with hydrazoic acid and substituted azides (Fig. 3.89).[150]

N-Substituted 4- or 5-fluoro-1,2,3-triazoles have been prepared by 1,3-cycloaddition reactions with azides. For example, reaction of perfluoropropadiene with phenylazide produces a mixture of regioisomeric 1,2,3-triazoles substituted with the phenyl, fluoro, and trifluoromethyl groups (Fig. 3.90). The major isomer was

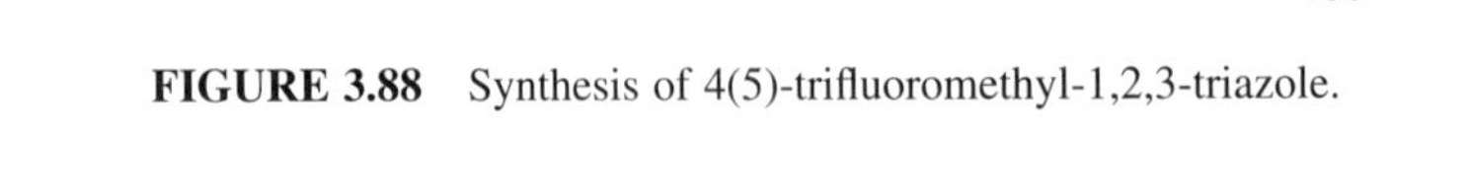

153

FIGURE 3.88 Synthesis of 4(5)-trifluoromethyl-1,2,3-triazole.

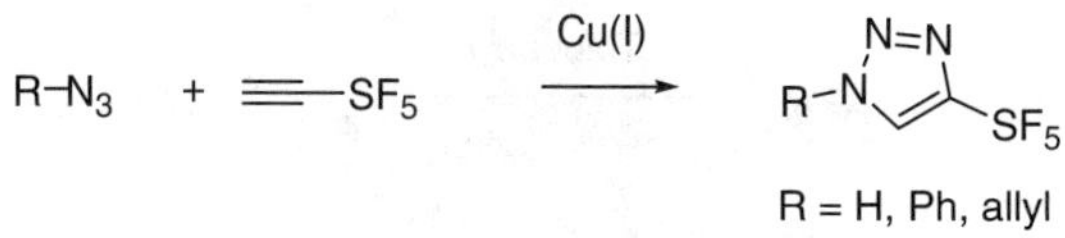

FIGURE 3.89 Synthesis of SF_5-substituted 1,2,3-triazoles.

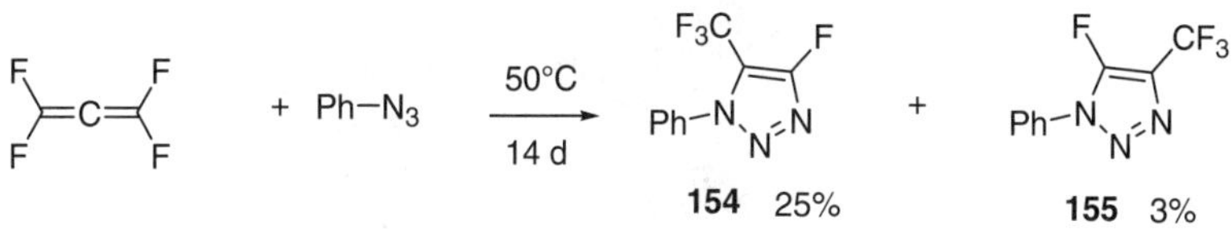

FIGURE 3.90 Synthesis of *N*-substituted 4- or 5-fluoro-1,2,3-tetrazoles by cycloaddition.

provisionally assigned the 4-fluoro-5-trifluoromethyl substitution pattern (**154**) based on theoretical grounds (highest occupied molecular orbital (HOMO)/lowest empty unoccupied molecular orbital (LUMO) interactions).[161] Reaction of phenylazide with pefluoropropyne mainly gave **155** (32%), the minor isomer obtained from the reaction with perfluoropropadiene.

Vinyl sulfones can function as acetylene equivalents through the elimination of sulfinic acid. Reck and coworkers prepared a series of vinyl sulfones for cycloaddition reactions with azides to prepare 1,2,3-triazole derivatives to be incorporated into new oxazolidinone antibacterial agents.[162] Included in the series was the cycloaddition of 1-fluoro-1-(phenylsulfonyl)ethylene with the oxizolidinone **156** to give, after elimination of phenylsulfonic acid, a 28% yield of a 1:7 mixture of the regioisomeric 4- and 5-fluorotriazoles (**157** and **158**) (Fig. 3.91).

1-Substituted-2-perfluoroalkyl propynes are convenient building blocks for the perfluoroalkyl-substituted 1,2,3-triazoles. For example, reaction of 1-aryl-3,3,3-trifluoropropynes with aryl azides produces good yields of the corresponding regioisomeric trifluoromethyl-substituted triazoles **159** and **160** in ratios of about 4:1 (Fig. 3.92).[163]

In their research on the synthesis of new reversed nucleosides, Miethchen and coworkers used masked fluorinated propynes to prepare perfluoroalkyl-substituted

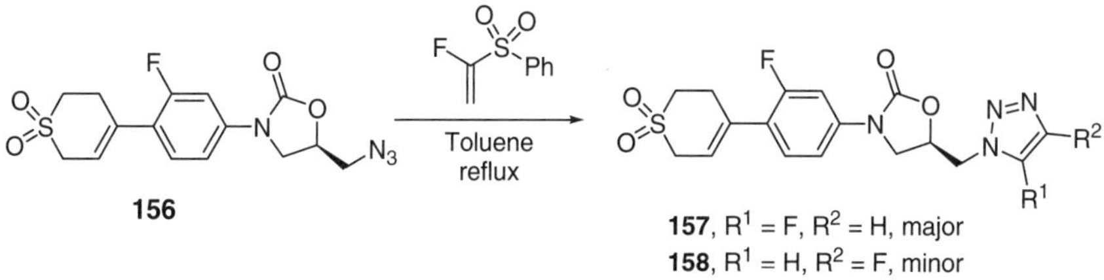

FIGURE 3.91 Fluorinated vinyl sulfones as fluoroacetylene equivalents in 1,2,3-triazole synthesis.

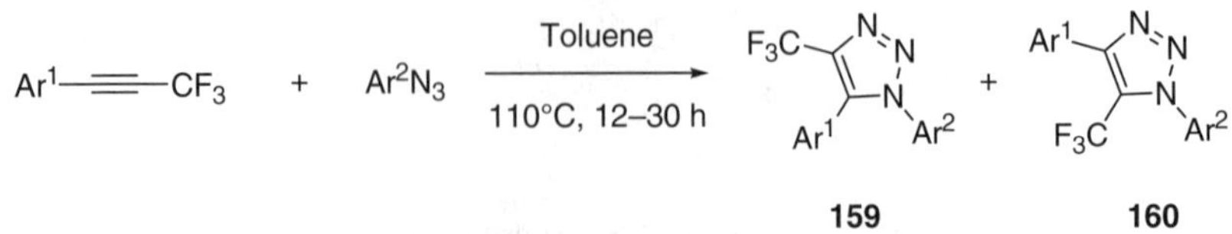

FIGURE 3.92 Synthesis of trifluoromethyl 1,2,3-triazoles by cycloaddition.

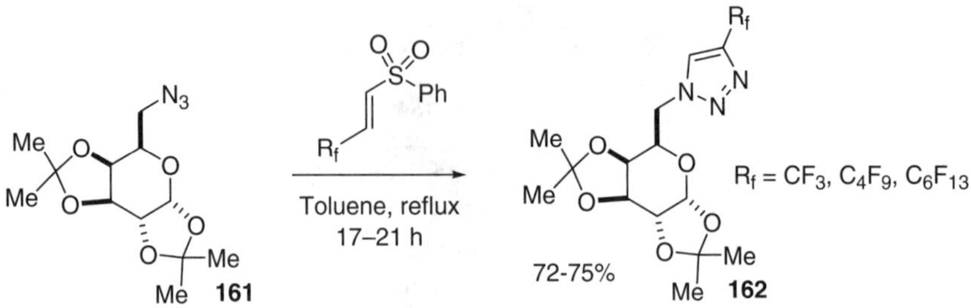

FIGURE 3.93 Synthesis of fluorinated triazole-containing reversed nucleosides by cycloaddition.

1,2,3-triazoles linked to sugars. It is noteworthy that the reactions produced only one of the two regioisomeric triazoles.[164] For example, regioselective reaction of the azide **161** with perfluorinated vinyl sulfones produced the 4-perfluoroalkyl-substituted 1,2,4-triazole **162**, a result of spontaneous loss of phenylsulfinic acid during the reaction (Fig. 3.93).

In related work, the same group prepared a series of trifluoromethyl-substituted 1,2,3-triazoles linked to D-galactose and D-gulose by the cycloaddition of 1-phenyl-2-trifluoromethyl acetylene to the carbohydrate-linked azide.[165] In this case, both regioisomeric triazoles were formed. The procedure is exemplified by the reaction with the galactose-derived azide **163** to give the carbohydrate-linked triazoles **164** and **165** (Fig. 3.94).

From these examples, the versatility of 1,3-cycloaddition reactions in the synthesis of 1,2,3-triazoles is apparent. Other strategies, however, are available. Greif and coworkers developed perfluoroalkyl-substituted β-chlorovinylaldehydes as new building blocks for a number of fluorinated heterocyclic systems such as thiazoles, pyridines, pyrazoles, and benzimidazoles.[166] Reaction of the chlorovinylaldehydes with sodium azide leads to the formation of moderate to good yields of 4-perfluor-

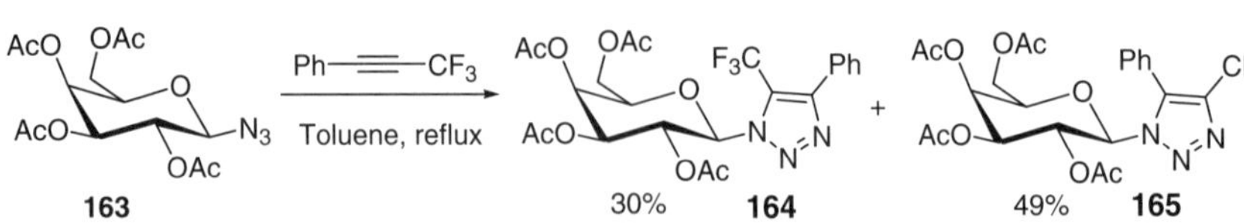

FIGURE 3.94 Synthesis of carbohydrate-linked triazoles by cycloaddition.

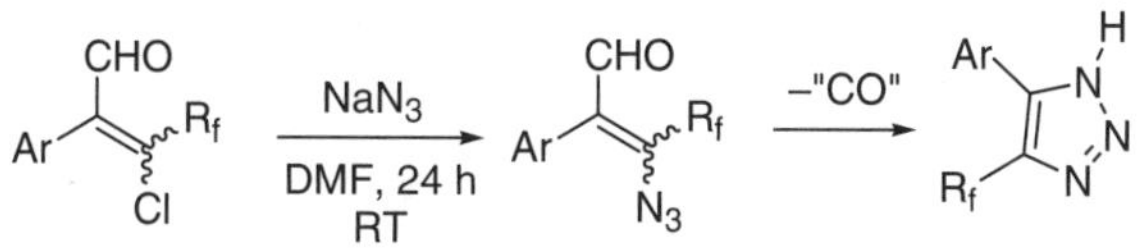

FIGURE 3.95 β-Chlorovinylaldehydes as building blocks for fluorinated 1,2,3-triazoles.

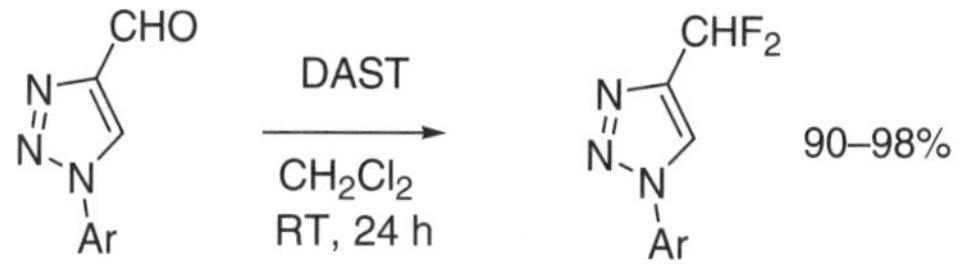

FIGURE 3.96 Synthesis of difluoromethyl-substituted triazoles as potential antitubercular agents.

oalkyl-5-aryl-1,2,3 triazoles. Electrocyclic ring closure of an initially formed azido vinyladehyde followed by loss of "CO" explains the formation of the triazoles (Fig.). Figure 3.95

As with the other aromatic and heteroaromatic systems, functional group transformations can also be used to produce fluorinated 1,2,3-triazoles. A series of 1-aryl-1,2,3-triazole-4-carbaldehydes were converted to the corresponding difluoromethyl-substituted triazoles by the action of DAST (Fig. 3.96).[167] The difluoromethylene derivatives as well as the precursor aldehydes were studied with respect to antitubercular activity.

3.5 FLUORINATED TETRAZOLES

Reports of simple ring-fluorinated tetrazoles appear to be limited. In one example, 1-benzyl-5-fluorotetrazole **166** was prepared by the action of KF on the corresponding chloro derivative.[168] This was used to prepare a series of *N*-(tetrazol-5-yl) azetidinones **167** by reaction with *N*-unsubstituted β-lactams (Fig. 3.97).

In contrast, there are many reports of trifluoromethyl- and other perfluoroalkyl-substituted tetrazoles prepared from fluorinated building blocks. For example, consistent with its high reactivity, trifluoroacetonitrile reacts exothermically with sodium azide to give sodium 5-trifluoromethyltetrazole **168** (Fig. 3.98).[169]

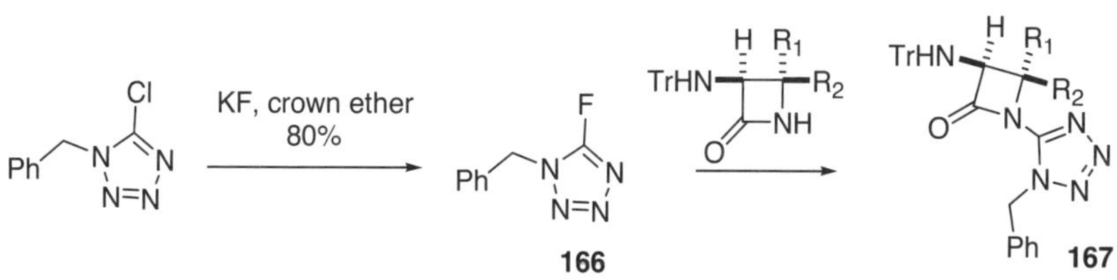

FIGURE 3.97 Synthesis and reactions of a 5-fluorotetrazole derivative.

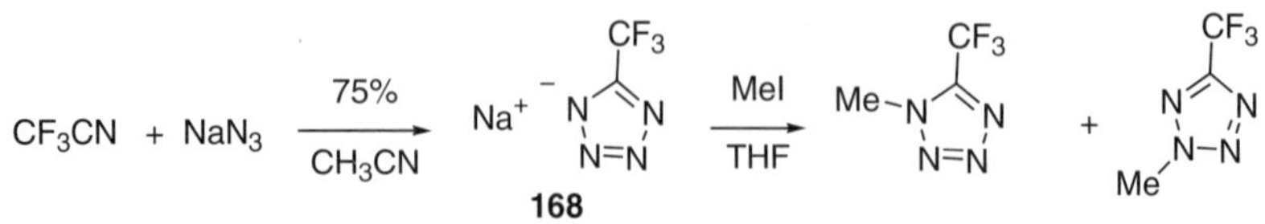

FIGURE 3.98 Synthesis and alkylation of 5-trifluoromethyltetrazolium anion.

5-Trifluoromethyltetrazole is a strong acid, with a pK_a of 1.14. Reaction of the anion with methyl iodide gives a 1:6 mixture of 1-methyl- and 2-methyl-5-trifluoromethyltetrazole.

The reaction of sodium azide with other perlfluoroalkylnitriles similarly leads to 5-pefluoroalkyltetrazoles.[170]

Reaction of imidoyl chlorides with azide ion produces tetrazoles through an imidoyl azide intermediate. Thus, *N*-substituted trifluoroacetimidoyl chlorides undergo facile nucleophilic displacement with azide ion and subsequently cyclize to give 1-substituted-5-trifloromethyltetrazoles (Fig. 3.99).[171] The acetimidoyl chlorides are readily prepared by refluxing a mixture of trifluoroacetic acid and primary amine in carbon tetrachloride in the presence of triethylamine and triphenylphosphine.

5-Trifluoromethyl-1-substituted tetrazoles in particular are employed in drug development. One of the many examples is the neurokinin-1 receptor antagonist **169** (GR205171) that is a potent and orally active antiemetic compound developed by Glaxo Wellcome.[172] The substituted tetrazole moiety was prepared through the acetimidoyl chloride (Fig. 3.100).

Another example of the use of a fluorine-containing tetrazole in drug development is found in the work of Taylor and coworkers. There is evidence that steroid sulfates

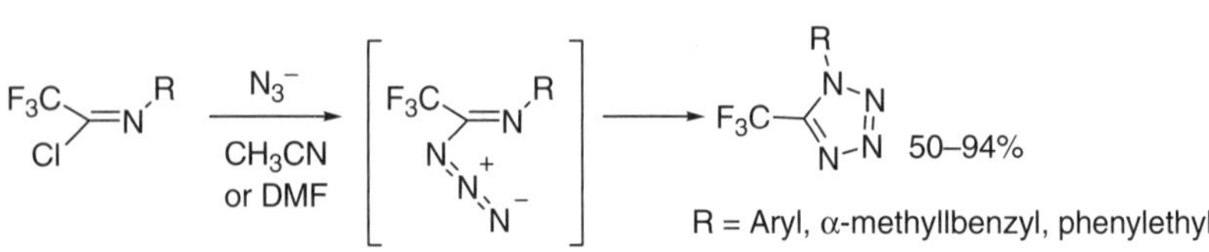

FIGURE 3.99 Synthesis of 1-substituted-5-trifluormethyl tetrazoles.

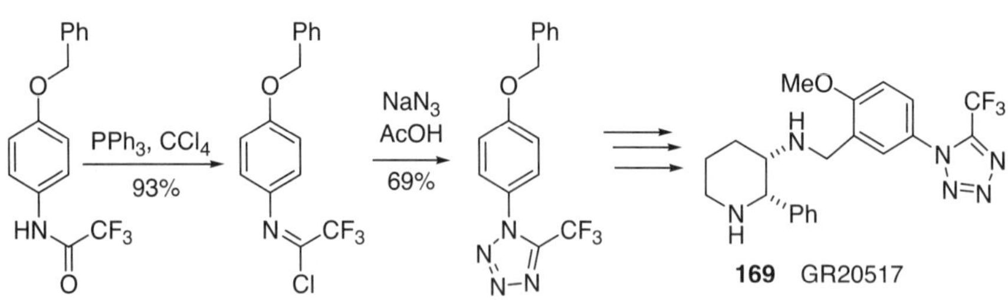

FIGURE 3.100 Synthesis of a trifluoromethyltetrazole-containing neurokinen receptor antagonist.

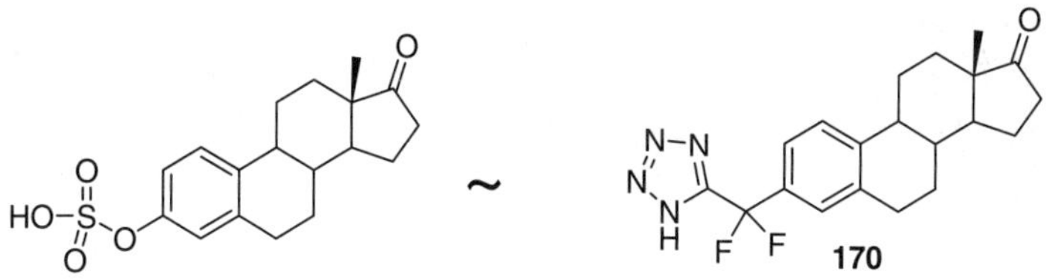

FIGURE 3.101 A fluorinated tetrazole as a mimic of a steroidal sulfate.

can function as a reservoir of steroids that produce estrogens in breast tumors. As part of a program to develop inhibitors of the steroid sulfatases responsible for the elaboration of the free steroids, Taylor and coworkers have prepared nonhydrolyzable isosteres of estrone and estradiol sulfates. The α,α-difluoromethylene tetrazole analogue **170** had an affinity for steroid sulfatase comparable to the natural substrate (Fig. 3.101).[173]

There are numerous reports in the patent literature of trifluoromethyl and other perfluoroalkyl-substituted tetrazoles, particularly with 1,5-subtitution, as pharmaceutical and agrochemical agents. Space does not allow a survey of this material here, but a detailed report of the application of fluorinated heterocyclic compounds in crop protection is given in another chapter of this book.[159]

3.6 PERFLUORINATED NITROGEN-CONTAINING HETEROCYCLES

Perfluorochemicals make up an important area of fluorine chemistry, in particular with respect to industrial applications.[174] Among these, perfluorinated nitrogen-containing heterocyclic compounds have received attention for a variety of applications, and this topic will be discussed briefly in this section.

3.6.1 *N*-Perfluoroalkylated Nitrogen Heterocycles

Sulfetrazone (**151**) and calfentrazone ethyl (**152**), briefly discussed above, are examples of azoles with a carbon-bearing fluorine directly attached to the ring nitrogen, a class of compound that has relatively received little attention. The reaction of five-membered aromatic heterocycles with tetrafluoroethylene and chlorotrifluoroethylene produced early examples of such compounds, including *N*-tetrafluoroethylpyrrole formed by the reaction of the potassium salt of pyrrole with tetrafluoroethylene.[175] Recently, Yagupolskii and coworkers have undertaken an extensive investigation of nitrogen-containing heterocycles possessing perfluorinated alkyl groups directly attached to nitrogen.[176–179]

Potassium salts of a set of heterocycles with different basicities, including imidazole, 2-methylebenzimidazle, 3,5-dimethylpyrazole, 1,2,4-triazole, and benzotriazole, were chosen for reactions with tetrafluoroethylene, chlorotrifluoroethylene, and 1,2-dichlorodifluoroethylene (Fig. 3.102).[179] Reactions with tetrafluoroethylene gave only *N*-tetrafluoroethyl derivatives, products of addition. (The reaction with the

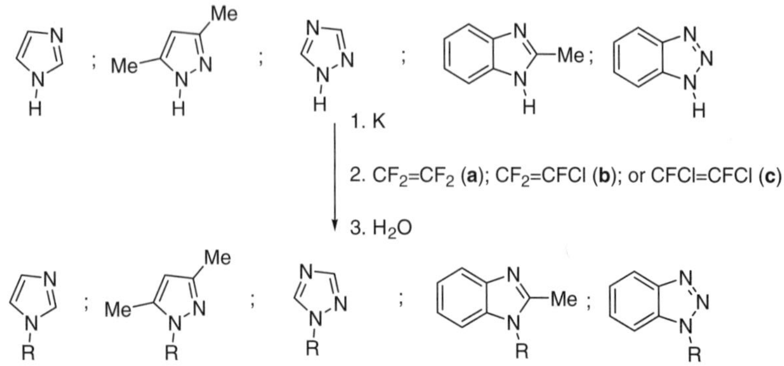

R =					
(**a**)	—CF_2CF_2H	—CF_2CF_2H	—CF_2CF_2H	—CF_2CF_2H	—CF_2CF_2H
(**b**)	—CF=CFCl —CF_2CFClH	—CF=CFCl —CF_2CFClH	—CF=CFCl —CF_2CFClH	—CF=CFCl —CF_2CFClH	—CF=CFCl —CF_2CFClH
(**c**)	—CF=CFCl	—CF=CFCl	—CF=CFCl	—CF=CFCl	

FIGURE 3.102 Reactions of nitrogen heterocycles with perhalogenated olefins.

highly basic imidazole did not require formation of the potassium salt.) Mixtures of substitution and addition products were formed from the reaction with chlorotrifluoroethylene, whereas reaction with tetrafluoroethylene gave only chlorodifluorovinyl products of substitution (no reaction with benzotriazole).

Attempts to prepare *N*-trifluorovinyl derivatives from chlorodifluoro products resulted in the surprising formation of *N*-(1,2,2,2)tetrafluoroethyl products (Fig. 3.103). A mechanism involving elimination of HF followed by displacement of chloride by fluoride and addition of fluoride to the 2-position was proposed, although no intermediate could be detected by NMR.

In the related work, nitrogen heterocycles, including, inter alia, imidazoles, pyrazoles, benzimidiazoles, and triazoles were converted to sodium or potassium salts and treated with Freon-113. Less-reactive amines required iodide catalysis.[176] In all cases, *N*-2,2-dichlorotrifluoro derivatives were formed in moderate to good yields (Fig. 3.104).

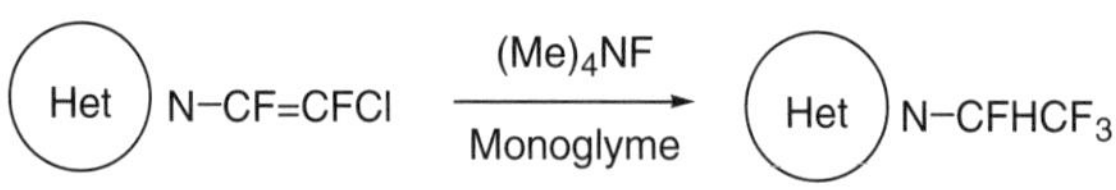

FIGURE 3.103 Reactions of *N*-chlorodifluorovinyl heterocycles with fluoride.

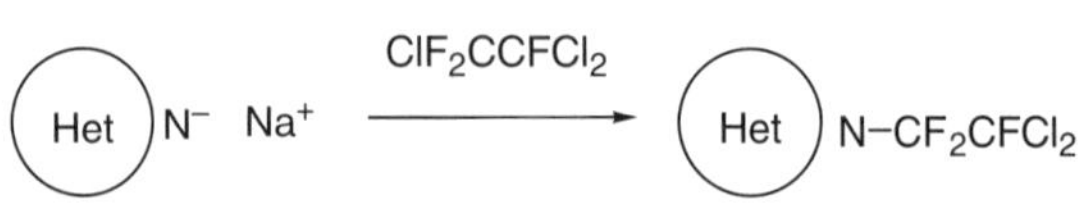

FIGURE 3.104 Reaction of nitrogen heterocycles with Freon-113.

To extend the utility of these reactions, chemical properties of the derived products were investigated.[178] *N*-Polyfluoroethylimidazoles and -pyrazoles were found to undergo electrophilic and nucleophilic substitution reactions, metalation reactions, and formation of quaternary salts as well as carbenes, all retaining the fluorine-containing groups. In another extension of this work, reactions of a similar series of nitrogen heterocycles with 1,2-dibromotetrafluoroethane gave the corresponding *N*-2-bromotetrafluoroethyl derivatives.[180] Reactions of the products with various sulfur nucleophiles extended the range of products accessible by this approach. A related study of nitration and halogenation reactions of *N*-difluoromethyl- and *N*-2-*H*-tetrafluoroethylpyrazoles further extended the inventory of products.[177]

The above are examples of research involving reaction of nitrogen-containing heterocyclic compounds with polyfluorinated hydrocarbons to produce *N*-perfluoroalkyl products. References to related work are found in the original papers.

3.6.2 Ring-Perfluorinated Pyrrolidines and Related Heterocycles

Electrochemical fluorination (ECF) is an effective procedure to replace C−H bonds with C−F bonds in an organic molecule to produce fluorocarbons for a host of uses, particularly in industrial applications. Included in medicinal applications of perfluorinated molecules are their use as inhalation anesthetics and in the development of oxygen carriers.[181]

Perfluorinated nitrogen heterocycles were among these compounds prepared as potential components of artificial blood compositions.[182]

A comprehensive review of this area would not be consistent with the focus of this chapter. However, an example of this work related to nitrogen heterocycles will be given. Thus, Abe and coworker have conducted a program to prepare nitrogen-containing perfluorocarboxylic acids as key intermediates to fluorochemicals for use as surfactants, water/oil repellents, and so on, which have the advantage of being soft-type (degradable) fluorochemicals owing to the presence of nitrogen. Included in a large series of (*N,N*-dialkylamino) acids and esters that were investigated were cyclic and acyclic derivatives related to glycine and alanine, as well as similarly substituted alcohols.[183–185] To illustrate this chemistry, the electrochemical fluorination of methyl esters of 3-dialkylamino acetic acid will be discussed, using pyrrolidine **170** as the dialkylamine.[183] Electrochemical fluorination in this case, as with other cyclic amino group-substituted carboxylic acid derivatives, led to products arising from C−C bond scission, C−N bond scission, as well as the desired perfluoroacid fluoride **171** (Fig. 3.105).

Electrochemical fluorination of (*N,N*-dialkylamino)alcohols led to similar products. Thus, compared to the above example, *N*-hydroxyethylpyrrolidine gave a similar product distribution of α- and γ-bond scission products and a 21.4% yield of the desired perfluoroacid fluoride **171** in addition to a small amount of perfluorinated *N*-ethylpyrrolidine.[185] The ease of preparation of the starting materials is one advantage of this procedure for producing these nitrogen-containing perfluorinated compounds.

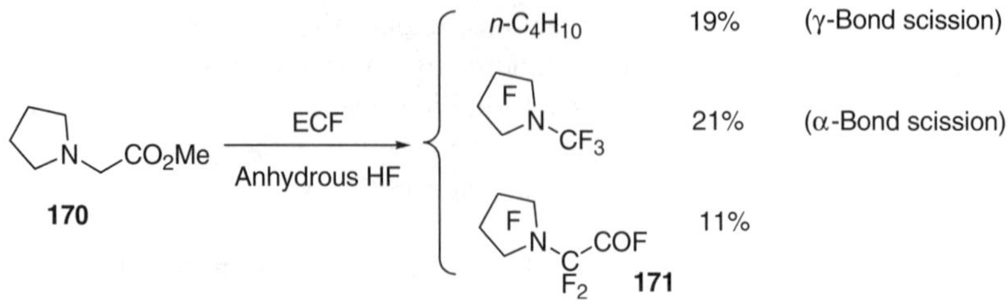

FIGURE 3.105 Electrochemical fluorination of pyrrolidino acetic acid methyl ester.

The above is just a very small sampling of the research involving perfluorinated molecules, an extremely important area of fluorine chemistry. However, since the focus of this chapter has been predominantly on selectively fluorinated molecules, only the above examples are included.

3.7 SUMMARY

The chemistry of fluorinated five-membered heterocyclic compounds containing nitrogen has been reviewed, with an emphasis on synthetic methods. The popularity of nitrogen-containing five-membered rings in drug development and for other uses is reflected in the large amount of published material related to their fluorinated analogues. Similarly, the presence in the class of two naturally occurring amino acids (proline and histidine) is the basis for a large body of literature related to their fluorinated analogues. No review such as this can be all-inclusive, but an attempt has been made to provide representative examples of synthetic methods and, to a lesser extent, applications of this important class of fluorinated compounds. Important examples may have been omitted due to space constraints and limitations of the author.

ACKNOWLEDGMENT

This work was supported by the intramural research funds of NIDDK.

REFERENCES

1. Giardina, G.; Dondio, G.; Grugni, M. Facile and efficient syntheses of novel (*S*)- and (*R*)-3-fluoropyrrolidines and 3,3-difluoropyrrolidine. *Synlett* **1995**, 55–57.
2. Goossens, F.; VanHoof, G.; De Meester, I.; Augustyns, K.; Borloo, M.; Tourwe, D.; Haemers, A.; Scharpé, S. Development and evaluation of peptide-based prolyl oligopeptidase inhibitors. Introduction of *N*-benzyloxycarbonyl-prolyl-3-fluoropyrrolidine as a lead in inhibitor design. *Eur. J. Biochem.* **1997**, *250*, 177–183.
3. Edmondson, S.D.; Mastracchio, A.; Mathvink, R.J.; He, J.; Harper, B.; Park, Y.-J.; Beconi, M.; Di Salvo, J.; Eiermann, G.J.; He, H.; Leiting, B.; Leone, J.F.; Levorse, D.A.; Lyons, K.; Patel, R.A.; Patel, S.B.; Petrov, A.; Scapin, G.; Shang, J.; Roy, R.S.; Smith, A.; Wu,

J.K.; Xu, S.; Zhu, B.; Thornberry, N.A.; Weber, A.E. (2*S*,3*S*)-3-Amino-4-(3,3-difluoropyrrolidin-1-yl)-*N*,*N*-dimethyl-4-oxo-2-(4-[1,2,4]triazolo[1,5-*a*]-pyridin-6-ylphenyl)butanamide: a selective α-amino amide dipeptidyl peptidase IV inhibitor for the treatment of type 2 diabetes. *J. Med. Chem.* **2006**, *49*, 3614–3627.

4. Xu, J.; Wei, L.; Mathvink, R.J.; Edmondson, S.D.; Eiermann, G.J.; He, H.; Leone, J.F.; Leiting, B.; Lyons, K.A.; Marsilio, F.; Patel, R.A.; Patel, S.B.; Petrov, A.; Scapin, G.; Wu, J.K.; Thornberry, N.A.; Weber, A.E. Discovery of potent, selective, and orally bioavailable oxadiazole-based dipeptidyl peptidase IV inhibitors. *Bioorg. Med. Chem. Lett.* **2006**, *16*, 5357–5377.
5. Caldwell, C.G.; Chen, P.; He, J.; Parmee, E.R.; Leiting, B.; Marsilio, F.; Patel, R.A.; Wu, J. K.; Eiermann, G.J.; Petrov, A.; He, H.; Lyons, K.A.; Thornberry, N.A.; Weber, A.E. Fluoropyrrolidine amides as dipeptidyl peptidase IV inhibitors. *Bioorg. Med. Chem. Lett.* **2004**, *14*, 1265–1268.
6. Hulin, B.; Cagral, S.; Lopaze, M.G.; Van Volkenburg, M.A.; Andrews, K.M.; Parker, J.C. New fluorinated pyrrolidine and azetidine amides as dipeptidyl peptidase inhibitors. *Bioorg. Med. Chem. Lett.* **2005**, *15*, 4770–4773.
7. Marson, C.M.; Melling, R.C. The first enantioselective syntheses of vicinal difluoropyrrolidines and the first catalytic synthesis mediated by the C_2 symmetry of a –CHFCHF– unit. *Chem. Commun.* **1998**, 1223–1224.
8. Chaudry, M.T.; Powers, G.A.; Stephens, R.; Tatlow, T.C. Lithium aluminum hydride reduction of polyfluorocyclic imides. *J. Chem. Soc.* **1964**, 874–876.
9. Roberts, R.D.; Spencer, T.A. Tetrafluoropyrrolidine. *Tetrahedron Lett.* **1978**, *29*, 2557–2558.
10. Inagaki, H.; Miyauchi, S.; Miyauchi, R.N.; Kawato, H.C.; Ohki, H.; Matsuhashi, N.; Kawakami, K.; Takahashi, H.; Takemura, M. Synthesis and structure–activity relationships of 5-amino-6-fluoro-1-[(1*R*,2*S*)-2-fluorocyclopropan-1-yl]-8-methylquinolonecarboxylic acid antibacterials having fluorinated 7-[(3*R*)-3-(1-aminocyclopropan-1-yl)-pyrrolidin-1-yl] substituents. *J. Med. Chem.* **2003**, *46*, 1005–1015.
11. Gottlieb, A.A.; Fujita, Y.; Undenfriend, S.; Witkop, B. Incorporation of *cis*- and *trans*-4-fluoro-L-prolines and hydroxylation of the *trans* isomer during collagen biosynthesis. *Biochemistry* **1965**, *4*, 2507–2513.
12. Shirota, F.N.; Nagasawa, H.T.; Elberling, J.A. Potential inhibitors of collagen biosynthesis. 4,4-Difluoro-L-proline and 4,4-dimethyl-DL-proline and their activation by prolyl-*t*-RNA ligase. *J. Med. Chem.* **1977**, *20*, 1176–1181.
13. Burger, K.; Rudolph, M.; Fehn, S.; Sewald, N. Synthesis of (2*S*)-4,4-difluoroproline and (2*S*,4*R*)-4-fluoroproline and their derivatives from (*S*)-aspartic acid. *J. Fluorine Chem.* **1994**, *66*, 87–90.
14. Demange, L.; Ménez, A.; Dugave, C. Practical synthesis of Boc and Fmoc protected 4-fluoro and 4-difluoroprolines from *trans*-4-hydroxyproline. *Tetrahedron Lett.* **1998**, *39*, 1169–1172.
15. Doi, M.; Nishi, Y.; Kiritoshi, N.; Iwata, T.; Nago, M.; Nakano, H.; Uchiyama, S.; Nakazawa, T.; Wakamiya, T.; Kobayashi, Y. Simple and efficient syntheses of Boc- and Fmoc-protected 4(*R*)- and 4(*S*)-fluoroproline solely from 4(*R*)-hydroxyproline. *Tetrahedron* **2002**, *58*, 8453–8459.
16. Shi, G.Q.; Cai, W.L. δ,ε-Unsaturated β,β-difluoro-α-keto esters: novel synthesis and utility as precursors of β,β-difluoro-α-amino acids. *J. Org. Chem.* **1985**, *60*, 6289–6295.

17. Thomas, K.M.; Naduthambi, D.; Tririya, G.; Zondlo, N.J. Proline editing: a divergent strategy for the synthesis of conformationally diverse peptides. *Org. Lett.* **2005**, *7*, 2397–2400.

18. Bonini, B.F.; Boschi, F.; Franchini, M.C.; Fochi, M.; Fini, F.; Mazzanti, A.; Ricci, A. First 1,3-dipolar cycloaddition of azomethine ylides with (*E*)-ethyl 3-fluoroacrylate: regio- and stereoselective synthesis of enantiopure fluorinated prolines. *Synlett* **2006**, 543–546.

19. Nadano, R.; Iwai, Y.; Mori, T.; Ichikawa, J. Divergent chemical syntheses of prolines bearing fluorinated one-carbon units at the 4-position via nucleophilic 5-*endo-trig* cyclizations. *J. Org. Chem.* **2006**, *71*, 8748–8754.

20. Di Lullo, G.A.; Sweeney, S.M.; Körkkö, J.; Ala-Kokko, L.; San Antonio, J.D. Mapping the ligand-binding sites and disease-associated mutations on the most abundant protein in the human, type I collagen. *J. Biol. Chem.* **2002**, *277*, 4223–4231.

21. Holmgren, S.K.; Bretscher, L.E.; Taylor, K.M.; Raines, R.T. A hyperstable collagen mimic. *Chem. Biol.* **1999**, *6*, 63–70.

22. Berg, R.A.; Prockop, D.J. The thermal transition of a non-hydroxylated form of collagen. Evidence for a role for hydroxyproline in stabilizing the triple helix of collagen. *Biochem. Biophys. Res. Commun.* **1973**, *52*, 115–120.

23. Bella, J.; Eaton, M.; Brodsky, B.; Berman, H.M. Crystal and molecular structure of a collagen-like peptide at 1.9 Å resolution. *Science* **1994**, *266*, 75–81.

24. Raines, R.T. Award address. 2005 Emil Thomas Kaiser award. *Protein Sci.* **2006**, *15*, 1219–1225, and references therein.

25. Bretscher, L.E.; Jenkins, C.L.; Taylor, K.M.; DeRider, M.L.; Raines, R.T. Conformational stability of collagen relies on a stereoelectronic effect. *J. Am. Chem. Soc.* **2001**, *123*, 777–778.

26. Hodges, J.A.; Raines, R.T. Stereoelectronic effects on collagen stability: the dichotomy of 4-fluoroproline diastereomers. *J. Am. Chem. Soc.* **2003**, *125*, 9262–9263.

27. Doi, M.; Nishi, Y.; Uchiyama, S.; Nishiuchi, Y.; Nakazawa, T.; Ohkubo, T.; Kobayashi, Y. Characterization of collagen model peptides containing 4-fluoroproline; (4(*S*)-fluoroproline-Pro-Gly)$_{10}$ forms a triple helix but (4(*R*)-fluoroproline-Pro-Gly)$_{10}$ does not. *J. Am. Chem. Soc.* **2003**, *125*, 9922–9923.

28. Shoulders, M.D.; Hodges, J.A.; Raines, R.T. Reciprocity of steric and stereoelectronic effects in the collagen triple helix. *J. Am. Chem. Soc.* **2006**, *128*, 8112–8113, and references therein.

29. Hodges, J.A.; Raines, R.T. Stereoelectronic and steric effects in the collagen triple helix: toward a code for strand association. *J. Am. Chem. Soc.* **2005**, *127*, 15923–15932, and references therein.

30. Kim, W.; Hardcastle, K.I.; Conticello, V.P. Fluoroproline flip-flop: regiochemical reversal of a stereoelectronic effect on peptide and protein structure. *Angew Chem., Int. Ed.* **2006**, *45*, 8141–8145.

31. O'Neil, I.A.; Thompson, S.; Kalindjian, S.B.; Jenkins, T.C. The synthesis and biological activity of *C*2-fluorinated pyrrolo[2,1-*c*][1,4]benzodiazepines. *Tetrahedron Lett.* **2003**, *44*, 7809–7812.

32. Kamal, A.; Shankaraiah, N.; Devaiah, V.; Reddy, K.L. An efficient solid-phase synthesis of biologically important DNA-interactive pyrrolo[2,1-*c*][1,4]benzodiazepine dimers (DSB-120) and their C2-fluorinated analogues. *Tetrahedron Lett.* **2006**, *47*, 6553–6556.

33. Kamal, A.; Reddy, P.S.M.M.; Reddy, D.R.; Laxman, E.; Murthy, Y.L.N. Synthesis of fluorinated analogues of SJG-136 and their DNA-binding potential. *Bioorg. Med. Chem. Lett.* **2004**, *14*, 5699–5702.

34. van der Ley, M. [^{18}F]Fluorine labeled aliphatic amino acids. *J. Label. Compd. Radiopharm.* **1983**, XX, 453–461.

35. Hamacher, K. Syntheses of N.C.A. *cis*- and *trans*-4-[^{18}F]fluoro-L-proline, radiotracers for PET-investigation of disordered matrix protein synthesis. *J. Label. Compd. Radiopharm.* **1999**, *42*, 1135–1144.

36. Mazza, S.M. Stereospecific, semi-automated, N.C.A. syntheses of *cis*-4-[^{18}F]fluoro-L-proline and *trans*-4-[^{18}F]fluoro-L-proline. *J. Label. Compd. Radiopharm.* **2000**, *43*, 1047–1058.

37. Wester, H.-J.; Herz, M.; Senekowitsch-Schmidtke, R.; Schwaiger, M.; Stöcklin, G.; Hamacher, K. Preclinical evaluation of 4-[^{18}F]fluoroprolines: diastereomeric effect on metabolism and uptake in mice. *Nucl. Med. Biol.* **1999**, *26*, 259–265.

38. Langen, K.-J.; Börner, A.R.; Müller-Mattheis, V.; Hamacher, K.; Herzog, H.; Ackermann, R.; Coenen, H.H. Uptake of *cis*-4-[^{18}F]fluoro-L-proline in urologic tumors. *J. Nucl. Med.* **2001**, *42*, 752–754.

39. Langen, K.-J.; Jarosch, M.; Hamacher, K.; Mühlensiepen, H.; Weber, F.; Floeth, F.; Pauleit, D.; Herzog, H.; Coenen, H.H. Imaging of gliomas with *cis*-4-[^{18}F]fluoro-L-proline. *Nucl. Med. Biol.* **2004**, *31*, 67–75.

40. Langen, K.-J.; Hamacher, K.; Bauer, D.; Bröer, S.; Pauleit, D.; Herzog, H.; Floeth, F.; Zilles, K.; Coenen, H.H. Preferred stereoselective transport of the D-isomer of *cis*-4-[^{18}F] fluoro-proline at the blood–brain barrier. *J. Cereb. Blood Flow Metab.* **2005**, *25*, 607–616.

41. Hudlický, M.; Merola, J.S. New stereospecific synthesis and X-ray diffraction structures of (−)-D-erythro- and (+)-L-threo-4-fluoroglutamic acid. *Tetrahedron Lett.* **1990**, *31*, 7403–7406.

42. Hart, B.P.; Coward, J.K. The synthesis of DL-3,3-difluoroglutamic acid from a 3-oxoprolinol derivative. *Tetrahedron Lett.* **1993**, *31*, 4917–4920.

43. McGuire, J.J.; Haile, W.H.; Bey, P.; Coward, J.K. DL-3,3-difluoroglutamate: an enhancer of folypolyglutamate elongation. *J. Biol. Chem.* **1990**, *265*, 14073–14079.

44. Konas, D.W.; Coward, J.K. Electrophilic fluorination of pyroglutamic acid derivatives: application of substrate-dependent reactivity and diastereoselectivity to the synthesis of optically active 4-fluoroglutamic acids. *J. Org. Chem.* **2001**, *66*, 8831–8842.

45. Qui, X.-L.; Qing, F.-L. Synthesis of *cis*-4-trifluoromethyl- and *cis*-4-difluoromethyl-L-pyroglutamic acids. *J. Org. Chem.* **2003**, *68*, 3614–3617.

46. Qiu, X.-L.; Meng, W.-D.; Qing, F.-L. Syntheses of 4-monofluoromethylenyl- and *cis*-4-monomethyl-L-pyroglutamic acids via a novel dehydrofluorination. *Tetrahedron* **2004**, *60*, 5201–5206.

47. Leroy, J.; Porhiel, E.; Bondon, A. Synthesis and characterization of partially β-fluorinated 5,10,15,20-tetraphenylporphyrins and some derivatives. *Tetrahedron* **2002**, *58*, 6713–6722.

48. Barnes, K.D.; Hu, Y.; Hunt, D.A. Electrophilic fluorination of a highly functionalized pyrrole. *Synth. Commun.* **1994**, *24*, 1749–1755.

49. Bondon, A.; Porhiel, E.; Pebay, C.; Thouin, L.; Leroy, J.; Moinet, C. Tetraarylporphyrin synthesis by electrochemical oxidation of porphyrinogens. *Electrochim. Acta* **2001**, *46*, 1899–1903.

50. Zhang, Y.; Shibatomi, K.; Yamamoto, H. Lewis acid catalyzed highly selective halogenation of aromatic compounds. *Synlett* **2005**, 2837–2842.

51. Leroy, J.; Wakselman, C. First access to 3,4-difluoro-1*H*-pyrrole. *Tetrahedron Lett.* **1994**, *35*, 8605–8608.

52. Woller, E.K.; Smirnov, V.V.; DiMagno, S.G. A straightforward synthesis of 3,4-difluoropyrrole. *J. Org. Chem.* **1998**, *63*, 5706–5707.

53. Buhr, G. Ringverengung von Azidocyclobutenen zu Cyclopropancarbonitrilen. *Chem. Ber.* **1973**, *106*, 3544–3558.

54. Onda, H.; Toi, H.; Aoyama, Y.; Ogoshi, H. Fluoropyrroles and tetrafluorporphyrins. *Tetrahedron Lett.* **1985**, *26*, 4221–4224.

55. Qiu, Z.-M.; Burton, D.J. A new approach to the synthesis of β-fluoropyrrole derivatives. *Tetrahedron Lett.* **1994**, *35*, 4319–4322.

56. Wang, Y.; Zhu, S. Convenient synthesis of polyfunctionalized β-fluoropyrroles from rhodium(II)-catalyzed intramolecular N–H insertion reactions. *Org. Lett.* **2003**, *5*, 745–748.

57. Chang, M.N.; Biftu, T.; Boulton, D.A.; Finke, P.E.; Hammond, M.L.; Pessolano, A.A.; Zambias, R.A.; Bailey, P.; Goldenberg, M.; Rackham, A. Syntheses and analgesic/antiinflammatory activities of novel 2-[5-aroyl pyrrolo]alkanoic acids. *Eur. J. Med. Chem.* **1986**, *21*, 363–369.

58. Wang, J.; Scott, A.I. Fluorination of pyrrolic compounds with xenon difluoride. *Tetrahedron Lett.* **1994**, *35*, 3679–3682.

59. Wang, J.; Scott, A.I. Synthesis of the 2-fluoro-11-hydroxy analog of porphobilinogen, a new suicide inhibitor of the enzyme porphobilinogen deaminase. *Tetrahedron* **1994**, *50*, 6181–6192.

60. Wang, J.; Scott, A.I. Fluoro-decarboxylation of pyrrolecarboxylic acids by F-TEDA-BF_4: a convenient general synthesis of fluoropyrroles. *J. Chem. Soc.* **1995**, 2399–2400.

61. Novikov, M.S.; Khlebnikov, A.F.; Sidorina, E.S.; Kostikov, R.R. A facile tandem carbene–ylide route to 2-fluoropyrrole derivatives. *J. Fluorine Chem.* **1998**, *90*, 117–119.

62. Novikov, M.S.; Khlebnikov, A.F.; Sidorina, E.S.; Kostikov, R.R. 1,3-Dipolar cycloaddition of azomethine ylides derived from imines and difluorocarbene to alkynes: a new active Pb-mediated approach to 2-fluoropyrrole derivatives. *J. Chem. Soc., Perkin Trans. 1* **2000**, 231–237.

63. Novikov, M.S.; Khlebnikov, A.F.; Shevchenko, M. A facile carbene route to 2-fluoro-2-pyrrolines via fluorinated azomethine ylides. *J. Fluorine Chem.* **2003**, *123*, 177–181.

64. Banks, R.E.; Khaffaff, S.N. Fluorocarbon derivatives of nitrogen. Part 18. Synthesis of fluorinated indolizines through reactions of pyridinium ethoxycarbonylmethylide or pyridinium phenacylide with perfluoropropene, perfluorobut-2-yne and 3,3,3-trifluoropropyne. *J. Fluorine Chem.* **1991**, *51*, 407–418, and references therein.

65. Wu, K.; Chen, Q.-Y. Synthesis of fluorinated indolizines and 4*H*-pyrrolo[1,2-*a*]benzimidazoles via 1,3-dipolar cycloaddition of fluoroalkenes to *N*-ylides. *Synthesis* **2003**, 35–40.

66. Zhang, X.-c.; Huang, W.-y. A one-step approach to 1-(fluoroalkyl)indolizine derivatives. *Synthesis* **1999**, 51–54.

67. Zhang, X.-c.; Huang, W.-y. Cycloaddition reactions of *N*-(cyanomethyl)pyridinium ylides with 2,2-dihydropolyfluoroalkanoates. *J. Fluorine Chem.* **1998**, *92*, 13–16.

68. Zhang, X.-c.; Huang, W.-y. A convenient synthesis of polyfluoroalkyl-substituted pyrazolo[1,5-*a*] pyridine, pyrrolo[1,2-*b*]pyridazine and indolizine derivatives. *J. Fluorine Chem.* **1998**, *87*, 57–64.

69. Zhang, X.-c.; Huang, W.-y. Reaction of 1-alkylbenzimidazolium 3-ylides with ethyl 2,2-dihydropolyfluoroalkanoates. *Tetrahedron* **1998**, *54*, 12465–12474.

70. Zhang, X.-c.; Huang, W-y. A new approach to pyrrolo[1,2-*a*]quinoxaline derivatives. *Tetrahedron Lett.* **1997**, *38*, 4827–4830.

71. Zhu, S.-z.; Qin, C.-y.; Wang, Y.-L.; Chu, Q.-l. Preparation of 1-trifluoroacetyl indolizines and their derivatives via the cycloaddition of pyridinium *N*-ylides with 4-4-ethoxy-1,1,1,-trifluorobut-3-en-one. *J. Fluorine Chem.* **1999**, *99*, 183–187.

72. Fang, X.; Wu, Y.-M.; Deng, J.; Wang, S.-W. Synthesis of monofluorinated indolizines and their derivatives by the 1,3-dipolar reaction of *N*-ylides with fluorinated vinyl tosylates. *Tetrahedron* **2004**, *60*, 5487–5493.

73. Novikov, M.S.; Khlebnikov, A.F.; Besedina, O.V.; Kostikov, R.R. The first example of intramolecular cycloaddition of carbene-derived azomethine ylides in a domino reaction of difluorocarbene with Schiff bases. *Tetrahedron Lett.* **2001**, *42*, 533–535.

74. Wang, X.; Seth, P.P.; Ranken, R.; Swayze, E.E.; Migawa, M.T. Synthesis and biological activity of 5-fluorotubercidin. *Nucleosides Nucleotides Nucleic Acids* **2004**, *23*, 161–170.

75. Reymond, J.-L.; Pinkerton, A.A.; Vogel, P. Total, asymmetric synthesis of (+)-castanospermine, (+)-6-deoxycastanospermine, and (+)-6-deoxy-6-fluorocastanospermine. *J. Org. Chem.* **1991**, *56*, 2128–2135.

76. Billig, M.J.; Baker, E.W. Monofluorodeuteroporphyrin IX dimethyl ester. *Chem. Ind.* **1969**, 654–655.

77. Naruta, Y.; Tani, F.; Maruyama, K. *meso*-Perfluorination of porphyrins with *N*-fluoropyridinium triflate. *Tetrahedron Lett.* **1992**, *33*, 1069–1072.

78. DiMagno, S.G.; Williams, R.A.; Therien, M.J. Facile synthesis of *meso*-tetrakis(perfluoroalkyl)porphyrins: spectroscopic properties and X-ray crystal structure of highly electron-deficient 5,10,15,20-tetrakis(heptafluoropropyl)porphyrin. *J. Org. Chem.* **1994**, *59*, 6943–6948.

79. DiMagno, S.G.; Wertsching, A.K.; Ross, C.R., II. Electronic consequences of nonplanar core conformations in electron-deficient porphyrins: the structure and spectroscopic properties of [5,10,15,20-tetrakis(heptafluoropropyl)-porphinato]cobalt(II). *J. Am. Chem. Soc.* **1995**, *117*, 8279–8280.

80. Suzuki, A.; Toi, H.; Aoyama, Y.; Ogoshi, H. Synthesis of 1-fluoro-1-demethylmesoporphyrin-IX. *Heterocycles* **1992**, *33*, 87–90.

81. Yamamoto, Y.; Nagao, S.; Hirai, Y.; Inose, T.; Terui, N.; Mita, H.; Suzuki, A. NMR investigation of the heme electronic structure in deoxymyoglobin possessing a fluorinated heme. *J. Biol. Inorg. Chem.* **2004**, *9*, 152–160.

82. Suzuki, A.; Tomizawa, T.; Hayashi, T.; Mizutani, T.; Ogoshi, H. Synthesis of ring-fluorinated porphyrins and reconstituted myoglobins with their iron complexes. *Bull. Chem. Soc. Jpn.* **1996**, *69*, 2923–2933.

83. Homma, M.; Aoyagi, K.; Aoyama, Y.; Ogoshi, H. Electron deficient porphyrins. 1. Tetrakis(trifluoromethyl)porphyrin and its metal complexes. *Tetrahedron Lett.* **1983**, *24*, 4343–4346.

84. Aoyagi, K.; Toi, H.; Aoyama, Y.; Ogoshi, H. Facile synthesis of perfluoroalkylporphyrins. Electron deficient porphyrins II. *Chem. Lett.* **1988**, 1891–1894.

85. Woller, E.K.; DiMagno, S.G. 2,3,7,8,12,13,17,18-Octafluoro-5,10,15,20-tetraarylporphyrins and their zinc complexes: first spectroscopic, electrochemical, and structural characterization of a perfluorinated tetraarylmetalloporphyrin. *J. Org. Chem.* **1997**, *62*, 1588–1593.

86. Leroy, J.; Bondon, A.; Toupet, L.; Rolando, C. 2,3,7,8,12,13,17,18-Octafluoro-5,10,15,20-tetraphenylporphyrin: first synthesis and X-ray crystal structure of the ZnII complex. *Chem. Eur. J.* **1997**, *3*, 1890–1893.

87. Sun, H.; Smirnov, V.V.; DiMagno, S.G. Slow electron transfer rates for fluorinated cobalt porphyrins: electronic and conformational factors modulating metalloporphyrin ET. *Inorg. Chem.* **2003**, *42*, 6032–6040.

88. Nelson, A.P.; DiMagno, S.G. Umpolung of a metal–carbon bond: a potential route to porphyrin-based methane functionalization catalysts. *J. Am. Chem. Soc.* **2000**, *122*, 8569–8570.

89. Che, C.-M.; Zhang, J.-L.; Zhang, R.; Huang, J.-S.; Lai, T.-S.; Tsui, W.-M.; Zhou, X.-G.; Zhou, Z.-Y.; Zhu, N.; Chang, C.K. Hydrocarbon oxidation by β-halogenated dioxoruthenium(VI) porphyrin complexes: effect of reduction potential ($Ru^{VI/V}$) and C—H bond–dissociation energy on rate constants. *Chem. Eur. J.* **2005**, *11*, 7040–7053, and references therein.

90. Porhiel, E.; Toupet, L.; Leroy, J.; Bondon, A. On the role of β-fluorine substitution on the extent of core deformation of porphyrin dications. *Tetrahedron Lett.* **2002**, *43*, 8293–8296.

91. Kumadaki, I.; Ando, A.; Omote, M. Synthesis of fluorinated analogs of porphyrins potentially useful for the diagnosis and therapy of cancer. *J. Fluorine Chem.* **2001**, *109*, 67–81.

92. Anzenbacher, P., Jr.; Try, A.C.; Miyaji, H.; Jursíková, K.; Lynch, V.M.; Marquez, M.; Sessler, J.L. Fluorinated calix[4]pyrrole and dipyrrolylquinoxaline: neutral anion receptors with augmented affinities and enhanced selectivities. *J. Am. Chem. Soc.* **2000**, *122*, 10268–10272, and references therein.

93. Sessler, J.L.; Cho, W.-S.; Gross, D.E.; Shriver, J.A.; Lynch, V.M.; Marquez, M. Anion binding studies of fluorinated expanded calixpyrroles. *J. Org. Chem.* **2005**, *70*, 5982–5986, and references therein.

94. Gross, Z.; Galili, N.; Saltsman, I. The first direct synthesis of corroles from pyrrole. *Angew Chem., Int. Ed.* **1999**, *38*, 1427–1429.

95. Gross, Z.; Galili, N.; Simkhovich, L.; Saltsman, I.; Botoshansky, M.; Bläser, D.; Boese, R.; Goldberg, I. Solvent-free condensation of pyrrole and pentafluorobenzaldehyde: a novel synthetic pathway to corrole and oligopyrromethenes. *Org. Lett.* **1999**, *1*, 599–602.

96. Steene, E.; Dey, A.; Ghosh, A. β-Octafluorocorroles. *J. Am. Chem. Soc.* **2003**, *125*, 16300–16309.

97. Liu, H.-Y.; Lai, T.-S.; Yeung, L.-L.; Chang, C.K. First synthesis of perfluorinated corrole and its Mn=O complex. *Org. Lett.* **2003**, *5*, 617–620.

98. Kirk, K.L.; Cohen, L.A. Photochemical decomposition of diazonium fluoroborates. Application to the synthesis of ring-fluorinated imidazoles. *J. Am. Chem. Soc.* **1971**, *93*, 3060–3061.

99. Kirk, K.L.; Cohen, L.A. Photochemistry of diazonium salts I. Synthesis of 4-fluoroimidazoles, 4-fluorohistamine, and 4-fluorohistidine. *J. Am. Chem. Soc.* **1973**, *95*, 4619–4624.

100. Kirk, K.L.; Nagai, W.; Cohen, L.A. Photochemistry of diazonium salts. II. Synthesis of 2-fluorohistidine and 2-fluorohistamine, and the halogen mobility of 2-fluoroimidazoles. *J. Am. Chem. Soc.* **1973**, *95*, 8389–8392.

101. Kirk, K.L.; Cohen, L.A. Biochemistry and pharmacology of ring-fluorinated imidazoles. In *Biochemistry of the Carbon–Fluorine Bond, ACS Symposium Series* No 28; Filler, R., Ed.; American Chemical Society, Washington D.C., 1976; pp. 23–36.

102. Nagai, W.; Kirk, K.L.; Cohen, L.A. Synthesis of 2-amino-L-histidine and 2-aminohistamine. *J. Org. Chem.* **1973**, *38*, 1971–1974.

103. Kirk, K.L.; Cohen, L.A. Photochemistry of diazonium salts. III. A new and facile synthesis of 4-fluoroimidazoles. *J. Org. Chem.* **1973**, *38*, 3647–3648.

104. Dolensky, B.; Takeuchi, Y.; Cohen, L.A.; Kirk, K.L. Synthesis of 4,5-difluoroimidazole. *J. Fluorine Chem.* **2001**, *107*, 147–148.

105. Takahashi, K.; Kirk, K.L.; Cohen, L.A. Photochemistry of diazonium salts. 5. Syntheses of ethyl 2,4-difluoroimidazole-5-carboxylate and related compounds. *J. Org. Chem.* **1984**, *49*, 1951–1954.

106. Pavlik, J.W.; Kurzweil, E.M. Phototransposition chemistry of 1-methylpyrazole. Deuterium, methyl, and fluorine substitution. *J. Org. Chem.* **1991**, *56*, 6313–6320.

107. Coad, E.C.; Liu, H.; Rasmussen, P.G. Synthesis and characterization of derivatives and dimers of 4,5-dicyanoimidazole. *Tetrahedron* **1999**, *55*, 2811–2826.

108. Sun, H.; DiMagno, S.G. Fluoride relay: a new concept for the rapid preparation of anhydrous nucleophilic fluoride salts from KF. *Chem. Commun.* **2007**, 528–529.

109. Burger, K.; Geith, G.; Hübel, D. Über eine neue Methode zur positionsselektiven Einführung von Trifluormethyl-Gruppen in Heteroaromaten, Teil 1. Synthese von trifluormethyl-substituierten 1,3-Azolen (oxazole, thiazole, imidazole). *Synthesis* **1988**, 189–194.

110. Burger, K.; Hübel, D.; Geith, K. Über eine neue Methode zur positionsselektiven Einführung von Trifluormethyl-Gruppen in Heteroaromaten, Teil 2. Nucleophile Substitution an 5-fluor-4-trifluormethyl-substituierten 1,3-Azolen. *Synthesis* **1988**, 194–198.

111. Bouchet, P.; Coquelet, C.; Elguero, J. Les fluoroazoles: synthèses, propriétés et applications. *Bull. Soc. Chim. Fr.* **1977**, 171–180, and thesis cited therein.

112. Bryce, M.R.; Chambers, R.D.; Mullins, S.T.; Parkin, A. Electrophilic fluorination of tin and mercury derivatives as a route to fluoroaromatics. *Bull. Chim. Soc. Fr.* **1986**, 930–932.

113. Baldwin, J.J.; Kasinger, P.A.; Novello, F.C.; Sprague, J.M. 4-Trifluoromethylimidazoles and 5-(4-pyridyl)-1,2,4-triazoles, new classes of xanthine oxidase inhibitors. *J. Med. Chem.* **1975**, *18*, 895–900.

114. Kimoto, H.; Kirk, K.L.; Cohen, L.A. A facile synthesis of 2-(trifluoromethyl)histamine and 2-(trifluoromethyl)histidine. *J. Org. Chem.* **1978**, *43*, 3403–3405.

115. Kimoto, H.; Fujii, S.; Cohen, L.A. Photochemical trifluoromethylation of some biologically significant imidazoles. *J. Org. Chem.* **1984**, *49*, 1060–1064.

116. Kimoto, H.; Fujii, S.; Cohen, L.A. Photochemical perfluorination of imidazoles. *J. Org. Chem.* **1982**, *47*, 2867–2872.

117. Barrio, J.R.; Namavari, M.; Phelps, M.E.; Satyamurthy, N. Elemental fluorine to 8-fluoropurines in one step. *J. Am. Chem. Soc.* **1996**, *118*, 10408–10411.

118. Barrio, J.R.; Namavari, M.; Keen, R.E.; Satyamurthy, N. The elusive 8-fluoroadenosine. *Tetrahedron Lett.* **1998**, *39*, 7231–7234.

119. Liu, J.; Barrio, J.R.; Satyamurthy, N. Kinetics and mechanism of the defluorination of 8-fluoropurine nucleosides in basic and acidic media. *J. Fluorine Chem.* **2006**, *127*, 1175–1187.

120. Namavari, M.; Barrio, J.R.; Toyokuni, T.; Gambhir, S.S.; Cherry, S.R.; Herschman, H.R.; Phelps, M.E.; Satyamurthy, N. Synthesis of 8-[^{18}F]fluoroguanine derivatives: *in vivo* probes for imaging gene expression with positron emission tomography. *Nucl. Med. Biol.* **2000**, *27*, 157–162.

121. Butora, G.; Schmitt, C.; Levorse, D.A.; Streckfuss, E.; Doss, G.A.; MacCoss, M. The elusive 8-fluoroadenosine: a simple non-enzymatic synthesis and characterization. *Tetrahedron* **2007**, *63*, 3782–3789.

122. Senanayake, C.H.; Hong, Y.; Xiang, T.; Wilkinson, H.S.; Bakale, R.P.; Jurgens, A. R.; Pippert, M.F.; Butler, H.T.; Wald, S.A. Properly tuned first fluoride-catalyzed TGME-mediated amination process for chloroimidazoles: inexpensive technology for antihistaminic norastemizole. *Tetrahedron Lett.* **1999**, *40*, 6875–6879.

123. Sun, H.; DiMagno, S.G. Room-temperature nucleophilic aromatic fluorination: experimental and theoretical studies. *Angew Chem., Int. Ed.* **2006**, *45*, 2720–2725.

124. Yeh, H.J.C.; Kirk, K.L.; Cohen, L.A.; Cohen, J.S. ^{19}F and ^{1}H nuclear magnetic resonance studies of ring-fluorinated imidazoles and histidines. *J. Chem. Soc., Perkin Trans. 2* **1975**, 928–934.

125. Takeuchi, Y.; Kirk, K.L.; Cohen, L.A. Imidazole cyclotrimers (trimidazoles), a novel heteroannular series. *J. Org. Chem.* **1979**, *44*, 4243–4246.

126. Jung, F.; Boucherot, D.; Delvare, C.; Olivier, A. Synthesis and structure–activity relationships of new cephalosporins with aminoimidazoles at C-7. Effect of the pK_a of the C-7 aminoimidazole on antibacterial spectrum and β-lactamase stability. *J. Antibiot.* **1993**, *46*, 992–1012.

127. Klein, D.C.; Kirk, K.L. 2-Fluoro-L-histidine: a histidine analog which inhibits enzyme induction. In *Biochemistry of the Carbon–Fluorine Bond, ACS Symposium Series*; Filler, R., Ed.; American Chemical Society, Washington D.C., 1976; pp. 37–56.

128. Creveling, C.R.; Kirk, K.L.; Highman, B. Effect of 2-fluorohistidine upon leukocytopoiesis in mice. *Res. Commun. Chem. Pathol. Pharmacol.* **1977**, *16*, 507–522.

129. Creveling, C.R.; Padgett, W.L.; McNeal, E.T.; Cohen, L.A.; Kirk, K.L. Incorporation of 2-fluorohistidine in murine protein *in vivo*. *Life Sci.* **1992**, *51*, 1197–1204.

130. Wimalasena, D.S.; Cramer, J.C.; Janowiak, B.E.; Juris, S.J.; Melnyk, R.A.; Anderson, D.E.; Kirk, K.L.; Collier, J.R.; Bann, J.G. Effect of 2-fluorohistidine labeling of the anthrax protective antigen on stability, pore formation, and translocation. *Biochemistry* **2007**, *46*, 14928–14936.

131. Penning, T.D.; Talley, J.J.; Bertenshaw, S.R.; Carter, J.S.; Collins, P.W.; Docter, S.; Graneto, M.J.; Lee, L.F.; Malecha, J.W.; Miyashiro, J.M.; Rogers, R.S.; Rogier, D.J.; Yu, S.S.; Anderson, G.D.; Burton, E.G.; Cogburn, J.N.; Gregory, S.A.; Koboldt, C.M.;

Perkins, W.E.; Siebert, K.; Veenhuizen, A.W.; Zhang, Y.Y.; Isakson, P.C. Synthesis and biological evaluation of the 1,5-diarylpyrazole class of cyclooxygenase-2 inhibitors: identification of 4-[5-(4-methylphenyl)-3-(trifluoromethyl)-1*H*-pyrazole-1-yl]benzenesulfonamide (SC-58635, celecoxib). *J. Med. Chem.* **1997**, *40*, 1347–1365.

132. Fabra, F.; Vilarrasa, J.; Coll, J. Fluoroazoles. III. Synthesis and ^{1}H and ^{19}F NMR spectra of 3-, 4-, and 5-fluoro-1-methylpyrazole. *J. Heterocycl. Chem.* **1978**, *15*, 1447–1449.
133. Fabra, F.; Fos, E.; Vilarrasa, ^{19}F-^{19}F Couplilng constants in the azole series. Synthesis of 3,4-, 4, 5-, and 3, 5-difluoropyrazoles. *Tetrahedron Lett.* **1979**, *34*, 3179–3180.
134. Makino, K.; Yoshioka, H. Selective fluorination of ethyl 1-methylpyrazole-4-carboxylates with poly(hydrogen fluoride)-amine complex under electrolytic anodic oxidation. *J. Fluorine Chem.* **1988**, *39*, 435–440.
135. Bumgardner, C.L.; Sloop, J.C. Ring-fluorinated pyrazoles. *J. Fluorine Chem.* **1992**, *56*, 141–146.
136. Molines, H.; Wakselman, C. Fluoromalonaldehyde bis(dialkyl acetals): synthesis by carbene condensation and transformation to dialkyl fluoromalonates and fluorinated heterocyclic compounds. *J. Org. Chem.* **1989**, *54*, 5618–5620.
137. Bouillon, J.-P.; Didier, B.; Dondy, B.; Doussot, P.; Plantier-Royon, R.; Portella, C. Efficient synthesis of 4-fluoro-5-(perfluoroalkyl)pyrazoles from organofluorosilicon building blocks. *Eur. J. Org. Chem.* **2001**, 187–192.
138. Ichikawa, J.; Kobayashi, M.; Noda, Y.; Yokata, N.; Amano, K.; Minami, T. Regiocontrolled syntheses of 3- or 5-fluorinated pyrazoles from 2,2-difluorovinyl ketones. *J. Org. Chem.* **1996**, *61*, 2763–2769.
139. Shi, X.; Ishihara, T.; Yamanaka, H.; Gupton, J.T. Reactions of β-fluorovinamidinium salt with bifunctional hetero nucleophiles. A new synthetic route to fluorinated heterocycles. *Tetrahedron Lett.* **1995**, *36*, 1527–1530.
140. Volle, J.-N.; Schlosser, M. Fluorine sacrificial cyclizations as an access to 5-fluoropyrazoles. *Eur. J. Org. Chem.* **2000**, 823–828.
141. Hanamoto, T.; Koga, Y.; Kido, E.; Kawanami, T.; Furuno, H.; Inanaga, J. Palladium catalyzed cross-coupling reaction of 5-tributylstannyl-4-fluoropyrazole. *Chem. Commun.* **2005**, 2041–2043.
142. Hanamoto, T.; Suetake, T.; Koga, Y.; Kawanami, T.; Furuno, H.; Inanaga, J. Synthesis and reactions of 5-tributylstannyl-4-fluoro-1*H*-pyrazoles. *Tetrahedron* **2007**, *63*, 5062–5070.
143. Wilkening, R.R.; Ratcliffe, R.W.; Fried, A.K.; Meng, D.; Sun, W.; Colwell, L.; Lambert, S.; Grennlee, M.; Nilsson, S.; Thorsell, A.; Mojena, M.; Tudela, C.; Frisch, K.; Chan, W.; Birzin, E.T.; Rohrer, S.P.; Hammond, M.L. Estrogen receptor β-subtype selective tetrahydrofluorenones: use of a fused pyrazole as phenol bioisostere. *Bioorg. Med. Chem. Lett.* **2006**, *16*, 3896–3901.
144. Ye, S.; Yoshida, S.; Frölich, R.; Haufe, G.; Kirk, K.L. Fluorinated phenylcyclopropylamines. Part 4: effects of aryl substituents and stereochemistry on the inhibition of monoamine oxidase by 1-aryl-2-fluoro-cyclopropylamines. *Bioorg. Med. Chem.* **2005**, *13*, 2489–2499.
145. Suda, M. Preparation and reactivity of an α-(difluoromethylene)-γ-lactone. *Tetrahedron Lett.* **1981**, *22*, 1421–1424.

146. Park, N.K.; Kim, B.T.; Moon, S.S.; Jeon, S.L.; Jeong, I.H. An efficient preparation of novel 5-fluoropyrazolin-3-one derivatives from α-trifluoromethylated α-arylacetates. *Tetrahedron* **2004**, *60*, 7943–7949.

147. Oh, L.M. Synthesis of celecoxib via 1,3-dipolar cycloaddition. *Tetrahedron Lett.* **2006**, *47*, 7943–7946.

148. Song, L.-p.; Chu, Q.-l.; Zhu, S.-z. Synthesis of fluorinated pyrazole derivatives from β-alkoxyvinyl trifluoroketones. *J. Fluorine Chem.* **2001**, *107*, 107–112.

149. Song, L.-p.; Zhu, S.-z. Regioselective synthesis of fluorinated pyrazole derivatives from trifluormethyl-1,3-diketone. *J. Fluorine Chem.* **2001**, *111*, 201–205.

150. Ye, C.; Gard, G.L.; Winter, R.W.; Syvret, R.G.; Twamley, B.; Shreeve, J.M. Synthesis of pentafluorosulfanylpyrazole and pentafluorosulfanyl-1,2,3-triazole and their derivatives as energetic materials by click chemistry. *Org. Lett.* **2007**, *9*, 3841–3844.

151. Naik, S.R.; Witkowski, J.T.; Robins, R.K. A novel route to 3(5)-fluoro-1,2,4-triazoles and 8-fluoropurines by displacement of the nitro group. *J. Org. Chem.* **1973**, *38*, 14353–4354.

152. Vilarrasa, J.; Galvez, C.; Calafell, M. Sintesis de fluoroazoles. *An. Quim.* **1975**, *71*, 631–633.

153. Zumbrunn, A. The first versatile synthesis of 1-alkyl-3-fluoro 1*H*-[1,2,4]-triazoles. *Synthesis* **1998**, 1357–1361.

154. Buscemi, S.; Pace, A.; Pibiri, I.; Vivona, N.; Spinelli, D. Fluorinated heterocyclic compounds. An expedient route to 4-perfluoroalkyl-1,2,4-triazoles via an unusual hydrazinolysis of 5-perfluoroalkyl-1,2,4-oxadiazoles: first examples of an ANRORC-like reaction in 1,2,4-oxadiazole derivatives. *J. Org. Chem.* **2003**, *68*, 605–608.

155. Funabike, K.; Noma, N.; Kuzuya, G.; Matusi, M.; Shibata, K. A direct and general synthesis of 5-substituted 3-trifluoromethyl-1,2,4-triazoles via the three component condensation reaction of ethyl trifluoroacetate, hydrazine and amidines. *J. Chem. Res. Synop.* **1999**, *300–301*, 1999, 1301–1308.

156. Xue, H.; Twamley, B.; Shreeve, J.M. The first 1-alkyl-3-perfluoroalkyl-4,5-dimethyl-1,2,4-triazolium salts. *J. Org. Chem.* **2004**, *69*, 1397–1400.

157. Pace, A.; Pibiri, I.; Buscemi, S.; Vivona, N. Photochemistry of fluorinated heterocyclic compounds. An expedient route for the synthesis of fluorinated 1,2,4-oxadizoles and 1,2,4-triazoles. *J. Org. Chem.* **2004**, *69*, 4108–4115.

158. Dayan, F.E.; Duke, S.O.; Weete, J.D.; Hancock, H.G. Selectivity and mode of action of carfentrazone-ethyl, a novel phenyl triazolinone herbicide. *Pestic. Sci.* **1997**, *51*, 65–73.

159. Hong, W. Agricultural products based on fluorinated heterocyclic compounds. In *Fluorinated Heterocyclic Compounds*; Petrov, V.A.,Ed.; Wiley-Blackwell: New York, 2008; p xx.

160. Crossman, J.M.; Haszledine, R.N.; Tipping, A.E. Organosilicon chemistry. Part X. Reaction of (diazomethyl)trimethylsilane with trifluoroacetonitrile and cyanaogen halides to give triazoles. *J. Chem. Soc.* **1973**, 483–486.

161. Blackwell, G.B.; Haszeldine, R.N.; Taylor, D.R. Polyhalogeno-allenes and -acetylenes. Part 15. Dipolar cycloadditions of *N*-phenylsydnone and aryl azides to perfluoropropadiene and perfluoropropyne. *J. Chem. Soc., Perkin Trans. 1* **1982**, 2207–2210.

162. Reck, F.; Zhou, F.; Girardot, M.; Kern, G.; Eyermann, C.J.; Hales, N.J.; Ramsay, R.R.; Gravestock, M.B. Identification of 4-substituted 1,2,3-trriazoles as novel oxazolidinone antibacterial agents with reduced activity against monoamine oxidase-A. *J. Med. Chem.* **2005**, *48*, 499–506.

163. Meazza, G.; Zanardi, G. Aryl trifluoromethyl-1,2,3-triazoles. *J. Fluorine Chem.* **1991**, *55*, 199–206.

164. Hager, C.; Miethchen, R.; Reinke, H. Organofluorine compounds and fluorinating agents. Part 26. New reversed nucleosides—perfluoroalkyl substituted 1,2,3-triazoles linked to D-galactose and D-altrose. *J. Fluorine Chem.* **2000**, *104*, 135–142.

165. Hager, C.; Miethchen, R.; Reinke, H. Organofluorine compounds and fluorinating agents. 27. New trifluoromethyl substituted 1,2,3-triazoles linked to D-galactose and D-gulose. *J. Prakt. Chem.* **2000**, *342*, 414–420.

166. Greif, D.; Eilitz, U.; Pulst, M.; Riedel, D.; Wecks, M. Perfluoralkylsubstituierte β-Chlorvinylaldehyde: eine neue Klasse von *building blocks* zur Synthese fluorierter Heterocyclen. *J. Fluorine Chem.* **1999**, *94*, 91–103.

167. Costa, M.S.; Boechat, N.; Rangel, E.A.; da Silva, F.d.C.; de Sousa A.M.T.; Rodrigues, C. R.; Castro, H.C.; Junior, I.N.; Lourenço, M.C.S.; Wardell, S.M.S.V.; Ferreira, V.F. Synthesis, tuberculosis inhibitory activity, and SAR study of *N*-substituted-phenyl-1,2,3-triazole derivatives. *Bioorg. Med. Chem.* **2006**, *14*, 8644–8653.

168. Klich, M.; Teutsch, G. Synthese de *N*-(tetrazol-5-yl) azetidin-2-ones. *Tetrahedron* **1986**, *42*, 2677–2684.

169. Norris, W.P. 5-Trifluoromethyltetrazole and its derivatives. *J. Org. Chem.* **1962**, *27*, 3248–3251.

170. Brown, H.C.; Kassal, R.J. 5-Perfluoroalkyltetrazoles. 1. Ring-opening reactions. *J. Org. Chem.* **1967**, *32*, 1871–1873.

171. Xiao, J.; Zhang, X.; Wang, D.; Yuan, C. Synthesis of trifluoromethyltetrazoles via building block strategy. *J. Fluorine Chem.* **1999**, *99*, 83–85.

172. Armour, D.R.; Chung, K.M.L.; Congreve, M.; Evans, B.; Guntrip, S.; Hubbard, T.; Kay, C.; Middlemiss, D.; Mordaunt, J.E.; Pegg, N.A.; Vinader, M.V.; Ward, P.; Watson, S.P. Tetrazole NK_1 receptor antagonists: the identification of an exceptionally potent orally active antiemetic compound. *Bioorg. Med. Chem. Lett.* **1996**, *6*, 1015–1020.

173. Lapierre, J.; Ahmed, V.; Chen, M.-J.; Ispahany, M.; Guillemette, J.G.; Taylor, S.D. The difluoromethylene group as a replacement for the labile oxygen in steroid sulfates: a new approach to steroid sulfatase inhibitors. *Bioorg. Med. Chem. Lett.* **2004**, *14*, 151–155.

174. Banks, R.E.; Smart, B.E.; Tatlow, J.C.,Eds.; *Organofluorine Chemistry. Principles and Commercial Applications*; Plenum Press: New York, **1994**.

175. England, D.C.; Melby, L.R.; Dietrich, M.A.; Lindsey, R.V., Jr. Nucleophilic reactions of fluoroolefins. *J. Am. Chem. Soc.* **1960**, *82*, 5116–5122.

176. Petko, K.I.; Yagupolskii, L.M. Interaction of the azole *N*-sodium derivatives with Freon-113: synthesis of *N*-(2,2-dichlorotrifluoroethyl)azoles. *Synth. Commun.* **2006**, *36*, 1967–1972.

177. Petko, K.I.; Sokolenko, T.M.; Yagupolskii, L.M. Chemical properties of *N*-difluoromethyl- and *N*-2-*H*-tetrafluoroethylypyrazoles. *Chem. Heterocycl. Comp.* **2006**, *42*, 1177–1184.

178. Yagupolskii, L.M.; Sokolenko, T.M.; Petko, K.I. Chemical properties of *N*-polyfluoroethylimidazole and -pyrrole derivatives. *Russ. Chem. Bull., Int. Ed.* **2005**, *54*, 796–802.

179. Rudyuk, V.V.; Fedyuk, D.V.; Yagupolskii, L.V. *N*-Polyfluoroethyl and *N*-2-chlorodifluorovinyl derivatives of azoles. *J. Fluorine Chem.* **2004**, *125*, 1465–1471.

180. Petko, K.I.; Sokolenko, T.M.; Bezdudny, A.V.; Yagupolskii, L.M. *N*-(2-Bromotetrafluoroethyl) derivatives of five-membered nitrogen-containing heterocycles. *J. Fluorine Chem.* **2005**, *126*, 1342–1346.

181. Kirk, K.L. *Biochemistry of the Elemental Halogens and Inorganic Halides*; Plenum Press: New York, 1991; pp 221–238.

182. Ono, T.; Inoue, Y.; Arakawa, Y.; Naito, Y.; Fukaya, C.; Yamanouchi, K.; Yokoyama, K. Synthesis of perflourochemicals as blood substitutes. Part IV. Electrochemical fluorination of *N*-cycloalkyl-pyrrolidines and -piperidines. *J. Fluorine Chem.* **1989**, *43*, 67–85.

183. Abe, T.; Hayashi, E.; Fukaya, H.; Baba, H. The electrochemical fluorination of nitrogen-containing carboxylic acids. Fluorination of methyl esters of cyclic amino-group substituted carboxylic acids. *J. Fluorine Chem.* **1990**, *50*, 173–176, and references therein.

184. Abe, T.; Hayashi, E.; Fukaya, H.; Hayakawa, Y.; Baba, H.; Ishikawa, S.; Asahino, K. The electrochemical fluorination of nitrogen-containing carboxylic acids. Fluorination of methyl esters of 3-dialkylamino propionic acids. *J. Fluorine Chem.* **1992**, *57*, 101–111, and references therein.

185. Abe, T.; Fukaya, H.; Hayashi, E.; Ono, T.; Nishida, M.; Soloshonok, I.; Okuhara, K. Electrochemical fluorination of (*N*,*N*-dialkylamino)alcohols. *J. Fluorine Chem.* **1999**, *97*, 229–237, and references therein.

4

FLUORINATED FIVE-MEMBERED HETEROCYCLES CONTAINING OXYGEN, SULFUR, SELENIUM, AND PHOSPHORUS

YURIY SHERMOLOVICH

Institute of Organic Chemistry, NAS of Ukraine, Murmanskaya 5, Kiev-94, 02094 Ukraine

The chemistry of fluorinated heterocyclic compounds shows intensive development. The analysis of presentations at the international conferences on fluorine chemistry for the past decade has shown that about 40% of papers are devoted to the heterocyclic compounds. The most obvious explanation of this fact consists in a high and various biological activity of fluorine-containing heterocyclic compounds, which triggers the investigation of the synthetic methods of these compounds. Other applications of fluorine-containing heterocycles include areas such as liquid dielectrics and crystals, high-temperature lubricants, complexones and extragents, and so on. The development of synthetic methods for obtaining the fluorinated heterocyclic compounds has been overviewed in several review articles.[1–4] The synthesis methods for nitrogen-containing fluorinated heterocycles are the most studied group. The synthesis of fluorine-containing five-membered heterocycles with the oxygen and sulfur atoms was partially covered in the reviews[5].

Fluorinated Heterocyclic Compounds: Synthesis, Chemistry, and Applications, Edited by Viacheslav A. Petrov

The present review includes data on the synthesis and chemical properties of the fluorine containing derivatives of furan, thiophene, selenophene, phospholene, and five-membered heterocycles with two or more oxygen and sulfur atoms published prior 2007. Every section starts with discussion on the synthetic methods for unsaturated heterocycles (furan, thiophene etc.) and is followed by synthesis and transformations of dihydro and tetrahydro derivatives.

Due to volume limitations this chapter covers only monocyclic heterocycles containing fluorine or fluoroalkyl substituents directly attached to heterocycle.

4.1 METHODS OF SYNTHESIS OF FIVE-MEMBERED HETEROCYCLES CONTAINING OXYGEN, SULFUR, SELENIUM, AND PHOSPHORUS ATOMS

4.1.1 Fluorinated Furan Derivatives

4.1.1.1 2-Fluorofuran Derivatives The furans are the most perspective class among heteroaromatic compounds common in nature.[6] The furan derivatives also are among commercially important pharmaceutical and flavor/fragrance products. However, till the beginning of the 1990s, only a few examples of the fluoro- or polyfluoroalkyl-substituted furans were reported. The synthesis of these compounds was extensively studied in the past 15 years.

In contrast to 2-chlorofuran,[7,8] 2-fluorofuran so far is not known. Several theoretical works were devoted to the investigation of fluorine effect on the stability and reactivity of this compound.[9–12] Furan ring is destabilized by π-electron donors and σ-electron acceptors. These effects are more pronounced for substituents in 2- than 3-position of the ring.[12] Fluorine substituent is known to be a strong σ-electron acceptor and a π-electron donor at the same time and the introduction of fluorine in positions 2 and 3 affects significantly the stability of the furan ring.

Electrophilic fluorination of the 2-alkyl-furan lithium derivative with *N*-fluorodibenzenesulfonimide leads to the formation of 2-fluoro-5-alkyl furane **1** in 60% yield.[13]

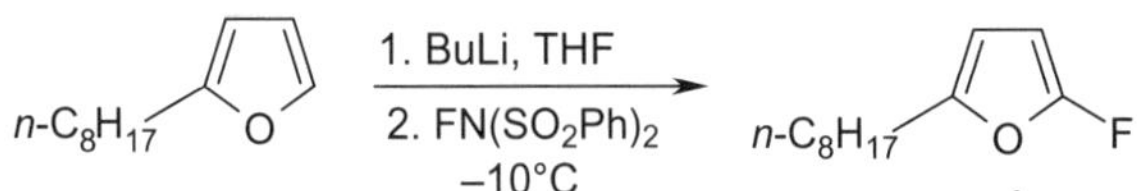

1

Forrest et al.[14] applied decarboxylation of furancarboxylic acids for the synthesis of 2-fluorofurans **2** and **3** using Selectluor® as fluorinating agent.

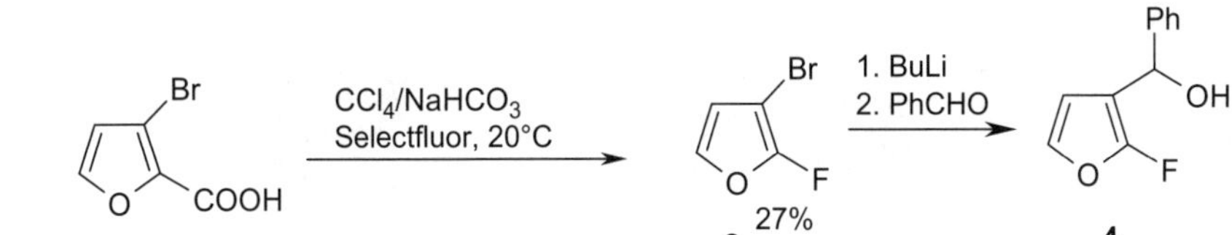

2 4

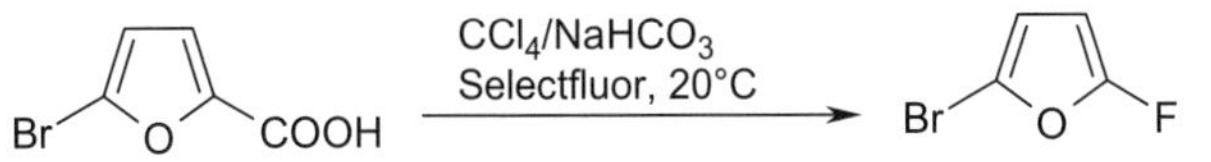

3

Both **2** and **3** were used for further transformations without isolation. Compound **2** was converted into secondary alcohol **4** by reaction with benzaldehyde.

Intramolecular nucleophylic cyclization of 1,1-difluoro-1-butenes bearing homoallylic hydroxyl group leads to the formation of 5-fluoro-2,3-dihydropyranes **5**.[15]

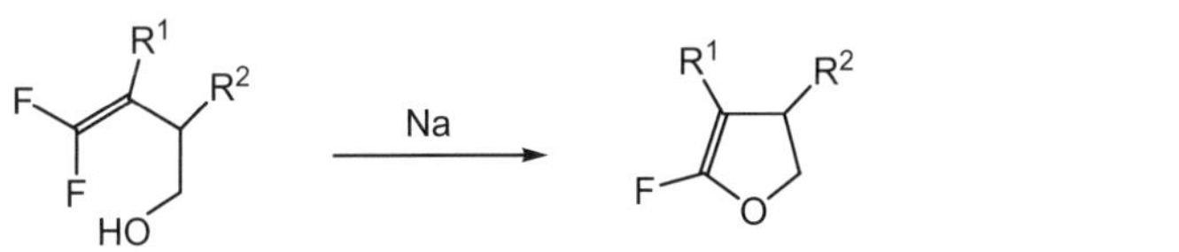

5

$R^1 = CH_2CH(Me)Ph$, $R^2 = Me$ (67%)

$R^1 = Bu$, $R^2 = CH_2Ph$ (62%)

As far as the synthesis of 2-fluorotetrahydrofuran derivatives are concerned, the basic method is the fluorination of tetrahydrofurans. 2-Fluorotetrahydrofuran was isolated in 80% yield by the anodic fluorination of THF in the presence of Et_3N 5HF.[16]

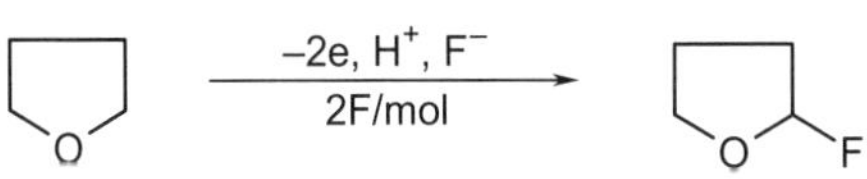

Anhydrous hydrogen fluoride or pyridinium poly(hydrogen fluoride), sulfur tetrafluoride, iodine fluoride, or N–F compounds can be used as fluorinating agents too. Glycosyl fluorides **6–8** are formed under the action of the hydrogen fluoride, on the protected furanosydes, in nitromethane/carboxylic acid anhydride mixture.[17,18]

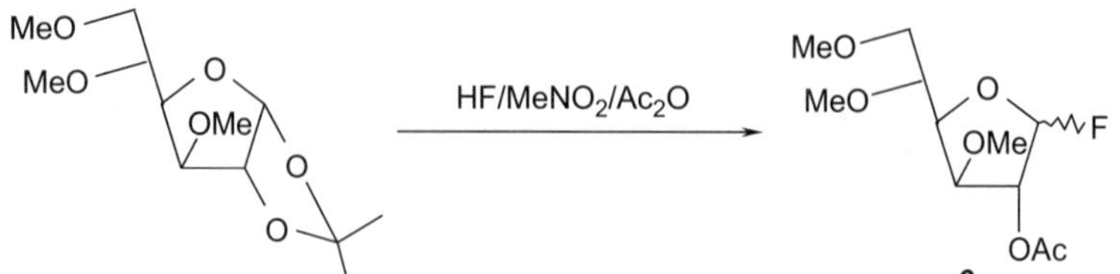

6

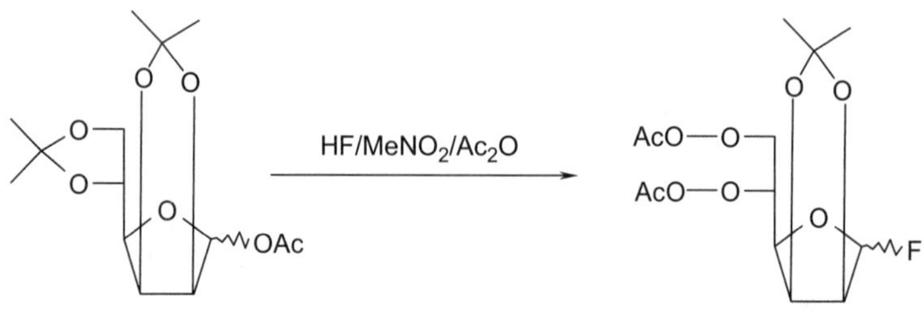

7

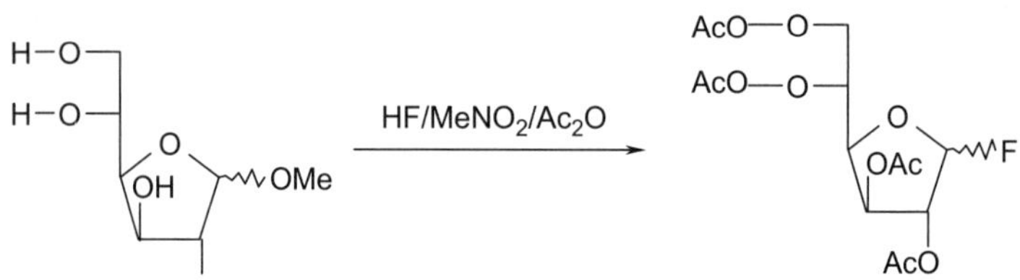

8

Prolonged treatment of tetra-*O*-benzoyl-α-D-xylofuranose with anhydrous hydrogen fluoride gave tri-*O*-benzoyl-α-D-lyxofuranosyl fluoride **9**.[19,20]

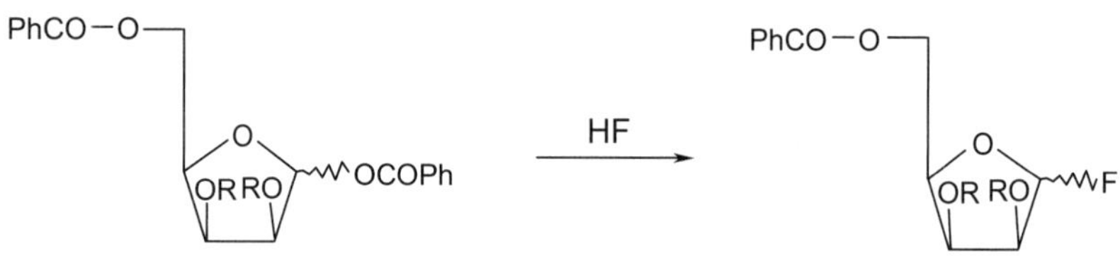

R = PhCO **9**

Pyridinium poly(hydrogen fluoride)[21] and iodine fluoride (generated "*in situ*" from silver fluoride and iodine[22,23]) were used for the synthesis of 2-fluorofuranosyl derivatives **10** and **11**.

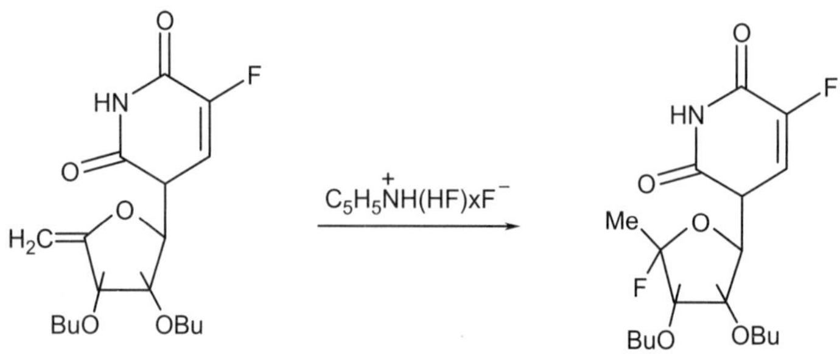

10

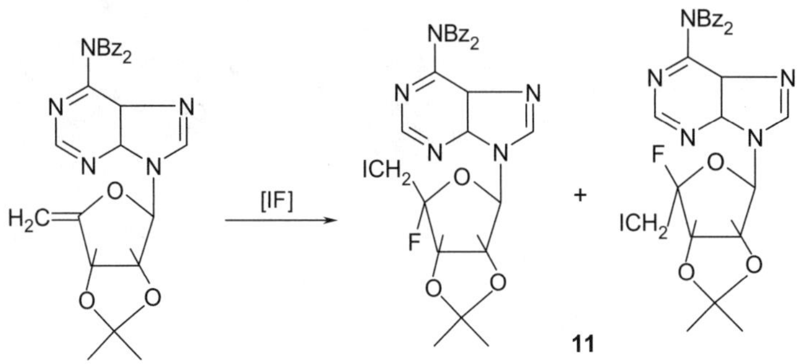

11

Isomers **11** were used for the synthesis of the natural antibiotic nucleocidin .[24]

4.1.1.2 3-Fluorofuran Derivatives In contrast to the corresponding chloro derivatives, unsubstituted 3-fluorofuran, -dihydrofurans, and 3-fluorotetrahydrofurans are not known.[25] 3-Fluoro-furfurole **12** was prepared by the electrophylic fluorination of carbanion generated from 3-bromo derivative.[13,26]

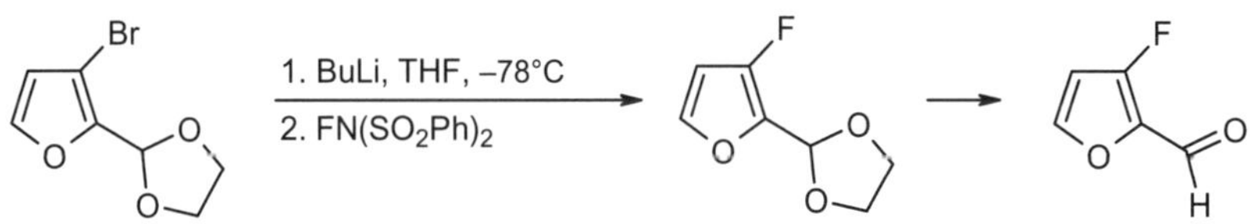

12

There are two known methods of 3-fluorofurans synthesis from the fluorine-containing synthones. Hydrolysis of *gem*-difluorocyclopropenyl acetales under acidic conditions gives 1-aryl-2-fluorofurans **13**.[27]

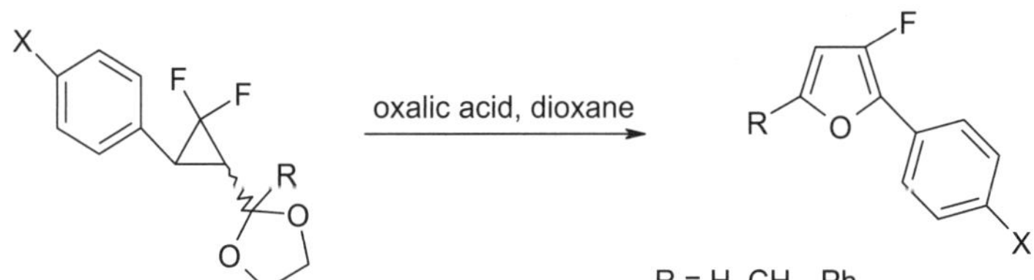

13

Regiocontrolled synthesis of 3-fluorofurans **14** is based on the Reformatsky reaction of phenylacetylenes **15** with aldehydes and subsequent base-promoted cyclization of the alcohol **16** into furan.[28]

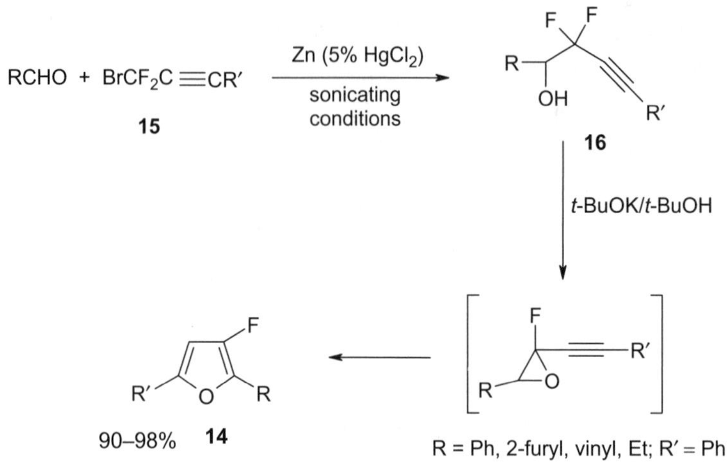

The main method for the synthesis of 3-fluorodihydrofuran and tetrahydrofuran derivatives is based on cyclization reactions. Cyclization of 2-fluorobut-2-enal **17** using triethylamine trihydrofluoride gave 3-fluoro-2,5-dihydrofuran **18**.[29]

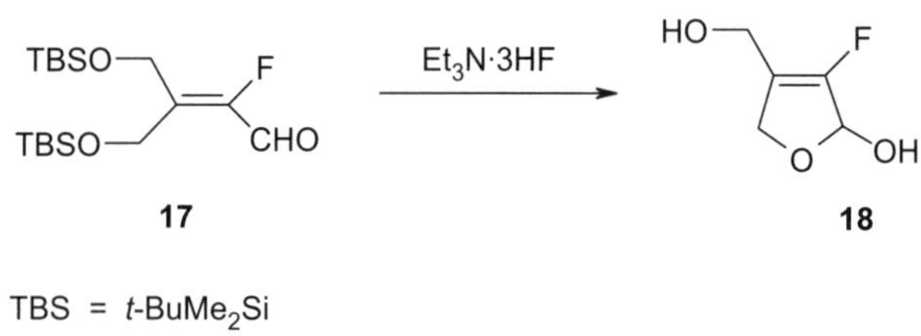

3-Fluoro-substituted derivatives of tetrahydrofuran were obtained by radical reaction of bromoacetals **19**.[30,31]

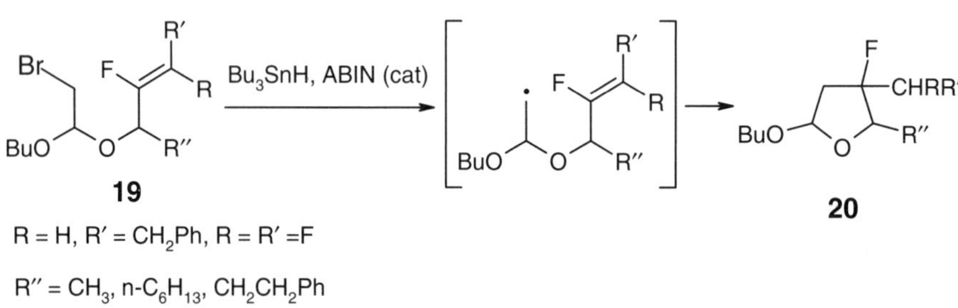

4.1.1.3 2-Perfluoroalkyl-Furan Derivatives The most common methods of synthesis of these compounds is perfluoroalkylation of furans, carried out using either radical, or ionic reactions. For example, the photolysis of furan/trifluoromethyliodide mixture yields of 2-trifluoromethylfurans in 40–50% after HI elimination.[32]

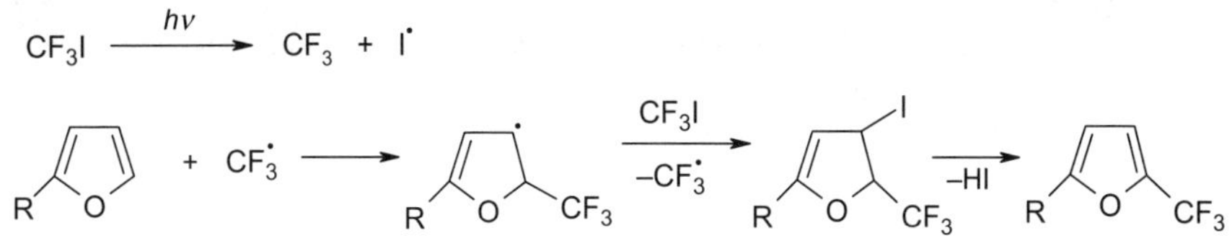

$R = H, C_4H_9$

Polyfluoroalkyl iodides react regioselectively with furan to form 2-polyfluoroalkyl derivatives[33] in the presence of catalytic amounts of tetrakis(triphenylphosphine) nickel at 60–80°C.

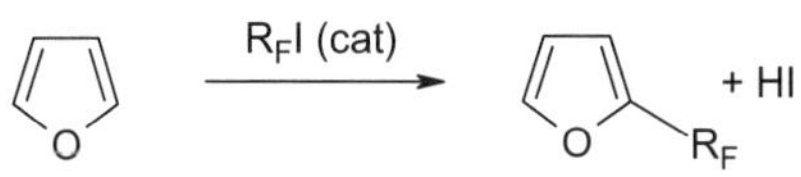

$R_F = Cl(CF_2)n \quad n = 4,6,8$

Photochemical or thermal (100°C) initiation of $(CF_3)_2Te$ in the presence of furan results in the formation of 2-trifluoromethylfuran, along with small quantities of trifluoromethylated dihydrofurans.[34]

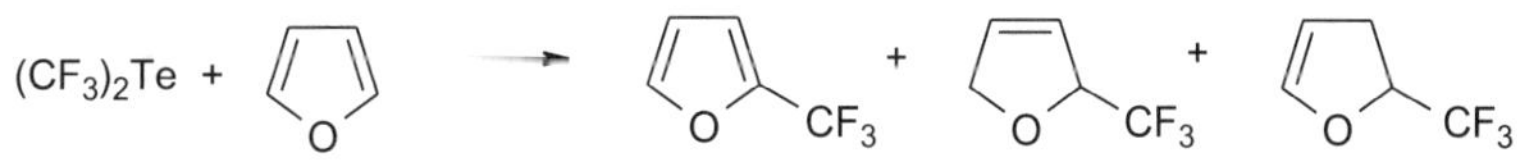

Fluorinated diacylperoxides were also used as the source of fluoroalkyl radicals. The electron-transfer mechanism was proposed for the reaction of compound **21** with furan.[35]

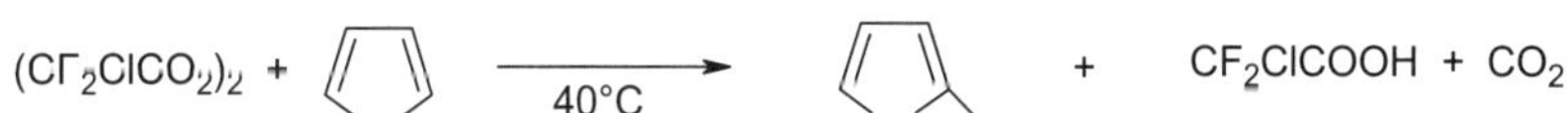

21

48%

$$\underset{\mathbf{21}}{(CF_2ClCO_2)_2} + HetH \longrightarrow (CF_2ClCO_2)_2^{\bullet-} + [HetH]^{\bullet+}$$

$$\downarrow$$

$$CO_2 + CF_2ClCOO^- + CF_2Cl^\bullet$$

$$CF_2Cl^\bullet + [HetH]^{\bullet+} \longrightarrow HetCF_2Cl + H^+$$

The reaction of perfluorohexanesulfonyl chloride with furan carried out in the presence of dichlorotris(triphenylphosphine)ruthenium (II) catalyst in a degassed sealed tube at 120°C, also leads to perfluoroalkylation products.[36] The redox reaction between the sulfonylchloride and catalyst affords anion radical, which cleaves homolitically to give perfluoroalkane sulfonyl radical, followed by extrusion of SO_2 with the formation of perfluoroalkyl radical and alkylation of the heterocycle. The subsequent hydrogen atom abstraction from intermediate heterocyclic radical by Ru^{III}-Cl species leads to the formation of product.[37]

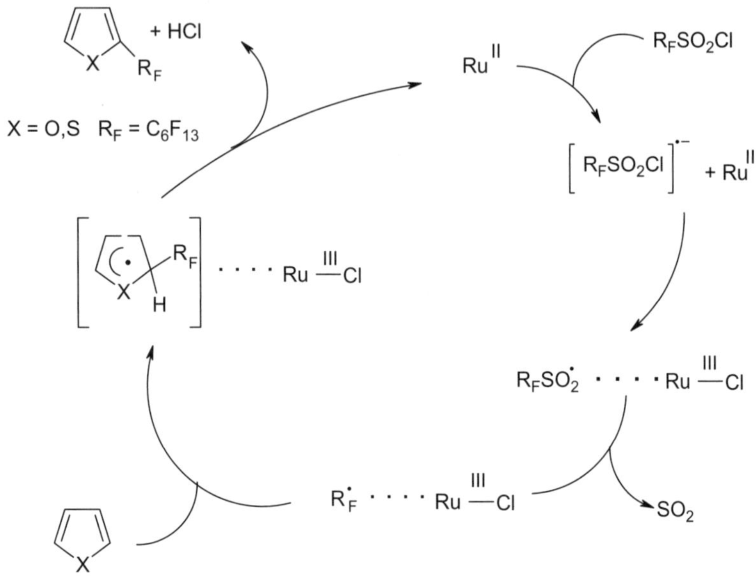

Fluorinated selenides[38–40] were also used as the source of polyfluoroalkyl radicals. Homolytic decomposition of **22** and **23** under UV irradiation in the presence of furan gives 2-polyfluoroalkyl furans.

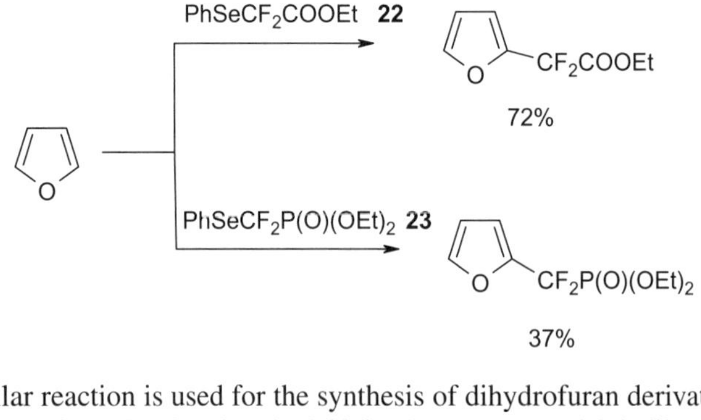

The similar reaction is used for the synthesis of dihydrofuran derivatives. Photolytic cleavage of the Se–CF_2 bond of **22** in the presence of 2,3-dihydrofuran, the phenylselenyl group transfer reaction proceeds quickly to provide **24**, and further photolysis of **24** results in the formation of the compound **25**.[39]

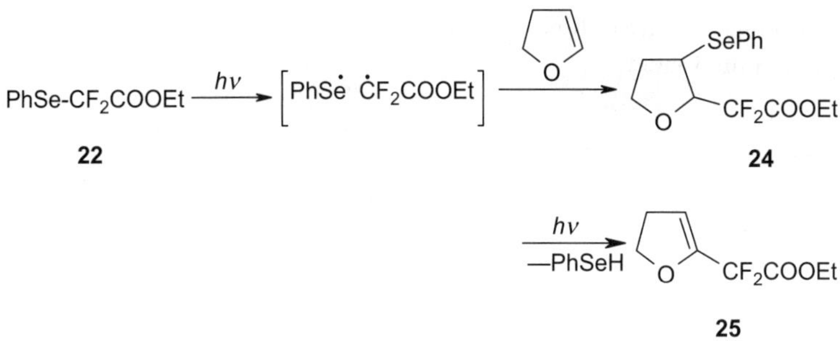

Trifluoromethyl group was introduced in furfurol by its reaction with trifluoroacetic acid and xenon difluoride. Xenon (II) trifluoroacetate intermediate undergoes decomposition under reaction conditions to yield carbon dioxide and trifluoromethyl radical, which reacts with furfurol giving the product.[41] The perfluoropropyl furan derivative **26** was prepared[42] using XeF_2/perfluorobutanoic acid mixture.

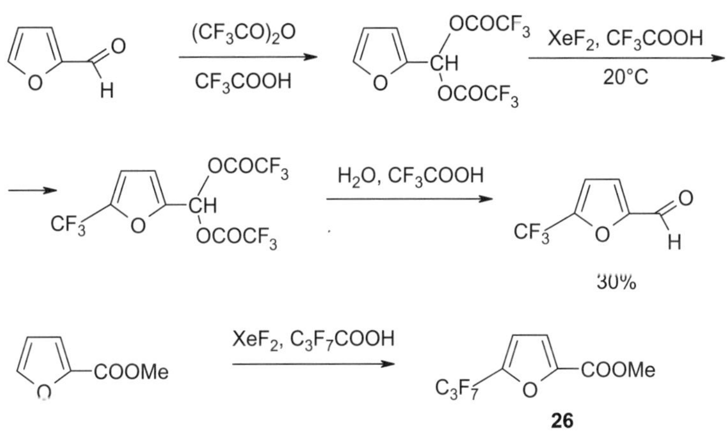

The only example of the cationic perfluoroalkylation of furan by trifluoromethyl cation (generated by radiolysis of CF_4) resulted in a mixture of 2- and 3-substituted furans.[43]

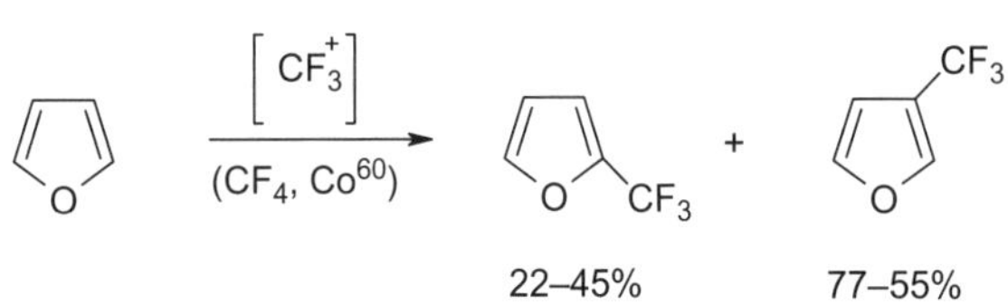

The reaction of perfluoroalkyl iodides, and bromofurans in the presence of copper metal in dimethyl sulfoxide, is a good example of anionic perfluoroalkylation process. Pefluoroalkylation of 5-bromofuroic acid produced a mixture of products, due to the fact that product **27** (or perhaps its copper salt) decomposes under reaction conditions.

So, if the longer reaction time is used, pure 2-monosubstituted furan **28** can be obtained in a high yield.[44]

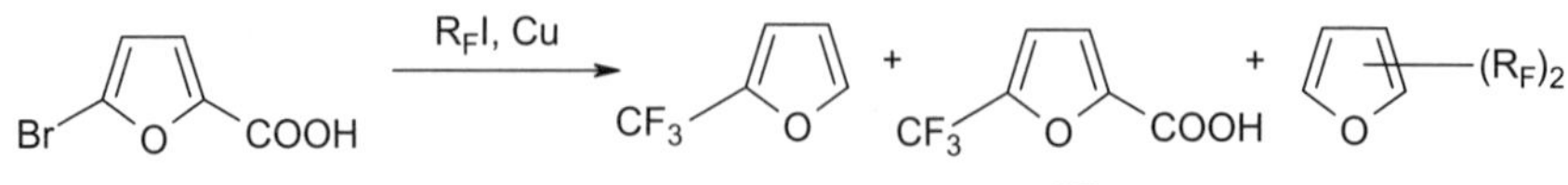

28 **27**

2,5-Furandicarboxylic acid fluorination by sulfur tetrafluoride in the presence of hydrogen fluoride leads to a mixture of products,[45,46] but decarboxylation of **27** in the presence of copper powder in quinoline gives 2-trifluoromethylfuran.

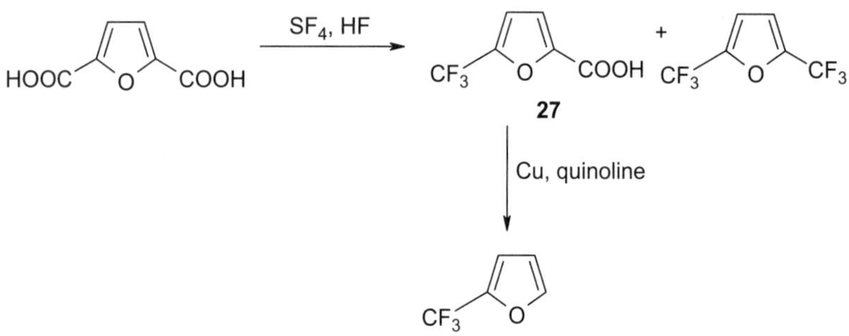

Bégué et al.[47] reported the preparation of substituted 2-trifluoromethylfurans through the iodocyclization of γ,δ- unsaturated ethyl trifluoroacetoacetate.

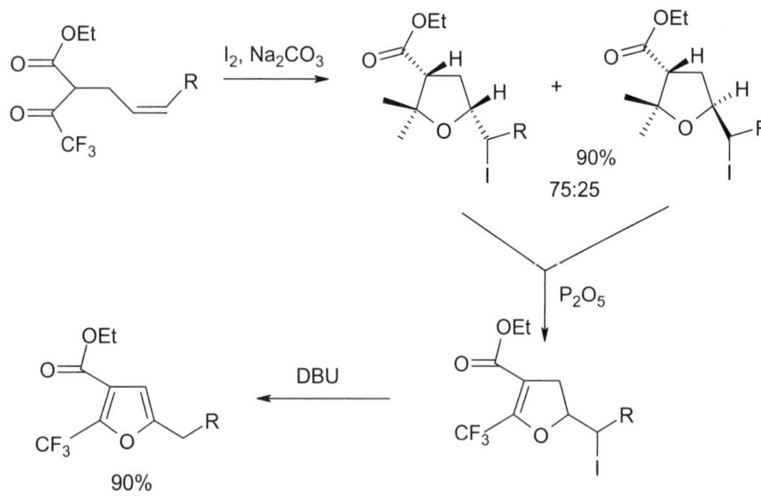

S-Ethyl-4-oxo-2-pentafluoroethynylpentanethioate **29** is a versatile intermediate for the synthesis of 2-trifluoromethylfurans.[48]

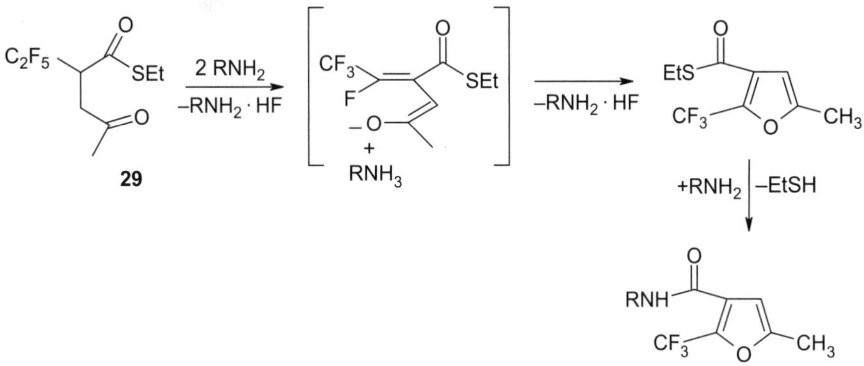

Rhodium (II) catalyzed 1,3-dipolar reactions of metal carbenoid **30** (derived from diazocompound **31**) leads to the formation of 2-perfluoroalkylfurans **32**.[49]

Flash vacuum thermolysis of β-keto-trimethylsilyl-enol ethers **33** gives furan **34** in good yield through an allenic intermediate **35**.[50]

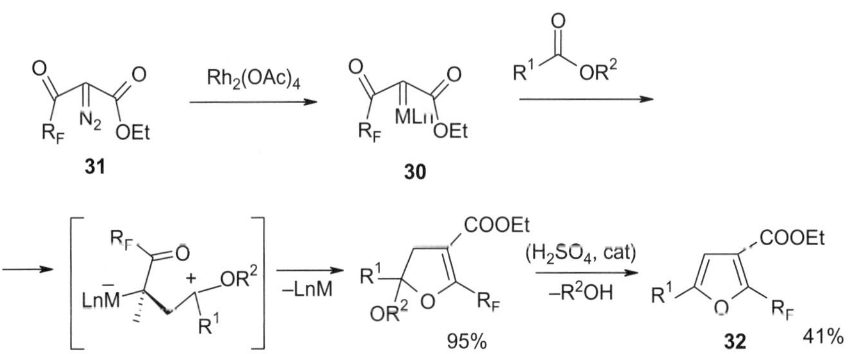

$R_F = CF_3, ClCF_2, BrCF_2, ClC_3F_6, C_5F_{11}$

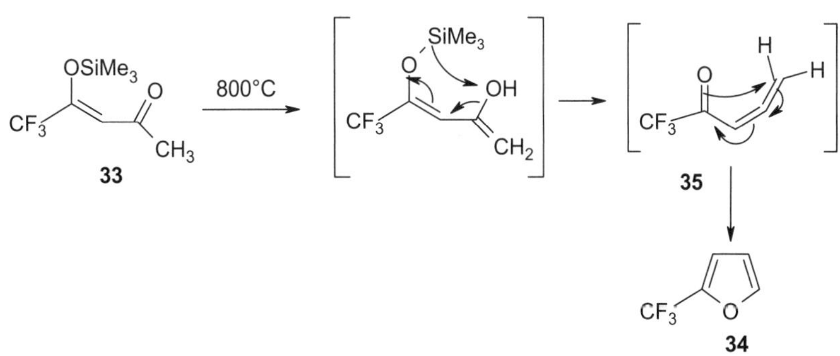

Cyclization of the unsaturated ketone **36** under the action of thiophenole, followed by aromatization of the cyclization product gives furan **37**.[51]

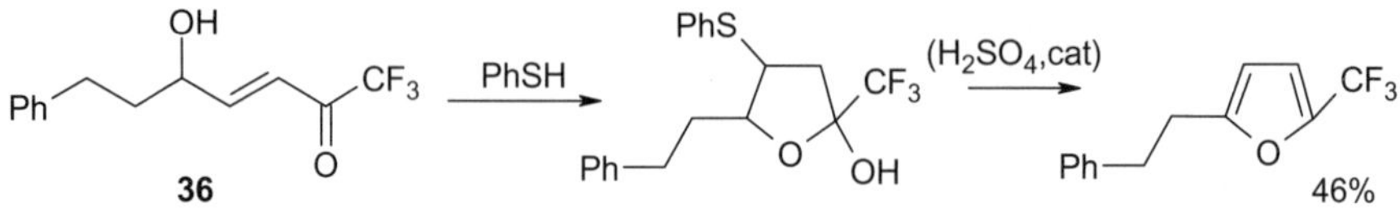

36 **37**

A series of 2-trifluoromethylfurans **38** was obtained by oxidation of ketones **39** using system $CF_3COOH–CH_2Cl_2–PbO_2$.[52]

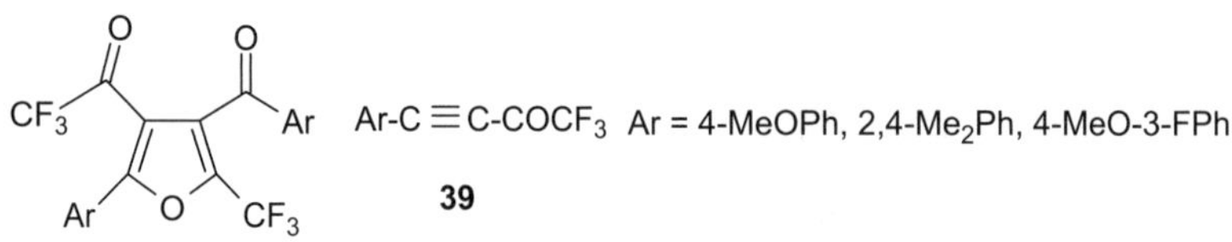

38 **39**

5-(ω-Iodoperfluoropropyl)furan-3(2H)-one **40** is formed in one step reaction of alcohol **41** with formic acid.[53]

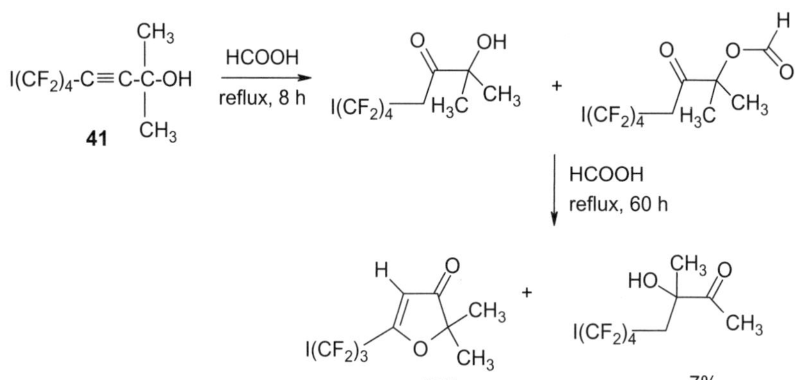

41 **40**

The methods of synthesis (direct perfluoroalkylation or building block approach) of perfluoroalkyl carbohydrates **42–44** with tetrahydrofuran fragments were described in the detailed review of Ref. 54

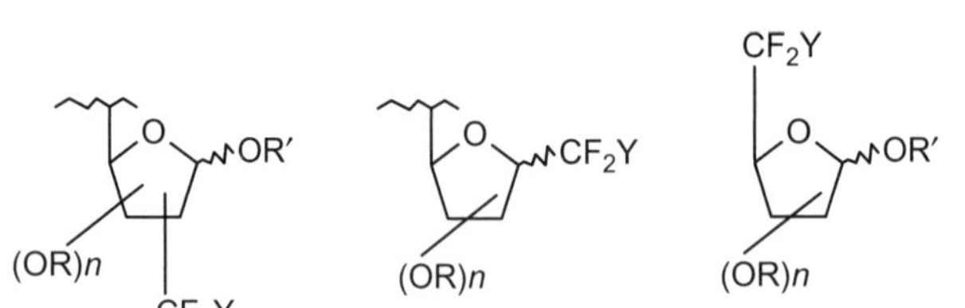

42 **43** **44**

$Y = F, C_nF_{2n+1}$

4.1.1.4 3-Perfluoroalkylfuran Derivatives Fluorination of 2,4-furandicarboxylic acid by sulfur tetrafluoride results in the mixture of products **45–47**.[45] The carboxylic group in position 4 is more reactive than in position 2 and the yield of the compound **46** higher than that of the compound **47**. Decarboxylation of the acid **46** in the presence of copper powder in quinoline gives 3-trifluoromethylfuran.

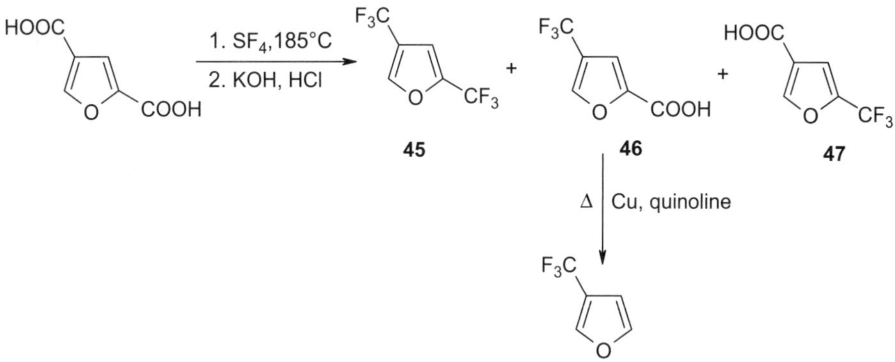

Often perfluoroalkylation of the heterocycle or cyclization of acyclic fluorine-containing synthones, are used for the synthesis of 3-fluoroalkylfuranes.

For example, 3-perfluoroalkyl-4-ethoxycarbonylfuranes **48** have been prepared from bicyclohepta-2,5-diene derivatives **49** by sequence involving cycloaddition of acetylene to tetracyclone **50** and retro Diels–Alder reaction of the cycloadduct.[55]

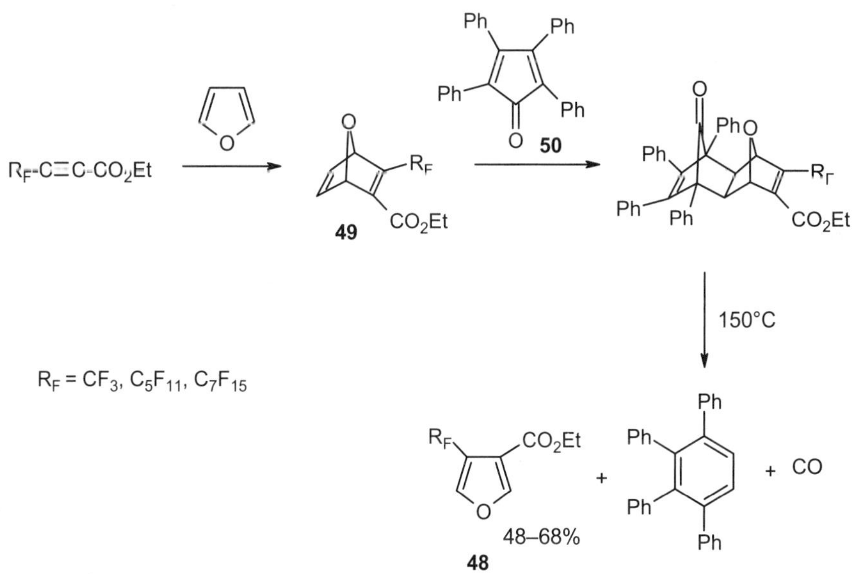

3-Trifluoromethyltetrahydrofuran was prepared by intramolecular cyclization of 4-chloro-2-trifluormethylbutanol.[56]

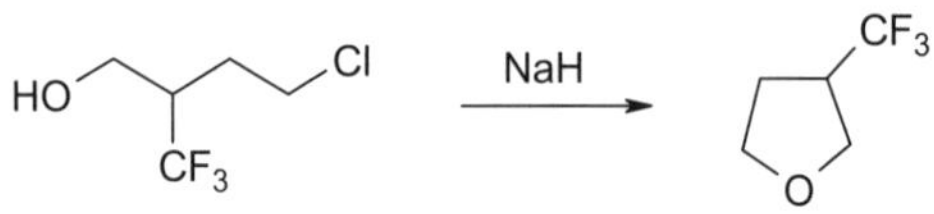

The reduction of the lacton **51** was used for the synthesis of 2-hydroxy-3-trifluoromethyltetrahydrofuran **52**.[57]

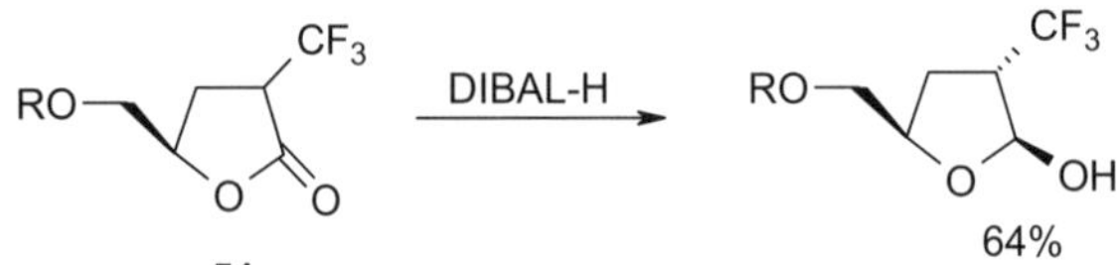

51

R = *t*-BuMe$_2$Si

64%

52

The derivatives of 3-perfluoroalkyltetrahydrofuran can be also prepared by radical addition to the C=C bond of dihydrofurans. Thus, irradiation of the mixture 4-[(phenylthio)difluoromethyl]-1,3-dioxalan-2-one and dihydrofurane by UV light gave the compound **53** in reasonable yield.[40]

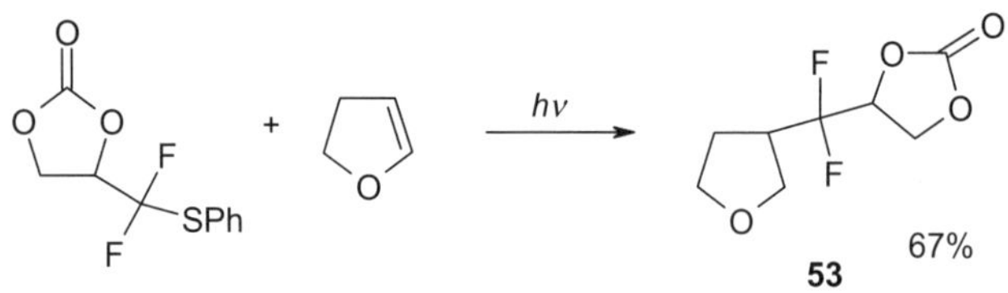

53 67%

Bromdifluoromethyl radical, generated from dibromdifluoromethane readily adds to C=C bond of dihydrofuran. The final product of this reaction, 2-hydroxy-3-bromdifluoromethyltetrahydrofuran **54** obviously forms by hydrolysis of 2-bromo-derivative **55** under reaction conditions.[58]

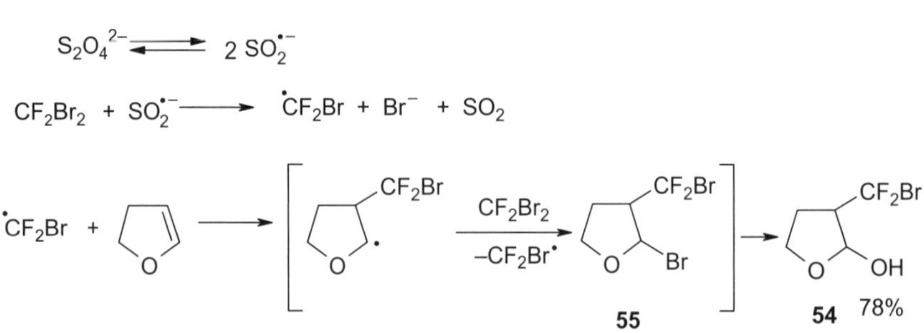

55 **54** 78%

The synthesis of sugars containing 3-perfluoroalkyltetrahydrofuran fragment was reviewed recently.[54] The preparation of 2,3-*trans*-substituted fluorinated tetrahydrofuran derivatives **56** from *S*-alkyl dithiocarbonates[59] should be mentioned among the most recent works. This strategy can be applicable to the synthesis of the new modified nucleosides.

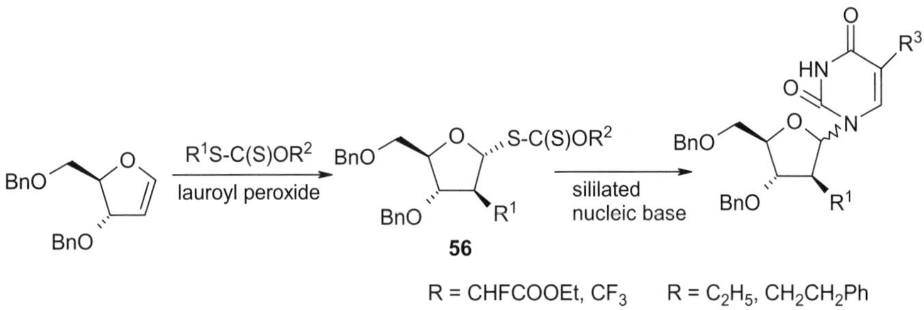

4.1.1.5 Furan Derivatives with Two Fluorine or Fluoroalkyl Substituents 2,5-Difluorofuran was isolated in low yield during fluorination of furan by $KCoF_4$[60] and 2,5-, 2,3-, and 3,4-difluorofurans were prepared by dehydrofluorination of the corresponding partially fluorinated oxolans or oxolens.[60]

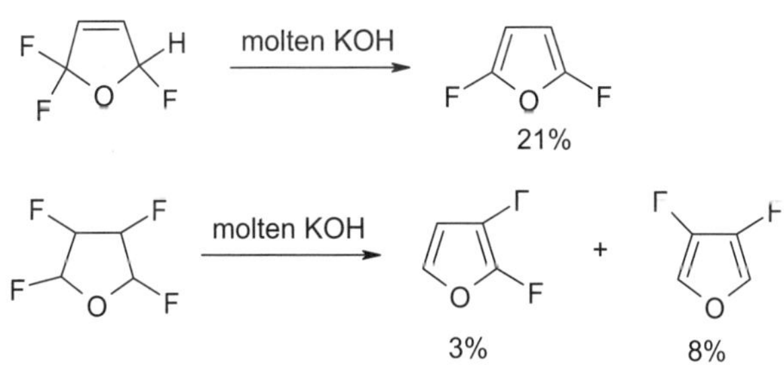

All these furans are unstable and polymerized rapidly and spontaneously. The 2,4-difluorofuran was not prepared so far.

The methods of synthesis for bis(perfluoroalkyl)furans are more developed. For example, 2-methyl-3,4-bis(trifluoromethyl)furan can be synthesized by the reaction of 2-methyl-furan-3,4-dicarboxylic with sulfur tetrafluoride.[61] However, the main synthetic methods rely on the use of unsaturated fluorine-containing acyclic compounds. So, 3-trifluoromethylfurans **57** were prepared from β,β-bis(trifluoromethyl)α,β-unsaturated ketones and tin(II) chloride.[62]

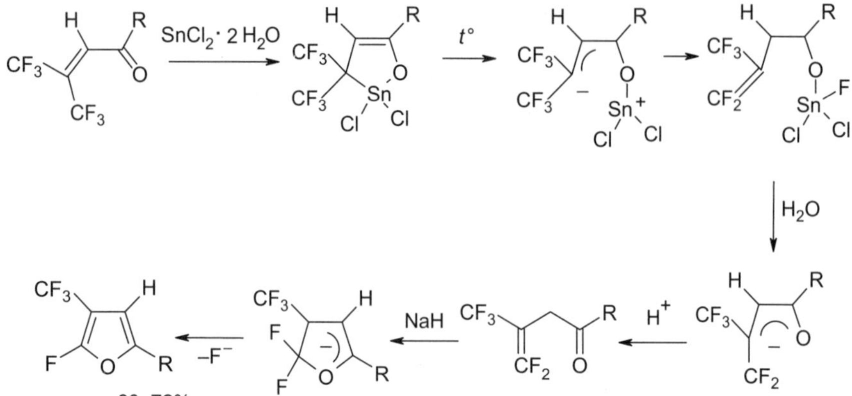

57 60–72%

R = Me, *p*-Tl, *p*-ClPh, *p*-FPh, 5-methyl-2-furyl, 2-thienyl

1-Perfluoroalkyl-1-fluoroalkenyl phosphates undergo sequence of transformations involving the dephosphorylation and fluoride ion catalyzed cyclocondensation to give 3,4-diperfluoroalkylated furans **58**.[63]

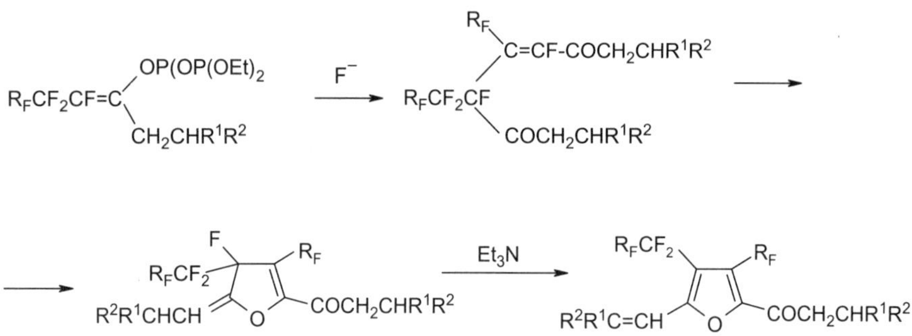

58 48–65%

$R_F = CF_3, CF_3(CF_2)_4$ R^1 = Me, Et, *i*-Pr, *n*-Bu R^2 = H, Me

Cycloadducts, formed by addition of hexafluorobut-2-yne to 2-methyl furan or 2-hydroxymethyl furan can be transformed into bis(trifluoromethyl)furan derivatives. For example, compounds **59** and **60** were prepared using this approach.[64]

The radical addition of propionaldehyde to hexafluoro-2-butyne under γ-ray irradiation leads to 4,5-bis(trifluoromethyl)octa-3,6-dione **61**, which was converted into 3,4-bis(trufluoromethyl)furan **62** by treatment with H_2SO_4.[65]

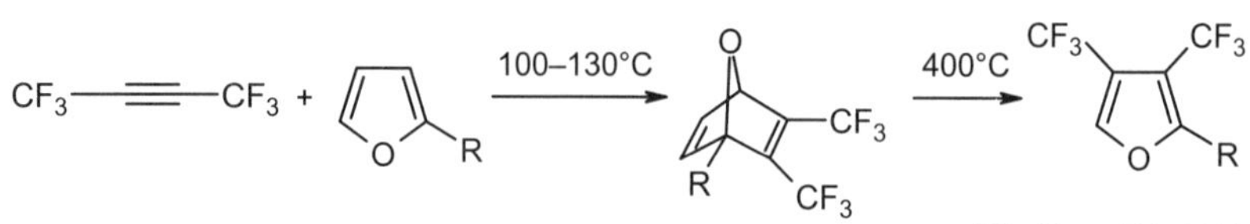

59 R = CH_3 96%
60 R = CH_2OH 60%

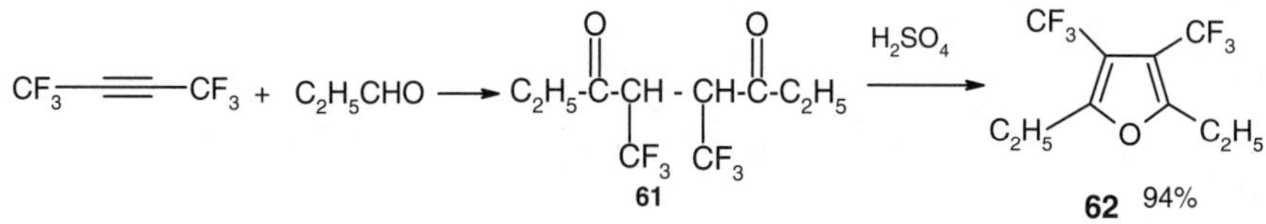

Fluorinated tetrahydrofuran derivatives are formed by cyclization of the saturated fluorine-containing aliphatic compounds.[66,67]

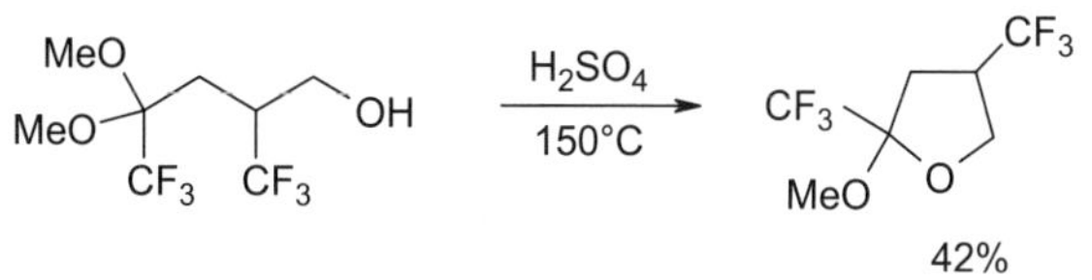

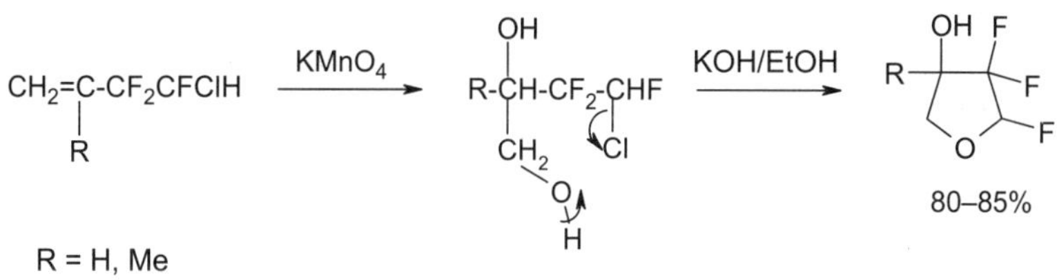

The tetrahydrofuran derivatives **63–65** are major products of the reactions of succinic acids with SF_4.[68]

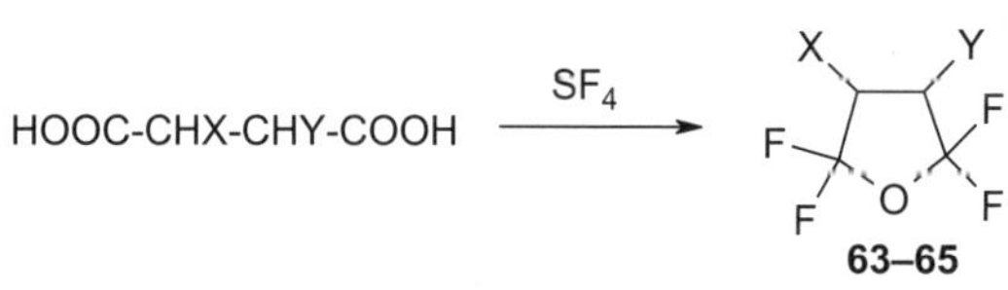

X = Y = H(**63**) ; X = H, Y = Cl (**64**); X = Y = Cl (**65**)

4.1.1.6 Furan Derivatives, Containing Three or Four Fluorine Atoms or Perfluoroalkyl Substituents Only a few furan derivatives with the named combination of substituents are known.

2,3,4-Tris(trifluoromethyl)furan **66** was prepared by the fluorination of acid **67** with sulfur tetrafluoride. The fluorination of acid **68** under similar conditions gives bicyclic product **69**.[69]

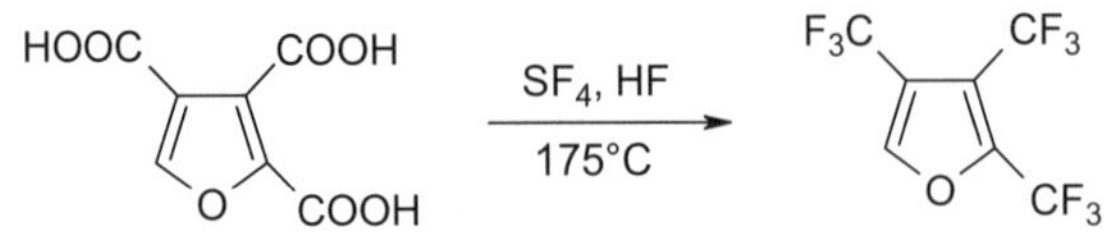

66 67

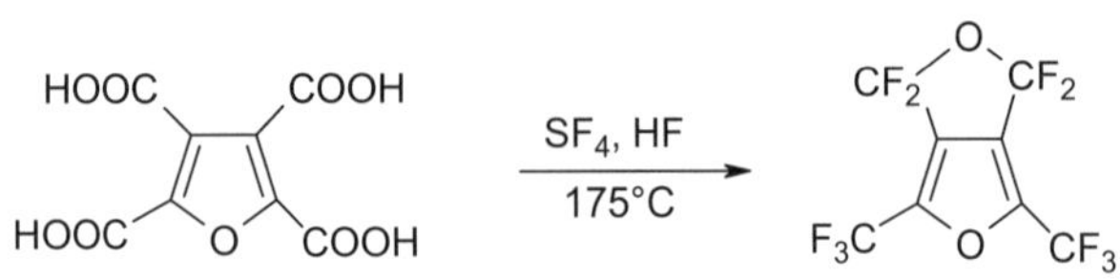

68 69

Several methods for the syntheses of this type of furan derivatives are based on the usage of unsaturated aliphatic fluorine-containing compounds. Gas-phase photolysis of the mixture of diazoketone **70** and hexafluoro-2 butyne yields the tetrakis(trifluoromethyl)furan as main product.[70] Ketocarbene intermediate **71** undergoes 1,3-addition to 2-butyne to give furan **72**.

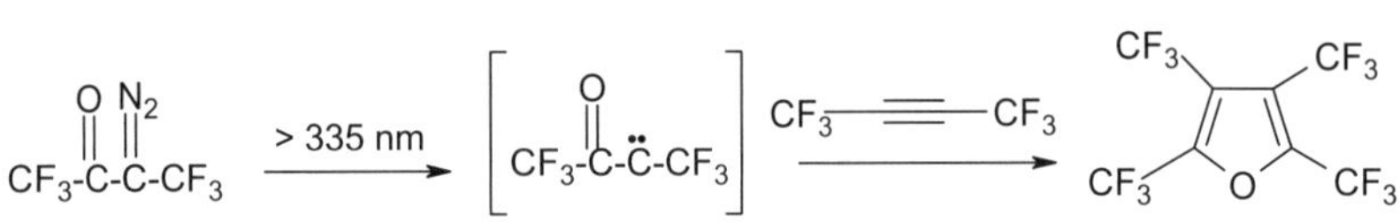

70 71 72

Reaction of perfluorodiene **73** with water leads to the formation of **72** in quantitative yield.[71]

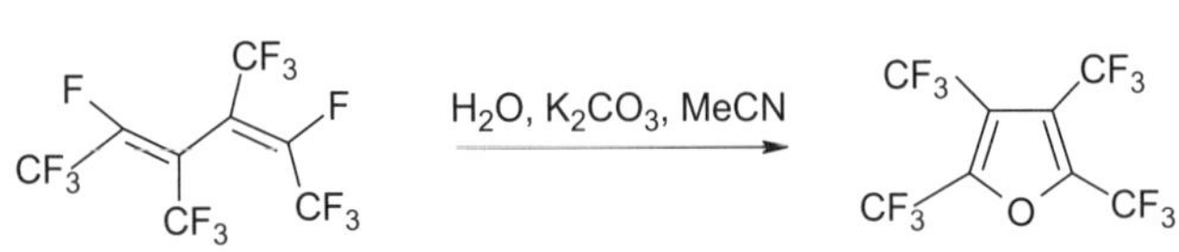

73 72

Free radical addition of acetaldehyde to perfluoro-3,4-dimethyl-3-hexene **74** under γ-irradiation followed by the pyrolysis of the product leads to furan **75**.[72]

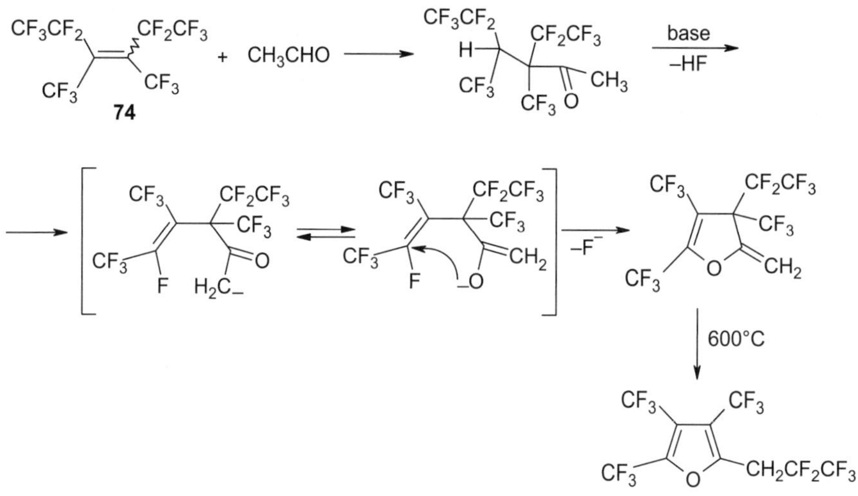

74 75

The derivatives of perfluorodihydrofurans are also prepared from the perfluorinated olefins. For example, the reaction of vinylketone **76** with difluorocarbene generated from hexafluoropropene epoxide at 225°C gave 4,5-dihydrofurane **77**.[73]

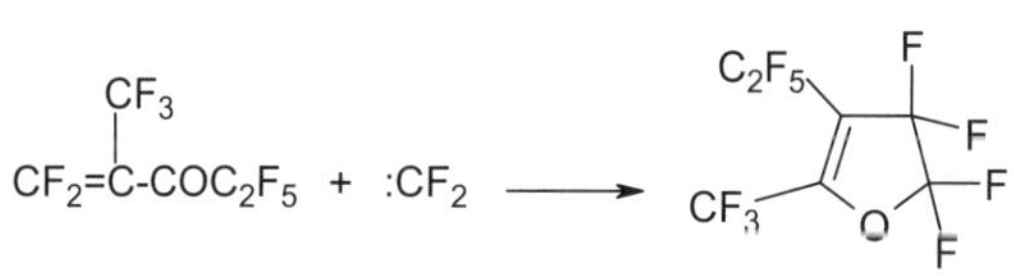

76 **77**

Perfluoro-4-ethyl-2,3,4,5-tetramethyl-4,5-dihydrofuran **78** was prepared from tetrafluoroethene pentamer. The possible reaction mechanism may involve the formation of biradical **79** by one-electron transfer from enolat-ion **80** to the CF_2 carbon that undergoes loss of fluoride ion followed by intramolecular cyclization of the biradical **79**.[74]

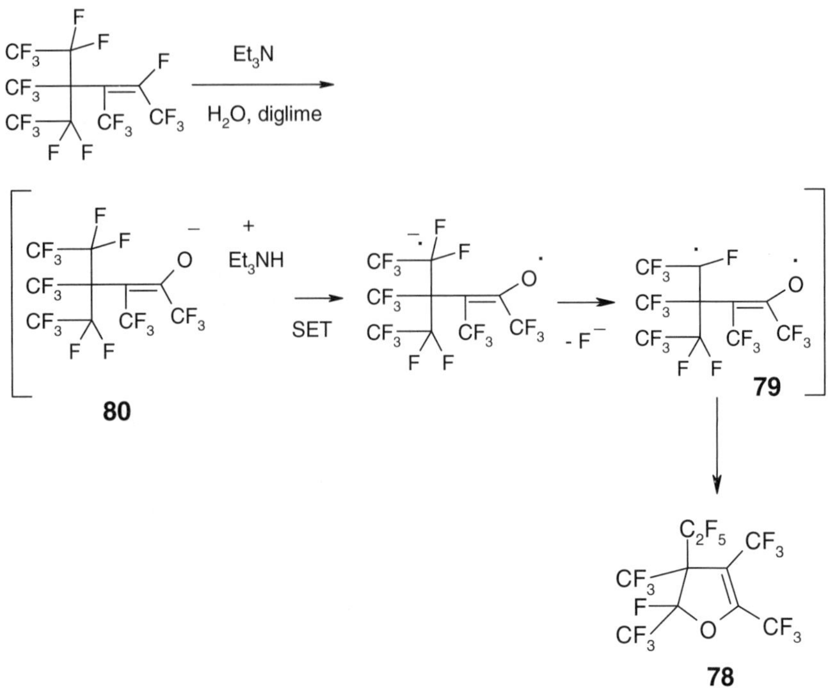

Perfluoro 2-oxolenes **81** and 3-oxolenes **82** are formed by heating 2-perfluorotetrahydrofurancarboxylic acid and (2-perfluorotetrahydrofuryl)acetic acid with K_2CO_3.[75,76]

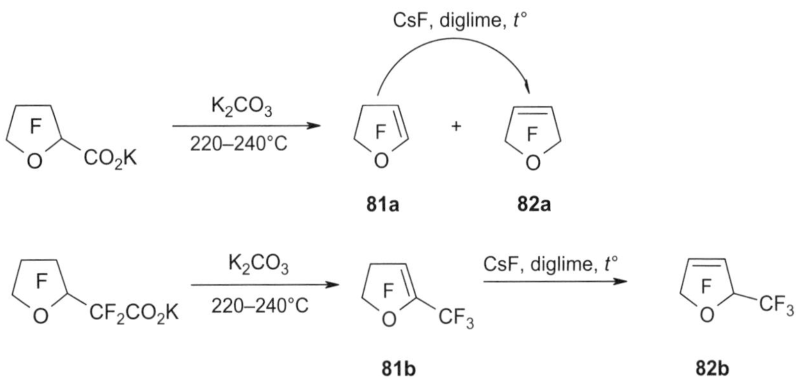

The interaction of 2-oxolenes **81a** and **81b** with cesium fluoride in diglyme results in isomerization into thermodynamically stable 3-oxolenes **82a** and **82b**.

The most studied are the methods for the synthesis of perfluorinated tetrahydrofuran derivatives.

Perfluorinated ketones, diketones, and α-oxides isomerize to the compounds of oxolane series[77] under the action of SbF_5. Electrophilic attack of SbF_5 on the carbonyl group results in generation of carbenium center and is followed by rearrangement of the secondary carbocation into the primary one and intramolecular alkylation of oxygen, resulting in the product.

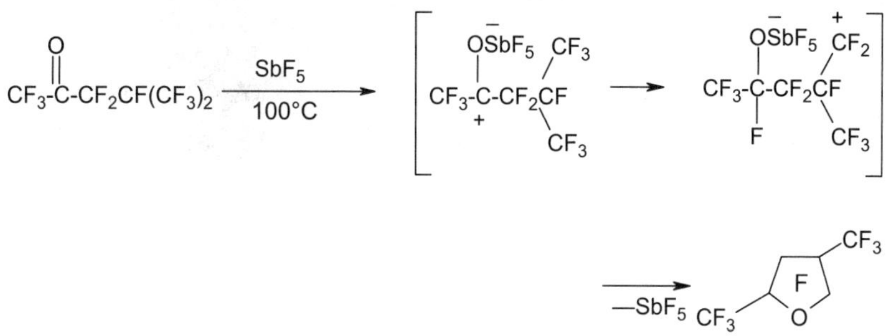

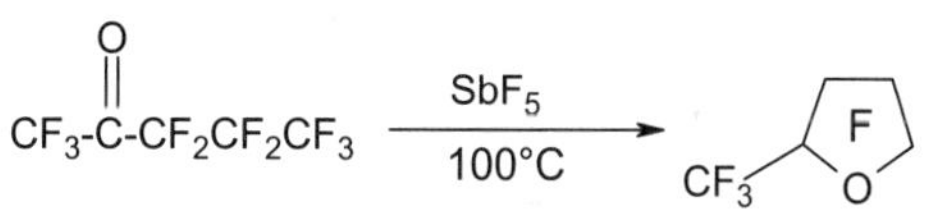

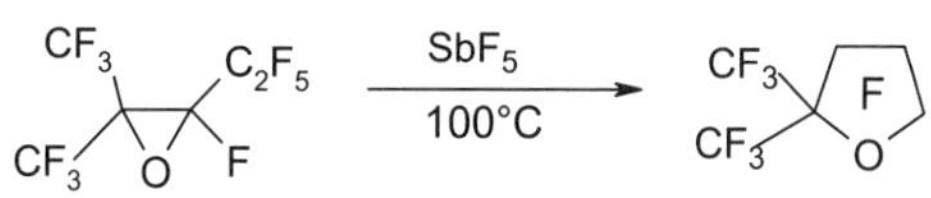

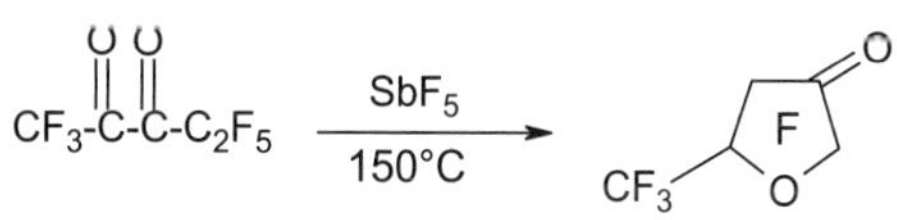

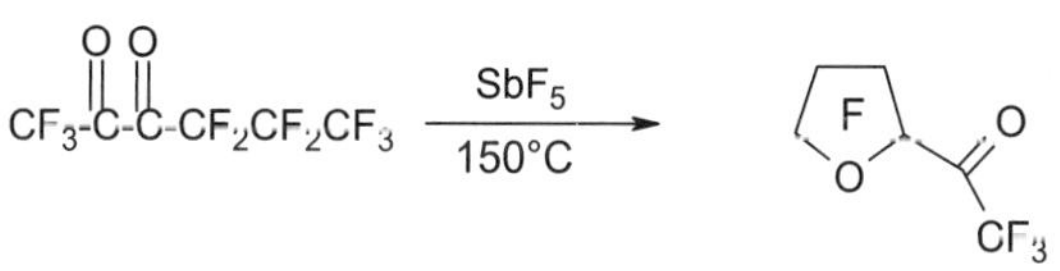

The most common method for the preparation of perfluorinated tetrahydrofurans is fluorination of tetrahydrofurans. Elemental fluorine[78,79] or high-valency metal fluorides (AgF_2, MnF_3, HgF_2, and CoF_3)[80] are usually used as fluorinating agents.[81]

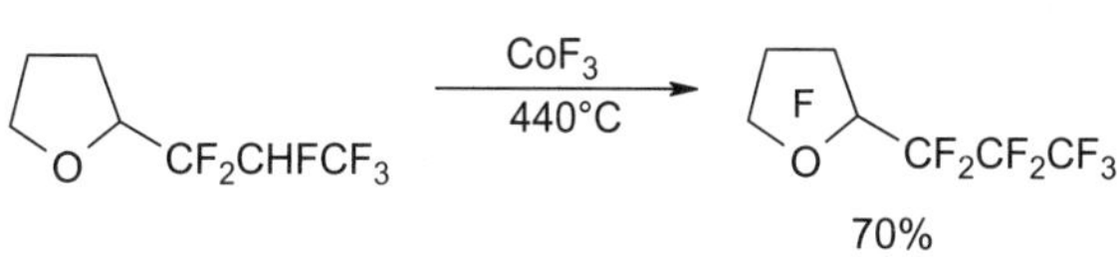

Electrochemical fluorination is the method of choice for the synthesis of perfluorinated tetrahydrofuranes. Tetrahydrofurans **83–87** were obtained with the yields 20–40% at electrochemical fluorination of the corresponding hydrocarbonic analogues in anhydrous hydrogen fluoride.[82]

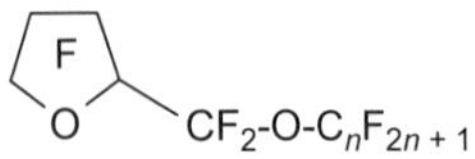

83

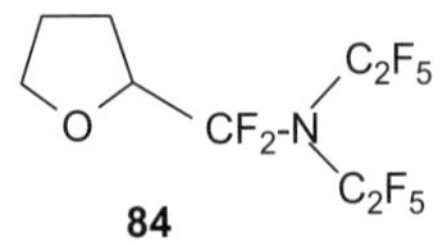

84

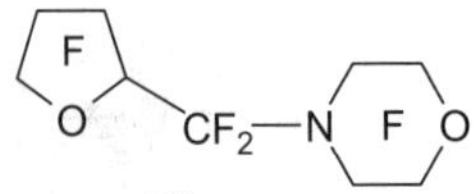

85

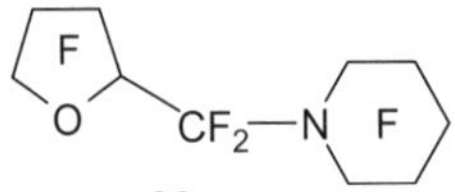

86

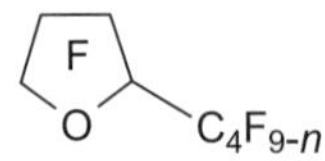

87

Electrochemical fluorination of the derivatives of tetrahydrofuran-2-yl-carboxylic acid and tetrahydrofuran-2-yl-methanol leads to the formation of the mixture of perfluorotetrahydrofurans **88–90**.[83]

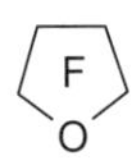

88

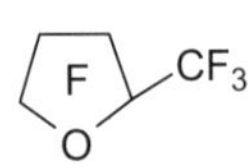

89

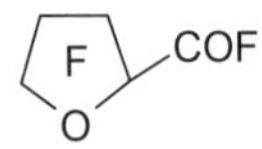

90

Electrochemical fluorination of the alkanoles of normal structure gives the mixture of five- and six-membered cyclic perfluoroethers with prevailing tetrahydrofuran derivatives.[84]

11.7%

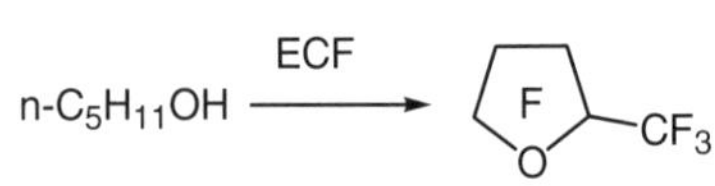

18.2%

10.3%

Electrochemical fluorination of α-alkyl-substituted carboxylic acid chlorides or esters produces the corresponding perfluorotetrahydrofuranes in about 20% yield, along with perfluoroalkanoyl fluorides.[85,86]

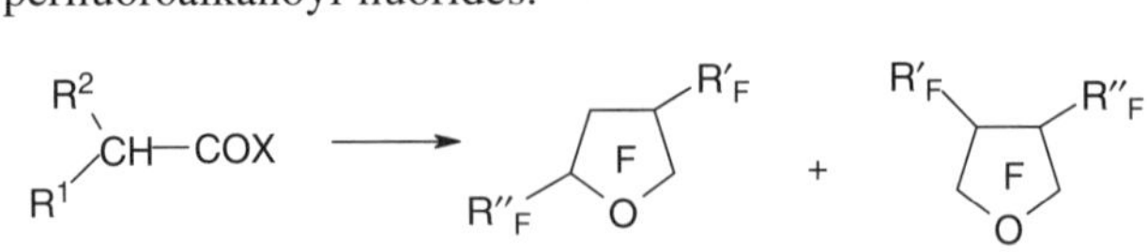

R^1 = *n*-Pr, *i*-Pr, *n*-Bu, *n*-Am $R'_F = CF_3, C_2F_5, C_3F_7, C_4F_9$

R^2 = Me, Et, *n*-Pr, *n*-Bu $R''_F = CF_3, C_2F_5, C_3F_7$

Additional information on the application of electrochemical fluorintaion for the syntheis of saturated heterocycles can be found in Chapters 9 and 10.

4.1.1.7 Fluorinated Derivatives of γ-Butyrolactone The fluorination of γ-butyrolactone by F_2 results in the formation of the mixture of α-,β-, and γ-fluorobutyrolactones **91–93**. β-F-Butyrolactone **92** and γ-F-butyrolactone **93** are major products.[87]

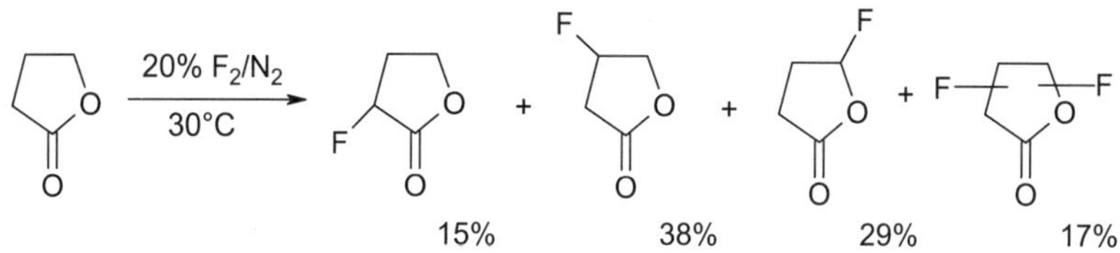

91 **92** **93**

Preparation of γ-fluorobutyrolactones **94** by the fluorination of γ-ketoacids with diethylaminosulfurtrifluoride (DAST) was reported back in 1984.[88]

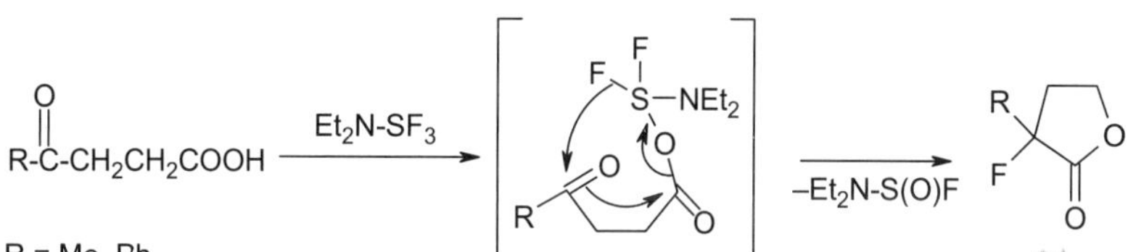

94

The difluoroiodotoluene is an effective reagent for the fluorination of α-phenelsulfanyl lactones **95**. The reaction involves fluoro-Pummerer rearrangment. A sequence involving conversion of α-fluorosulfide **95′** into sulfoxide and elimination step was used for the synthesis of 3-fluoro-2(5H) furanones **96**.[89]

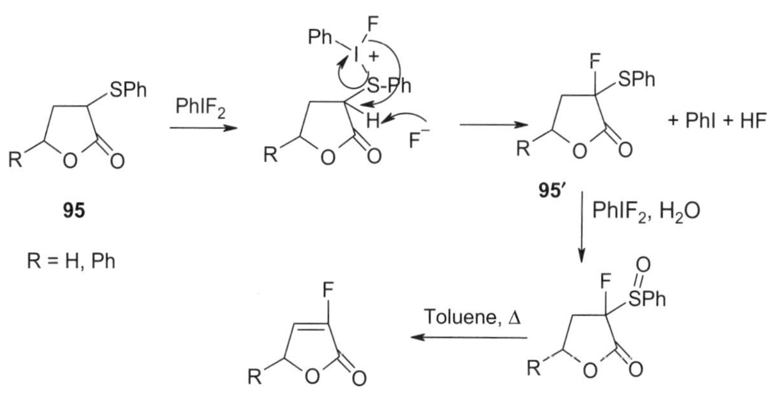

96

γ-Polyfluoroalkylbutyrolactones are among the most studied the polyfluorinated lactones. The most common non methods of synthesis are based on the cyclization of the fluorine-containing saturated[90–93] or unsaturated oxy-acids.[94–96]

For example, acid-catalyzed cyclization of 2,4-dihydroxyesters **97** gives lactones **98** as a mixture of diastereomeres.[97]

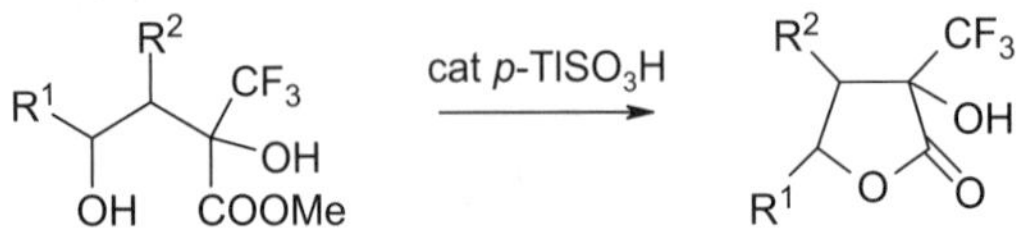

97 **98**

R^1 = Ph, R^2 = H R^1 = Me, R^2 = H

Upon treatment of γ-hydroxy-α-fluoro-α-trifluoromethyl carboxamides **99** by an organoaluminum reagent a single diastereomer of α-trifluoromethyl lactones **100** forms.[93]

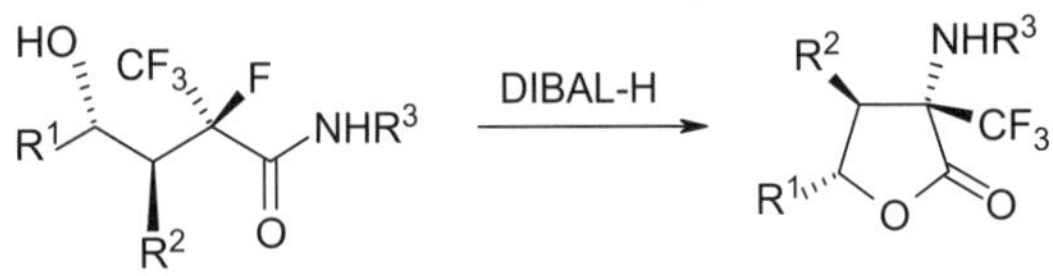

99 **100**

R^1 = H, Me, *i*-Bu, *n*-Bu; R^2 = H, Me, CH_2Ts; R^3 = Bn, Ph, *p*-MeOPh

γ-Hydroxycarboxylic acids **101** formed as an intermediate during the reduction of unsaturated ketones **102** undergo cyclization, giving the lactones **103** as a mixture of diastereomeres.[91]

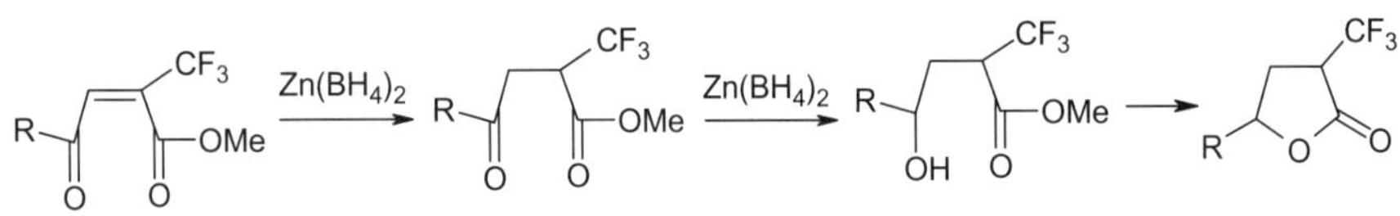

102 **101** **103**

R = Me, Ph

A series of *trans*- and *cis*-γ-lactones **104** with high optical purity have been prepared from the lactonization of (*s*)-(−)-2-(trifluoromethyl)-4-pentenoic acid under acidic conditions.[94,96]

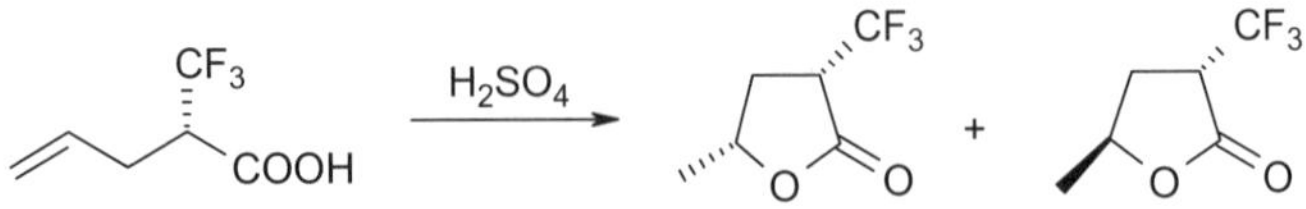

104

Seemingly, this reaction is general for the substituted pentenoic acids.[95]

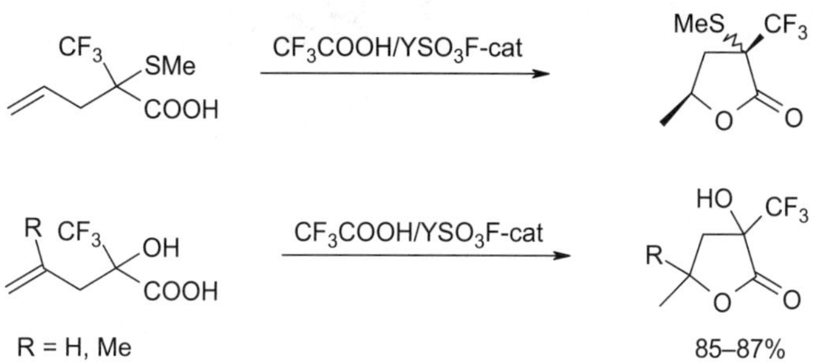

Iodolactonization of enantiomerically pure α-methoxy-β-trifluoromethyl-γ,δ-unsaturated carboxylic acid derivatives was reported as a highly stereoselective synthesis method of the γ-lactones.[98]

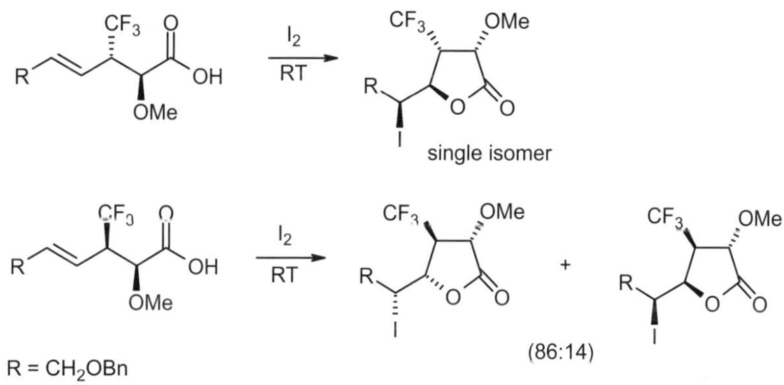

Similarly, iodolactonization of 2-fluoro-2-trifluoromethyl-4-alkenamides **105** proceeds stereoselectively and results in formation of a diastereomeric mixture of lactones **106** and **107**.[99]

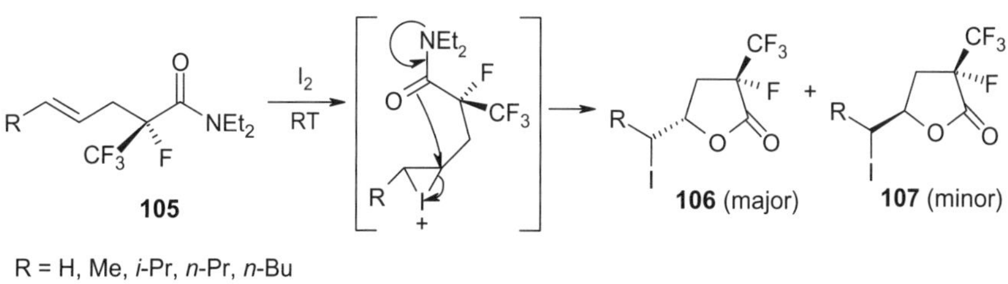

β-Polyfluoroalkyl-γ-butyrolactones **108** and **109** were prepared by sulfoxide-directed lactonization of β-fluoroalkyl vinyl sulfoxides **110**.[100]

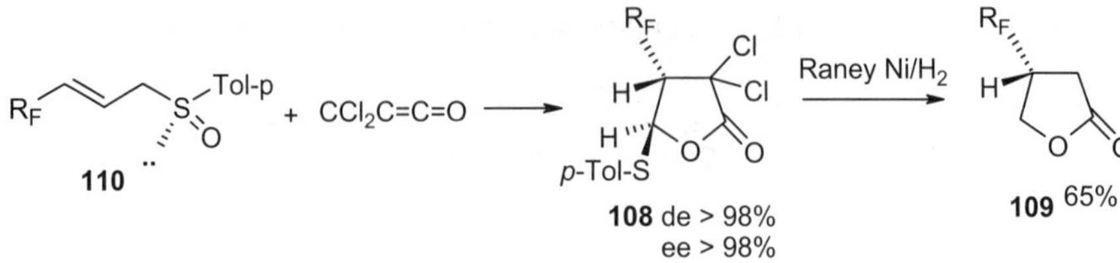

4.1.2 Fluorinated Thiophenes

4.1.2.1 2-Fluorothiophenes 2-Fluorothiophene was prepared first time in low yield by the fluorination of 2-iodothiophene by antimony trifluoride.[101]

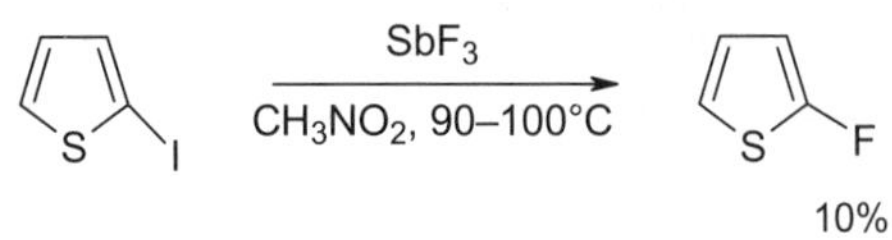

2-Fluorothiophenecarbonic acid **111** was synthesized by the hydrolysis of the 2-nitro-5-cyanthiophene.[102]

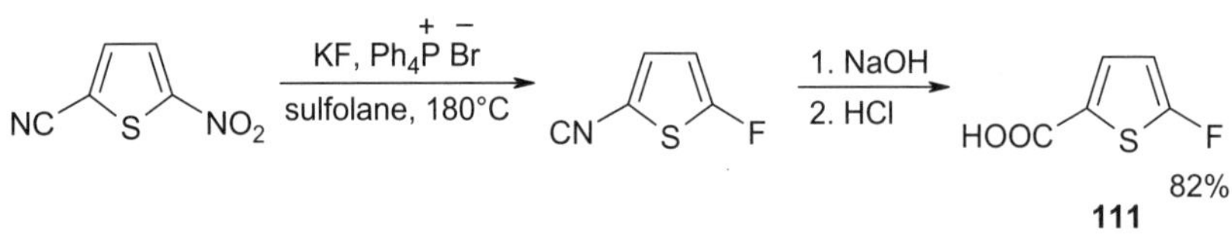

Electrophilic fluorination of the thiophene derivatives was used for preparation of 2-fluorothiophenes but fluorination of thiophene by F_2/He mixture leads to a mixture of two isomers, with 2-fluorothiophene being the major product.[103]

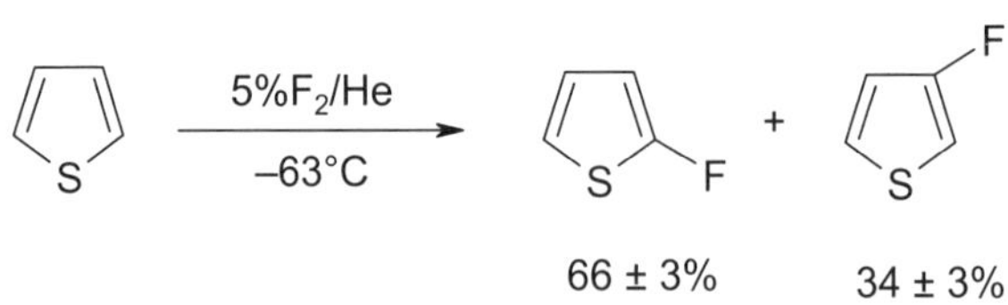

2-Fluorothiophene and 2-fluoro-5-methylthiophene were prepared by the exothermic reaction of perchloryl fluoride with the corresponding organolithium heterocyclic compounds in anhydrous ether.[104]

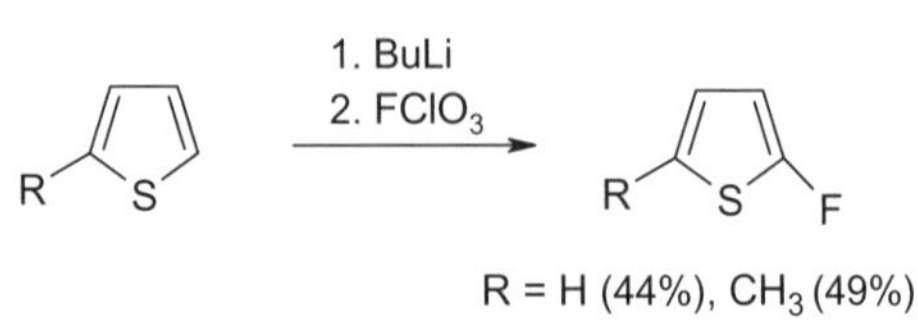

N-Fluorodibenzensulfonimide was successfully applied for the preparation of 2-fluorothiophenes **112** and **113**[13,105] and *N*-fluoroquinuclidinium fluoride[106] was used in the synthesis of 2-fluorothiophene.

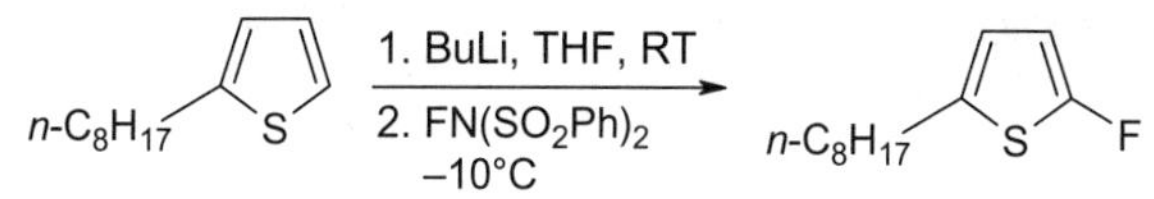

112 70%

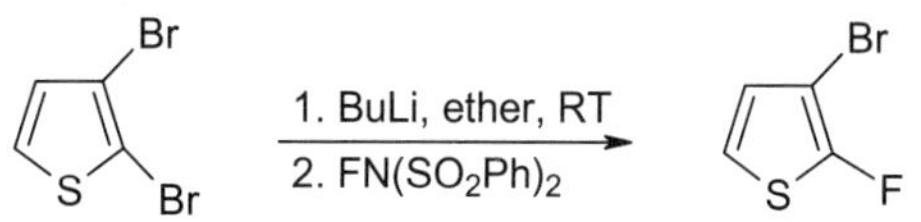

113

The synthetic methods for 2-fluorothiophenes with the usage of acyclic synthones are less studied. For example, 5-fluoro-2,3-dihydrothiophene **114** was prepared by cyclization of 1,1-difluoro-1-butenes.[15]

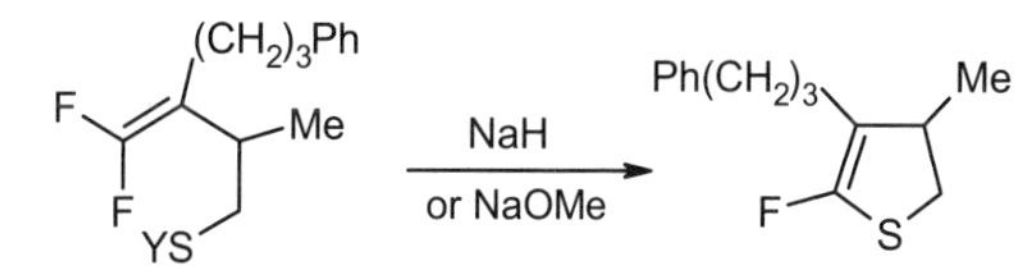

Y = H (76%); Ac (69%)

114

4.1.2.2 3-Fluorothiophene Derivatives 3-Fluorothiophene **115** was obtained by a multistep synthesis from 3-chlorothiophene.[107] The substitution of chlorine by fluorine under the action of cesium fluoride is possible only if electron-withdrawing substituent is present in 2-position of 3-chlorothiophene.

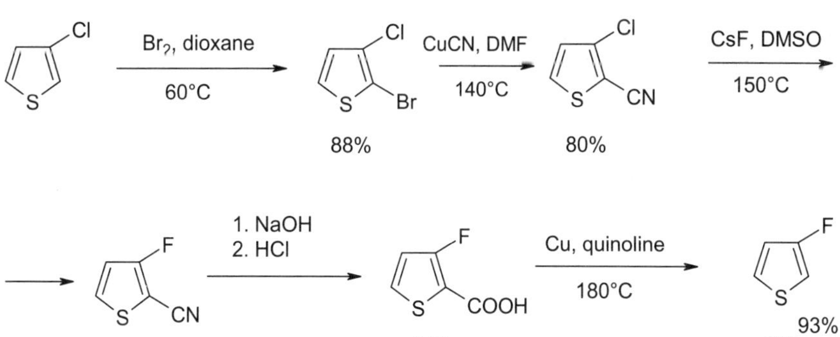

115

Substituted 3-fluorothiophenes were synthesized using a Balz–Schieman reaction. Diazotization of 3-aminothiophene **116** provided the corresponding diazonium hexafluorophosphate in an excellent yield. Thermolysis of this compound leads to the formation of 3-fluorothiophene **117**.[108,109]

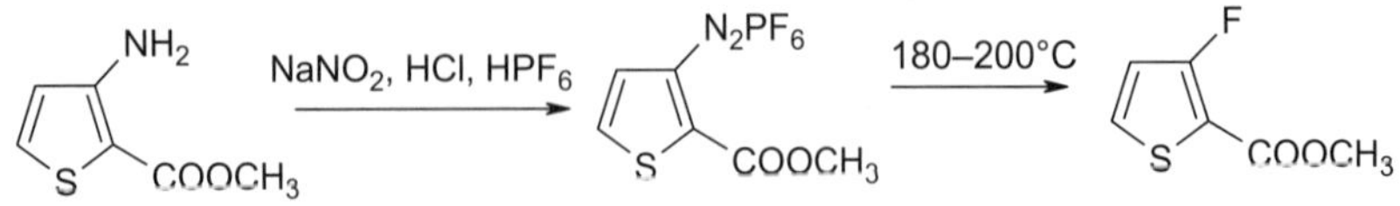

116 100% **117** 47%

The 4-fluoro isomer **118** was also prepared using this approach.[108]

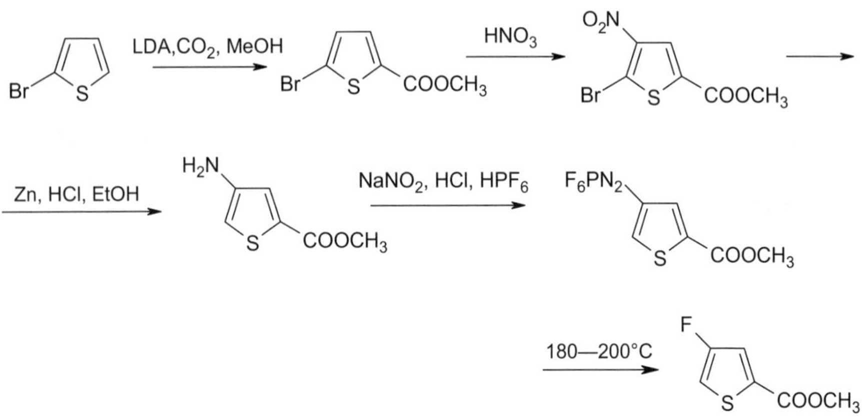

118

The methods of electophilic fluorination of thiophenes are more perspective for the synthesis of 3-fluorothiophenes. For example, thiophene fluorination by F_2 affords 3-fluorothiophene in 34% yield,[103] but the use of other electrophilic fluorinating agents gives much better results.

So, the treatment of 3-bromothiophenes by butyllithium and N-fluorodibenzene-sulfonimide results in formation of 3-fluorothiophenes **119–121** in 60–78% yield.[103,105,110]

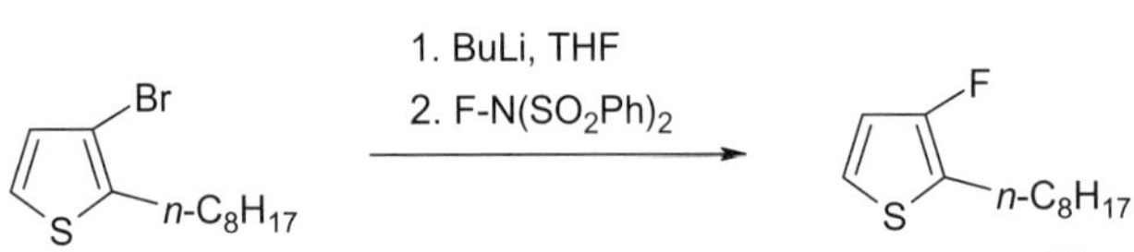

119 60%

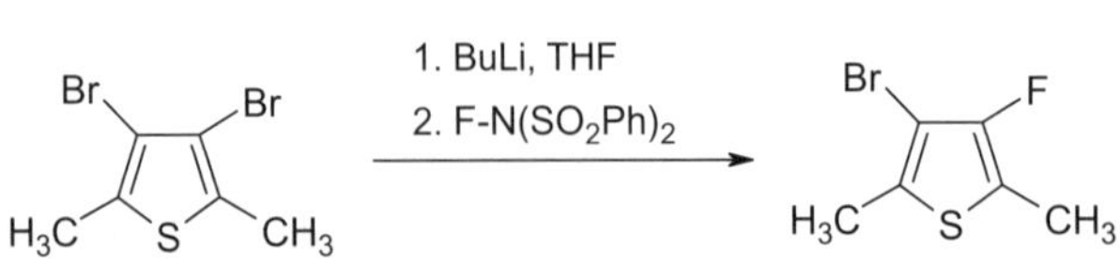

120

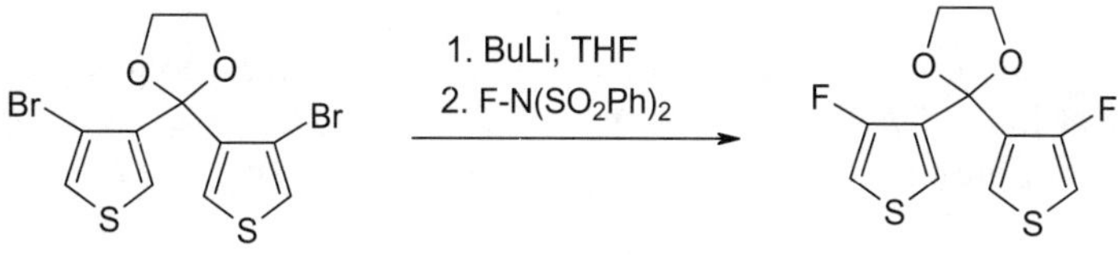

121 78%

2,5-Dihydrothiophendioxide **122** adds ClF (generated from HF and *N*-chlorosuccinimide) giving tetrahydrothiophene-1,1-dioxide **123** used as a starting material for the synthesis of 3-chloro-4-fluorothiophene-1,1-dioxide **124**. Compound **124** can be used in organic synthesis as fluorodiene.[111]

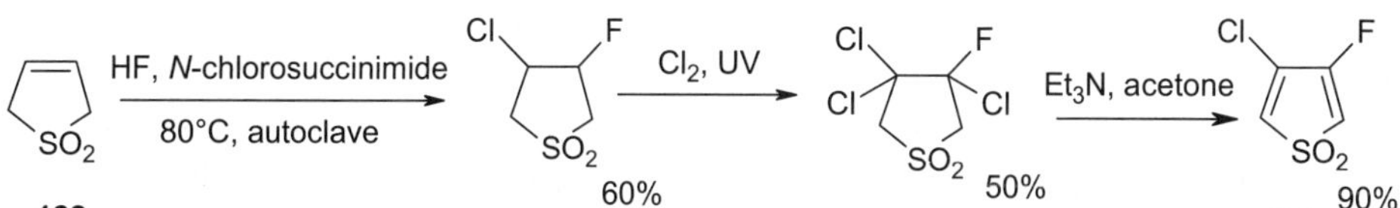

122 **123** **124**

Oxidative deselenenylation of compound **125** by the hydrogen peroxide oxidation affords the dihydrothiophen-1,1-dioxide **126**, exclusively.[112]

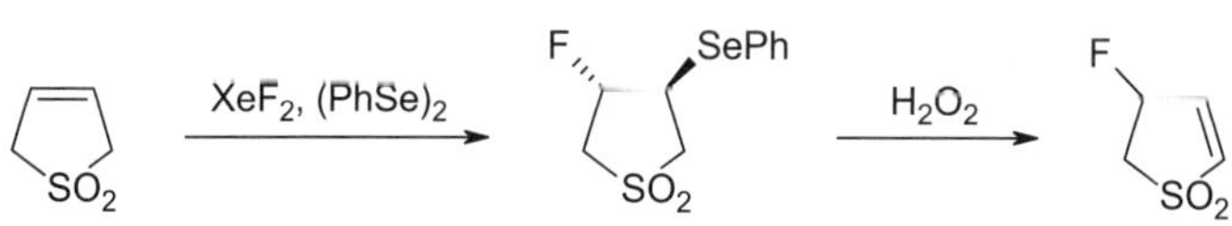

125 **126**

Fluorination of 4-thiofuranose hydroxyl-containing derivatives by dialkylamino sulfur trifluoride was successfully applied for the synthesis of 2-fluoro-4-thiosugars **127**,**128**.[113,114] The reactions proceed with complete retention of C–F carbon atom configuration.

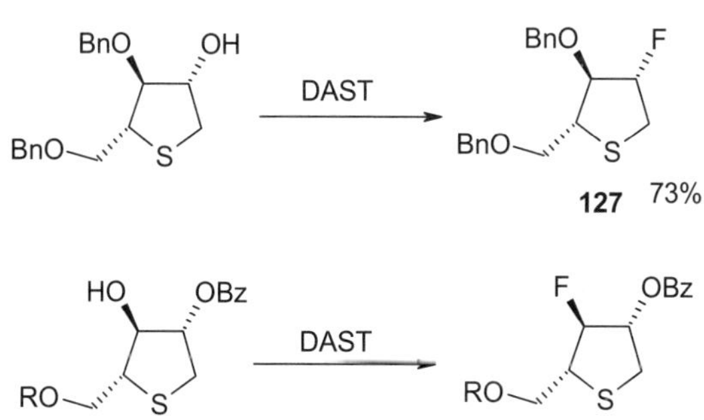

127 73%

128

The key intermediate of the synthesis 2-fluoro derivative **130** was prepared by treatment of **129** by DAST. The reaction proceeds in an unusual way with retention of configuration, presumably due to the intermediate formation of an episulfonium ion.[221]

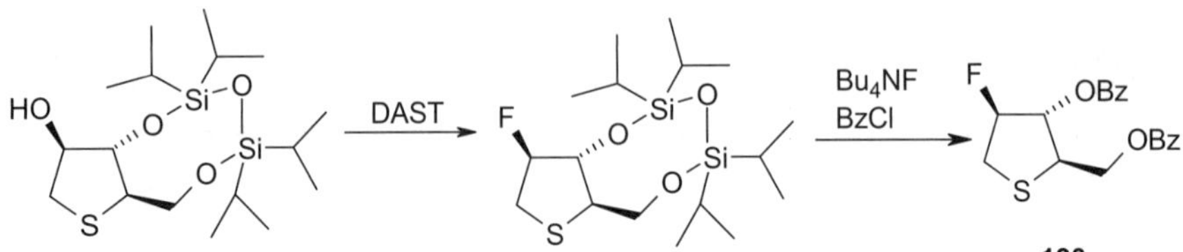

129 **130**

4.1.2.3 2-Perfluoroalkylthiophenes The methods for 2-perfluoroalkylthiophenes synthesis are similar to the methods used for the preparation of 2-perfluoroalkylfurans and based on the perfuloroalkylation reactions.

Polyfluoroalkyliodides in the presence of catalytic amounts of tetrakis(triphenylphosphine)nickel,[33] and perfluoroalkyl chlorides in the presence of sodium dithionite[115] were used for the generation of perfluoroalkyl radicals that were reacted with thiophene.

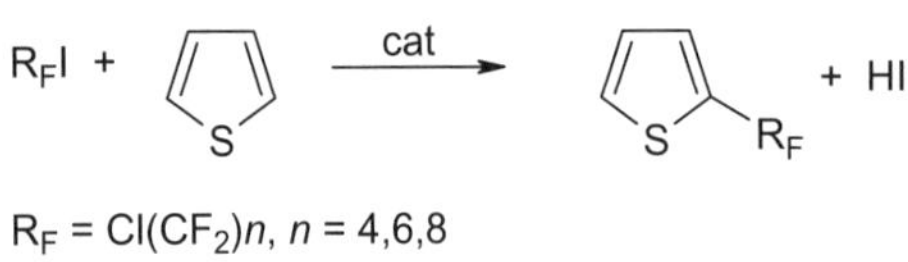

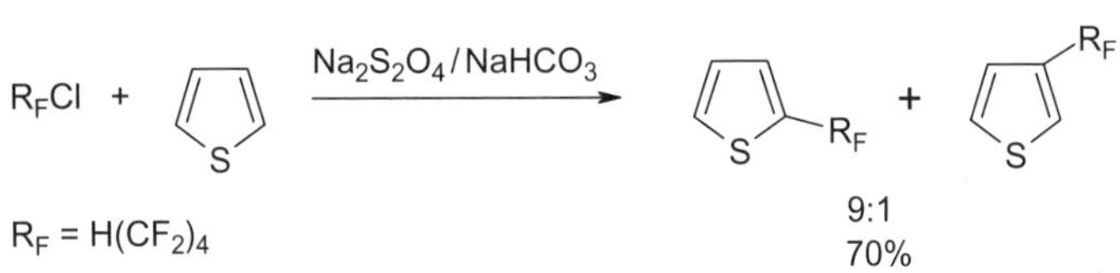

The highest yields of 2-perfluoroalkylthiophenes were obtained when bis(perfluoroalkanoyl)peroxides were used as the source of perfluoroalkyl radicals.[35,116,117] Mechanistically reactions with thiophenes (similarly to furans) are considered to be initiated by one-electron transfers from the substrates to peroxide.

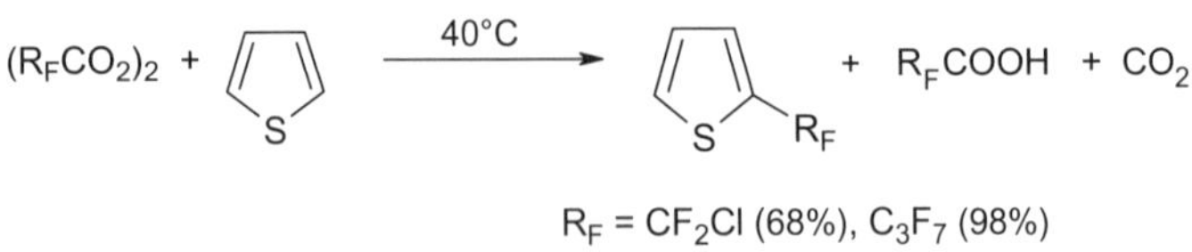

α-Substituted thiophenes can also be prepared in a high yields using this reation.[116]

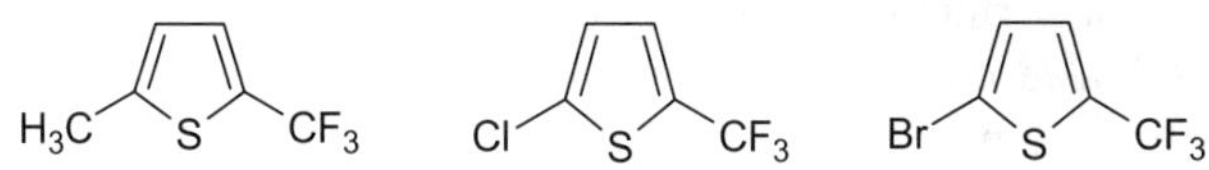

92% 78% 64%

In reaction of 2-Me- and 2-(chloro)bromothiopenes with bis(perfluorobutyryl) peroxides heptafluoropropyl group was introduced into a 5-position. In the analogous reaction of 3-bromothiophene, a heptafluoropropyl group was introduced into both 2- and 5-position in the yields 49% and 30%, respectively.[117]

Similar to furan, thiophene can be perfluoroalkylated in position 2 by the radicals, generated from perfluoroalkane sulfonyl chloride by ruthenium (II) catalyst.[36,37]

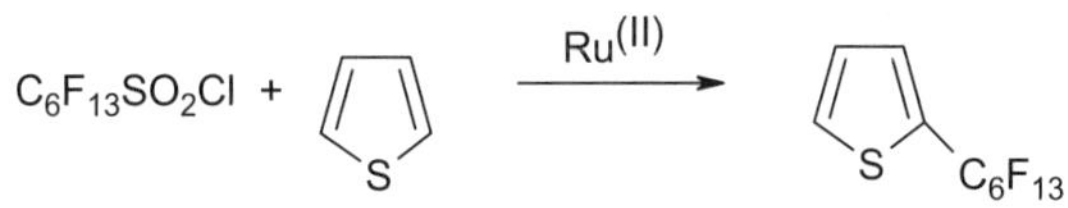

Once again, selenydes **22** and **23** react with thiophene under photolysis, forming 2-substituted thiophenes **70** and **71**.[38,118]

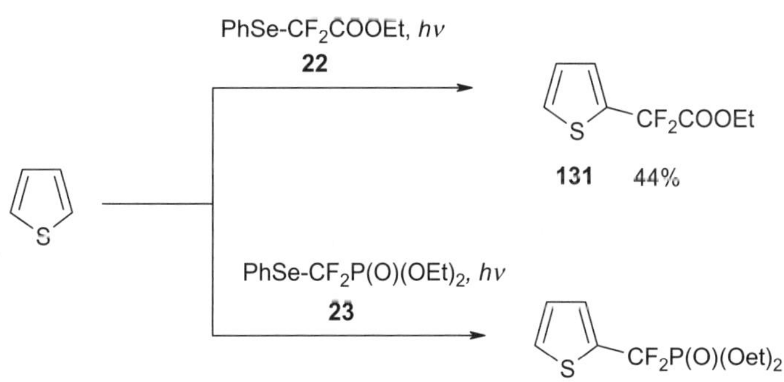

131 44%

132 37%

As in furan case, there is only one communication about cationic trifluoromethylation of thiophene by CF_3^+ generated from CF_4 (^{60}Co, γ-irradiation).[43]

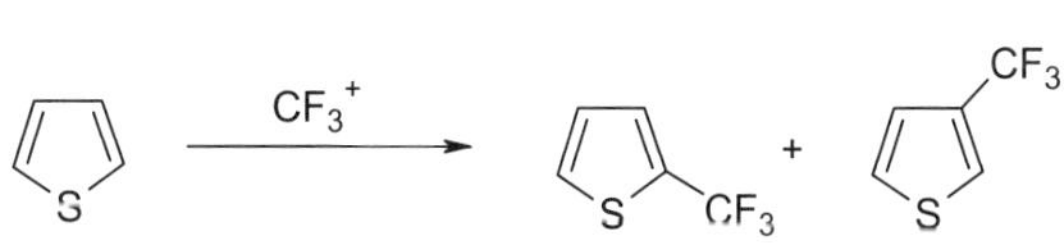

56–62% 36–41%

The more general approach to fluoroalkylthiophenes is anionic perfluoroalkylation of 2-bromthiophenes with perfluoroalkyliodides in the presence of copper.[44,119–122] The mechanism of perfluoroalkylation reaction consists of the steps involving the formation of solvated complex of fluoroalkylcopper, coordination of heterocyclic halide and ligands exchange at copper center.[122]

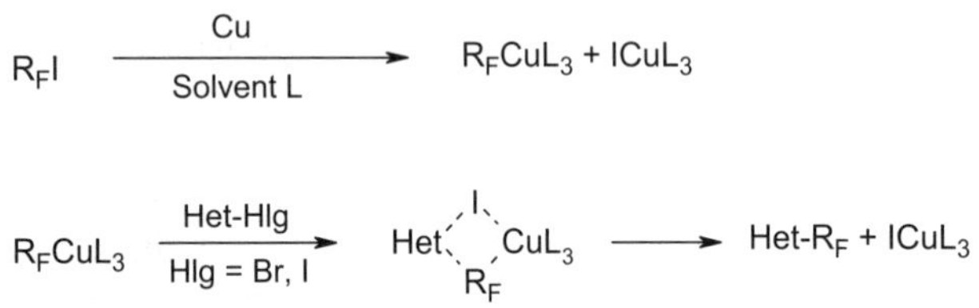

The thiophenes **133–136** were obtained by this method.[119–122]

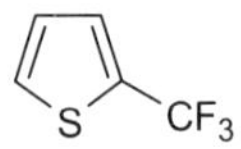

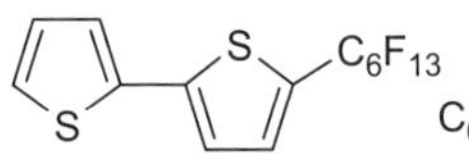

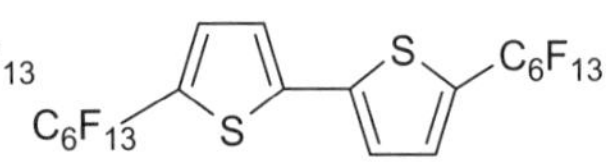

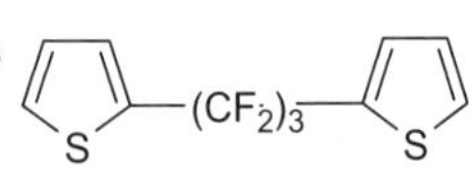

133 **134** **135** **136**

Sodium trifluoroacetate in the presence of copper (I) iodide is also used for introduction of trifluoromethyl group in 2-iodothiophen. The proposed mechanism of the formation of 2-trifluoromethyl thiophene (**72**) is based on CF_3CuI^- intermediate.[123]

Fluorination of 2,2-thiophendicarboxylic acid with sulfur tetrafluoride in the presence of HF gives the mixture of mono- and bistrifluoromethylated products and is analogous to the reaction of furanes.[46]

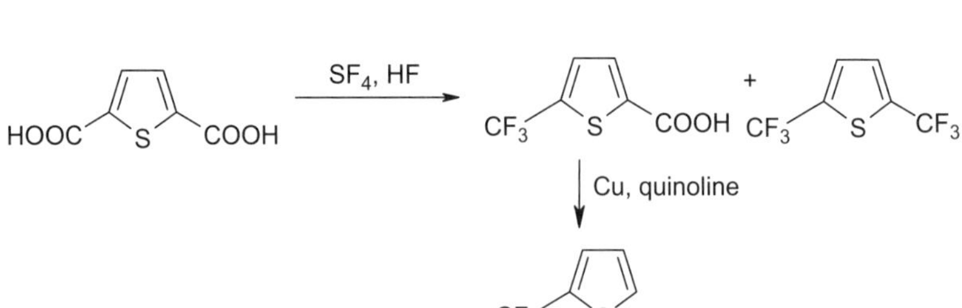

72

The new α,α-difluoroamides **137** result from the reactions of α-ketoacid **138** with Deoxofluor® or DAST.[124]

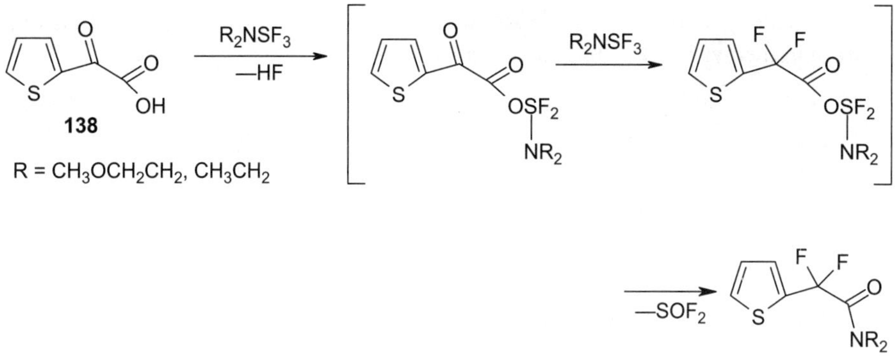

138

R = $CH_3OCH_2CH_2$, CH_3CH_2

137

Deoxofluor reacts with glyoxal hydrate **139** to form in good yield polyfluoroethers **140** as a mixture of meso and racemic (1:1).[125,126]

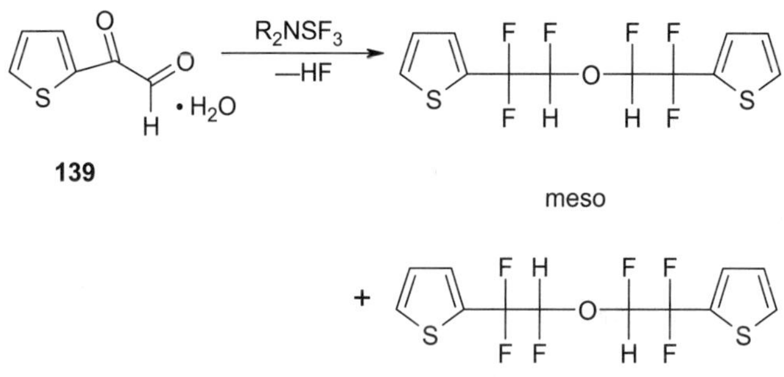

139

meso

140

racemic

The reaction of 1,1-dihydropolyfluoroalkylketones **141** or ethers **142** with methyl 2-sulfanylacetate giving thiophenecarboxylates **143** and **144**,[127] probably is the most promising synthetic approach for the preparation of 2-perfluoroalkylthiophenes.

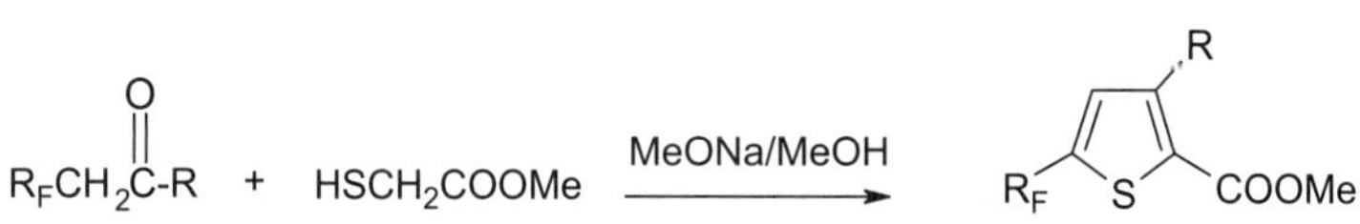

141

143 54–76%

R_F = CF_3CF_2, CF_3CBr_2, $Cl(CF_2)_4$, n-C_4F_9

R = H, Me, t-Bu

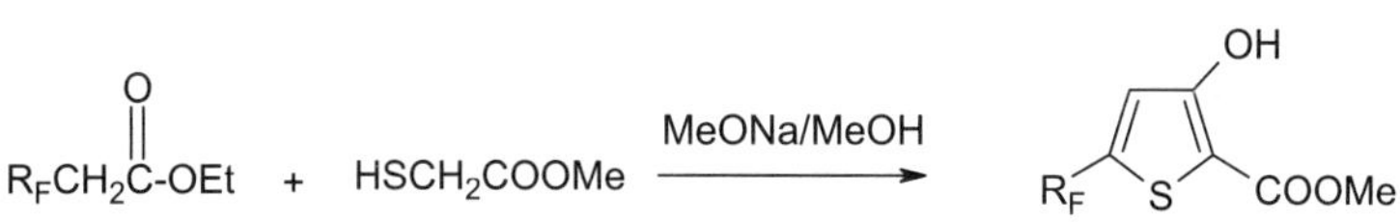

142

144 71–80%

R_F = CF_3CF_2, CF_3CBr_2, $ClCF_2CF_2$, $Cl(CF_2)_4$, $PhCH{=}CHCF_2CF_2$

Vinylaldehyde **145** was also used as starting material for the preparation of 2-trifluoromethyl-3,5-diarylthiophenes **146**.[128]

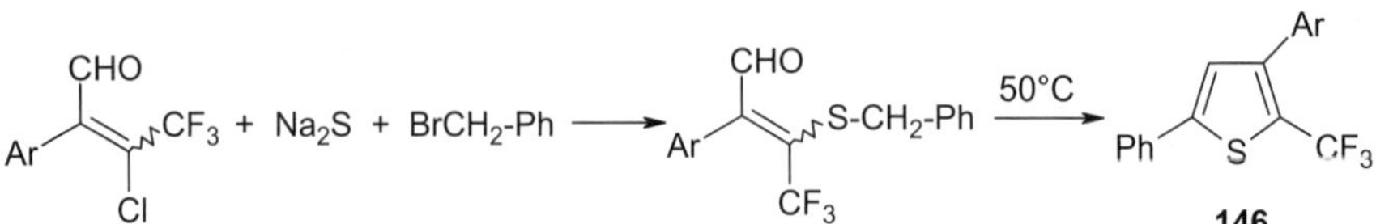

145 **146**

Ar = Ph, p-NO_2-Ph

4.1.2.4 3-Perfluoroalkylthiophenes 3-Trifluoromethylthiophene and 3-octafluorobutylthiophene are formed along with the 2-substituted isomers in the radical[129] or cationinc[43] perfluoroalkylation of thiophene (see section 3 of this chapter). The anionic perfluoroalkylation of 3-bromine(iodine)thiophenes by perfluoroalkyl halides in the presence of copper gives the mixture of 3- and 2-perfluoroalkyl isomers with prevailing 3-isomer.[121,130,131]

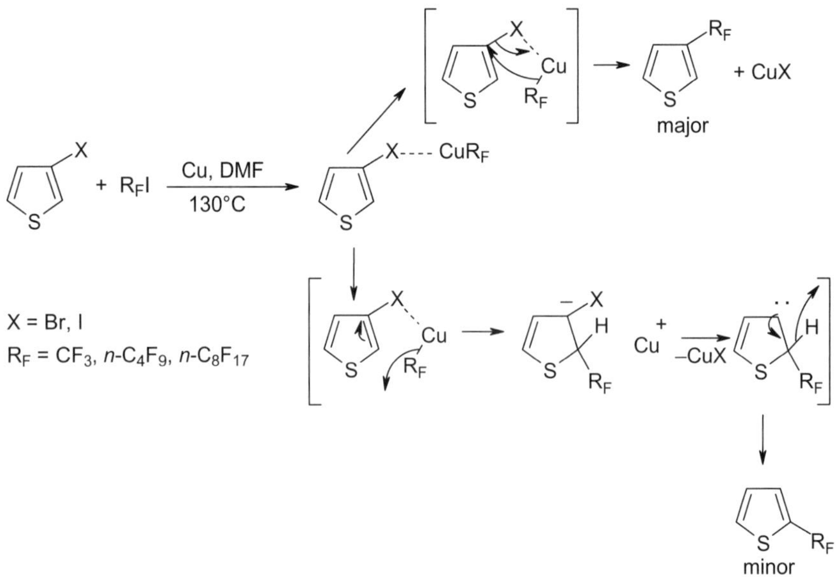

Direct attack of R_FCu on C(3) carbon leads to the most favorable product, probably by a concerted mechanism. The ratio of two isomers is defined by the nature of the halogen in thiophene ring. For example, in the case of the reaction of 3-bromothiophene and n-perfluorooctyl iodide, 18% of a rearranged product (2-isomer) was formed, whereas in the reaction of 3-iodothiophene, <4% of 2-perfluoro-n-octylthiophene was formed. The possible reason for 2-isomer formation may be addition of perfluoroalkyl anion to C(2)–C(3) double bond with formation of carbene, which undergoes isomerization into 2-fluoroalkylthiophene.[130]

1,3-Dipolar cycloaddition of mesoionic 1,3-dithiolium-4-olanes **147** and alkynes **148** should be mentioned among other methods of synthesis the 3-perfluoroalkylthiophenes involving acyclic starting compounds. Various triaryl-3-trifluoromethylthiophenes **149** and **150** were synthesized using this method.[131]

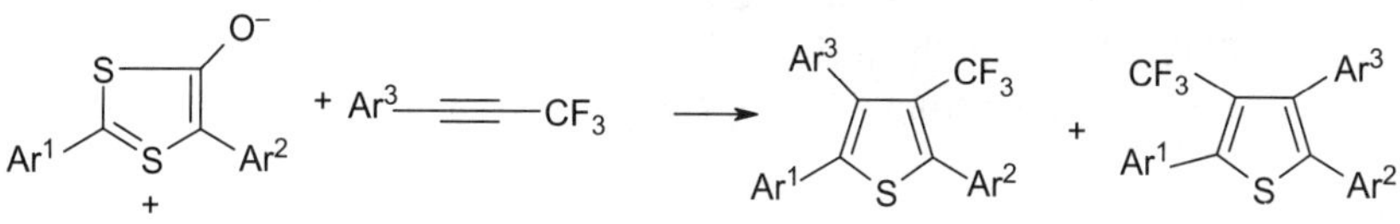

147 **148** **149** **150**

Ar^1	Ar^2	Ar^3	**149/150**
Ph	4-ClPh	4-ClPh	91/9
Ph	4-ClPh	4-MeOPh	93/7
Ph	4-ClPh	4-NO_2Ph	89/11

The derivatives of fluorine-containing 3-(methoxycarbonylthio) acrylic acids easily undergo base catalyzed cyclization giving 3-hydroxy(amino)-4-trifluoromethylthiophenes.[132]

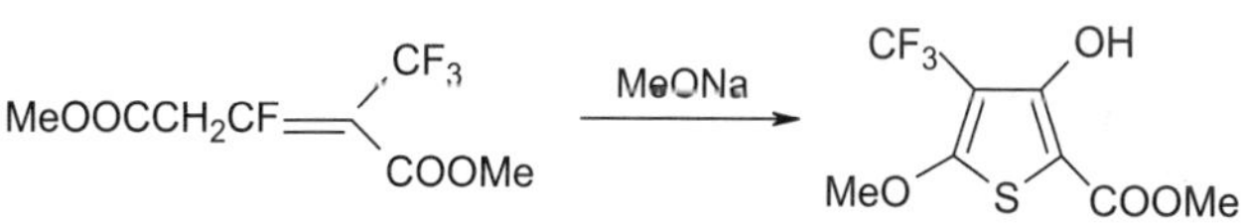

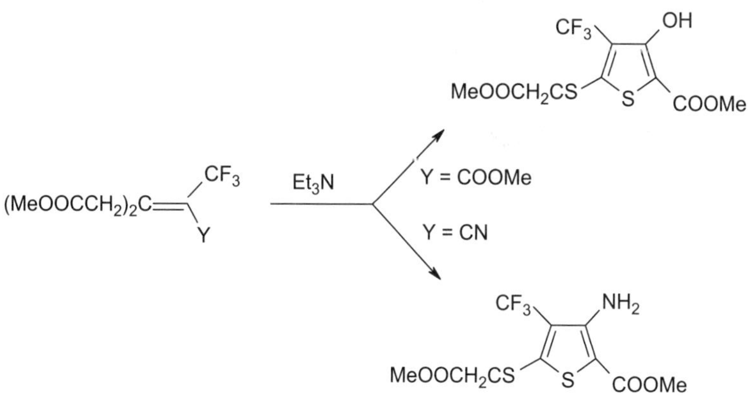

4.1.2.5 Thiophenes Containing Two Fluorinated Substituents 2,5-Difluorothiophene was obtained from 2,2,5,5-tetrafluoro-3-thiolen **151**. Compound **151** is a major product in the fluorination of thiophene over potassium tetrafluorocobaltate.[133,134]

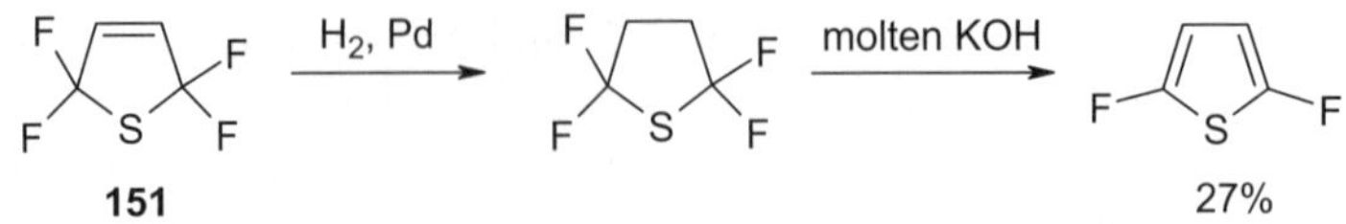

The other isomeric unsubstituted difluorothiophenes are not known. 2,5-Bis (trimethylsilyl)-3,4-difluorothiophene **152** was prepared by electrophilic fluorination of Li derivative, obtained from dibromo compound **153**.[135]

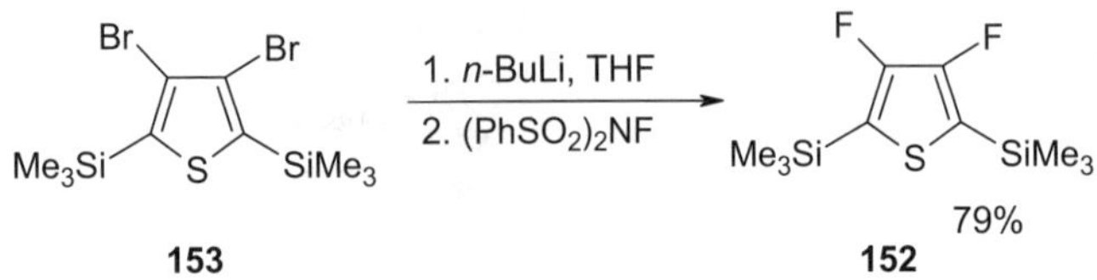

As in case of furan derivatives, the methods for the synthesis of bis(polyfluoroalkyl thiophenes are more studied. Fluorination of 2,5-thiophendicarboxylic acid with sulfur tetrafluoride leads to the formation of 2,5-bis(trifluoromethyl)thiophen.[46] Fluorination of perchlorinated thiophen **154** with AgF gives thiophene **155**[136] and replacement of the β-chlorine atoms by fluorine could not be achieved in this reaction.

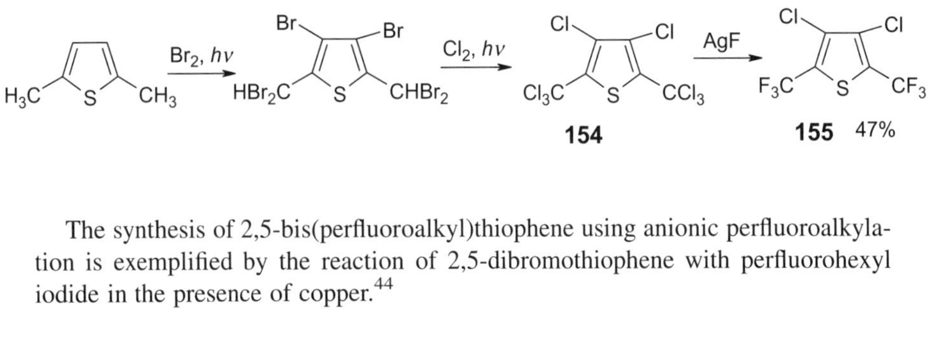

The synthesis of 2,5-bis(perfluoroalkyl)thiophene using anionic perfluoroalkylation is exemplified by the reaction of 2,5-dibromothiophene with perfluorohexyl iodide in the presence of copper.[44]

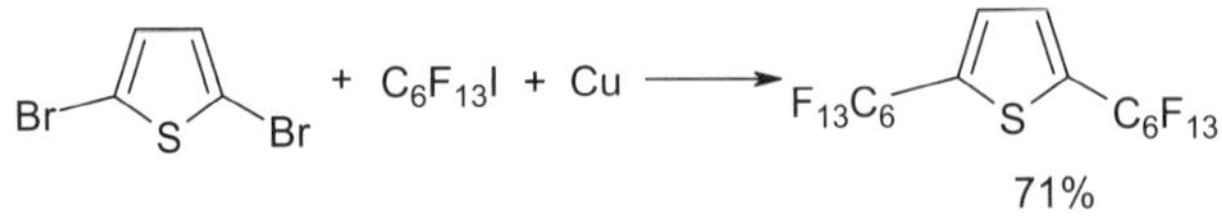

There are several examples on the synthesis of di-substituted thiophenes from the fluorine-containing acyclic unsaturated compounds. For example, 2-fluoro-3-trifluromethylthiophenes **157** were prepared starting from 3-trifluoromethylbut-3-en-1-one **156** and phosphorus pentasulfide.[137,138]

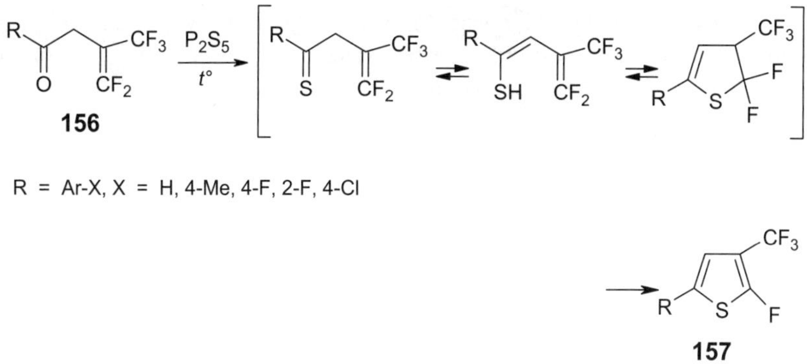

R = Ar-X, X = H, 4-Me, 4-F, 2-F, 4-Cl

2,3-Bis(trifluoromethyl)thiophene was prepared in low yield by heating bis(trifluoromethyl)-1,2-dithietene **158** with acetylene.[139]

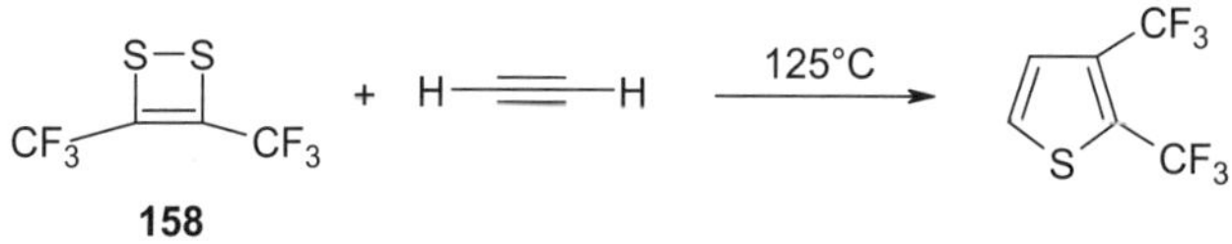

Octafluorotetrahydrothiophene **159** was prepared from sulfur and tetrafluoroethylene in ~15% yield.[140]

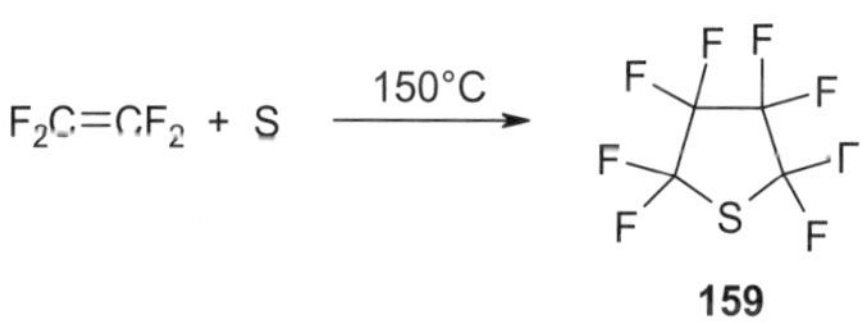

Modification of this reaction by the addition of hydrocarbon olefin has been found to be a general preparative method for synthesis of partially fluorinated tetrahydrothiophenes.[141]

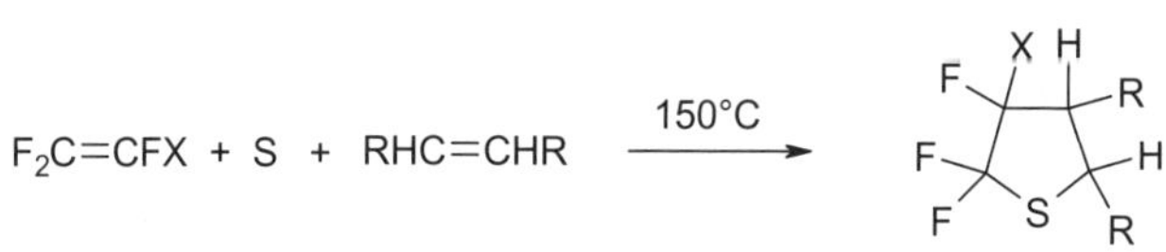

X = F, Cl, Br

about 60%

R = H, CH_3, CN, $COOCH_3$, $OCOCH_3$, OCH_2CH_3, CH_2OH, Cl, CF_3, Ph

4.1.2.6 *Fluorinated Tetrasusbstituted Thiophenes*

Tetrafluorothiophene **160** was prepared by dechlorination of 2,3,4,5-tetrachlorotetrafluorotetrahydrothiophene.[142]

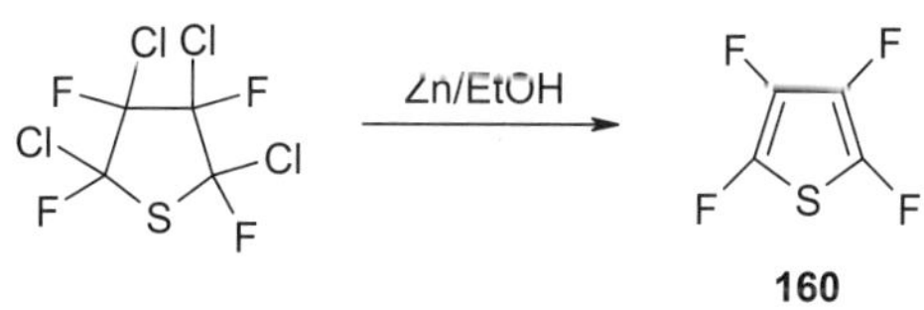

Tetrakis(trifluoromethyl)thiophene **161** was synthesized by the reaction of perfluorodiene with potassium sulfide[71] or by exhaustive trifluoromethylation of tetraiodothiophene using trifluoromethyl copper.[222]

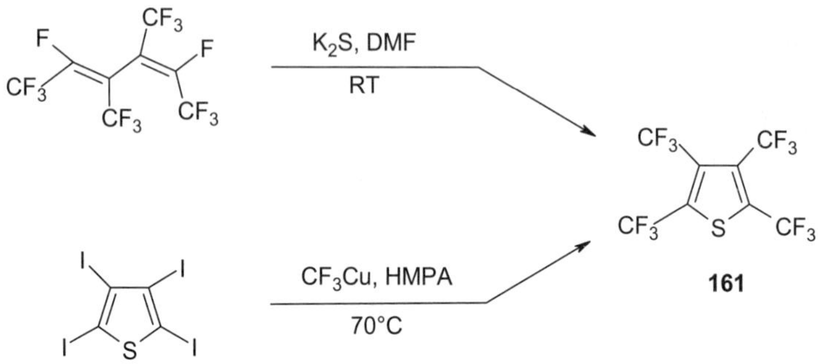

Sulfur and tetrafluoroethylene were found to react in the presence of iodine at 250–300°C under pressure to form octafluorotetrahydrothiophene **159**.[140]

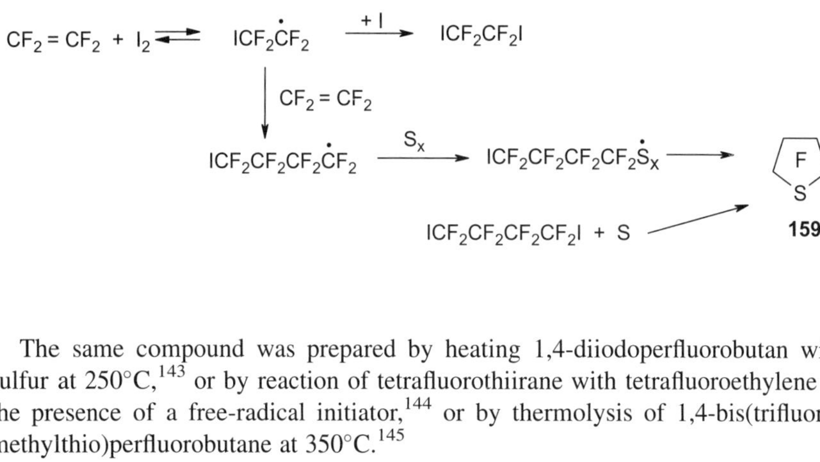

The same compound was prepared by heating 1,4-diiodoperfluorobutan with sulfur at 250°C,[143] or by reaction of tetrafluorothiirane with tetrafluoroethylene in the presence of a free-radical initiator,[144] or by thermolysis of 1,4-bis(trifluoromethylthio)perfluorobutane at 350°C.[145]

The reaction of diiodide with phosphorus pentasulfide instead of sulfur gives dithiolactone **162**.[146]

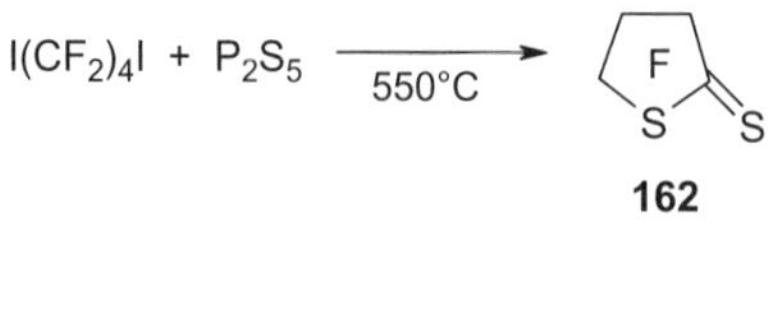

Partially fluorinated tetrahydrothiophenes **164** and **165** are isolated by fluorination of 3*H*,4*H*-tetrafluoro-dihydrothiophene **163** over $KCoF_4$.[147]

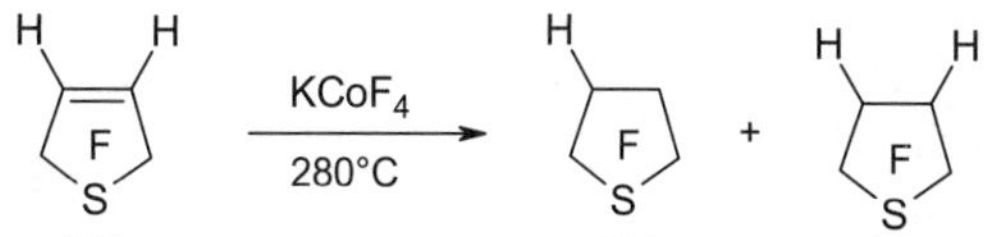

163 **164** **165**

Electrochemical fluorination, widely used for preparation of perfluorinated terahydrofuranes can not be applied for the synthesis of tetrahydrothiophene derivatives due to oxidation of the sulfur. Electrochemical fluorination of tetrahydrothiophene and 2-methyltetrahydrothiophene resulted in low yield formation of the products **166** and **167**, respectively.[148]

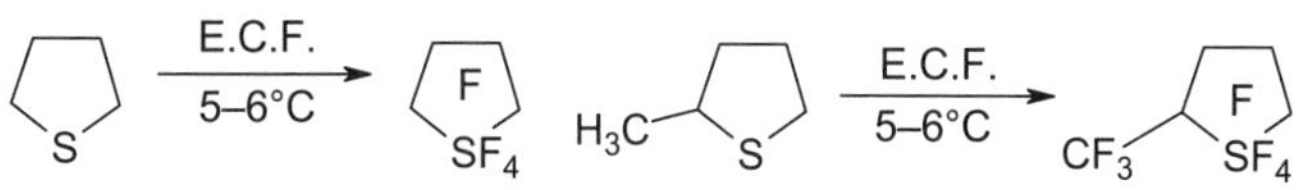

166 **167**

4.1.3 Fluorinated Derivatives of Selenophene

2-Fluoro- and 2,5-difluoroselenophenes were prepared by reaction of the corresponding selenophene lithium derivatives with FClO3 in 73% and 11% yield, respectively.[149] The first polyfluorinated derivatives of selenophene-octafluoroselenonane **168** and octafluoro-1,4-diselenane- were isolated in low yields in the reaction of selenium, tetrafluoroethylene, and iodine.[140] Selenonane **169** was prepared in low yield by reaction of equimolar amounts of selenium, tetrafluoroethylene, and propylene in the presence of iodine.[141]

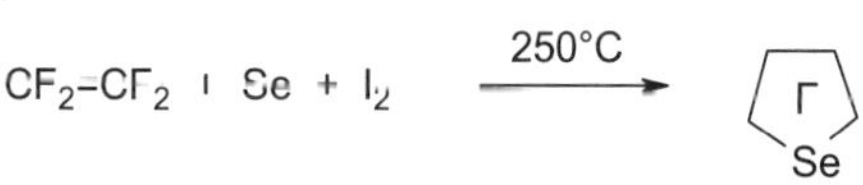

168 10%

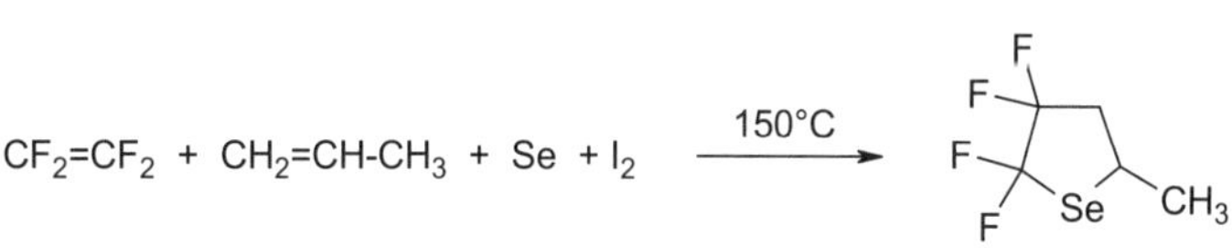

169 8%

4.1.4 Fluorine-Containing Five-Membered Heterocycles with More Than One Heteroatom

4.1.4.1 1,2 and 1,3-Dioxolanes Cyclic peroxide **170** was obtained in 2% yield along with the other products by the direct fluorination of 1-hydroxy-3-trichloroacetoxypropane.[150] This compound is more stable than peroxide **171**[151] and can be handled at room temperature.

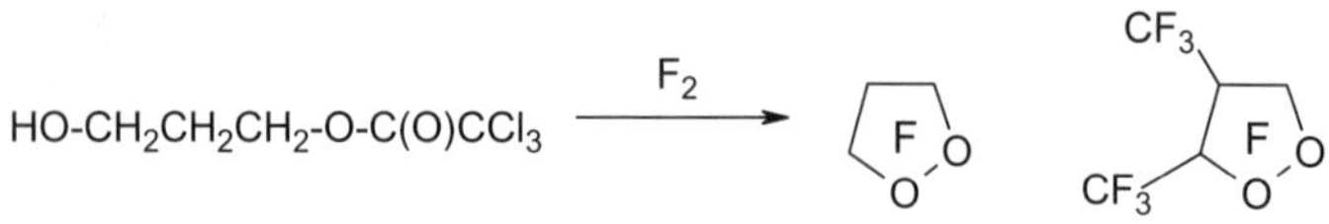

170 171

1,3-Dioxolanes are more studied. As a rule, these compounds are obtained from the fluorinated olefins, nitriles, or hexafluoroacetone etc.[152–155,223]

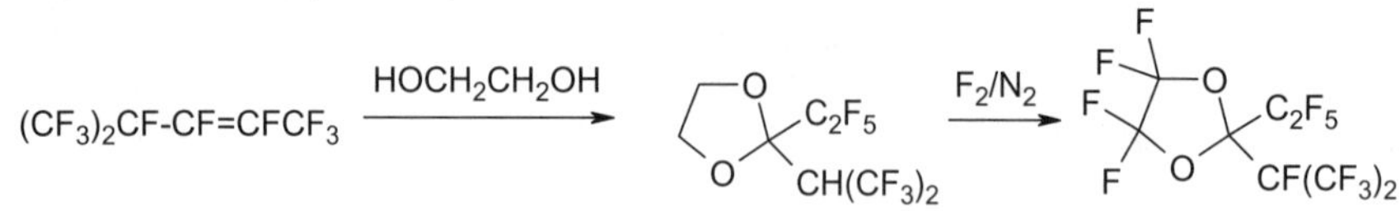

172 173 69%

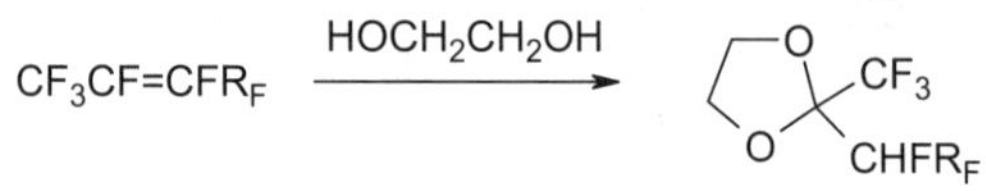

174

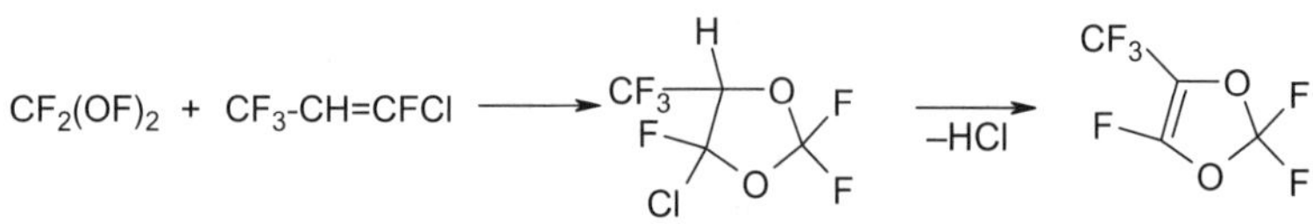

175 72% 176 61%

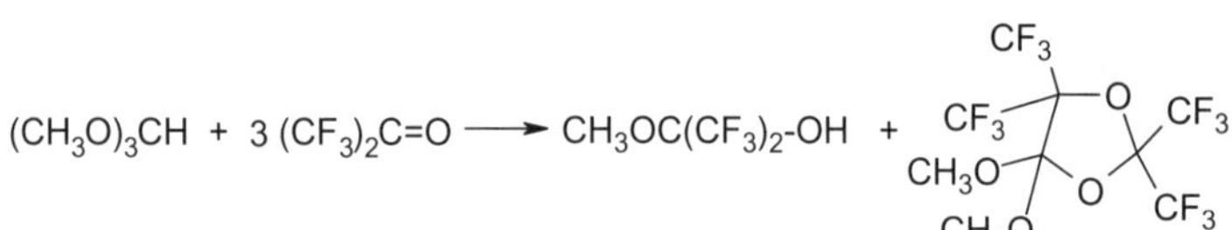

79%

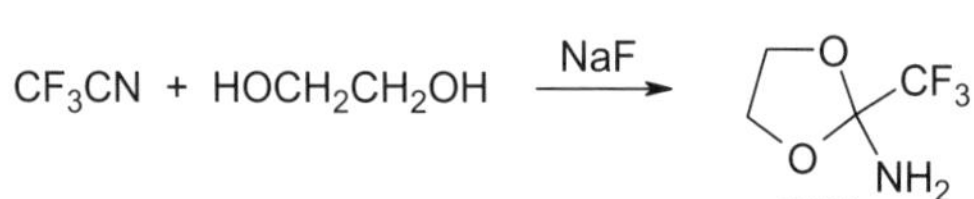

85%

4.1.4.2 1,3-Oxathiolanes, 1,2-Dithiolanes, 1,3-Dithiolanes, and 1,2,3-Trithiolanes Only three types of fluorinated 1,3-oxathiolanes derivatives are known: **177**,[156,157] **178**,[158] and **179**.[144]

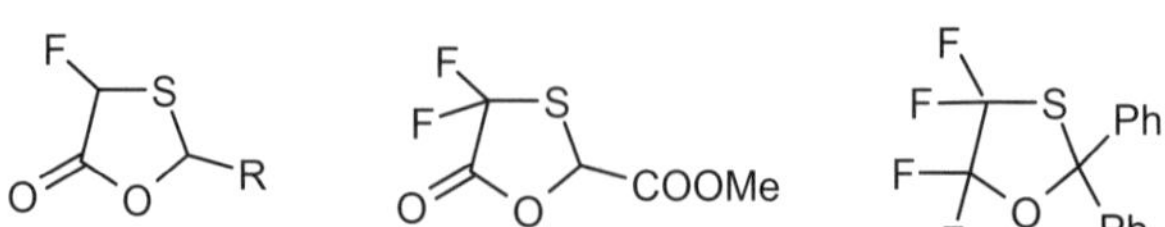

177 178 179

Five-membered fluorine-containing heterocyclic compounds with two or three sulfur atoms are more extended. A ring expansion reaction of thietanes into 1,2-dithiolanes involving heating with sulfur was used.[159]

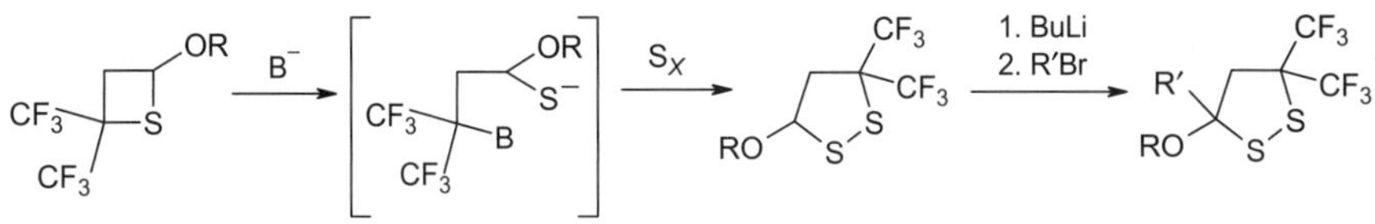

3,5-Bis(trifluoromethyl)-1,2-dithiolan **180** was prepared by heating 1,1,1-trifluorothioacetone with sulfur.[160] The action of the thionazing agents on polyfluorinated unsaturated ketones or β-ketoethers results in 1,2-dithioles **181**,[161] **182**,[162] and **183**.[163]

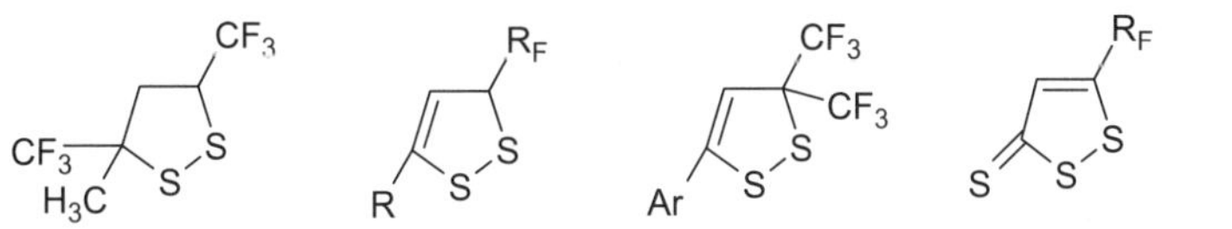

180 **181** **182** **183**

R = Alk, Ar; $R_F = CF_3, C_4F_9, H(CF_2)_6$ $R_F = HCF_2, CF_3, C_4F_9$

The 1,2 dithiolane **184** was obtained by the chlorination of 1,2-dithiole-3-thone **185**.[164]

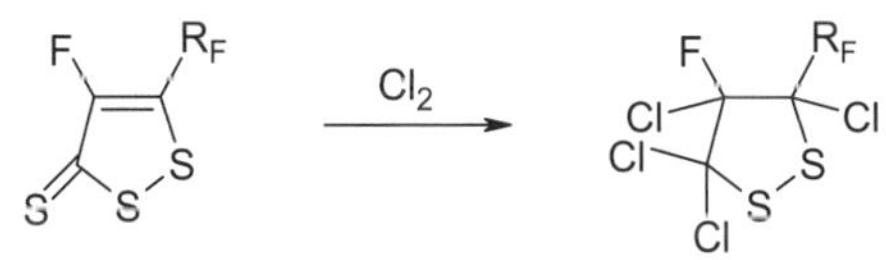

185 **184**

$R_F = HCF_2CF_2$

Fluorine-containing 1,3-dithiolanes can be prepared by the reaction of trifluoromethylketones or trifluoroacetoaldehyde with ethanedithiole in the presence of boron trifluoride.[165]

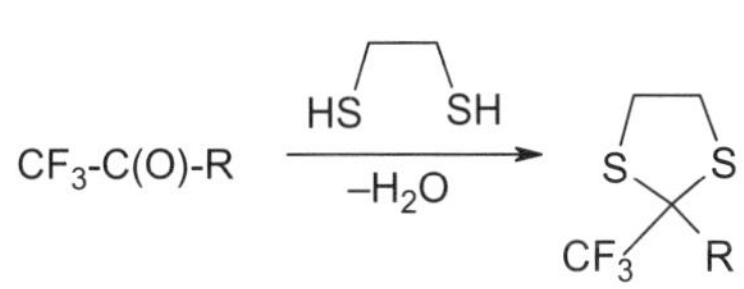

R = H, Me, Ph

The addition of ethanedithiole to polyfluoro-2-alkynoic acids gives the 2-carboxymethyl-2-polyfluoroalkyl-1,3-dithiolanes **186**.[166]

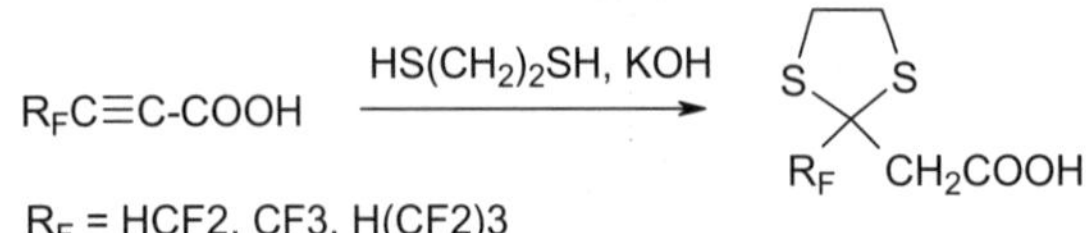

186

Compounds **187** and **188** were prepared by the fluorination of ethylene trithiocarbonate[167] and reaction of hexafluorothioacetone with diazomethane,[168] correspondingly.

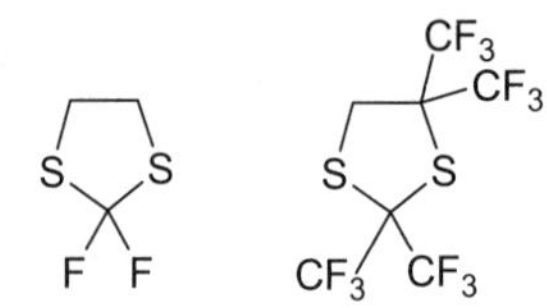

187 **188**

A number of 4,5-bis(trifluoromethyl)1,3-dithioles was obtained by the reactions of hexafluoro-2-butyne with the derivatives of dithiocarboxylic acids.[169–171]

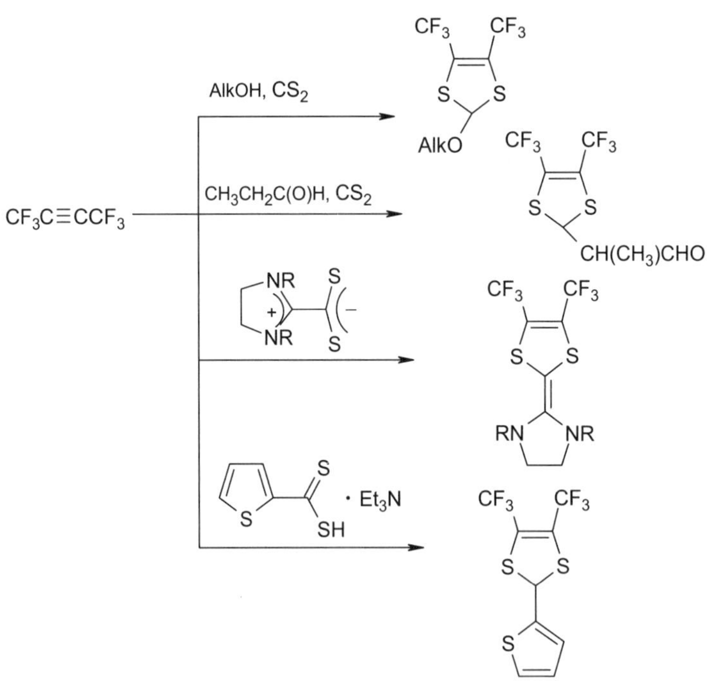

Tetrafluoro-1,2,3-trithiolane **189** was synthesized by the reaction of sulfur with tetrafluoroethene.[172] At elevated temperature, tetrafluoroethylene reacts with the complex $S_4(Sb_2F_{11})_2$ with the formation of perfluoro-3,5-dimethyl-3,5-diethyl-1,2,4-trithiolane **190**.[173]

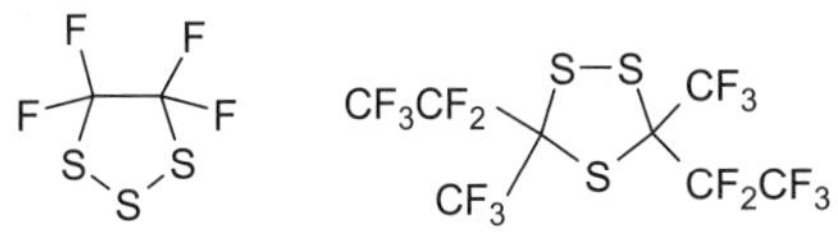

189 **190**

4.1.5 Fluorinated Phospholane Derivatives

The rich chemistry of phospholes, phospholenes, and phospholanes is described in several reviews.[174,175]

However, the information on fluorine-containing derivatives of these materials is very limited. P–F Phosphoranes **193** and **194**, resulted from the fluorination of diphosphine disulfide **192** or from the reaction of 1-methyl-methylenephosphorane **191** with hydrogen fluoride, were among the first representatives of the fluorine-containing phospholanes.[176]

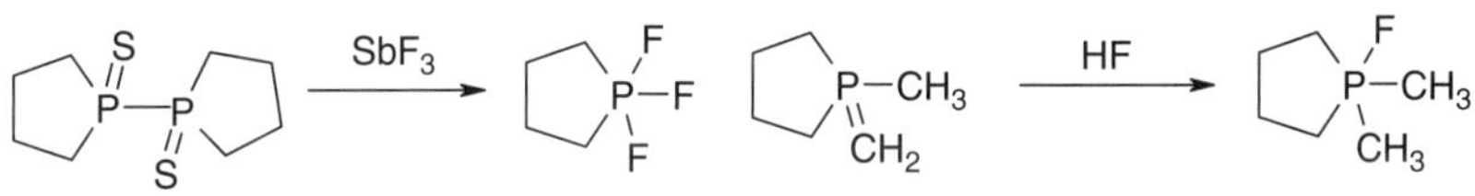

192 **193** **191** **194**

3-Phospholenes **195** with P–F bond are prepared by the reaction of 1,3-dienes with fluorophosphates.[177]

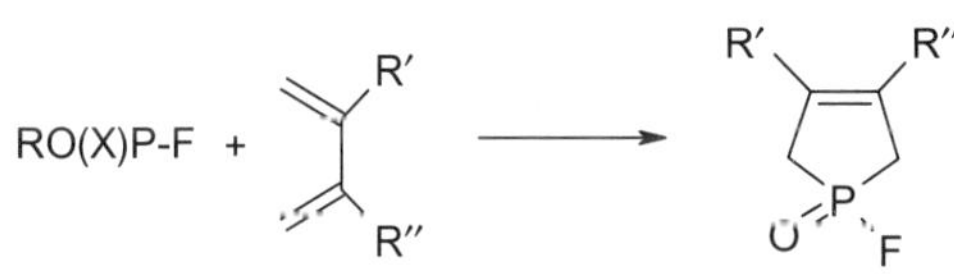

195

X = RO, F; R′ = H, Me; R″ = H, Me

Phospholen-2 **196** was obtained by the fluorination of 1,1,1-trichloro-3-methyl-2-phospholen.[178]

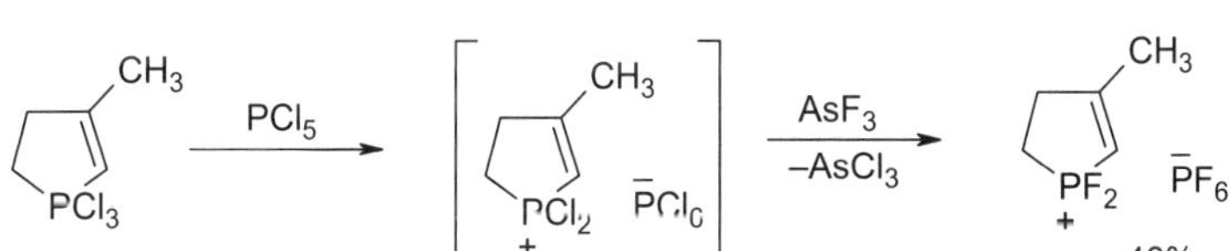

196 48%

The first phospholane **197**, containing fluorine atoms at the cycle carbons, was obtained by the reaction of tetrafluoroethylene with phosphorus in the presence of iodine.[140] The hydrolysis of **197** proceeds with ring opening leading to the corresponding phosphonic acid.

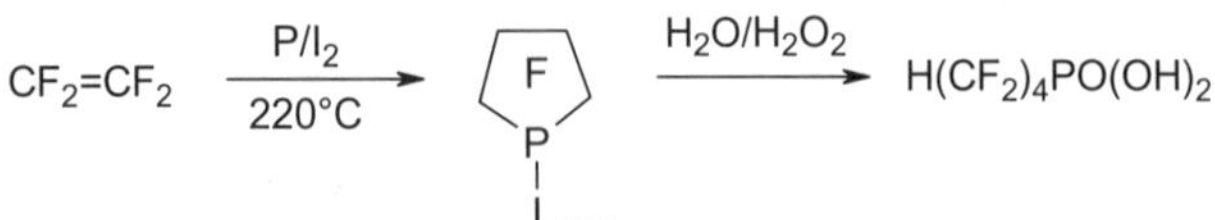

197

Unsubstituted α-fluoro- and β-fluoro-phospholes are unknown although their possible electronic structure was studied theoretically.[179] Tetrakis(trifluoromethyl)-phosphole **198** was obtained in thermal reaction of hexafluorobut-2-yne with methyl(triphenylphosphine) gold.[180]

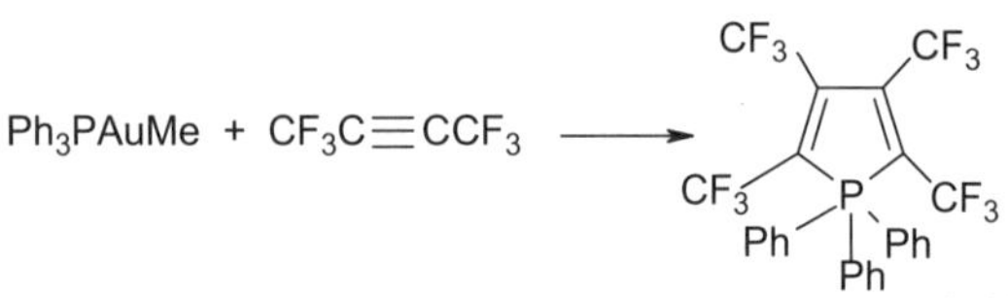

31%

198

1,2-Oxophospholanes **199** were obtained by cyclization of hydroxyl-containing α-fluorineohisohinic acids.[181,182] Oxaphospholanes **200** were prepared by the reaction of 4,4,4-trifluoro-3-hydroxy-1-phenylbutane-1-one with dichlorophosphines,[183] and oxaphospholanes **202** by the ring closure reaction of phosphonates **201**.[184]

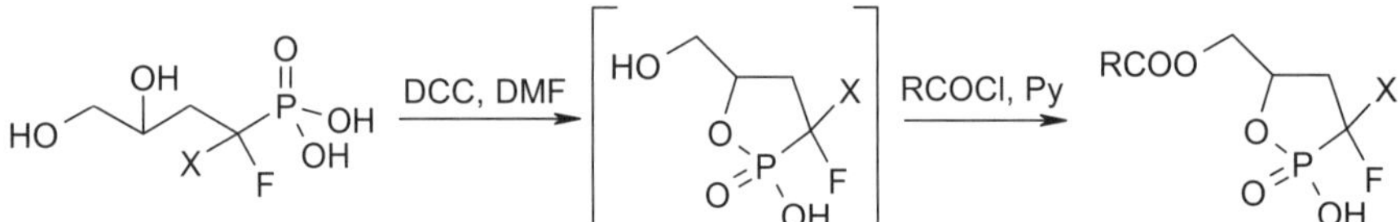

X = H, F; R = $C_{17}H_{33}$

199

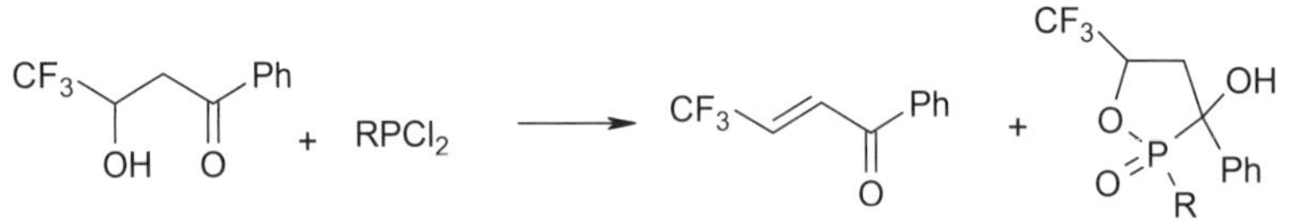

200

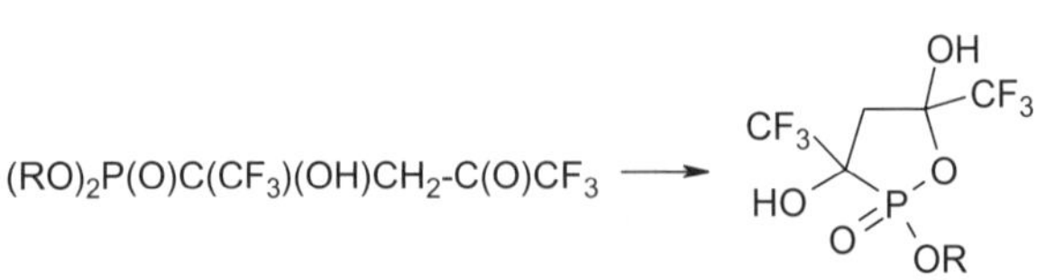

201 **202**

1,3,2-dioxaphospholanes and phospholenes **203** and **204**, containing pentacoordinated phosphorus atom, represent the most numerous group of fluoro-containing five-membered heterocycles with phosphorus atom. Main methods of their synthesis are the reactions of the compounds of three-valent three-coordinated phosphorus with fluorine-containing carbonyl compounds. The structure and chemical properties of these compounds were studied in details.[185–189]

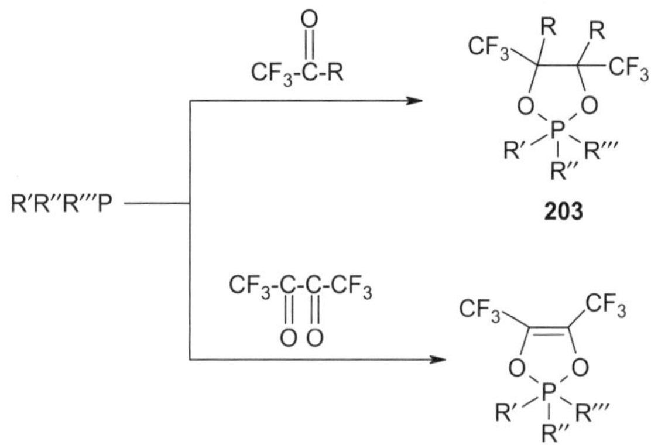

203

204

4.2 NMR SPECTRA AND STRUCTURE OF THE FLUORINE-CONTAINING FURAN AND THIOPHENE DERIVATIVES

Detailed analys of the ^{1}H and ^{19}F NMR spectra of furanosyl fluorides **205** was reported and can be found in ref.[190]

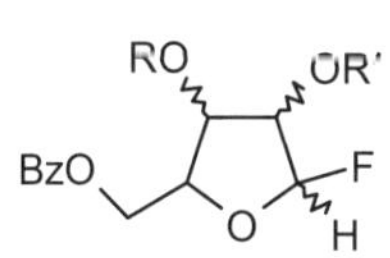

R = Ac, R′ = Bz
R = Bz, R′ = Ac
R = R′ = Bz

205

With the help of ^{1}H-^{19}F heteronuclear decoupling experiments relatively large ^{4}J coupling have been detected between *trans*-fluorine and H-4 substituents (5.5–7.9 Hz). These and other coupling constants give possibility to establish the ring conformation of the anomeric tri-*O*-benzoyl-D-ribofuranosyl fluorides.

The consideration of polyfluorofurans[60] NMR ^{19}F spectra allows to represent the fields of chemical fluorine nuclei shifts in 2- and 3-positions of furan cycle as −111–−137 and −145–−196 ppm, respectively. Chemical shifts of fluorine nuclei at C=C bond for perfluoro-2-oxolenes[76] lie between -105 ppm (C(2)–F) and -200 ppm (C(3)–F), and chemical shifts of fluorine atoms at sp^3 carbon atoms between -112 (C(4)–F) and -90 (C(5)–F ppm, respectively. Resonances of fluorines at C=C carbon bonds in perfluoro-3-oxolenes lie in the field −120–−150 ppm, but for fluorines connected to sp^3 carbon are shifted upfield (−55–−77 ppm).[60,76] Additional data on ^{19}F NMR spectra for perfluorooxolanes can be found in Refs 78, 191, 192. The NOESY and COSY NMR ^{19}F spectra have been used to assign all of the fluorine resonances in oxolane **206**.[193]

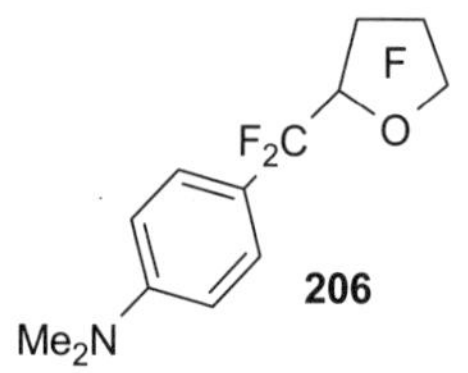

206

Chemical shifts of the fluorine nuclei in 2- and 5-positions of perfluorooxolanes located in −69–−89 ppm range, but much broader range of chemical shifts is observed for 3- and 4-isomers (−86–−210 ppm). On the other hand, chemical shifts of the CF_3 group, connected with furan cycle, varies insignificantly (−57–−68 ppm).[70]

According to the single crystal X-ray diffraction data, fluorine-containing oxolene cycle in spirocyclic compound **207** is almost flat.[194]

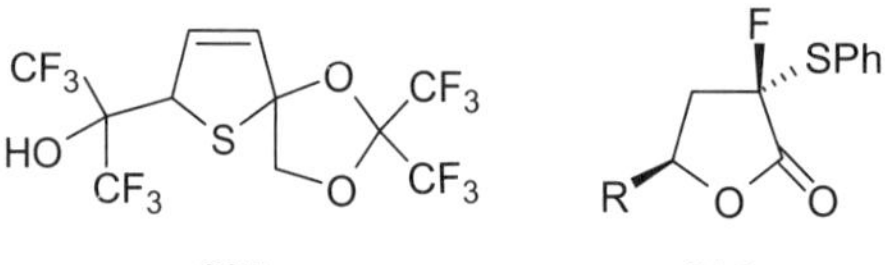

207 **208**

According to X-ray in compound **208** fluorine substituent is located in *syn* position relative to the *S*-phenyl group and phenylsulfanyl group occupies a pseudoaxial position.[89]

The derivatives of 3-*F*-thiophene-2-carboxylic acid,[195] 4-*F*-thiophene-2-carboxylic acid,[196] and 5-*F*-thiophene-2-carboxylic acid[197] were well characterized by ^{1}H, ^{13}C, and ^{19}F NMR spectroscopy. ^{1}H NMR was used to demonstrate the existence of a strong intramolecular hydrogen bond in the molecule of 3-*F*-thiophene-2-carboxylic acid amide **209**, that leads to the significant increase of the energy of rotation around C–N bond, which is comparable to the value reported for the compound **210**.[195]

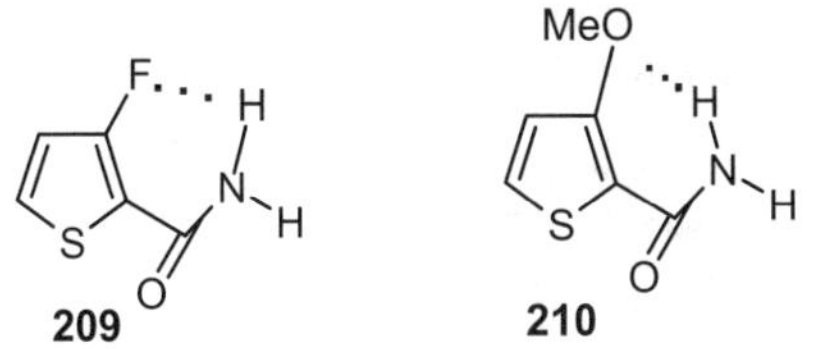

209 E = 14.4 kcal/mol

210 E = 15.2 kcal/mol

The electronic and thermochemical properties of the model monofluorinated phospholes **211–213** were studied using quantum chemical calculations.[179]

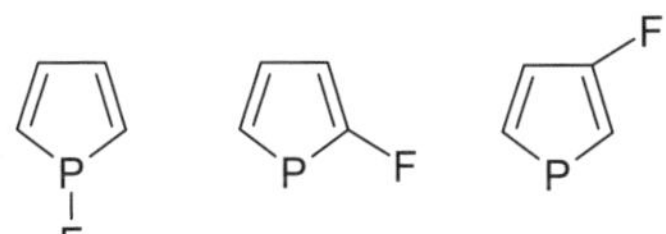

211 **212** **213**

The influence of the position of fluorine substituent on the ionization energy, the frontier orbital energy gap, the site of protonation and the basicity of phospholes were studied using this technique. Remarkably, compound **211** has the lowest LUMO–HOMO energy gap, which is reduced by almost 1 eV. A similar decrease could be expected for electron affinities of these compounds.

4.3 CHEMICAL PROPERTIES OF FLUORINATED FURANS AND THIOPHENES

Introduction of the fluorine atoms or perfluoroalkyl substituents to the molecules of furan or thiophene results in significant change of their chemical properties compared to hydrocarbon analogues.

4.3.1 Transformation Involving Heterocycle C-H Fragment

2-Lithio-3,4-bis(trifluoromethyl)furan easily prepared from 3,4-bis(trifluoromethyl) furan and BuLi was used for the synthesis of a wide variety of derivatives by the reactions with appropriate electrofile.[198]

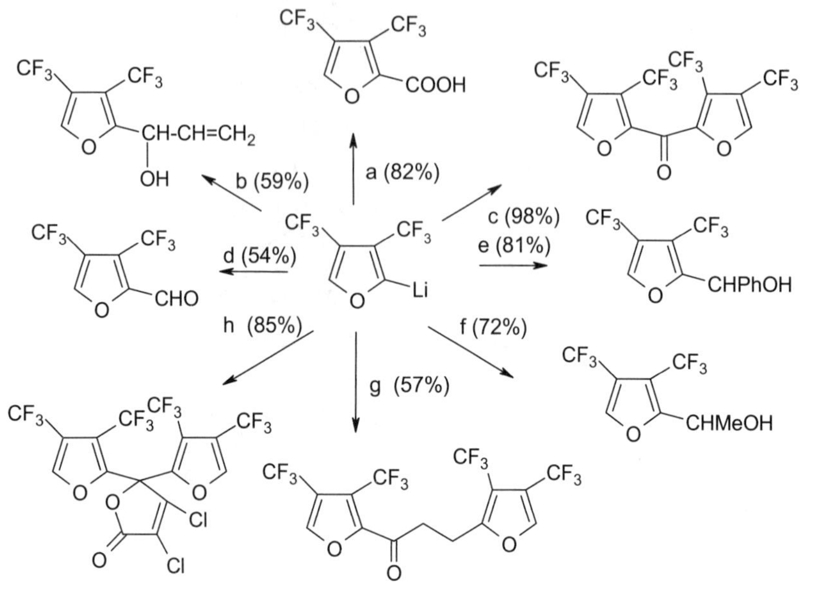

a, CO_2; b, CH_2=CHCHO; c, $ClCO_2Et$; d, DMF, H_2O; e, PhCHO; f, MeCHO;

g, CH_2=$CHCO_2Me$; h,

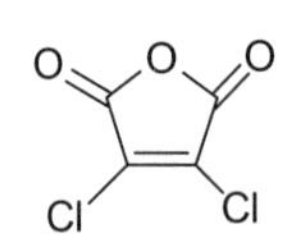

Using a similar procedure, ethynylfuran **214** and ethynylthiophene **215** were prepared.[46]

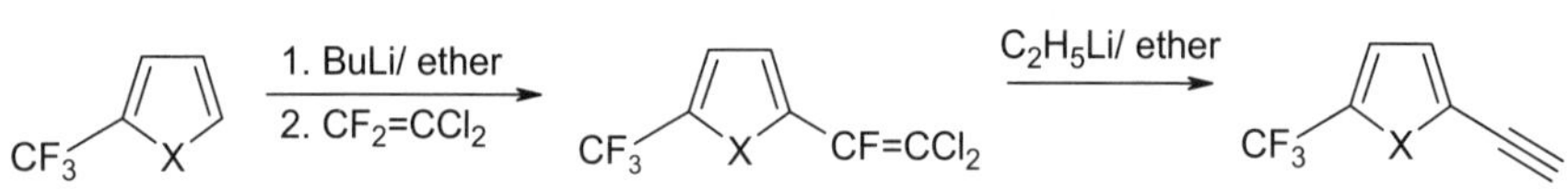

X = O, **214**
X = S, **215**

The synthesis of polythiophenes with alkyl and perfluoroalkyl substituents[135,199] is also based on the ability of fluorine-containing thiophenes with α-hydrogen to react with organometallic compounds.

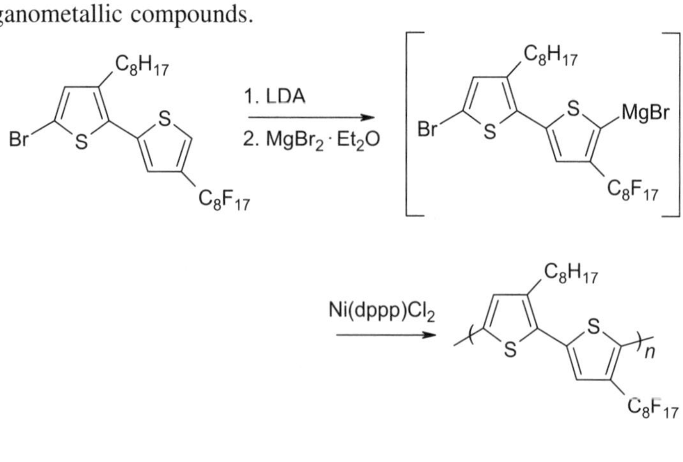

The presence of two trifluoromethyl groups in the molecule increases the acidity of α-proton of 3,4-bis(trifluoromethyl)furans. In the reaction of 2-ethynyl-3,4-bis(trifluoromethyl)furan **216** with ketones **217** both propyn-1-ols **218** and triarylmethanols **219** were obtained and converted into the corresponding salts by treatment with perchloric acid.[200]

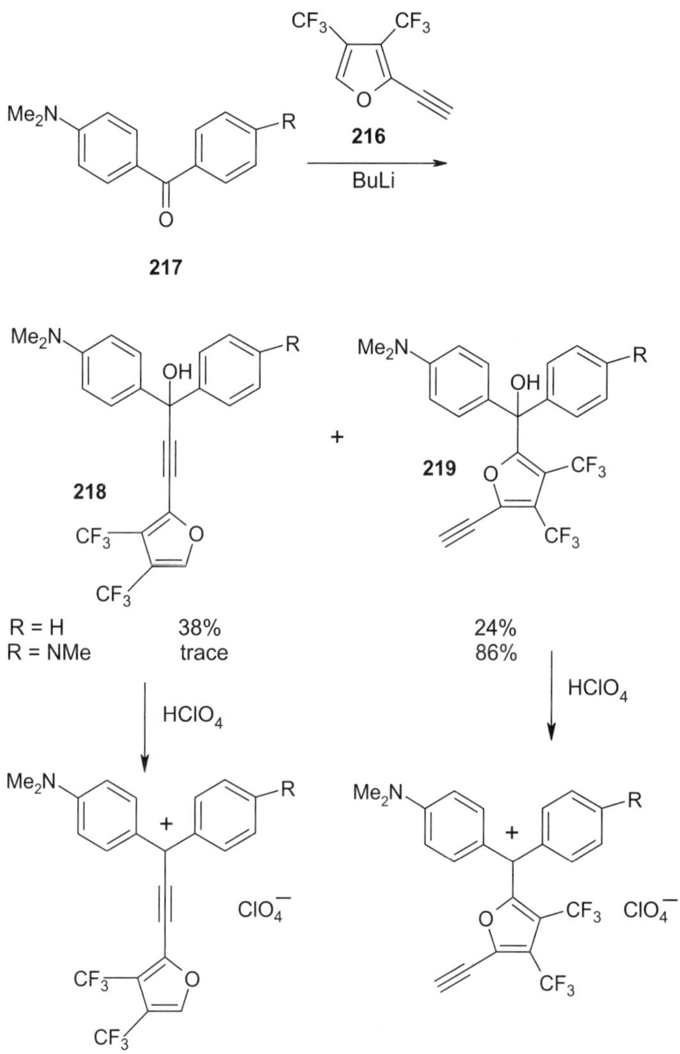

These results show that the acidity of hydrogen of the furan ring with two trifluoromethyl groups in a molecule is comparable to that of acetylenic hydrogen.

Exhaustive chlorination of 2-(2*H*-hexafluoropropyl)-oxalan at 70–210°C under UV irradiation gave pentachloro-3-oxolen **220**, quantitatively.[201]

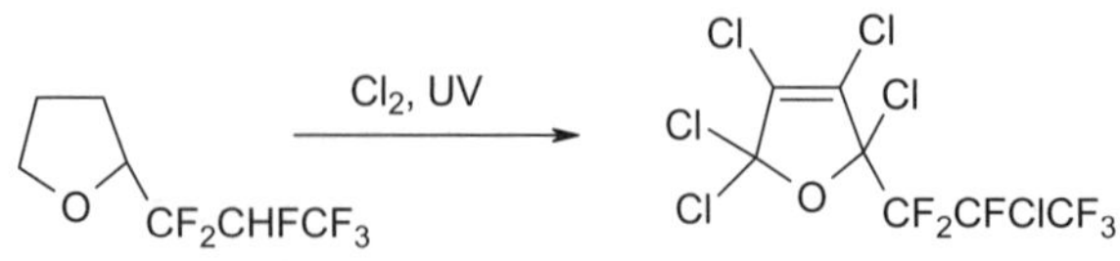

220

The KOH action on the partially fluorinated tetrahydrofurans results in dehydrofluorination. For example, 4*H*/3*H*-2-difluoromethyl-pentafluotetrahydrofuran **221** gives dihydrofurans **222** and **223**.[202]

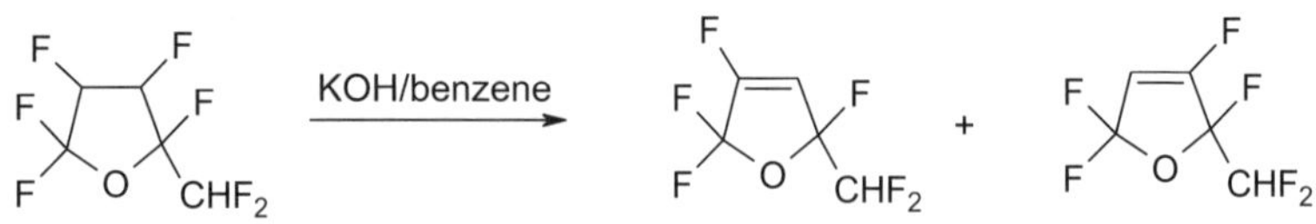

221 **222** 58% **223** 42%

4.3.2 Transformation of Heterocyclic C-Halogen Fragment

The reactions of α-halo-substituted fluorine-containing furans with nucleophilic reagents[203–205] are the most studied reactions proceeding with participation of halogen atoms. A fluorine atom in 2-fluoro-3-trifluoromethyl-5-aryl-furans **224** is replaced smoothly by various nucleophiles.[205] The similar reactions were reported for 2-fluoro-3-trifluoromethyl-5-aryl-thiophenes.[162]

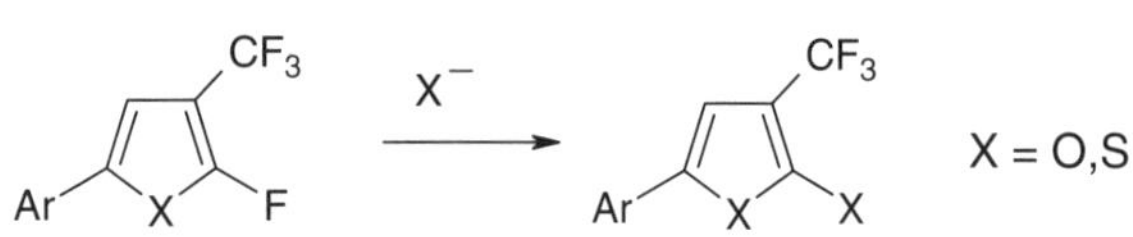

224

X^- = Ph,H,RO,RS,CN,R_2N

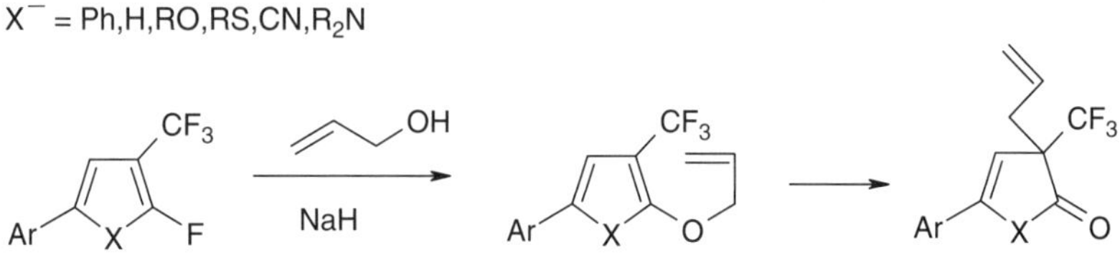

X = O, X = S **225**

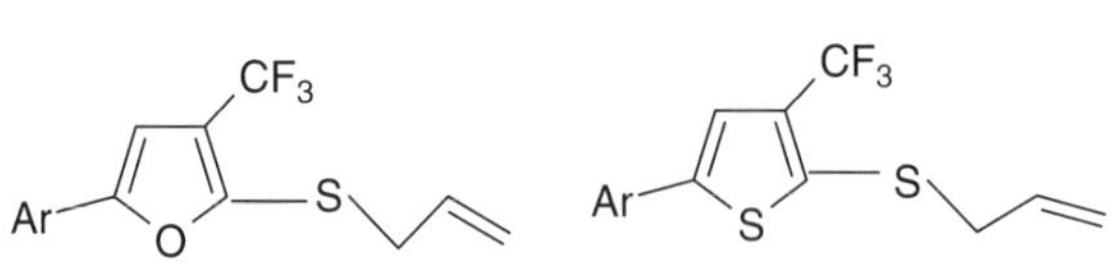

226 **227**

The substitution of the α-fluorine was used for preparation of trifluoromethyl-substituted butenolides **225**[206] and corresponding thiophene derivatives[207] via

Claisen rearrangement. On contrary, thioethers **226** and **227** synthesized by the reaction of the corresponding furan or thiophene with allyl mercaptane are thermally stable and no evidence of thio-Claisen rearrangement of these compounds was found.[207] The fluorine atom of 2-fluoro-3-trifluoromethylfurans and thiophenes can be replaced upon nucleophilic substitution with benzyl alcohols. Compounds **228** are susceptible to [1,3]- and [1,5]-benzyl group migration.[204]

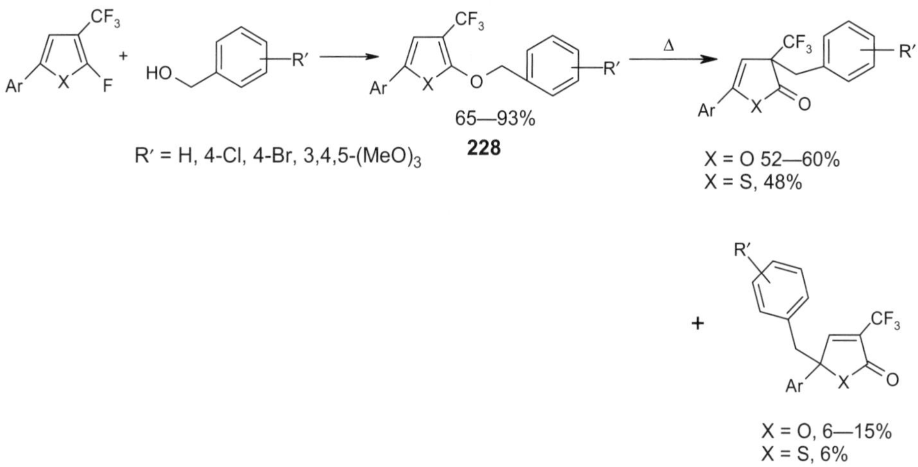

Fluorinated derivatives of tetrahydrofurans **229** and **230** were used to obtain the analogues of pyrimidinenucleosides[208] and aromatic β-C-nucleosides **231**.[209]

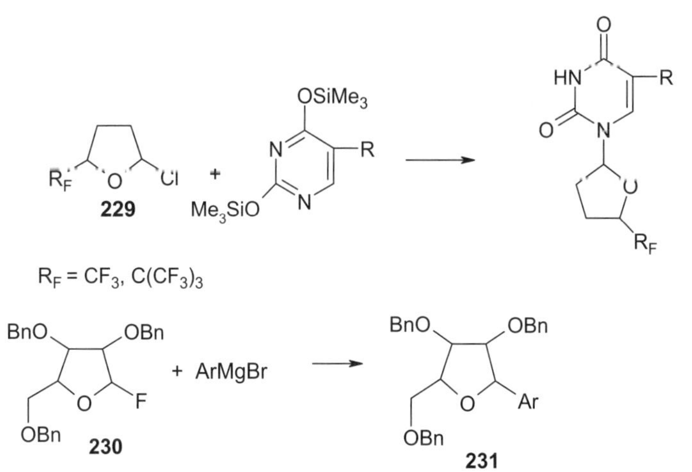

The substitution of the fluorine α-atoms by chlorine occurs at elevated temperature in the reaction of perfluoro-2,4-dialkyloxolanes with anhydrous aluminum chloride.[210]

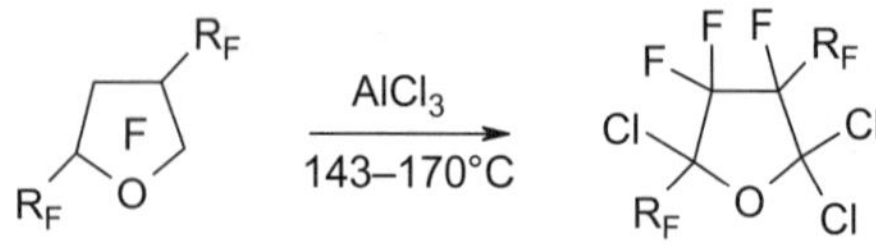

4.3.3 The Transformation Involving Endocyclic double Bond

The syntheses of perfluorinated epoxides by oxidation of perfluorodihydrofurans[211] with sodium hypochlorite and Diels–Alder reaction of 2,5-dialkyl-3,4-bis(trifluoromethyl)furan with hexafluoro-2-butyne[212] should me mentioned among the reactions proceeding with participation of endocyclic C=C bonds.

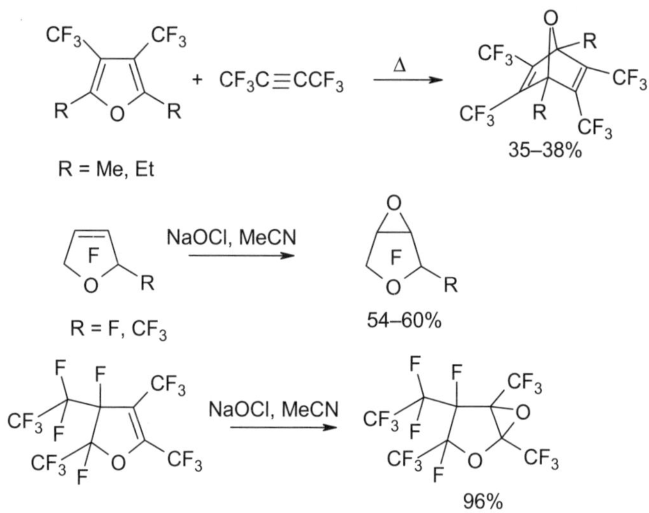

The last reaction was studied in details[213] using 3-fluorofuran-2(5*H*)-one **232** and 3,4-difluorofuran-2(5*H*)-ones **233** as dienophiles and phenylisobenzofuran or cyclopentadiene as dienes.

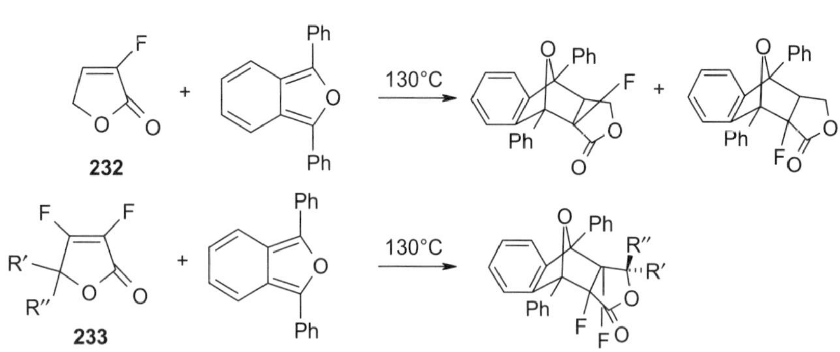

R′ = Me, R″ = H; R′ = $BrCH_2CH_2CH_2$, R″ = H; R′ = R″ = Me

Compound **232** gave *exo* cycloadducts, along with smaller amount of the *endo* isomer in reaction with both dienes. Reactions of **233** with diphenylisobenzofuran led exclusively to the *exo*-isomers, in contrast to the *endo* diastereoselectivity observed for nonfluorinated α,β-unsaturated carbonyl compounds.

3-Chloro-4-fluorothiophene-1,1-dioxide **234**, a new synthetically useful fluorodiene was prepared from commercially available 3-sulfolene. This compound reacts with different types of dienophiles: acetylenes, alkenes, furans quinine, and antracene giving the 3-fluoro-4-chloro-substituted arenes or cyclic chlorofluorodienes.[214]

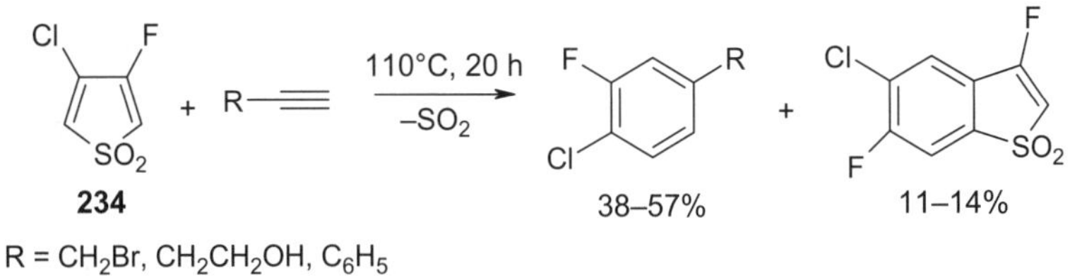

4.3.4 Transformation of Furan or Thiophene Cycle

There are several reactions, proceeding with transformation of the fluorine-containing furan cycle. Photochemical transformations of perfluoro-tri and tetra-methyl furans under irradiation by high pressure Hg lamp lead to cyclopropenylcarbonyl derivatives **235**.[203] No Dewar furan derivatives **236** were detected.

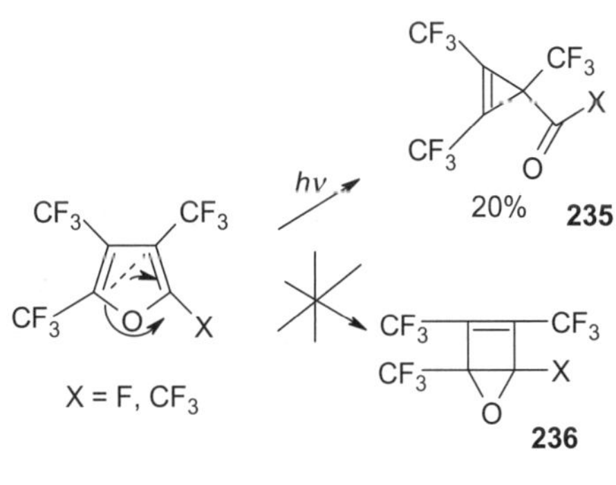

At the same time, a Dewar type isomer **237** was obtained by the photoreaction of tetrakis(trifluoromethyl)thiophene[215,216] and tetrakis(2-chlorotetrafluoroethyl)thiophene.[217] Gas-phase irradiation of 2,3-bis(trifluoromethyl)thiophene **238** by a low pressure Hg lamp gave a mixture of Dewar thiophenes **239**.[218] In contrast to **238**, irradiation of 2,5-bis(trifluoromethyl)thiophene **240** gave only 2,4-bis(trifluoromethyl)thiophene.[218]

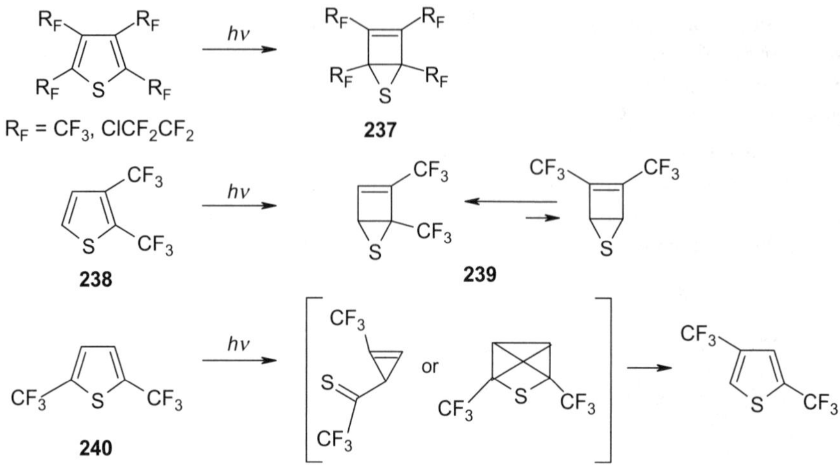

$R_F = CF_3, ClCF_2CF_2$

237 **238** **239** **240**

2-Ethoxy-3-perfluoroalkyl-tetrahydrofuran **241** reacted with α,ω-dithiols to give the corresponding fluorinated cyclic dithioacetales **242**.[219]

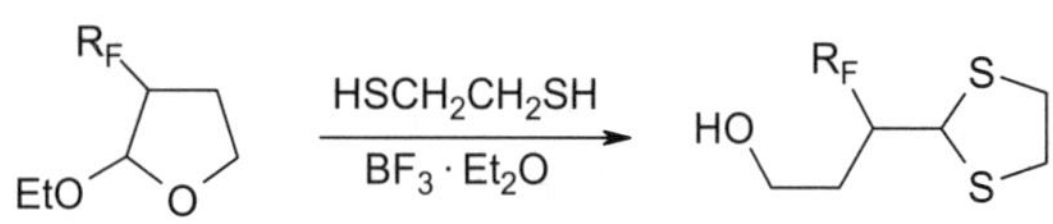

241 **242**

$R_F = Cl(CF_2)_4$

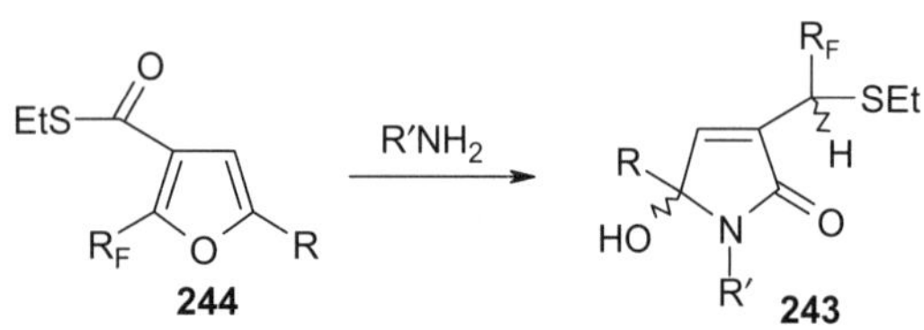

244 **243**

R′ = Bn, nC_5H_{11}, *i*-Pr, Ph, C_5H_4N, $(CH_2)_2OH$, $CHPhCH_2OH$, (CH_2)

$R_F = CF_3, H(CF_2)_2$

The original method for α,β-unsaturated lactams **243** obtained from 2-trifluoromethylfurans **244** was proposed.[220]

The presented data confirm that the chemistry of the five-membered fluorine-containing O-, S-, Se-, P-heterocycles are under intensive study. Alongside, it should be mentioned that many of the considered heterocycle types and reactions are represented only by a few examples. That is why investigation of fluorine-containing heterocycles is still the issue of the day for the modern organofluorine chemistry.

REFERENCES

1. Filer, R.; Kobayashi, Y. *Biomedical Aspects of Fluorine Chemistry*, Kodansha Elsevier Biomedical Press: Tokyo, New York, 1982.
2. Filer, R.; Kobayashi, Y.; Yagupolskii, L.M. *Studies in Organic Chemistry. 48. Organofluorine Compounds in Medicinal Chemistry and Biomedical Applications*, Elsevier: Amsterdam, 1993.
3. Dolbier, W.D. In *Chemistry of Organic Fluorine Compounds. II. A Critical Review. ACS Monograph*; Hudlicky, M.; Pavlath, A.E. Eds.; American Chemical Society: Washington, DC; Vol. *187*, 1995; p 797.
4. Furin, G.G. Fluorine-containing heterocycles. Part. IV. Electrochemical fluorination of heterocyclic compounds. *Adv. Heterocycl. Chem.* **2006**, *90*, 239–310.
5. Burger, K.; Wucherpfenning, U.; Brunner, E. Fluoro heterocycles with five-membered rings. *Adv. Heterocycl. Chem.* **1994**, *60*, 2–64.
6. Faulkner, D.J. Interesting aspects of marine natural products chemistry. *Tetrahedron* **1977**, *33*, 1421–1443.
7. Shepard, A.F.; Wilson, N.R.; Johnson, J.R. The simple halogen derivatives of furan. *J. Am. Chem. Soc.* **1930**, *52*, 2083–2086.
8. Nazarova, Z.N.; Babaev, Yu.A.; Nataliina, T.N. Investigation of the chemistry of 5-haloidfurans. *Zhurn. Obsch. Khim.* **1963**, *33*, 1431–1436.
9. Gorb, L.G.; Morozova, I.M.; Belenkii, L.I.; Abronin, I.A. Activity and selectivity in the reactions of electrophilic substitution of five-membered heterocyclic compounds. *Izv. AN SSSR, Ser. Khim.* **1983**, *4*, 828–835.
10. Pieniazek, S.N.; Houk, K.H. The origin of the halogen effect on reactivity and reversibility of Diels–Alder cycloadditions involving furan. *Angew. Chem., Int. Ed.* **2006**, *45*, 1442–1445.
11. John, I. G.; Radow, L. Conformations, stabilities, and charge distributions in 2- and 3-monosubstituted furans. An *ab initio* molecular orbital study. *J. Am. Chem. Soc.* **1978**, *100*, 3981–3991.
12. Novak, J. *Ab initio* thermochemistry of some halofurans and halothiophenes. *J. Org. Chem.* **2001**, *66*, 9041–9042.
13. Dvornikova, E.; Bechcicka, M.; Kamieńska-Trela, K.; Krówczyński, A. Synthesis and NMR studies of 2- and 3-fluorosubstituted five-membered heterocycles. *J. Fluorine Chem.* **2003**, *124*, 159–168.
14. Forrest, A.K.; O'Hanlon, P.J. The preparation and lithiation of bromofluorofurans via a novel fluorodecarboxylation. *Tetrahedron Lett.* **1995**, *36*, 2117–2118.
15. Ichikawa, J.; Wada, Y.; Fujiwara, M.; Sakoda, K. The nucleophilic 5-*endo-trig* cyclization of 1,1-difluoro-1-alkenes: ring-fluorinated hetero- and carbocycle synthesis and remarkable effect of the vinylic fluorines on the disfavored process. *Synthesis* **2002**, *13*, 1917–1936.
16. Hasegawa, M.; Ishii, H.; Fuchigami, T. Electroorganic synthesis under solvent-free conditions. Highly regioselective anodic monofluorination of cyclic ethers, lactones, and a cyclic carbonate. *Tetrahedron Lett.* **2002**, *43*, 1503–1505.
17. Miethchen, R.; Gabriel, T. Von Glycosofluoriden zu chiralen Kronenethern. *Chem. Ber.* **1993**, *126*, 115–123.
18. Miethchen, R.; Gabriel, T. Organofluorine compounds and fluorinating agents. Part II. Glycosyl fluorides from acetal-protected sugars. *J. Fluorine Chem.* **1994**, *67*, 11–15.

19. Bock, K.; Pedersen, C. Reaction of sugar esters with hydrogen fluoride. XIV. rearrangement of D-xylose and D-lyxose derivatives. *Acta Chem. Scand.* **1976**, *B30*, 727–732.
20. Bock, K.; Pedersen, C. Reaction of sugar esters with hydrogen fluoride. X. Derivatives of D-glucofuranose and D-mannofuranose. *Acta Chem. Scand.* **1972**, *26*, 2360–2366.
21. Ajmera, S.; Bapat, A.R.; Stephanian, E.; Danenberg, P.V. Synthesis and interaction with uridine phosphorylase of 5′-deoxy-4′,5-difluorouridine, a new prodrug of 5-fluorouracyl. *J. Med. Chem.* **1988**, *31*, 1094–1098.
22. Jenkins, I.D.; Verheyden, P.H.; Moffatt, J.G. 4′-Substituted nucleosides. 2. Synthesis of the antibiotic nucleocidin. *J. Am. Chem. Soc.* **1976**, *98*, 3346–3357.
23. Owen, G.R.; Verheyden, J.P.H.; Moffatt, J.G. 4′-Substituted nucleosides. 3. Synthesis of some 4′-fluorouridine derivatives. *J. Org. Chem.* **1976**, *41*, 3010–3017.
24. Morton, G.O.; Lancaster, J.E.; VanLear, G.E.; Fulmor, W. The structure of nucleocidin. III. (A new structure). *J. Am. Chem. Soc.* **1969**, *91*, 1535–1537.
25. Nguen, D.L.; Schlosser, M. Eine einfache Synthese von Rosenfuran und verwandten Verbindungen. *Helv. Chim. Acta* **1977**, *60*, 2085–2088.
26. Cantrell, A.S.; Engelhardt, P.; Högberg, M.; Jaskunas, S.R.; Johansson, N.G.; Jordan, C. L.; Kangasmetsä, J.; Kinnick, M.D.; Lind, P.; Morin, J.M.; Muesing, M.A.; Noreén, R.; Oberg, B.; Pranc, P.; Sahlberg, C.; Ternansky, R.J.; Vasileff, R.T.; Vrang, L.; West, S.J.; Zhang, H. Phenethylthiazolylthiourea (PETT) compounds as a new class of HIV-1 reverse transcriptase inhibitors. 2. Synthesis and further structure-activity relationship studies of PETT analogs. *J. Med. Chem.* **1996**, *39*, 4261–4274.
27. Xu, W.; Chen, Q.Y. A novel approach of cycloaddition of difluorocarbene to α,β-unsaturated aldehydes and ketones: synthesis of *gem*-difluorocyclopropyl ketones and -2-fluorofurans. *Org. Biomol. Chem.* **2003**, *1*, 1151–1156.
28. Sham, H.L.; Betebenner, D.A. A new and concise synthesis of 3-fluoro-2,5-disubstituted furans. *J. Chem. Soc., Chem. Commun.* **1991**, 1134–1135.
29. Moon, H.R.; Kim, H.O.; Jeong, L.S. Synthesis of 2′,3′-didehydro-2′,3′-dideoxy-2′-fluoro apionucleosides as potential antiviral agents. *J. Chem. Soc., Perkin Trans. 1* **2002**, 1800–1804.
30. Morikawa, T.; Nishiwaki, T.; Iitaka, Y.; Kobayashi, Y. Radical cyclization to fluorinated double bonds: 5-*exo* ring. closure of bromoacetals derived from fluoroallyl alcohols. *Tetrahedron Lett.* **1987**, *28*, 671–674.
31. Morikawa, T.; Nishiwaki, T.; Kobayashi, Y. Radical cyclization to the trifluoromethyl-substituted double bond: regioselectivity and tandem cyclization. *Tetrahedron Lett.* **1989**, *30*, 2407–2410.
32. Girard, Y.; Atkinson, J.G.; Bélanger, P.C.; Fuentes, J.J.; Rokach, J.; Rooney, C.S.; Remy, D.C.; Hunt, C.A. Synthesis, chemistry, and photochemical substitutions of 6,11-dihydro-5*H*-pyrrolo[2,1-*b*][3]benzazepin-11-ones. *J. Org. Chem.* **1983**, *48*, 3220–3234.
33. Zhou, Q.L.; Huang, Y.Z. Direct fluoroalkylation of aromatic compounds catalyzed by tetrakis(triphenylphosphine)nickel. *J. Fluorine Chem.* **1989**, *43*, 385–392.
34. Naumann, D.; Kischkewitz, J. Trifluoromethylierungsreactionen von CF_3I, $Te(CF_3)_2$, $Sb(CF_3)_3$, $Hg(CF_3)_2$, und $Cd(CF_3)_2{\cdot}D$ mit furan, thiophen, pyrrol und *p*-benzochinon. *J. Fluorine Chem.* **1990**, *46*, 265–281.
35. Yoshida, M.; Morinaga, Y.; Ueda, M.; Kamigata, N.; Oyoda, M. Conventient preparation of (chlorodifluoromethyl)arenes. Useful precursors for aryldifluoromethyl radicals. *Chem. Lett.* **1992**, 227–230.

36. Kamigata, N.; Ohtsuka, T.; Shimizu, T. Regioselective perfluoroalkylation of heteroaromatic compounds by perfluoroalkanesulfonyl chloride in the presence of a ruthenium(II) complex. *Phosphorus, Sulfur Silicon Relat. Elem.* **1994**, *95–96*, 491–492.

37. Kamigata, N.; Ohtsuka, T.; Fukushima, T.; Yoshida, M.; Shimizu, T. Direct perfluoroalkylation of aromatic and heteroaromatic compounds with perfluoroalkanesulfonyl chlorides catalysed by a ruthenium(II) phosphine Complex. *J. Chem. Soc., Perkin Trans. 1* **1994**, 1339–1346.

38. Murakami, S.; Kim, S.; Ishii, H.; Fuchigami, T. Aromatic substitution with photochemically generated difluoromethyl radicals bearing electron-withdrawing group. *Synlett* **2004**, *5*, 815–818.

39. Murakami, S.; Ishii, H.; Tajima, T.; Fuchigami, T. Photochemical substitution of olefins and aromatic compounds with difluoromethyl radicals bearing ester and phosphonate groups. *Tetrahedron* **2006**, *62*, 3761–3769.

40. Murakami, S.; Fuchigami, T. Synthesis of precursors of *gem*-difluorodiols and amino alkohols using electrochemical and photochemical reactions. *Synlett* **2006**, *7*, 1015–1020.

41. Popkov, S.V.; Kuzenkov, A.V. Study of trifluoromethyl-substituted heteroaromatic aldehydes. *Izv. Akad. Nauk, Ser. Khim.* **2005**, *7*, 1623–1625.

42. Tanabe, Y.; Matsuo, N.; Ohno, N. Direct perfluoroalkylation including trifluoromethylation of aromatics with perfluoro carboxylic acids mediated by xenon difluoride. *J. Org. Chem.* **1988**, *53*, 4582–4585.

43. Bucci, R.; Laguzzi, G.; Pompili, M.L.; Speranza, M. Gas-phase heteroaromatic substitution. 12. Reaction of free trifluoromethyl cation with simple five-membered heteroarenes in the gas phase. *J. Am. Chem. Soc.* **1991**, *113*, 4544–4550.

44. Chen, G.J. Polufluoroalkylation of bromoheterocyclic compounds via perfluoroalkylcopper intermediates. *J. Fluorine Chem.* **1990**, *46*, 137–159.

45. Lyalin, V.V.; Grigorash, R.V.; Alekseeva, L.A.; Yagupolskii, L.M. Fluorination of furan tetracarboxylic acid with sulfur tetrafluoride. *Zhurn. Org. Khim.* **1975**, *11*, 460–461.

46. Nishida, M.; Fujii, S.; Aoki, T.; Hayakawa, Y.; Muramatsu, H.; Morita, T. Synthesis and polimerization of ethynylthiophenes and ethynylfurans containing trifluoromethyl groups. *J. Fluorine Chem.* **1990**, *46*, 445–459.

47. Bégué, J.P.; Bonnet-Delpon, D.; Dogbeavou, R.; Ourévitch, M. Trifluoromethylated furans via iodocyclisation of γ-unsaturated ethyl trifluoroacetoacetates. *J. Chem. Soc., Perkin Trans. 1* **1993**, 2787–2791.

48. Bouillon, J.P.; Hénin, B.; Huot, J.F.; Portella, C. 1,1-Bis(ethylsulfanyl) perfluorobut-1-ene as starting material for the synthesis of substituted 2-trifluoromethylfurans and -pyrroles. *Eur. J. Org. Chem.* **2002**, 1556–1561.

49. Haas, A.; Klare, C.; Krächter, H.U. (Perhalogenmethyltio)heterocyclen. XIX. Darstellung und Eigenschaften von Tetrakis(trifluormethylthio)thiophen, Tetrakis(trifluormethylthio-1,2-dithiin, 2,3,4-Tris(Trifluormethylthio)-5-trifluormethylthiophen und 1-Chlor-terakis (trifluormethylthio)pyrrol. *J. Heterocycl. Chem.* **1985**, *22*, 1631–1633.

50. Jullien, J.; Pechine, J.M.; Perez, F.; Piade, J.J. Flash vacuum thermolysis of β-keto-trimethylsilyl-enol-ethers. Synthesis of allenic and furanic derivatives. *Tertrahedron* **1982**, *38*, 1413–1416.

51. Linderman, R.J.; Jamois, E.A.; Tennyson, S.D. Synthesis of and analysis of thiol additions to β-alkyl-α, β-unsaturated trifluoromethyl ketones. *J. Org. Chem.* **1994**, *59*, 957–962.

52. Aristov, S.A.; Vasil'ev, A.V.; Rudenko, A.P. Oxidation of aryl ethynyl ketones in the system CF_3CO_2H–CH_2Cl_2–PbO_2. *Zhurn. Org. Khim.* **2006**, *42*, 74–80.

53. Amato, C.; Calas, P. A novel leiparvarin analogue possessing a fluorinated segment. *Synthesis* **2003**, *17*, 2709–2712.

54. Plantier-Royon, R.; Portella, C. *C*-Difluoromethylene-containing, *C*-trifluoromethyl and *C*-perfluoroalkyl carbohydrates. *Carbohydr. Res.* **2000**, *327*, 119–146.

55. Nezis, A.; Fayn, J.; Cambon, A. Obtention de 3-*F*-alkyl-4-ethoxycarbonyl furanes. *J. Fluorine Chem.* **1991**, *53*, 297–299.

56. Kubota, T.; Aoyagi, R.; Sando, H.; Kawasumi, M.; Tanaka, T. Preparation of trifluoromethylated compounds by anodic oxidation of 3-hydroxy-2-trifluoromethylpropionic. *Acid. Chem. Lett.* **1987**, 1435–1438.

57. Zhang, X.; Qing, F.L.; Yu, Y. Synthesis of 2′,3′-dideoxy-2′-trifluoromethylnucleosides from α-trifluoromethyl-α,β-unsaturated ester. *J. Org. Chem.* **2000**, *65*, 7075–7082.

58. Wu, F.H.; Huang, B.N.; Lu, L.; Huang, W.Y. Addition reactions of dibromodifluoromethane promoted by sulfinatodehalogenation reagents. *J. Fluorine Chem.* **1996**, *80*, 91–94.

59. Jean-Baptiste, L.; Yemets, S.; Legay, R.; Lequeux, T. Synthesis of 2,3-*trans* disubstituted tetrahydrofurans through sequential xanthate radical addition–substitution reactions. *J. Org. Chem.* **2006**, *71*, 2352–2359.

60. Burdon, J.; Chivers, G.E.; Tatlow, J.C. Highly fluorinated heterocycles. Part III. Preparation and reactions of some polyfluorofurans. *J. Chem. Soc. C* **1970**, 2146–2151.

61. Lyalin, V.V.; Grigorash, R.V.; Alekseeva, L.A.; Yagupolskii, L.M. Interaction of the derivatives of furan 3,4-dicarboxylic acid with sulfur tetrafluoride in hydrogen fluoride media. *Zhurn. Org. Khim.* **1984**, *20*, 846–849.

62. Burger, K.; Helmreich, B. Synthesis of 3-trifluoromethylfurans from β,β-bis(trifluoromethyl) α,β-unsaturated ketones and tin(II) chloride. *J. Chem. Soc., Chem. Commun.* **1992**, 348–349.

63. Kuroboshi, M.; Shinozaki, T.; Ishihara, T.; Ando, T. A convenient synthesis of fluorine-containing highly-substituted furans through new fluoride ion-catalyzed reaction of 1-alkyl-*F*-1-alkenyl phosphates. *Chem. Lett.* **1987**, 1621–1622.

64. Chambers, R.D.; Roche, A.J.; Rock, M.H. Polyhalogenated heterocyclic compounds. Part 41. Cycloaddition reactions involving hexafluorobut-2-yne and 1,1,1,2,4,4,4-heptafluorobut-2-ene. *J. Chem. Soc., Perkin Trans. 1* **1996**, 1095–1100.

65. Nishida, M.; Hayakawa, Y.; Matsui, M.; Shibata, K.; Muramatsu, H. Synthesis of 2,5-diethyl-3,4-bis(trifluoromethyl)furan and its derivatives. *J. Heterocycl. Chem.* **1991**, *28*, 225–229.

66. Kubota, T.; Ishii, T.; Minamikawa, H.; Yamaguchi, S.; Tanaka, T. Preparation of trifluoromethylated compounds via anodic oxidation of 4,4,4-trifluoro-3,3-dimethoxybutyric. *Acid. Chem. Lett.* **1988**, 1987–1990.

67. Dědek, V.; Liška, F.; Fikar, J.; Chvátal, Z.; Pohořelský, L. Synthesis of fluorochloroalkanols and their cyclisation to fluorinated oxolanes. *Collect. Czech. Chem. Commun.* **1975**, *40*, 1008–1020.

68. Dmovski, W.; Koliński, R.A. Reactions of sulfur tetrafluoride with calboxylic acids. Part IV. Reactions with alkane–dicarboxylic acids. *Polish J. Chem.* **1978**, *52*, 71–85.

69. Lyalin, V.V.; Grigorash, R.V.; Alekseeva, L.A.; Yagupolskii, L.M. Mono-and bis(trifluoromethyl) furans. *Zhurn. Org. Khim.* **1975b**, *11*, 1086–1090.

70. Mahaffy, P.G.; Visser, D.; Torres, M.; Bourdelande, J.L.; Strausz, O.P. The gas-phase photolysis of perfluoro α-diazo ketones: furan formation and the involvement of transient oxirenes. *J. Org. Chem.* **1987**, *52*, 2680–2684.
71. Briscoe, M.W.; Chambers, R.D.; Mullins, S.J.; Nakamura, T.; Drakesmith, F.G. A synthesis of novel perfluorodienes. *J. Chem. Soc., Chem. Commun.* **1990**, 1127–1128.
72. Bartlett, S.; Chambers, R.D.; Kelly, N.M. A novel furan synthesis. *Tetrahedron Lett.* **1980**, *21*, 1891–1892.
73. England, D.C. Fluoroketenes. 10. Synthesis and chemistry of a perfluoroacylketene and a related perfluorovinyl ketone. *J. Org. Chem.* **1981**, *46*, 147–153.
74. Coe, P.L.; Ray, N.C. Reactions of tetrafluoroethene oligomers. Part XIII. Reactions of a perfluorinated dihydrofuran; perfluoro-4-ethyl-2,3,4,5-tetramethyl-4,5-dihydrofuran. *J. Fluorine Chem.* **1991**, *53*, 15–31.
75. Zapevalova, T.B.; Plashkin, V.S.; Selischev, B.N.; Bil'dinov, K.N.; Scherbakova, M.S. Synthesis and properties of perfluorooxolan-2-carboxylic acid. *Zhurn. Org. Khim.* **1977**, *13*, 2573–2574.
76. Filyakova, T.I.; Kodess, M.I.; Zapevalov, A.Y.; Saloutin, V.I. Synthesis of perfluoro- and 2-trifluoromethyl pentafluorooxolenes and epoxides on their base. *Zhurn. Org. Khim.* **2003**, *39*, 1073–1078.
77. Chepik, S.D.; Cherstkov, V.F.; Mysov, E.I.; Aerov, A.F.; Galakhov, M.V.; Sterlin, S.R.; German, L.S. Electrophilic isomerization of fluoroaliphatic oxygen-containing compounds. *Izv. Akad. Nauk SSSR, Ser. Khim.* **1991**, *11*, 2611–2618.
78. Modena, S.; Calini, P.; Gregorio, G.; Moggi, G. Synthesis of perfluoroethers by liquid phase fluorination. *J. Fluorine Chem.* **1988**, *40*, 349–357.
79. Chambers, R.D.; Spink, R.C.H. Microreactors for elemental fluorine. *Chem. Commun.* **1999**, 883–884.
80. Chambers, R.D.; Grievson, B.; Drakesmith, F.G.; Powell, R.L. Free radical chemistry. Part 5. A new approach to the synthesis of perfluorinated ethers. *J. Fluorine Chem.* **1985**, *29*, 323–339.
81. Takubo, S.; Sekiya, A. Monofluorination of fluorinated oxolanes with high-valency metal fluorides. *J. Fluorine Chem.* **1998**, *87*, 105–109.
82. Ryabinin, N.A.; Kolenko, I.P.; Lundin, B.N. Haloidfurans. II. Electochemical fluorination of α-substituted tetrahydrofurans. *Zhurn. Obsch. Khim.* **1967**, *37*, 1229–1232.
83. Abe, T.; Tamura, M.; Sekiya, A. Electrochemical fluorination of several esters derived from oxolane-2-yl-carboxylic acid, oxolane-2-yl-methanol and oxane-2-yl-methanol. *J. Fluorine Chem.* **2005**, *126*, 325–332.
84. Abe, T.; Nagase, S.; Baba, H. The electrochemical fluorination of aliphatic primary monohydric alkohols and aldehydes. *Bull. Chem. Soc. Jpn.* **1976**, *49*, 1888–1892.
85. Abe, T.; Kodaira, K.; Baba, H.; Nagase, S. The electrochemical fluorination of α-alkyl-substituted carboxylic acids. *J. Fluorine Chem.* **1978**, *12*, 1–25.
86. Abe, T.; Baba, H.; Okuhara, K.; Fukaya, H. Electrochemical fluorination of several methyl and/or ethyl esters of morpholino-substituted carboxylic acids. *J. Fluorine Chem.* **2001**, *111*, 115–128.
87. Sasaki, Y.; Ebara, R.; Nanbu, N.; Takehara, M.; Ue, M. Direct fluorination of γ-butyrolactone. *J. Fluorine Chem.* **2001**, *108*, 117–120
88. Patrick, T.B.; Poon, Y.F. Preparation of fluorolactones from the reaction of γ-Ketoacids with diethylaminosulfur trifluoride. *Tetrahedron Lett.* **1984**, *25*, 1019–1022.

89. Motherwell, W.B.; Greaney, M.F.; Tocher, D.A. Fluorination of α-phenylsulfanyl esters using difluoroiodotoluene. *J. Chem. Soc., Perkin Trans. 1* **2002**, *24*, 2809–2815.

90. Osipov, S.N.; Kolomiets, A.F.; Fokin, A.V. [2 + 4] Cycloaddition reactions of trifluoroacetylimine of trifluoropyruvate. *Izv. Akad. Nauk SSSR, Ser. Khim.* **1991**, 1130–1133.

91. Paleček, J.; Kvíčala, J.; Paleta, O. Fluorinated butanolides and butenolides Part 9. Synthesis of 2-(trifluoromethyl)butan-4-olides by Wittig reaction using methyl 3,3,3-trifluoropyruvate. *J. Fluorine Chem.* **2002**, *113*, 177–183.

92. Haddad, M.; Molines, H.; Wakselman, C. Preparation of β-Trifluoromethyl-γ-butyrolactones via Claisen Rearrangement. *Synthesis* **1994**, 167–169.

93. Ogu, K.; Matsumoto, S.; Akazome, M.; Ogura, K. Novel Synthesis of α-Trifluoromethylated α-Amino Acid Derivatives from γ-Hydroxy-α-fluoro-α-trifluoromethyl Carboxamides. *Org. Lett.* **2005**, *7*, 589–592.

94. Watanabe, S.; Sakai, Y.; Takeda, M.; Kitazume, T.; Yamazaki, T. Synthesis of optically active γ-lactones and dopants for ferroelectric liquid crystals possessing a trifluoromethyl group. *J. Fluorine Chem.* **1994**, *67*, 149–152.

95. Andreev, V.G.; Sizov, A.Y.; Kolomiets, A.F. 2-(Methylthio) pentafluoro-propene. *Izv. Akad. Nauk, Ser. Khim.* **1994**, *1*, 80–84.

96. Andreev, V.G.; Kolomiets, A.F.; Fokin, A.V. Synthesis and properties of 2-trifluoromethyl-substituted pent-4-en-, pent-3,4-dienic acids and their esters. *Izv. Akad. Nauk SSSR, Ser. Khim.* **1991**, *12*, 2805–2809.

97. Paleta, O.; Paleček, J.; Dolenský, B. Fluorinated butanolides and butenolides. Part 8. 2-(Trifluoromethyl)butan-4-olides by synthesis from methyl 3,3,3-trifluoropyruvate as building block. *J. Fluorine Chem.* **2001**, *111*, 175–184.

98. Konno, T.; Kitazume, T. Synthesis of trifluoromethylated compounds with four consecutive asymmetric centers via sequential [3,3]-Ireland–Claisen rearrangement and iodolactonization. *Tetrahedron: Asymmetry* **1997**, *8*, 223–230.

99. Ogu, K.; Akazome, M.; Ogura, K. A novel reaction of allylic alkohols with hexafluoropropene-diethylamine adduct (PPDA) to form 2-fluoro-2-trifluoromethyl-4-alkenamide. *J. Fluorine Chem.* **2003**, *124*, 69–80.

100. Bravo, P.; Arnone, A.; Bandiera, P.; Bruché, L.; Ohashi, Y.; Ono, T.; Sekine, A.; Zanda, M. Stereoselective synthesis of enantiomerically pure β-fluoroalkyl γ-butyrolactones by sulfoxide-directed lactonization. *Eur. J. Org. Chem.* **1999**, 111–115.

101. VanVleck, R.T. Preparation of fluorothiophene. *J. Am. Chem. Soc.* **1949**, *71*, 3256–3257.

102. Chambers, R.J.; Marfat, A. An improved synthesis of 5-fluorothiophene-2-carboxylic acid. *Synth. Commun.* **2000**, *30*, 3629–3632.

103. Cerichelli, G.; Crestoni, M.E.; Fornarini, S. The fluorination of simple five-membered heteroatomic compounds with elemental fluorine. *Gazz. Chim. Ital.* **1990**, *120*, 749–755.

104. Schuetz, R.D.; Taft, D.D.; O'Brien, J.P.; Shea, J.L.; Mork, H.M. Fluorinated heterocycles. *J. Org. Chem.* **1963**, *28*, 1420–1421.

105. Higashiguchi, K.; Matsuda, K.; Asano, Y.; Murakami, A.; Nakamura, S.; Irie, M. Photochromism of dithienylethenes containing fluorinated thiophene rings. *Eur. J. Org. Chem.* **2005**, 91–97.

106. Banks, R.E.; Du Boisson, R.A.; Tsiliopoulus, E. Electrophilic fluorinatuion with *N*-fluoroquinuclidinium fluoride. *J. Fluorine Chem.* **1986**, *32*, 461–466.

107. Kassmi, A.E.; Fache, F.; Lemaire, M. A convenient synthesis of 3-fluorothiophene. *Synth. Commun.* **1994**, *24*, 95–101.

108. Kiryanov, A.A.; Seed, A.J.; Sampson, P. Ring fluorinated thiophenes: applications to liquid crystal synthesis. *Tetrahedron Lett.* **2001**, *42*, 8797–8800.

109. Kobarfard, F.; Kauffman, J.M.; Boyko, W.J. Attempted synthesis of aminofluorothiophenes. *J. Heterocycl. Chem.* **1999**, *36*, 1247–1251.

110. Chonan, T.; Takahashi, K. The synthesis of difluoro and dimethyl derivatives of 2,6-bis (dicyanomethylene)-2,2-dihydro-4*H*-cyclopenta[2,1-*b*:3,4-*b*′]-dithiophen-4-one (CPDT-TCNQ) and the conducting properties of the metallic salts based on the dimethyl derivative. *Bull. Chem. Soc. Jpn.* **2004**, *77*, 1487–1497.

111. Dmowski, W.; Manko, V.A.; Nowak, I. 3-Chloro-4-fluorothiophene-1,1-dioxide. A new synthetically useful fluorodiene. *J. Fluorine Chem.* **1998**, *88*, 143–151.

112. Uneyama, K.; Kanai, M. Benzeneselenyl fluoride equivalent in sity generated with XeF_2–$(PhSe)_2$ in CH_2Cl_2 for fluoroselenenylation of olephines. *Tetrahedron Lett.* **1990**, *31*, 3583–3586.

113. Jeong, L.S.; Moon, H.R.; Yoo, S.J.; Lee, S.N.; Chun, M.W.; Lim, Y.H. Ring contraction and rearrangement of 4-thiofuranose derivatives observed during DAST fluorination in the synthesis of L-2-"up"fluoro-4′-thiothymidine. *Tetrahedron Lett.* **1998**, *39*, 5201–5204.

114. Jeong, L.S.; Yoo, S.J.; Moon, H.R.; Kim, Y.H.; Chun, M.W. Participation of sulfur occurred during Mitsunobu reaction: synthesis of novel isodideoxythionucleosides. *J. Chem. Soc., Perkin Trans. 1* **1998**, 3325–3326.

115. Cowell, A.B.; Tamborski, C. Fluoroalkylation of aromatic compounds. *J. Fluorine Chem.* **1981**, *17*, 345–356.

116. Yoshida, M.; Sasage, S.; Kamigata, N.; Sawada, H.; Nakayama, M. Nucleophilic displacement of bis(perfluoroalkanoyl) peroxides with perfluoroalkanoate. *Bull. Chem. Soc. Jpn.* **1989**, *62*, 2416–2418.

117. Yoshida, M.; Yoshida, T.; Kamigata, N.; Kobayashi, M. Heptafluoropropylation of various substituted thiophenes with bis(heptafluorobutyryl) peroxide. Preparation of 3-heptafluoropropylthiophene. *Bull. Chem. Soc. Jpn.* **1988**, *61*, 3549–3552.

118. Murakami, S.; Ishii, H.; Fuchigami, T. Electrosynthesis of ethyl α,α-difluoro-α-(phenylseleno)acetate and its photochemical synthetic application. *J. Fluorine Chem.* **2004a**, *125*, 609–614.

119. Facchetti, A.; Yoon, M.H.; Stern, C.L.; Hutchison, G.R.; Ratner, M.A.; Marks, T.J. Building blocks for *N*-type molecular and polymeric electronics. Perfluoroalkyl- versus alkyl functionalized oligothiophenes (*n*Ts; $n = 2$–6). Systematic synthesis, spectroscopy, electrochemistry, and solid-state organization. *J. Am. Chem. Soc.* **2004**, *126*, 13480–13501.

120. Facchetti, A.; Deng, Y.; Wang, A.; Koide, Y.; Sirringhaus, H.; Marks, T.J.; Friend, R.H. Tuning the Semiconducting Properties of sexithiophene by α,ω-substitution-Diperfluorohexylsexthiophene: The First Sexthiophene for Thin-Film Transistors. *Angew. Chem., Int. Ed.* **2000**, *39*, 4547–4551.

121. Paratian, J.M.; Sibille, S.; Périchon, J. An efficient electrochemical trifluoromethylation of aromatic halides with bromotrifluoromethane and a sacrificial copper anode. *J. Chem. Soc., Chem. Commun.* **1992**, 53–54.

122. McLoughlin, V.C.R.; Thrower, J. A route to fluoroalkyl-substituted aromatic compounds involving fluoroalkylcooper intermediates. *Tetrahedron* **1969**, *25*, 5921–5940.

123. Carr, G.E.; Chambers, R.D.; Holmes, T.F. Sodium perfluoroalkane carboxylates as sources of perfluoroalkyl groups. *J. Chem. Soc., Perkin Trans. 1* **1988**, 921–926.

124. Singh, R.P.; Shreeve, J.M. One-pot route to new α,α-difluoroamides and α-ketoamides. *J. Org. Chem.* **2003**, *68*, 6063–6065.

125. Singh, R.P.; Twamley, B.; Shreeve, J.M. Polyfluoroether derivatives via nucleophilic fluorination of glyoxal hydrates with deoxofluor. *J. Org. Chem.* **2002**, *67*, 1918–1924.

126. Singh, R.P.; Shreeve, J.M. Concentration-dependent reactions of deoxofluor with arylglyoxal hydrates: a new route to polyfluoro ethers. *Org. Lett.* **2001**, *3*, 2713–2715.

127. Guan, H.P.; Luo, B.H.; Hu, C.M. An effective synthesis of 2-trifluoromethyl- or 2-(1,1-difluoroalkyl)thiophenes. *Synthesis* **1997**, 461–464.

128. Greif, D.; Eilitz, U.; Pulst, M.; Riedel, D.; Wecks, M. Perfluoroalkylsubstituierte β-chlorvinylalgehyde: eine neue Klasse von *buildings blocs* zur Synthese fluorierter Heterocyclen. *J. Fluorine Chem.* **1999**, *94*, 91–103.

129. Huang, X.T.; Long, Z.Y.; Chen, Q.Y. Fluoroalkylation of aromatic compounds with per (poly)fluoroalkyl chlorides initiated by sodium dithionite in DMSO. *J. Fluorine Chem.* **2001**, *111*, 107–113.

130. Leroy, J.; Rubinstein, M.; Wakselman, C. Cooper-mediated perfluoroalkylation of halogenothiophenes. *J. Fluorine Chem.* **1985**, *27*, 291–298.

131. Meazza, G.; Zanardi, G.; Guglielmetti, G.; Piccardi, P. Synthesis of 2,3,5-triaryl-4-trifluoromethyl thiophenes. *J. Fluorine Chem.* **1997**, *82*, 175–180.

132. Kovregin, A.N.; Sizov, A.Y.; Ermolov, A.F. Synthesis of substituted thiophenes based on fluoro-containing vinylsulfides. *Izv. Akad. Nauk, Ser. Khim.* **2002**, *6*, 945–947.

133. Burdon, J.; Parsons, I. W. Highly fluorinated heterocycles. Part XIV. The conversion of 2,2,5,5-tetrafluoro-3-thiolen into 2,5-difluorothiophen: an apparent example of the thermal elimination of molecular fluorine. *J. Fluorine Chem.* **1979**, *13*, 159–162.

134. Burdon, J.; Campbell, J.G.; Parsons, J.W.; Tatlow, J.C. Highly fluorinated heterocycles. Part V. The preparation and some reactions of tetrafluorothiophen and some polyfluorothiophens. *J. Chem. Soc. C.* **1971**, 352–355.

135. Sakamoto, Y.; Komatsu, S.; Suzuki, T. Tetradecafluorosexithiophene: The first perfluorinated oligothiophene. *J. Am. Chem. Soc.* **2001**, *123*, 4643–4644.

136. Ulrich, H.; Kober, E.; Rätz, R.; Schroeder, H.; Grundmann, C. Synthesis of polyfluorinated heterocycles by indirect fluorination with silver fluorides. V. Fluorothiophene. *J. Org. Chem.* **1962**, *27*, 2593–2595.

137. Burger, K.; Helmreich, B.; Jendrewski, O. 3-Trifluoromethylthiophens from hexafluoroacetone. *J. Fluorine Chem.* **1994a**, *66*, 13–14.

138. Burger, K.; Helmreich, B.; Popkova, V.Y.; German, L.S. Synthesis of 2-fluoro-3-trifluoromethylthiophenes and 3-trifluoromethylthiophenes from Hexafluoroacetone. *Heterocycles* **1994b**, *39*, 819–832.

139. Krespan, C.G.; McKusick, B.C. Bis-(polyfluoroalkyl)-acetylenes. V. Addition of bis-(trifluoromethyl)-1,2-dithietene to olefins and acetylenes. *J. Am. Chem. Soc.* **1961**, *83*, 3438–3440.

140. Krespan, C.G.; Langkammerer, C.M. Fluorinated heterocyclic derivatives of sulfur, selenium and phosphorus. *J. Org. Chem.* **1962**, *27*, 3584–3587.

141. Krespan, C.G. Fluorinated thiolanes and a selenonane. *J. Org. Chem.* **1962**, *27*, 3588–3594.

142. Burdon, J.; Campbell, J.C.; Parsons, I.W.; Tatlow, J.C. Tetrafluorothiophene. *Chem. Commun.* **1969**, *1*, 27–28.

143. Tiers, G.V.D. Fluorocarbon sulfides. II. Synthesis and physical properties of thiaperfluorocyclopentane. *J. Org. Chem.* **1961**, *26*, 2538–2539.

144. Brasen, W.R.; Cripps, H.N.; Bottomley, C.G.; Farlow, M.W.; Krespan, C.G. Tetrafluorothiirane. *J. Org. Chem.* **1965**, *30*, 4188–4193.

145. Dear, R.E.A.; Gilbert, E.E. Telomerization of bis(trifluoromethyl)disulfide with polyfluoro-olefins. *J. Fluorine Chem.* **1974**, *4*, 107–110.

146. Middleton, W.J.; Howard, E.G.; Sharkey, W.H. Fluorothiocarbonyl compounds. I. Preparation of thio ketones, thioacyl halides, and thio esters. *J. Org. Chem.* **1965**, *30*, 1375–1384.

147. Burdon, J.; Parsons, I.W.; Shommakhi, A. The replacement of vinylic hydrogen by fluorine. *J. Fluorine Chem.* **1982**, *20*, 357–363.

148. Abe, T.; Nagase, S.; Baba, H. The electrochemical fluorination of dithiols and cyclic sulfides. *Bull. Chem. Soc. Jpn.* **1973**, *46*, 3845–3848.

149. Gronowitz, S.; Johnson, I.; Hornfeldt, A.B. Fluorinated heterocycles. *Chemica Scripta* **1973**, *3*, 94–96.

150. Prager, J.H. 1,3-Bis(fluoroxy)perfluoropropane and other oxygen-containing compounds by direct fluorination. *J. Org. Chem.* **1966**, *31*, 392–394.

151. Talbott, R.L. Fluorocarbon peroxides. Synthesis and characterization of *cis-* and *trans-*perfluoro-3,5-dimethyl-1,2-dioxolane. *J. Org. Chem.* **1965**, *30*, 1429–1431.

152. Moldavski, D.D.; Furin, G.G. Polyfluorination of organic compounds. *Zhurn. Obsch. Khimii.* **1996**, *66*, 1995–2002.

153. Russo, A.; Navarini, W. Perfluoro-4-methyl-1,3-dioxole: a new monomer for high-T_g amorphous fluoropolymers. *J. Fluorine Chem.* **2004**, *125*, 73–78.

154. Braum, R.A. Cyclic ethers from hexafluoroacetone. *J. Am. Chem. Soc.* **1965**, *87*, 5516–5517.

155. Hung, M.H.; Resnik, P.R. Thermal rearrangement of fluorinated dioxoles. *J. Am. Chem. Soc.* **1990**, *112*, 9671–9672.

156. Higashiya, S.; Narizuka, S.; Konno, A.; Maeda, T.; Momota, K.; Fuchigami, T. Electrolytic partial fluorination of organic compounds. 30. Drastric improvement of anodic monofluorination of 2-substituted 1,3-oxothiolan-5-ones using the novel fluorine source Et_4N 4HF. *J. Org. Chem.* **1999**, *64*, 133–137.

157. Fuchigami, T. Selective anodic monofluorination of sulfur-containing heterocycles: Potent applications towards pharmaceuticals. *Phosphorus, Sulfur Silicon Relat. Elem.* **1997**, *120–121*, 343–344.

158. Gouault, S.; Pommelet, J.C.; Lequeux, T. First synthesis of 4,4-difluoro-1,3-oxathiolan-5-one. *Synlett* **2002**, *6*, 996–997.

159. Kitazume, T.; Otaka, T.; Takei, R.; Ishikawa, N. Preparation and reaction of 4-alkoxy-2,2-bis(trifluoromethyl)-thietanes and 5-alkoxy-3,3-bis(trifluoromethyl)dithiolanes. *Bull. Chem. Soc. Jpn.* **1976**, *49*, 2491–2494.

160. Popkova, V.Y.; Bekker, R.A.; Misov, E.I.; Galahov, M.V.; Knuniantz, I.L. Fluorine-containing unsymmetrical disulfides. *Izv. AN SSSR, Ser. Khim.* **1985**, *12*, 2740–2746.

161. Pashkevich, K.I.; Ratner, V.G.; Latipov, R.R. β-Polyfluoroalkyl-containing α,β-unsaturated ketones. *Izv. AN SSSR, Ser. Khim.* **1987**, *4*, 921–924.

162. Burger, K.; Helmreich, B. Synthesis of 2-fluoro-3-trifluoromethylthiophenes and 3-trifluoromethylthiophenes from hexafluoroacetone. *Heterocycles* **1994**, *39*, 819–832.

163. Pashkevich, K.I.; Saloutin, V.I.; Bobrov, M.B.; Yufit, D.S.; Struchkov, Y.T. Nucleophilic tionation of polyfluorinated β-dicarbonylic compounds. *Zhurn. Org. Khim.* **1988**, *24*, 2397–2402.

164. Fesun, I.N.; Timoshenko, V.M.; Chernega, A.N.; Shermolovich, Y.G. Chlorination and oxidation reactions of 4-fluoro-5-tetrafluoroethyl-1,2-dithuiol-3-thine. *Zhurn. Org. Khim.* **2006**, *42*, 130–136.

165. Schuler, B.; Sundermeyer, W. Die thermolyse oxidierter 1,3-dithiolane: ein neuer syntheseweg fur thiocarbonyl-verbindungen und sulfine. *Chem. Ber.* **1990**, *123*, 177–184.

166. Yamanaka, H.; Tamura, K.; Funabiki, K.; Fukunishi, K.; Ishihara, T. Reactions of polyfluoro-2-alkynoic acids with bifunctional hetero nucleophiles leading to polyfluoroalkylated heterocycles. *J. Fluorine Chem.* **1992**, *57*, 177–190.

167. Harder, R.J.; Smith, W.C. Chemistry of sulfur tetrafluoride. VI. Fluorination of thiocarbonyl compounds. *J. Am. Chem. Soc.* **1961**, *83*, 3422–3424.

168. Middleton, W.J.; Sharkey, W.H. Fluorothiocarbonyl compounds. II. Reactions of hexafluorothioacetone. *J. Org. Chem.* **1965**, *30*, 1384–1390.

169. Hartzler, H.D. 1,3-Dithiolium carbenes from acetylenes and carbon disulfide. *J. Am. Chem. Soc.* **1973**, *95*, 4379–4387.

170. Krasuski, W.; Nikolauas, D.; Regitz, M. CS_2-Dipole aus elektronenreichen Olephinen–Herstellung und cycloadditionreactionen. *Lieb. Ann. Chem.* **1982**, 1451–1465.

171. Petrov, M.L.; Turbanova, E.C.; Kuzmina NYa, Stepanova N.P.; Petrov, A.A. Interaction of dithio-anions with hexafluorobutyne. *Zhurn. Org. Khimi.* **1984**, *20*, 2627–2628.

172. Krespan C.G.; Brasen W.R. Fluorinated cyclic polysulfides and thei polymers. *J. Org. Chem.* **1962**, *27*, 3995–3999.

173. Desjardins, C.D.; Passmore, J. Reaction of the fluoroantimony(V) and fluoroarsenic (V) salts of the polyatomic cations of sulphur with tetrafluoroethylene and hexafluoropropylene. *Can. J. Chem.* **1977**, *55*, 3136–3146.

174. Hughes, A.N.In *Handbook of Organophosphorus Chemistry*. Engel, R.,Ed.; Marcel Dekker: New York, **1992**; p 483.

175. Mathey, F. *Phospholes in Science of Synthesis*. In Maas, G.Ed.; Thieme: Stuttrart, **2001**; p 533.

176. Schmidbaur, H.; Holl, P. Ube rein spirocyclisches Pentaalkylphosphoran und Alkoxytetraalkylphosphoran, sowie uber ein monocyclisches kovalentes Fluorotetraalkylphosphoran. *Z. Anorg. Allg. Chem.* **1979**, *458*, 249–256.

177. Razumov, N.A.; Zubtsova, L.I.; Petrov, A.A. Phosphoris-containing heterocycles. *Zhurn. Obsch. Khimi.* **1970**, *40*, 2563–2565.

178. Fridland, S.V.; Miftahov, M.N. Interaction of the phosphorus pentachloride and isoprene adducts with arsenic(III) fluoride. *Zhurn. Obsch. Khimi.* **1988**, *58*, 41–46.

179. Delaere, D.; Pham-tran, N.N.; Minh Tho, Nguen; Remarkable influence of fluorine substitution on electronic and thermochemical properties of phospholes. *Chem. Phys. Lett.* **2004**, *383*, 138–142.

180. Mitchell, C.M.; Stone, F.G.A. Reaction of low-valent metal complexes with fluorocarbons. XXII. Tertiary phosphine gold complexes. *J. Chem. Soc., Dalton.* **1972**, 102–107.

181. Yong Xu.; Jiang C.; Tsukahara, R.; Fujiwara, Y.; Togyi, C.; Prestwich, C.D. Phosphonothioate and fluoromethylene phosphonate analogues of cyclic phosphatidic acid. *J. Med. Chem.* **2006**, *49*, 5309–5315.

182. Murano, T.; Yuassa, Y.; Koboyakawa, H.; Yokomatsu, T.; Shibuya, S. Synthesis of acyclic nucleotide analogues possessing a difluoromethylene phosphonyl group at the side chain. *Tetrahedron* **2003**, *59*, 10223–10230.

183. Ratner, V.G.; Lork, E.; Pashkevich, K.I.; Roschenthaler, G.V. Trifluoromethylated 1,2λ^5-oxaphospholanes via λ^5 –P-tricyclic phosphoranes from 4,4,4-trifluoro-3-hydroxy-1-phenylbutane-1-one and dichloriphosphines. *J. Fluorine Chem.* **1997**, *85*, 129–133.

184. France, R.; Heine, J.; Roschenthaler, G.V. Phosphites and hexafluoroacetone-hexafluoroacetylacetone: a comparison. *Phosphorus, Sulfur Silicon Relat. Elem.* **1990**, *49/50*, 377–380.

185. Sarma, R.; Ramirez, F.; McKeever, B.; Marechek, J.F.; Prasad, V.A. Interconversion of compounds with penta- and hexacoordinated phosphorus. *Phosphorus Sulfur* **1979**, *5*, 323–330.

186. Brown, R.K.; Day, R.O.; Husebye, S.; Holmes, R.R. Sterically oriented structures of perfluoromethyl-substituted spirophosphoranes centered between the trigonal bipyramid and the rectangular pyramid. *Inorg. Chem.* **1978**, *17*, 3276–3285.

187. Polezhaeva, N.A.; Cherkasov, R.A. Reactivity of 1,3,2-diheterophosphacyclanes with pentacoordinated phosphorus atom. *Uspehi. Khimi.* **1985**, *54*, 1899–1939.

188. Swamy, K.C.K.; Burton, S.D.; Holmes, J.M.; Day, R.O.; Holmes, R.R. Cyclic oxyphosphoranes. *Phosphorus Sulfur* **1990**, *53*, 437–455.

189. Mironov, V.F.; Gubauidullin, A.T.; Konovalova, I.V.; Litvinov, I.A.; Burnaeva, L.M.; Zyablikova, T.A.; Romanov, S.V.; Iovleev, R.A. Reaction of 2-*R*-4-oxo-5,6-benzo-1,3,2-dioxaphosphorinanov with methyl-2-oxo-3,3,3-trifluoropropionate. *Zhurn. Obsch. Khimii.* **2000**, *70*, 1812–1832.

190. Hall, L.D.; Steiner, P.R.; Pedersen, C. Studies of specifically fluorinated carbohydrates. Part VI. Some pentofuranosyl fluorides. *Can. J. Chem.* **1970**, *48*, 1155–1165.

191. Abe, T.; Baba, H.; Soloshonok, I. Electrochemical fluorination of several 1,4-bis [(methoxycarbonyl)alkyl] substituted piperazines. *J. Fluorine Chem.* **2001**, *108*, 215–228.

192. Burdon, J.; Coe, P.L.; Smith, J.A.; Tatlow, J.C. Highly fluorinated heterocycles. Part XVIII. 2-Difluoromethyl- and 2-trifluoromethyl- hexa- and penta-fluoro-oxolans. *J. Fluorine Chem.* **1991**, *51*, 179–196.

193. Battiste, J.L.; Jing, N.; Newmark, R.A. 2D $^{19}F/^{19}F$ NOESY for the assignment of NMR spectra of fluorochemicals. *J. Fluorine Chem.* **2004**, *125*, 1331–1337.

194. Kosnikov, A.Yu.; Lindemay, S.V.; Struchkov, Yu.T.; Gluhovtsev, V.G.; Illin, Yu.D. Structure of 2,2-bis(trifluoromethyl)-7,7-bis(trifluoromethyl)oxymethyl-1,3,6-trioxaspiro[4,4]nonene-8. Izv. AN. SSSR, Ser. Khim. **1990**, *1*, 64–67.

195. Bottini, A.; Fisichella, S.; Occipinti, S.; Arnone, C.; Consiglio, G.; Spinelli, D. ^1H Nuclear magnetic resonance spectra of some 3-and 5- substituted thiophen-2-carboxamides. *J. Chem. Soc., Perkin Trans. 2* **1983**, 459–460.

196. Andrès, D.F.; Laurent, E.G.; Marquet, B.S. Reaction of thioglycolate with α-fluoro-β-(phenylthio)enones (or -enals): synthesis of substituted α-carboxy-γ-fluorothiophenes. *Tetrahedron Lett.* **1997**, *38*, 1049–1052.

197. Nato, R.; Lamartina, L.; Arnone, C.; Spinelli, D. An analysis of ^{13}C nuclear magnetic resonance substituent chemical shifts in 3-substituted thiophene-2-carboxylic and 2-substituted benzoic acids by linear free energy relationships. *J. Chem. Soc., Perkin Trans. 2* **1988**, 887–892.

198. Abubakar, A.B.; Booth, B.L.; Tipping, A.E. Reaction of 2-lithio-3,4-bis(trifluoromethyl) furan with electrophiles. *J. Fluorine Chem.* **1991**, *55*, 189–198.

199. Li, L.; Collard, D.M. Tuning the electronic structure of conjugated polymers with fluoroalkyl substitution: alternating alkyl/perfluoroalkyl-substituted polythiophene. *Macromolecules* **2005**, *38*, 372–378.

200. Muramatsu, H.; Okumura, A.; Shibata, K.; Matsui, M. Synthesis, absorption spectra, and photostability of triarylmethane dye ethynylogues containing trifluoromethyl group(s). *Chem. Ber.* **1994**, *127*, 1627–1632.

201. Dmowski, W.; Voellnagel-Neugebauer, H. Preparation of some highly halogenated derivatives of furan. *J. Fluorine Chem.* **1981**, *18*, 223–241.

202. Burdon, J.; Coe, P.L.; Smith, J.A.; Tatlow, J.C. Highly fluorinated heterocycles. Part XIX. Dehydrofluorination studies on 2-difluoromethyl- and 2-trifluoromethyl-hexa- and -pentafluoro-oxolans. *J. Fluorine Chem.* **1991**, *52*, 165–184.

203. Chambers, R.D.; Lindley, A.A. Photochemical transformations of perfluoro-tri- and -tetra-methylfurans. *J. Fluorine Chem.* **1978**, *12*, 337–340.

204. Burger, K.; Fuchs, A.; Hennig, L.; Helmreich, B.; Greif, D. Domino reactions with 2-fluoro-3-trifluoromethylfurans and -thiophenes. *Monatsh. Chem.* **2001a**, *132*, 929–945.

205. Burger, K.; Helmreich, B. Ein neues Verfahren zur positionsselektiven Einführung von Trifluormethylgruppen in Heteroaromaten: Synthese von 3-3-Trifluormethylfuranen. *J. Prakt. Chem.* **1992**, *334*, 311–316.

206. Burger, K.; Fuchs, A.; Hennig, L.; Helmreich, B. Low temperature domino reactions. A ready access to trifluoromethyl substituted butenolides and their thioanalogues. *Tetrahedron Lett.* **2001**, *42*, 1657–1659.

207. Burger, K.; Hennig, L.; Fuchs, A.; Greif, D.; Spengler, J.; Albericio, F. Domino reactions with fluorinated five-membered heterocycles: synthesis of trifluoromethyl substituted butenolides and γ-ketoacids. *Monatsh. Chem.* **2005**, *136*, 1763–1779.

208. Yagupolskii, L.M.; Vechirko, E.P.; Kondratenko, N.V.; Liepinsh, E.E.; Shpaer, E.G.; Zhuk, R.A. 2-Trifluoromethyl- and 2-perfluoro tert.butyl tetrahydrofuranes and analogous of pirimidinnucleozides on their base. *Zhurn. Org. Khim.* **1981**, *17*, 186–191.

209. Yokoyama, M.; Toyoshima, H.; Shimizu, M.; Mito, J.; Togo, H. Simple synthesis of aromatic β-*C*-nucleosides via coupling of aryl grignard reagents with sugar fluorides. *Synthesis* **1998**, 409–412.

210. Abe, T.; Nagase, S. The reaction of perfluoro-2,4-dialkyloxolanes with anhydrous aluminum chloride. *J. Fluorine Chem.* **1978**, *12*, 359–379.

211. Coe, P.L.; Ray, N.C. Tetrafluorethene oligomers. Part 14. Reactions of perfluoro-3-methylpent-2-ene (tetrafluoroethylene trimer) with nucleophiles. *J. Fluorine Chem.* **1998**, *88*, 169–178.

212. Nishida, M.; Hayakawa, Y.; Matsui, M.; Shibata, K.; Muramatsu, H. Synthesis and photochemical reaction of 1,4-dialkyl-7-oxa-2,3,5,6-tetrakis(trifluoromethyl) bicyclohepta-2,5-diene. *J. Heterocycl. Chem.* **1992**, *29*, 113–116.

213. Hajduch, J.; Paleta, O.; Kvicala, J.; Haufe, G. Fluorinated furan-2(5H)ones: reactivity and stereoselectivity in Diels–Alder reactions. *Eur. J. Org. Chem.* **2007**, 5101–5111.

214. Nowak, I.; Rogers, L.M.; Rogers, R.D.; Thrasher, J.S. The effect of fluorine on the diastereoselectivity of the addition of α-oxyradicals to 3-fluoro-2,3-dihydro-1*H*-λ^6-thiophene-1,1-dione. *J. Fluorine Chem.* **1999**, *99*, 73–81.

215. Wiebe, H.A.; Braslavsky, S.; Heichlen, J. The photolysis of tetra(trifluoromethyl) thiophene vapor. *Can. J. Chem.* **1972**, *50*, 2721–2724.

216. Kobayashi, Y.; Kumadaki, I.; Ohsawa, A.; Sekine, Y. Derivation of oxahomocubanes from dewar thiophene. *Tetrahedron Lett.* **1974**, 2841–2844.

217. Smith, J.R.; Lemal, D.M. Synthetic approaches to a fluorinated tetrahedrane. *J. Fluorine Chem.* **2000**, *102*, 323–332.

218. Kobayashi, Y.; Kawada, K.; Ando, A.; Kumadaki, I. Photoisomerization of bis (trifluoromethyl)thiophenes. *Tetrahedron Lett.* **1984**, *25*, 1917–1920.

219. Guan, H.P.; Luo, B.H.; Wang, Q.F.; Hu, C.M. Preparation and reaction of 3,3,3-trifluoropropiondithioacetals as trifluoromethyl-containing building blocks. *J. Chem. Soc., Perkin Trans. 1* **1998**, 279–281.

220. Bouillon, J.P.; Tinant, B.; Nuzillard, J.M.; Portella, C. Synthesis of new 3-(1-ethylsulfanyl-2-perfluoroalkyl)-5-hydroxy-5-methyl 5-phenyl)-1,5-dihydro-pyrrol-2-ones starting from γ-keto thioesters and imines. *Synthesis* **2004**, 711–721.

221. Yoshimura, Y.; Endo, M.; Sakata, S. An alternative Synthesis of antineoplastic 4-thiocytidine analoque 4-thioFAC. *Terahedron Lett.* **1999**, *40*,1937–1940.

222. Burton, D.; Yang, Z.Y.; Fluorinated Organometallics: Perfluoroalkyl and Functionalized Perfluoroalkyl Organometallic Reagents in Organic Synthesis. *Tetrahedron.* **1992**, *48*, 189–275.

223. Glemser, O.; Shreeve, J.M. Fluoride Ion Induced Cyclization of Trifluoroacetonitrile with Oxygen- and Sulfur(II)-Containing Nucleophiles. *Inorg. Chem.* **1979**, *18*, 2319–2321.

5

SYNTHESIS OF FLUORINATED SUGARS FROM FLUORINE-CONTAINING SYNTHONS

Xiao-Long Qiu[1] and Feng-Ling Qing[1,2]

[1] *Key Laboratory of Organofluorine Chemistry, Shanghai Institute of Organic Chemistry, Chinese Academy of Science, 354 Fenglin Lu, Shanghai 200032, China*

[2] *College of Chemistry, Chemical Engineering and Biotechnology, Donghua University, 2999 North Renmin Lu, Shanghai 201620, China*

5.1 INTRODUCTION

It is well known that the introduction of fluorine into biological molecules often results in significant changes in their chemical, physical, and biological properties.[1–3] For example, research has clearly demonstrated the important effects of fluorine substitution on the inter- and intramolecular forces, which affect binding of ligands, and thus introduce receptors subtype selectivity, at cholinergic and adrenergic receptors.[4–6] Also as an important tool, fluorine substitution has a profound effect on drug disposition, in terms of distribution, drug clearance, route(s), and extent of drug metabolism.[7] In the past several decades, the noteworthy increase in the utilization of fluorine-containing compounds in fluorinated materials, fluorinated amino acids, fluorinated sugars, fluorinated steroids and fluorinated nucleosides unambiguously illustrated the significant impact that fluorine(s) have made on all the aspects of modern life.

Up to date, a lots of fluorinated compounds were synthesized and biologically evaluated. Among these fluorinated compounds, fluorinated sugars have recently attracted more and more attentions from organic and medicinal chemists. It should be noted that just one fluorine-containing carbohydrate derivative has been isolated from

Fluorinated Heterocyclic Compounds: Synthesis, Chemistry, and Applications, Edited by Viacheslav A. Petrov

an organism—the 4′-fluoro-5′-*O*-sulfamoyladenoside, nucleocidin.[8] Generally speaking, two major methodologies were developed to synthesize fluorinated sugars. One method featured the introduction of fluorine atom(s) into modified sugar derivatives via fluorinating reagents (for example, DAST, TBAF, etc.). In some cases, this method suffered from low regioselectivities, low stereoselectivities, and even low yields. Alternatively, the other strategy involved the synthesis of fluorinated sugars from the some versatile fluorinated building blocks in a straightforward fashion. So far, the topic "Fluorinated Sugars" has been reviewed by several groups[9–15] and most of these reviews only focused on the synthesis of fluorinated sugars utilizing fluorinating reagents. Described herein is our mini review about synthesis of fluorinated sugars starting from fluorine-containing synthons. This review is grouped into two types: synthesis fluorinated furanoses or pyranoses and synthesis of fluorinated azasugars.

5.2 SYNTHESIS OF FLUORINATED FURANOSES OR PYRANOSES

The Reformatsky reactions of fluorinated synthons, difluoroacetate **1**, with aldimine was utilized to prepare the 2,3-dideoxy-2,2-difluoro-3-amino sugars in the Kobayashi group.[16] This group found that reaction of the Zn reagent, *in situ* formed by treating iododifluoroacetate or bromodifluoroacetate with Zn powder, with aldimines **2** and **3** smoothly provided the diastereoisomers, difluoro-β-lactam **4a–b** and **9a–b** in 71% and 65% yield, respectively (Scheme 5.1). Conversion of the lactam **4a** to the diacetate form **5** of 2,3-dideoxy-3-amino-2,2-difluorosugar was realized by a three-step procedure including reduction with DIBAL-H followed by deprotection and acetylation. Treatment of the isomer **4b** with $KOH/Et_2O/H_2O$ followed by CH_2N_2-mediated methylation gave the β-amino ester **6**, from which the diacetate form **7** of L-2,3,6-trideoxy-3-*N*-benzylamino-2,2-difluorogulopyranoside was obtained in 66% yield. Using the similar reaction conditions, the difluoro sugars **10** and **11** were also prepared as α,β anomeric mixture from the difluoro-β-lactam **9a** and **9b**, respectively. Interestingly, the key intermediate **6** could also be obtained with moderate *syn* selectivity (*syn*/*anti* = 3.7:1) by reaction of the **2** with fluorinated building block, ketene silyl acetal **8**.

In 1993, Taguchi and coworkers synthesized the dihydropyrones **13a–d** from the fluorinated building block, 2,4-dialkoxy-1,1-difluoro-1,3-diene **12** using a Lewis acid-catalyzed hetero Diels–Alder reaction. Treatment of dihydropyrone **13a** with $TsOH/CH_2Cl_2$ followed by Luche reduction and alcoholysis gave the racemic 2,4-dideoxy-4,4-difluorosugars (±)-**14a–b** (Scheme 5.2).[17]

Several years ago, the Percy group performed the synthesis of some 4,4-difluorosugars via a direct sequence involving ring-closing metathesis (RCM) and indium-mediated difluoroallylation using *gem*-difluorinated synthon, 1-bromo-1,1-difluoropropene **15**.[18] Percy et al. found that indium-mediated additions of fluorinated synthon **15** to aldehydes **16** gave the homoallylic difluoroalcohols **17a–f** in moderate to high yields (Scheme 5.3). Allylation of **17e** delivered the ether **18** in quantitative yield, which was subjected to RCM to provide the compound **19** in 97% yield.

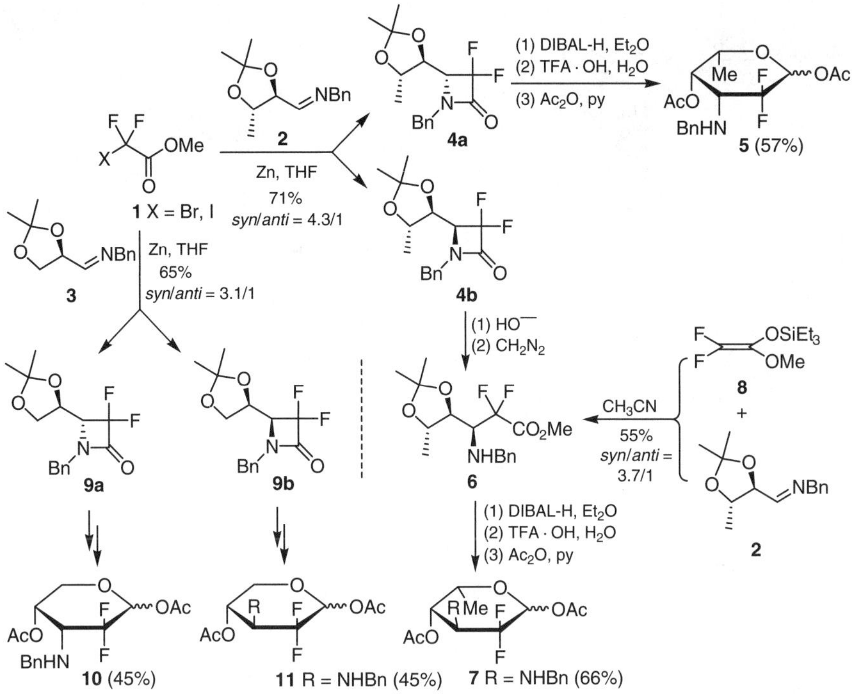

SCHEME 5.1

Dihydroxylation of **19** with OsO_4/NMO occurred without discrimination between the alkene faces to afford the fluoro sugars (±)-**20** and (±)-**21** as a 1:1 mixture of separable *cis* and *trans* isomers. Synthesis of the 4,4-difluoro-4-deoxy-β-DL-ribo-pyranside derivatives (±)-**23e–f** and 4,4-difluoro-4-deoxy-β-DL-lyxo-pyranside derivatives (±)-**24e–f** were realized via transacetalisation of the difluoro homoallylic alcohols **17e–f** followed by RCM and dihydroxylation, respectively.

Very recently, 3-deoxy-3,3-difluoro-D-ribohexose **32** was also conveniently prepared starting from the *gem*-difluoromethylenated synthon **15** in the Qing group.[19] As the first step, indium-mediated reaction of **15** with 1-(*R*)-glyceraldehyde acetonide **25** gave the versatile intermediate **26** with *anti* isomer as major product (*anti*/*syn* = 7.7:1)

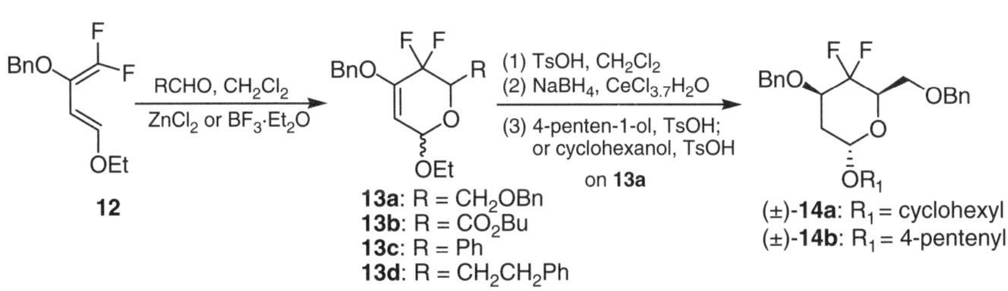

SCHEME 5.2

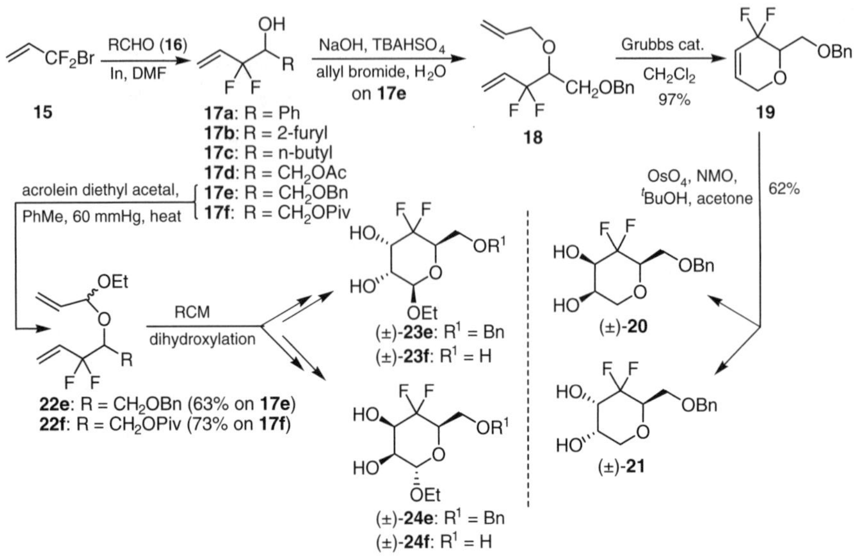

SCHEME 5.3

(Scheme 5.4). Conversion of *gem*-difluorohomoallyl alcohol **26** to the separable diastereoisomers **27a** and **27b** was accomplished by the kinetic resolution (NaH/BnBr/TBAI) and subsequent Os-catalyzed dihydroxylation. Disilylation of isomer **27b** furnished the compound **28** in 94% yield. $SnCl_2$-mediated removal of the

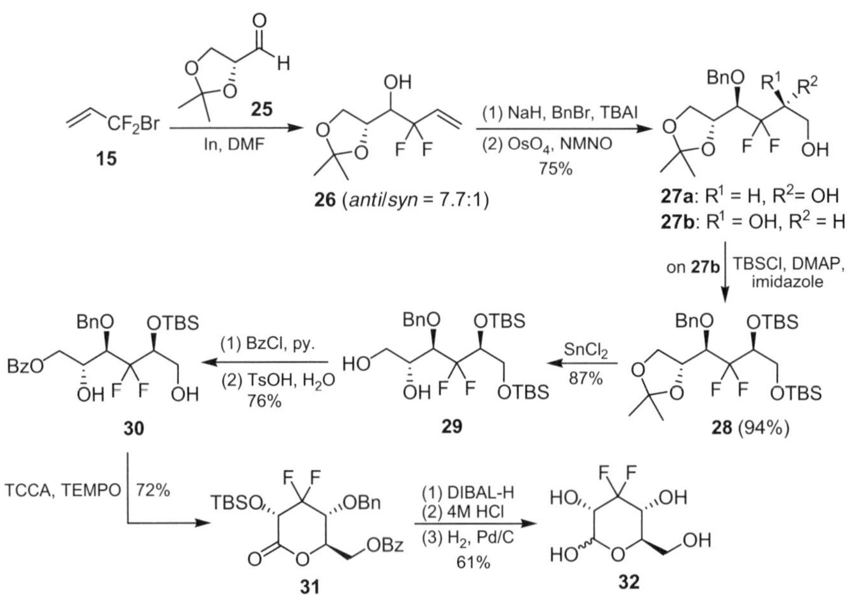

SCHEME 5.4

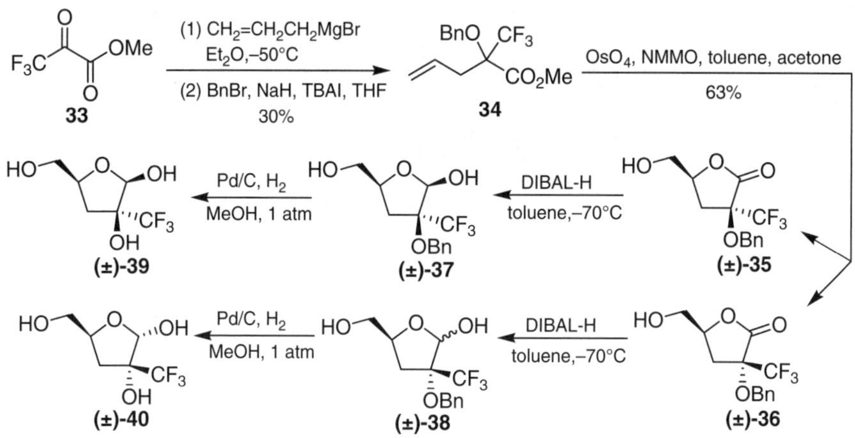

SCHEME 5.5

isopropylidene ketal in **28** afforded the diol **29** in 87% yield. Selective benzoylation of **29** followed by removal of the primary TBS group gave the 1,5-diol **30** in medium yield. Oxidation of **30** using the excess TCCA/TEMPO provided lactone **31** in 72% yield. Reduction of compound **31** with DIBAL-H followed by removal of all the protecting groups afforded the target fluoro sugar **32** in 61% yield over three steps.

In 1994, Burger and coworkers described the synthesis of racemic 2-*C*-trifluoromethyl-substituted 3-deoxypentoses (±)-**39** and (±)-**40**.[20] Their synthesis commenced with the Grignard reaction of methyl trifluoropyruvate **33** with allyl magnesium bromide (Scheme 5.5). The resultant product was further subjected to *O*-benzylation to furnish the ester **34** in 30% yield over two steps. Subsequent dihydroxylation with OsO_4 yielded the simultaneous lactonization products (±)-**35** and (±)-**36** in 63% yield and in 1:1 ratio. DIBAL-H mediated reduction of chromatographically separated diastereomeric lactones (±)-**35** and (±)-**36** gave the lactols (±)-**37** and (±)-**38**, respectively. After hydrogenolytic removal of the benzyl groups, the (±)-3-deoxy-2-*C*-trifluoromethylarabinose β-**39** and (±)-3-deoxy-2-*C*-trifluoromethylribose α-**40** was accessed, respectively.

Several years later, same group carried out the synthesis of racemic 2-*C*-trifluoromethyl ribose (±)-**46** and racemic 2-*C*-trifluoromethyl arabinose (±)-**47** starting from the same building block **33** (Scheme 5.6).[21] The synthesis featured the four-step reaction sequence. The first important step involved the diastereoselective allyl ether addition to **33** leading to separable diastereomeric compounds (±)-**42** and (±)-**43** were obtained in high yield. Diastereoselective dihydroxylation of the C=C double bond in (±)-**42** and (±)-**43** with potassium osmate followed by simultaneous lactonization afforded lactone (±)-**44** and (±)-**45** in medium yields with very high diastereomeric excess (>99% de). SMEAH reduction (also known as Red-Al reduction) of compound (±)-**44** and (±)-**45** delivered the target sugars (±)-**46** and (±)-**47** in 70% and 64% yield, respectively. It should be noted that the carbohydrates (±)-**46** and (±)-**47** exhibited unexpected the anomeric stability. That is, 2-*C*-trifluoromethyl ribose could crystallize to give exclusively α-furanose (±)-**46**,

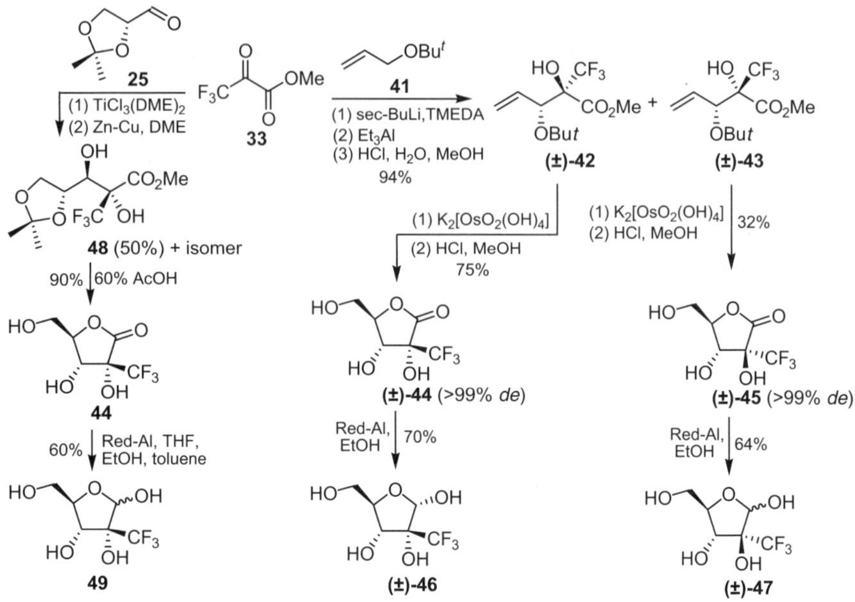

SCHEME 5.6

while 2-*C*-trifluoromethyl arabinose (±)-**47** was only obtained as an oily mixture of both anomers. Stereoselective synthesis of 2-*C*-trifluoromethyl ribose **49** was built on the addition of the synthon **33** to the chiral pool 1-(*R*)-glyceraldehyde acetonide **25** and the major diasteroisomeric product **48** was generated in 50% yield. Removal of isopropylidene ketal of **48** in the presence of AcOH afforded lactone **44** in high yield, which was further subjected to Red-Al reduction to give the target fluoro sugar, diastereomerically pure 2-*C*-trifluoromethyl ribose **49**.

Trifluoroacetylated vinyl ethers of type **50** are versatile building blocks for synthesizing trifluoromethyl-containing compounds. In 1989, Hojo et al. pioneered the facile hetero-Diels–Alder reaction of heterodienes **50** with a variety of vinyl ethers **51**. The cycloadditions proceeded smoothly under thermal conditions to give approximately 1:1 mixture of endo and exo cycloadducts (±)-**52** and (±)-**53** in good yield (Scheme 5.7).[22] Based on Hojo's results, the Greuter group[23] and the Brooker group[24] further found that using aluminium- or titanium-derived Lewis acids ($TiCl_4$, $ZnCl_2$, Et_2AlCl, $TiCl_2(OPr^i)$, etc.) as catalysts, these reactions afforded the *cis* isomer (±)-**52** as the major products, which could be converted to the racemic 2,6-dideoxy-6,6,6-trifluorinated sugar derivatives (±)-**55** and (±)-**56** by hydroboration-oxidation and removal of benzyl group subsequence. In addition, Greuter et al. also prepared the enantiomercially pure trifluoromethylated sugar (+)-**57** by conversion of (±)-**54** to the camphanic acid derivative, followed by separation of the diastereomeric esters and saponification.

The synthesis of trifluoromethylated branched sugars (±)-**66** and (±)-**67** were performed by the Kobayashi group starting from the 4-trifluoromethyl-2(5*H*)-furanose **58**.[25] The synthesis embarked on the conversion of trifluoromethylated

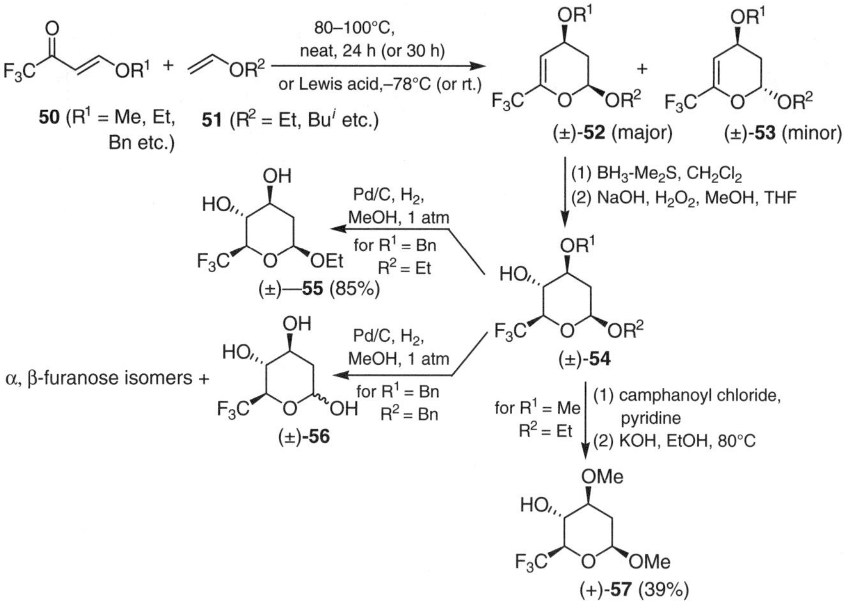

SCHEME 5.7

building unit **58** to 2-trialkylsilyloxy-4-trifluoromethylfuran **59** via treatment with $R_3SiCl/Et_3N/ZnCl_2$ (cat) (Scheme 5.8). $BF_3 \cdot OEt_2$-promoted aldol reaction of the silyloxyfuran **59** with aldehyde **60** yielded the γ-adduct ($\pm$)-**61** as the single diastereoisomer in 74% yield. After protection of the hydroxyl group with MOM group, the resultant lactone ($\pm$)-**62** was reduced to afford the hemiacetal ($\pm$)-**63** in

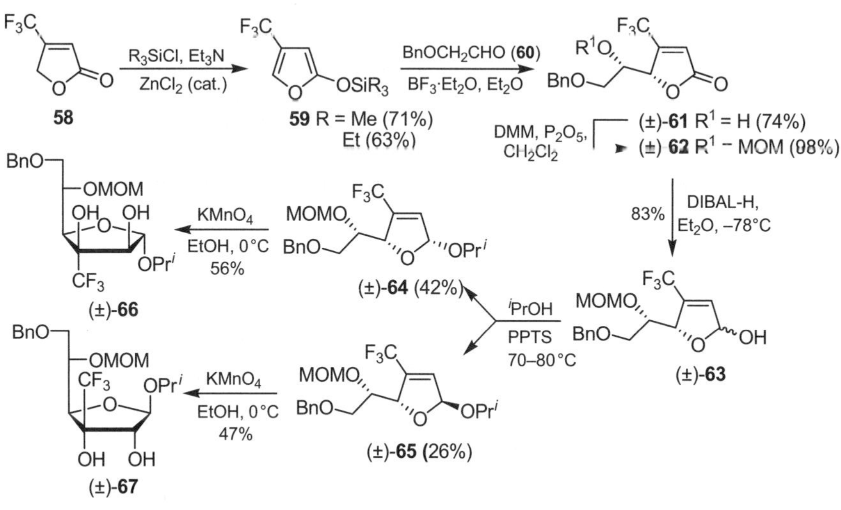

SCHEME 5.8

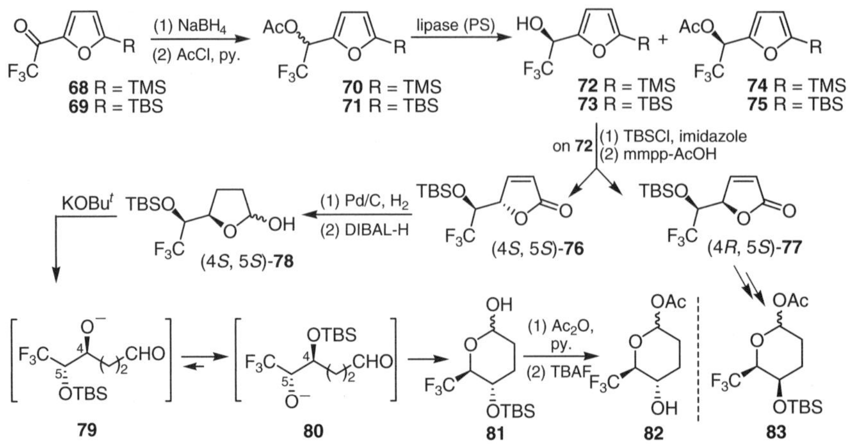

SCHEME 5.9

83% yield. Isopropylation of (±)-**63** generated two separable diastereoisomers (±)-**64** and (±)-**65** in a ratio of 1.6:1. Introduction of the vicinal diol function into (±)-**64** and (±)-**65** was accomplished through dihydroxylation with $KMnO_4$ in EtOH, and the (±)-3-trifluoromethylhexofuranoside α-gulofuranoside derivative **66** and (±)-3-trifluoromethylhexofuranoside β-talofuranoside derivative **67** were isolated in 56% and 47% yield, respectively.

Starting from furyl ketones **68–69**, the Kitazume group fulfilled the preparation of optically active 2-butenolides **76–77** bearing a trifluoromethyl group using enzymatic optical resolution and the resultant chiral 2-butenolides were selectively transformed into 6,6,6-trifluororhodinose and amicetose via base-promoted 1,2-migration of TBS moiety in a highly efficient manner (Scheme 5.9).[26] This synthetic pathway was based on the excellent resolutions of the esters **70** and **71** with lipase PS and the chiral alcohols **72** and **73** with (*S*)-configuration were isolated in 39% and 48% yield with 98% and >99% ee, respectively. Further protection of the furanol **72** with TBSCl and MMPP-mediated oxidation resulted in readily separable diastereoisomers (4*S*,5*S*)-**76** and (4*R*,5*S*)-**77**. After hydrogenation and reduction of (4*S*,5*S*)-**76**, the resultant furanose **78** was treated with *t*-BuOK to afford the rearranged pyranose **81**. This key step involved the TBS migration from the oxygen at C5 in intermediate **79** to C4 in the intermediate **80**. Acetylation of the pyranose **81** followed by TBAF-promoted removal of silyl group gave the 6,6,6-trifluoro analogue of D-amicetose **82**. In a similar fashion, the 6,6,6-trifluoro D-rhodinose derivative **83** was also prepared.

One year later, Kitazume et al. also described the synthesis of 6-deoxy-6,6,6-trifluoro-D-mannose and D-allose using intermediate (4*S*,5*S*)-**76** (Scheme 5.10).[27] Thus, oxidation of **76** with potassium permanganate in the presence of catalytic 18-crown-6 provided the diol **84** as the only isomer in 42% yield. Protection of diol moiety in **84** as its acetonide followed by DIBAL-H reduction afforded the lactol **85**, which was subjected to the *t*-BuOK promoted isomerization to give the desired fluoro sugar **86** in 35% yield. The precursor **87** of fluoro sugar **89** was obtained by reduction

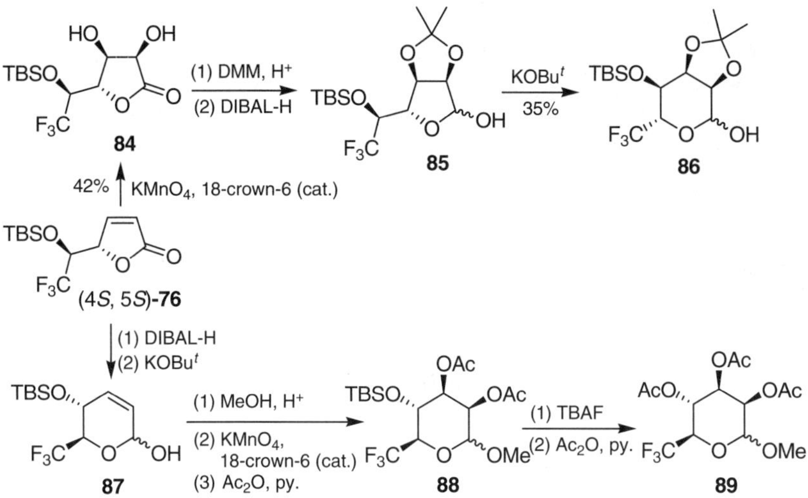

SCHEME 5.10

of **76** with DIBAL-H and subsequent *t*-BuOK promoted isomerization. After derivatization of **87** into methyl glycoside, permanganate oxidation led to a stereochemically pure diol, which was further acetylated to the diacetate **88**. Desilylation of **88** followed by acetylation gave the trifluoromethylated sugar **89** in 90% yield.

Starting from the *trans*-1-benzyloxy-4,4,4-trifluoro-2-butene **90**, the Qing group prepared the β-L-fucofuranose **97** and β-L-4,6-dideoxylohexopyranose **100** in high stereoselectivity and in a straightforward fashion.[28] The Sharpless AD reaction of **90** provided the chiral diol **91** in 95% yield. Silylation of **91** followed by hydrogenolytic debenzylation yielded the alcohol **92** in quantitative yield (Scheme 5.11). Oxidation of **92** with TCCA/TEMPO and subsequent Horner–Wadsworth–Emmons (HWE) reaction gave the ester **93** in 82% yield. Another Sharpless AD reaction on **93** catalyzed by AD-mix-α generated the diol **94** in 97% yield. After disilylation, reduction of the resultant tetra-TBS ester **95** afforded the alcohol **96**, which was transformed to the trifluoromethylated sugar **97** by mean of oxidation followed by deprotection and acetylation sequence. One key step of to the synthesis fluorinated sugar **100** was the regioselective ring opening of trifluoromethylated cyclic sulfate **98**. The synthetic steps of the preparation of target molecular **100** from compound **99** were same as those used in synthesizing the 6-deoxy-6,6,6-trifluorosugar **97** from diol **91**.

5.3 SYNTHESIS OF FLUORINATED AZASUGARS

In 1994, Kitazume et al. described the highly stereoselective synthesis of 1,6-dideoxy-6,6-difluoroazasugar analogues starting from the difluorinated building block, difluoro-acetaldehyde ethyl hemiacetal **101** (Scheme 5.12).[29] Their basic strategy was based on the concept that chiral 6-difluoromethyl-5,6-dihydro-4-pyridone **105**

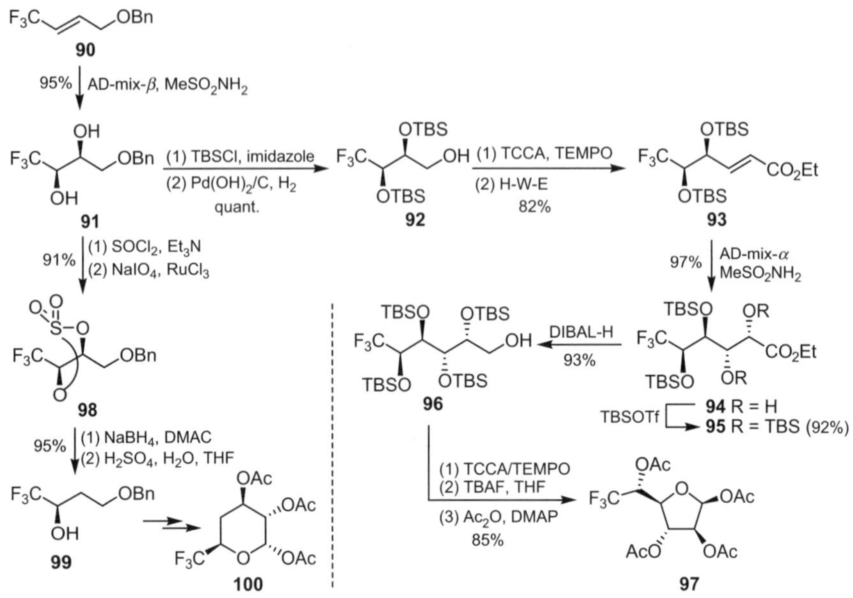

SCHEME 5.11

and **106** were accessed through the aza-Diels–Alder reaction of 1-methoxy-3-[(trimethylsilyl)oxyl]-1, 3-butadiene **104** and chiral (αR)-*N*-(2, 2-difluoroethylidene)(α-methylbenzyl)amine **103**, which was derived from the addition of (*R*)-(α-methylbenzyl)amine **102** to fluorinated semiketal **101**. $BF_3 \cdot Et_2O$ was found to be a most effective Lewis acid catalyst to achieve the highly diastereoselective addition. Stereocontrolled hydroxylation of pyridone **105** and **106** were carried out via treatment with $NaN(TMS)_2$/*N*-phenylsulfonyl 2-phenyloxazolidine, and the compounds **107** and **108** were obtained after protection with TBSCl, respectively. Epimerization of the pyridone **107** was realized via exposure to LDA/THF and epimer **109** was afforded. Transformation of **109** to the 1,6-dideoxy-6,6-difluoroazasugar derivative **110** was accomplished by treatment with BH_3 followed by 10% $NaOH/H_2O_2$ and acid-promoted desilylation. Starting from the pyridone **107** and **108**, 1,6-dideoxy-6,6-difluoroazasugar derivative **111** and **112** were also synthesized using the same reaction conditions used for preparation of **110** from **109**.

Recently, a novel series of *gem*-4,4-difluoromethylenated azasugars, D-1,4,6-trideoxy-4,4-difluoronojirimycin **120**, and L-1,4,6-trideoxy-4,4-difluoronojirimycin **122**, were stereoselectively synthesized starting from the trifluoroethanol **113** (Scheme 5.13).[30] The key step was highly diastereoselective construction of the piperidine ring via reductive amination. Firstly, the *O*-allyl ether **114** was prepared in three steps and a sigmatropic rearrangement of **114** proceeded well by treatment with LDA/THF at −78°C to give the intermediate **115** in 32% yield over four steps. Exposure of allylic alcohol **115** to $MsCl/Et_3N$ provided the mesylate **116** in 96% yield, which was subjected to Pd(0)-catalyzed regioselective allylic substitution with NaN_3 to deliver the azide **117**. Conversion of **117** to the *N*-Cbz-amine **118** was

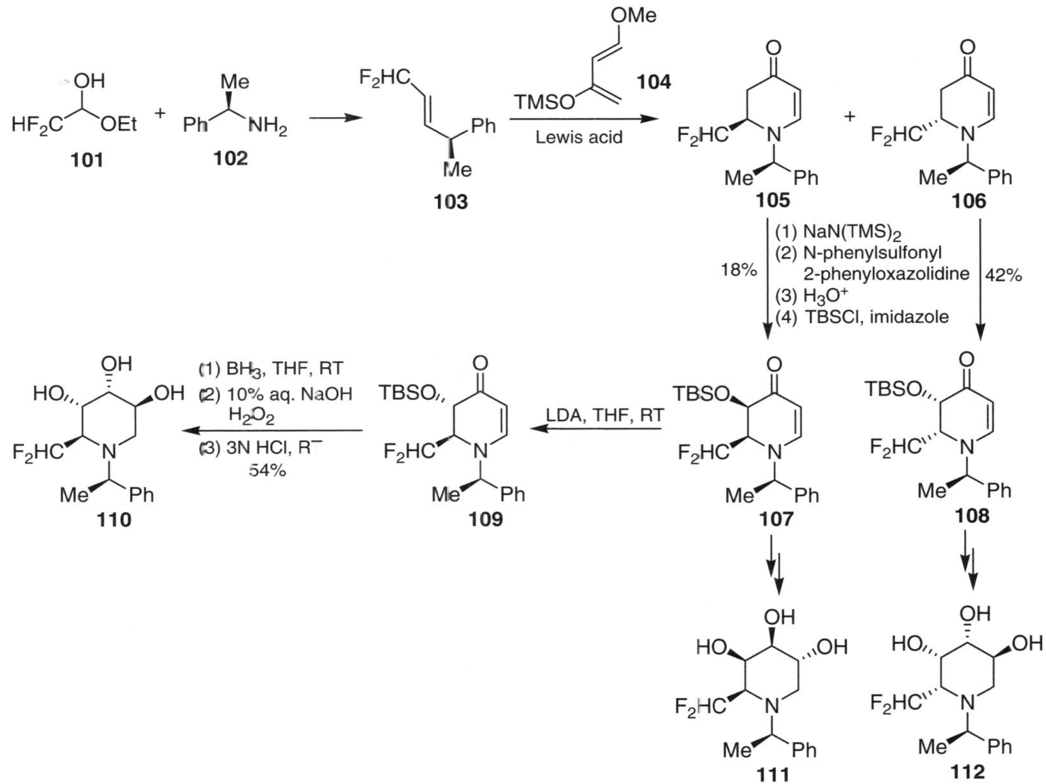

HF2C
OH
OEt
101
Me
Ph
NH2
102
F2HC
Ph
Me
103
OMe
TMSO
104
Lewis acid
105
106
(1) NaN(TMS)2
(2) N-phenylsulfonyl 2-phenyloxazolidine
(3) H3O+
(4) TBSCl, imidazole
18%
42%
TBSO
107
108
LDA, THF, RT
109
(1) BH3, THF, RT
(2) 10% aq. NaOH H2O2
(3) 3N HCl, RT
54%
110
111
112

SCHEME 5.12

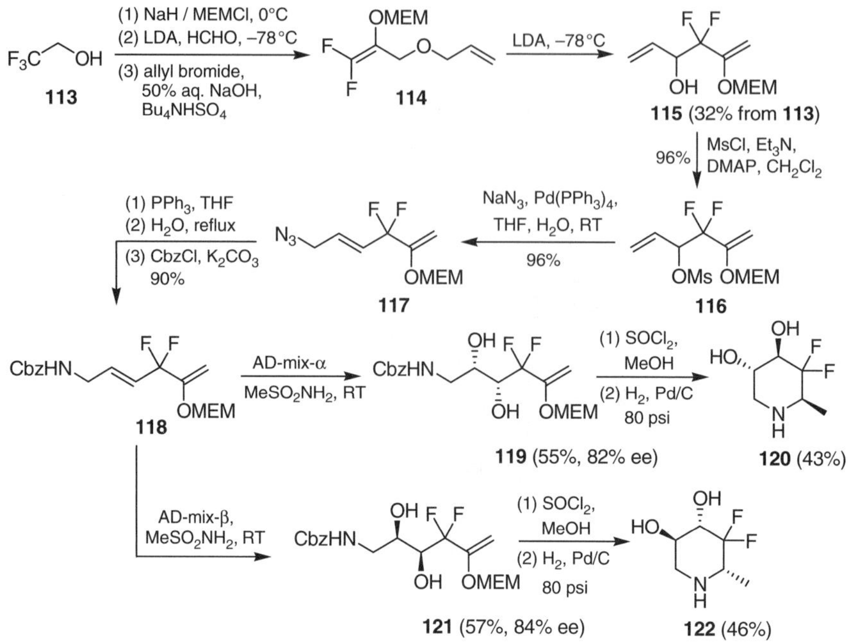

SCHEME 5.13

accomplished via treatment with PPh_3/THF followed by hydrolysis and addition of CbzCl. Using $(DHQ)_2$PHAL and $(DHDQ)_2$PHAL as the ligands, asymmetric AD reaction of **118** furnished the diols **119** and **121** in 82% and 84% ee, respectively. Finally, the iminosugars **120** and **122** were provided by means of removal of MEM group and highly diastereoselective hydrogenation.

Starting from the intermediates **27a–b**, three optically pure azasugars, D-1,4-dideoxy-4,4-difluoromannonojirimycin **128**, L-1,4-dideoxy-4,4-difluorogulonojirimycin **129** and D-1,4-dideoxy-4,4-difluoronojirimycin **133**, were synthesized in a straightforward fashion (Scheme 5.14).[31] Selective benzoylation of **27a** gave the benzoate **123** in high yield, which was further converted to the azide **124** via treatment with Tf_2O/pyridine followed by NaN_3/DMF. HOAc-mediated removal of isopropylidene ketal in **124** generated the diol **125** in 96% yield. After selective mesylation of **125** (81% yield), reduction of the resultant azide **126** with PPh_3 and subsequent treatment with CbzCl afforded the carbamate **127** in 82% yield. The fluorinated sugar **128** was obtained through one-step hydrogenolytic removal of Bn and Cbz groups followed by debenzoylation with NH_3/MeOH. Following the same procedure, fluorinated iminosugar **129** was also prepared from **27b**. The synthesis of fluorinated azasugar **133** required the reversion of the C2 center in **125** and the inversion was achieved via AcOK/AcOH mediated S_N2 nucleophilic substitution of mesylate **130**, which was delivered in 82% yield via selective silylation of diol **125** followed by mesylation. The removal of acetyl groups of compound **131** with HCl/MeOH provided the diol **132** in 97% yield, which was converted to the target fluorinated

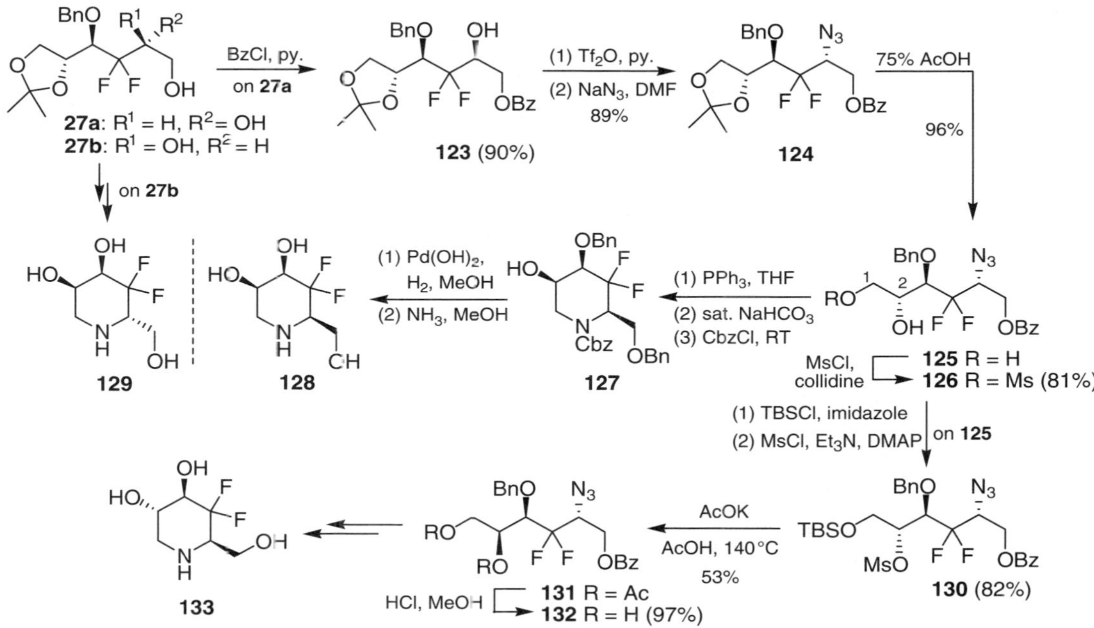

BzCl, py.
on 27a
(1) Tf2O, py.
(2) NaN3, DMF
89%
75% AcOH
96%
27a: R1 = H, R2 = OH
27b: R1 = OH, R2 = H
123 (90%)
124
on 27b
(1) Pd(OH)2,
H2, MeOH
(2) NH3, MeOH
(1) PPh3, THF
(2) sat. NaHCO3
(3) CbzCl, RT
129
128
127
MsCl,
collidine
125 R = H
126 R = Ms (81%)
(1) TBSCl, imidazole
(2) MsCl, Et3N, DMAP
on 125
AcOK
AcOH, 140°C
53%
130 (82%)
HCl, MeOH
131 R = Ac
132 R = H (97%)
133

SCHEME 5.14

azasugar **133** just using the similar procedure described for synthesis of iminosugar **128** from diol **125**.

ACKNOWLEDGMENTS

The former and current colleagues, who contributed to the synthesis of the fluorinated sugars at Shanghai Institute of Organic Chemistry, are greatly acknowledged. The funding support in this area came from the National Natural Science Foundation of China and Shanghai Municipal Scientific Committee.

REFERENCES

1. Banks, R.E.; Smart, B.E.; Tatlow, J.C. *Organofluorine Chemistry: Principles and Commerical Applications*; Plenum Press, New York, 1994.
2. Ojima, I.; McCarthy, J.R.; Welch, J.T. *Biomedical Frontiers of Fluorine Chemistry*, ACS Symposium Series 639, ACS, Washington DC, Oxford University Press, 1996.
3. Soloshonok, V.A. Fluorine-Containing Synthons, ACS Symposium Series 911, ACS, Washington DC, Oxford University Press: 2005.
4. Bravo, P.; Resnati, G.; Angeli, P.; Frigerio, M.; Viani, F.; Arnone, A.; Marucci, G.; Cantalanmessa, F. Synthesis and pharmacological evaluation of enantiomerically pure 4-deoxy-4-fluoromuscarines. *J. Med. Chem.* **1992**, *35*, 3102–3110.
5. Lu, S.F.; Herbert, B.; Haufe, G.; Laue, K.W.; Padgett, W.L.; Oshunleti, O.; Daly, J.W.; Kirk, K.L. Syntheses of (*R*)- and (*S*)-2- and 6-fluoronorepinephrine and (*R*)- and (*S*)-2- and 6-fluoroepinephrine: effect of stereochemistry on fluorine-induced adrenergic selectivities. *J. Med. Chem.* **2000**, *43*, 1611–1619.
6. Tewson, T.J.; Stekhova, S.; Kinsey, B.; Chen, L.; Wiens, L.; Wiens, L.; Barber, R. Synthesis and biodistribution of *R*- and *S*-isomers of [^{18}F]-fluoropropranolol, a lipophilic ligand for the β-adrenergic receptor. *Nucl. Med. Biol.* **1999**, *26*, 891–896.
7. Park, B.K.; Kitteringham, N.R. Effects of fluorine substitution on drug metabolism: pharmacological and toxicological implications. *Drug Metab. Rev.* **1994**, *26*, 605–643.
8. O'Hagan, D.; Harper, D.B. Fluorine-containing natural products. *J. Fluorine Chem.* **1999**, *100*, 127–133.
9. Penglis, A.A.E. Fluorinated carbohydrates. *Adv. Carbohydr. Chem. Biochem.* **1981**, *38*, 195–285.
10. Card, P.J. Synthesis of fluorinated carbohydrates. *J. Carbohydr. Chem.* **1985**, *4*, 451–487.
11. Tsuchiya, T. Chemistry and developments of fluorinated carbohydrates. *Adv. Carbohydr. Chem. Biochem.* **1990**, *48*, 91–277.
12. Novo, B.; Resnati, G. Soloshonok, V.A. Ed.; *Enantiocontrolled Synthesis of Fluoro-Organic Compounds*, John Wiley & Sons, Inc.: Chichester, UK, 1999; p 349.
13. Dax, K.; Albert, M.; Ortner, J.; Paul, B.J. Synthesis of deoxyfluoro sugars from carbohydrate precursors. *Carbohy. Res.* **2000**, *327*, 47–86.
14. Plantier-Royon, R.; Portella, C. *C*-Difluoromethylene-containing, *C*-trifluoromethyl and *C*-perfluoroalkyl carbohydrates. Synthesis by carbohydrate transformation or building block methods. *Carbohydr. Res.* **2000**, *327*, 119–146.

15. Miethchen, R. Modified natural substances—fluorinated and fluoroalkylated monosaccharides and inositols. *J. Fluorine Chem.* **2004**, *125*, 895–901.

16. Taguchi, T.; Kitagawa, O.; Suda, Y.; Ohkawa, S.; Hashimoto, A.; Iitaka, Y.; Kobayashi, Y. Synthesis of 3,3-difluoro-2-azetidinones and 2,3-dideoxy-2,2-difluoro-3-amino-sugars through the reformatsky reaction of difluoroacetate with imine. *Tetrahedron Lett.* **1988**, *29*, 5291–5294.

17. Taguchi, T.; Kodama, Y.; Kanazawa, M. Synthetic approach to 4-deoxy-4,4-difluoropyranosides via cycloaddition of 2,4-dialkoxy-1,1-difluoro-1,3-dienes with aldehydes. *Carbohydr. Res.* **1993**, *249*, 243–252.

18. Audouard, C.; Fawcett, J.; Griffiths, G.A.; Percy, J.M.; Pintat, S.; Smith, C.A. Synthesis of 4,4-difluoroglycosides using ring-closing metathesis. *Org. Biolo. Chem.* **2004**, *2*, 528–541.

19. Xu, X.H.; You, Z.W.; Zhang, X.; Qing, F.L. Synthesis of 3-deoxy-3,3-difluoro-D-ribohexose from gem-difluorohomoallyl alcohol. *J. Fluorine Chem.* **2007**, *128*, 535–539.

20. Wucherpfennig, U.; Logothetis, T.A.; Eilitz, U.; Burger, K. 2-*C*-Trifluoromethyl substituted 3-deoxypentoses. *Tetrahedron* **1996**, *52*, 143–148.

21. Logothetis, T.A.; Eilitz, U.; Hiller, W.; Burger, K. 2-*C*-Trifluoromethyl substituted pentoses. *Tetrahedron* **1998**, *54*, 14023–14030.

22. Hojo, M.; Masuda, R.; Okada, D. A facile synthesis of 2,4-dialkoxy-, 2-alkoxy-4-phenoxy- and 2,4-diphenoxy-6-trifluoromethyl-3,4-dihydro-2*H*-pyrans. Hetero–Diels–Alder reactions of *trans*-β-trifluoroacetylvinyl ethers with various vinyl ethers. *Synthesis* **1989**, 215–216.

23. Differding, E.; Frick, W.; Lang, R.W.; Martin, P.; Schmit, C.; Veenstra, S.; Greuter, H. Fluorinated heterocycles: target in the search for bioactive compounds and tools for their preparation. *Bull. Soc. Chim. Belg.* **1990**, *99*, 647–671.

24. Hayman, C.M.; Larsen, D.S.; Brooker, S. A facile synthesis of 2,6-dideoxy-6,6,6-trifluorinated carbohydrate analogues. *Aus. J. Chem.* **1998**, *51*, 545–553.

25. Kawada, K.; Kitagawa, O.; Taguchi, T.; Hanzawa, Y.; Kobayashi, Y.; Iitaka, Y. Studies on organic fluorine compounds XLII. synthesis of trifluoromethylated sugars through Aldol reaction of 2-trimethylsilyloxy-4-trifluoromethylfuran. *Chem. Pharm. Bull.* **1985**, *33*, 4216–4222.

26. Yamazaki, T.; Mizutani, K.; Takeda, M.; Kitazume, T. Chiral trifluoromethylated 2-butenolides for the construction of 6-deoxy-6,6,6-trifluorosugars. *J. Chem. Soc., Chem. Commun.* **1992**, 55–57.

27. Yamazaki, T.; Mizutani, K.; Kitazume, T. Preparation of 6-deoxy-6,6,6-trifluoro-D-mannose and D-allose from enzymatically resolved 2-butenolides. *Tetrahedron: Asymmetry* **1993**, *4*, 1059–1062.

28. Wang, B.L.; Yu, F.; Qiu, X.L.; Jiang, Z.X.; Qing, F.L. Synthesis of trifluoromethylated analogues of β-L-fucofuranose and β-L-4,6-dideoxyxylohexopyranose. *J. Fluorine Chem.* **2006**, *127*, 580–587.

29. Kitazume, T.; Murata, K.; Okabe, A.; Takahashi, Y.; Yamazaki, T. A highly stereocontrolled synthetic approach to 1,6-dideoxy-6,6-difluoroazasugar derivatives. *Tetrahedron: Asymmetry* **1994**, *5*, 1029–1040.

30. Wang, R.W.; Qing, F.L. Highly stereocontrolled synthesis of *gem*-difluoromethylenated azasugars: D- and L-1,4,6-trideoxy-4,4-difluoronojirimycin. *Org. Lett.* **2005**, *7*, 2189–2192.

31. Wang, R.W.; Qiu, X.L., Bols, M., Ortega-Caballero, F.; Qing, F.L. Synthesis and biological evaluation of glycosidase inhibitors: *gem*-difluoromethylenated nojirimycin analogues. *J. Med. Chem.* **2006**, *49*, 2989–2997.

6

SYNTHESIS OF RING-FLUORINATED PYRIDINES

Anatoliy M. Shestopalov[1], Alexander A. Shestopalov[2], Lyudmila A. Rodinovskaya[1], and Anna V. Gromova[3]

[1] *N.D. Zelinsky Institute of Organic Chemistry, RAS, 47 Leninsky Prospect Moscow 119991, Russia*
[2] *Department of Chemistry, Duke University, Durham, NC 27708-0346, USA*
[3] *UNC Eshelman School of Pharmacy, Division of Medicinal Chemistry and Natural Products, The University of North Carolina, Chapel Hill, NC 27599-7360, USA*

6.1 INTRODUCTION

An arising interest toward fluoropyridines is explained by their interesting and unusual physical, chemical, and biological properties owing to the presence of the strong electron-withdrawing substituent(s) in the aromatic ring. Fluoropyridines have reduced basicity and are usually less reactive than their chlorinated and brominated analogues. A selective synthesis of fluoropyridines remains a challenging problem. Here a synthetic method for preparation of 2-, 3-, 4-fluoropyridines and di- and polyfluoropyridines are reviewed, along with some synthetic routes toward ^{18}*F-substituted* pyridines, which present a special interest as potential imaging agents for various biological applications.

6.2 SYNTHESIS OF 2-FLUOROPYRIDINES

6.2.1 Direct Fluorination of Pyridines

Direct fluorination of pyridine or its derivatives with fluorine in the inert atmosphere (F_2/N_2 mixture) is not regioselective and usually gives 2-fluoropyridines in low yield.

Fluorinated Heterocyclic Compounds: Synthesis, Chemistry, and Applications, Edited by Viacheslav A. Petrov

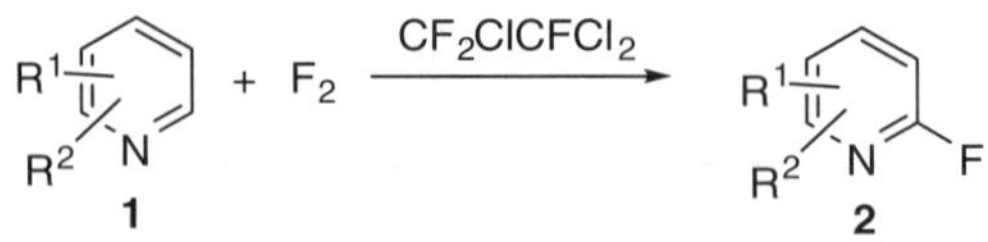

R^1, R^2 = Me, Et, Bz, i-Pr, Cl, COOMe, Ac

SCHEME 6.1

This phenomenon can be explained by low selectivity of fluorination process leading to the formation of hard to separate mixtures of difluoro- and polyfluoropyridines.

In the late 1980s, Van Der Puy reported a direct fluorination of substituted pyridines **1** by the fluorine–nitrogen mixture, producing substituted 2-fluoropyridines **2** in low to moderate yields. The reaction was conducted at −25 to 0°C by passing F_2/N_2 mixture through the 0.1 molar solutions of pyridine **1** in CFC-113 (Scheme 6.1, Table 6.1).[1,2]

Another interesting protocol utilized fluorine–iodine mixtures as fluorinated agents to prepare 2-fluoropyridines **4** (Table 6.2).[3] This regioselective reaction was

TABLE 6.1 Preparation of the Substituted 2-Fluoropyridines 2 by Direct Fluorination of Pyridines

Pyridine **1**	Fluoropyridine **2**	Yield, % (Literature)
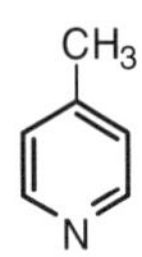	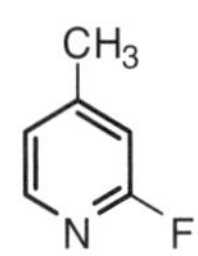	24 (Ref. 1)
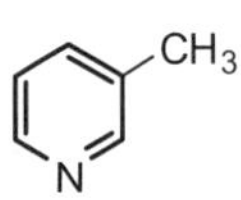	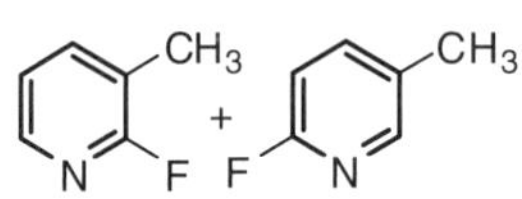	43 (total) (Ref. 1)
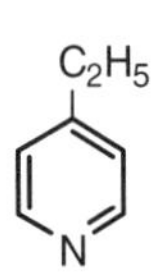	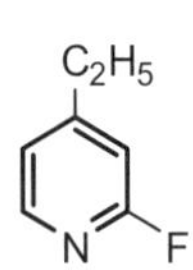	32 (Ref. 1)
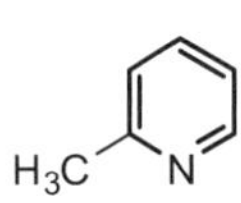	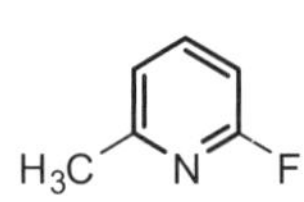	6 (Ref. 1)
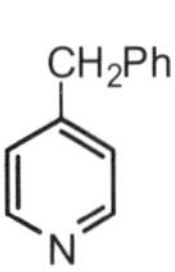	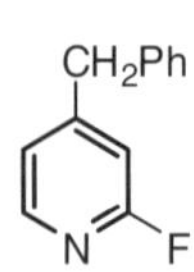	25 (Refs 1,2)

TABLE 6.1 (*Continued*)

Pyridine **1**	Fluoropyridine **2**	Yield, % (Literature)
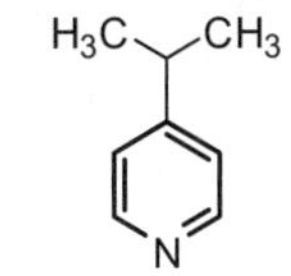	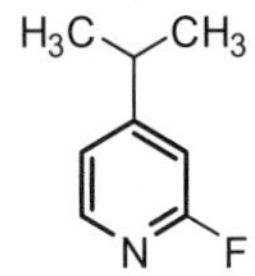	47 (Refs 1,2)
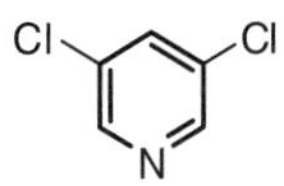	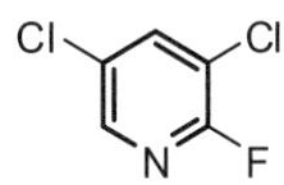	8 (Ref. 1)
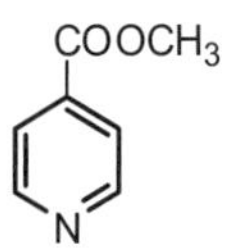	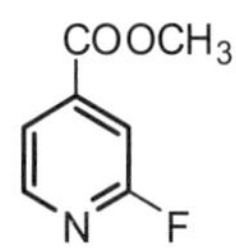	46 (Ref. 2)
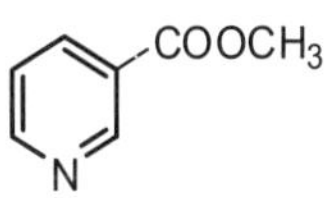	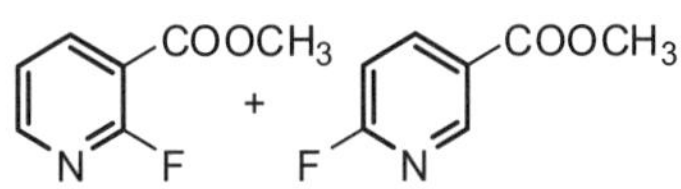	36 (total) (Ref. 2)
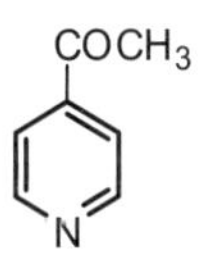	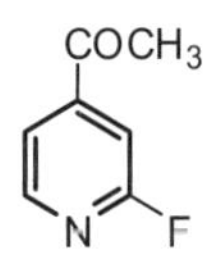	61 (Ref. 1)
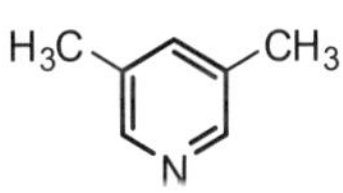	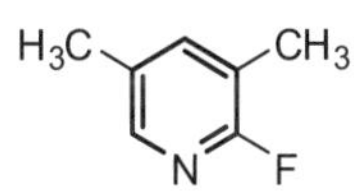	37 (Ref. 2)

performed by passing a 10% F_2 in nitrogen into the solution of the pyridine, iodine, and triethylamine in CFC-113 (Scheme 6.2).

2-Chloropyridines in this reaction along with the expected products also form products of a halogen exchange (Table 6.2).

TABLE 6.2 Yield of the Substituted 2-Fluoropyridines 4

Pyridine	Fluoropyridine	R^1	R^2	Yield % (Conversion %)
3a	**4a**	H	H	56 (59)
3b	**4b**	Et	H	54 (78)
3c	**4c** + **4a**	H	Cl	**4a** 14 **4c** 70 (61)
3d	**4d** + **4a**	H	Br	**4a** 30 **4d** 59 (100)

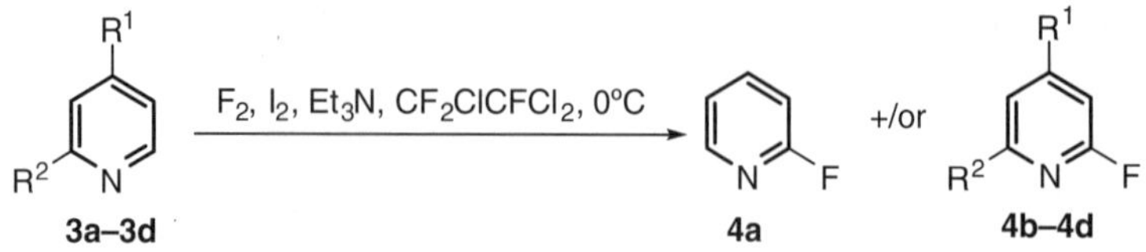

SCHEME 6.2

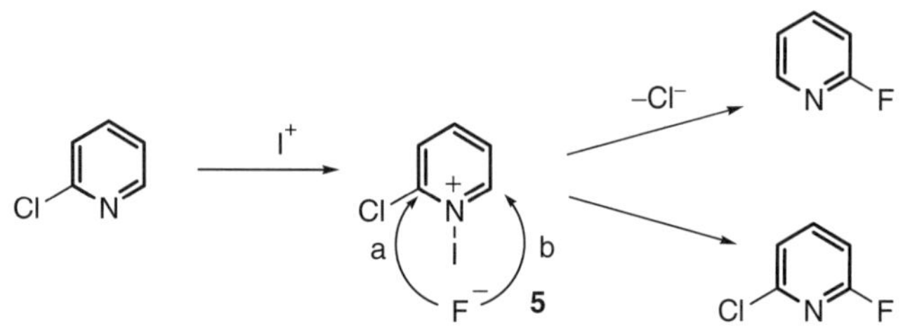

SCHEME 6.3

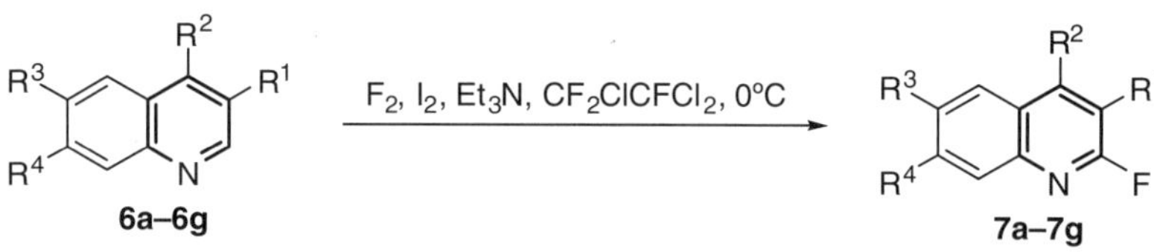

SCHEME 6.4

The high regioselectivity of this reaction is the result of the formation of pyridine–iodine complexes **5**, in which electrophilicity of the α-carbons in pyridine ring is significantly amplified, which in its turn governs the formation of 2-fluoropyridines over the other isomers (Scheme 6.3).

Under the same conditions, quinolines and quinoxalines also undergo regioselective fluorination and give monofluorinated products **7a–7g** in high yields[3] (Scheme 6.4, Table 6.3).

TABLE 6.3 Yield of the Substituted 2-Fluoroquinolines 7

Quinoline	Fluoroquinoline	R^1	R^2	R^3	R^4	Yield % (Conversion %)
7a	**7a**	H	H	H	H	54 (77)
7b	**7b**	Br	H	H	H	85 (56)
7c	**7c**	H	Cl	H	H	90 (76)
7d	**7d**	H	Me	H	H	49 (58)
7e	**7e**	H	H	Cl	H	82 (81)
7f	**7f**	H	Cl	H	Cl	88 (69)
7g	**7g**	H	Cl	H	CF_3	84 (74)

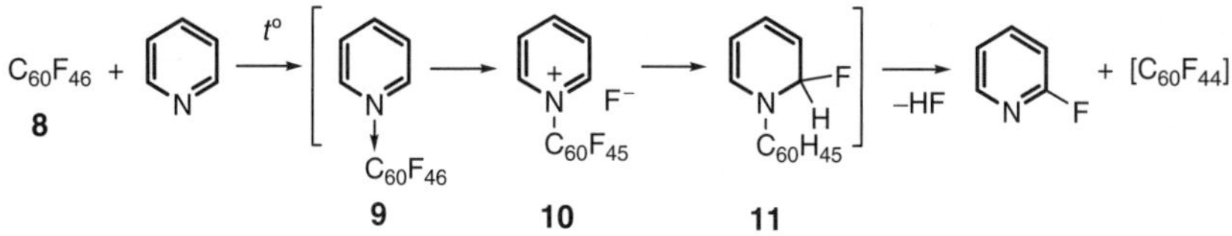

SCHEME 6.5

Gakh and coworker developed a regioselective method for fluorination of pyridines using fluorofullerenes.[4] This reaction proceeds as an ionic process and can be described as a sequence of *N*-addition followed by cine substitution (Scheme 6.5).[4–7]

6.2.2 Utilization of *N*-Fluoropyridinium Salts for the Synthesis of 2-Fluoropyridines

N-Fluoropyridinium salts **12**, efficient precursors in the synthesis of substituted 2-fluoropyridines, can be conveniently prepared in good yields by the reaction of the corresponding pyridine with F_2/N_2 in the presence of strong acid.[8] *N*-Fluoropyridinium tetrafluoroborates, hexafluoroantimonates, or hexafluorophosphates (**12**, X = BF_4, SbF_6, and PF_6) upon treatment with a base undergo an exothermic reaction selectively forming 2-fluoropyridines in moderate to high yield (Scheme 6.6, Table 6.4).[9]

As it can be seen from Table 6.4, the reaction yields depend on the media's basicity and in a stronger degree on the presence of substituents in the pyridine ring. In addition, it was demonstrated that the yields of compounds **13** using ammonium fluoride as a base without a solvent were identical to the yields of **13** using Et_3N. Based on experimental data it was suggested that the fluorine substituent in product **13** arrives from counter anion (BF_4^-, SbF_6^-, or PF_6^-).[9]

Compound **13** can be obtained in one-pot process by reacting the corresponding pyridines with F_2/N_2 mixture followed by the subsequent treatment with Et_3N.[9] However, the yields of the fluorinated pyridines obtained by this protocol are significantly lower (22–35%).

The mechanism of this reaction was discussed in several publications.[8,9] It was demonstrated that under workup with triethylamine in CH_2Cl_2 or CH_2Br_2, triflate salt **14** gives a mixture of three compounds: 2-halopyridine **15**, compound **16**, and 2-fluoropyridine (Scheme 6.7).[8] Similarly, it was demonstrated that salts **12** give 2-diethylaminopyridines, 2-phenylaminopyridines, or 2-(2-furyl and 3-furyl) pyridines when they are reacted with Et_2NH, benzene, or furan.

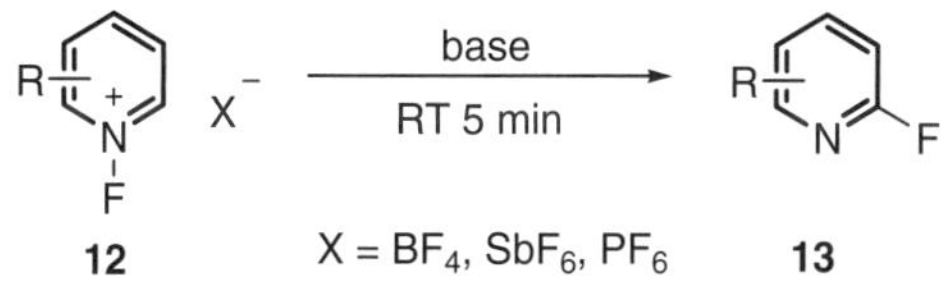

SCHEME 6.6

TABLE 6.4 Preparation of 2-Fluoropyridines 13 from *N*-Fluoropyridinium Salts 12[9]

R	X	Base (equiv)	Yield %
H	BF_4	Et_3N (1)	66
H	BF_4	Et_3N (3)	73
H	BF_4	Et_3N (10)	79
H	BF_4	$n\text{-}Bu^4N^+F^-$ (2.6)	80
H	SbF_6	Et_3N (10)	78
H	BF_4	KF (9) (7 days, 40°C)	26
H	PF_6	Et_3N (10)	74
4-Me	BF_4	Et_3N (10)	80
3,5-$(Me)_2$	BF_4	Et_3N (10)	87
3,5-$(Me)_2$	BF_4	Py (10)	30
4-*t*-Bu	BF_4	Et_3N (10)	91
2-MeO	BF_4	Et_3N (10)	75
2-MeO	BF_4	Py (10)	10
3,5-*bis*(CF_3)	BF_4	Et_3N (10)	99
3-CN	BF_4	Et_3N (10)	51
3-CN	BF_4	Py (10)	49
4-NO_2	BF_4	Et_3N (10)	21
4-NO_2	BF_4	Py (10)	31

It was proposed that under basic conditions, salt **14** undergoes heterolytic C^2–H bond cleavage with formation of carbene **17** ↔ **18**, which in its turn eliminates F^- to give cation **19** ↔ **20**. A subsequent reaction of **19** ↔ **20** with nucleophiles or π-electron-containing molecules gives above-mentioned products (Scheme 6.8).

Some chemical transformations of intermediate **17** lead to 2-substituted pyridines **15** and **21–24** are shown in Scheme 6.9.[8,10]

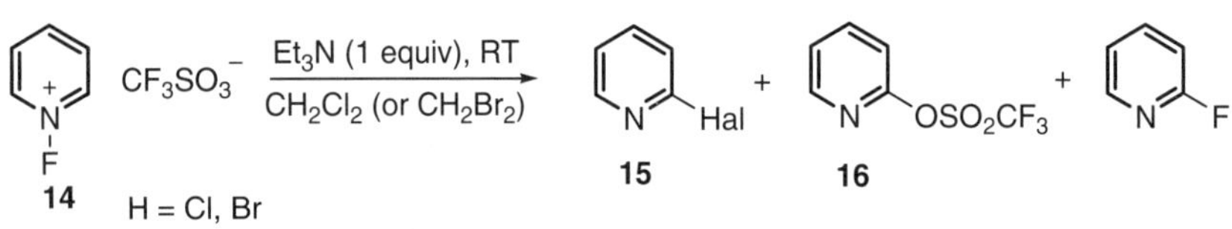

SCHEME 6.7

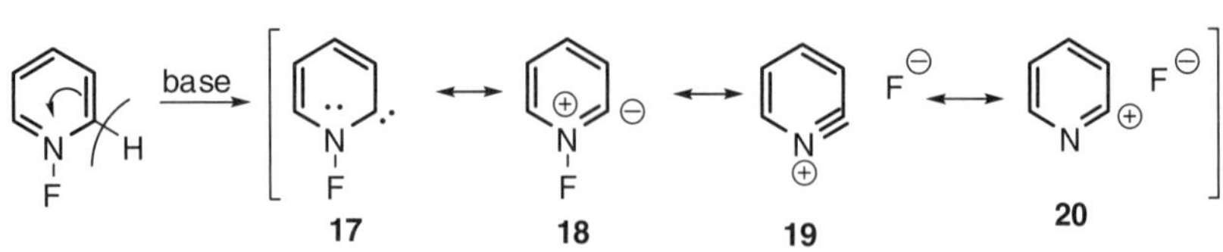

SCHEME 6.8

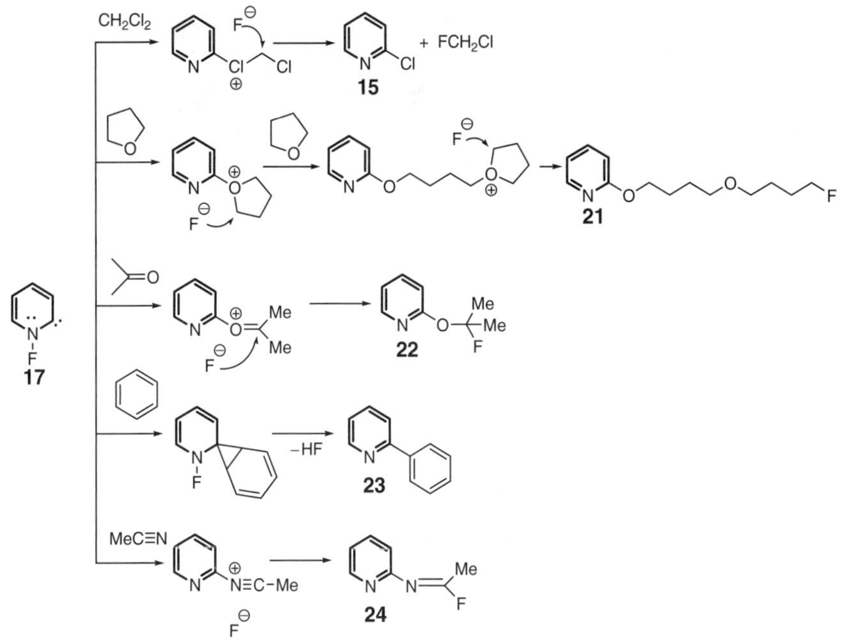

SCHEME 6.9

Direct fluorination of pyridine can also be carried out using $CsSO_4F$ as a source of fluorine. It was shown that pyridine readily reacts with $CsSO_4F$ at room temperature producing a mixture of different products (2-fluoro-, 2-fluorosulfonate- and 2-chloro- or 2-alkoxypyridines).[11]

6.2.3 Synthesis of 2-Fluoropyridines from 2-Aminopyridines

One of the typical examples of the Baltz–Schiemann reaction is synthesis of fluoro-substituted pyridines **26** from aminopyridines **25** (Scheme 6.10).[12] In this variation, the Baltz–Schiemann reaction is most often used for the synthesis of 2-fluoropyridines.[13] In the first step, a diazonium tetrafluoroborate is generated from 2-aminopyridine, $NaNO_2$, and solution of HF and BF_3 (HBF_4), while subsequent thermal decomposition of the diazonium salt leads to formation of 2-fluoropyridines. In this chapter, we present several literature examples of synthesis of 2-fluoropyridines and examples of specific use of the Baltz–Schiemann reaction for preparation of biologically active derivatives of 2-fluoropyridines.

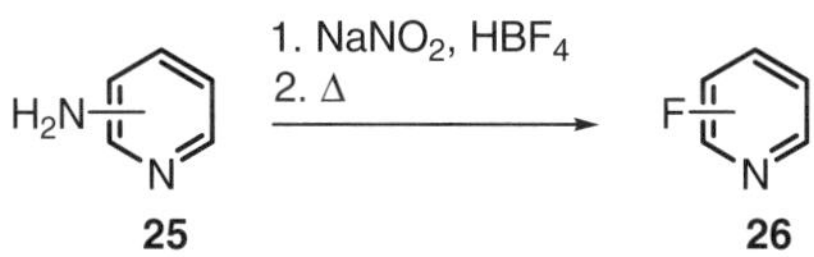

SCHEME 6.10

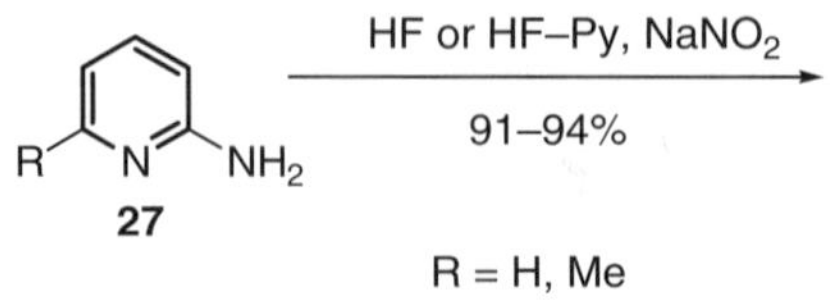

SCHEME 6.11

Several variations of the Baltz–Schiemann reaction allow synthesis of fluorinated pyridines in almost quantitative yields. For example, 2-fluoropyridines **28** were prepared in 91–94% yields by diazotation of 2-aminopyridines **27** with sodium nitrite in anhydrous HF or HF–pyridine complex (Scheme 6.11).[14]

Substituted 2-fluoro-5-fluoroalkoxypyridines **30** were prepared in good to high yields by diazotation of substituted 2-aminopyridines **29** with $NaNO_2$ in HF (Scheme 6.12). Subsequently, they were used as starting materials for the synthesis of herbicides and insecticides.[15]

3-Hydroxy-2-fluoropyridine (**32**) was prepared from 2-amino-3-hydroxypyridine (**31**) by diazotation with $NaNO_2$ in HBF_4 solution[16] (Scheme 6.13). Subsequently, compound **32** was used for the preparation of 2-fluoro-3-[2(*S*)-2-azetidinylmethoxy] pyridine (**33**), a closely related analogue of the high-affinity nicotinic ligand A-85380.

Synthesis of *exo*-2-(2′-fluoro-substituted 5′-pyridinyl)-7-azabicyclo[2.2.1] heptanes **35**, novel nicotinic receptor antagonists, was based on diazotation reaction of corresponding 2′-aminopyridines **34** using HF–pyridine complex (Scheme 6.14).[17–20]

2-Halopyridine-5-yl-boronic acids **38** and esters **39** that are useful reagents for the Suzuki coupling reaction were prepared from the corresponding 2-fluoro-5-iodopyridine **37** according to the Scheme 6.15.[21] Compound **37** was prepared by the diazotation reaction of 2-amino-5-iodopyridine (**36**) with $NaNO_2$ in HBF_4.

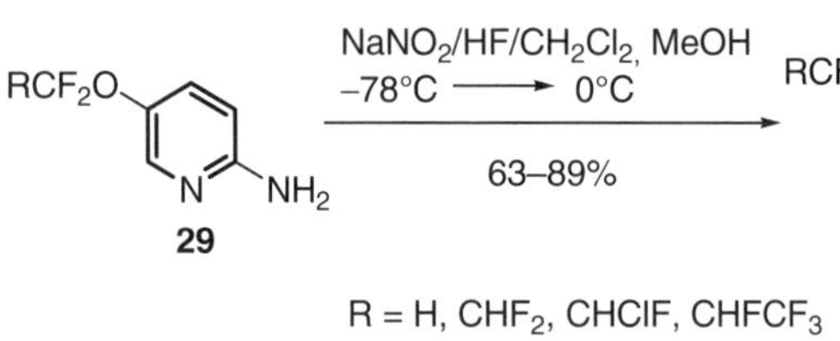

SCHEME 6.12

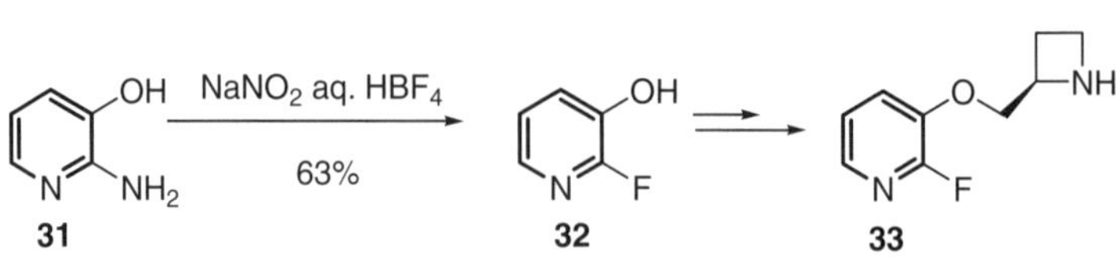

SCHEME 6.13

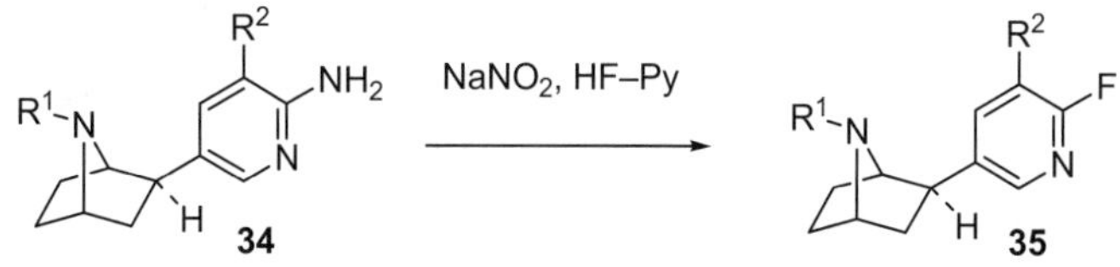

SCHEME 6.14

2-Fluoro-5-methylpyridine, which is often used for the synthesis of herbicides, can be also prepared by the Baltz–Schiemann reaction from the corresponding 2-amino-5-methylpyridine.[22]

Nucleoside analogues can be used to investigate a variety of enzyme–substrate interactions, including polymerase dNTP recognition or protein–DNA targeting. They can also be incorporated into nucleic acid sequences using conventional synthesis protocols to explore the structural and functional aspects of DNA or RNA. In one class of DNA analogues, –F replaces the carbonyls and $-CH_3$ replaces the exocyclic amino groups in the nucleobase heterocycle yielding a hydrophobic isoster of the natural nucleoside with the desired molecular shape.[23–25] Substituted 2-fluoropyridines were recently used in the synthesis of pyridine C-nucleosides as analogues of the natural nucleosides dC and dU.[26] Commercially available 2,6-diaminopyridine (**40**) was used as the starting material for these synthesis. Compound **40** was fist transformed into the 2,6-diamino-3-iododpyridine (**41**), which was acylated and then converted into 6-amino-2-fluoro-3-iodopyridine (**44**), which was transformed into 6-(4-nitrophenyldimethoxy)-2-fluoro-3-iodopyridine **45** (Scheme 6.16). Both **44** and **45** were used for the synthesis of nucleosides **46** and **47**.[26]

6.2.4 Nucleophilic Substitution in 2-Fluoro-Substituted Pyridines

Pyridines containing leaving groups (Hal, R_3N^+, SO_2R, and NO_2) in position 2 are often used in nucleophilic substitution reactions as starting materials for preparation of 2-fluoropyridines. Typical nucleophiles that are most often used in these syntheses

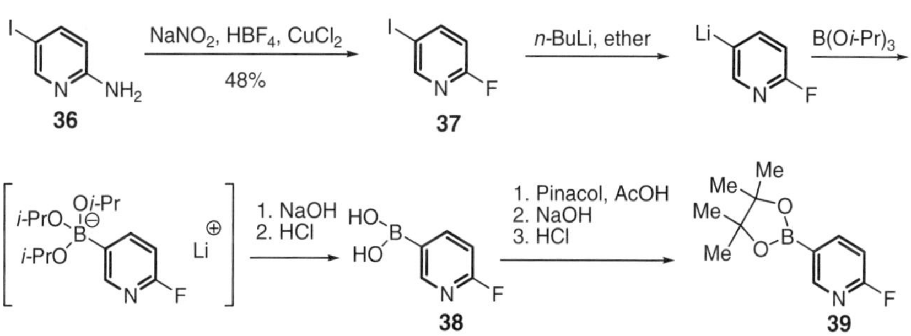

SCHEME 6.15

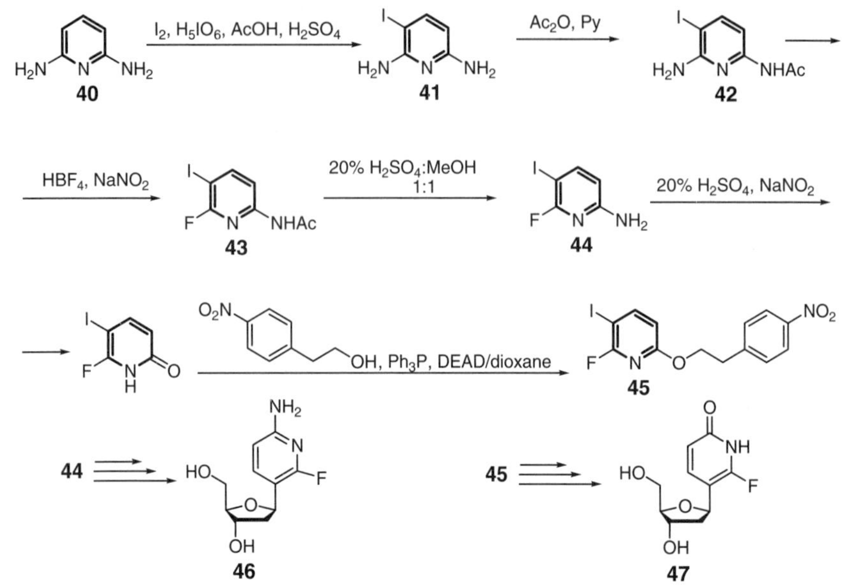

SCHEME 6.16

include fluorides of alkaline metals, hydrofluoric acid, tetrabutylammonium fluoride, and fluoroboric acid. Although this method allows preparation of 2-fluoropyridines in good yields, its main disadvantages include a set of special demands toward fluorine-producing reagents, which, if not otherwise met, will significantly reduce the yield of the final products. In majority of all cases, these reactions must be conducted in a dry aprotic solvents (DMSO, DMF, and THF) with fluoride source introduced as a fine dry powder (normally due to its low solubility in these solvents), since the hydration significantly reduces the nucleophilicity of fluoride anion. Dry environment for these reactions is dictated by a very high solvolysis ratio of the fluoride anion in water, which in its turn significantly increases its steric hindrance and reduces its nucleophilicity. However, in some cases, high reactivity of the fluoride anion in water-organic solvent two-phase system can be maintained, for example, using crown ethers.[27] Recently, it was shown that bulky *tert*-butanol as a solvent in nucleophilic substitution reactions gives only partially shielded solvates with fluoride anion and actually increases fluoride anion reactivity.[28]

It was shown that 2-pyridines **48** containing chlorine substituent in position 3 can be selectively converted into 2-fluoropyridines **49** by treatment with KF.[29] The reactions were conducted at elevated temperature (100–200°C) producing final pyridines **49** in 14–94% yields (Table 6.5).

Phase-transfer catalysis (PTC) is one of the most important methods for the enhancement of reactivity of inorganic reagents having poor solubility in organic solvents. As such, PTC is often used for reactions involving potassium fluoride, which has very low solubility in aprotic solvents. Both types of phase-transfer catalysts, quaternary ammonium salts and 18-crown-6, have been used to accelerate

TABLE 6.5 Preparation of the Substituted 2-fluoropyridines 49 from 2-pyridines 48

48a–48f **49a–49f**

Compounds	X	Y	R	Temp. °C	Compounds	Y	R	Yield %
48a	Cl	Cl	Cl	200	**49a**	Cl	Cl	76.6
48b	Cl	H	Cl	200	**49b**	H	Cl	72.4
48c	Cl	H	CH_3	200	**49c**	H	CH_3	33
48d	Cl	Cl	CH_3	200	**49d**	Cl	CH_3	69.4
48e	Cl	H	CF_3	200	**49e**	H	CF_3	83–94
48f	Br	H	NO_2	100	**49f**	H	NO_2	14

such reactions. However, recent studies on the reactivity of KF-quaternary ammonium salt systems have indicated some problems associated with their use in fluorination reactions, notably the very low extraction coefficient of F^- and the low thermal stability of these catalysts in the presence of F^-. On the other hand, tetraarylphosphonium fluorides show high thermal stability and strong $P^+ \cdots F^-$ interactions in the complexes with KF. Recently, KF-tetrarylphosphonium salts were described as useful new systems for the nucleophilic fluorinations. As an example, 2-chloro-5-nitropyridine **50** was completely converted into 2-fluoro-3-nitropyridine **51** using anhydrous KF and KF-Ph_4PBr in acetonitrile at 120°C.[30] The reaction yielded fluorinated pyridine in high yield, whereas in the reaction with anhydrous KF in DMF at 120°C, compound **51** was obtained from **50** in 78% yield only (Scheme 6.17).[31]

Fluorinated pyridines are often used as starting materials for the preparation of potent radioligands for *in vivo* imaging by positron emission tomography (PET). For example, the nucleophilic fluorination was used to prepare ^{18}F-containing ligand for imaging of central nicotinic acetylcholine receptors. In this work, [^{18}F]FK–K_{222} complex was prepared from cyclotron-produced [^{18}F] fluoride ion, and then was used to obtain 2-[^{18}F]fluoropyridine from its standard precursors (Scheme 6.18).[32]

Another highly potent radioligand 2-[^{18}F]fluoro-3-[2(*S*)-2-azetidinylmethoxy] pyridine **54** was similarly prepared in two steps from the Boc-protected nitro precursor **52** (Scheme 6.19).[16]

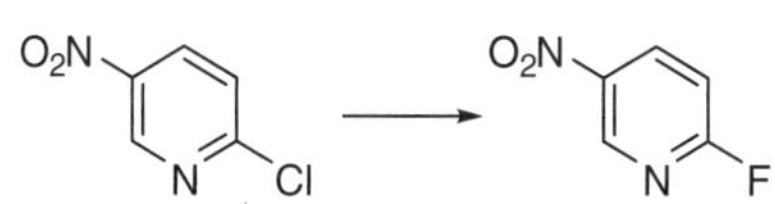

50 **51**

SCHEME 6.17

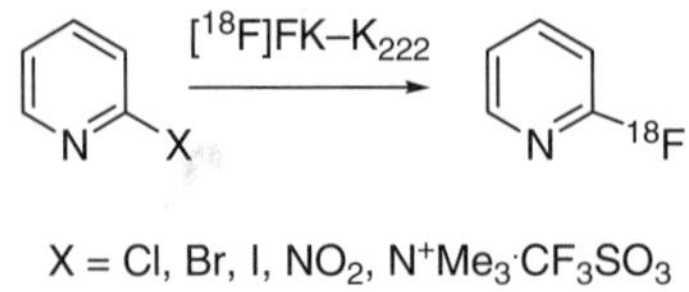

SCHEME 6.18

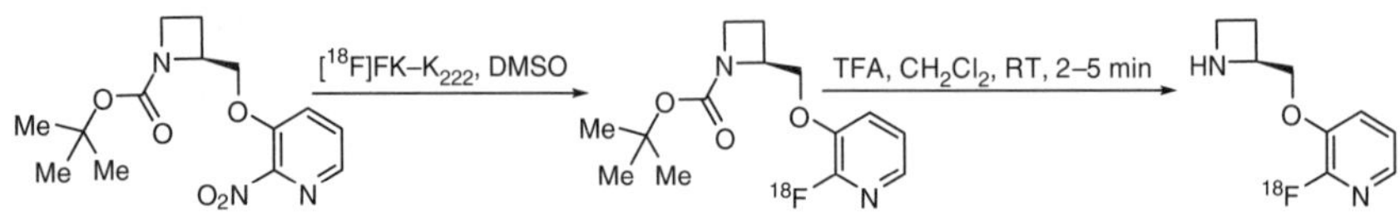

52 53 54

SCHEME 6.19

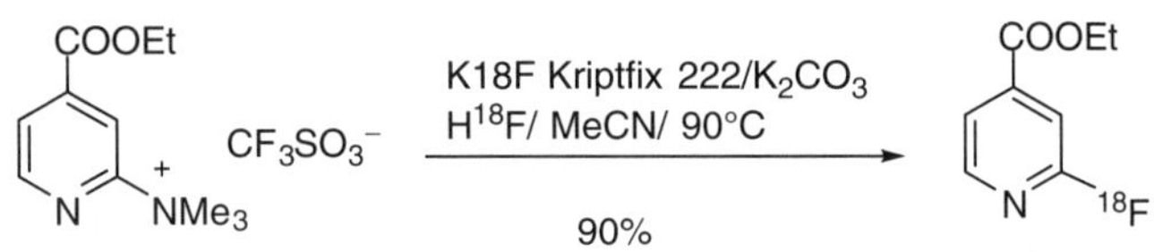

55 56

SCHEME 6.20

Similar strategy was employed for the preparation of 6-chloro-3[(2-(*S*)-azetidinyl) methoxy]-5-(2-[^{18}F]fluoropyridine, a novel potential radioligand for studying exothalamic nicotylcholine receptors by PET,[33] and ethyl ester of 2-[^{18}F]fluoroisonicotinic acid **56**, which was used in the synthesis of pyridinecarbohydrazide folates (Scheme 6.20).[34,35] Compound **56** was obtained in high yield from trimethylammonium salt **55** using [^{18}F]FK–K_{222} complex in a mixture of [^{18}F]FH–acetonitrile at 90°C.

Similarly, 2-amino-6-(2-[^{18}F]fluoropyridine-4-ylmethoxy)-9-(octyl-β-D-glucosyl)-purine **57**, a novel radioligand for PET studies of the O6-methylguanine-DNA methyltransferase activity in brain tumors, was prepared in several steps from the corresponding 2-chloropyridine and dry 2.2.2 Kryptofix/[^{18}F]fluoride complex in DMF as a solvent.[36]

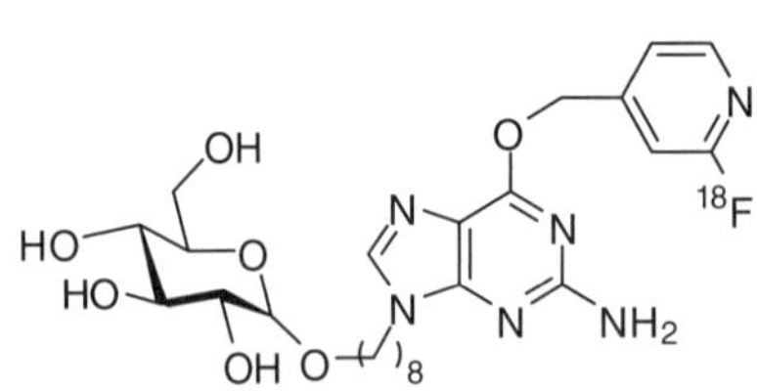

57

Fluoropyridyl derivatives of [3,2-*c*]pyrazollocorticosteroids **58** and **59** show excellent anti-inflammatory activity, which is considerably better than the activity of potent glucocorticoid dexamethasone. In this case, fluorination of the precursors was achieved using a nucleophilic substitution reaction, which was also used for radiolabeling with the positron-emitting isotope ^{18}F. It was proposed that considering its superior biological activity and adaptability for facile radiosynthesis, compound **59** has the potential for imaging of glucocorticoid receptor containing tissues using PET.[37,38] Initially, thermal activation was used to accelerate the exchange of substituents in compounds **58** on ^{18}F[37]; however, taking into account a short half-life of the ^{18}F ($t_{1/2} = 110$ min), microwave irradiation was used later on to speed up the substitution of chlorine by ^{18}F in **58**.[38]

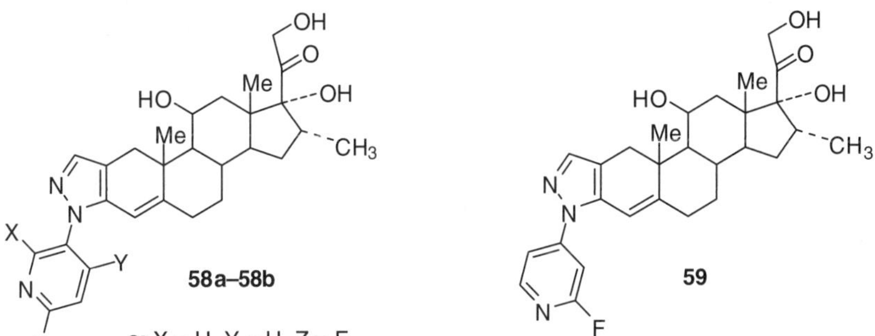

58a–58b

a: X = H, Y = H, Z = F
b: X = F, Y = H, Z = H
c: X = H, Y = F, Z = H

59

Tetrabutylammonium fluoride (TBAF) is one of the most efficient reagents in the nucleophilic fluorination reactions. It was demonstrated that TBAF can be successfully used to substitute nitro group in pyridines **60** (Scheme 6.21).[39] The developed technique allows direct synthesis of 2-fluoropyridines from 2-nitropyridines.

2-Fluoropyridines and substituted fluorodiazines were successfully prepared from the corresponding chlorazines using triethylamine tris(hydrogen fluoride) system as a selective nucleophilic fluorinating reagent.[40]

Pentasubstituted 2-fluoropyridines **66** were prepared by the reaction of perfluoroalkene **65** with *N*-silyl-1-azaallyl anion **64**, generated by coupling of a functional silane **62** and aryl/alkyl nitrile **63** using *n*-BuLi in tetrahydrofurane[41] (Scheme 6.22).

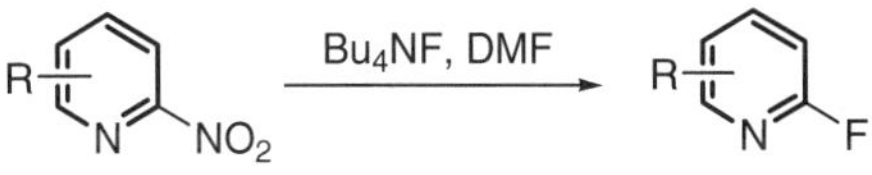

R = 3-Br, 5-Br, 3-OEt, 3-(1-OCH$_2$-3-NH$_2$-C$_6$H$_4$)

SCHEME 6.21

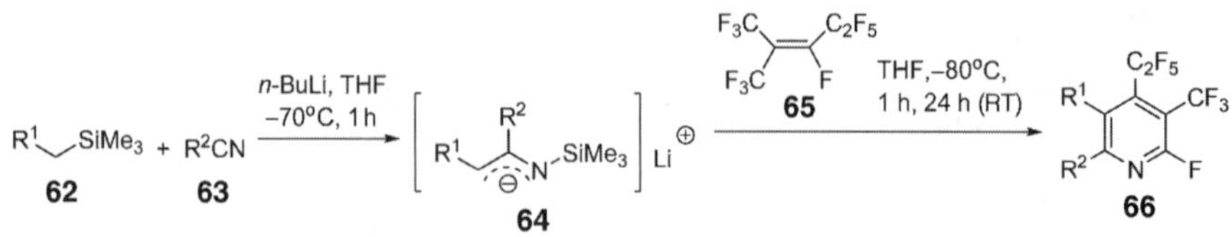

SCHEME 6.22

6.3 SYNTHESIS OF 3-FLUOROPYRIDINES

6.3.1 Direct Fluorination and Other Methods

In comparison to 2-fluoropyridines, synthetic methods leading to 3-fluoropyridines are less developed, most likely due to the decreased reactivity of the C3 atom in the pyridine cycle.

Several examples of electrophilic substitution reactions at the C3-position of the pyridium cycle can be found in literature. For example, direct fluorination of 2,4,6-trimethylpyridine **67**, in which all activated positions are occupied with methyl groups, results in the formation of *N*-fluoro-2,4,6-trimethylpyridiniumtriflate **68**,[42] which on treatment with triethylamine gives substituted 3-fluoropyridine **69** (Scheme 6.23).[42]

3-Deaza-3-fluoropurine ribonucleoside **72** was obtained in several steps from **70**. Fluorine substituent was introduced into the ribonucleoside of hydroxyimidazole **70** using DAST (Et_2NSF_3). This step was followed by the ring closure resulting in the formation of 3-fluoropyridinone **72**[43] (Scheme 6.24).

N-Azinylpyridinium *N*-aminides **73** have been shown to be convenient precursors for the regioselective synthesis of 3-fluoro-2-aminopyridines **76**[44] (Scheme 6.25). First, compound **73** was treated with xenon difluoride, which resulted in the regioselective introduction of fluorine in 3 of substituted pyridine ring leading to compounds **74**, which were further alkylated to give compounds **75** and finally, the radical reduction was used to form 3-fluoropyridines **76**.

Substituted 3-fluoropyridine **78** was synthesized by the regioselective reaction of 2,3-difluoro-5-chloropyridine **77** with hydrazine hydrate followed by the treatment with $CuSO_4$[45] (Scheme 6.26).

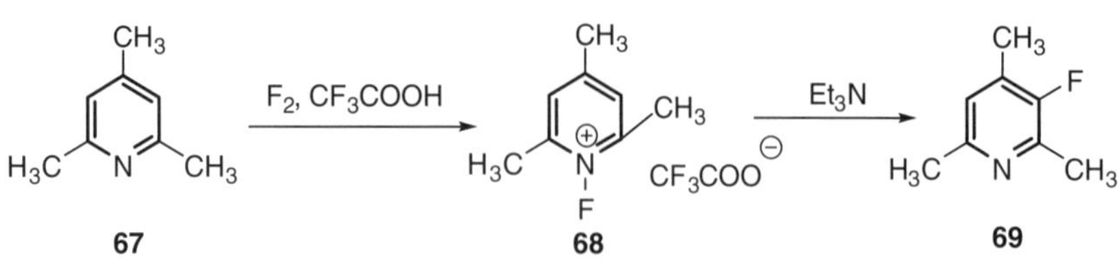

SCHEME 6.23

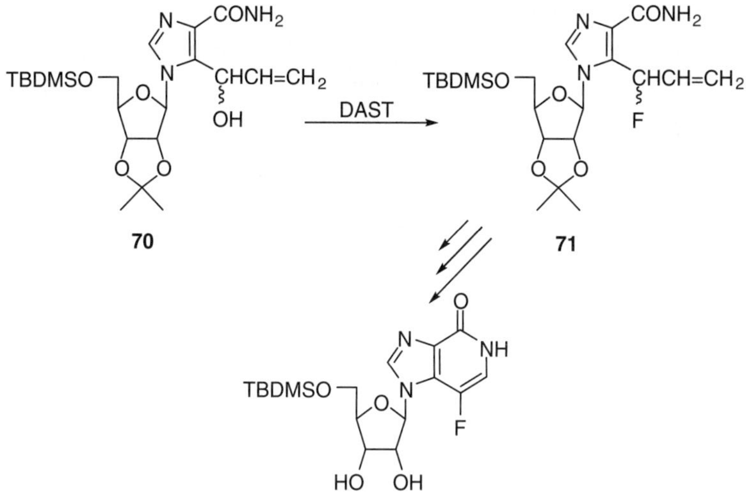

70 71 72

SCHEME 6.24

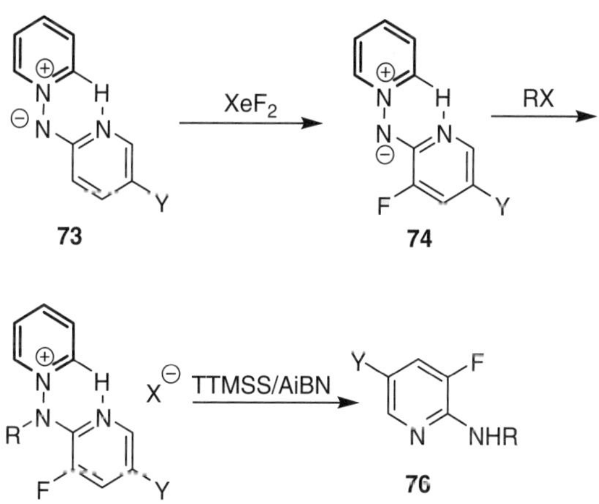

73 74 75 76

TTMSS - tris(trimethylsillyl)silane
AiBN - azoizobutyronitrile
Y = Cl, Br; R = H, CH_3, C_2H_5, *n*-C_3H_7; X = Hal

SCHEME 6.25

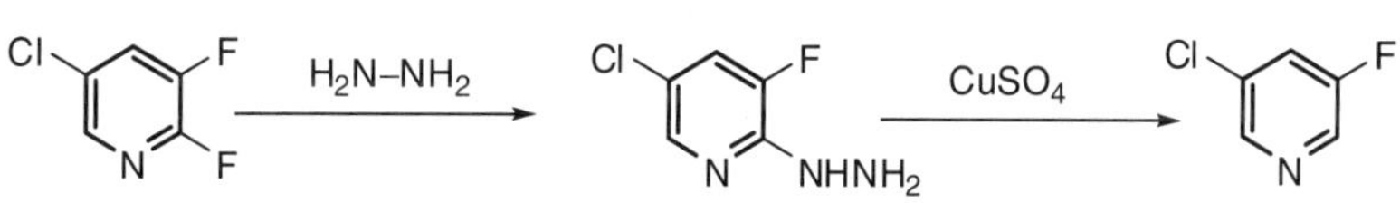

77 78

SCHEME 6.26

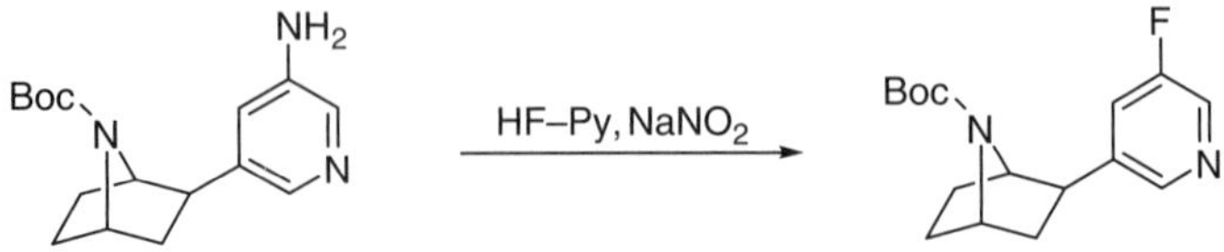

79

SCHEME 6.27

6.3.2 Baltz–Schiemann Reaction in the Synthesis of 3-Fluoropyridines

The Baltz–Schiemann reaction is the most often used method for the synthesis of 3-fluoropyridines. This method utilizes readily accessible 3-nitropyridines as the precursors, since they can be readily reduced into amines and then used in the Baltz–Schiemann reaction. In this section, selected examples applied for the synthesis of practically important compounds are given. For example, the Baltz–Schiemann reaction was used for the synthesis of fluoro-substituted epibatidine analogue **79** (epibatidine is a high-affinity nonselective ligand for nicotinic cholinergic receptor (nAChRs))[46] (Scheme 6.27).

3-Deoxy-3-fluoropyridoxamine 5′-phosphate **86** (a coenzyme B_6 analogue) was also synthesized using the Baltz–Schiemann reaction[47] (Scheme 6.28). First, substituted pyridine **80** was nitrated to form 3-nitropyridine **81**, which was subsequently treated with PCl_5 to form 2-chloro-5-nitropyridine **82**. It was then reduced in two steps to form 3-aminopyridine **84**, was converted into 3-fluoropyridine **85** by the Baltz–Schiemann reaction, and afterward was transformed into 3-deoxy-3-fluoropyridoxamine 5′-phosphate (F-PMP) **86**.

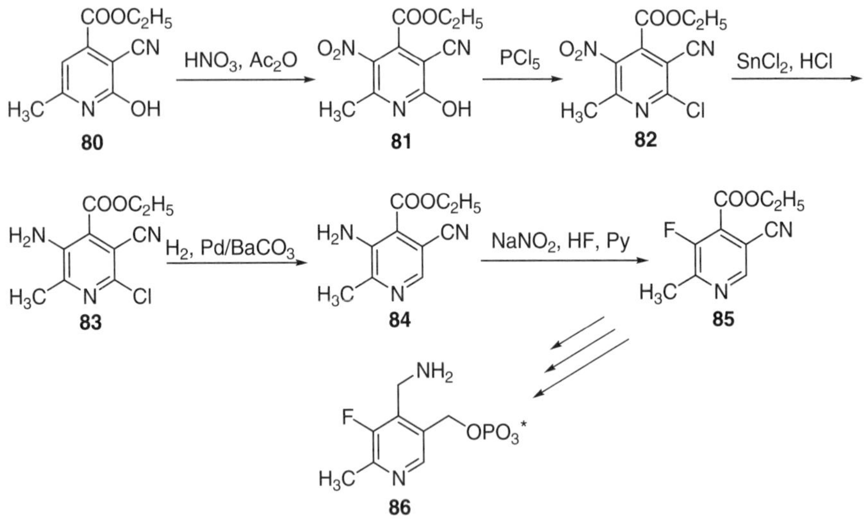

SCHEME 6.28

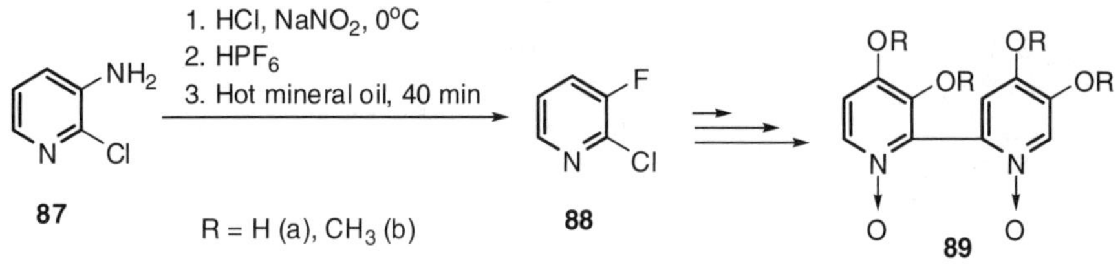

SCHEME 6.29

Similarly, the Baltz–Schiemann reaction was also used for the synthesis of 3-fluoropyridine-4-carboxylate, which can also be prepared by the nucleophilic substitution from 3-nitro-4-ethylcarboxypyridine.[48] 2-Chloro-3-fluoropyridine **88**, the key intermediate in the synthesis of orellanine **89** (toxin from *Cortinarius orellanus* Fries mushroom), was prepared starting from 3-amino-2-chloropyridine **87**[49] (Scheme 6.29).

6.3.3 Nucleophilic Substitution Reactions in the Synthesis of 3-Fluoropyridines

The nucleophilic substitution reactions leading to 3-fluoropyridines are rare. Although 2-amino (or butylthio)-3-aminopyridines do not react with TBAF,[39] the introduction of the electron-withdrawing group in position 2 of the pyridine ring in some cases makes possible such transformations. For example, 2-cyano-3-nitropyridine reacts with TBAF forming 2-cyano-3-fluoropyridine in 64% yield.[39] Similar transformations were reported for 3-substituted-4-carbethoxypyridines, which also undergo nucleophilic substitution at the position 3 of pyridine ring.[48]

6.4 SYNTHESIS OF 4-FLUOROPYRIDINES

In general, the reactivity of the pyridine ring in nucleophilic substitution reaction decreases in the row C2 > C4 > C3. Consequently, more synthetic routes are reported for 4-fluoropyridines compared to 3-fluoropyridines. Pyridines can form cationic complexes with electrophiles resulting in activation of heterocyclic ring toward nucleophilic substitution. On the other hand, pyridines have significantly reduced reactivity toward electrophiles and typically undergo electrophilic substitution reactions in the present of strong Lewis acids selectively in the position 3.[50]

6.4.1 Electrochemical Fluorination in the Synthesis of 4-Fluoropyridines

It was demonstrated that 4-fluoropyridine can be synthesized by electrochemical fluorination of pyridine on the platinum anode at constant voltage in acetonitrile solution containing $Et_3N{\cdot}3HF$ as supporting electrolytes and fluorine sources.[51]

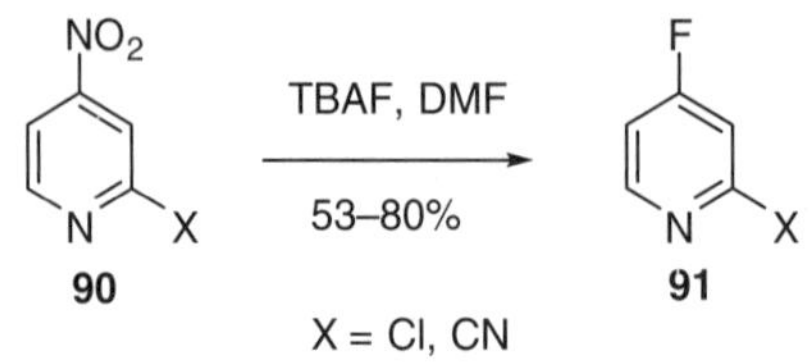

SCHEME 6.30

6.4.2 Nucleophilic Substitution Reactions in the Synthesis of 4-Fluoropyridines

Usually 4-fluoropyridines are synthesized from their nucleofuge-containing precursors by the nucleophilic substitution reaction. For example, 4-nitropyridines **90** react with TBAF in DMF with the formation of substituted 4-fluoropyridines **91**[39] (Scheme 6.30). This reaction is highly regioselective despite the presence of relatively good leaving group (Cl or CN) in position 2 of pyridine.

Radiolabeled 4-[^{18}F]fluoropyridine can be synthesized by no-carrier-added nucleophilic aromatic substitution with K[^{18}F]F–K_{222}.[52] In another instance, the nucleophilic substitution reaction was also employed for the synthesis of steroids-containing 4-fluoropyridine motif,[37,38] and for the synthesis of 4-fluoropyridines annulated with pyrrole (azoindoles).[53,54] Substantial difference in the reactivity of the pyridinium ring toward nucleophilic substitution in 5-iodo-2,4-difluorpyridine was effectively used for the preparation of 4-fluoropyridines **93** and **94** using difluoropyridine **92** as starting material[55] (Scheme 6.31).

6.4.3 Baltz–Schiemann Reaction in the Synthesis of 4-Fluoropyridines

The Baltz–Schiemann reaction can also be used for the synthesis of 4-fluoropyridine derivatives.[12,13,17–20] For example, it was successfully applied to the synthesis of

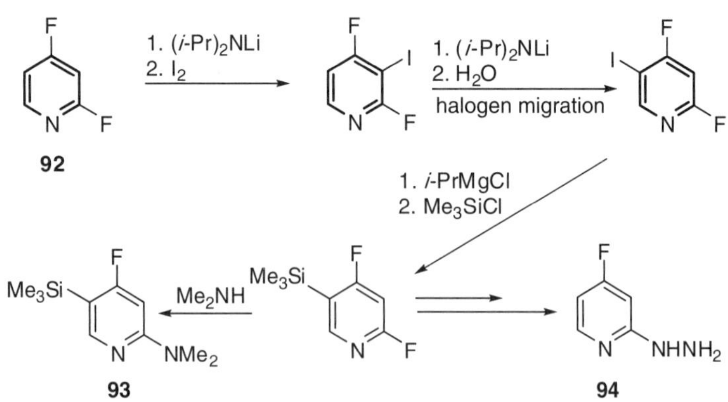

SCHEME 6.31

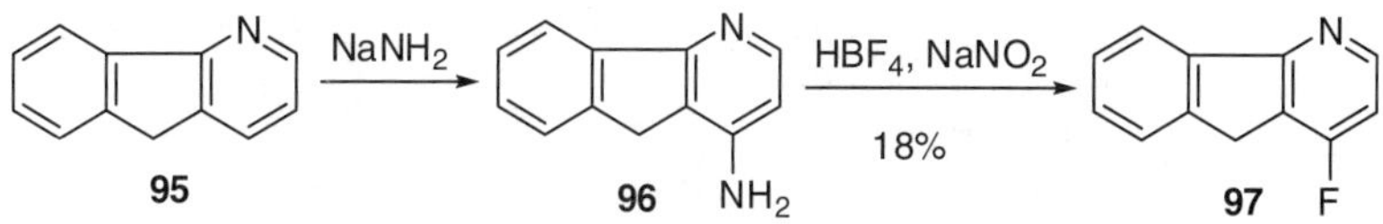

SCHEME 6.32

4-fluorofluorene.[56] First, 1-amino-4-azafluorene **96** was synthesized by amination of 4-azafluorene **95** using the Chichibabin reaction and then was converted into 1-fluoro-4-azafluoren **97** in 18% yield (Scheme 6.32).

6.5 SYNTHESIS OF DI- AND POLYFLUOROPYRIDINES

In many cases, di- and polyfluoropyridines can be prepared using the same reactions for preparation of monofluorinated analogues. The degree of fluorination in some cases can be controlled, however, often leading to mixtures of polyfluorinated compounds. Some polyfluoropyridines can be reduced back to di- or monofluoropyridines, which can be successfully used for a selective synthesis of these compounds.

6.5.1 Synthesis of Difluoropyridines

Pentafluoro- and tetrafluoropyridines, which are usually prepared from pentachloropyridine using Halex process, can be used as the starting materials for the synthesis of difluoropyridines.[45] For example, it was demonstrated that pentafluoropyridine can be utilized in the synthesis of substituted 3,5-difluoropyridines, which were investigated as new antithrombotic drugs.[57,58] However, one of the most commonly used reaction for the synthesis of difluoropyridines is a selective reduction of polyhalogenated pyridines.[45] For example, chlorodifluoropyridines **98**, **100**, and **102** can be reduced to the corresponding difluoropyridines **99**, **101**, and **103** using palladium on carbon, ammonium formate, and 80% acetic acid (Scheme 6.33). The described reaction is highly selective and only chlorine atom is getting reduced.

Similarly, a catalytic hydrogenation of 3-chloro-4,5-difluoropyridine **104** provided the mixture of 3,4-difluoropyridine **105** along with small amount of 3-fluoropyridine (ratio 95:3)[45] (Scheme 6.34).

Other possible synthetic route leading to difluororpyridines such as **103**, **105**, and **108** is based on the reductive deamination of difluoropyridine hydrazines in the presence of $CuSO_4$ often combined with the removal of SiR_3 group[45] (Scheme 6.35).

3,4-Difluoropyridine can be synthesized by the nucleophilic substitution of chlorine in 4-chloro-3-fluoropyridine with KF,[45] while, 2,5-difluoropyridine can be prepared by deamination reaction of 2-hydrazino-3,6-difluropyridine in

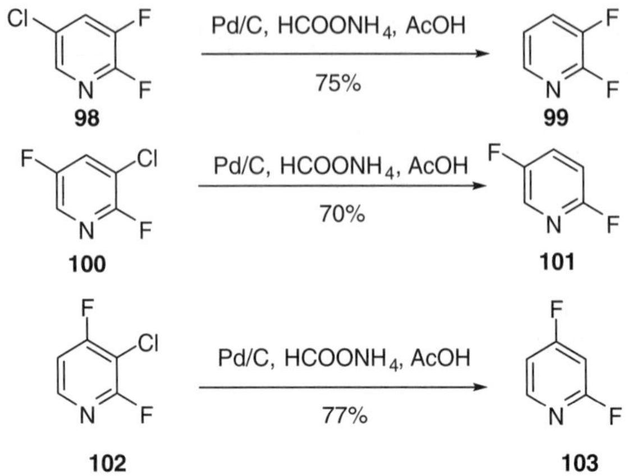

SCHEME 6.33

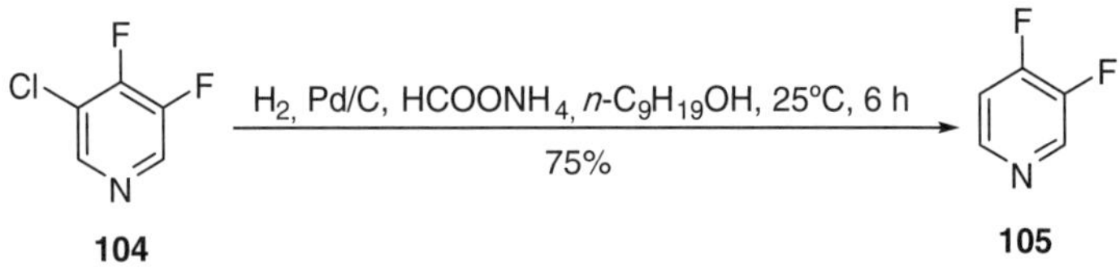

SCHEME 6.34

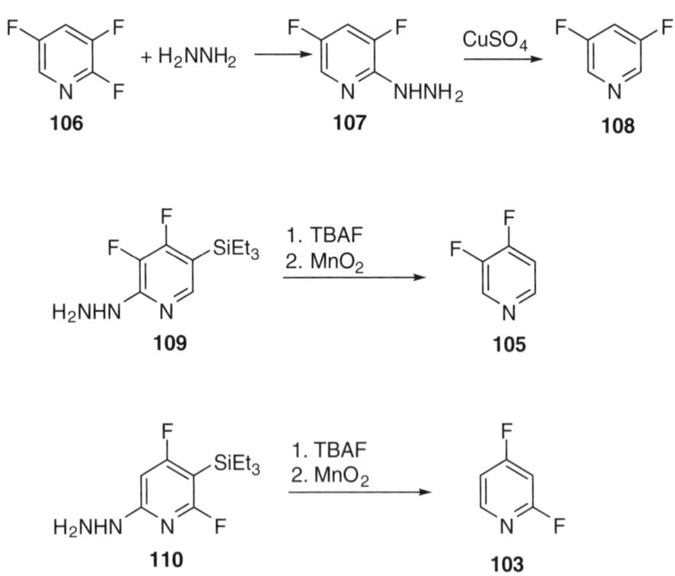

SCHEME 6.35

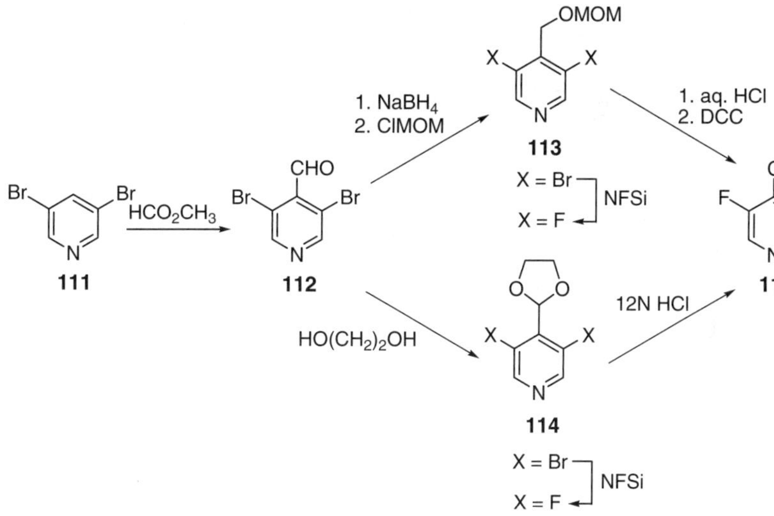

SCHEME 6.36

the presence of NaOH.[45]Scheme 6.36 shows an interesting synthesis of 2,6-difluoropyridine **115**.[59] This compound was obtained from 3,5-dibromo-4-formylpyridine **112** by electrophilic fluorination of its protected forms **113** or **114** by *N*-fluoro-benzenesulfonimide (NFSi).

As it was mentioned, substituted difluoropyridines can be used for the synthesis of monofluorinated pyridines. For example, various difluoropyridines were converted into monofluoropyridyl carboxylic acids[60,61] and hydrazines[56,62] by the reaction with the corresponding nucleophilic reagents.

6.5.2 Synthesis of Trifluoropyridines and Polyfluoropyridines

Normally, trifluoropyridines are prepared by the reduction or nucleophilic substitution of perhalogened pyridines.[45] However, the reaction of the corresponding 2,3,5-trichloropyridine with KF (sulfolane, dimethylpropyleneurea, 220°C, 16 h) resulted only in partial fluorination and formation of 2,3-difluoro-5-chloropyridine.[45] Attempts to prepare 2,3,5-trifluoropyridine from the corresponding trichloropyridine were unsuccessful. Pentachloropyridine **116** was used as the starting material in the reaction with KF, first producing dichlorotrifluoropyridine **117**. At higher temperature, this compound was converted into 3-chlorotetrafluoropyridine **118** and then pentafluoropyridine **119**[45] (Scheme 6.37, also see Chapter 8).

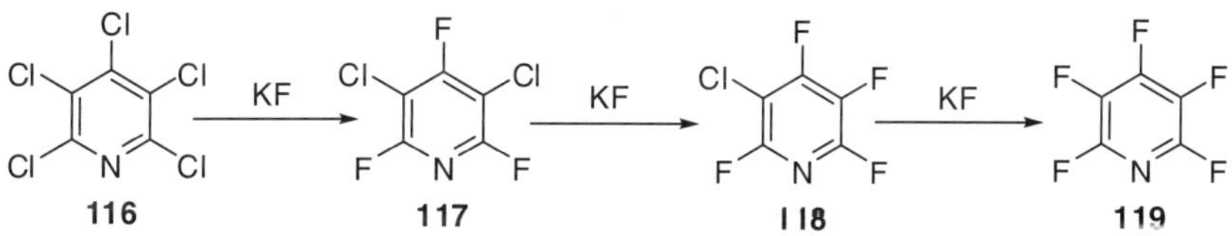

SCHEME 6.37

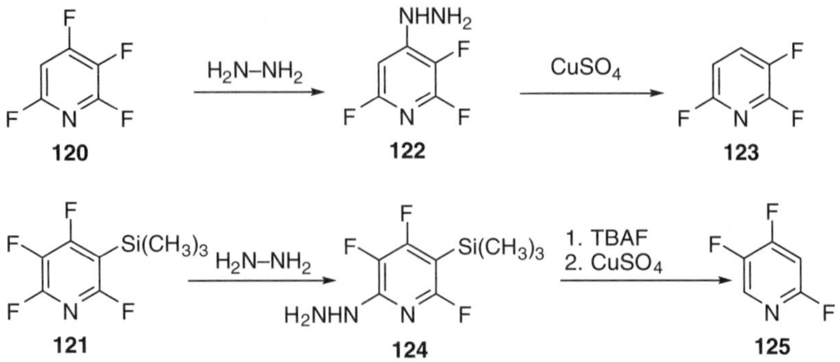

SCHEME 6.38

Tetrafluoropyridines **120** and **121** were used in the reduction reactions for the selective synthesis of 2,3,6-trifluoropyridine **123** or 2,4,5-trifluoropridine **125**[45] (Scheme 6.38).

Scheme 6.39 shows methods of synthesis of tri- and tetrafluoropyridines **130**, **131**, and **135** from corresponding di- and trifluoropyridines **99**, **108**, and **132**. The starting material is first lithiated by *n*-BuLi and then transformed into chlorofluoropyridines **128**, **129**, and **134**. The last step of the synthesis is based on Halex exchange reaction using spray-dried KF in anhydrous DMSO to give the corresponding polyfluorinated pyridines **130**, **131**, and **135**[45] (Scheme 6.39).

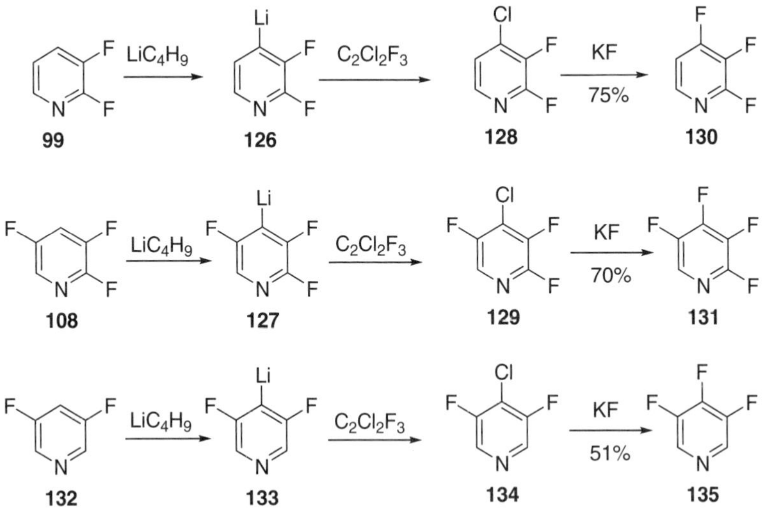

SCHEME 6.39

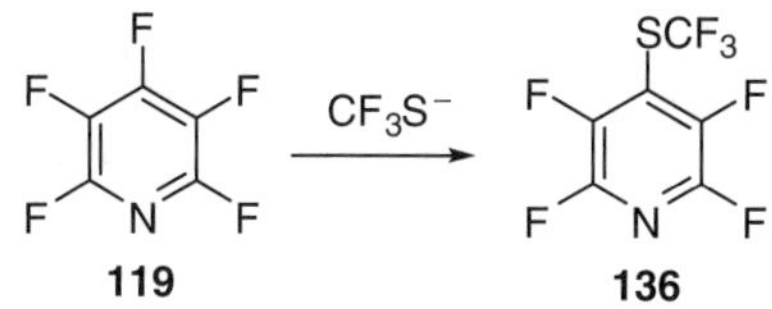

SCHEME 6.40

Mixtures of polyfluorinated pyridines can be obtained from the corresponding pyridines by fluorination with tetrafluorocobaltates(III),[63] however, this reaction has a low selectivity. For example, the reaction mixture derived from the reaction of pyridine with $KCoF_4$ at 220°C is reported to contain more than seven fluoropyridines, two fluoro-2-azahexenes, three azahexadienes, and two fluoro-*N*-methylpyrrolidines. Four fluorinated products were isolated from a fluorination of pyridine by CoF_3 at 150°C: a 2-azahexene, two *N*-methylpyrrolidines, and 4*H*-nona-fluoropiperidine.[64]

2,3,5,6-Tetrafluoro-4-trifluoromethylthiopyridine **136** was prepared in high yield by the reaction of pentafluoropyridine **119** with the CF_3S^- anion, generated from $F_2C{=}S$ or its trimer, and cesium fluoride at −15°C[65] (Scheme 6.40).

When the trimer was used as a precursor of the CF_3S^- anion, compound **136** reacted further at 20°C to give a mixture of polysubstituted pyridines **137–139**[65] (Scheme 6.41).

Syntheses of tri- and tetrafluoro-substituted pyridiniumtrifluoromethylsulfoxides and their hydroxyl(methoxy)-substituted analogues were previously reported in the same reference.[65] Reactions of pentafluoropyridine with pentafluoro- and 4-nitrophenols were also studied.[66] It was found that the first substitution occurs at the C4-position of pentafluoropyridine, subsequently giving a mixture of di- and triphenoxy-substituted fluoropyridines.

Polyfluopyridines are versatile fluorinated building blocks. They were used as starting materials in the synthesis of fluorinated chinolonic acids, which show antibacterial activity. Examples of bactericidal mono- and difluoro-substituted 1,8-naphtiridines **140** and **141** prepared from pentfluoropyridine are shown below.[67,68]

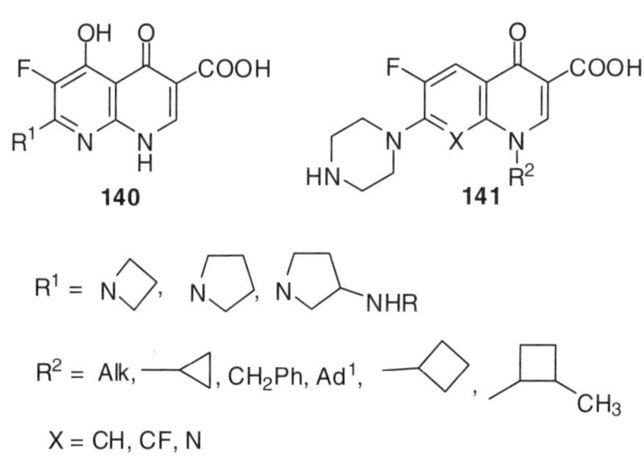

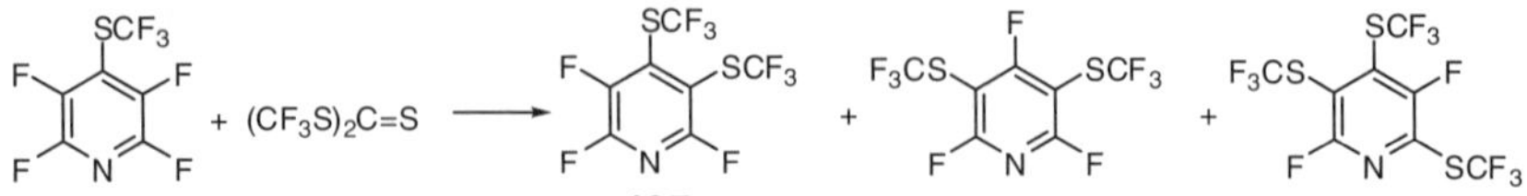

SCHEME 6.41

Additional information on the reactivity of perfluorinated aromatic heterocyclic compounds can be found in Chapter 8.

6.6 CONCLUSION

Many fluorinated pyridines have found some practical applications, especially as drugs and potential drug candidates. Therefore, the development of the selective synthetic methods for the preparation of these compounds as well as the advances in new highly efficient fluorinating techniques still remains an important task of organic chemistry.

REFERENCES

1. Van Der Puy, M.; Eibeck, R.E. *US Patent* 4,786,733, **1988**; *Chem. Abstr.* **1989**, *110*, 212617w.
2. Van Der Puy, M. Direct fluorination of substituted pyridines. *Tetrahedron Lett.* **1987**, *28*(3), 255–258.
3. Chambers, R.D.; Parsons, M.; Sandford, G.; Skinner, C.J.; Atherton, M.J.; Moilliet, J.S. Elemental fluorine. Part 10. Selective fluorination of pyridine, quinoline and quinoxaline derivatives with fluorine–iodine mixtures. *J. Chem. Soc., Perkin. Trans. 1* **1999**, 803–810.
4. Gakh, A.A.; Tuluman, A.A.; Adcock, J.L.; Compton, R.N. Highly fluorinated fullerenes as oxidizers and fluorinating agents. *Tetrahedron Lett.* **1993**, *34*(45), 7167–7170.
5. Gakh, A.A.; Romaniko, S.V.; Ugrak, B.I.; Fainzilberg, A.A. *N*-fluorination with cesium fluoroxysulfate. *Tetrahedron* **1991**, *47*(35),7447–7458.
6. Gakh, A.A.; Nikishin, K.G.; Kagramanov, N.D.; Semenov, V.V. Fluorine nitrate (FONO2): an electrophilic fluorinating agent. *Izv. Acad. Nauk SSSR, Ser. Khim.* **1991**, 2403.
7. Gakh, A.A.; Kiselyov, A.S.; Semenov, V.V. Hetarylation of C-nucleophiles: a new course for the reaction of *N*-fluoropyridiniun salts with carbanions. *Tetrahedron Lett.* **1990**, *31*(50), 7379–7382.
8. Umemoto, T.; Tomizawa, G. Base-initiated reactions of *n*-fluoropyridinium salts; a novel cyclic carbene proposed as a reactive species. *Tetrahedron Lett.* **1987**, *28*(24), 2705–2708.
9. Umemoto, T.; Tomizawa, G. Preparation of 2-fluoropyridines via base-induced decomposition of *N*-fluoropyridinium salt. *J. Org. Chem.* **1989**, *54*(7),1726–1731.

10. Kiselyov, A.S.; Strekowski, L. Synthesis of 2-substituted pyridines by the reaction of *N*-fluoropyridinium fluoride with trimethylsilyl derivatives. *J. Org. Chem.* **1993**, *58* (16),4476–4478.

11. Stavber, S.; Zupan, M. Reactions of caesium fluoroxysulphate with pyridine. *Tetrahedron Lett.* **1990**, *31*(5),775–776.

12. The Merck Index. Organic Name Reactions. ONR-94; Merck & Co., Inc.: Whitehouse Station, NJ, USA; **2001**.

13. Brel, V.K.; Pirguliyev, N.S.; Zefirov, N.S. Chemistry of xenon derivatives. Synthesis and chemical properties. *Russ. Chem. Rev.* **2001**, *70*(3),231–264.

14. Fukuhara, T.; Yoneda, N.; Suzuki, A. Facile preparation of fluoropyridines from aminopyridines via diazotization and fluorodediazoniation in HF or HF-pyridine solutions. *J. Fluorine Chem.* **1988**, *38*, 435–438.

15. Fuss, A.; Koch, V. Chemistry of 3-hydroxypyridine. Part 4. Synthesis of 2- and 2,3-substituted 5[fluoro(chloro)alkoxy]pyridines via 5-hydroxy-2-(4-nitrophenylazo)pyridines. *Synthesis* **1990**, *8*, 681–685.

16. Dolle, F.; Valette, H.; Bottlaender, M.; Hinnen, F.; Vaufrey, F.; Guenther, I.; Crouzel, C. Synthesis of 2-[^{18}F]fluoro-3-[2(*S*)-2-azetidinylmethoxy]pyridine, a highly potent radioligand for *in vivo* imaging central nicotinic acetylcholine receptors. *J. Label. Compd. Radiopharm.* **1998**, *41*, 451–463.

17. Carroll, F.I.; Lee, J.R.; Navarro, H.A.; Brieaddy, L.E.; Abraham, P.; Damaj, M.I.; Martin, B.R. Synthesis, nicotinic acetylcholine receptor binding, and antinociceptive properties of 2-*exo*-2-(2′-substituted-3′-phenyl-5′-pyridinyl)-7-azabicyclo [2.2.1]-heptanes. Novel nicotinic antagonist. *J. Med. Chem.* **2001**, *44*(24),4039–4041.

18. Carroll, F.L.; Liang, F.; Navarro, H.A.; Brieaddy, L.E.; Abraham, P.; Damaj, M.I.; Martin, B.R. Synthesis, nicotinic acetylcholine receptor binding, and antinociceptive properties of 2-*exo*-2-(2′-substituted 5′-pyridinyl)-7-azabicyclo [2.2.1]heptanes. Epibatidine analogues. *J. Med. Chem.* *44*(13),2229–2237.

19. Carroll, F.I.; Lee, J.R.; Navarro, H.A.; Ma, W.; Brieaddy, L.E.; Abraham, P.; Damaj, M.I.; Martin, B.R. Synthesis, nicotinic acetylcholine receptor binding, and antinociceptive properties of 2-*exo*-2-(2′,3′-disubstituted 5′-pyridinyl)-7-azabicyclo [2.2.1]heptanes: epibatidine analogues. *J. Med. Chem.* **2002**, *45*(21),4755–4761.

20. Carroll, F.I., Ware, R.; Brieaddy, L.E.; Navarro, H.A.; Damaj, M.I.; Martin, B.R. Synthesis, nicotinic acetylcholine receptor binding, and antinociceptive properties of 2′-fluoro-3′-(substituted phenyl)deschloroepibatidine analogues. Novel nicotinic antagonist. *J. Med. Chem.* **2004**, *47*(18),4588–4594.

21. Bouillon, A.; Lancelot, J.-C.; Collot, V.; Bovy, P.R.; Rault, S. Synthesis of novel halopyridinylboronic acids and esters. Part 1. 6-Halopyridin-3-yl-boronic acids and esters. *Tetrahedron* **2002**, *58*, 2885–2890.

22. Liu, Y.; Zhao, Q.; Wang, Q.; Li, H.; Huang, R.; Li, Y. Synthesis and herbicidal activity of 2-cyano-3-(2-fluoro-5-pyridyl)methylaminoacrylates. *J. Fluorine Chem.* **2005**, *126*, 345–348.

23. Dzantiev, L.; Alekseyev, Y.O.; Morales, J.C.; Kool, E.T.; Romano, L.J. Significance of nucleobase shape complementarity and hydrogen bonding in the formation and stability of the closed polymerase–DNA complex. *Biochemistry* **2001**, *40*(10), 3215–3221.

24. Moran, S.; Ren, R.XF.; Rumney, S.; Kool, E.T. Difluorotoluene, a nonpolar isostere for thymine, codes specifically and efficiently for adenine in DNA replication. *J. Am. Chem. Soc.* **1997**, *119*(8),2056–2057.

25. Matray, T.J.; Kool, E.T. Selective and stable DNA base pairing without hydrogen bonds. *J. Am. Chem. Soc.* **1998**, *120*(24),6191–6192.

26. Sun, Z.; Ahmed, S.; McLaughlin, L.W. Syntheses of pyridine C-nucleosides as analogues of the natural nucleosides dC and dU. *J. Org. Chem.* **2006**, *71*(7),2922–2925.

27. Kim, D.W.; Jeong, H.-J.; Lim, S.T.; Sohn, M.-H.; Katzenellenbogen, J.K.; Chi, D.Y. Facile nucleophilic fluorination reactions using *tert*-alcohols as a reaction medium: significantly enhanced reactivity of alkali metal fluorides and improved selectivity. *J. Org. Chem.* **2008**, *73*(3),957–962.

28. Adams, D.B.; Kilcoyne, J.P.; MacBride, J.A.; Hugh, M.M. Biphenylenes and heterocyclic analogs of biphenylene. Part 7. Preparation of hexafluoro-1,8-diazadiphenylene and comparison of the site of nucleophilic substitution with frontier orbital predictions. *J. Chem. Res. Synop.* **1990**, *6*, 172–173.

29. Mutterer, F.; Weis, C.D. Halogenierte pyridine V. Fluorierte und bromierte pyridinverbindungen. *Helv. Chim. Acta* **1976**, *59*(23–24),229–235.

30. Clark, J.H.; Macquarrie, D.J. The synthesis of organofluorine compounds using potassium fluoride–tetraphenylphosphonium bromide systems. *Tetrahedron Lett.* **1987**, *28*(1), 111–114.

31. Finger, G.C.; Starr, L.D. Aromatic fluorine compounds. IX. 2-Fluoropyridines. *J. Am. Chem. Soc.* **1959**, *81*, 2674–2675.

32. Dolci, L.; Dolle, F.; Jubeau, S.; Vaufrey, F.; Crouzel, C. 2-[^{18}F]Fluoropyridines by no-carrier-added nucleophilic aromatic substitution with [^{18}F]F K–K_{222}: a comparative study. *J. Label. Compd. Radiopharm.* **1999**, *42*, 975–985.

33. Zhang, Y.; Horti, A.G. Synthesis of 6-chloro-3-((2-(*S*)-azetidinyl)methoxy)-5-(2-[^{18}F] fluoropyridin-4-yl)pyridine ([^{18}F]NIDA 522131), a novel potential radioligand for studying extrathalamic nicotinic acetylcholine receptors by PET. *J. Label. Compd. Radiopharm.* **2004**, *47*, 947–952.

34. Jammaz, I.A.; Al-Otaibi, B.; Okarvi, S.; Amartey, J. Novel synthesis of [^{18}F]-fluorobenzene and pyridinecarbohydrazide-folates as potential PET radiopharmaceuticals. *J. Label. Compd. Radiopharm.* **2006**, *49*, 125–137.

35. Amartey, J.K.; Al-Jammaz, I.; Al-Otaibi, B.; Esguerra, C. Novel synthesis of 2-[^{18}F]-fluoroisonicotinic acid hydrazide and initial biological evaluation. *Nucl. Med. Biol.* **2002**, *29*, 817–823.

36. Schirrmacher, R.; Mühlhausen, U.; Wängler, B.; Schirrmacher, E.; Reinhard, J.; Nagel, G.; Kaina, B.; Piel, M.; Wießler, M.; Rösch, F. Synthesis of 2-amino-6-(2-[^{18}F]fluoropyridine-4-ylmethoxy)-9-(octyl-β-D-glucosyl)-purine: a novel radioligand for positron emission tomography studies of the O^6-methylguanine-DNA methyltransferase (MGMT) status of tumour tissue. *Tetrahedron Lett.* **2002**, *43*, 6301–6304.

37. Hoyte, R.M.; Zhang, J.; Lerum, R.; Oluyemi, A.; Persaud, P.; O'Connor, C.; Labaree, D.C.; Hochberg, R.B. Synthesis of halogen-substituted pyridyl and pyrimidyl derivatives of

[3,2-c]pyrazolo corticosteroids: strategies for the development of glucocorticoid receptor mediated imaging agents. *J. Med. Chem.* **2002**, *45*(24),5397–5405.

38. Kahn, M.G.C.; Konde, E.; Dossou, F.; Labaree, D.C.; Hochberg, R.B.; Hoyte, R.M. Microwave-enhanced nucleophilic fluorination in the synthesis of fluoropyridyl derivatives of [3,2-*c*]pyrazolo-corticosteroids, potential glucocorticoid receptor-mediated imaging agents. *Bioorg. Med. Chem. Lett.* **2006**, *16*, 3454–3458.
39. Kuduk, S.D.; DiPardo, R.M.; Bock, M.G. Tetrabutylammonium salt induced denitration of nitropyridines: synthesis of fluoro-, hydroxy-, and methoxypyridines. *Org Lett.* **2005**, *7* (4),577–579.
40. Darabantu, M.; Lequeux, T.; Pommelet, J.-C.; Ple, N.; Turck, A. Selective fluorination by halogen exchange of chlorodiazines and chloropyridines promoted by the 'proton sponge'–triethylamine tris(hydrogen fluoride) system. *Tetrahedron* **2001**, *57*, 739–750.
41. Suzuki, H.; Sakai, N.; Iwahara, R.; Fujiwaka, T.; Satoh, M.; Kakehi, A.; Konakahara, T. Novel synthesis of 7-fluoro-8-(trifluoromethyl)-1*H*-1,6-naphthyridin-4-one derivatives: intermolecular cyclization of an *N*-silyl-1-azaallyl anion with perfluoroalkene and subsequent intramolecular skeletal transformation of the resulting pentasubstituted pyridines. *J. Org. Chem.* **2007**, *72*(15),5878–5881.
42. Umemoto, T.; Harasawa, K.; Tomizawa, G.; Kawada, K.; Tomita, K. Syntheses and properties of *N*-fluoropyridinium salts. *Bull. Chem. Soc. Jpn.* **1991**, *64*, 1081–1092.
43. Minakawa, N.; Kojima, N.; Matsuda, A. Nucleosides and nucleotides. 184. Synthesis and conformational investigation of anti-fixed 3-deaza-3-halopurine ribonucleosides. *J. Org. Chem.* **1999**, *64*, 7158–7172.
44. Viedma, A.G.; Martinez-Barrasa, V.; Burgos, C.; Izquierdo, M.L.; Alvarez-Builla, J. *N*-Azinylpyridinium *N*-aminides: intermediates for the regioselective synthesis of 3-fluoro-2-aminopyridine derivatives. *J. Org. Chem.* **1999**, *64*, 1007–1010.
45. Bobbio, C.; Rausis, T.; Schlosser, M. Removal of fluorine from and introduction of fluorine into polyhalopyridines: an exercise in nucleophilic hetarenic substitution. *Chem. Eur. J.* **2005**, *11*, 1903–1910.
46. Carroll, F.I.; Ma, W.; Yokota, Y.; Lee, J.R.; Brieaddy, L.E.; Navarro, H.A.; Damaj, M.I.; Martin, B.R. Synthesis, nicotinic acetylcholine receptor binding, and antinociceptive properties of 3′-substituted deschloroepibatidine analogues. Novel nicotinic antagonists. *J. Med. Chem.* **2005**, *48*, 1221–1228.
47. Pieper, P.A.; Yang, D.; Zhou, H.; Liu, H. 3-Deoxy-3-fluoropyridoxamine 5′-phosphate: synthesis and chemical and biological properties of a coenzyme B6 analog. *J. Am. Chem. Soc.* **1997**, *119*(8),1809–1817.
48. Tjosaas, F.; Fiksdahl, A. A simple synthetic route to methyl 3-fluoropyridine-4-carboxylate by nucleophilic aromatic substitution. *Molecules* **2006**, *11*, 130–133.
49. Dehmlow, E.V.; Schulz, H.-J. Synthesis of orellanine. The lethal poison of a toadstool. *Tetrahedron Lett.* **1985**, *26*, 4903–4906.
50. Joule, J.A.; Mills, K.; *Heterocyclic Chemistry*, 4th ed.; Blackwell Science Limited: Oxford, 2000.
51. Fang, B.; Tao, H.; Kan, X.; Shang, Y. Selective electrochemical synthesis of 4-fluoropyridine using Et_3N 3HF. *Heterocycl. Commun.* **2004**, *10*(4–5),305–306.

52. Karramkam, M.; Hinnen, F.; Vaufrey, F.; Dolle, F. 2-, 3- and 4-[^{18}F]Fluoropyridine by no-carrier-added nucleophilic aromatic substitution with K[^{18}F]F–K_{222}: a comparative study. *J. Label. Compd. Radiopharm.* **2003**, *46*, 979–992.

53. Thibault, C.; L'Heureux A, Bhide, R.S.; Ruel, R. Concise and efficient synthesis of 4-fluoro-1*H*-pyrrolo [2,3-*b*]pyridine. *Org. Lett.* **2003**, *5*(26),5023–5025.

54. Caldwell, J.J.; Cheung, K.M.; Collins, I. Synthesis of 4-(cyclic dialkylamino)-7-azaindoles by microwave heating of 4-halo-7-azaindoles and cyclic secondary amines. *Tetrahedron Lett.* **2007**, *48*, 1527–1529.

55. Schlosser, M.; Bobbio, C.; Rausis, T. Regiochemically flexible substitutions of di-, tri-, and tetrahalopyridines: the trialkylsilyl trick. *J. Org. Chem.* **2005**, *70*, 2494–2502.

56. Varlamov, A.V.; Levov, A.N.; Toze, F.; Chernyshev, A.I.; Davydov, V.V.; Ryabov, M.A.; Egorova, O.A. Amination of 4-azafluorene under chichibabin reaction conditions. Some chemical transformations of 1-amino-4-azafluorene. *Chem. Heterocycl. Compd.* **2002**, *38*(12),1484–1490.

57. Kohrt, J.T.; Filipski, K.J.; Cody, W.L.; Cai, C.; Dudley, D.A.; Van Huis, C.A.; Willardsen, J.A.; Rapundalo, S.T.; Saiya-Cork, K.; Leadley, R.J.; Narasimhan, L.S.; Zhang, E.; Whitlow, M.; Adler, M.; McLean, K.; Chou, Y.-L.; McKnight, C.; Arnaiz, D.O.; Shaw, K.J.; Light, D.R.; Edmunds, J.J. The discovery of fluoropyridine-based inhibitors of the factor VIIa/TF complex. *Bioorg. Med. Chem. Lett.* **2005**, *15*, 4752–4756.

58. Kohrt, J.T.; Filipski, K.J.; Cody, W.L.; Cai, C.; Dudley, D.A.; Van Huis, C.A.; Willardsen, J.A.; Narasimhan, L.S.; Zhang, E.; Rapundalo, S.T.; Saiya-Cork, K.; Leadley, R.J.; Edmunds, J.J. The discovery of fluoropyridine-based inhibitors of the factor VIIa/TF complex: Part 2. *Bioorg. Med. Chem. Lett.* **2006**, *16*, 1060–1064.

59. Ko, Y.-J.; Park, K.-B.; Shim, S.-B.; Shin, J.-H. Synthesis of difluorinated pyridinecarboxaldehyde via electrophilic fluorination. *J. Fluorine Chem.* **2006**, *127*, 755–759.

60. Bobbio, C.; Schlosser, M. Regiochemical flexibility: the optional functionalization of 2,3,5-trihalopyridines at the 4- or 6-position. *Eur. J. Org. Chem.* **2001**, 4533–4536.

61. Bobbio, C.; Schlosser, M. Selective functionalization of 2-fluoropyridine, 2,3-difluoropyridine, and 2,5-difluoropyridine at each vacant position. *J. Org. Chem.* **2005**, *70*, 3039–3045.

62. Schlosser, M.; Rausis, T.; Bobbio, C. Rerouting nucleophilic substitution from the 4-position to the 2- or 6-position of 2,4-dihalopyridines and 2,4,6-trihalopyridines: the solution to a long-standing problem. *Org. Lett.* **2005**, *7*(1),127–129.

63. Plevey, R.G.; Rendell, R.W.; Tatlow, J.C. Fluorinations with complex metal fluorides. Part 6. Fluorination of pyridine and related compounds with caesium tetrafluorocobaltate (III). *J. Fluorine Chem.* **1982**, *21*, 159–169.

64. Coe, P.L.; Holton, A.G.; Tatlow, J.C. Fluorinations with potassium tetrafluorocobaltate (III). Part VII. Further investigations on the fluorination of pyridine. *J. Fluorine Chem.* **1982**, *21*, 171–189.

65. Dmowski, W.; Haas, A. Trifluoromethanethiolate ion. Part 2. Nucleophilic substitution in pentafluoropyridine. Synthesis and characteristics of trifluoromethylthio and trifluoromethylsulphonyl derivatives. *J. Chem. Soc., Perkin Trans. 1* **1987**, 2119–2124.

66. Aksenov, V.V.; Vlasov, V.M.; Yakobson, G.G. Interaction of pentafluoropyridine with 4-nitrophenol and pentafluorophenol in the presence of potassium fluoride and 18-crown-6-ether. *J. Fluorine Chem.* **1982**, *20*, 439–458.

67. Hansen, T.M.; Gu, Y.-G.; Rehm, T.M.; Dandliker, P.J.; Chovan, L.E.; Bui, M.H.; Nilius, A. M.; Beutel, B.A. Synthesis and antibacterial activity of 5-methoxy- and 5-hydroxy-6-fluoro-1,8-naphthyridone-3-carboxylic acid derivatives. *Bioorg. Med. Chem. Lett.* **2005**, *15*, 2716–2719.

68. Bouzard, D.; Cesare, P.; Essiz, M.; Jacquet, J.P.; Remuzon, P.; Weber, A.; Oki, T.; Masuyoshit, M. Fluoronaphthyridines and quinolones as antibacterial agents. 1. Synthesis and structure–activity relationships of new 1-substituted derivatives. *J. Med. Chem.* **1989**, *32*, 537–542.

7

SYNTHESIS AND CHEMICAL TRANSFORMATION OF SIX-MEMBERED AROMATIC HETEROCYCLES CONTAINING PERFLUOROALKYL GROUPS

VIACHESLAV A. PETROV
DuPont Central Research and Development, Experimental Station, P.O. Box 80500 Wilmington, DE 19880-0500, USA

7.1 INTRODUCTION

Majority of known perfluoroalkylated aromatic heterocyclic compounds contain one or several nitrogen atom(s) in the aromatic ring. Examples of fluoroalkylated aromatic six-membered heterocycles containing atoms, such as phosphorous or antimony are extremely rare and this is the reason why this chapter is mostly limited to preparation and chemical transformations of *nitrogen*-containing aromatic heterocycles, providing limited number of references to only few known examples of trifluoromethyl *s*-1,3,5,2,4,6-triazatriphosphinines.

Methods of preparation of perfluoroalkyl aromatic heterocyclic compounds can be divided into *nonregioselective* and *regioselective*. The first group of synthetic methods is limited mostly to radical perfluoroalkylation of hydrocarbon heteroaromatic precursors. This group of reactions is reviewed in the first part of this chapter. More numerous regioselective methods of the introduction of fluoroalkyl groups are summarized in second part. This group of methods in turn can be divided into several subgroups, based on the reagents or chemical transformations used in the synthesis.

Fluorinated Heterocyclic Compounds: Synthesis, Chemistry, and Applications, Edited by Viacheslav A. Petrov

Methods based on electrophilic reactions form a relatively small group, due to the fact that aromatic heterocyclic compounds have relatively low reactivity toward electrophiles. Rather limited in number, this group of methods is quite important from practical point of view, since many of commercially available fluoroalkyl heterocycles are prepared using these transformations.

Various cyclization processes traditionally used in the synthesis of hydrocarbon heterocycles and it is not surprising that this methodology is popular for the preparation of fluoroalkyl analogues. These methods are included in second subgroup of synthetic transformations used for the preparation of fluoroalkylated heterocycles. Usually, these are multicomponent reactions and one of the building blocks contains a fluoroalkyl group, which ends up into the synthesized heterocycle.

The third and biggest subgroup of regioselective methods of synthesis perfluoroalkyl aromatic heterocycles is based on nucleophilic reactions and usually required activation of hydrocarbon heterocycle either by introduction in the ring of an additional heteroatom such as nitrogen, oxygen, or halogen or can be achieved through the alkylation, acylation, or oxidation of the nitrogen of heterocycle. The reagents used for the introduction of R_F-group typically are a good nucleophiles, such as perfluoroalkyl derivatives of copper, lithium, magnesium, or silicon.

Due to the fact that the chemical behavior of perfluoroalkylated heterocycles is often similar to that of their hydrocarbon counterparts, the section on chemical transformations of this chapter is mostly focused on reactions that differ perfluoroalkylated heterocycles from their hydrocarbon analogues.

7.2 NONREGIOSELECTIVE SYNTHESIS OF AROMATIC PERFLUOROALKYL HETEROCYCLES

Although the nonregioselective methods of the synthesis of perfluoroalkyl heterocycles are mostly limited to radical reactions, the sources of polyfluorinated radicals and methods of their generations can vary a lot.

Owing to their relatively low thermal stability, perfluoroacylperoxides $[R_FC(O)O]_2$ undergo thermal decomposition generating perfluoroalkyl radicals.[1] Although $[R_FC(O)O]_2$ were successfully used for the perfluoroalkylation furane, thiophene, and pyrrole, the reaction of perfluoroacylperoxides with pyridine does not result in perfluoroalkylation, due to side reactions caused by interaction of the peroxides with a lone electron pair of nitrogen. Sterically hindered pyridines, however, were reported to give perfluoroalkylated products. For example, compound **1** reacts with peroxide **2** giving isomeric pyridines **3** and **4** in low yield (Fig. 7.1).[1]

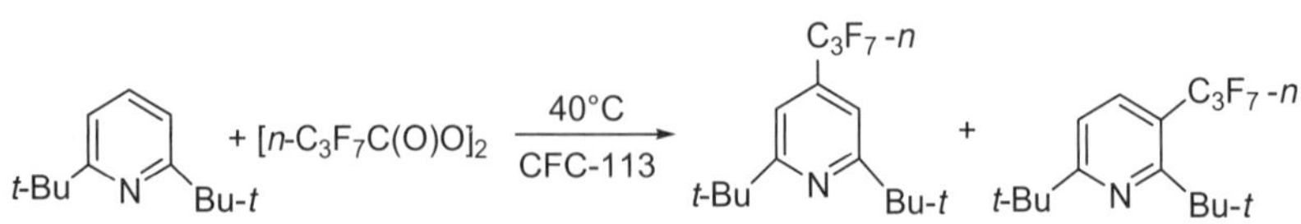

1 2 3, 16% 4, 8%

Figure 7.1 Reaction of acylperoxide **2** with pyridine **1**.

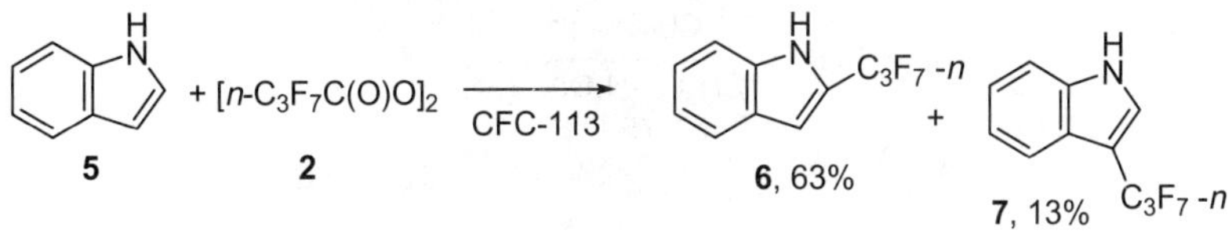

Figure 7.2 Perfluoroalkylation of indole using perfluoroacyl peroxide **2**.

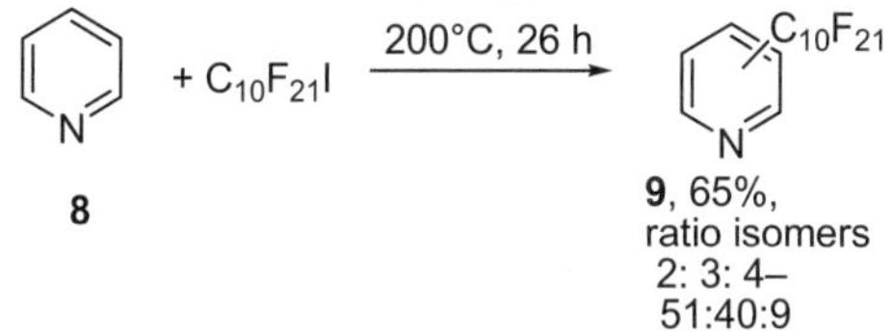

Figure 7.3 Thermal perfluoroalkylation of pyridine using $C_{10}F_{21}I$.

Although the reaction of indol **5** and peroxide **2** proceeds under mild conditions, it is nonregioselective resulting in the mixture of two isomers **6** and **7** (Fig. 7.2).

Perfluoroalkyl iodides were demonstrated to be efficient reagents for perfluoroalkylation of pyridine (**8**) at elevated temperature (Fig. 7.3).[2]

Monoperfluoroalkyl pyridines **9** form as a mixture of regioisomers, with significant predominance of 2 and 3 isomers.[2]

Perfluoroalkylation of pyridines can be carried out under significantly milder conditions if perfluoroalkyl iodide or bromide reacts with aromatic substrate in the presence of reducing agent, such as $HOCH_2SO_2Na$ (rongalite) in $CH_3CN–H_2O$ mixture at 70–75°C (Fig. 7.4).[3,4]

It should be pointed out that this process is not limited to perfluoroalkyl iodides and the corresponding bromides (but not chlorides) also function as perfluoralkylating agents.[3,4] The proposed reaction mechanism involves single electron transfer from rongalite to perfluoroalkyl iodide (bromide), generation of $R_F^{\bullet}$ through the decomposition of initially formed $[R_FX]^{-\bullet}$ and consecutive reaction of $R_F^{\bullet}$ with heterocyclic substrate.[3]

The ratio of 2, 3, and 4 isomers formed in the reaction of pyridine (**8**) with polyfluoroalkyl bromides and iodides (~5:5:1) is approximately constant in all theses reactions and is consistent with free radical mechanism of the process.[3] Quinoline and

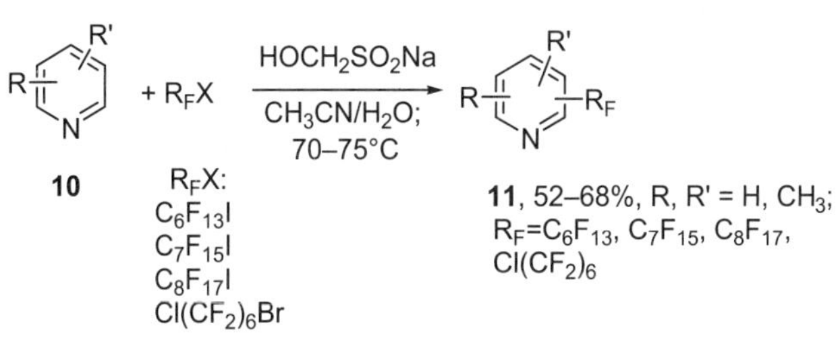

Figure 7.4 Redox perfluoroalkylation of pyridines using R_FX/rongalite system.

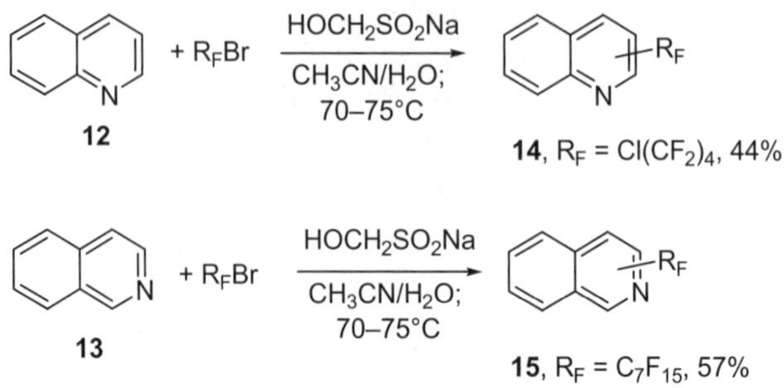

Figure 7.5 Redox fluoroalkylation of quinoline and isoquinoline.

isoquinoline also give a mixture of isomeric monopolyfluoroalkylated products in moderate yields (Fig. 7.5), but pyrazine and imidazole were reported to be inactive in this process.[3]

7.3 REGIOSELECTIVE METHODS OF SYNTHESIS OF AROMATIC POLYFLUOROALKYL HETEROCYCLIC COMPOUNDS

Regioselective methods of preparation of aromatic six-membered polyfluoroalkyl heterocycles, based on the type of the reaction used for the synthesis, can be divided in three major groups:

(a) electrophilic and radical
(b) cyclization processes
(c) nucleophilic.

7.3.1 Electrophilic and Radical Reactions

First group of methods is limited mostly to electrophilic processes. On relatively small, multigram laboratory scale trifluoromethyl pyridines **16–18** and isoquinoline **19** can be readily prepared by the reaction of the corresponding aromatic acids with SF_4 in HF solvent (Fig. 7.6).[5]

Porwisiak and Dmowski prepared a number of bis- and tris-trifluoromethyl pyridines **20–24** using this process (Fig. 7.7).[6]

Using this method, all isomers of bis(trifluoromethyl)pyridine[7] along with 2-, 3-, and 4-trifluoromethyl-quinolines were synthesized by the Kobayashi research group.[8] Anhydrous HF is required for the synthesis of CF_3-pyridines, since in the absence of HF the reaction of carboxylic acid and SF_4 has a tendency to stop at acyl fluoride stage.[5] In general, the method is reliable and gives trifluoromethylated heteroaromatics in acceptable yields, however, the necessity to carry out the reaction under pressure using aggressive and toxic SF_4 and HF significantly limits its application.

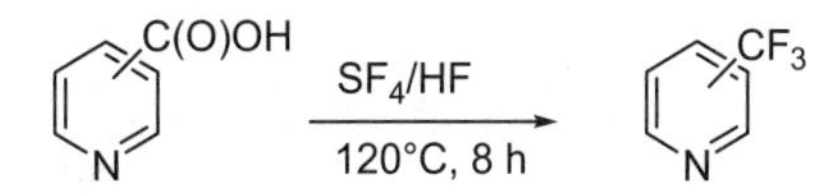

16, 2-CF_3, 53%
17, 3-CF_3, 25%
18, 4-CF_3, 57%

19, 72%

Figure 7.6 Fluorination of heteroaromatic carboxylic acids using SF_4 and HF.

Interesting, but rather exotic variation of the synthesis of trifluoromethyl pyridines involves the reaction of the corresponding nicotinic acids with MoF_6 at 190–200°C, which gives pyridines **16**, **18**, and **22** in 60–80% yield.[9]

High temperature gas-phase chlorofluorination of picolines[10,11] is a commercial process used for the preparation of an important intermediates for a variety of agricultural and pharmaceutical products—trifluoromethylpyridines **17**, **25–27**. The details of synthesis and synthetic applications of intermediates **17**, **25–27** are given in Chapter 11 and Refs 10–12.

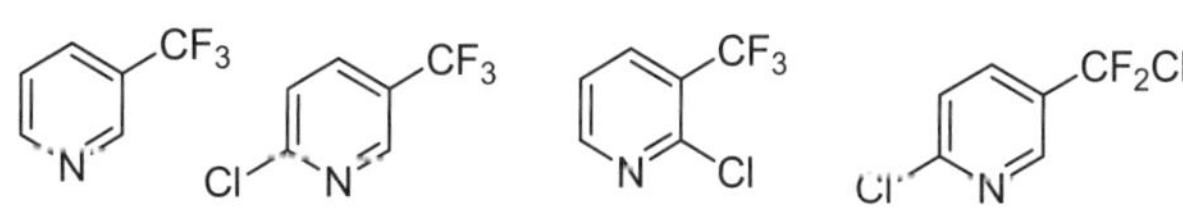

17 **25** **26** **27**

Compound **25** was produced on commercial scale by ICI company using chlorofluorination technology since the early 1980s.[12]

Various trifluoromethylated quinolines also can be prepared in good yield by the liquid-phase fluorination of the corresponding trichloromethyl- derivatives using anhydrous HF and antimony pentafluoride (chloride) as a catalyst.[13]

Due to a low regioselectivity of the process examples of selective introduction of fluoroalkyl group using radical reactions are rare and limited to special cases, such as reaction of **28** reported by Yoshida et al. (Fig. 7.8).[1]

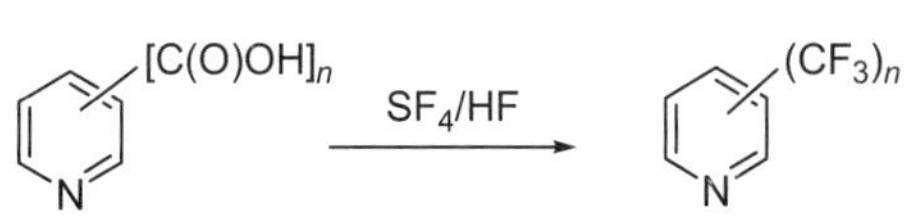

20, n = 2, 2,5-CF_3
21, n = 2, 2,4-CF_3
22, n = 2, 2,6-CF_3
23, n = 2, 3,5-CF_3
24, n = 3, 2,4,6-CF_3

Figure 7.7 Synthesis of poly(trifluoromethyl) pyridines using SF_4/HF system.

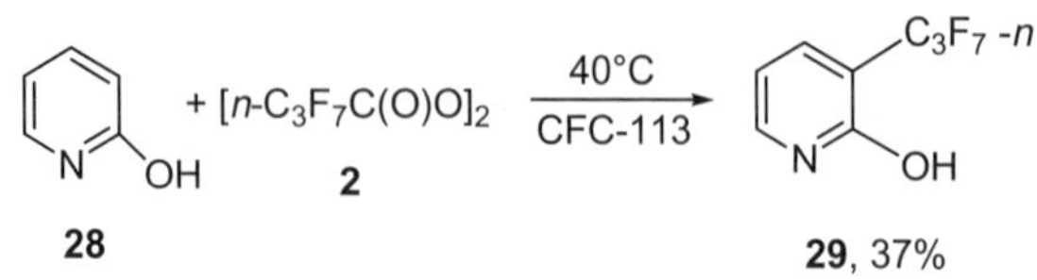

Figure 7.8 Selective radical perfluoralkylation 2-hydroxypyridine (**28**).

7.3.2 Cyclization Processes

Major types of cyclization reactions used for the synthesis of aromatic heterocycles containing fluoroalkyl group include (but, not limited to) cycloadditions, nucleophilic, and electrophilic cyclizations.

2-Trifluoromethylpyridine (**16**) can be prepared in a flow system at elevated temperature by the hetero-Diels–Alder reaction between butadiene-1,3 and CF_3CN (Fig. 7.9).[14]

This reaction is general and other perfluorinated nitriles R_FCN ($R_F = CF_3$, $ClCF_2$, FCl_2C, C_2F_5, n-C_3F_7) under similar conditions give the corresponding 2-fluoroalkyl pyridines in high yields.[15]

Nucleophilic cyclization reactions are widely used for the synthesis fluoroalkyl heterocycles. This process relies on the use of fluorinated building block for the introduction of perfluoroalkyl group into heterocycle. For example, heating of compound **30** with urea or guanidine results in the formation of the corresponding pyrimidines **31** and **32** in moderate yield (Fig. 7.10).

Interestingly, cyclizations in the presence of HCl proceeds at ambient temperature giving compounds **31**–**33** in 60–75% yield.[16] Pyrimidine **34** can be prepared in low yield by the reaction of **30** with NH_4Cl in formamide at elevated temperature (Fig. 7.10).[16]

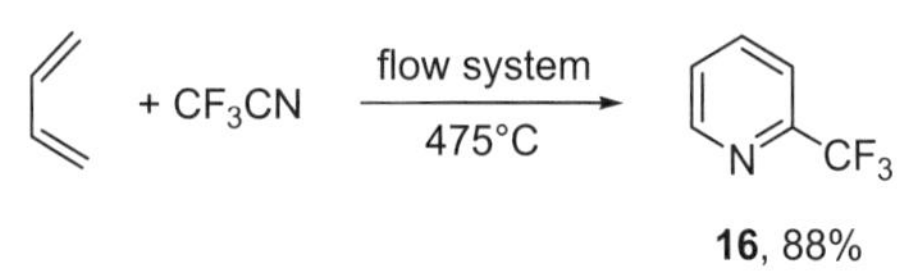

Figure 7.9 [4 + 2] cycloaddition reaction of butadiene-1,3 and CF_3CN.

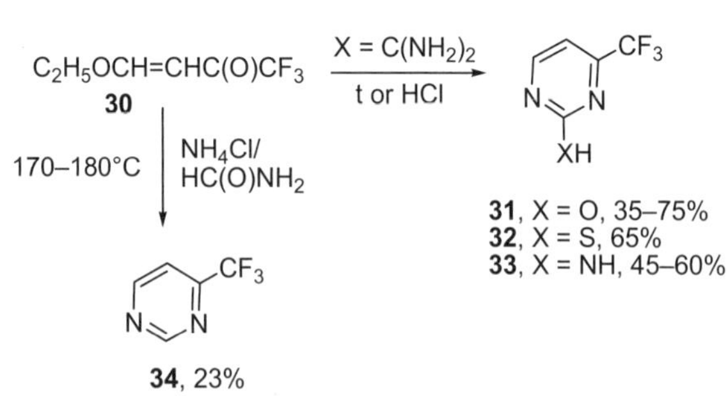

Figure 7.10 Synthesis of 4-trifluoromethylpyrimidines using cyclization reactions.

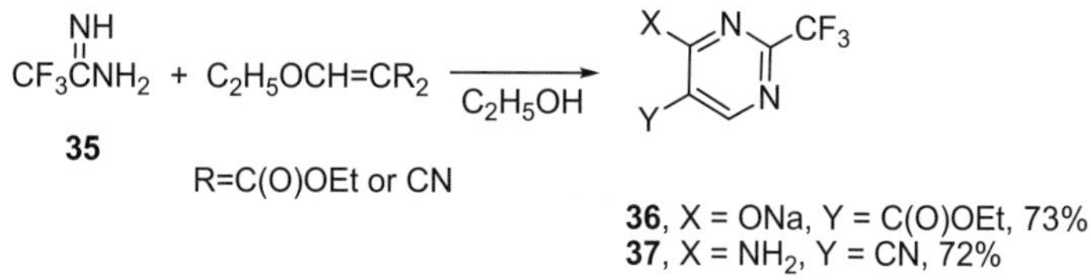

Figure 7.11 Trifluoroacetamidine based synthesis of 2-trifluoromethylpyrimidines.

Amidines of perfluorocarboxylic acids are excellent synthons for the preparation of fluoroalkyl pyrimidines. The reaction of **35** with diethyl ethoxyethylene malonate or malononitrile results in the formation of the corresponding 2-trifluoromethylpyrimidines **36** and **37** in high yield (Fig. 7.11).[17]

Recently, it was demonstrated that imines of polyfluorinated β-diketones **38**, **40**, and **43** could serve as a valuable building blocks for the preparation of polyfluorinated pyridines and pyrimidines-containing multiple fluoroalkyl groups.[18–20] The reaction of **38** with acetone gives pyridine **39**,[20] while homologous imine **40** reacts with acetone forming a mixture two isomeric pyridines **41** and **42** with significant predominance of 2-C_2F_5 isomer (Fig. 7.12). The reaction between **38** and butan-2-one also leads to a mixture of two isomeric pyridines.[20]

The mechanism of the condensation involves the formation (after elimination of ammonia) of the corresponding 1,2- and 1,5-dihydropyrimidines intermediates, which are further converted into the pyridines **39**, **41**, and **42** after water elimination.[20]

The synthesis of 2-arylpyrimidines carrying fluoroalkyl groups in 4- and 6-position can be achieved through base-catalyzed condensation of fluorinated imines **38**, **40**, and **43** with the corresponding aromatic aldehydes **44** (Fig. 7.13).[19]

The condensation of imine **38** with anhydrides or chlorides of perfluorocarboxylic acids provides access to pyrimidines **46** bearing three perfluoroalkyl groups (Fig. 7.14).[18]

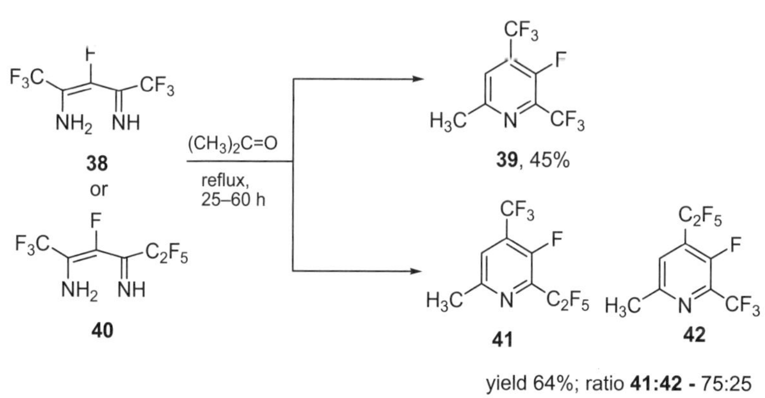

Figure 7.12 Synthesis of polyfluorinated pyridines by condensation of imines **38** and **40** with ketones.

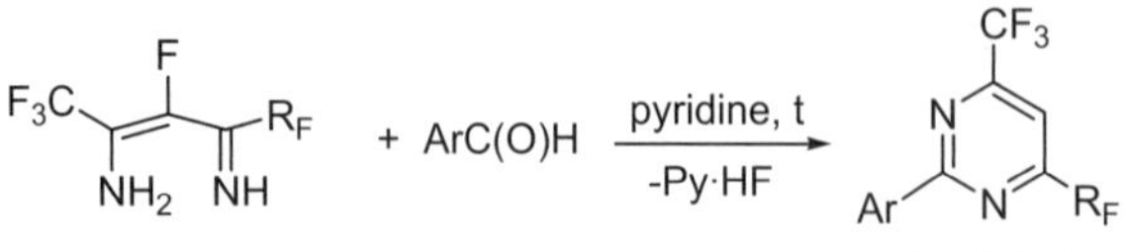

38, $R_F = CF_3$
40, $R_F = C_2F_5$
43, $R_F = n\text{-}C_3F_7$

44

45, 73-97%
$R_F = CF_3; C_2F_5; n\text{-}C_3F_7$
Ar = C_6H_5; HO-C_6H_4; MeO-C_6H_4, $Me_2N\text{-}C_6H_4$

Figure 7.13 Synthesis of fluorinated 2-arylpyrimidines.

Cyclocondensation processes are often used for the synthesis of various fluoroalkyl quinolines and isoquinolines.[21–29] A variety of fluorinated building blocks can be used as starting material in this processes. The regioselectivity of reported by Linderman and Kirollos acid-catalyzed cyclization of trifluoromethyl-containing aryl enamines **47** is extremely sensitive the reaction conditions. For example, the cyclization of enamines **47** under action of polyphosphoric acid leads to the formation of 4-CF_3-quinolines **48** and in case of enamine **47** (Ar = 3,5−CH_3O−C_6H_3) leads to selective formation of 4-CF_3 isomer **48** (X, Y = CH_3O, 93%). On the other hand, the reaction of 3,5-dimethoxyaniline with compound **49** in $CF_3C(O)OH$ leads to selective formation of 2-CF_3-quinoline **50** (Fig. 7.15).[21]

Selective formation of aromatic heterocycles containing 4-CF_3-pyridine unit under action of $TiCl_4$ on RCH=C[C(O)CF_3]$_2$ (R = ArNH−) was reported by Soufayne et al.,[23] however, in detailed investigation carried out by the Schlosser research group,[24,25] it was shown that the cyclization of enamines **51** under action of P(O)Cl_3 leads to regioselective formation of 2-CF_3-quinolines **52** and **53** (Fig. 7.16).

The synthesis 2-R_F-4-hydroxyqunolines reported by Froissard et al. is based on the cyclization of fluorinated enamines **54** under action of polyphosphoric acid at elevated temperature providing access to variety of 2-fluoroalkyl quinolines **55** (Fig. 7.17).[26]

This reaction is general and it was applied by the Schlosser group for the regioselective synthesis of various 2-R_F-6-alkyl quinolines ($R_F = C_2F_5$, $n\text{-}C_3F_7$; R = H, D, CH_3, C_2H_5).[25]

Polyfluorinated aldehydes or their hemiketals were successfully used for the synthesis of 2-R_F quinolines. Recently reported reaction of aldehydes **56** involves intermediate formation of unsaturated aldehyde **57**, which further reacts with aniline giving quinolines **52** and **58** in moderate to high yields (Fig. 7.18).[28]

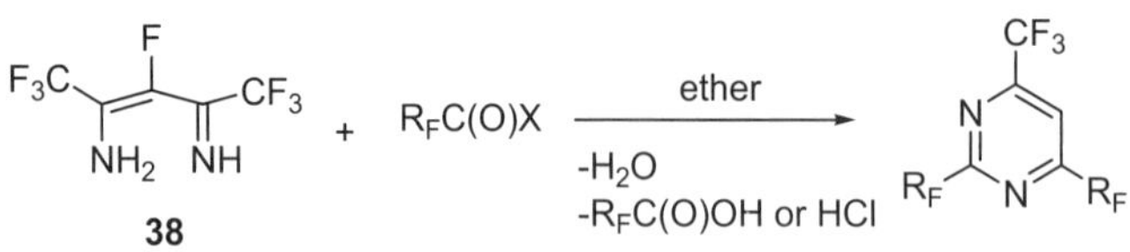

38

46, 71–88%
$R_F = CF_3; C_2F_5; n\text{-}C_3F_7$

Figure 7.14 Synthesis of pyrimidines **46**.

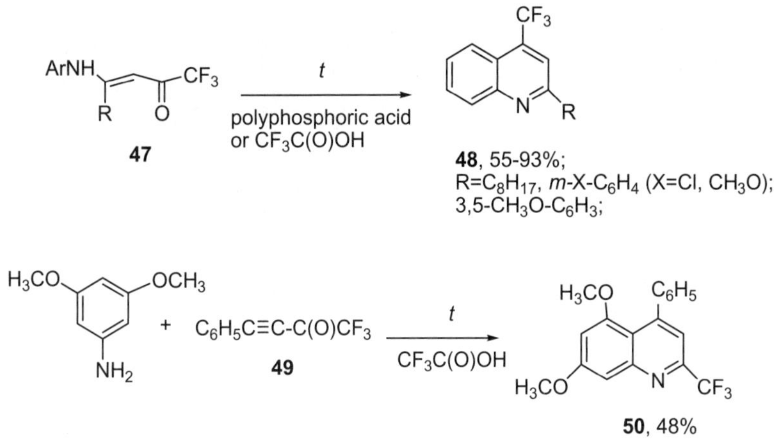

Figure 7.15 Preparation of trifluoromethylated quinolines **48** and **50**.

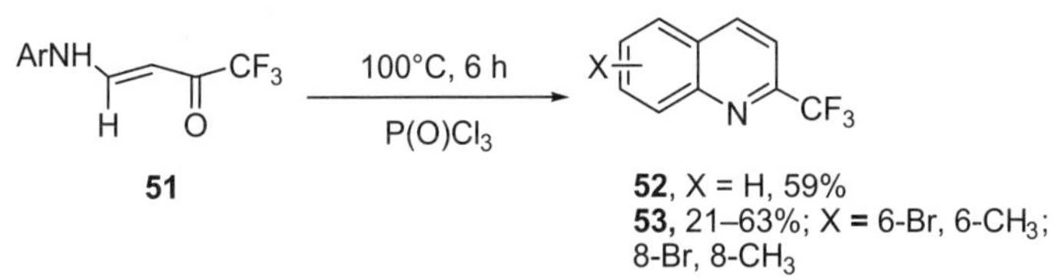

Figure 7.16 $P(O)Cl_3$ promoted synthesis of 2-trifluoromethyl quinolines.

This method is based on readily available starting materials, is highly regioselective and provides access to a wide variety of 2-fluoroalkyl quinolines. Recently reported variation of the synthesis of 2-R_F-quinolines involves the reaction of *o*-amino styrenes with hemiketals or hydrates of perfluorinated ketones in the presence of $(CH_3)_3SiCl$/Py as dehydrating agent.[29] The process is regioselective although it leads to the formation the mixture 2-R_F- quinoline and 2-R_F-1,2-dihydroquinolines ($R_F = CF_3$, C_2F_5, CF_2Cl); however, the latter can be converted into the corresponding quinoline by oxidation with air.

Another two-step procedure for synthesis of 1-R_F-isoquinolines was reported by Pastor and Cambon.[27] The cyclization step involves dehydration of amides **59** by

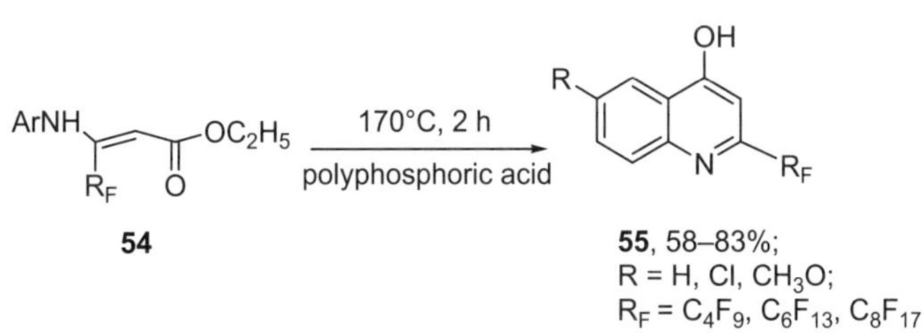

Figure 7.17 Synthesis of 2-R_F-4-hydroxyquinolines.

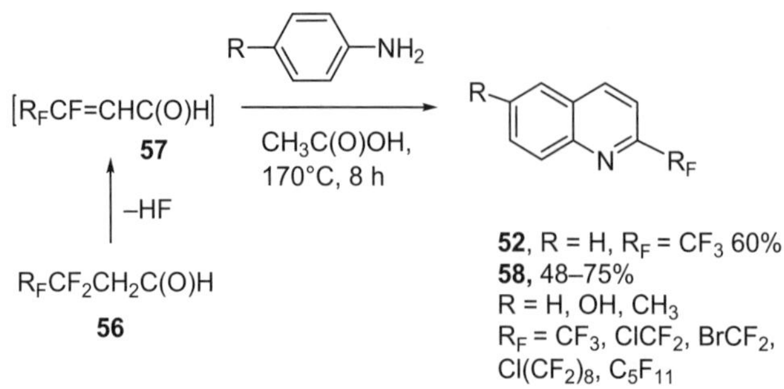

Figure 7.18 Synthesis of 2-fluoroalkylquinolines by reaction of fluorinated aldehydes and anilines.

P_2O_5 leading to dihydroisoquinolines **60** and is followed by aromatization step (Fig. 7.19).

Perfluoroalkyl-*s*-triazines can be prepared by thermal trimerization of the corresponding nitriles in low yield.[30] However, the trimerization can be significantly accelerated by HCl. For example, the reaction of CF_3CN in the presence of HCl catalyst proceeds at ambient temperature and gives 2,4,6-tris(trifluoromethyl)-1,3,5-triazine in >90% yield.[31] Some antimony (V) derivatives, also were found to be effective catalysts for trimerization of perfluorinated nitriles at ambient temperature. For example, 2,4,6-tris(perfluoropentyl)-1,3,5-triazine was obtained in 85% yield using $(C_4H_9)_3Sb(OH)_2$ as the catalyst.[32] Triethylamine/water system at 50–150°C and high pressure (1000–14,000 kg/cm^2) was demonstrated to be an effective catalyst for trimerization of perfluorinated nitriles, containing bulky, branched perfluoroalkyl groups.[33] Reported in 1957, thermal trimerization of amidines perfluorcarboxylic acids proceeding with the elimination of ammonia, still remains an attractive route for

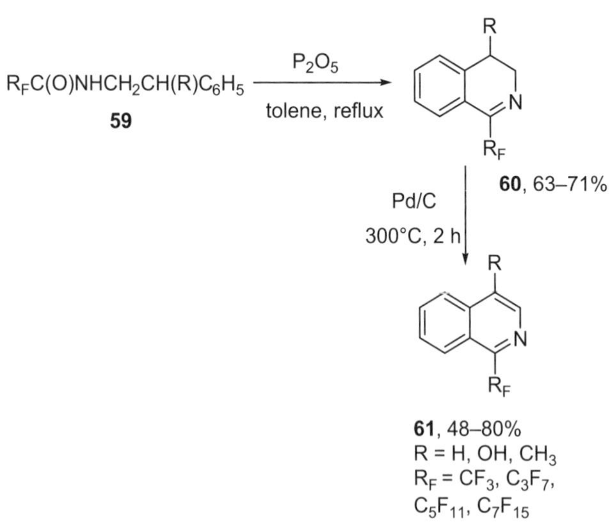

Figure 7.19 Two step synthesis of 1-fluoroalkylisoquinolines.

the preparation of symmetrical triazines, since it proceeds under relatively mild conditions, giving triazines in acceptable yields.[30] Methods of the synthesis of *s*-triazines based on direct introduction of perfluoroalkyl groups into cyanuric chloride or fluoride are reviewed in next section of this chapter.

7.3.3 Nucleophilic Perfluoroalkylation of Aromatic Heterocycles

The introduction of perfluoroalkyl group cannot be carried out using direct reaction of perfluoroalkyl carbanion with aromatic hydrocarbon heterocycle due to not sufficient electron deficiency of the heterocycles. All know nucleophilic fluoroalkylation procedures (with very few exceptions) require either modification of perfluoroalkyl nucleophiles or activation of heteroaromatic system. An excellent example of the first synthetic approach is the use of perfluoroalkyl copper derivatives for the introduction of R_F group into halogenated heterocycles. Being a variation of the Ullmann reaction, this methodology was first introduced by McLoughlin and Thrower.[34] The reaction of perfluoroalkyl iodides with aromatic halogenated compounds in the presence of copper powder in polar solvent results in the introduction of perfluoroalkyl group into heteroaromatic ring. The mechanism of this process[34] involves oxidative addition of R_FI to copper with the formation of solvent stabilized copper (I) intermediate, which is followed by its oxidative insertion into C–Hal bond of the aromatic halide and reductive elimination step leading to the formation of the coupling product (Fig. 7.20).

Since the perfluoroalkyl copper intermediates are stabilized by interaction with the solvent, the choice of the media for this reactions is very important. Usually these reactions are carried out either in polar solvents, such as $HC(O)NMe_2$ (DMF), $CH_3C(O)NMe_2$ (DMAA), *N*-methylpyrrolidone (NMP), $(CH_3)_2S{=}O$ (DMSO), $(MeO)_3P{=}O$. Sometimes, nonpolar solvents (such as C_6F_6), can be also used, however, in the presence of sufficient amount of cosolvent, such as DMF or DMSO.

Due to availability of perfluoroalkyl iodides and high regioselectivity of the process, this reaction was used for the synthesis of a wide variety of perfluoroalkyl heterocyclic compounds.

Perfluoroalkyl cooper reagent also can be generated *in situ* through the reaction of perfluoroalkyl anions (or their synthetic equivalents) with Cu(I) salts. First part of this section contains a brief review of major types of reagents used for copper mediated

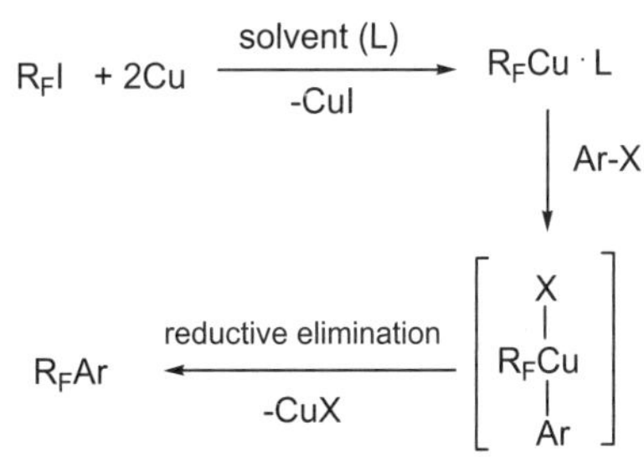

Figure 7.20 Mechanism of cooper mediated perfluoroalkylation of heteroaromatic compounds.

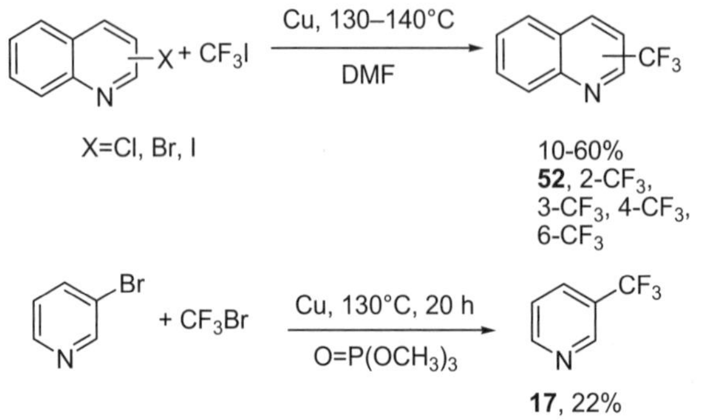

Figure 7.21 Trifluoromethylation of heterocycles using CF_3X/Cu system.

perfluoroalkylations and it is followed by review of nucleophilic reactions, which require activation of heterocyclic substrate.

Trifluoromethylation reactions of heteroaromatic compounds represent a majority of reported nucleophilic perfluoroalkylations. For example, trifluoromethylation of halogenated quinolines, isoquinolines, or pyridines carried out under pressure using CF_3X/Cu powder system (X = I or Br) in DMF or $(MeO)_3P{=}O$ solvent affords the corresponding trifluoromethylated heterocycles in 22–60% yield (Fig. 7.21).[35,36]

One of disadvantages of this process is the necessity to run the reaction under pressure due to gaseous nature of CF_3I and CF_3Br. Yagupolskii et al. used transmetalation reaction between $(CF_3)_2Hg$ and Cu for the generation of CF_3Cu (NMP or DMAA solvents, 140°C), which was further involved into reaction with 2-iodopyridine to give 2-CF_3-pyridine (**17**) in 74% yield.[37] A year later, another interesting modification of trifluoromethylation procedure was reported.[38] The reaction of 2-Br-pyridine with sodium trifluoroacetate and CuI in NMP at 140–160°C was reported to give 2-CF_3-pyridine in 41% yield. This process was expanded and studied in great details by the Chambers research group and a number of trifluoromethylated and pentafluoroethylated heterocycles, such as **20**, **25**, **62**, and **63** was prepared in moderate to high yields (Fig. 7.22).[39]

A procedure for simultaneous preparation of CF_3- and C_2F_5-containing heterocycles developed by Clark et al.[40,41] is based on discovered by Burton reaction of CF_2Br_2 and cooper leading through disproportionation mechanism to CF_3Cu.[42] While the interaction of generated in this process CF_3Cu with highly electrophilic substrates, such as 2-chloro-3-nitropyridine (**64**) results in selective formation of trifluoromethylated product,[41] the reaction with less active substrates leads to the formation of substantial amount of C_2F_5-containing heterocycles, as the result of the reaction of heterocyclic substrate with C_2F_5Cu relatively slowly forming from CF_3Cu under reaction conditions (Fig. 7.23).[42]

Recently reported by Cottet and Schlosser regioselective synthesis of trifluoromethyl pyridines and -quinolines[43,44] is based on the trifluoromethylation of iodoheteroarynes using $CF_3Si(CH_3)_3$/KF/CuI system originally applied by Urata and Fuchikami for trifluromethylation of aryl halides.[45] Trifluoromethyl copper in this

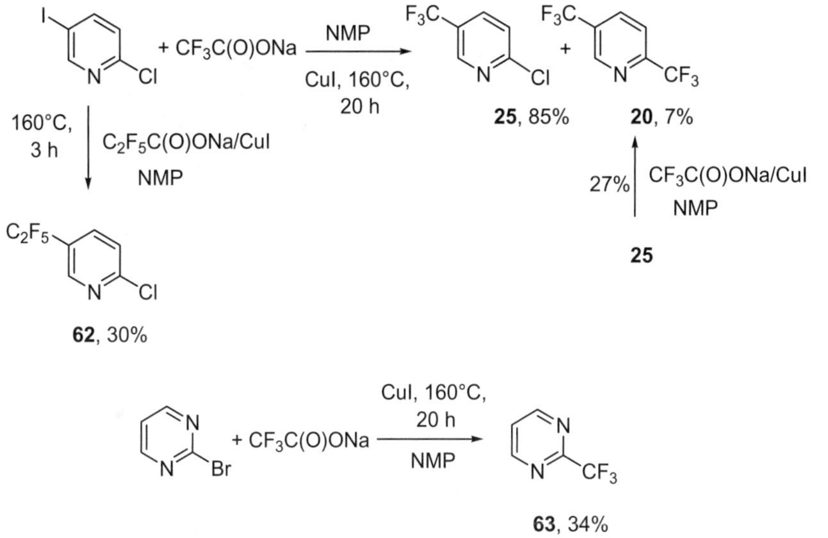

Figure 7.22 Perfluoroalkylation of halogenated heterocyclic compounds using $R_FC(O)ONa$/CuI system.

process is generated from $CF_3Si(CH_3)_3$ and CuX (X = Cl, Br, I, or CN) in the presence of alkali metal fluoride. The synthesis of trifluoromethyl-pyridines and -quinolines is carried out in DMF or NMP solvent at 25–50°C, leading to moderate to high yield formation of heterocycles **68–70** (Fig. 7.24).[43,44]

Iodoheteroarenes have significantly higher reactivity toward CF_3Cu and usually replacement of iodo-substituent by CF_3 group proceeds exclusively, providing access to chloro- and bromo-trifluoromethylated heterocycles, which can be further functionalized. It should be pointed out that this regiospecific method is based on commercial available starting materials, does not involve handling gaseous and highly toxic materials and is relatively easy to scale up in a laboratory.

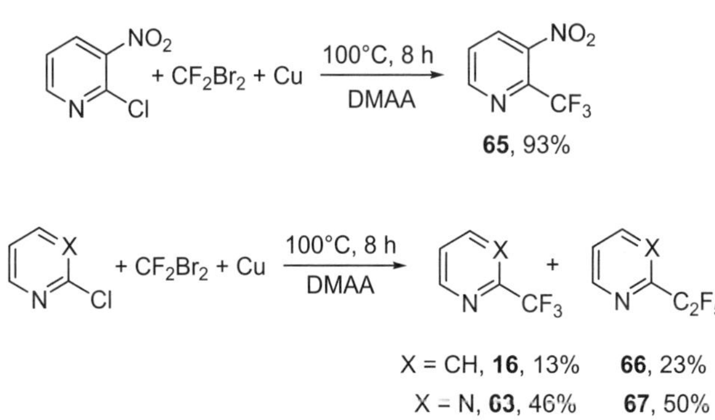

Figure 7.23 Perfluoroalkylation of aromatic heterocycles using CF_2Br_2/Cu system.

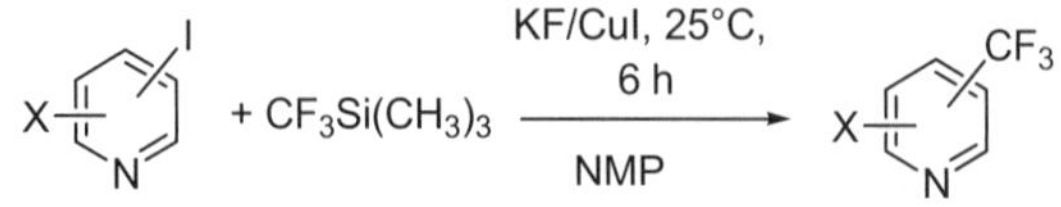

68, 29–79%,
X = Cl, Br

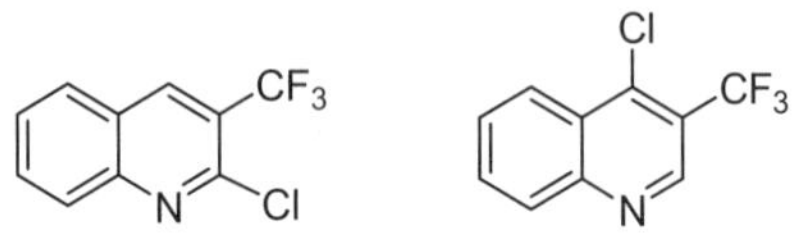

69, 80% **70**, 79%

Figure 7.24 Preparation of trifluoromethylated heterocycles using $CF_3Si(CH_3)_3$/CuI/KF system.

First example of *perfluoroalkylation* of heteroarenes is the reaction of 3-iodopyridine by 1,3-diiodohexafluoropropane resulting in the formation of derivative **71** reported in 1969 (Fig. 7.25).[34]

Due to the availability of perfluoroalkyl iodides, this method quickly became the method of choice for the regioselective synthesis of perfluoroalkylated heterocycles. Chen and Tamborski used this reaction for the preparation of variety perfluoroalkylpyridines,[46] azines,[47] and triazines.[48] For example, pyridines **72–73** and pyrimidine **74** were synthesized by addition of perfluorohexyl iodide to the agitated mixture of the corresponding heteroaryl bromide, cooper bronze powder in DMSO solvent under nitrogen at elevated temperature (Fig. 7.26).[46]

Later on, it was found that in case of diazenes, the corresponding chlorides can be used as starting materials for regioselective preparation of perfluoroalkyl derivatives **75–77**.[47] The reaction can be carried out in the presence of 2,2′-bipyridyl catalyst, in C_6F_6 and DMSO (or DMF) cosolvent (Fig. 7.27).

Cyanuric chloride under similar conditions gave the corresponding *s*-triazines in low yield (34–37%).[48] However, in case of electron-deficient cyanuric fluoride it was shown that nucleophilic perfluoralkylation can be carried out by fluoride anion catalyzed reaction with perfluoralkyltrimethylsilanes. The corresponding *s*-triazines **78** are formed in 41–77% yield under mild conditions (Fig. 7.28).[48]

In some instances, perfluorinated olefins can be used for the introduction of perfluoroalkyl group into highly electrophilic heterocycles. For example, reported in patent literature nucleophilic reaction of $C_2F_5^-$ (generated from tetrafluoroethylene

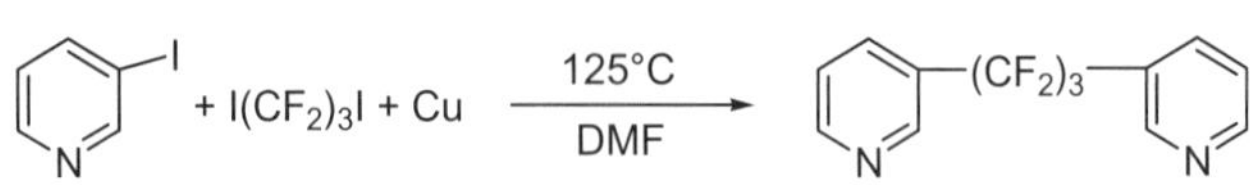

71, 52%

Figure 7.25 Copper mediated coupling of 1,3-diiodohexafluoropropane and 3-iodopyridine.

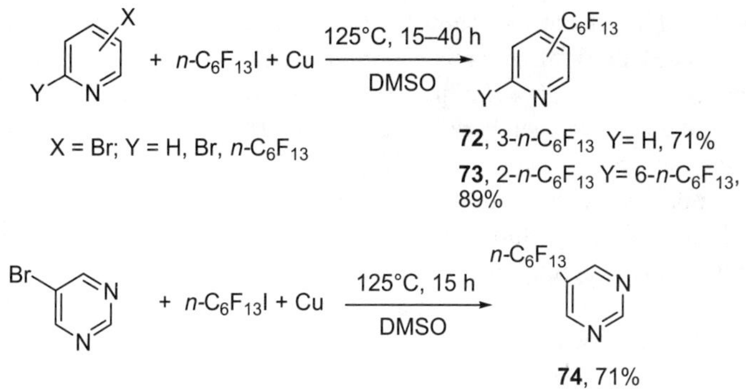

Figure 7.26 Perfluoroalkylation of bromoheteroarenes using perfluorohexyl iodide.

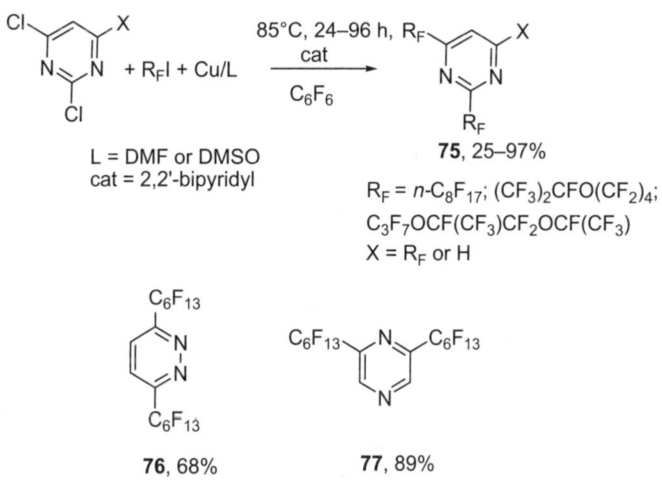

Figure 7.27 Preparation of perfluoroalkyldiazines using the copper mediated perfluoroalkylation reaction.

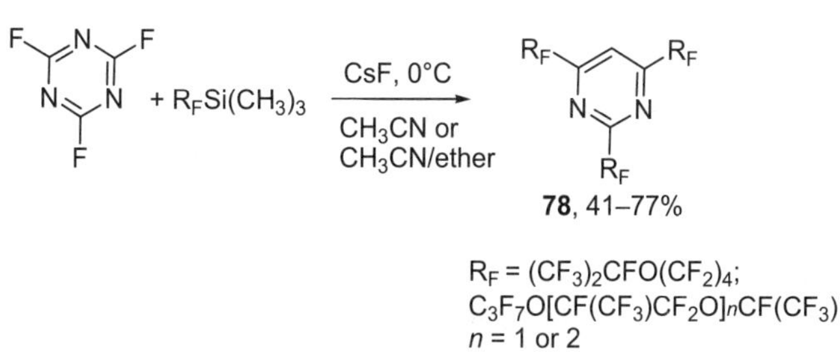

Figure 7.28 Nucleophilic perfluoroalkylation of cyanuric fluoride.

in the presence of metal fluoride catalyst) with cyanuric fluoride leads to the formation of tris(pentafluoroethyl)-*s*-triazine in up to 44% yield.[49]

Fluorinated *s*-triazatriphosphinines represent group of heterocyclic aromatic compounds, in which six-membered heterocyclic aromatic ring is constructed from nitrogen and phosphorous (V) atoms. Despite the fact that *s*-triazatriphosphinines-containing fluorinated alkoxy groups (readily prepared by the reaction of $[NPCl_2]_3$ with various derivatives of fluorinated alcohols) are well known,[50–54] phosphinines carrying perfluoroalkyl groups at phosphorous are still rare. For example, 2,2,4,4,6,6-hexakis(trifluoromethyl)-1,3,5,2,4,6-triazatriphosphinine (**79**) was recently prepared in 90% yield through the reaction of 2,2,4,4,6,6- hexafluoro-1,3,5,2,4,6-triazatriphosphinine $[NPF_2]_3$ with Ruppert–Prakash reagent.[55]

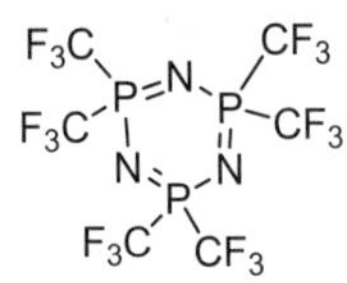

79, 90%

Several examples, of *s*-triazatriphosphinines containing alkyl and trifluoromethyl groups attached to phosphorous are described in Ref. 56.

As it was mentioned above, the activation of heteroarene toward nucleophilic attack is another methodology, often applied for the synthesis of perfluoroalkyl heteroarenes. Uno et al. discovered that pyridine oxide reacts with generated *in situ* at low temperature perfluorohexyl lithium giving 2-perfluorohexylpyridine in low yield.[57] However, the addition of boron trifluoride etherate facilitates perfluoralkylation through the activation of heterocycle. For example, quinoline and 2-alkylquinolines were react with variety of perfluoroalkyl lithium leading to selective formation of the corresponding dihydroquinolines **80**.[57] Compound **80** ($R = H$, $R_F = C_6F_{13}$) at ambient temperature undergoes oxidation by atmospheric oxygen to give isoquinolines **81** (Fig. 7.29).

Isoquinoline under similar conditions gives 1-perfluorohexyl-1,2-dihydroisoquinoline, which undergoes oxidation during chromatographic purification forming isoquinolines **82** and **83** (Fig. 7.29). Additional examples of preparation of 2-perfluoroalkyl-4-hydroxyisoquinolines are described in Ref. 58. Both pyridazine and pyrimidine undergo perfluoroalkylation giving dihydro derivatives **84** and **85** in reasonable yields, but pyrazine reaction under similar conditions results in the formation of complex mixture of products.[57]

2-(1,1,1,2-Tetrafluoroethyl)pyridines can be prepared using an unusual reaction of the pyridine oxides with electrophilic fluoroolefins, such as hexafluoropropene (HFP). This reaction discovered by Mailey and Ocone,[59] was further expanded by the Haszeldine group[60] and recently was extensively studied by Makosza et al.[61,62] The reaction between heterocyclic *N*-oxides and HFP rapidly proceeds in DMF at ambient temperature and atmospheric pressure, resulting in the formation of the corresponding 2-(1,1,1,2-tetrafluoroethyl)- heterocycles **86–88** (Fig. 7.30).[61]

While the reaction of quinoline, isoquinoline, and 4-substituted pyridines results in the formation of one product, oxides of 3-substituted pyridines produce a mixture of

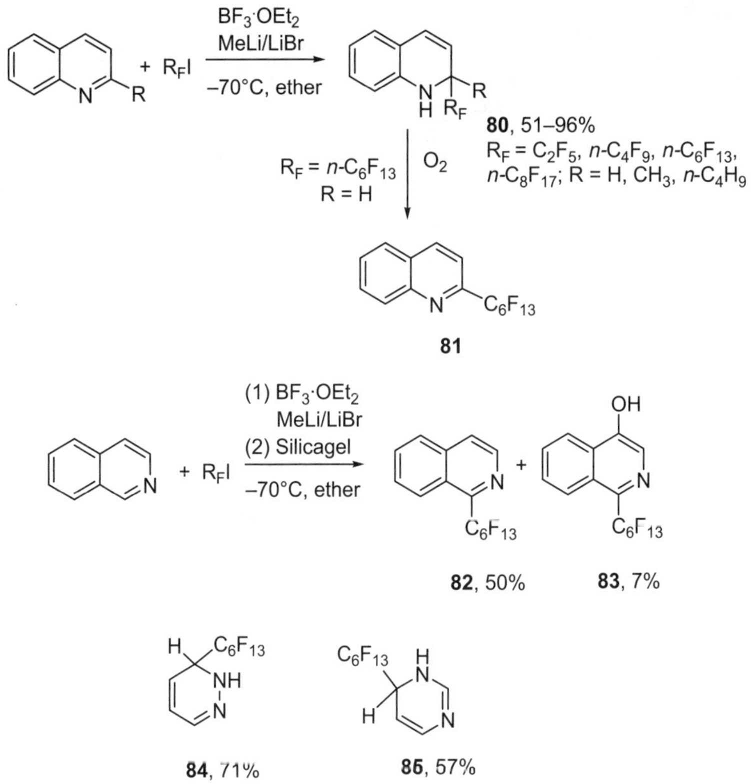

Figure 7.29 Boron trifluoride promoted perfluoroalkylation of nitrogen containing aromatic heterocycles.

two isomers **88a** and **88b** (Fig. 7.30).[59–61] The reaction mechanism involves the formation of intermediate **89**, which under reaction conditions undergoes ring opening forming **90**. Elimination of $C(O)F_2$ and proton migration leads to the formation of two isomeric pyridines **88a** and **88b**. The formation of cyclic intermediate similar to **89** was experimentally observed in the reaction of quinoline oxide with HFP.[61]

Recently, Makosza et al. reported that the reaction of different pyridine and quininolin oxides HFP carried out in the presence of methanol, amines, or thiols leads to the formation of the corresponding functional derivatives **91–93** (Fig. 7.31).[62]

It should be pointed out that this reaction is of general nature and it can be used for the synthesis of different types heteroarenes, including derivatives **94–95** obtained in the reaction of quinoline oxide with chlorotrifluoroethylene and 2-*H*-pentafluoropropene, respectively (Fig. 7.31).[62]

Makosza et al. recently reported two-step procedure for selective introduction of CF_3 or $(CF_3)_2CF$ group into heteroaromatic ring, using Reissert type reaction between *N*-acylated heterocycles and the corresponding perfluoroalkyl anions.

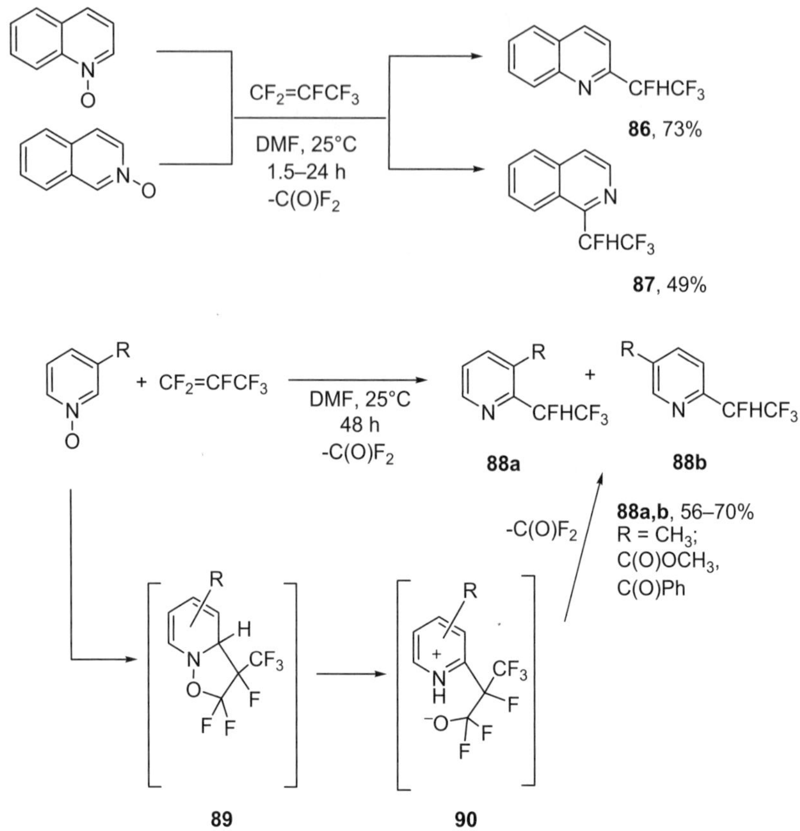

Figure 7.30 Mechanism of the reaction pyridine oxides with hexafluoropropene.

Due to the low solubility of KF in CH_2Cl_2 the trifluoromethylation step (using $CF_3Si(CH_3)_3$/KF) is carried out using Ph_3SnF as a phase-transfer catalyst (Fig. 7.32).[63]

The aromatization of dihydro heterocycles was carried out using cerium ammonium nitrate (CAN) at ambient temperature and results in trifluoromethylated heterocycles in 18–68% yield (Fig. 7.32). While trifluoromethylation of 3-picoline is regioselective giving 5-methyl isomer in 41% yield, the reaction of other 3-substituted pyridines leads to the mixture of – 3 and – 5 regioisomers.

Same methodology is applicable for the synthesis of heteroarenes containing $(CF_3)_2CF$ substituent.[64] Since hexafluoropropene readily forms $(CF_3)_2CF^-$ under action of dry KF or CsF, the first step involves the reaction of HFP and the corresponding pyridinium salt in presence of KF, leading to the corresponding dihydropyridines **96** in 72–83% yield (Fig. 7.33). The solvent has a pronounced effect on the regioselectivity of this reaction and high selectivity in favor of 2 isomer (up to 11:1 ratio for – 2 and – 4 isomers at 72% yield) can be achieved in solvents with

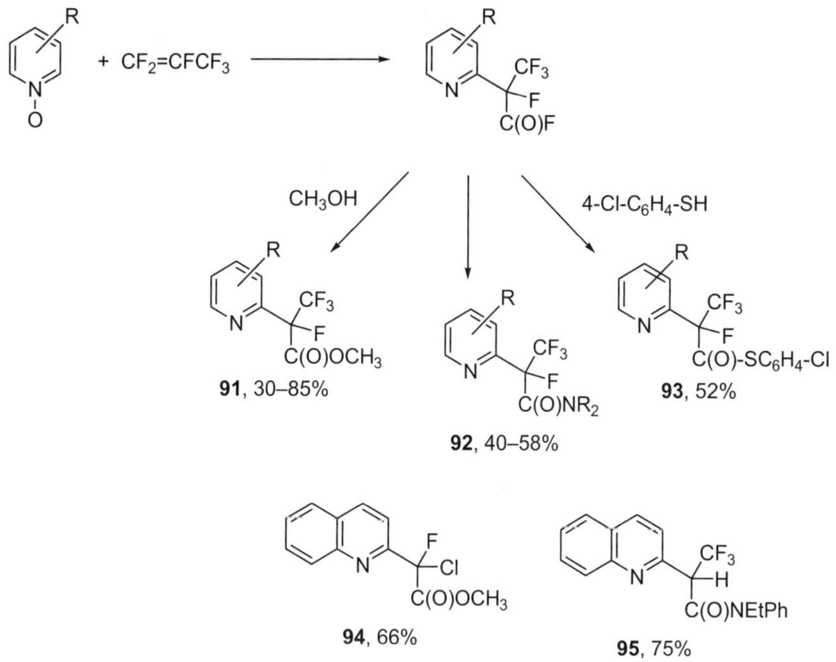

Figure 7.31 Synthesis of 2-heteroaryl derivatives of fluorinated acids.

low polarity, such as 1,2-dimethoxyethane.[64] The oxidation of dihydropyridines **96** by CAN results in high yield formation of the corresponding pyridines **97**.

It should be pointed out that the only known example of pyridine containing branched substituent at C-2 [2-(1,1,1,3,3,3-hexafluoro-*iso*-propyl)pyridine] was first prepared in 15% yield by reaction of pyridine oxide with $(CF_3)_2C{=}CFC_2F_5$[60] and later was synthesized through the reaction of $(CF_3)_2C{=}S{=}O$ and pyridine oxide in 70% yield.[65]

7.4 REACTIONS OF PERFLUOROALKYL HETEROCYCLES

In general perfluoroalkylated aromatic heterocycles chemically are similar to their hydrocarbon counterparts and participate in all reactions typical for aromatic heterocycles. On the other hand, due to the presence of R_F group these materials undergo some unique chemical transformations, having no analogy in chemistry of hydrocarbon heterocycles. This section is focused mostly on chemical transformations typical for perfluoroalkylated heterocycles.

The basicity of trifluoromethylated pyridines differs significantly depending on the position of CF_3-group in the cycle. Timperley et al. recently demonstrated that both 3- and 4-trifluoromethyl pyridines in reaction with CH_3I at ~70°C form the corresponding pyridinium salts **98** and **99** in 72 and 39% yield. However,

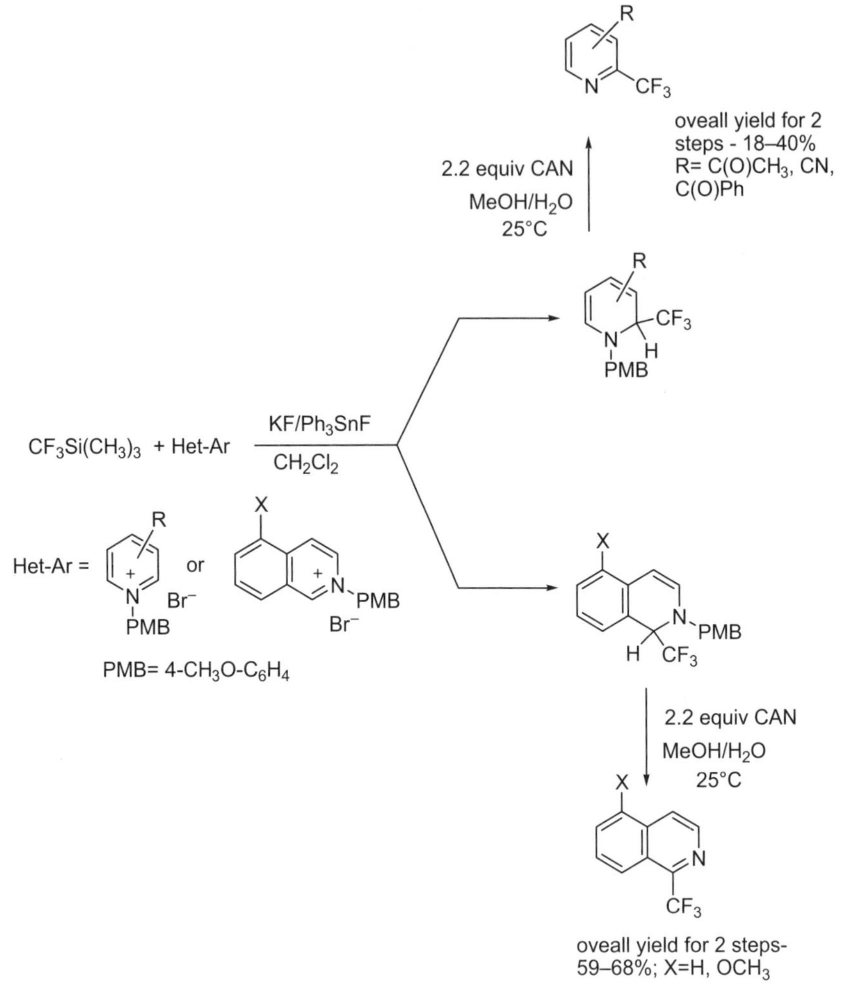

Figure 7.32 Two step trifluoromethylation process using $CF_3Si(CH_3)_3$/KF system.

2-CF_3-pyridine does not undergo methylation under similar conditions even after 595 h! (Fig. 7.34).[66]

Low basicity of **16** is likely to be caused by combination of electron withdrawing properties and steric bulk of CF_3 group located in α-position to nitrogen.[66] Since 2-fluropyridine also failed to give methylation product in reaction with CH_3I in boiling THF after 65 h,[66] it is clear that pK_a value of **16** should close to pK_a of 2-fluoropyridine (pK_a = −0.44 *vs.* 5.17 and 2.97 for pyridine and 3-fluoropyridine, respectively [67]).

On the other hand, 2-fluoroalkyl pyridines undergo oxidation under action of peroxyacids giving the corresponding oxides **100** in high yield (Fig. 7.35).[60]

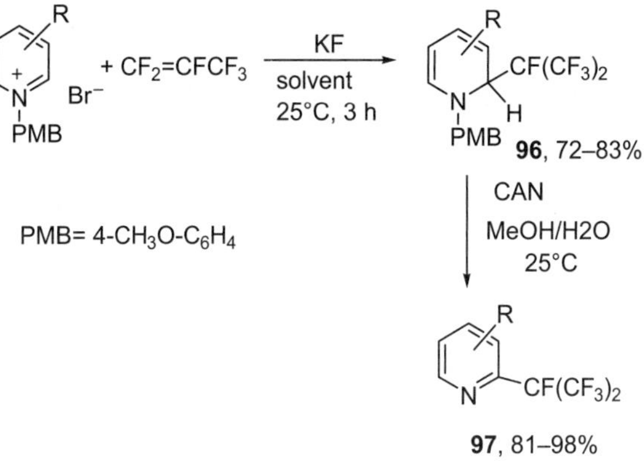

Figure 7.33 Two-step synthesis of 2-(heptafluoro-*i*-propyl)pyridines.

Kobayashi et al. reported the synthesis of 2-, 3- and 4- trifluoromethyl- quinoline oxides,[8] having quite different chemical properties. While in the reaction 2-CF_3-quinoline with $P(O)Cl_3$ gave 4-chloro-2-CF_3-quinoline, 3- and 4-CF_3 isomers selectively produce the corresponding 2-chloro- derivatives.[8]

2-Tetrafluoroethyl-substituted heterocycles can serve as starting materials for the synthesis of other fluorinated compounds. For example, Mailey reported the conversion of 2-(1,1,1,2-tetrafluoroethyl)pyridine into 2-trifluorovinylpyridine through dehydrofluorination over NaF bed at elevated temperature.[68] The Yagupol'skii group developed the synthesis of 2 CH_2F- derivatives by treatment of 2-(1,1,1,2-tetrafluoroethyl) pyridine or quinoline with MeONa in methanol, followed by acidic hydrolysis-decarboxylation of the corresponding orthoformate HetAr–CFHC$(OCH_3)_3$.[69]

The reaction of heterocycles containing CF_3 group in α-position to nitrogen with strong nucleophiles may result in a nucleophilic displacement of trifluoromethyl group. For example, 2-CF_3-pyridine, quinoline, and isoquinoline react with sodium amide with the formation of the corresponding 2-NH_2 derivatives in 88%, 69.5%, and 57% yield, respectively.[70] The mechanism of this process shown in Fig. 7.36 involves

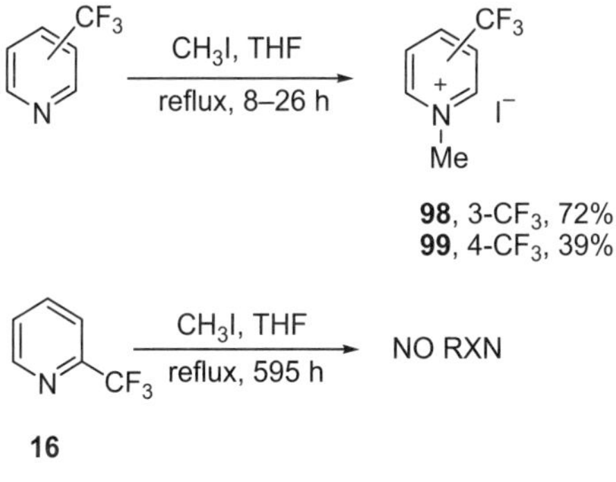

Figure 7.34 Alkylation of isomeric trifluoromethyl pyridines.

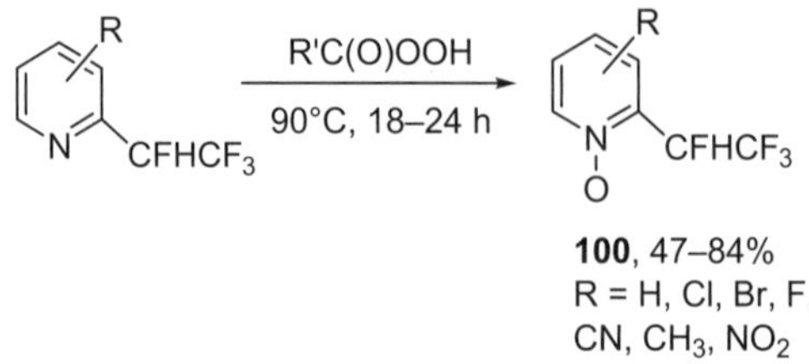

Figure 7.35 Synthesis of *N*-oxides of 2-(tetrafluoroethyl)pyridines.

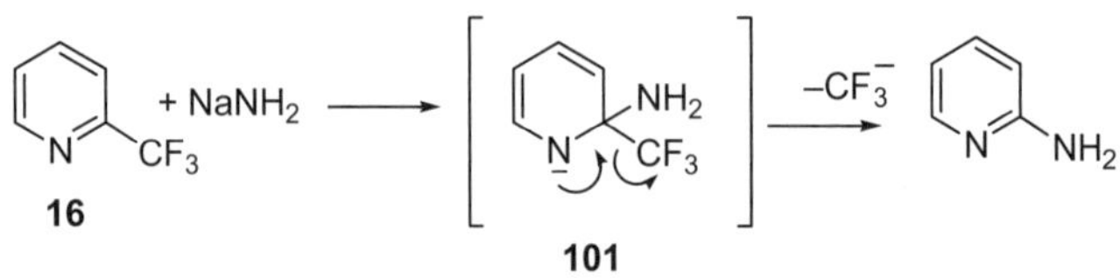

Figure 7.36 Mechanism of reaction of pyridine **16** with $NaNH_2$.

the attack of NH_2^- on the carbon bearing CF_3 group with the intermediate formation of anion **101**, followed by irreversible extrusion of trifluoromethyl anion.[70]

The reaction of 2-, 4-, and 6-CF_3 quinolines with $LiAlH_4$ leads to complete reduction of CF_3 group resulting in the corresponding heterocycles **102–104** (Fig. 7.37).[71]

The mechanism of nucleophilic reduction involves the formation of an intermediate containing exocyclic $-C{=}CF_2$ double bond and its further conversion into $-CH_3$ group through the sequential hydrogenation/HF elimination processe.[71] Interestingly, 3-CF_3 quinoline reacts with $LiAlH_4$ with the formation of dihydroquinoline **105** and the product of similar structure **106** was isolated in the reaction of 3-CF_3-pyridine with butyl lithium.[72]

Detailed study of the reactions of trifluoromethylated pyridines with alkyl lithium reagents revealed an interesting behavior of these materials.[6] While pyridines

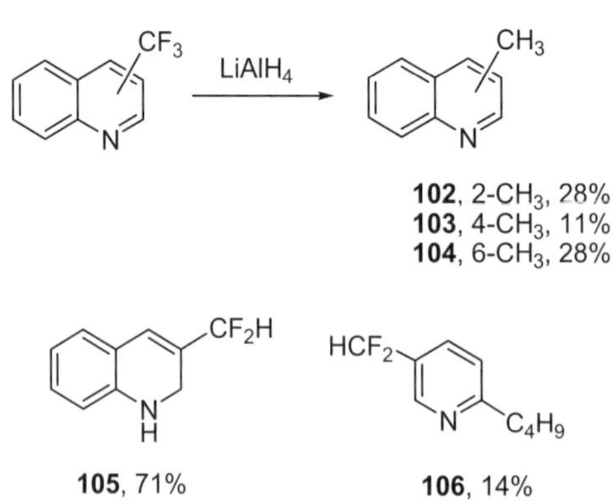

Figure 7.37 Reaction of trifluoromethylated quinolines with $LiAlH_4$.

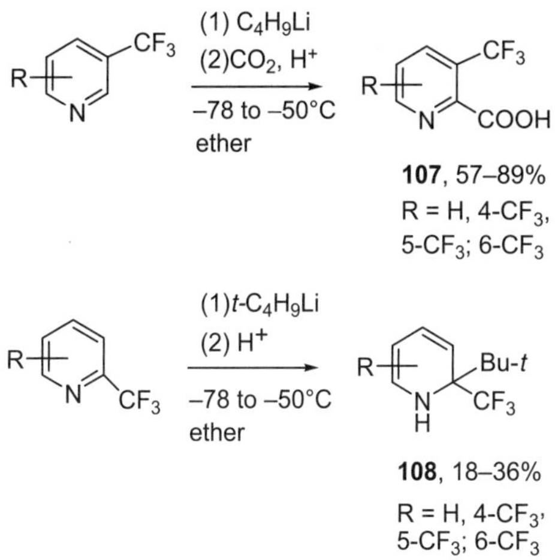

Figure 7.38 Reaction of trifluoromethylated pyridines with alkyllithium reagents.

containing 3-CF_3– group undergo smooth and highly regioselective lithiation at C-2 position, producing the corresponding acids **107** upon reaction with CO_2 (Fig. 7.38), 2-CF_3– isomers in reaction with *t*-butyl lithium give 1,2-dihydropyridines **108** in low yield (Fig. 7.38).

Regioselective metalation of a wide variety of trifluoromethylated pyridines and quinolines was exhaustively studied by the Schlosser research group.[44,72–74] It was demonstrated that regioselectivity of direct lithiation of fluorinated heterocycles depends significantly on reagents, reaction conditions and type of solvent. For example, the lithiation of 2-CF_3-pyridine using lithium 2,2,6,6-tetramethylpiperidine (LiTMP) in THF at −75°C results in exclusive formation of acid **109**, while the reaction with BuLi in the presence of $Et_2NCH_2CH_2OLi$ in less polar ether leads to selective formation of isomeric acid **110** (Fig. 7.39).[72]

The presence of halogen X (X = F, Cl, Br, or I) in the molecule of trifluoromethylated heterocycle opens an additional possibility for further regioselective transformations. Functiolization of these materials can be achieved using metal–halogen exchange (for X = Cl, Br, or I) followed by the reaction of lithiated intermediate with different electrophiles.[43,75]

Numerous transformations of perfluoroalkylated heterocycles are not limited to metalation reactions and there are many other transformations, of heterocyclic

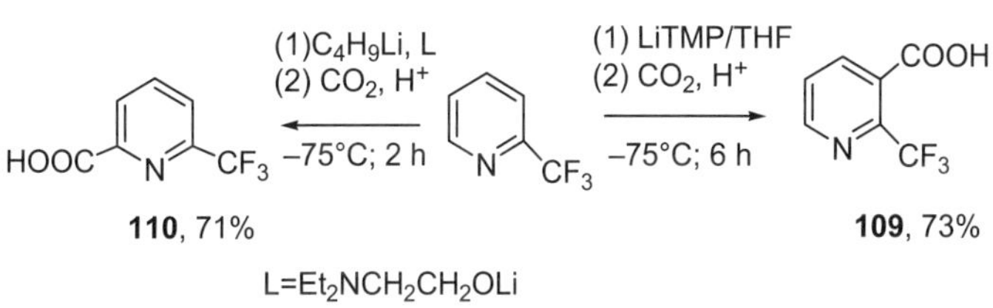

Figure 7.39 Regioselective introduction of functional group into 2-CF_3-pyridine.

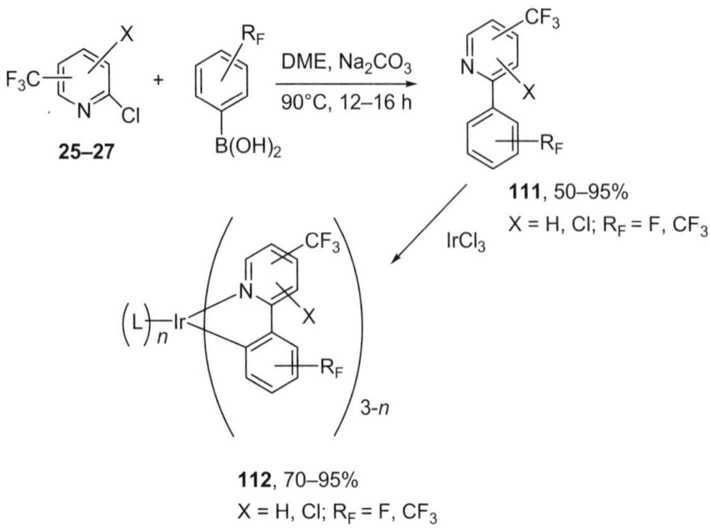

Figure 7.40 Synthesis of fluorinated 2-(aryl)pyridines **111** and iridium complexes **112**.

compounds, which are not mentioned in this chapter. As a representative example of potentially important from practical point of view transformations, the synthesis of fluorinated 2-(aryl)pyridines can be mentioned. Fluorinated 2-(aryl)pyridines are important precursors for the synthesis of highly efficient electroluminescent iridium complexes **112**.[76–78] These materials can be prepared using traditional Suzuki cross coupling of commercially available 2-chloropyridines, such as **25–27** and aryl boronic acids (Fig. 7.40).[76–79] The coupling carried out in refluxing monoglyme/water mixture, using potassium carbonate as a base and tetrakis(triphenylphosphine) palladium catalyst leads to formation of fluorinated 2-(aryl)pyridines **111** in high yield. The reaction conditions for cyclometalation step may vary, depending on type of synthesized iridium complex. Tris-cyclometalated complexes **112** (L = **111**) are usually prepared using reaction of $IrCl_3$ with excess of pyridines **111** under solvent free conditions in the presence of silver trifluoroacetate.[76,77]

ACKNOWLEDGMENT

Author thanks Dr Roman Larichev for valuable comments and proofreading the manuscript.

REFERENCES

1. Yoshida, M.; Yoshida, T.; Kobayashi, M.; Kamigata, N. Perfluoroalkylations of nitrogen-containing heteroaromatic compounds with bis(perfluoroalkanoyl) peroxides. *J. Chem. Soc., Perkin Trans. 1* **1989**, (5), 909–914.

2. Cowell, A.B.; Tamborski, C. Fluoroalkylation of aromatic compounds. *J. Fluorine Chem.* **1981**, *17*(4), 345–356.
3. Haung, B.; Liu, J. Sulfinatodehalogenation. XXVI. Perfluoroalkylation of nitrogen-containing heteroaromatic compounds with perfluoroalkyl halides and Rongalite. *J. Fluorine Chem.* **1993**, *64*(1–2), 37–46.
4. Huang, B.; Liu, J. A facile method for the perfluoroalkylation of pyridine and its derivatives. *Tetrahedron Lett.* **1990**, *31*(19), 2711–2712.
5. Raasch, M.S. Chemistry of sulfur tetrafluoride. IX. Reaction with amino acids in hydrogen fluoride. *J. Org. Chem.* **1962**, *27*, 1406–1409.
6. Porwisiak, J.; Dmowski, W. Reactions of (trifluoromethyl)pyridines with alkyllithium reagents. Directing effects of the trifluoromethyl groups. *Tetrahedron* **1994**, *50*(42), 12259–12266.
7. Kobayashi, Y.; Chinen, E. Organic fluorine compounds. I. Syntheses and infrared absorption of bis(trifluoromethyl)pyridines. *Chem. Pharm. Bull.* **1967**, *15*(12), 1896–1900.
8. Kobayashi, Y.; Kumadaki, I.; Taguchi, S. Organic fluorine compounds. VI. Preparations and reactions of (trifluoromethyl)quinoline *N*-oxide. *Chem. Pharm. Bull.* **1969**, *17*(11), 2335–2339.
9. Shustov, L.D.; Nikolenko, L.N.; Senchenkova, T.M. Reaction of molybdenum hexafluoride with carboxylic acids. *Zh. Obshch. Khim.* **1983**, *53*(1), 103–105.
10. Haga T.; Tsujii T.; Hayashi K.; Kimura F.; Sakashita N.; Fujikawa, K. *Synthesis and Chemistry of Agrochemicals II.* In ACS Symposium Series 443, Baker D.R.; F.J.G.; Moberg W.K. Ed.; American Chemical Society: Washington, DC, 1991; pp 107–119.
11. Nishiyama, R.; Fujikawa, K.; Yokomichi, I.; Tsujii, Y.; Nishimura, S. Trifluoromethylpyridines. (Shihara Sangyo Kaisha, Ltd., Japan.) *European Patent* EP42696, **1981**.
12. Powell R. Fluorine chemistry: the ICI Legacy. In *Fluorine Chemistry at the Millennium. Fascinated by Fluorine*, Banks, R.E. Ed.; Elsevier: New York, 2000; pp 376–377.
13. Salamon, Z.; Imre, I.; Sebestyen, M. Fluoromethylquinoline derivatives (Alkaloida Vegyeszeti Gyar, Hung). *European Patent* EP 111329, **1984**.
14. Jarvie, J.M.S., Fitzgerald, W.E.; Janz, G.J. 2-Trifluoromethylpyridine. *J. Am. Chem. Soc.* **1956**, *78*, 978–979.
15. Janz, G.J.; Monahan, A.R. Synthesis of pyridines from butadiene and cyanogen like molecules. *J. Org. Chem.* **1964**, *29*(3), 569–571.
16. Gerus, I.I.; Vdovenko, S.I.; Gorbunova, M.G.; Kukhar, V.P. 4-(Trifluoromethyl) pyrimidines. *Khim. Geterotsikl. Soedin.* **1991**, (4), 502–511.
17. Barone, J.A.; Peters, E.; Tieckelmann, H. 2-Trifluoromethylpyrimidines. *J. Org. Chem.* **1959**, *24*, 198–200.
18. Petrova, O.E.; Kurykin, M.A.; Gorlov, D.V. Synthesis of 5-fluoro-2,4,6-tris (perfluoroalkyl)pyrimidines. *Russ. Chem. Bull.* **1999**, *48*(11), 2169–2170.
19. Petrova, O.E.; Kurykin, M.A.; Kagramanova, E.M. Condensation of polyfluorinated beta-diimines with aromatic aldehydes. *Russ. Chem. Bull.* **2002**, *51*(4), 712–714.
20. Petrova, O.E.; Kurykin, M.A.; Mysov, E.I. Reaction of polyfluorinated beta-diimines with ketones. A novel method for the synthesis of fluorinated pyridines. *Chem. Heterocycl. Compd.* **2004**, *40*(8), 1024–1030.

21. Linderman, R.J.; Kirollos, K.S. Regioselective synthesis of trifluoromethyl-substituted quinolines from (trifluoroacetyl)acetylenes. *Tetrahedron Lett.* **1990**, *31*(19), 2689–2692.
22. Gerus, I.I.; Kropachev, L.V.; Gorbunova, M.G.; Ilchenko, A.Y.; Kukhar, V.P. Synthesis of trifluoromethyl-containing quinolines from beta-(*N*-arylamino)vinyltrifluoromethyl ketones. *Ukr. Khim. Zh.* **1993**, *59*(4), 408–411.
23. Soufyane, M.; Mirand, C.; Levy, J. Synthesis of some fluorinated nitrogen heterocycles from (diethylaminomethylene)hexafluoroacetylacetone (DAMFA). *Tetrahedron Lett.* **1993**, *34*(48), 7737–7740.
24. Keller, H.; Schlosser, M. 2-(Trifluoromethyl)quinolines from anilines: a novel mode of isomerization and cyclization. *Tetrahedron* **1996**, *52*(13), 4637–4644.
25. Schlosser, M.; Keller, H.; Sumida, S.-I.; Yang, J. How 2-anilinovinyl perfluoroalkyl ketones can be mechanistically correlated with their cyclization products 2-(perfluoroalkyl)quinolines. *Tetrahedron Lett.* **1997**, *38*(49), 8523–8526.
26. Froissard, J.; Greiner, J.; Pastor, R.; Cambon, A. Synthesis of 2-(perfluoroalkyl)-4-hydroxyquinolines. *J. Fluorine Chem.* **1981**, *17*(3), 249–263.
27. Pastor, R.; Cambon, A. Synthesis of perfluoroalkyl-substituted isoquinolines. *J. Fluorine Chem.* **1979**, *13*(4), 279–296.
28. Wang, Q.-F.; Mao, Y.-Y.; Qin, C.-Y.; Zhu, S.-Z.; Hu, C.-M. A novel synthesis of 2-fluoroalkyl quinolines. *Monatsh. Chem.* **2000**, *131*(1), 55–63.
29. Yanai, H.; Mimura, H.; Kawada, K.; Taguchi, T. Convenient synthesis of fluorinated quinoline, 1,2-dihydroquinoline, and 1,2,3,4-tetrahydroquinoline derivatives. *Tetrahedron* **2007**, *63*(10), 2153–2160.
30. Reilly, W.L.; Brown, H.C. Reactions of the perfluoronitriles. II. Syntheses of 2,4,6-tris (perfluoroalkyl)-1,3,5-triazines. *J. Org. Chem.* **1957**, *22*, 698–700.
31. Bissell, E.R.; Spengler, R.E. Fluorine-containing nitrogen compounds. II. Trimerization of trifluoroacetonitrile. *J. Org. Chem.* **1959**, *24*, 1147–1148.
32. Zollinger, J.L.Catalyzed nitrile trimerization. (Minnesota Mining and Manufacturing Co.) *US Patent* 3,470,176, **1969**.
33. Tolmacheva, G.M.; Zhulin, V.M.; Krukovskii, S.P.; Ponomarenko, V.A. Trimerization of perfluorooxaalkanoic acid nitriles. *Izv. Akad. Nauk SSSR, Ser. Khim.* **1976**, (6), 1424–1426.
34. McLoughlin, V.C.R.; Thrower, J. Route to fluoroalkyl-substituted aromatic compounds involving fluoroalkylcopper intermediates. *Tetrahedron* **1969**, *25*(24), 5921–5940.
35. Kobayashi, Y.; Kumadaki, I.; Ohsawa, A.; Yamada, T. Trifluoromethylation with bromotrifluoromethane. *Chem. Pharm. Bull.* **1972**, *20*(8), 1839.
36. Kobayashi, Y.; Kumadaki, I. Trifluoromethylation of aromatic compounds. *Tetrahedron Lett.* **1969**, (47), 4095–4096.
37. Kondratenko, N.V.; Vechirko, E.P.; Yagupolskii, L.M. A convenient method for the introduction of trifluoromethyl groups into organic compounds. *Synthesis* **1980**, (11), 932–933.
38. Matsui, K.; Tobita, E.; Ando, M.; Kondo, K. A convenient trifluoromethylation of aromatic halides with sodium trifluoroacetate. *Chem. Lett.* **1981**, (12), 1719–1720.
39. Carr, G.E.; Chambers, R.D.; Holmes, T.F.; Parker, D.G. Sodium perfluoroalkanecarboxylates as sources of perfluoroalkyl groups. *J. Chem. Soc., Perkin Trans. 1* **1988**, (4), 921–926.

40. Clark, J.H.; McClinton, M.A.; Blade, R.J. The direct trifluoromethylation of aryl chlorides using Burton's reagent. *J. Chem. Soc., Chem. Commun.* **1988**, (10), 638–639.
41. Clark, J.H.; Denness, J.E.; McClinton, M.A.; Wynd, A.J. The trifluoromethylation of chloroaromatics using the copper-dibromodifluoromethane-dialkylamide reaction system. *J. Fluorine Chem.* **1990**, *50*(3), 411–426.
42. Wiemers, D.M.; Burton, D.J. Pregeneration, spectroscopic detection and chemical reactivity of (trifluoromethyl)copper, an elusive and complex species. *J. Am. Chem. Soc.* **1986**, *108*(4), 832–834.
43. Cottet, F.; Schlosser, M. Trifluoromethyl-substituted pyridines through displacement of iodine by in situ generated (trifluoromethyl)copper. *Eur. J. Org. Chem.* **2002**, (2), 327–330.
44. Cottet, F.; Marull, M.; Lefebvre, O.; Schlosser, M. Recommendable routes to trifluoromethyl-substituted pyridine and quinolinecarboxylic acids. *Eur. J. Org. Chem.* **2003**, (8), 1559–1568.
45. Urata, H.; Fuchikami, T. A novel and convenient method for trifluoromethylation of organic halides using CF3SiR′3/KF/Cu(I) system. *Tetrahedron Lett.* **1991**, *32*(1), 91–94.
46. Chen, G.J.; Tamborski, C. Polyfluoroalkylation of bromoheterocyclic compounds via perfluoroalkylcopper intermediates. *J. Fluorine Chem.* **1990**, *46*(1), 137–159.
47. Chen, G.J.; Chen, L.S. Perfluoroalkylations and perfluorooxaalkylations. Part 3. Chloro-substituted diazines as substrates in copper-mediated cross-coupling. *J. Fluorine Chem.* **1995**, *73*(1), 113–119.
48. Chen, G.J.; Chen, L.S. New methods for the preparation of perfluoroalkyl- and perfluoroalkylether-*s*-triazines. *J. Fluorine Chem.* **1998**, *89*(2), 217–221.
49. Deem, W.R. Reaction of perfluoro-1,3,5-triazine with perfluoroalkenes. (Imperial Chemical Industries, Ltd.) *British Patent* GB 1148676, **1969**.
50. Schmutz, J.L.; Allcock, H.R. Phosphorus-nitrogen compounds. XXIII. Reaction of sodium 2,2,2-trifluoroethoxide with hexachlorocyclotriphosphazene. *Inorg. Chem.* **1975**, *14*(10), 2433–2438.
51. Horn, H.G.; Kolkmann, F. Synthesis and properties of some cyclotriphosphazene derivatives. *Chem. Ztg.* **1981**, *105*(7–8), 213–216.
52. Elias, A.J.; Kirchmeier, R.L.; Shreeve, J.n.M. Polyfluoroalkoxy and aryloxy cyclic phosphazenes: an alternative synthetic route to substitution reactions using siloxanes in the presence of fluoride ion catalysts. *Inorg. Chem.* **1994**, *33*(13), 2727–2734
53. Allcock, H.R.; Smeltz, L.A. Phosphorus-nitrogen compounds. 25. Substituent exchange and carbon–oxygen bond cleavage with aryloxycyclophosphazenes. *J. Am. Chem. Soc.* **1976**, *98*(14), 4143–4149.
54. Allcock, H.R.; Schmutz, J.L.; Kosydar, K.M. A new route for poly(organophosphazene) synthesis. Polymerization, copolymerization, and ring–ring equilibration of trifluoroethoxy- and chloro-substituted cyclotriphosphazenes. *Macromolecules* **1978**, *11*(1), 179–186.
55. Singh, R.P.; Vij, A.; Kirchmeier, R.L.; Shreeve, J.n.M. A novel synthesis of hexakis (trifluoromethyl)cyclotriphosphazene. Single-crystal X-ray structures of N3P3(CF3)6 and N3P3F6. *Inorg. Chem.* **2000**, *39*(2), 375–377.
56. Karthikeyan, S.; Neilson, R.H. Synthesis of Some *P*-Trifluoromethyl-Substituted (Silylamino)phosphines, *N*-Silylphosphoranimines, and Phosphazenes. *Inorg. Chem.* **1999**, *38*(9), 2079–2084.

57. Uno, H.; Okada, S.; Suzuki, H. Preparation of perfluoroalkyl azaarenes with a perfluoroalkyllithium-boron trifluoride system. *Tetrahedron* **1991**, *47*(32), 6231–6242.

58. Uno, H.; Okada, S.; Suzuki, H. A novel method for the synthesis of 4-isoquinolinols. *J. Heterocycl. Chem.* **1991**, *28*(2), 341–346.

59. Mailey, E.A.; Ocone, L.R. Fluoroalkylpyridines. A novel rearrangement. *J. Org. Chem.* **1968**, *33*(8), 3343–3344.

60. Banks, R.E.; Haszeldine, R.N.; Robinson, J.M. Reactions of pyridine 1-oxides, 5-methylpyrimidine 1-oxide, and quinoline 1-oxide with perfluoropropene and perfluoro (2-methylpent-2-ene): synthesis of 2-(1,2,2,2-tetrafluoroethyl)pyridines or -pyrimidines and their *N*-oxides and of 2,2,3-trifluoro-2,3-dihydro-3-trifluoromethylfuro[3,2-b] pyridine. *J. Chem. Soc., Perkin Trans. 1* **1976**, (11), 1226–1230.

61. Loska, R.; Makosza, M. Simple method for the introduction of tetrafluoroethyl substituents into nitrogen heterocycles. *Mendeleev Commun.* **2006**, (3), 161–163.

62. Loska, R.; Makosza, M. New synthesis of 2-heteroarylperfluoropropionic acids derivatives by reaction of azine *N*-oxides with hexafluoropropene. *Chem. Eur. J.* **2008**, *14*(8), 2577–2589.

63. Loska, R.; Majcher, M.; Makosza, M. Synthesis of trifluoromethylated azines via nucleophilic oxidative substitution of hydrogen by trifluoromethyl carbanions. *J. Org. Chem.* **2007**, *72*(15), 5574–5580.

64. Loska, R.; Makosza, M. Synthesis of perfluoroalkyl-substituted azines via nucleophilic substitution of hydrogen with perfluoroisopropyl carbanions. *J. Org. Chem.* **2007**, *72*(4), 1354–1365.

65. Schwab, M.; Sundermeyer, W. Bis(trifluoromethyl)sulfine, $(CF_3)_2C{:}SO$. Addition reactions with thiols, amine *N*-oxides, ethers, and 1,3-dipoles. *Chem. Ber.* **1988**, *121* (1), 75–79.

66. Timperley, C.M.; Bird, M.; Heard, S.C.; Notman, S.; Read, R.W.; Tattersall, J.E.H.; Turner, S.R. Fluorinated pyridine derivatives. *J. Fluorine Chem.* **2005**, *126*(8), 1160–1165.

67. Brown, H.C.; McDaniel, D.H. The base strengths and ultraviolet absorption spectra of the 2- and 3-monohalopyridines. *J. Am. Chem. Soc.* **1955**, *77*, 3752–3755.

68. Mailey, E.A. 2-(Trifluorovinyl)pyridine from 2-(1,2,2,2-tetrafluoroethyl)pyridine. (to Pennwalt Corp.) US Patent US 3541102, **1970**.

69. Bogolyubskii, A.V.; Skrynnikova, A.A.; Il'chenko, A.Y.; Popov, V.I.; Yagupol'skii, L.M. Synthesis of monofluorinated compounds from 1,2,2,2-tetrafluoroethyl compounds. *Zh. Org. Khim.* **1988**, *24*(5), 1082–1087.

70. Kobayashi, Y.; Kumadaki, I.; Taguchi, S.; Hanzawa, Y. Novel reactions of trifluoromethyl group in heterocycles. *Tetrahedron Lett.* **1970**, (45), 3901–3902.

71. Kobayashi, Y.; Kumadaki, I.; Taguchi, S. Organic fluorine compounds. IX. Reactions of (trifluoromethyl)quinolines with metal hydrides. *Chem. Pharm. Bull.* **1972**, *20*(4), 823–826.

72. Schlosser, M.; Marull, M. The direct metalation and subsequent functionalization of trifluoromethyl-substituted pyridines and quinolines. *Eur. J. Org. Chem.* **2003**, (8), 1569–1575.

73. Cottet, F.; Schlosser, M. Three chloro(trifluoromethyl)pyridines as model substrates for regioexhaustive functionalization. *Eur. J. Org. Chem.* **2004**, (18), 3793–3798.

74. Cottet, F.; Marull, M.; Mongin, F.; Espinosa, D.; Schlosser, M. Further metalations and functionalizations of chloro-, bromo- and iodo(trifluoromethyl)pyridines. *Synthesis* **2004**, (10), 1619–1624.

75. Mongin, F.; Tognini, A.; Cottet, F.; Schlosser, M. Halogen shuffling in pyridines: site selective electrophilic substitutions of 2-chloro-6-(trifluoromethyl)pyridine. *Tetrahedron Lett.* **1998**, *39*(13), 1749–1752.

76. Grushin, V.V.; Herron, N.; LeCloux, D.D.; Marshall, W.J.; Petrov, V.A.; Wang, Y. New, efficient electroluminescent materials based on organometallic Ir complexes. *Chem. Commun.* **2001**, (16), 1494–1495.

77. Grushin, V.V.; Thorn, D.L.; Marshall, W.J.; Petrov, V.A. New applications of electrophilic aromatic C–H activation with metal trifluoroacetates. *ACS Symp. Ser.* **2004**, *885* 393–406.

78. Ionkin, A.S.; Wang, Y.; Marshall, W.J.; Petrov, V.A. Synthesis, structural characterization, and initial electroluminescent properties of bis-cycloiridiated complexes of 2-(3,5-bis (trifluoromethyl)phenyl)-4-methylpyridine. *J. Organomet. Chem.* **2007**, *692*(22), 4809–4827.

79. Wang, Y.; Herron, N.; Grushin, V.V.; LeCloux, D.; Petrov, V. Highly efficient electroluminescent materials based on fluorinated organometallic iridium compounds. *Appl. Phys. Lett.* **2001**, *79*(4), 449–451.

8

PERFLUORINATED SIX-MEMBERED AROMATIC HETEROCYCLES CONTAINING ONE OR MORE HETEROATOM

Graham Sandford

Department of Chemistry, Durham University, South Road, Durham DH1 3LE, UK

8.1 INTRODUCTION

Perfluorinated systems are molecules in which all of the hydrogen atoms have been replaced by fluorine and so, in this chapter, we are concerned with the chemistry of perfluorinated heteroaromatic systems containing one or more nitrogen atoms such as pentafluoropyridine, tetrafluoropyrimidine, and trifluoro-*s*-triazine. Since the first realistic synthesis of pentafluoropyridine was reported in the 1960s,[1] the chemistry of perfluorinated heteroaromatic systems has developed extensively and been discussed in several detailed reviews[2–4] and in more general monographs.[5–8]

The exploration of the chemistry of entirely synthetic perfluorinated heteroaromatic systems has led to the discovery of some remarkable new chemistry. In this discussion, general features of perfluoroheteroaromatic chemistry are presented with the aim of highlighting the unique range of chemistries possible for these fascinating systems and new reactions that extend our understanding and knowledge of organic chemistry in general. Of course, this discussion can only provide a starting point to the study of some of the unique chemistry of these systems.

Perfluoroheteroaromatic compounds are either colorless liquids or white solids and, apart from having relatively high volatilities, no special handling procedures are required for their use in synthetic chemistry. The boiling points of the perfluorinated

---

Fluorinated Heterocyclic Compounds: Synthesis, Chemistry, and Applications, Edited by Viacheslav A. Petrov

TABLE 8.1 Comparison of Boiling Points of Perfluoroheteroaromatic Systems with the Corresponding Hydrocarbon Systems

	Boiling Point (°C)	Boiling Point of Perfluorinated Compound (°C)
Pyridine	115.5	84
Quinoline	238	205
Isoquinoline	243	212
Pyridazine	208	117
Pyrimidine	124	83
Pyrazine	115	54

heterocycles are, in general, somewhat lower than those for the corresponding parent hydrocarbons (see Table 8.1) and this is attributed to the much lower intermolecular forces and the very low basicities of the fluorocarbon systems that compensate for the increase in mass upon replacing hydrogen by fluorine.[9]

All the perfluoroheteroaromatic systems are very weak bases and, for instance, superacids are required to protonate pentafluoropyridine. Relative base strengths of the perfluorinated heteroaromatic systems have been determined by NMR competition experiments[10] and the major influence is that of the fluorine atoms *ortho* to ring nitrogen that significantly decrease the basicity of the system (see Fig. 8.1).

8.2 SYNTHESIS OF PERFLUORINATED SIX-MEMBERED AROMATIC HETEROCYCLES

The standard method for the synthesis of perfluorinated heteroaromatic compounds is most commonly by halogen exchange of the corresponding perchlorinated system using molten potassium fluoride in an autoclave at very high temperature (300–480°C) in the absence of solvent.[2,3] In many cases, the synthesis of the perchlorinated precursors is the most difficult part of the synthetic sequence and this is usually accomplished by chlorination of the parent hydrocarbon by using a

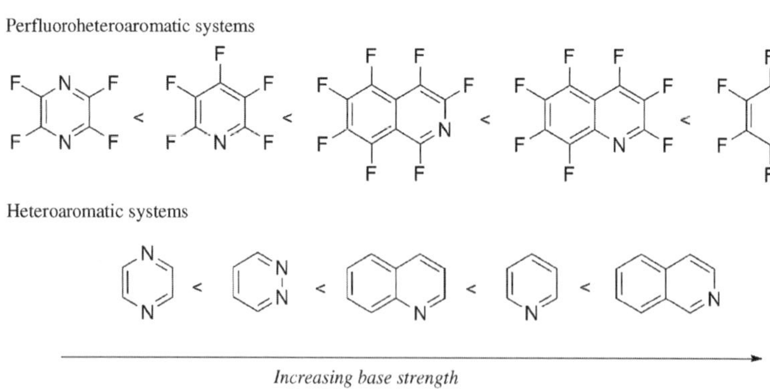

FIGURE 8.1 Basicity of perfluoroheteroaromatic systems.

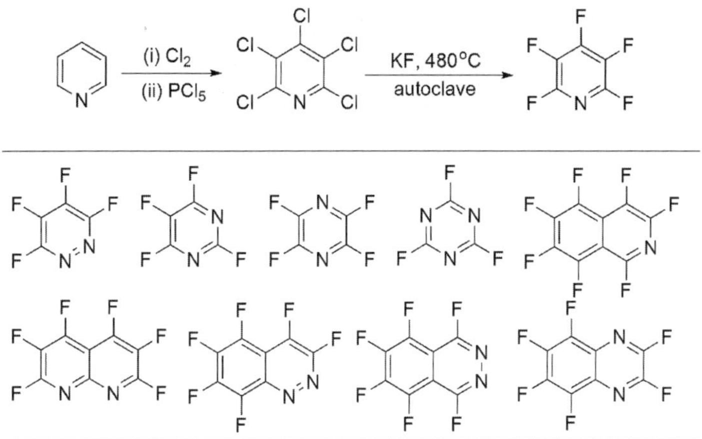

FIGURE 8.2 Synthesis of perfluoroheteroaromatic compounds.

combination of chlorine gas and phosphorous pentachloride. The synthesis of pentafluoropyidine is shown in Fig. 8.2 to illustrate the synthetic strategy along with a range of perfluoroheteroaromatic systems that have been synthesized using this general methodology.

However, among these systems, only pentafluoropyridine and tetrafluoropyridazine are currently commercially available.

8.3 REACTIONS OF PERFLUORINATED SIX-MEMBERED AROMATIC HETEROCYCLES

The chemistry of perfluoroheteroaromatic systems[2,3] is dominated by reactions with nucleophilic species because the presence of several highly electronegative fluorine atoms attached to the ring make the heterocycles very susceptible toward nucleophilic attack. Nevertheless, several reactions with electrophilic and radical species have been reported, although few in number.

In the following discussion, reactions of each perfluoroaromatic heterocyclic system are discussed in turn after a more general description of the mechanism of nucleophilic aromatic substitution that underpins the majority of the chemistry presented here.

8.3.1 Nucleophilic Aromatic Substitution: Mechanism

Nucleophilic aromatic substitution reactions follow the well-established two-step addition–elimination mechanism[2,3] via a Meisenheimer intermediate (Fig. 8.3). Indeed, reaction of fluoride ion with trifluoro-*s*-triazine, gives the corresponding perfluorocarbanion system that has been directly observed by ^{19}F NMR spectroscopy,[11] supporting this mechanistic rationale. This reactivity has been termed "mirror-image" chemistry, which contrasts the very well-known chemistry of

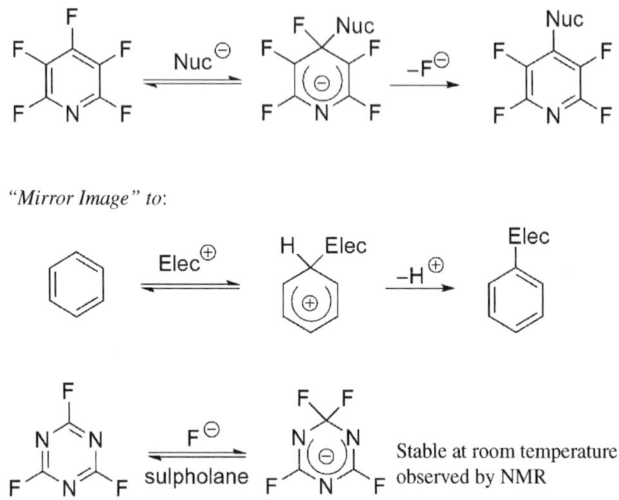

FIGURE 8.3 Nucleophilic aromatic substitution (S_NAr) mechanism and perfluorocarbanions observed by ^{19}F NMR.

aromatic hydrocarbons that undergo electrophilic substitution via carbocationic Wheland intermediates.

Extensive discussions regarding the factors that affect the regioselectivity of such nucleophilic aromatic substitution processes have been published in the course of developing the chemistry of pentafluoropyridine and related heteroaromatic systems.[2,3]

Comparison of rates of reactivity[12–14] of a series of perfluorinated heteroaromatic systems with nucleophiles shows that the ring nitrogen exerts the dominating influence and activates *ortho* and *para* positions due to the stabilizing effect of charge located on nitrogen in the intermediate Meisenheimer complex. Reactions of 2,4,6-trifluoropyridine[14] show that the relative rate of nucleophilic attack at the 4- and 2-positions is in the ratio 3:1, indicating that substitution at the *para* positions is slightly favored (Fig. 8.4). The fact that pentafluoropyridine reacts exclusively with

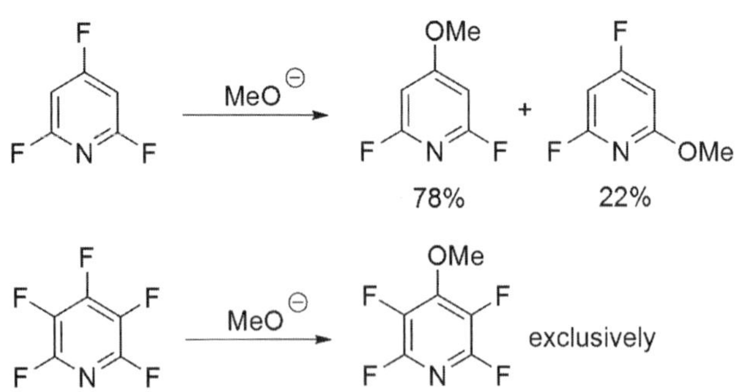

FIGURE 8.4 Regioselectivity of S_NAr processes.

nucleophiles at the 4-position, therefore, indicates that the fluorine atoms attached to the heteroaryl ring play an important role in effecting the regioselectivity of the nucleophilic substitution processes in these systems.

Kinetic studies[3,12–14] (Fig. 8.5) have been used to ascertain the separate activating influences of fluorine attached to aromatic rings and it was concluded that fluorine

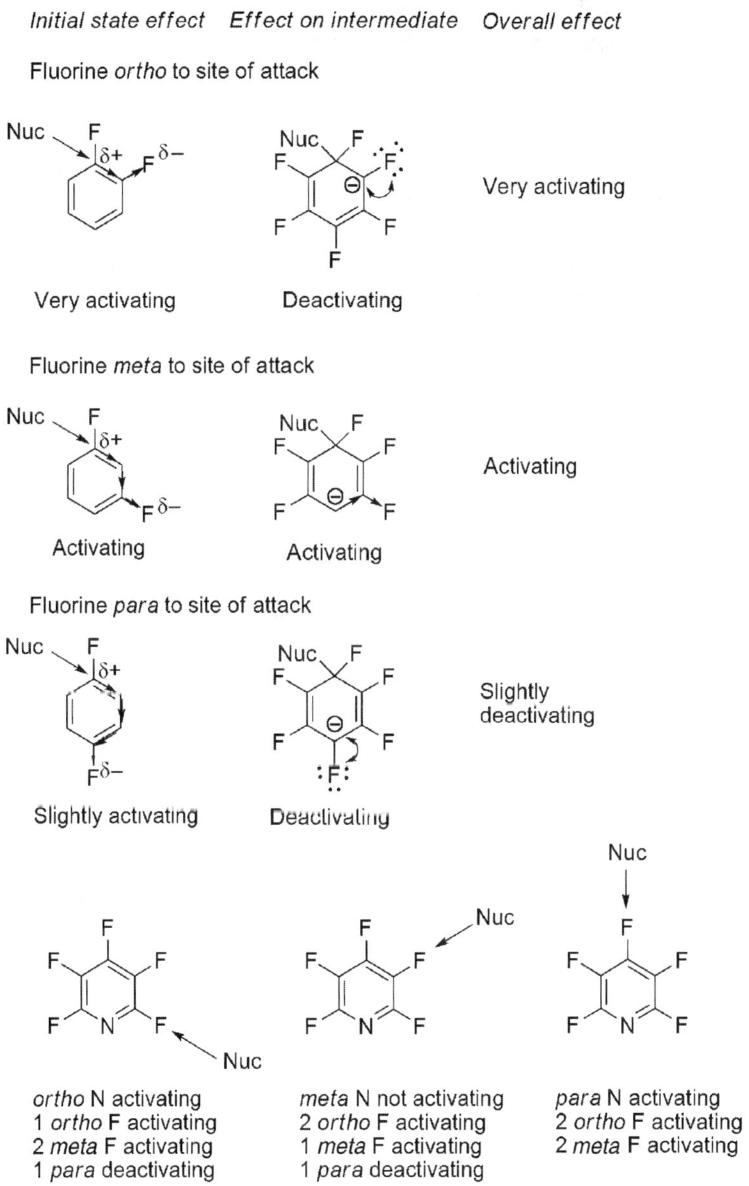

FIGURE 8.5 Effects of fluorine on nucleophilic aromatic substitution processes.

located at sites *ortho* and *meta* to the site of nucleophilic attack is strongly activating whereas fluorine that is *para* is slightly deactivating. A consideration of carbanion stabilities explains the deactivating effect of *para* fluorine. In planar carbanions, such as in the Meisenheimer intermediates for heteraryl systems, repulsion between the fluorine atom lone pairs and the negative charge are maximized and, therefore, overall destabilizing. For fluorine atoms *meta* to the site of attack, the electronegative fluorine atom inductively stabilizes the negatively charged intermediate thereby overall activating the system. We would expect fluorine atoms *ortho* to the site of attack to have a similar effect to the *para* fluorine atoms if the stability of the Meisenheimer intermediates were the only factors to consider, but experimentally determined kinetic measurements show that this is not the case. Here, *ortho* fluorine atoms influence the site of attack in the initial state, making the carbon–fluorine bond under attack more electron deficient in nature by inductive withdrawal. Consequently, nucleophilic aromatic substitution in perfluoroheteroaromatic systems occurs at sites that are preferentially *para* to ring nitrogen and also at sites that maximize the number of activating *ortho* and *meta* fluorine atoms while minimizing the number of fluorine atoms *para* to the site of attack. This situation is shown for pentafluoropyridine where the 4-position is clearly favored over the 2- and 3-sites and this is consistent with experimental observations.

These ideas have been expanded to predict and explain the orientation of nucleophilic aromatic substitution processes for a variety of perfluorinated heteroaromatic systems[3] and the sites of nucleophilic substitution are indicated in Fig. 8.6 for a number of systems studied. Specific examples that confirm this mechanistic analysis are given in subsequent appropriate sections below.

8.3.1.1 Pentafluoropyridine Reactions of pentafluoropyridine with nucleophiles proceed very readily and, in the vast majority of cases, substitution of fluorine located at the 4-position is achieved regioselectively.[2,3] Reactions involving a very wide range of oxygen, nitrogen, sulfur, and carbon-centered nucleophiles have been reported and representative examples of such processes[15–18,19] are indicated in Table 8.2. Of

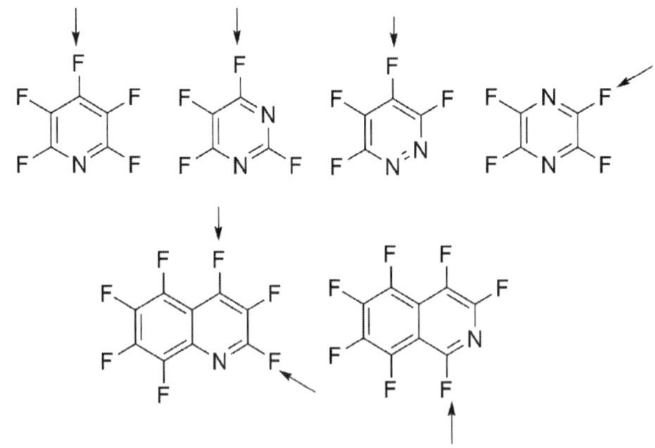

FIGURE 8.6 Orientation of S_NAr processes in perfluoroheteroaromatic systems.

TABLE 8.2 Reactions of Pentafluoropyridine with Monofunctional Nucleophiles

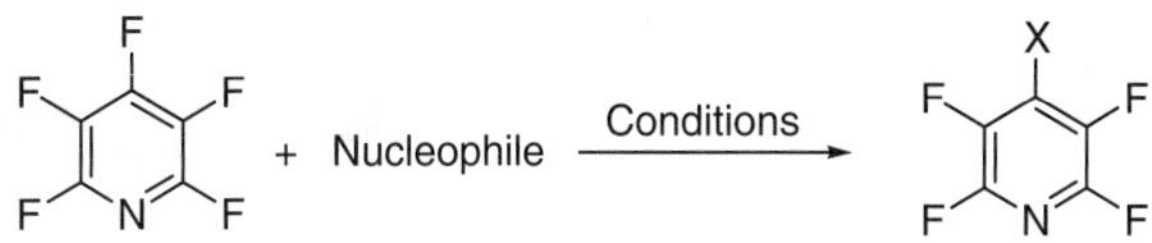

Nucleophile	X	Reference
$LiAlH_4$	H	15
MeONa	OMe	15
Me_2NH	Me_2N	15
BuLi	Bu	18
NH_2OH	NH–OH	19
PhSNa	PhS	16
NaI	I	17

course, each derivative can be used as the starting material in a wide range of synthetic transformations.

The great majority of nucleophilic substitution reactions of pentafluoropyridine occur at the 4-position, but there are exceptions. Keto-oximate salts give high proportions of the 2-substituted isomer (Fig. 8.7) and this is postulated to be due to the directing effect of a complex formed between the pyridine ring and the incoming salt.[20]

2,3,5,6-Tetrafluoropyridine derivatives, synthesized by nucleophilic substitution reactions of pentafluoropyridine are, of course, electron-deficient aromatic ring systems and further nucleophilic substitution can occur. However, few studies exploring the effects of the 4-substituent upon the regiochemistry of further nucleophilic substitution processes have been reported and examples of reactions of various 2,3,5,6-tetrafluoropyridine derivatives with diethylamine are shown in Table 8.3 to illustrate the synthetic possibilities.[21,22]

In general, the outcome of the substitution reactions of most of the tetrafluoropyridine derivatives (X = Et_2N, EtO, H, Br, $CF(CF_3)_2$, SO_2Ph) reflect the dominating *ortho* directing effect of ring nitrogen. However, a 4-pyridinol salt forms from the corresponding 4-methoxy pyridine derivative, due to nucleophilic displacement of the methyl group by the amine nucleophile, rather than ring fluorine displacement because, here, the electron-deficient pyridine group acts as a good leaving group

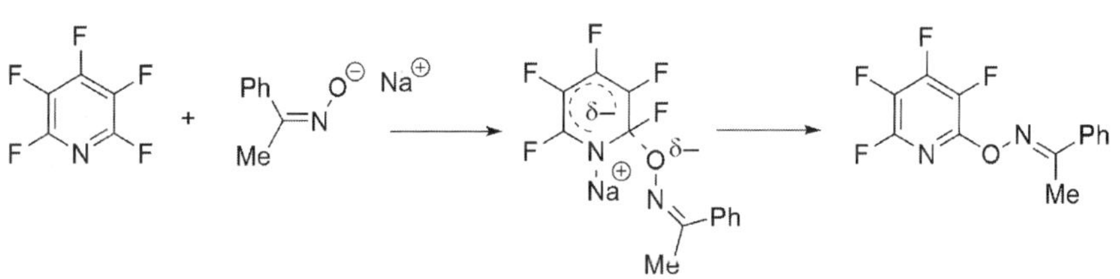

FIGURE 8.7 Reaction of ketooximates with pentafluoropyridine.

TABLE 8.3 Reactions of 2,3,5,6-Tetrafluoropyridine Derivatives with Diethylamine

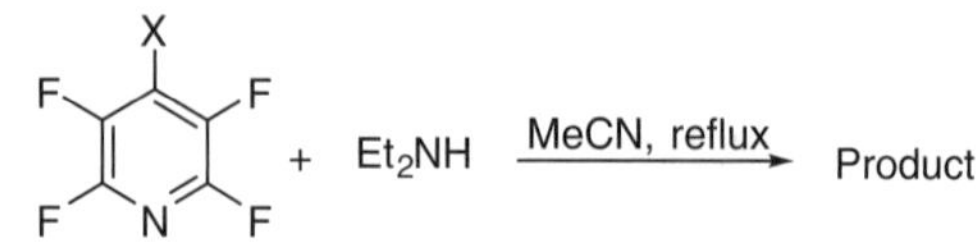

X	Product (Yield)
NEt_2	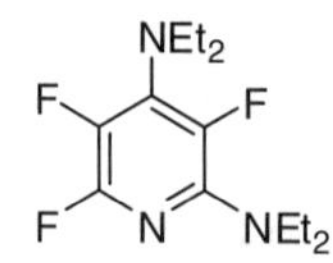 63%
OEt	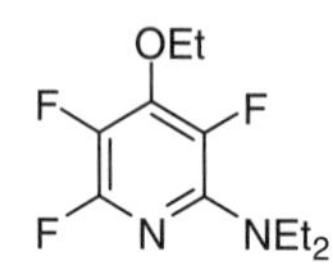 63%
Br	 61%
$CF(CF_3)_2$	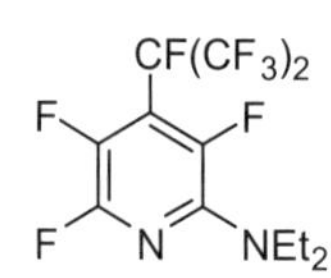 61%

(Fig. 8.8). The 4-nitropyridine derivative gives a mixture of products (Fig. 8.8) arising from substitution of fluorine *ortho* to nitrogen and the nitro group due to the high lability of nitro groups in nucleophilic aromatic substitution processes.[22]

In general, therefore, reactions of 2,3,5,6-tetrafluoroyridine derivatives with nucleophiles give products arising from substitution of fluorine *ortho* to ring nitrogen although this situation can be affected by competing substitution processes dependent on the substituent and the nucleophilic species.

Recently, the reactivity profile of pentafluoropyridine, where reaction at the 4-position is followed by substitution at the 2- and 6-positions selectively, has been used by medicinal chemists for the synthesis of small arrays of biologically active

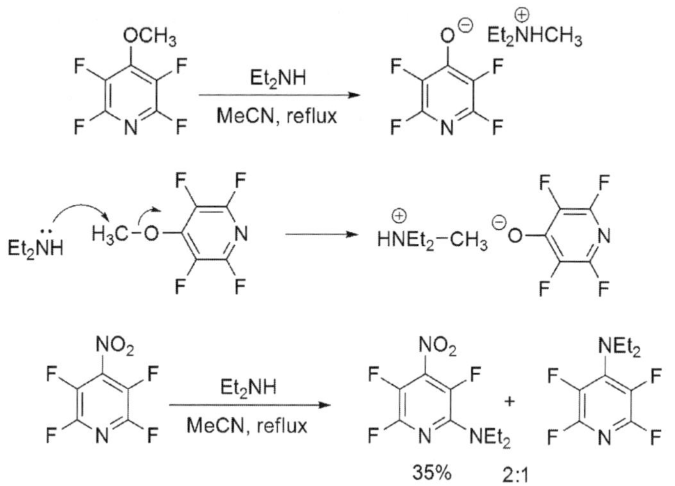

FIGURE 8.8 Reactions of 4-methoxy- and 4-nitro-tetrafluoropyridine.

pyridine systems (Table 8.4) by sequential, regioselective nucleophilic substitution of the 4-, 2-, and 6-positions, respectively.[23,24]

This sequential substitution methodology, therefore, offers great scope for the synthesis of a wide range of highly substituted pyridine derivatives that are of continuing interest to the life science and materials industries.

The order of nucleophilic substitution for the displacement of fluorine atoms in pentafluoropyridine, as outlined above, is generally in the order $4 > 2 > 3$, but this reactivity can be altered by reaction with bifunctional nucleophiles.[25,26] Substitution of the 4-position may be followed by attack at the adjacent 3-position due to the geometric constraints of the system as outlined in Fig. 8.9. Similarly, tetrafluoropyridine derivatives bearing substituents at the 4-position react with appropriate difunctional nucleophiles to give polyfunctional annelated systems.[22,27]

A variety of ring-fused systems have been prepared by reaction of pentafluoropyridine and various tetrafluoropyridine systems with difunctional nitrogen nucleophiles. For example, tetrahydropyrido[2,3-*b*]pyrazine[22,26,27] and imidazopyridine[28] systems can be prepared by reaction of pentafluoropyridine and appropriate tetrafluoropyridine systems with suitable diamines. The [5,6] and [6,6]-ring-fused systems are also useful substrates for further nucleophilic substitution processes and, consequently, act as versatile scaffolds for the construction of a range of functionalized annelated systems[26,27] (Fig. 8.10). Of course, such scaffolds are of great interest to the life science industries where access to novel heterocyclic skeletal diversity is a major factor driving the discovery of new chemical entities in lead generation.

The annelated systems described above arise from intramolecular attack by two nucleophilic centers upon the pentafluoropyridine unit. In contrast, reaction of an excess of pentafluoropyridine with similar difunctional nitrogen nucleophiles gives access to systems in which two pyridine units are linked by a "bridging unit" which, upon reaction with a further equivalent of difunctional nucleophile, release

TABLE 8.4 Biologically Active Polysubstituted Systems Synthesized from Pentafluoropyridine

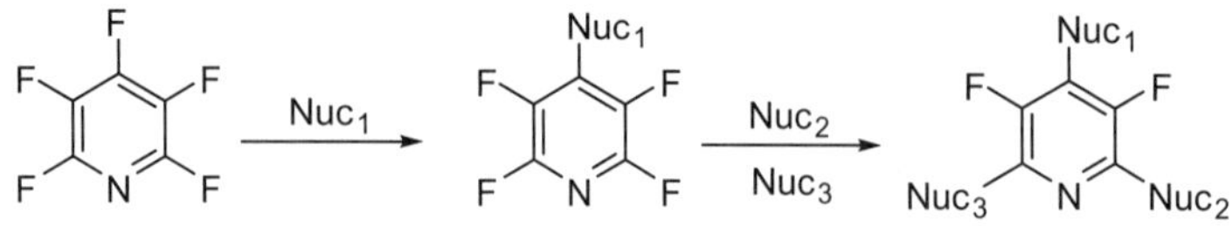

Nuc_1	Nuc_2	Nuc_3	Product
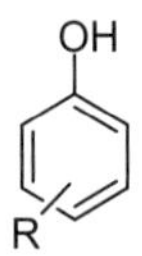	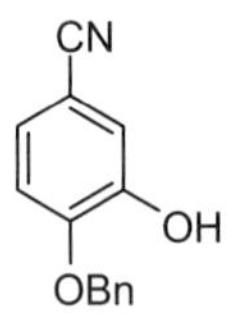	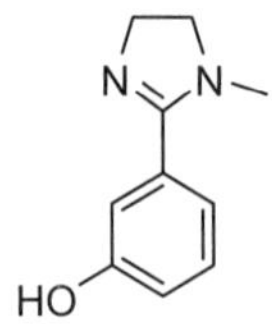	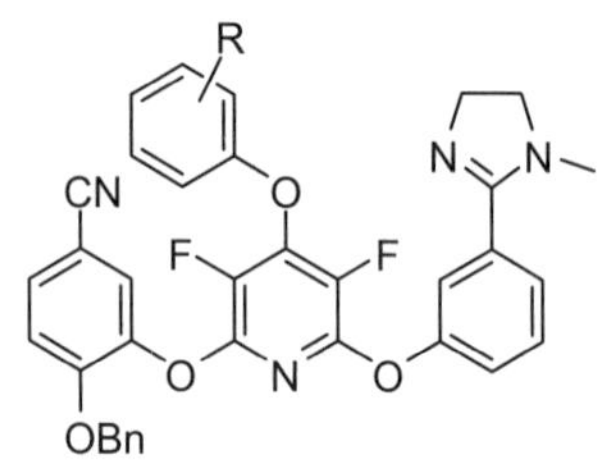
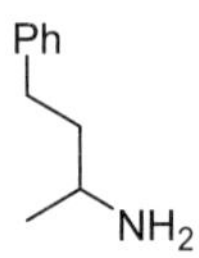	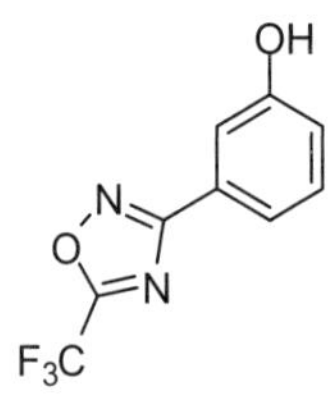	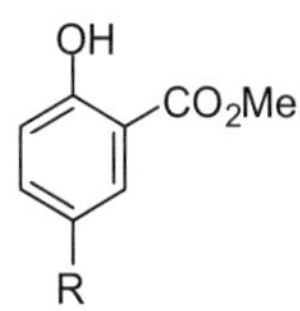	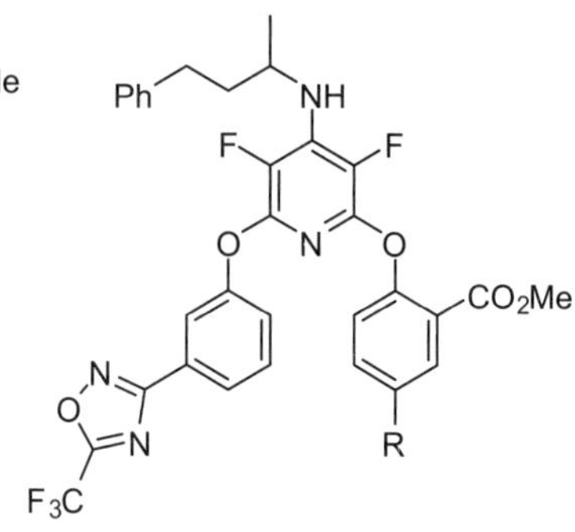
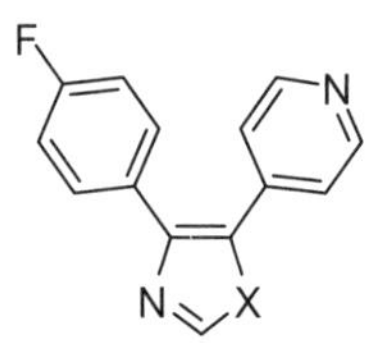	NH_3	NH_3	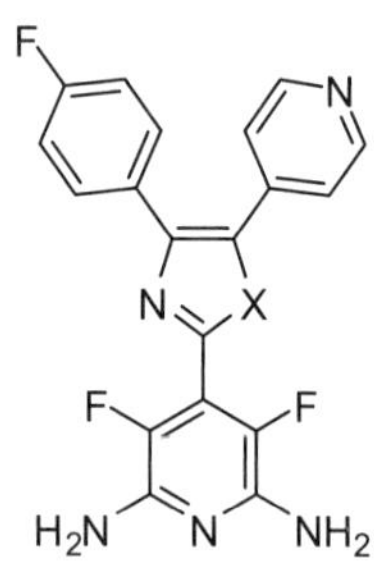

corresponding macrocyclic systems (Fig. 8.11).[29,30] This strategy offers, in principle, a general route to many structurally diverse macrocycles with specific function that is dependent upon, for example, the structural variation and functionality of the exocyclic ring substituents.

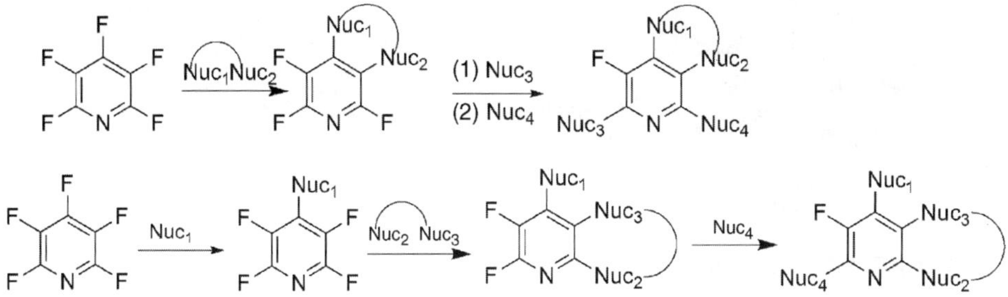

FIGURE 8.9 Possible synthetic routes to polyfunctional, ring-fused pyridine analogues.

For example, perfluoro-4-*iso*-propylpyridine was found to be an excellent building block for macrocycle synthesis.[29] Nucleophilic substitution reaction with di-oxyanions, prepared *in situ* from bis-trimethylsilyl derivatives of appropriate alkyl and aryl diols and catalytic quantities of fluoride ion, firstly give bridged systems, in

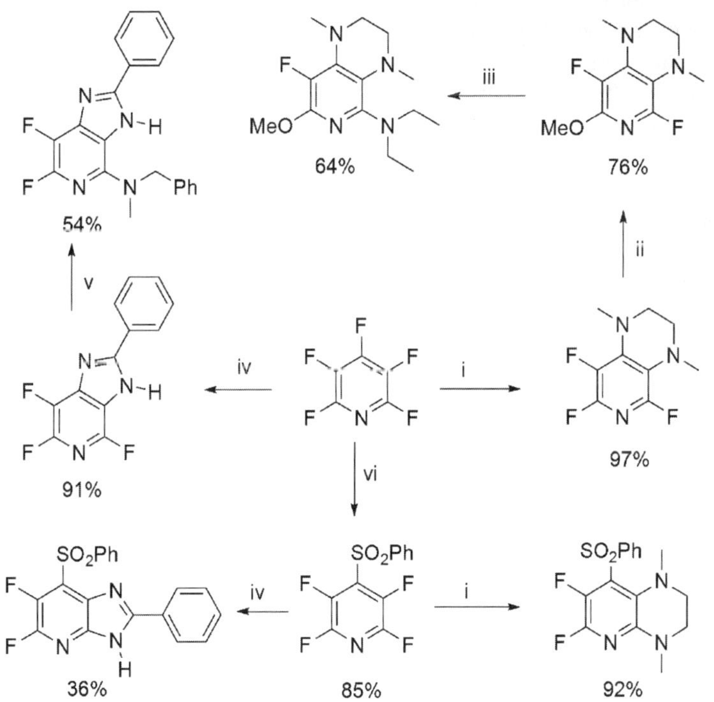

Reagents and conditions: i, $CH_3NHCH_2CH_2NHCH_3$, $NaHCO_3$, MeCN, reflux; ii, MeONa, MeOH, reflux, 2 d; iii, Et_2NLi, THF, reflux; iv, (a) $PhC{=}NHNH_2$, $NaHCO_3$, MeCN, reflux, (b) LDA, THF, rt; v, $PhCH_2NHMe$, THF, 150°C, microwave; vi, $PhSO_2Na$, DMF, 140°C;

FIGURE 8.10 Polysubstituted ring-fused systems from pentafluoropyridine.

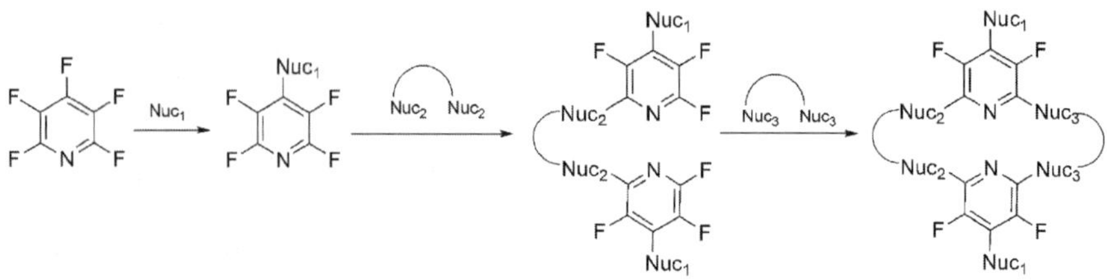

FIGURE 8.11 General strategy for the synthesis of macrocyclic derivatives from pentafluoropyridine.

which two pyridine subunits are connected by a polyether chain. Ring closure to the corresponding macrocycles, is readily achieved by addition of a further equivalent of dinucleophile (Fig. 8.12).

Reactions of perfluorinated alkenes, such as hexafluoropropene, with fluoride ion give perfluoroalkylcarbanions which can act as nucleophiles in S_NAr reactions with perfluoroheteroaromatic systems[7,31,32] (Fig. 8.13). These reactions are another example of "mirror-image" chemistry and reflect well-known Friedel–Crafts reactions of hydrocarbon systems that proceed by reaction of the corresponding electrophile and carbocationic intermediates. Polysubstitution processes are possible and, indeed, all five fluorine atoms may be replaced upon reaction with an excess of tetrafluoroethylene and fluoride ion.[33]

Various factors are important in determining the outcome of polysubstitution processes of this kind after two perfluoroalkyl groups have been introduced at the 4- and 2-positions, respectively. Further substitution processes are influenced by the

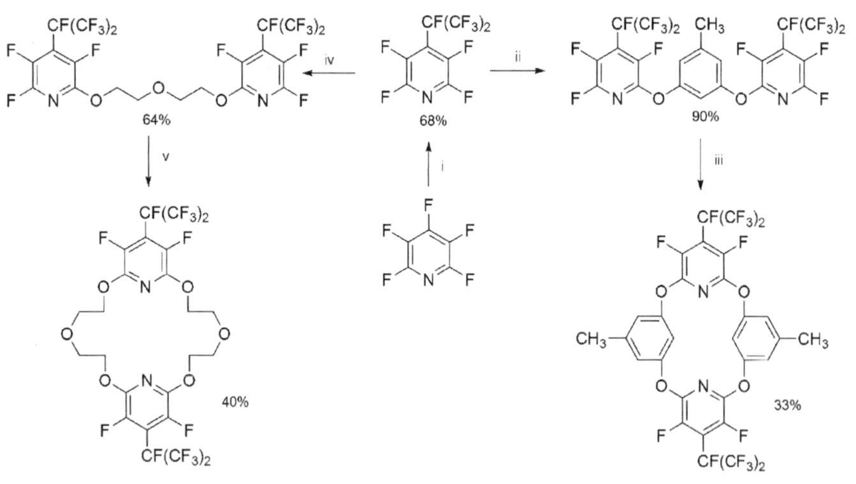

Reagents and conditions: i, CF_2=CF-CF_3, TDAE, 60°C, 16 h; ii, CH_3-$C_6H_3(OSiMe_3)_2$, CsF, monoglyme, reflux, 4 d; iii, CH_3-$C_6H_3(OSiMe_3)_2$, CsF, monoglyme, reflux, 4 d; iv, $Me_3SiOCH_2CH_2OCH_2CH_2OSiMe_3$, CsF, monoglyme, reflux, 2 d; v, $Me_3SiOCH_2CH_2OCH_2CH_2OSiMe_3$, CsF, monoglyme, reflux, 4 d.

FIGURE 8.12 Macrocycles from pentafluoropyridine.

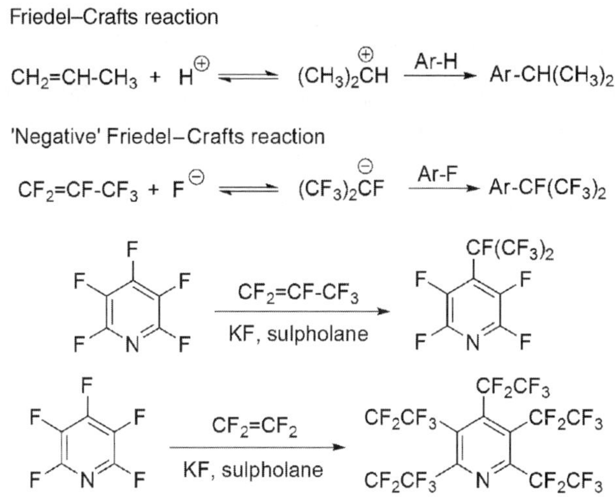

FIGURE 8.13 "Mirror-image" chemistry, negative Friedel–Crafts reactions.

perfluoroalkyl groups, which influence the site of attack by electronic and steric factors and the reversibility of these reactions in the presence of fluoride ion (Fig. 8.14). This leads to competition between the formation of kinetic and thermodynamic products where, for instance, the reaction of the 2,4-disubstituted system gives the 2,4,5-trisubstituted kinetic product at low temperature but the less sterically crowded more thermodynamically stable 2,4,6-trisubstituted product at higher temperatures.[34]

8.3.1.2 Tetrafluoropyrimidine Perfluorinated diazines (pyrimidine, pyrazine, and pyridazine) are typically 1000 times more reactive toward nucleophiles than pentafluoropyridine. Various reactions of tetrafluoropyrimidine with a small range of nucleophiles have been reported[2,3,7] (Fig. 8.15) and, in all cases, nucleophilic substitution occurs selectively at the 4-position, consistent with the mechanistic principles discussed above.

Sequential nucleophilic substitution reactions involving tetrafluoropyrimidine and a range of nucleophiles would, in principle, lead to the synthesis of many novel polyfunctional pyrimidine derivatives. Highly functionalized pyrimidine derivatives are of great importance to the life science industries and, indeed, many pyrimidine derivatives have been used for various medicinal applications.[35,36] However, the use of tetrafluoropyrimidine as a starting material for the synthesis of a range of

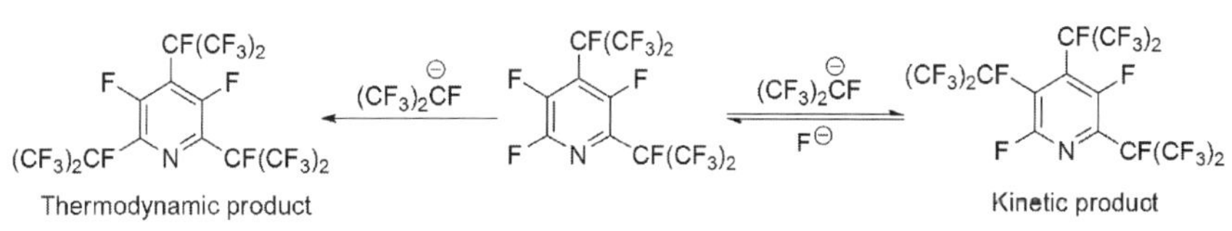

FIGURE 8.14 Polyperfluoroalkylation processes.

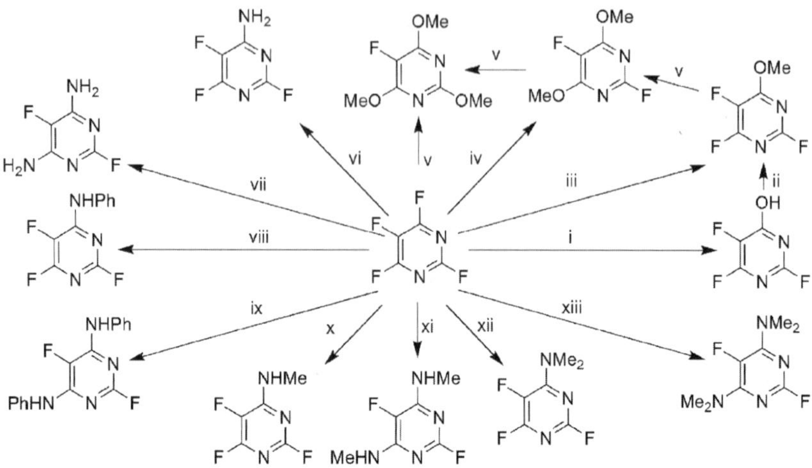

Reagents and conditions: i, H_2O/THF, rt; ii, CH_2N_2/ether, 20°C; iii, MeOH/Na_2CO_3, rt; iv, MeOH/MeONa 0°C; v, MeOH/MeONa, reflux; vi, NH_3 (aq), rt; vii, NH_3 (aq), 60°C; viii, $PhNH_2$/Na_2CO_3/THF, 15°C; ix, $PhNH_2$/THF, reflux; x, $MeNH_2$ (aq), 0–20°C; xi, $MeNH_2$ (aq)/DMF; xii, Me_2NH (aq), 0–20°C; xiii, Me_2NH (aq)/DMF, 60°C.

FIGURE 8.15 Reactions of tetrafluoropyrimidine with nucleophiles.

polysubstituted systems has not been developed to any great extent and, indeed, only very limited number of reports concerning sequential polysubstitution reactions of tetrafluoropyrimidine with nucleophiles have been published.[2,3] Recently, a systematic study of reactions of 4-aminotrifluoropyrimidine derivatives (Table 8.5) allowed the synthesis of a small array of 5-fluoro-trisubstituted pyrimidine derivatives[37] where the second and third substitution processes occurred selectively at the 6- and 2-positions, respectively.

8.3.1.3 ***Tetrafluoropyrazine*** Reactions of tetrafluoropyrazine with nucleophiles occur readily[2,3,7] (Fig. 8.16) and, of course, there are no issues regarding regioselectivity of the first nucleophilic substitution process due to the symmetry of this system. The reduced reactivity of tetrafluoropyrazine compared to the other perfluorinated diazines reflects the absence of highly activated sites *para* to ring nitrogen.[17]

The regiochemistry of the reaction of trifluoropyrazine derivatives with nucleophiles is influenced by the nature of the substituent as well as the presence of the remaining fluorine atoms. If the substituent is either an alkoxy or amino group, the site of attack is generally *ortho* to the substituent, although steric effects can also influence the outcome of this reaction. In contrast, when the substituent is an alkyl group or chlorine, the site of attack is *para* to the substituent.

Reactions of tetrafluoropyrazine with appropriate poly-hydroxylated systems has allowed the synthesis of various velcraplexes (Fig. 8.17) due to the availability of a pair of vicinal fluorines that can undergo nucleophilic substitution to form several nine-membered rings in a one-pot process.[38]

TABLE 8.5 Tetrafluoropyrimidine as a Core Scaffold

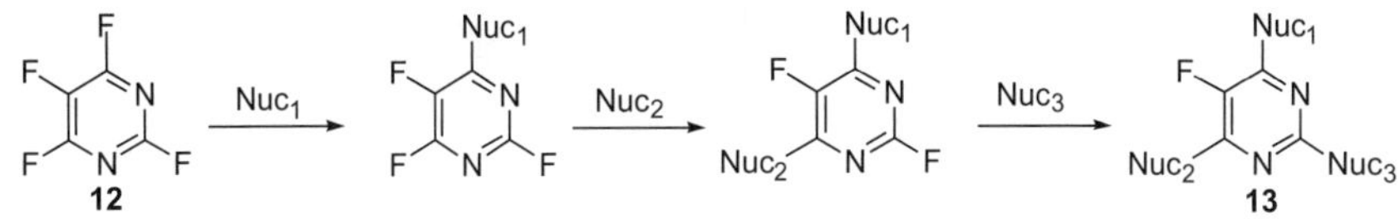

First nucleophile

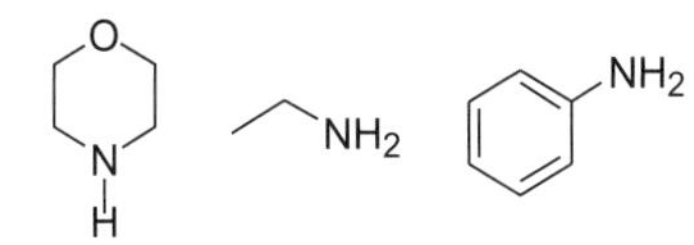

Second nucleophile

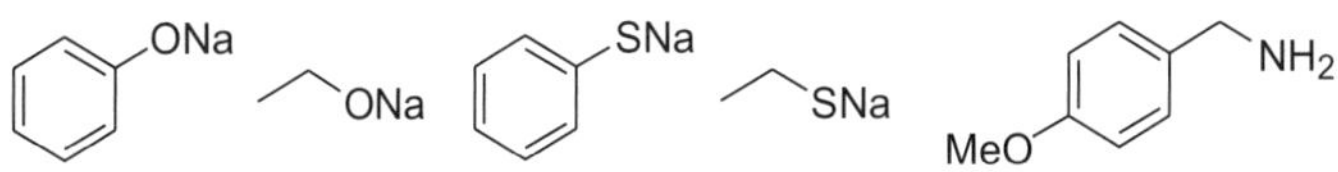

Third nucleophile

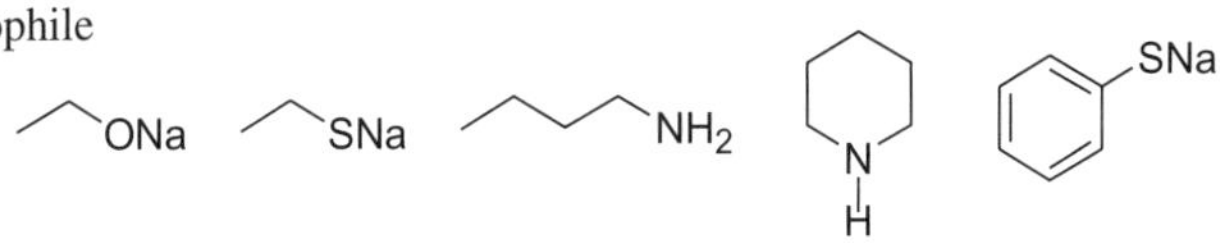

Examples of trisubstituted systems **13** synthesized

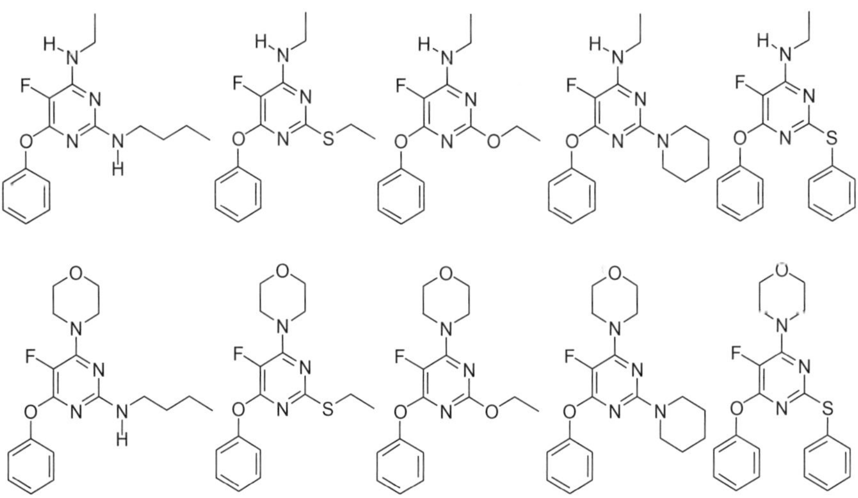

8.3.1.4 Tetrafluoropyridazine Tetrafluoropyridazine is very reactive toward nucleophiles and a short series of reactions have been reported[2,3,7] (Fig. 8.18). Substitution occurs selectively at positions *para* to ring nitrogen and polysubstitution is relatively simple. This diazine is the most basic of the

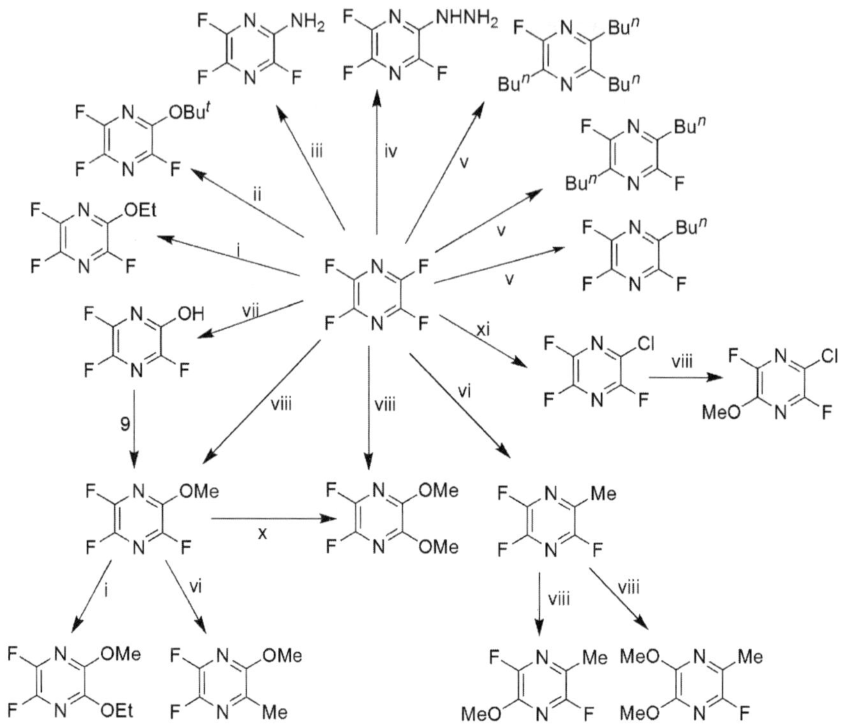

Reagents and conditions: i, NaOEt/EtOH; ii, $KOBu^t/Bu^tOH/Et_2O$; iii, NH_3 (aq); iv, N_2H_4, H_2O/EtOH; v, Bu^nLi/Et_2O; vi, $MeLi/Et_2O$; vii, KOH/Bu^tOH; viii, NaOMe/MeOH; ix, CH_2N_2/Et_2O; x, $MeOH/H_2SO_4$; xi, (a) N_2H_4,H_2O/ETOH, (b) $CuCl_2$-HCl(aq).

FIGURE 8.16 Reactions of tetrafluoropyrazine.

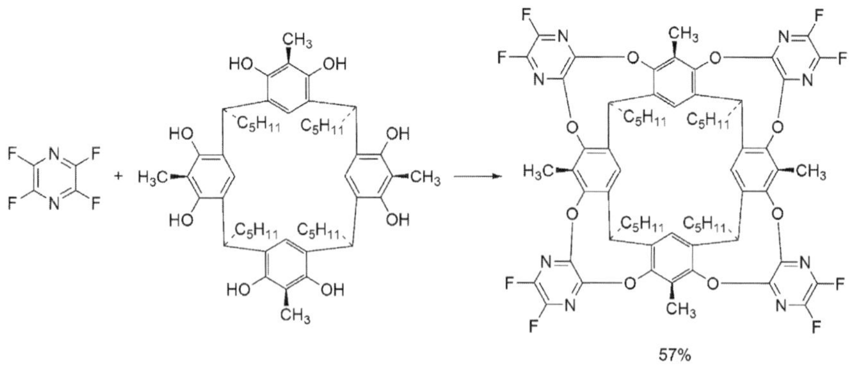

FIGURE 8.17 Synthesis of velcraplexes.

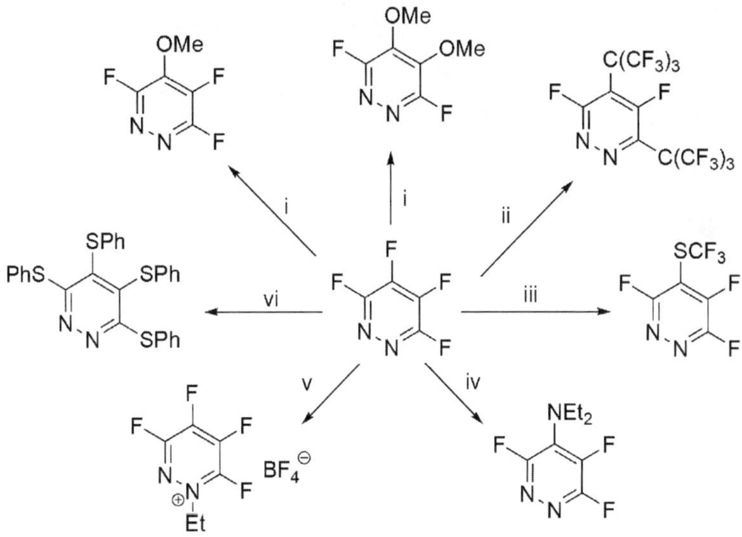

Reagents and conditions: i, NaOMe, MeOH; ii, CF_2=$(CF_3)_2$, CsF; iii, CF_2S, CsF; iv, $NHEt_2$, NMP; v, Et_3O^+ BF_4^-, CH_2Cl_2; vi, NaSPh, NMP

FIGURE 8.18 Reactions of tetrafluoropyridazine.

tetrafluorodiazines and alkylation of ring nitrogen is possible if a strong alkylating agent is used.

8.3.1.5 ***Trifluorotriazines*** As would be expected, the trifluorotriazines are highly activated toward nucleophilic attack[2,3,7] and some examples are shown in Fig. 8.19. Indeed, control of the reactions for such highly reactive systems can be difficult to achieve and polysubstitution usually competes.[39,40,41]

8.3.2 Synthesis of Valence Bond Isomers

Perfluoroalkylheteroaromatic systems can be used to prepare a range of particularly interesting and remarkably stable valence bond isomers upon irradiation.[7] Many of the valence bond isomers formed have been isolated and fully characterized, reflecting the inductive stabilizing influence of the perfluoroalkyl groups on small ring systems. Examples of the Dewar benzene and prismane systems that have been synthesized by photolysis of appropriate perfluoroalkyl heteroaromatic systems are given in Fig. 8.20.[3]

8.3.3 Reactions with Electrophiles

As would be expected, perfluoroheteroaromatic derivatives are very weak bases and, indeed, it is very difficult to observe protonation of pentafluoropyridine even

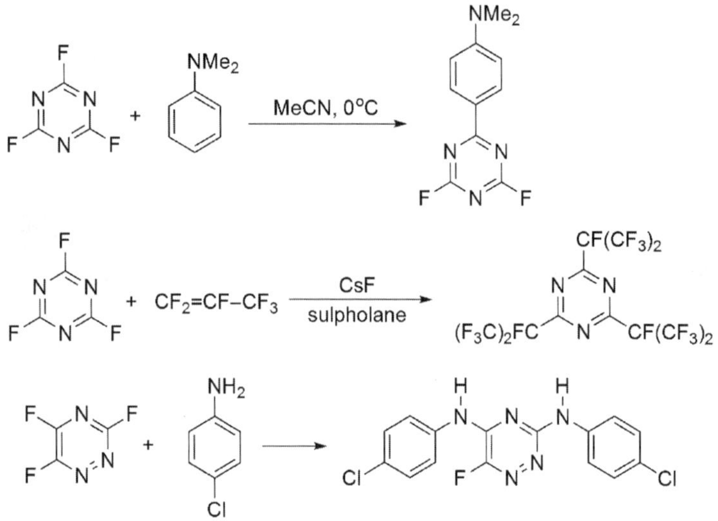

FIGURE 8.19 Reactions of trifluorotriazines.

by superacids. However, some substitution reactions can be induced by the presence of strong Lewis acids in highly acidic conditions (Fig. 8.21). Polybromination of various perfluoroheteraryl systems can be achieved by reactions involving anhydrous hydrogen bromide gas and aluminium tribromide in an autoclave at

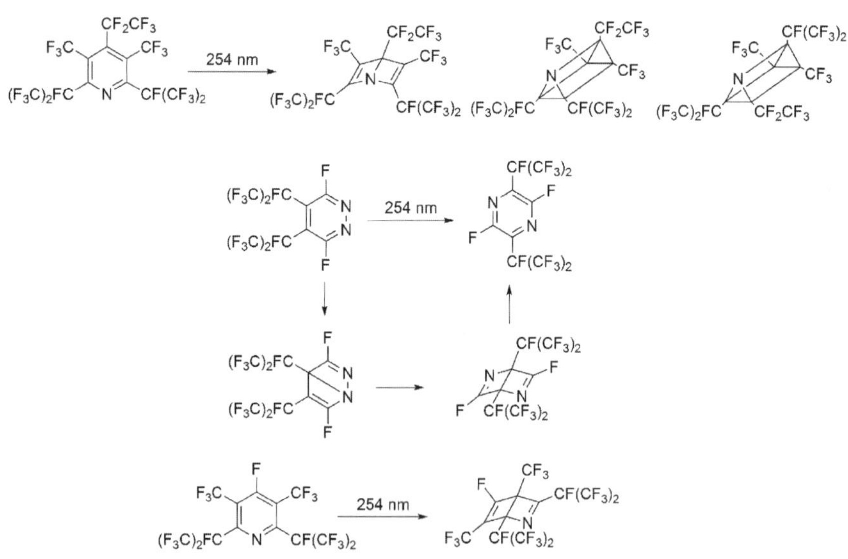

FIGURE 8.20 Valence bond isomers.

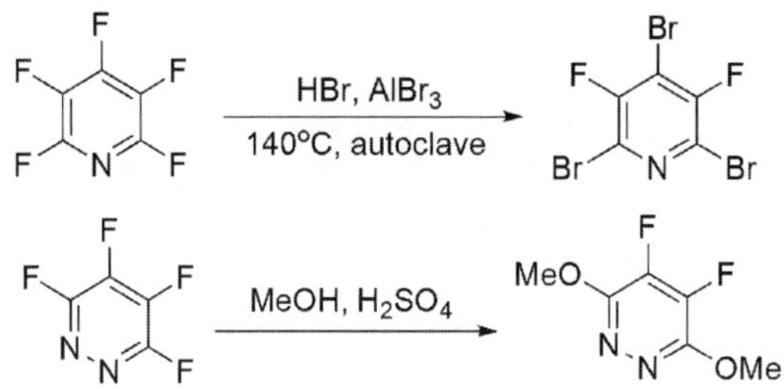

FIGURE 8.21 Reactions promoted by strong acids.

high temperature.[42] Reactions of the most basic diazine, tetrafluoropyridazine, with nucleophiles in strongly acidic conditions give products arising from substitution of fluorine at positions *ortho* to ring nitrogen, in contrast to those reactions carried out under basic conditions.[43]

8.3.4 Addition Reactions and Radical Processes

Photochemical addition reactions involving [2 + 2] additions between electron-deficient heteraryl rings and electron-rich alkenes and alkynes have been established.[44,45] Reactions of carbon-centered radicals with pentafluoropyridine gives 4-substituted products, reflecting the nucleophilic character of the radical species[46] (Fig. 8.22).

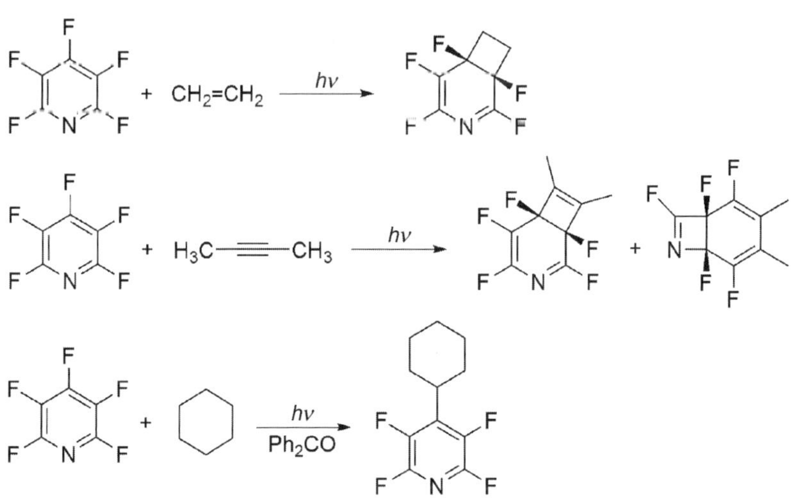

FIGURE 8.22 Addition reactions and radical processes.

REFERENCES

1. Chambers, R.D.; Hutchinson, J.; Musgrave, W.K.R. The preparation of fluoro-, chloro- and chlorofluorohydro-pyridines. *J. Chem. Soc.* **1964**, 3573–3576.
2. Brooke, G.M. The preparation and properties of polyfluoro aromatic and heteroaromatic compounds. *J. Fluorine Chem.* **1997**, *86*, 1–76.
3. Chambers, R.D.; Sargent, C.R. Polyfluoroheteroaromatic compounds. *Adv. Heterocycl. Chem.* **1981**, *28*, 1–71.
4. Silvester, M.J. Recent advances in perfluoroheteroaromatic chemistry. *Adv. Heterocycl. Chem.* **1994**, *59*, 1–37.
5. Baasner, B.; Hagemann, H.; Tatlow, J.C. (Eds.) *Houben–Weyl Organo-Fluorine Compounds*, Thieme: Stuttgart, 2000.
6. Banks, R.E.; Smart, B.E.; Tatlow, J.C. *Organofluorine Chemistry. Principles and Commercial Applications*, Plenum: New York, 1994.
7. Chambers, R.D. *Fluorine in Organic Chemistry, Blackwell*: Oxford, 2004.
8. Uneyama, K. *Organofluorine Chemistry*, Blackwell: Oxford, 2006.
9. Weast, R.C. *Handbook of Chemistry and Physics*, Chemical Rubber Publishing Company: Cleveland, OH, 1975.
10. Bell, S.L.; Chambers, R.D.; Musgrave, W.K.R.; Thorpe, I.G. Relative base strengths of some polyfluoroaryl–nitrogen heterocyclic systems. *J. Fluorine Chem.* **1971**, *1*, 51–57.
11. Chambers, R.D.; Philpot, P.D.; Russell, P.L. Anionic *s*-complexes from polyfluorinated *s*-triazines. *J. Chem. Soc., Perkin Trans. 1* **1977**, 1605–1608.
12. Chambers, R.D.; Close, D.; Musgrave, W.K.R.; Waterhouse, J.S.; Williams, D.L.H. Orienting effects of chlorine substituents in nucleophilic aromatic substitution. *J. Chem. Soc., Perkin Trans. 2* **1977**, 1774–1778.
13. Chambers, R.D.; Close, D.; Williams, D.L.H. Variation in activating influence of halogen substituents in nucleophilic aromatic substitution. *J. Chem. Soc., Perkin Trans. 2* **1980**, 778–780.
14. Chambers, R.D.; Waterhouse, J.S.; Williams, D.L.H. Activating effects of fluorine in polyfluoropyridines in reactions with ammonia. *J. Chem. Soc., Perkin Trans. 2* **1977**, 585–588.
15. Banks, R.E.; Burgess, J.E.; Cheng, W.M.; Haszeldine, R.N. Nucleophilic substitution in pentafluoropyridine: the preparation and properties of some 4-substituted 2,3,5,6-tetrafluoropyridines. *J. Chem. Soc.* **1965**, 575–581.
16. Banks, R.E.; Haszeldine, R.N.; Karsa, D.R.; Rickett, F.E.; Young, I.M. Synthesis of tetrafluoropyridine-4-sulphonic acid and related sulphur-containing derivatives of pentafluoropyridine. *J. Chem. Soc. C* **1969**, 1660–1662.
17. Banks, R.E.; Haszeldine, R.N.; Philips, E.; Young, I.M. Synthesis and reactions of 2,3,5,6-tetrafluoro-4-iodopyridine. *J. Chem. Soc. C* **1967**, 2091–2095.
18. Chambers, R.D.; Iddon, B.; Musgrave, W.K.R. 2,3,5,6-Tetrafluoro-4-methylpyridine and related compounds. *Tetrahedron* **1968**, *24*, 877–885.
19. Miller, A.O.; Furin, G.G. *N*-(Polyfluoroaryl)-hydroxylamines. Synthesis and properties. *J. Fluorine Chem.* **1987**, *36*, 247–272.

20. Banks, R.E.; Jondi, W.; Tipping, A.E. SNAr displacement of fluorine from pentafluoropyridine by sodium oximates: unprecedented substitution patterns. *J. Chem. Soc., Chem. Commun.* **1989**, 1268–1269.

21. Chambers, R.D.; Hassan, M.A.; Hoskin, P.R.; Kenwright, A.; Richmond, P.; Sandford, G. Reactions of perfluoro-isopropyl-pyridine with oxygen, nitrogen and carbon nucleophiles. *J. Fluorine Chem.* **2001**, *111*, 135–146.

22. Hargreaves, C.A.; Sandford, G.; Slater, R.; Yufit, D.; Howard, J.A.K.; Vong, A. Synthesis of tetrahydropyrido[2,3-*b*]pyrazine scaffolds from 2,3,5,6-tetrafluoropyridine derivatives. *Tetrahedron* **2007**, *63*, 5204–5211.

23. Ng, H.P.; Buckman, B.O.; Eagen, K.A.; Guilford, W.J.; Kochanny, M.J.; Mohan, R.; Shaw, K.J.; Wu, S.C.; Lentz, D.; Liang, A.; Trinh, D.; Ho, E.; Smith, D.; Subramanyam, B.; Vergona, R.; Walters, J.; White, K.A.; Sullivan, M.E.; Morrissey, M.M.; Phillips, G.B. Design, synthesis, and biological activity of novel factor Xa inhibitors, 4-aryloxy substituents of 2,6-diphenoxypyridines. *Bioorg. Med. Chem.* **2002**, *10*, 657–666.

24. Revesz, L.; Di Padova, F.E.; Buhl, T.; Feifel, R.; Gram, H.; Hiestand, P.; Manning, U.; Wolf, R.; Zimmerlin, A.G. SAR of 2,6-diamino-3,5-difluoropyridinyl substituted heterocycles as novel p38 MAP kinase inhibitors. *Bioorg. Med. Chem. Lett.* **2002**, *12*, 2109–2112.

25. Kolchina, E.F.; Gerasimova, T.N. Cyclization of polyfluorinated 2-amino- and 2-acetamidodiaryl ethers into phenoxazine derivatives. *Izv. Akad. Nauk SSSR* **1990**, *4*, 846–849.

26. Sandford, G.; Slater, R.; Yufit, D.S.; Howard, J.A.K.; Vong, A. Tetrahydro-pyrido[3,4*b*] pyrazine scaffolds from pentafluoropyridine. *J. Org. Chem.* **2005**, *70*, 7208–7216.

27. Baron, A.; Sandford, G.; Slater, R.; Yufit, D.S.; Howard, J.A.K.; Vong, A. Polyfunctional tetrahydro-pyrido[2,3*b*]pyrazine scaffolds from 4-phenylsulfonyl-tetrafluoropyridine. *J. Org. Chem.* **2005**, *70*, 9377–9381.

28. Cartwright, M.W.; Sandford, G.; Yufit, D.S.; Howard, J.A.K.; Christopher, J. Imidazopyridine scaffolds from perfluorinated pyridine derivatives. *Tetrahedron* **2007**, *63*, 7027–7035.

29. Chambers, R.D.; Hoskin, P.R.; Kenwright, A.R.; Khalil, A.; Richmond, P.; Sandford, G.; Yufit, D.S.; Howard, J.A.K. Macrocycles from pentafluoropyridine. *Org. Biomol. Chem.* **2003**, *1*, 2137–2147.

30. Sandford, G. Macrocycles from perhalogenated heterocycles. *Chem. Eur. J.* **2003**, *9*, 1464–1469.

31. Chambers, R.D.; Hutchinson, J.; Philpot, P.D. Reactions of perfluoro-4-vinylpyridine. *J. Fluorine Chem.* **1977b**, *9*, 15–23.

32. Drayton, C.J.; Flowers, W.T.; Haszeldine, R.N. Fluoride-initiated reactions of hexafluoropropene with pentafluoropyridine and tetrafluoropyridine-4-carbonitrile. *J. Chem. Soc., Perkin Trans. 1* **1975**, 1029–1034.

33. Barlow, M.G.; Haszeldine, R.N.; Dingwall, J.G. Valence-bond isomers of pentakis (pentafluoroethyl)pyridine. *J. Chem. Soc., Perkin Trans. 1* **1973**, 1542–1545.

34. Bell, S.L.; Chambers, R.D.; Gribble, M.Y.; Maslakiewicz, J.R. Syntheses involving octafluoroisobutene. *J. Chem. Soc., Perkin Trans. 1* **1973**, 1716–1720.

35. Katritzky, A.R.; Rees, C.W. *Comprehensive Heterocyclic Chemistry*, Pergamon Press: Oxford, 1984.

36. Pozharskii, A.F.; Soldantenkov, A.T.; Katritzky, A.R. *Heterocycles in Life and Society*, John Wiley & Sons, Inc.: New York, 1997.
37. Parks, E.L.; Sandford, G. Polysubstitution of tetrafluoropyrimidine. Unpublished results. **2007**.
38. Cram, D.J.; Choi, H.J.; Bryant, J.A.; Knobler, C.B. Solvophobic and entropic driving forces for forming velcraplexes which are four-fold, lock-key dimers in organic media. *J. Am. Chem. Soc.* **1992**, *114*, 7748–7765.
39. Barlow, M.G.; Haszeldine, R.N.; Simon, C.; Simpkin, D.J. Preparation and some nucleophilic substitution reactions of trifluoro-1,2,4-triazine. *J. Chem. Soc., Perkin Trans. 1* **1982**, 1251–1254.
40. Chambers, R.D.; Korn, S.R.; Sandford, G. Tertiary aromatic amines as carbon nucleophiles with activated perfluorinated aromatic compounds. *Tetrahedron* **1992**, *48*, 7939–7950.
41. Dressler, R.L.; Young, J.A. Catalytic perfluoroalkylation of halotriazines. *J. Org. Chem.* **1967**, *32*, 2004–2005.
42. Chambers, R.D.; Hall, C.W.; Hutchinson, J.; Millar, R.W. Fluorinated nitrogen heterocycles with unusual substitution patterns. *J. Chem. Soc., Perkin Trans 1* **1998**, 1705–1713.
43. Chambers, R.D.; McBride, J.A.H.; Musgrave, W.K.R. Formation and nucleophilic substitution of polyfluoropyridazinium cations. *J. Chem. Soc. C* **1968**, 2989–2994.
44. Barlow, M.G.; Brown, D.E.; Haszeldine, R.N. The photochemical addition of ethylene to pentafluoropyridine: formation of 1:1- and 2:1-adducts. *J. Chem. Soc., Perkin Trans 1* **1978**, 363–365.
45. Barlow, M.G.; Brown, D.E.; Haszeldine, R.N. Photochemical addition of but-2-yne to pentafluoropyridine to give 1:1 and 2:1 adducts. *J. Fluorine Chem.* **1982**, *20*, 745–750.
46. Sket, B.; Zupan, M.; Zupancic, N.; Pahor, B. Photosubstitution reactions on aromatic and heteroaromatic rings evidence for addition and substitution mechanism. *Tetrahedron* **1991**, *47*, 5029–5042.

9

PERFLUORINATED NONAROMATIC HETEROCYCLES

VIACHESLAV A. PETROV
DuPont Central Research and Development, Experimental Station, P.O. Box 80500, Wilmington, DE 19880-0500, USA

9.1 INTRODUCTION

This chapter is one of the first attempts of systematic review of the synthetic methods and chemistry of nonaromatic saturated and unsaturated perfluorinated heterocycles. It summarizes data on the synthesis and chemical transformation of perfluorinated six- and to some extent five- membered nonaromatic heterocyclic compounds containing oxygen, nitrogen, sulfur, selenium, and phosphorous. It should be pointed out that the review is not all-inclusive one, but it is an attempt to provide representative examples of synthetic methods and chemistry of perfluorinated nonaromatic heterocyclic materials.

Perfluorinated heterocyclic materials can be considered as being a "mirror" image of hydrocarbon heterocyclic analogues. The replacement of all hydrogens by fluorines results in very different and often unpredictable reactivity of these materials. Despite the fact that the saturated perfluorinated heterocycles expected to be chemically inert, this group of material actually has an interesting and unusual chemistry.

The review is organized according to the compounds class. Synthetic methods for the preparation of *saturated* heterocycles containing oxygen, nitrogen, sulfur (II), selenium (II), phosphorous (III), sulfur in higher oxidation state, and finally mixed sulfur–oxygen and sulfur–nitrogen heterocycles are reviewed in the first part of the chapter. The discussion on chemistry of saturated fluorinated heterocycles containing

---

Fluorinated Heterocyclic Compounds: Synthesis, Chemistry, and Applications, Edited by Viacheslav A. Petrov

oxygen, sulfur, and nitrogen is given in the second part of the chapter and is followed by review of synthesis and chemical transformations of *unsaturated* fluorinated cyclic compounds, such as dioxins, cyclic imidoyl fluorides, and azadienes. The unusual order of presentation may be counterintuitive, however, it is justified by the fact that saturated cyclic compounds are often used as a starting material for the synthesis of unsaturated nonaromatic heterocycles.

9.2 PREPRATION AND REACTIONS OF SATURATED PERFLUORINATED HETEROCYCLES CONTAINING OXYGEN, NITROGEN, SULFUR, SELENIUM, AND PHOSPHOROUS

9.2.1 Preparation of Saturated Perfluorinated Heterocycles Containing Oxygen, Nitrogen, Sulfur, Selenium, and Phosphorous

9.2.1.1 Fluorinated Oxygen Containing Heterocycles Electrochemical fluorination (ECF) is the method of choice for the preparation of perfluorinated saturated heterocycles such as cyclic ethers or amines. Discovered by J.H. Simmons and later on commercialized by 3M Company, ECF quickly became "working horse" process for the preparation of wide variety of perfluorinated materials. For example, *F*-oxane and *F*-tetrahydrofurane are prepared in 42 and 35% yield by ECF of the parent hydrocarbon materials.[1,2] The formation of perfluorinated cyclic ethers in ECF of the corresponding noncyclic carboxylic acids or acyl chlorides was first reported in patent literature[3,4] and later this method was extended for the preparation of large number of perfluorinated tetrahydrofuranes. For example, the electrolysis of $C_5H_{11}C(O)Cl$ in HF was reported to produce a mixture of *F*-2-ethyltetrahydrofurane and *F*-2-methyloxane in (32.3% total yield, ratio 1:1.6), however, *F*-2-methyloxane forms in this process only as minor product.[5] On the other hand, the fluorination of 2-alkyloxanes using ECF leads to perfluorinated oxanes in higher yield (23–30%).[5,6] Typical distribution of reaction products in fluorination of 2-alkyloxanes is shown below (Fig. 9.1).

Electrochemical fluorination of tetrahydrofuran carboxylic acids derivatives proceeds with preferential formation of *F*-tetrahydrofurane (**5**), however, in case of furans containing carbonyl group in the side chain, an interesting formation of spiro-ethers **7** is observed.[7] *F*-Oxanes **8** and **9** are prepared in low yield by ECF of the corresponding oxane derivatives[7] (Fig. 9.2).

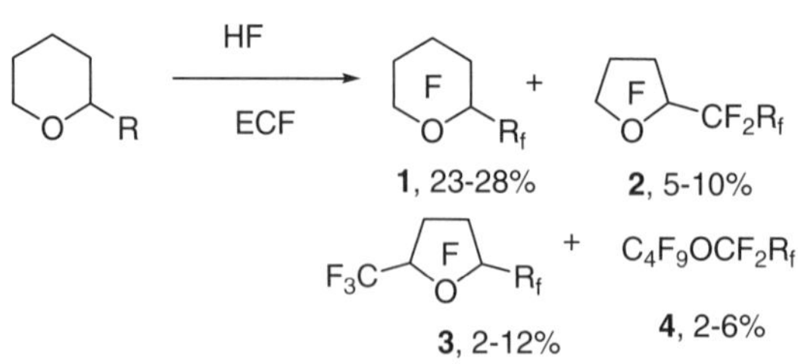

FIGURE 9.1 Products distribution in electrochemical fluorination of 2-alkyloxanes.

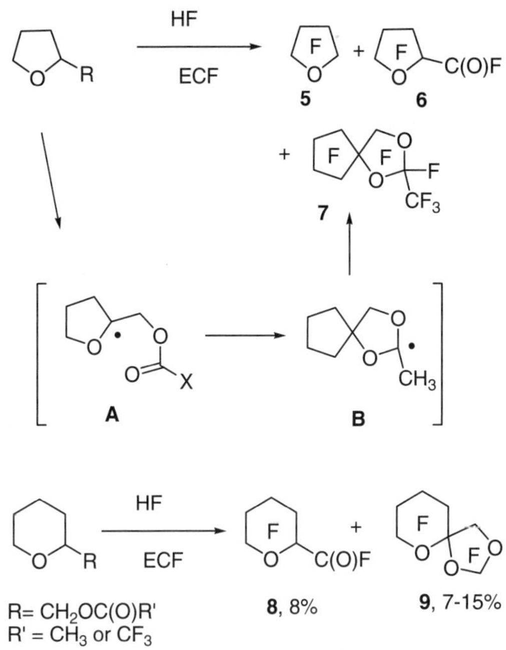

FIGURE 9.2 Electrochemical fluorination of cyclic ethers derivatives.

It is believed that the process involving the formation of intermediate **A** and its intramolecular cyclization with producing radical **B** is responsible for the formation of bicyclic ethers **7** and **9** (Fig. 9.2). Similar mechanism was suggested to explain the formation of bicyclic and spiro-ethers **10–13**, obtained from the corresponding esters containing five- and six-membered alicyclic fragment.[8]

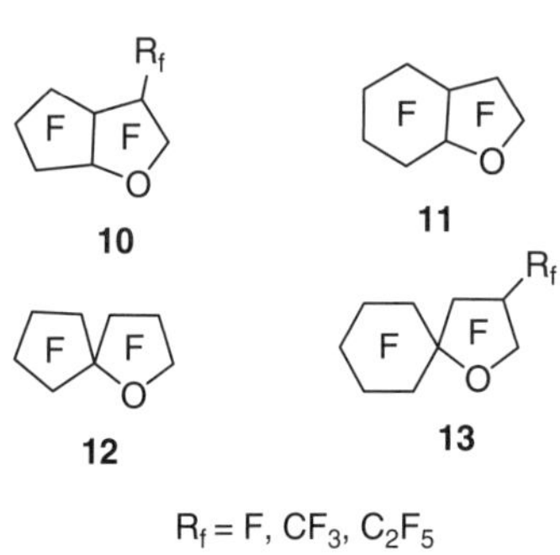

Structurally similar bicyclic ethers **15** were prepared in 10–23% yield in ECF of esters of α-cyclohexenylcarboxylic acids **14**[9,10] (Fig. 9.3).

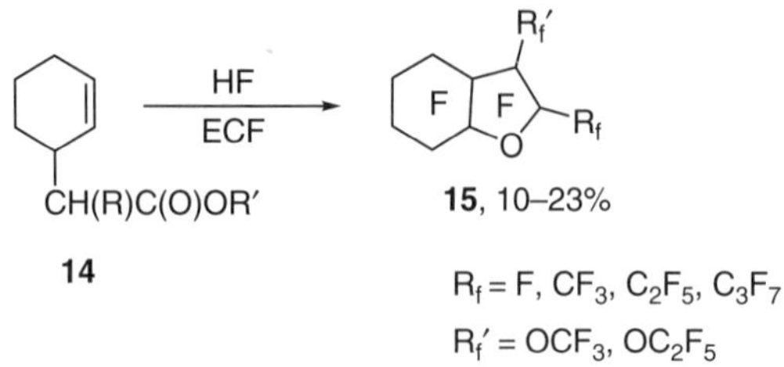

FIGURE 9.3 Electrochemical fluorination of esters of α-cyclohexenylcarboxylic acids **14**.

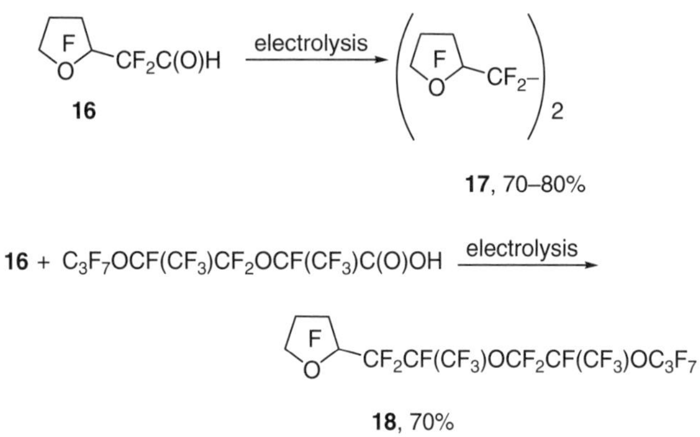

FIGURE 9.4 Homo- and cross- Kolbe-coupling of cyclic perfluorinated acids.

Interesting variation of the Kolbe reaction reported by Sokolov et al.[11,12] was used for the synthesis of *F*-1,2-di-tetrahydofurane **17** (Fig. 9.4).

It is worth of mentioning that the cross-coupling of **16** and acid derived from hexafluoropropene oxide trimer leads to clean formation of **18** in relatively high yield (Fig. 9.4). More data on the synthesis of five-membered perfluorinated ethers can be found in Chapter 4.

Sulfur tetrafluoride and F_2 gas are among other fluorinating agents used for the synthesis of fluorinated ethers. For example, *F*-oxane was reported to form in the reaction of anhydride of *F*-glutaric acid or *F*-glutaryl fluoride with excess of SF_4.[13] Perfluorinated spiro-ethers **19** were prepared in high yield by exhaustive fluorination of the parent hydrocarbons using elemental fluorine.[14]

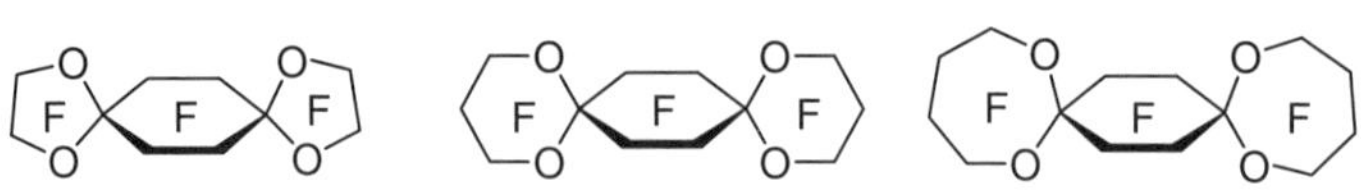

19

Perfluorinated lactones are usually synthesized by selective hydrolysis of polyfluorinated cyclic ethers. Since perfluorinated ethers are resistant to hydrolysis, usually cyclic ethers, containing CCl_2 fragment are used as feedstock. Tiers was

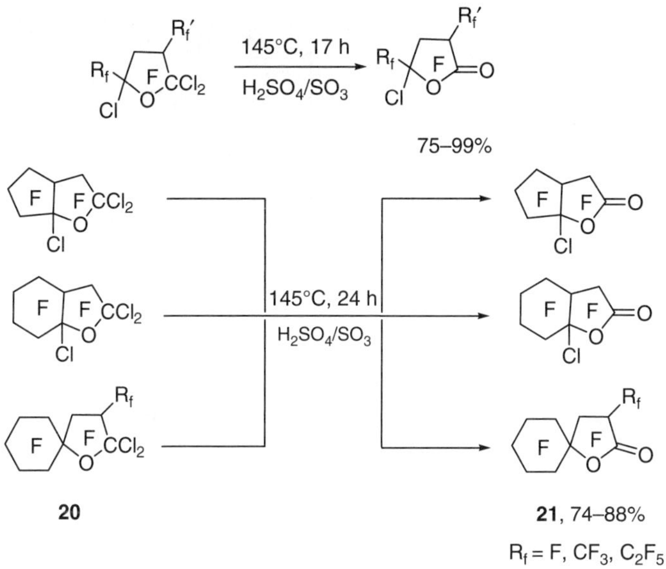

FIGURE 9.5 Synthesis of polyhalogenated lactones acidic hydrolysis of chlorofluoroethers **20**.

first to demonstrate that polyfluorinated cyclic α,α,α-trichloro ethers undergo selective hydrolysis of CCl_2 group under the action of fuming sulfuric acid.[15] Later, this reaction was adopted for the hydrolysis of polyfluorinated 2,4-dialkyltetrahydrofuranes.[16] The reaction of cyclic chlorofluoroethers **20** results in the formation of lactones **21** in 74–88% yield (Fig. 9.5).[10]

Perfluorinated lactones can also be prepared by electrophilic cyclization of polyfluorinated carbonyl compounds. The preparation of *F*-butyrolactone (**22**), reported by Hauptschtein in 1952, is based on the reaction of silver *F*-glutarate with iodine at elevated temperature.[17] This material can be also prepared in moderate yield (12–53%) using reaction of $I(CF_2)_4I$ with fuming H_2SO_4.[18] Significantly higher yields of **22** (up to 80%) are reported for the reaction of $I(CF_2)_3C(O)X$ (X = F or OCH_3) with 30% fuming H_2SO_4.[18] The reaction of $I(CF_2)_4C(O)F$ under similar conditions leads to the formation of the six-membered lactone **23** (Fig. 9.6).[19]

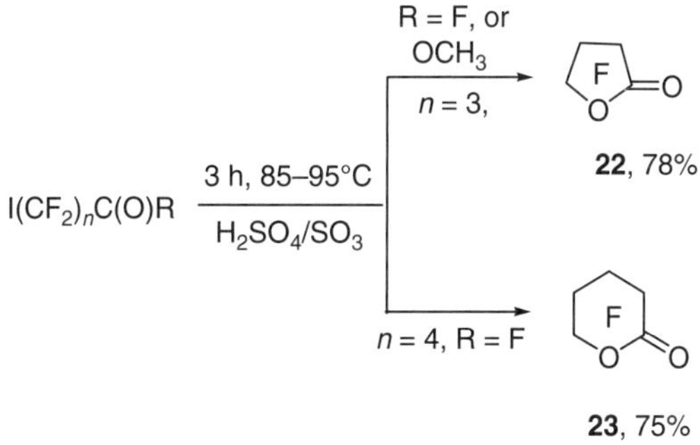

FIGURE 9.6 Synthesis of perfluorinated lactones **22** and **23**.

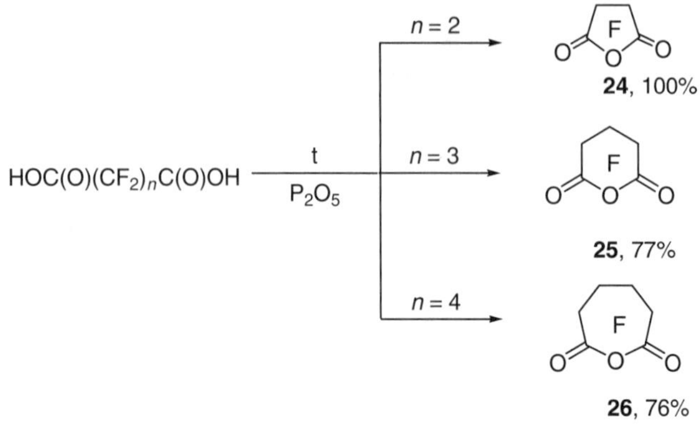

FIGURE 9.7 Preparation of perfluorinated cyclic anhydrides **24–26**.

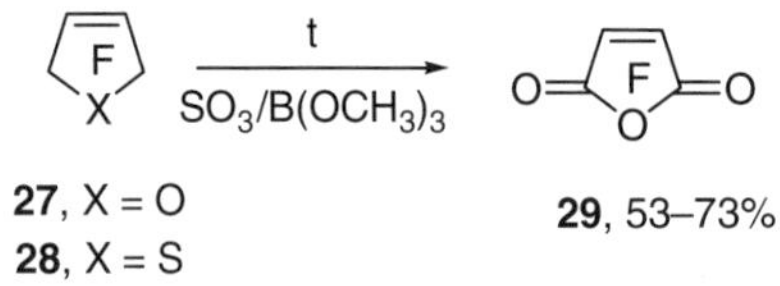

FIGURE 9.8 Synthesis of *F*-maleic anhydride.

Perfluorinated cyclic anhydrides usually are prepared by cyclization of the corresponding perfluorinated dicarboxylic acids using P_2O_5 as dehydration agent. Cyclic anhydrides of *F*-succinic (**24**),[20] *F*-glutaric (**25**),[21] and *F*-adipic acids (**26**)[22] are synthesized in high yield using this protocol (Fig. 9.7).

As it was mentioned above, perfluorinated ethers are resistant toward hydrolysis, however, activated CF_2 group can be converted into carbonyl using SO_3 and boron catalyst.[23] For example, both *F*-dihydrofurane **27** and *F*-dihydrothiophene **28** were converted into **29** using the reaction shown in Fig. 9.8.[23,24]

9.2.1.2 Preparation of Perfluorinated Nitrogen and Nitrogen–Oxygen Containing Heterocycles A wide range of perfluorinated amines was prepared in last 60 years. Electrochemical fluorination is the main method used for the preparation of variety saturated perfluorinated heterocycles. Despite the low yields of perfluorinated materials, the process is reliable and can be used for the preparation of perfluorinated materials. Detailed reports on the use of ECF for the preparation of fluorinated materials can be found in several reviews published in last 40 years.[25–29] Major types of perfluorinated *N*-fluoroamines prepared by fluorination of cyclic secondary amines are given below.

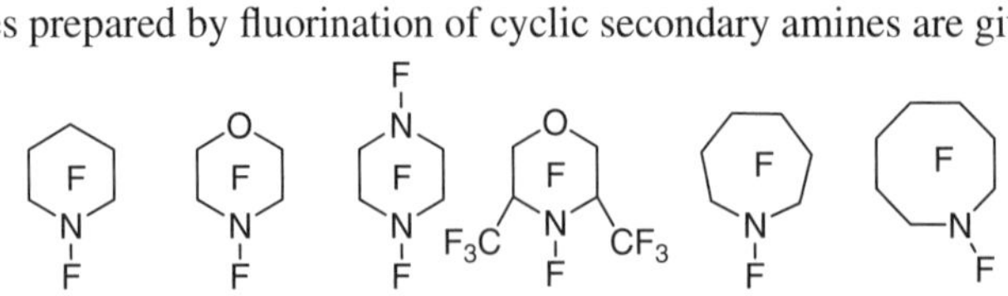

30 **31** **32** **33** **34** **35**

Perfluorinated secondary amines such as perfluoro-*N*-fluoropiperidine (**30**), perfluoro-*N*-fluoromorpholine (**31**),[30,31] or perfluoro-*N*,*N*-difluoropiperazine (**32**)[32] were prepared by ECF of the hydrocarbon feedstock in low to moderate yield. Recently, Lin and Lagow prepared compounds **32–35**, in 60–95% yield using elemental fluorine.[33]

A large number of perfluorinated heterocyclic compounds was prepared (mostly using ECF technology) in 1970–1990 during the quest for the best materials to be used as oxygen carriers in perfluorocarbon emulsions, so-called "blood substitutes" (for detailed review on this effort, see Ref. 34). The list of synthesized materials includes perfluorinated *N*-alkyl(cycloalkyl) pyrrolidines,[35] oxazolidines,[36] piperidines and morpholines, and larger saturated heterocycles.[37–43] Major types of perfluorinated heterocyclic materials are shown below.

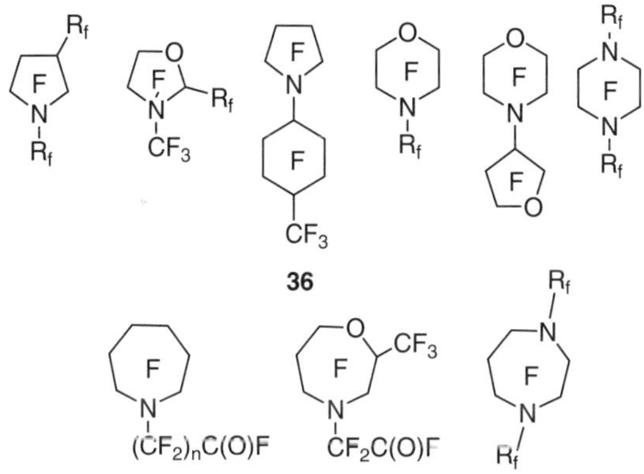

36

37 38 39

R_f=CF_3, $CF_2C(O)F$

In the Soviet Union, compound **36** was used as a component of in the emulsion, called "Perftoran®," which successfully went through clinical trails.[34]

Gas-phase fluorination of hydrocarbon heterocycles using CoF_3 as fluorinating agent was successfully applied for the preparation of some perfluorinated heterocyclic compounds[44] (see Fig. 9.9).

Among other methods, leading to the formation of saturated perfluorinated five- and six-membered oxygen and nitrogen heterocycles, having limited application, but

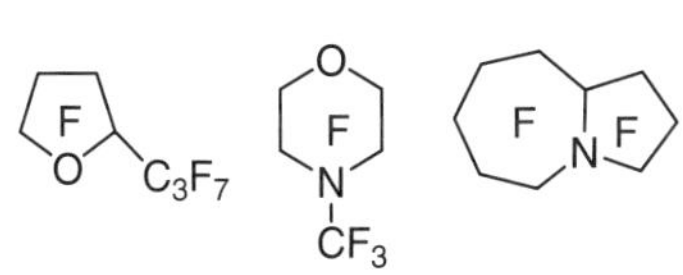

FIGURE 9.9 Examples of perfluorinated heterocycles prepared using CoF_3 as fluorinating agent.

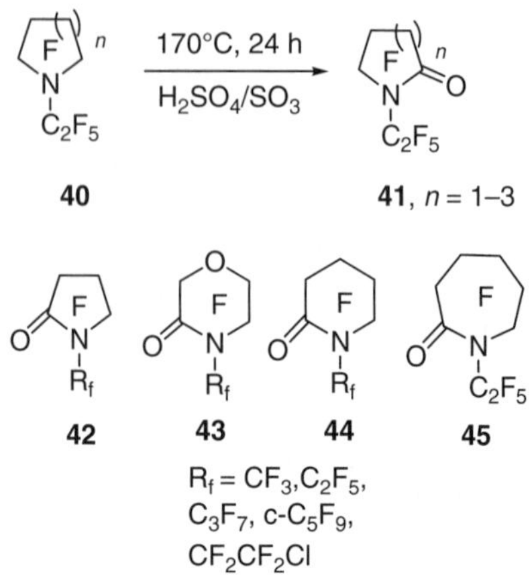

FIGURE 9.10 Perfluorinated lactams prepared by hydrolysis of perfluorinated cyclic amines.

not involving fluorination are dimerization of *F*-2-methyloxaziridine into *F*-2, 5-dimethyl-1,4,2,5-dioxadizine catalyzed by SbF_5,[45] Lewis acid-catalyzed conversion of *F*-2-methyl -1,2-oxazetidine into oxazines,[46] and trimerization of *F*-2-azapropene.[47,48]

Due to the combination of lower electronegativity and the presence of unshared electron pair on nitrogen atom, perfluorinated cyclic amines are more reactive toward electrophiles, compared to ethers. As it is demonstrated by Hayashi et al., cyclic perfluorinated amines undergo selective, high-yield conversion into the corresponding lactams **41–45** upon treatment with fuming H_2SO_4[49,50] (Fig. 9.10).

The proposed mechanism of the reaction involves insertion of SO_3 into α-C–F bond of the heterocycle, followed by decomposition of intermediate fluorosulfate with the formation of lactam. Experimentally observed[50] higher reactivity of derivatives having $-CF_2CF_2Cl$ substituent compared to materials carrying C_2F_5 group, correlates well with lower electronegativity of $-CF_2CF_2Cl$ substituent.

Similar to cyclic anhydrides, imides of perfluorinated acids usually are prepared using derivatives of perfluorinated bis-carboxylic acids as starting material. For example, imides of *F*-succinic acid (**46**),[51,52] *F*-glutaric (**47**),[52,53] and *F*-adipic (**48**)[54] acids were synthesized through the reactions of either amido derivatives or cyclic anhydrides.

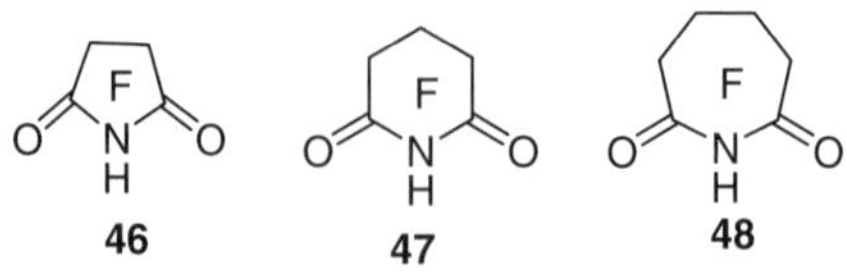

N-Fluoro imides **49** and **50** can be conveniently prepared by fluorination of **46–47** using XeF_2[55] (Fig. 9.11).

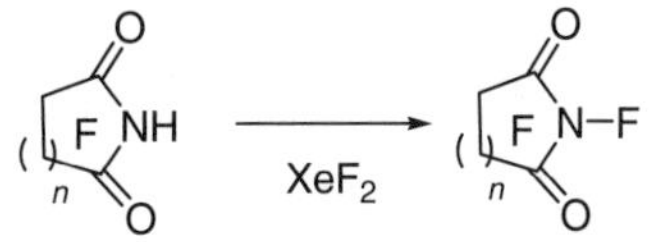

46, 47, 49, 55–65%, 50, 50–60%

FIGURE 9.11 Preparation of N–F imides **49** and **50**.

9.2.1.3 Preparation of Saturated Sulfur-, Selenium-, and Phosphorous-Containing Perfluorinated Heterocycles Number of perfluorinated five- and six-membered heterocycles containing sulfur, selenium, and phosphorous can be prepared through an unusual reaction of fluorinated olefins and sulfur, the process that was discovered by C. Krespan in early 1960s. For example, the reaction of tetrafluoroethylene (TFE) gas with boiling sulfur at atmospheric pressure results in the formation of *F*-1,2,3-trithiolane (**51**) and *F*-1,2,3,4-tetrathiane (**52**).[56] It is noteworthy, that only trace of *F*-1,4-dithiane (**53**) is formed under this conditions, however, prolonged heating of **51** or **52** causes elimination of elemental sulfur and formation of thermodynamic product **53**[56] (Fig. 9.12).

When the reaction of TFE and sulfur is carried out in a closed system by continuous addition of TFE to molted sulfur, compound **53** (contaminated by trace of dithiolane **54**) can be prepared in 61% yield. The formation of **54** can be suppressed by using CS_2 as a solvent[57] (Fig. 9.13).

Perfluorinated thiolanes **55** and **56** were synthesized in low yield by the reaction of sulfur with TFE at 150°C (in CCl_4) or hexafluoropropene (HFP, neat) at 300°C[57] (Fig. 9.14).

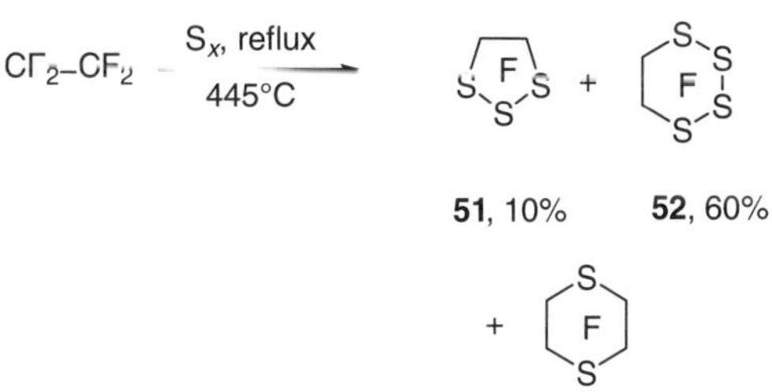

51, 10% 52, 60% 53, trace

FIGURE 9.12 Tetrafluoroethylene based synthesis of cyclic perfluorinated polysulfides.

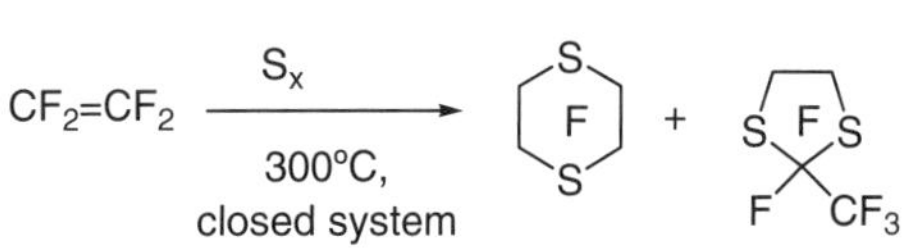

53, 61% 54, 1-3%

FIGURE 9.13 Synthesis of *F*-1,4-dithiane (**53**).

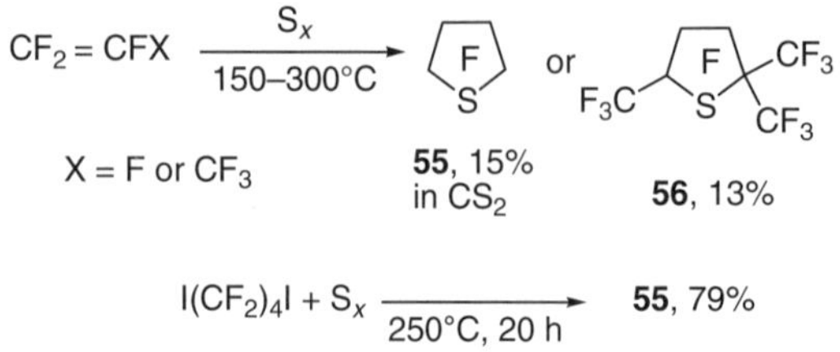

FIGURE 9.14 Synthesis of *F*-thiolanes **55** or **56**.

According to Tiers,[58] compound **55** can be synthesized in much better yield (up to 79%) by the reaction of $I(CF_2)_4I$ and sulfur at an elevated temperature (250°C, 20 h).

The proposed mechanism of thermal reaction of TFE and sulfur involves the formation biradical intermediate $\bullet SCF_2CF_2\bullet$ and consecutive reaction with TFE leading to the formation **55**. Cyclodimerization of $\bullet SCF_2CF_2\bullet$ leads to compound **53**. The intermediacy of $\bullet SCF_2CF_2\bullet$ in all these reactions was supported by interception of this biradical by olefins $CH_2{=}CHR$ (R = H, CH_3, CN, C_6H_5, OR, C(O) OCH_3, CF_3), resulting in the formation of the corresponding partially fluorinated thiolanes in 15–56% yield.[59] Sulfur in this three-component process can be replaced by selenium. In this case, the corresponding partially fluorinated selenolanes were isolated.[59]

Known examples of saturated perfluorinated selenium-containing heterocycles are limited to *F*-selenolane (**57**) and *F*-1,4-diselenane (**58**) prepared in <10% yield by the reaction of TFE with Se at 250°C.[57]

The corresponding *F*-1-iodophosphalane (**59**) and *F*-1,4-diiodophosphane (**60**) (Fig. 9.15) were prepared in similar reaction involving red phosphorus, iodine, and TFE.[57] In contrast to the synthesis of perfluorinated S(II), Se(II), and P(III) heterocycles, all known methods for the synthesis of perfluorinated S(IV) and S(VI) derivatives involve the reaction of S(II) derivatives with an fluorooxidant. For example, Abe and Shreeve reported the preparation of thiolanes derivatives **61**–**64** through the fluorination of the corresponding S(II) or S(IV) derivatives by ClF[60] (Fig. 9.16).

The sufoxide **65** and sulfone **66** are prepared by the hydrolysis of fluorides **61** and **62**.[60] Compound **62** was also prepared in low yield by ECF of sulfolene.[61]

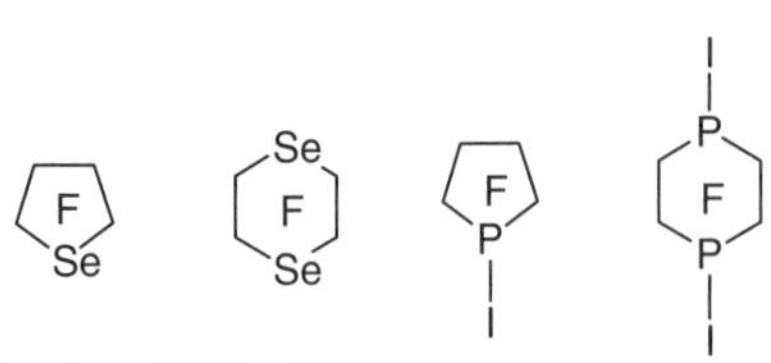

57, 10% **58**, trace **59**, 4% **60**, 27%

FIGURE 9.15 Known examples of saturated perfluorinated selenium and phosphorus heterocycles.

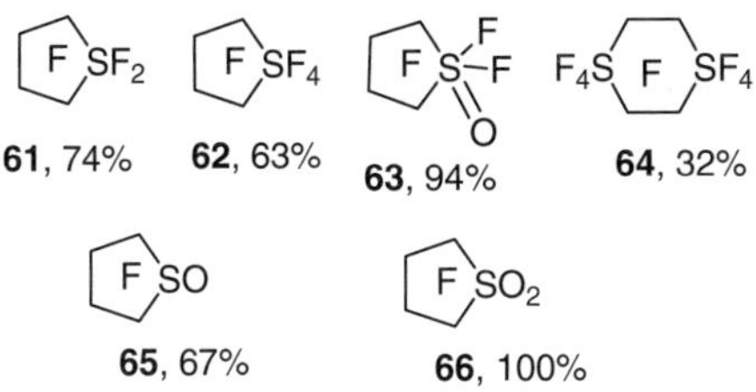

FIGURE 9.16 Polyfluorinated heterocycles containing sulfur in higher oxidation states.

ECF can be used for the preparation of sulfur–nitrogen-containing heterocycles. For example, Simmons, and later Mitsch, reported the preparation of compound **67** by ECF. [62,63]

The yields of sulfur-containing heterocycles are significantly higher in the process, which is based on the exhaustive fluorination using diluted fluorine gas at low temperature with gradual increase of temperature and fluorine concentration. Compounds **62**, **67**, and **68** are prepared in 51–80% using this process (Fig. 9.17).[64] In contrast to ECF, the yield of the fluorinated products are significantly higher. However, this process is lengthy and the synthesis may take up to several days.[64]

It should be pointed out that the fluorination process is tolerant to some functional groups. For example, the direct synthesis of **66** and **69** was carried out starting from the corresponding hydrocarbon sulfones.

Perfluorinated disulfonic acids are readily available through the hydrolysis of the corresponding disulfonyl fluorides, which are prepared by ECF. Several cyclic derivatives of disulfonic acid are known. The preparation of cyclic anhydrides of the perfluorinated disulfonic acid is disclosed in patent literature.[65] The process includes the hydrolysis of the corresponding sulfonyl fluorides by KOH, acidification of the salt and the cyclization of diacid under the action of P_2O_5 (Fig. 9.18).

Anhydrides **70–73** are clear liquids, fuming in moist air and rapidly undergoing the hydrolysis.[65]

Cyclic imides of perfluorinated disulfonic acids **74–76**, reported by Koshar in 1982[66,67] and later studied by DesMarteau group,[68] are prepared by action of anhydrous ammonia on disulfonyl fluorides, followed by acidification of ammonium salt of imide by concentrated H_2SO_4 (Fig. 9.19).

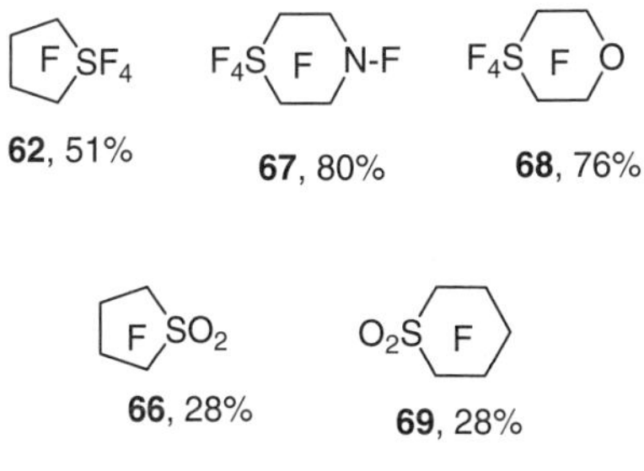

FIGURE 9.17 Synthesis of perfluorinated S(VI) heterocycles using F_2.

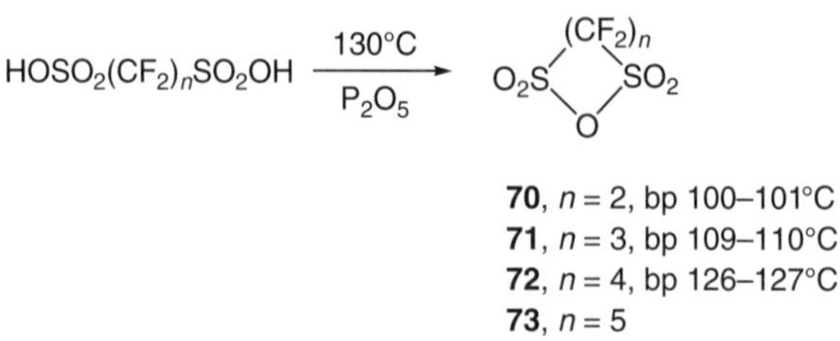

FIGURE 9.18 Preparation of cyclic anhydrides of perfluorinated disulfonic acids.

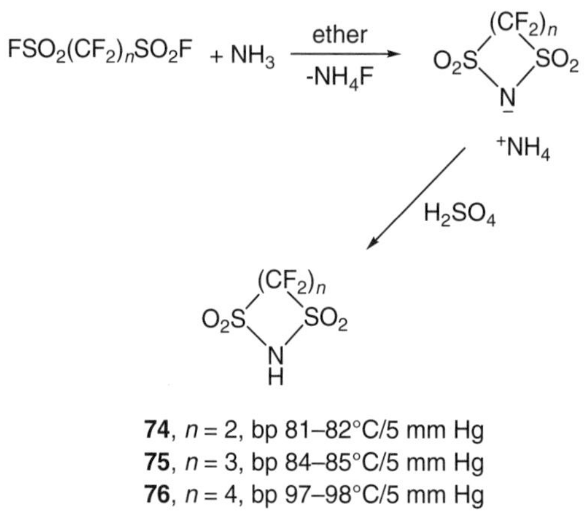

FIGURE 9.19 The Koshar synthesis of imides **74–76**.

Similar to other bis(perfluoralkylsulfonyl)imides, known to be very strong protic acids (currently $(C_4F_9SO_2)_2NH$ holds the record as the strongest acid in gas phase[69,70]), cyclic imides form stable salts with variety of counter cations, such as ammonium, potassium, sodium, lithium, etc. Lithium salts of **74–76** were patented as conductive salts for nonaqueous electrolytes of lithium batteries.[71] The DesMarteau group reported synthesis and isolation of stable benzenediazonium salt containing anion of imide **75** and unusual reactions of this material.[72]

In 1997, DesMarteau's group reported the synthesis of the corresponding N–F imides, including cyclic imides **77–79**[73] (Fig. 9.20).

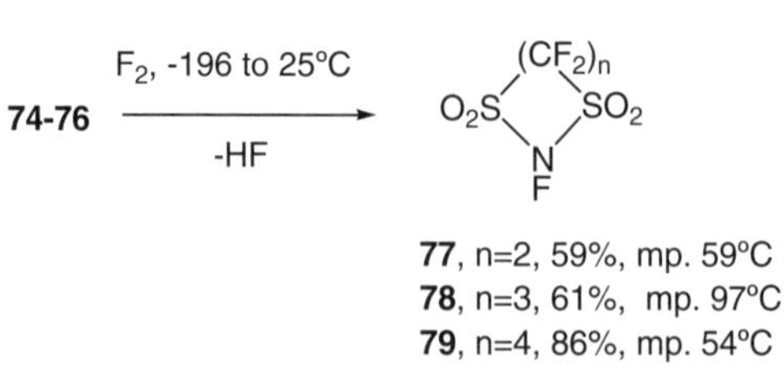

FIGURE 9.20 Preparation of cyclic perfluorinated *N*-fluoroimides **77–79**.

9.2.2 Chemical Transformations of Saturated Perfluorinated Heterocycles

9.2.2.1 Reactions of Saturated Perfluorinated Heterocycles Due to the strong electron withdrawing effect of perfluorinated framework, saturated ethers or amines are extremely week bases and they do not form stable compounds as a result of protonation or alkylation of heteroatom. Unshared electron pair(s) of heteroatom, however, manifests itself in different manner—through the activation of α-fluorine substituents.

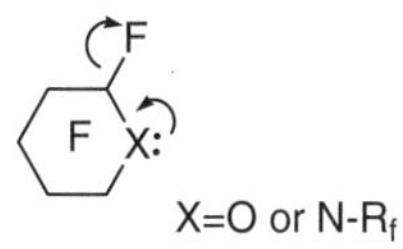

Redistribution of the electron density in perfluorinated heterocycles leads to the significant increase in the "basicity" of the fluorine substituent, resulting in an unusual reactivity toward strong electrophiles. This effect makes possible reactions, which are virtually unknown for cyclic hydrocarbons, such as reactions of perfluorinated ethers or amines with strong Lewis ($AlCl_3$, SbF_5, SO_3) or protic acids. Short summaries on these reactions can be found in two reviews.[24,74]

The ability of perfluorinated ethers to react with anhydrous $AlCl_3$ at elevated temperatures was discovered by Tiers.[15,75] While the reaction of alicyclic perfluorinated ethers results in the cleavage with the formation of acyl chloride and chlorofluoroalkanes,[15] perfluorinated α-alkyl tetrahydrofuranes undergo replacement of all three α-fluorines with the formation of cyclic α,α,α-trichloro derivatives.[75] Later, this reaction was extensively studied by Abe (Fig. 9.21).[5,10,16,36]

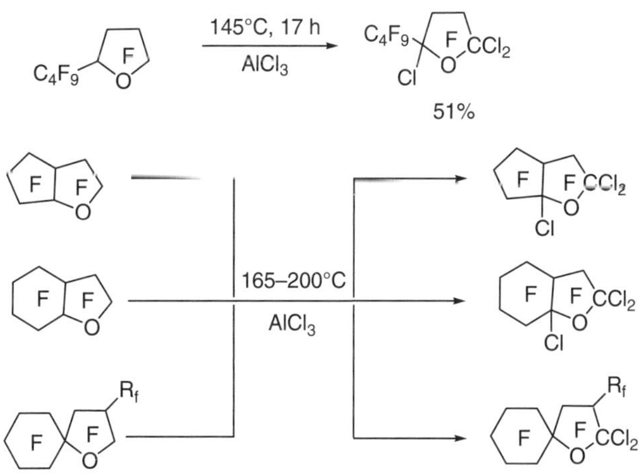

20, 39–71%

$R_f = F, CF_3, C_2F_5$

FIGURE 9.21 Reaction of perfluorinated cyclic ethers with $AlCl_3$.

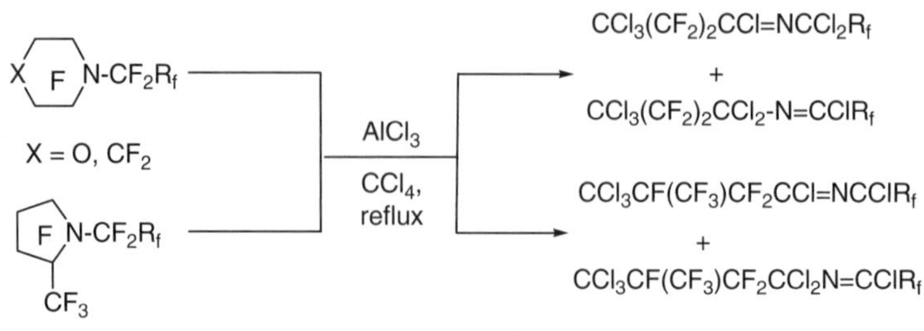

80, 60–80%

FIGURE 9.22 Ring opening of cyclic perfluorinated amines under the action of $AlCl_3$.

As it was shown by Sokolov et al., perfluorinated amines are more reactive toward $AlCl_3$. The reaction of perfluorinated cyclic amines results in the ring opening and is accompanied by fluorine–chlorine exchange, leading to noncyclic isomeric imidoyl chlorides **80** (Fig. 9.22).[74,76–78]

In perfluorinated piperidines or morpholines, carrying *F*-tetrahydrofurane substituent at nitrogen, heterocycle containing nitrogen has higher reactivity toward $AlCl_3$ and undergoes cleavage, while oxygen-containing fragment stays intact (see Ref. 24 and references therein).

The reaction of perfluorinated cyclic amines with strong Lewis acids, such as SbF_5[79,80] or aluminum chlorofluoride (ACF),[24,81] at elevated temperature results in rapid cleavage of amine, leading to the formation of cyclic perfluorinated imidoyl fluorides (Fig. 9.23). In contrast to $AlCl_3$ reactions, this process is catalytic. The outcome of the reaction depends on the structure of amine. For example, the reaction of *N*-alkyl piperidines[80] or morpholines[81] leads to the elimination of perfluoroalkane

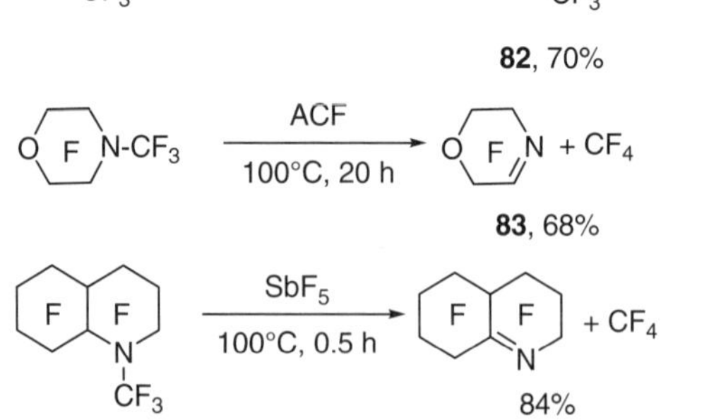

FIGURE 9.23 Catalytic cleavage of cyclic amines under the action of strong Lewis acids.

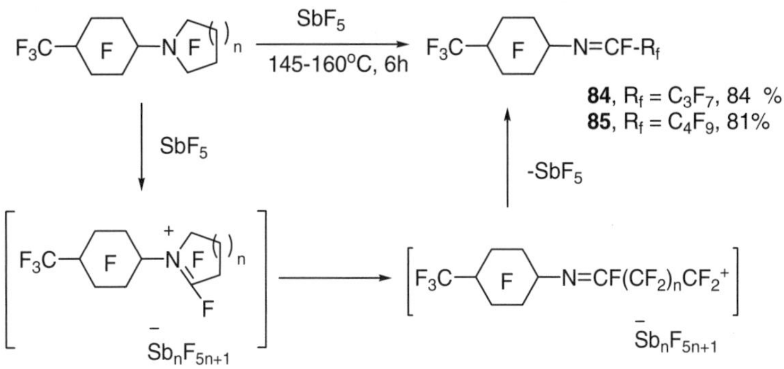

FIGURE 9.24 Catalytic isomerization of cyclic amines by SbF_5.

with the formation of cyclic imidoyl fluorides **81–84** (Fig. 9.23). It should be pointed out that the amines bearing CF_3-group at nitrogen are significantly more reactive toward SbF_5 and undergo cleavage at lower temperature, compared to the analogues with longer perfluorinated alkyl groups attached to nitrogen.[80]

Cyclic perfluorinated amines bearing cycloalkyl group undergo isomerization with the ring opening of heterocycle unit and the formation of the corresponding imidoyl fluorides **84** and **85**, bearing alicyclic substituent of nitrogen (Fig. 9.24).[79]

The proposed mechanism of the reaction involves abstraction by Lewis acid of F^- from activated by nitrogen α-CF_2 group of heterocycle, leading to the formation of cyclic immonium cation, which is followed by its ring opening and the formation of imidoyl fluoride **84** or **85** (Fig. 9.24).[24,79]

Other examples of electrophilic reactions of perfluorinated heterocyclic compounds include mentioned earlier hydrolysis of cyclic chlorinated ethers **20** (Fig. 9.5), unsaturated ether **27** (Fig. 9.8), and cyclic amines leading to fluorinated lactams (Fig. 9.10).

Due to the presence of N–F bond, the cyclic perfluorinated *N*-fluoro amines have an interesting reactivity. N–F bond in *F*-piperidine (**30**), *F*-morpholine (**31**), and *F*-piperazine (**32**) is relatively weak and easily dissociate under UV irradiation leading the formation of the corresponding nitrogen centered radicals, which were intercepted by reaction with *F*-cyclobutene (**86**), oxygen, bromine, or tetrafluoroethylene, resulting in the corresponding derivatives **87–90**[32,82,83] (Fig. 9.25).

The ability of *N*-fluoro amines to undergo reductive defluorination was discovered by Mitsch in 1965.[84] The treatment of amines **30**, **31**, and **67** by ferrocene results in the formation of the corresponding cyclic imidoyl fluorides **81**, **83**, and **91**[63,84] (Fig. 9.26).

This process has a high selectivity for N–F bond and SF_4 group in compound **67** does not undergo reduction under reaction conditions.[63] The ability of *N*-fluoro amines selectively convert triphenyl- phospine, arsine, and stibine into the corresponding difluorides (reported in 1967 by Haszeldine group[85]) was utilized in an effective laboratory scale synthesis of imidoyl fluorides **81** and **83** involving the reaction of amines **30** and **31** with triphenylphosphine.[86]

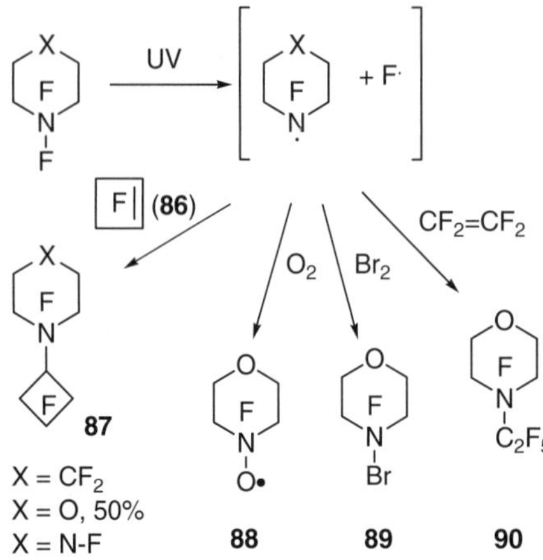

FIGURE 9.25 Photochemical reactions of cyclic *N*-fluoro amines.

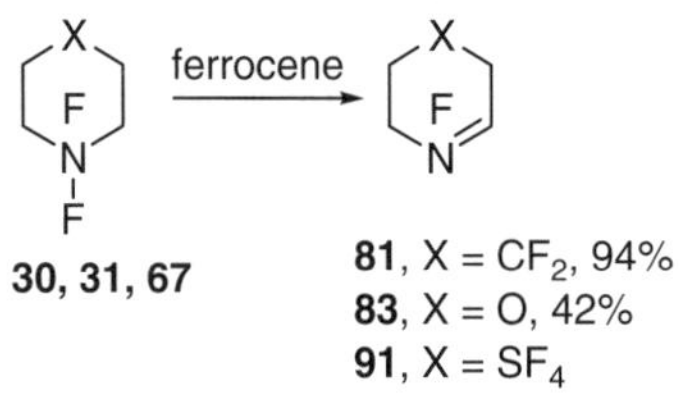

FIGURE 9.26 Reductive defluorination of cyclic perfluorinated *N*-fluoroamines.

The mechanism of the reductive defluorination of perfluorinated N−F amines may involve a relatively low laying LUMO of N−F bond. The injection of an electron on this orbital leads to the formation of radical anion **D**, which can further lose fluoride anion, forming radical **E**. The reduction of **E** to azaanion **F** followed by second fluoride anion elimination may lead to the corresponding cyclic imidoyl fluoride as a result of reductive defluorination (Scheme 9.1).

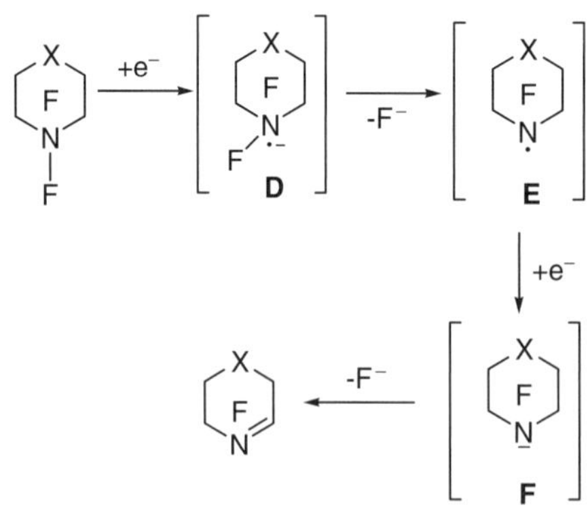

SCHEME 9.1 Mechanism of reductive defluorination of cyclic *N*-fluoro amines.

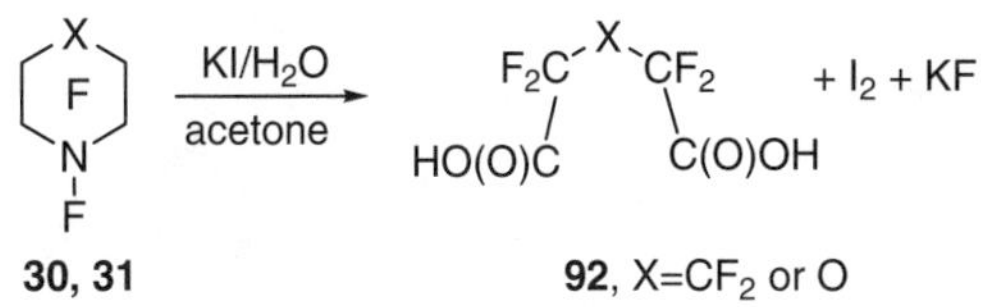

FIGURE 9.27 Reductive hydrolysis of *N*-fluoro amines **30** and **31**.

It should be pointed out that this mechanism is quite general and is likely to be predominant pathway in fluorination reactions involving N–F reagents.

Compounds **30** and **31** are inert to the actions of either base (KOH/H_2O, 80°C) or acids (HCl/H_2O, 80°C), but both react rapidly with KI in aqueous acetone with quantitative formation of I_2 and the corresponding diacids **92**[87] (Fig. 9.27).

Anhydrous HI selectively reduces N–F bond of cyclic amines, converting **30** into 1-*H*-octafluoropiperidine (**93**) and compound **31** into a mixture of 1-*H*-hexafluoromorpholine (**94**) and imine **83**, respectively[88] (Fig. 9.28).

The formation of **83** in this reaction is not surprising, since secondary amines $(R_f)_2NH$ are known to be prone to HF elimination, which can be caused by such week base as a glass. Obviously, the formation of diacids **92** in reaction with KI/H_2O/acetone (or diethyl ester of perfluoroglutaric acid in the reaction of **31** with ethanol[31]) is the result of ring opening hydrolysis of the corresponding imines **81** or **83**, formed as intermediates in this process (for detailed mechanism see Ref. 87). It should be pointed out that the conversion of amine **31** into imidoyl fluoride **81** at elevated temperature in the presence of Fe or Pt metals was also reported.[31]

Compound **30** can be used as fluorinating agent, formally transferring "F^+" to substrates such as phenols,[89] anilines,[90] and various carbanions[86,91,92] including $(CF_3)_2CF^-$[93] (Fig. 9.29).

One of the drawbacks of amine **30** as fluorinating agent is a side reactions of highly reactive imidoyl fluoride **81**, formed as by-product in fluorination process. This problem can be eliminated by the use of N–F amine, which in the course of the reaction will be converted into stable, unreactive azaanion. Indeed, higher yields of fluoride **96** were reported in the reaction of fluorinating reagents **97** and **98** with salt **95**.[86]

Perfluorinated sulfonyl N–fluoro imides were demonstrated to be by far better fluorinating agents compared to perfluorinated *N*-fluoro amines, due to a combination

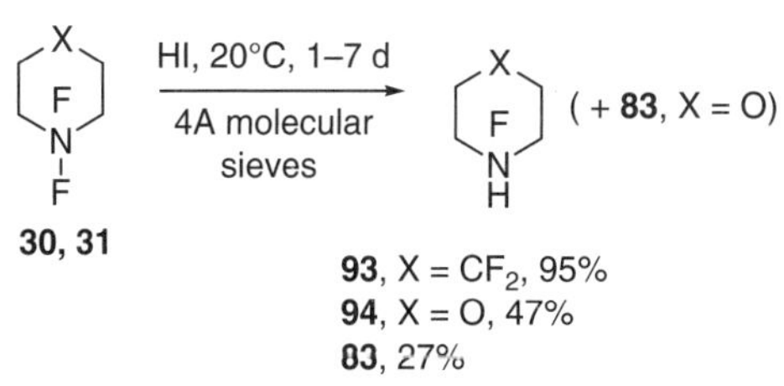

FIGURE 9.28 Reduction of cyclic *N*-fluoroamines by anhydrous HI.

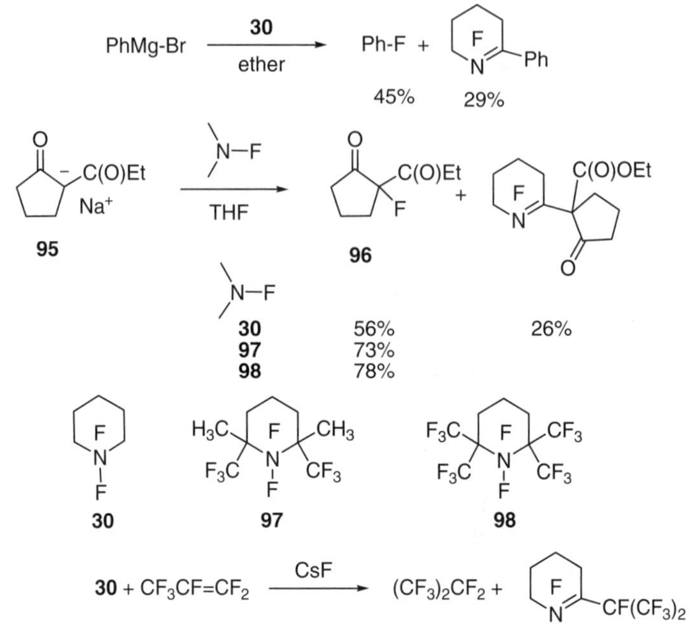

FIGURE 9.29 Cyclic *N*-fluoroamines as fluorinating agents.

of significantly lower reduction potential[92] and extremely high stability (or rather "unreactivity") of the corresponding azaanion or imide forming in this process. Cyclic sulfonyl imines **77–79** were reported to be excellent fluorinating agents for electron-rich aromatic compounds.[73] It is likely that the mechanism of fluorination of organic substrates by perfluorinated sulfonyl imides (including cyclic imides) involves single electron transfer (SET) process leading to the formation of radical anion, similar to **E** (see Scheme 9.1). Although, at this point S_N2 type ("fluorophilic")[94] mechanism cannot be ruled out for these processes at this point. More information on different types of *N*-fluoro compounds, their use in organic synthesis and the mechanism of fluorination can be found in excellent review article.[94]

9.2.2.2 Chemical Transformations of Saturated Heterocyclic Compounds Containing C=O or SO_2 Groups The reactivity of this class of compounds is governed by the functional groups such as C=O or SO_2. For example, the lactone **22**,[17,21,95] cyclic anhydrides of carboxylic acids **24** or **25**, [21,51] or sulfonic acids **70–73**[65,96] readily undergo reactions with a variety of nucleophiles, such as ammonia, amines, thioles, metal chlorides, and fluorides producing the corresponding derivatives of diacids. Some representative examples are shown in Fig. 9.30.

It should be pointed out that often these reactions are highly selective and result in nonsymmetrical derivatives of diacids, providing access to interesting and valuable materials, such as **99–103**.

Difluoromaleic anhydride (**29**) also undergoes ring opening in reactions with nitrogen and oxygen nucleophiles, giving monoamides and monoesters.[97] On the

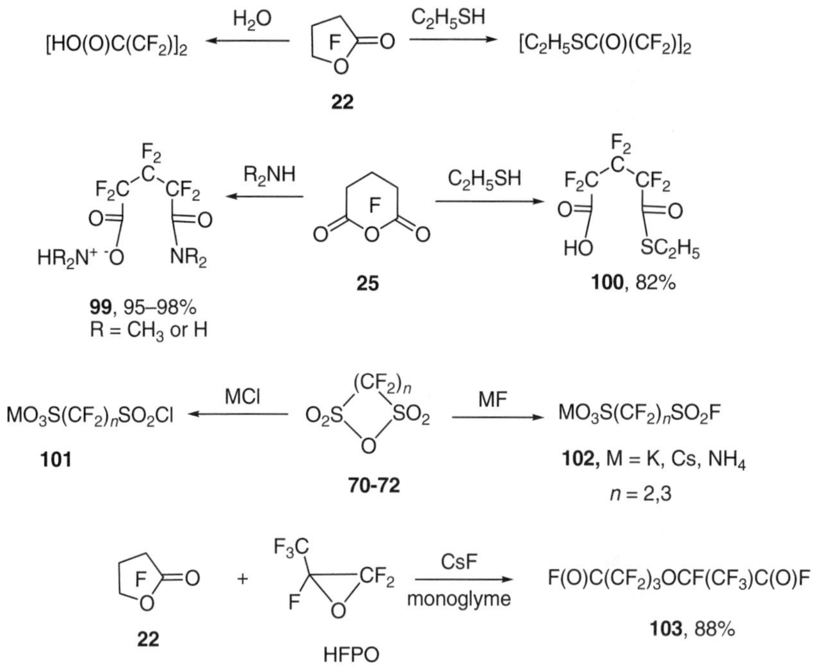

FIGURE 9.30 Nucleophilic ring opening reactions of cyclic esters.

other hand, the presence of reactive and electron deficient double bond makes this compound an excellent dienophile for Diels–Alder reactions. It undergoes the cycloaddition with anthracene and furan (Fig. 9.31).[97] The last reaction is stereoselective, leading to *endo* isomer of the corresponding cycloadduct, which was hydrolyzed into the corresponding *exo*, *exo*-diacids **104**.

In 1976, it was reported that the treatment of cyclic perfluoro-*N*-alkylamines with SO_3, followed by hydrolysis leads to the formation of 1:1 mixture of bis- and

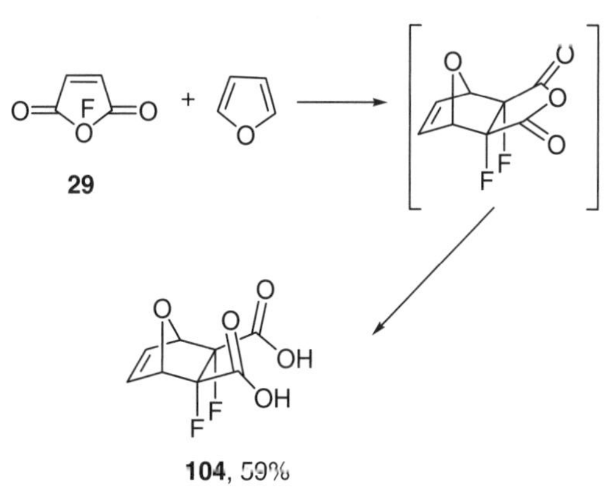

FIGURE 9.31 Diels–Alder reaction of **29** and furan.

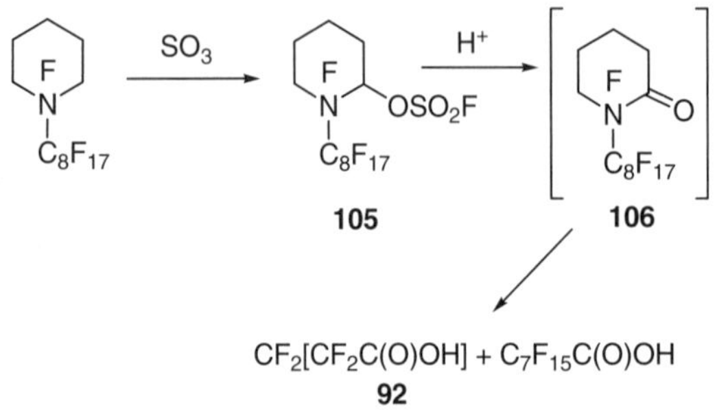

FIGURE 9.32 Hydrolysis of cyclic perfluoro-*N*-alkylamines.

monocarboxylic acids, derived from the hydrolysis of the heterocycle and side chain of cyclic amines, respectively.[98] For example, the reaction of *F*-*N*-octylpiperidine results in the mixture of **92** (X = CF_2) and perfluoroctanoic acid (Fig. 9.32).

The reaction proceeds through the insertion of SO_3 into α-C−F bond of amine leading to stable derivative **105**. The hydrolysis step is likely to involve the formation of the lactam **106**, which undergoes ring opening and hydrolysis giving mixture of acid and diacid. Indeed, the treatment of isolated flurosulfato derivatives similar to **105** by alcohols was reported to give the corresponding esters of perfluorcarboxylic acid.[98] The formation of imino ester **107** observed in controlled methanolisis of lactam **41** (Fig. 9.33) by Hayashi et al.[49] is an additional evidence in favor of this mechanism.

Imides of perfluorinated carboxylic acids also undergo the ring opening in the reactions with nucleophiles, giving derivatives of dicarboxylic acids. However, there are a number of processes proceeding with the preservation of heterocycle, such as reaction of silver salt of imide *F*-succinic acid **46** with diazomethane to produce *N*-Me derivative.[52]

The ability to form stable salts of under the action of base is an important property of imides of perfluorocarboxylic acids. Henne and Zimmer used generated "*in situ*" silver salts of **46** and **47** for the preparation of the corresponding *N*-bromo derivatives,[52] a potent source of positive bromine.[52,99] Young et al. used the reaction of *N*-bromoglutarimide with benzoyl bromide for the synthesis of *N*-benzoyl-*F*-glutarylimide.[100] The mentioned above (Fig. 9.11) reaction of imides of *F*-succinic and *F*-glutaric acids with XeF_2 is a convenient and relatively simple route for the preparation of *N*-fluoro compounds **50** and **51**.

The reactions involving perfluorinated cyclic sulfides are scarce and known examples limited to the preparation of **61** and **62** by fluorination of **55** and chemical

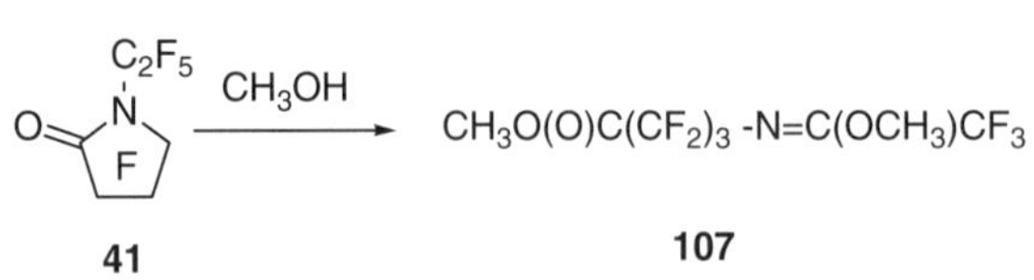

FIGURE 9.33 Controlled methanolisis of lactam **41**.

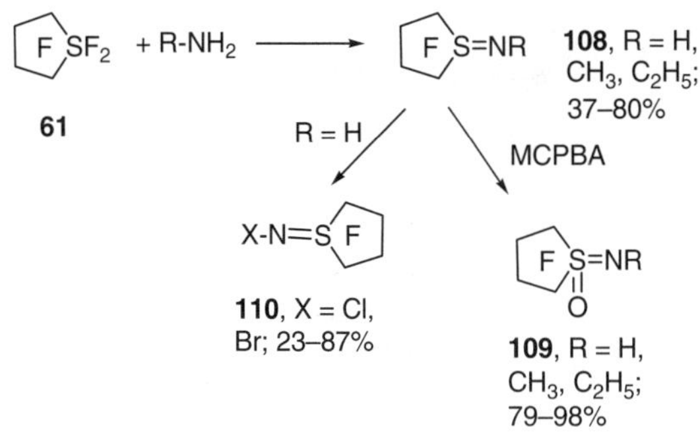

FIGURE 9.34 Reactions of difluorosulfurane **61**.

transformations of difluorsulfurane **61**,[101,102] including reactions with aliphatic amines, oxidation of the products **108** into **109** and synthesis of *N*-haloimines **110** (Fig. 9.34).

9.3 SYNTHESIS AND REACTIONS OF UNSATURATED PERFLUORINATED OXYGEN- AND NITROGEN-CONTAINING HETEROCYCLES

9.3.1 Synthesis of Unsaturated Perfluorinated Oxygen- and Nitrogen-Containing Heterocycles

This group of perfluorinated compounds is relatively small and limited mostly to nitrogen-containing heterocycles. There are very few known examples of oxygen-containing unsaturated heterocycles. Tatlow et al. reported the synthesis of fluorinated dioxenes **111–113** by dehydrofluorination of the corresponding monohydro-*F*-dioxanes (isolated from reaction mixture derived from fluorination of 1,4-dioxane over CoF_3).[103,104]

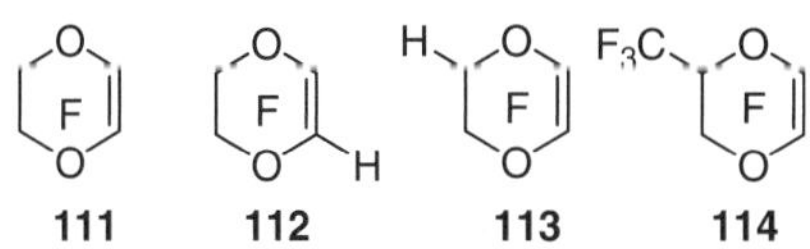

In 1991, Krespan reported the preparation of **111** using the sequence, including photochemical chlorination of 1,4-dioxane, chlorine–fluorine exchange, and dechlorination of 1,2-dichlorohexafluoro-1,4-dioxane.[105] Compound **114** was prepared using similar procedure.

Very interesting thermal rearrangement of *F*-3,4-methyl-2,3,4,5-dioxahexane was reported by the Chambers group. On heating this material undergoes ring expansion giving *F*-2,3,5,6-metyl-2,3-dihydrodioxine isolated in 92% yield[106] (Fig. 9.35).

Several methods are available for the preparation of perfluorinated cyclic imidoyl fluorides. For example, reductive defluorination of cyclic *N*-fluoro compounds

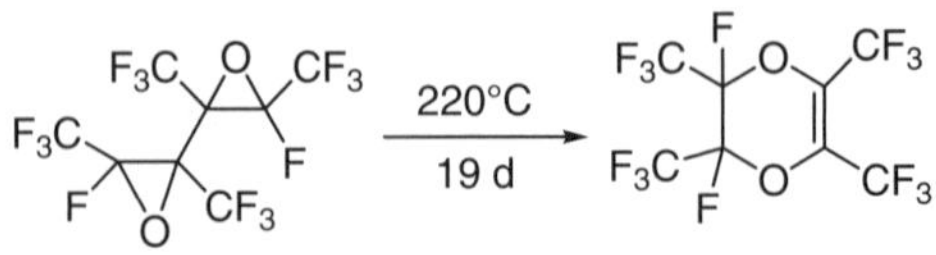

FIGURE 9.35 Isomerization of *F*-2-methyl-3-(3-methyloxiran-2-yl)oxirane.

(Fig. 9.26) by ferrocene or PPh_3 was used for the preparation of imidoyl fluorides **81**, **83**, and **91**. The cleavage of perfluorinated tertiary amines under the action of SbF_5 (see Fig. 9.23) is another convenient method for the synthesis of cyclic imines, such as **81–83**.

It should be pointed out that perfluorinated aromatic heterocyclic compounds is another valuable feedstock for the synthesis of unsaturated fluorinated heterocycles. For example, fluorination of *F*-pyridine over CoF_3 at 120°C results in saturation of both C=C, leading to **81** as principal product, along with smaller amount of *F*-1-azacyclohexadiene-1,3.[44,107] Fluorination of substituted pyridine **115** over CoF_3 gives diene **116** in a high yield.[108]

The gas-phase fluorination of substituted pyrimidine **117** ($R_f = CF(CF_3)_2$) over CoF_3 leads to the selective and high-yield formation of azadiene **118**, however, unsubstituted *F*-pyrimidine under similar conditions forms the dimer **119**. On the other hand, the reaction of isomeric *F*-pyridazine with CoF_3 is straightforward leading to azadiene **120** (Fig. 9.36). Detailed discussion on the radical cation mechanism

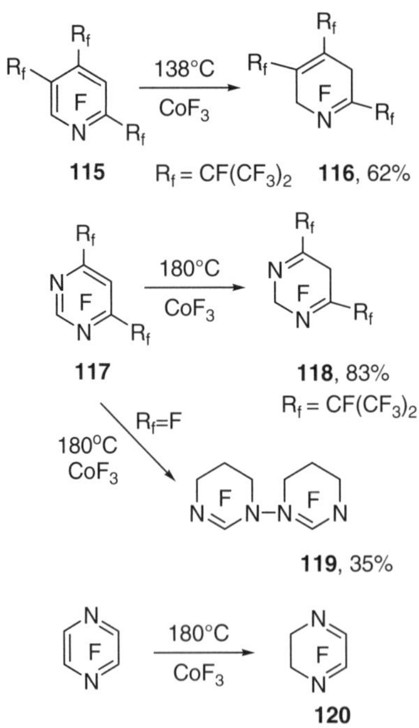

FIGURE 9.36 Fluorination of perfluoroaromatic compounds over CoF_3.

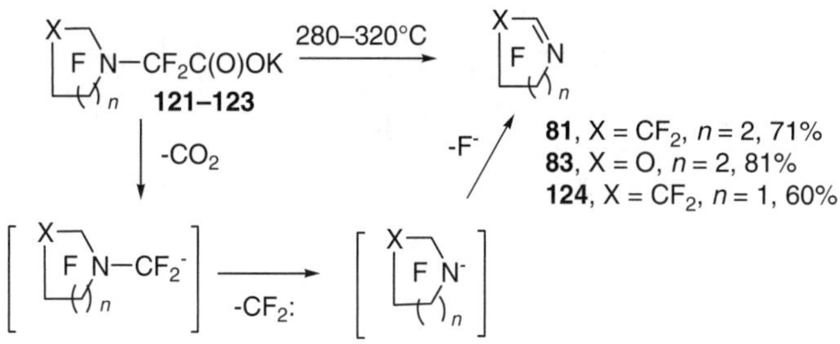

FIGURE 9.37 Synthesis of perfluorinated cyclic imines by pyrolysis of N-heteroaryl acetic acids.

of the fluorination of nitrogen-containing heteroaromatic compounds, which defines the direction of fluorination can be found in the original paper.[107]

Cyclic imines can be also prepared using "traditional" chlorination–fluorination approach. Ulrich et al. prepared imines **81** in a good yield by reacting imides **47** with PCl_5, followed by the treatment of chlorinated products with AgF.[109]

Recently reported method of preparation of perfluorinated cyclic imines is based on an unusual decarboxylation reaction of fluorinated acetic acid derivatives containing perfluorinated heterocyclic substituent. Pyrolysis of potassium salts **121–123** leads to imines **81**, **83**, and **124** in 60–82% yield (Fig. 9.37).[110]

The mechanism of this process, involves extrusion of CO_2 with the formation of intermediate carbanion, which further eliminates difluorcarbene giving cyclic azaanion and its conversion into final product through the elimination of fluoride anion (Fig. 9.37).

9.3.2 Reactions of Unsaturated Perfluorinated Nitrogen-Containing Heterocycles

The presence of highly electrophilic C=N bond defines the reactivity of this class of compounds. Similar to other perfluorinated imidoyl fluorides, cyclic analogues are sensitive to the action of nucleophiles. For example, compounds **81** and **124** react with metal fluoride anion in polar solvents with the formation of the salt **125**, which can react with a strong electrophiles (Fig. 9.38).

The treatment of **125** ($n = 1$) with Cl_2 or MeI[112] leads to compounds **126** and **127** in decent yields[111] and the alkylation of **125** ($n = 2$) with fluorinated alkyl trifluorosulfonates results in the formation of amines **128** (Fig. 9.38).[113] Isomeric azadienes **129** and **130** both produce stable unsaturated azaanion **131** upon treatment with CsF in polar media. The reaction of **129** with MeI leads to the formation of unsaturated heterocycle **132** and the treatment with boron trifluoride results in the mixture of dienes **129** and **130**[114] (Fig. 9.39).

It is noteworthy, that in contrast to noncyclic perfluorinated imidoyl fluorides $R_fN{=}CFR'_f$, cyclic imines **124**[113] and **81** undergo the dimerization upon treatment with fluoride anion[111] (Fig. 9.40).

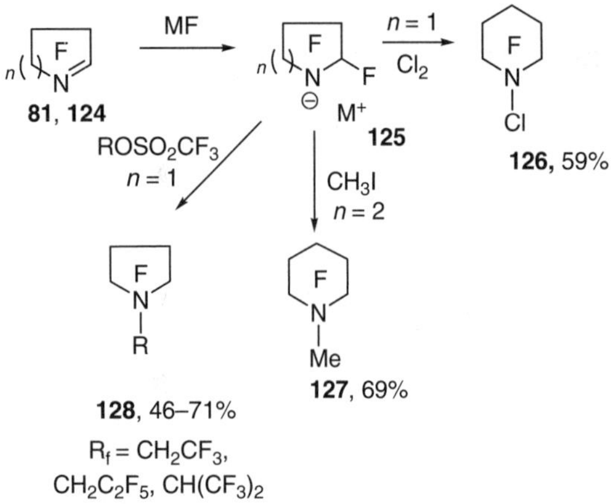

FIGURE 9.38 Reactions involving cyclic azaanions.

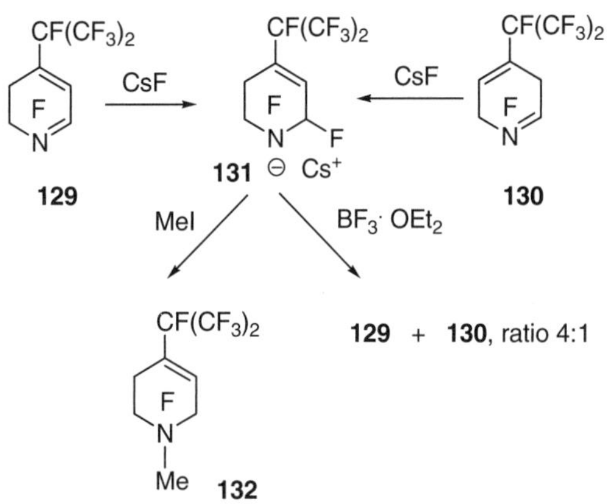

FIGURE 9.39 Generation and reactions of cesium salt **131**.

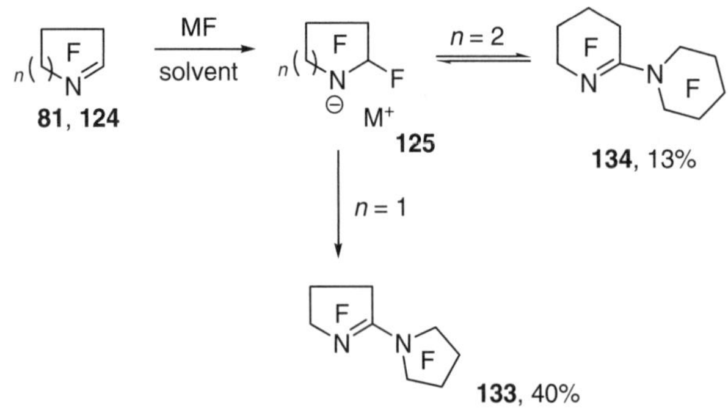

FIGURE 9.40 Dimerization of cyclic perfluorinated imidoyl fluorides.

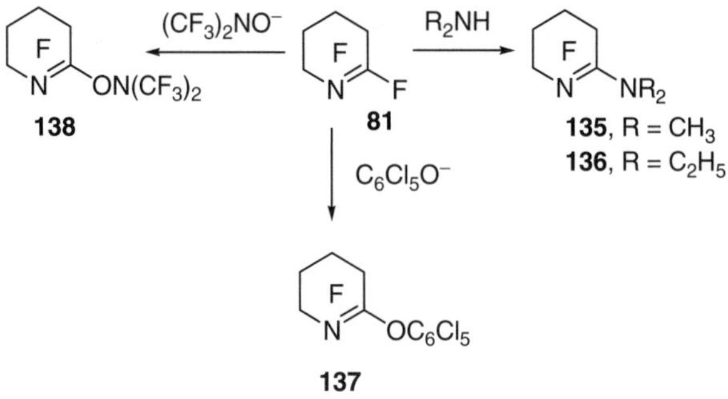

FIGURE 9.41 Nucleophilic reactions of imidoyl fluoride **81**.

The process involves intermediate formation of the salt of azaanion **125**, and the reaction of **125** with starting material resulting in the replacement of vinylic fluorine with the formation of dimers **133** and **134**.

The reaction of with nucleophilic reagents is the most studied group of chemical transformations involving cyclic imidoyl fluorides. Some representative reactions of compound **81** with secondary amines, pentachlorophenoxide and $(CF_3)_2NONa$ leading to products **135–138** are shown below (Fig. 9.41).[115]

The reaction of **81** with sodium azide results in the formation of compound **139**, which exists in equilibrium with cyclic tetrazole **140**[116] (Fig. 9.42).

The reaction of **81** with diazomethane is another example of nucleophilic cycloaddition process. Elimination of nitrogen from initially formed triazole **142** leads to aziridine **141**.[117] Vinylic fluorine in **81** can be replaced by chlorine using the reaction with $ClSi(CH_3)_3$/KF. Alternatively, chloride **143** can be prepared by the reaction of **81** with $AlCl_3$.[115]

Alkylation of cyclic azaanions is also well studied. It can be exemplified by the reactions of **81** reported by Banks group.[118] The fluoride ion-catalyzed reaction of

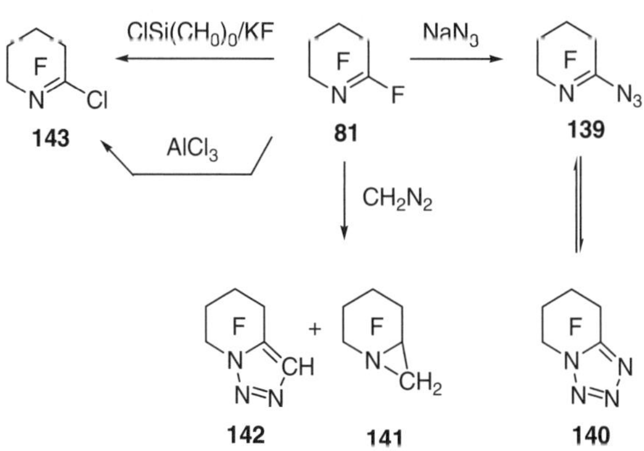

FIGURE 9.42 Reactions of **81** with sodium azide and diazomethane.

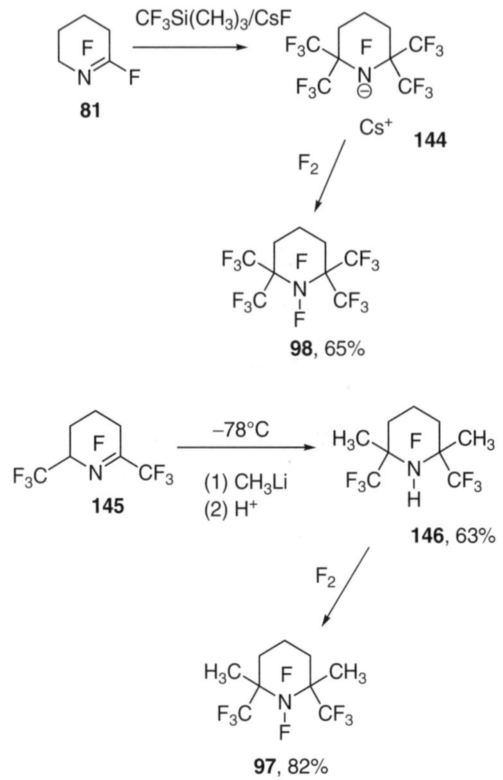

FIGURE 9.43 Reaction of **81** with carbon nucleophiles.

81 with $CF_3Si(CH_3)_3$ leads to complete replacement of all α-fluorines in **81** and the formation of stable cesium salt **144**, which was converted into compound **98** by the reaction with diluted fluorine gas (Fig. 9.43).

Imine **145** is converted into amine **146** by the reaction with excess of CH_3Li, followed by hydrolysis step. Consecutive fluorination of **146** using F_2/N_2 mixture results in *N*-fluoroamine **97**.[118]

The list of other carbon nucleophiles used in the reactions with **81**, **83**, and **124** includes $C_6F_5Si(CH_3)_3$,[119,120] $CF_3Si(CH_3)_3$,[121] and $R_fCH_2OSi(CH_3)_3$.[113,122]

Available data on chemical transformations of perfluorinated cyclic dienes are limited. Dienes **129** and **130** were reported to undergo interconversion under the action of CsF (Fig. 9.39)[114] and compound **119**—an interesting isomerization into **147** under the action of CsF[123] (Fig. 9.44).

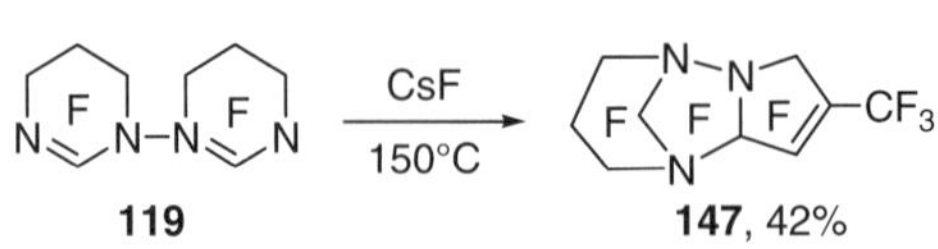

FIGURE 9.44 Nucleophilic isomerization of **119**.

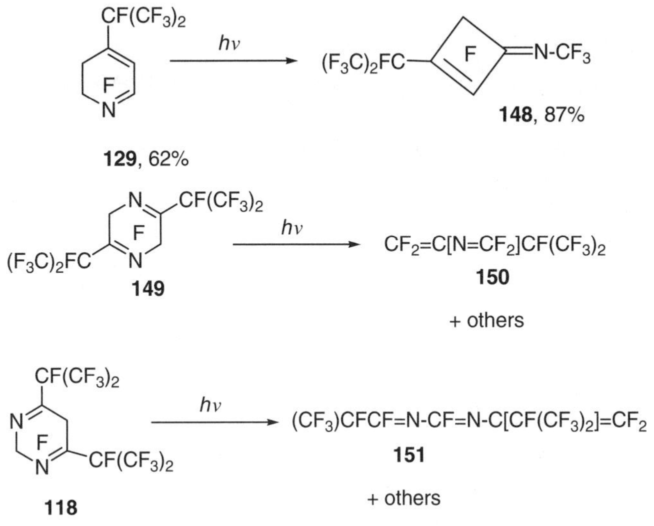

FIGURE 9.45 Photochemical isomerization of perfluorinated cyclic azadienes.

In sharp contrast to heterocycles, containing only one double bond, polyfluorinated cyclic azadienes undergo photochemical isomerizations. The outcome of the reaction depends on the structure of starting material. For example, irradiation of **129** results in high-yield formation of cyclobutene **148**[124] (Fig. 9.45).

Dienes **149** and **118** under UV irradiation undergo ring opening with the formation of the corresponding azadienes **150** and **151** as major products however, both **118** and **130** remain unchanged under similar conditions.[124]

9.4 CONCLUSION

A wide variety of perfluorinated cyclic materials has been prepared and a large body of experimental material on the synthesis and chemistry of perfluorinated nonaromatic heterocyclic was accumulated. Despite the fact that the saturated perfluorinated heterocycles are expected to be chemically inert, this group of materials actually have an interesting and unusual chemistry, which can be exemplified by reductive defluorination of cyclic perfluorinated N−F amines under the action of mild reducing agents or cleavage of cyclic amines under the action of strong acids.

On the other hand, the synthesis and especially the chemistry of sulfur-, selenium-, and phosphorous-containing heterocycles is the least studied and underdeveloped sector of perfluorinated heterocycles and there is no doubt that new materials and interesting chemistries yet to be discovered in area.

REFERENCES

1. Kauck, E.A.; Simons, J.H. Fluorocarbon tertiary amines. (Minnesota Mining and Manufacturing Co.) *US Patent* 2,616,927, **1952**.
2. Kauck, E.A.; Simons, J.H. Cyclic perfluoroalkylene oxides. (Minnesota Mining and Manufacturing Co.) *US Patent* 2,594,272, **1952**.
3. Sander, M.; Helfrich, F.; Blochl, W. Electrolytic preparation of 5- and 6-membered cyclic perfluoroalkyl ethers. *German Patent* DE 1069639, **1959**.
4. Organic cyclic perfluoroethers having 5 or 6-membered rings. *British Patent* GB 862538, **1961**.
5. Abe, T.; Hayashi, E.; Baba, H.; Kodaira, K.; Nagase, S. Fluorination of 2-alkyl-substituted oxanes. The synthesis and purification of perfluoro(2-alkyl-substituted oxane)s. *J. Fluorine Chem.* **1980**, *15*(5), 353.
6. Abe, T.; Nagase, S. The electrochemical fluorination of 2-methyloxane, oxepane, 2-chloromethyloxane and 2-cholоromethyloxolane. *J. Fluorine Chem.* **1979**, *13*(6), 519.
7. Abe, T.; Tamura, M.; Sekiya, A. Electrochemical fluorination of several esters derived from oxolane-2-yl-carboxylic acid, oxolane-2-yl-methanol and oxane-2-yl-methanol. *J. Fluorine Chem.* **2005**, *126*(3), 325.
8. Abe, T.; Baba, H.; Hayashi, E.; Nagase, S. Synthesis of perfluorobicyclic ethers [1]. The electrochemical fluorination of cycloalkyl-substituted carboxylic acids. *J. Fluorine Chem.* **1983**, *23*(2), 123.
9. Abe, T.; Hayashi, E.; Baba, H.; Nagase, S. Synthesis of perfluorobicyclic ethers [2]. The electrochemical fluorination of [α]-cyclohexenyl-substituted carboxylic acid derivatives. *J. Fluorine Chem.* **1984**, *25*(4), 419.
10. Abe, T.; Hayashi, E.; Baba, H.; Nagase, S. The reaction of perfluorobicyclic ethers and perfluorospiro-ethers with anhydrous aluminum chloride. *J. Fluorine Chem.* **1984**, *26*(3), 295.
11. Chechina, O.N.; Sobolev, V.P.; Zerzelidi, M.K.; Snegirev, V.F.; Goncharenko, A.A.; Utarbaev, G.K. Anodic cross coupling involving perfluorotetrahydro-α-furylacetic acid. *Elektrokhimiya* **1991**, *27*(7), 914–917.
12. Levin, A.I.; Sokolov, S.V.; Chechina, O.N.; Bil'dinov, K.I. Kolbe reaction in a series of perfluorinated acids. *Zh. Obshch. Khim.* **1969**, *39*(2), 440–442.
13. De Pasquale, R.J. Preparation of highly fluorinated ethers. *J. Org. Chem.* **1973**, *38*(17), 3025–3030.
14. Lagow, R.J.; Bierschenk, T.R.; Juhlke, T.J.; Kawa, H. A new synthetic procedure for the preparation and manufacture of perfluoropolyethers. In *Synthetic Fluorine Chemistry*; Olah, G.A.; Chambers, R.D.; Prakash, G.K.S., Eds.; John Wiley & Sons, Inc.: New York, 1992; pp 97–126.
15. Tiers, G.V.D. The chemistry of perfluoro ethers. I. Substitution of α-fluorine by chlorine: the α,α,α′-trichloro perfluoro ethers. *J. Am. Chem. Soc.* **1955**, *77*, 4837–4840.
16. Abe, T.; Nagase, S. The reaction of perfluoro-2,4-dialkyloxolanes with anhydrous aluminum chloride. *J. Fluorine Chem.* **1978**, *12*(5), 359.
17. Hauptschein, M.; Stokes, C.S.; Grosse, A.V. The properties and reactions of perfluorobutyrolactone. *J. Am. Chem. Soc.* **1952**, *74*, 1974–1976.
18. Kumai, S.; Samejima, S.; Yamabe, M. Preparation of perfluoro-γ-butyrolactone. *Asahi Garasu Kenkyu Hokoku* **1979**, *29*(2), 105–109.

19. Yamabe, M.; Munekata, S.; Kumai, S.; Samejima, S. Perfluorolactones. (Asahi Glass Co., Ltd., Japan.) *German Patent* DE 2642824, **1977**.

20. Banks, R.E.; Birchall, J.M.; Clarke, T.; Haszeldine, R.N.; Stevenson, M.J.; Iserson, H. Fluoro olefins. VI. A convenient synthesis of trifluoroacryloyl fluoride. *J. Chem. Soc. C* **1968** (3), 266–268.

21. Hauptschein, M.; Stokes, C.S.; Nodiff, E.A. Thiol esters of perfluorocarboxylic acids. *J. Am. Chem. Soc.* **1952**, *74*, 4005–4010.

22. Malichenko, N.A.; Yagupol'skii, L.M. Delta-(2-Benzothiazolyl)perfluorovaleric acid and its derivatives. *Zh. Obshch. Khim.* **1969**, *39*(8), 1751–1755.

23. Krespan, C.G. Abbreviated route to difluoromaleic anhydride. *J. Fluorine Chem.* **1990**, *48*, 339–343.

24. Krespan, C.G.; Petrov, V.A. The chemistry of highly fluorinated carbocations. *Chem. Rev.* **1996**, *96*(8), 3269–3301.

25. Nagase, S. Electrochemical fluorination. *Fluorine Chem. Rev.* **1967**, *1*(1), 77–106.

26. Banks, R.E.; Barlow, M.G. (Eds.) *Specialist Periodical Report: Fluorocarbon and Related Chemistry*, Vol. 3, 1976; pp 491.

27. Banks, R.E.; Barlow, M.G. (Eds.) *Fluorocarbon and Related Chemistry*, Vol. 2. 1974; p 491.

28. Banks, R.E.; Barlow, M.G. *Fluorocarbon and Related Chemistry,* Vol. 1: *A Review of the Literature Published During 1969 and 1970*. **1971**; p 307.

29. Drakesmith, F.G. Electrofluorination of organic compounds. In *Top. Curr. Chem.* **1997**, *193*, 197–242.

30. Simmons, T.C.; Hoffmann, F.W.; Beck, R.B.; Holler, H.V.; Katz, T.; Koshar, R.J.; Larsen, E.R.; Mulvaney, J.E.; Paulson, K.E.; Rogers, F.E.; Singleton, B.; Sparks, R.E. Fluorocarbon derivatives. II. Cyclic nitrides. *J. Am. Chem. Soc.* **1957**, *79*, 3429–3432.

31. Banks, R.E.; Cheng, W.M.; Haszeldine, R.N. Heterocyclic polyfluoro compounds. II. Reactions of undecafluoropiperidine. The preparation of perfluoro-2,3,4,5-tetrahydropyridine and perfluoro-(1-methylpyrrolidine). *J. Chem. Soc.* **1962**, *3407*-16.

32. Banks, R.E.; Carson, P.A.; Haszeldine, R.N. *N*-Fluoro compounds. VI. Preparation and some reactions of perfluoro-*N,N'*-difluoropiperazine. Crosslinking of a nitroso rubber. *J. Chem. Soc., Perkin Trans. 1* **1973**, (10), 1111–1114.

33. Lin, W.H.; Lagow, R.J. The synthesis of perfluoro highly branched heterocyclic fluorine compounds by direct fluorination. *J. Fluorine Chem.* **1990**, *50*(1), 15–30.

34. Riess, J.G., Fluorocarbon emulsions—designing an efficient shuttle service for the respiratory gases—the so-called blood substitutes. In *Fluorine Chemistry at the Millennium. Fascinated by Fluorine*; Banks, R.E., (Ed.); Elsevier: New York, 2000; pp 385–431.

35. Abe, T.; Fukaya, H.; Hayashi, E.; Hayakawa, Y.; Nishida, M.; Baba, H. The electrochemical fluorination of N-containing carboxylic acids. Part 4. Fluorination of methyl 3-(dialkylamino)isobutyrates and methyl 3-(dialkylamino)-*n*-butyrates. *J. Fluorine Chem.* **1994**, *66*(2), 193–202.

36. Abe, T.; Hayashi, E. The synthesis of perfluoro (*N,N*-dialykylcarbamoyl fluorides) by the reaction of perfluoro (*N,N*-dialkylmethylamines) with oleum. *J. Fluorine Chem.* **1989**, *45*(2), 293.

37. Munekata, S.; Yamabe, M.; Kaneko, I.; Jitsugiri, Y. Iodoacyl fluoride containing an ICF_2CF_2O group from tetrafluoroethylene. (Asahi Glass Co., Ltd., Japan.) FR 2463115, **1981**.

38. Mazalov, S.A.; Gerasimov, S.I.; Sokolov, S.V.; Zolotavin, V.L. Methods of preparation and properties of organic fluorine compounds. I. The electrochemical method of preparation of perfluoro tertiary heterocyclic amines. *Zh. Obshch. Khim.* **1965**, *35*(3), 485–489.

39. Abe, T.; Hayashi, E.; Fukaya, H.; Baba, H. The electrochemical fluorination of nitrogen-containing carboxylic acids. Fluorination of methyl esters of cyclic amino-group substituted carboxylic acids. *J. Fluorine Chem.* **1990**, *50*(2), 173.

40. Abe, T.; Fukaya, H.; Hayashi, E.; Hayakawa, Y.; Nishida, M.; Baba, H. The electrochemical fluorination of N-containing carboxylic acids. Part 4. Fluorination of methyl 3-dialkylamino-isobutyrates and methyl 3-dialkylamino-*n*-butyrates. *J. Fluorine Chem.* **1994**, *66*(2), 193.

41. Abe, T.; Fukaya, H.; Ono, T.; Hayashi, E.; Soloshonok, I.; Okuhara, K. Electrochemical fluorination of N-containing carboxylic acids. Part 5. Fluorination of the methyl esters of *cis*-2,6-dimethylmorpholino-group substituted carboxylic acids. *J. Fluorine Chem.* **1998**, *87*(2), 193.

42. Abe, T.; Pandey, S.K.; Baba, H. Electrochemical fluorination of hexahydroazepine, methylpiperidines and methyl 1-hexahydroazepine-acetate: the preparation of *F*-(1-hexahydroazepine-acetyl fluoride) and its derivatives. *J. Fluorine Chem.* **2000**, *105*(1),149.

43. Abe, T.; Baba, H.; Fukaya, H.; Tamura, M.; Sekiya, A. Simons electrochemical fluorination of substituted homopiperazines(hexahydro-1,4-diazepines) and piperazines. *J. Fluorine Chem.* **2003**, *119*(1), 27.

44. Chambers, R. *Fluorine in Organic Chemistry*. Blackwell Publishing; CRC Press: **2004**; pp 32–33.

45. Lam, W.Y.; DesMarteau, D.D. Unusual cycloaddition reactions with 2-(trifluoromethyl)-3,3-difluorooxaziridine. *J. Am. Chem. Soc.* **1982**, *104*(14), 4034–4035.

46. Malacrida, A.; Desmarteau, D.D. Fluorinated heterocyclic compounds and process for preparing them. (Ausimont S.r.l., Italy.) *European Patent* EP 89-112384. **1990**.

47. Pawelke, G.; Buerger, H. Formation of the perfluoro-2,6-dimethyl-2,4,6-triazahepta-3,4-dienium ion by trimerization of perfluoroazapropene. *J. Fluorine Chem.* **1984**, *26*(3), 321–330.

48. Knunyants, I.L.; Gontar, A.F.; Vinogradov, A.S. Electrophilic condensation of perfluoro-2-azapropene. *Izv. Akad. Nauk SSSR, Ser. Khim.* **1983** (2), 476–478.

49. Hayashi, E.; Abe, T.; Nagase, S. The reaction of perfluoro (*N*-alkyl cyclic amines) with oleum. The formation and characterization of perfluorolactams. *J. Fluorine Chem.* **1988**, *41*(2), 213–225.

50. Hayashi, E.; Abe, T.; Nagase, S. Formation of perfluoro(*N*-alkyl lactams) by the reaction of perfluoro(*N*-alkyl cyclic amines) with fuming sulfuric acid. *Chem. Lett.* **1985** (3), 375–376.

51. Sankina, L.V.; Kostikin, L.I.; Ginsburg, V.A. Interaction of cyclic anhydrides of perfluorodicarboxylic acids with nucleophilic agents. *Zh. Org. Khim.* **1972**, *8*(6), 1330–1331.

52. Henne, A.L.; Zimmer, W.F. Positive bromine from perfluorinated *N*-bromoimides. *J. Am. Chem. Soc.* **1951**, *73*, 1103–1104.

53. McBee, E.T.; Wiseman, P.A. Polyfluoro aliphatic diamines. *US Patent* 2515246, **1950**.

54. Fedorova, G.B.; Dolgopol'skii, I.M.; Suvorova, M.A.; Gubanov, V.A.; Gracheva, P.E. Bifunctional derivatives of perfluorocarboxylic acids. IV. Synthesis of ω-cyano-substituted chlorides of perfluoro carboxylic acids. *Zh. Org. Khim.* **1973**, *9*(11), 2211–2214.

55. Yagupol'skii, Y.L.; Savina, T.I. Reaction of perfluorodicarboxylic acid imides with xenon difluoride. *Zh. Org. Khim.* **1981**, *17*(6), 1330–1331.

56. Krespan, C.G.; Brasen, W.R. Fluorinated cyclic polysulfides and their polymers. *J. Org. Chem.* **1962**, *27*, 3995–4000.

57. Krespan, C.G.; Langkammerer, C.M. Fluorinated heterocyclic derivatives of sulfur, selenium, and phosphorus. *J. Org. Chem.* **1962**, *27*, 3584–3587.

58. Tiers, G.V.D. Fluorocarbon sulfides. *J. Org. Chem.* **1961**, *26*, 2538–2539.

59. Krespan, C.G. Fluorinated thiolanes and a selenolane. *J. Org. Chem.* **1962**, *27*, 3588–3594.

60. Abe, T.; Shreeve, J.n.M. Perfluorotetramethylenesulfur difluoride and its derivatives. Perfluoro-1,3-dithietane octafluoride and perfluoro-1,4-dithiane octafluoride. *J. Fluorine Chem.* **1973**, *3*(1), 17.

61. Geisler, K.; Koemm, U.; Metzinger, H.G. Partially nonfluorinated compounds in the electrochemical fluorination of sulfolene. *J. Fluorine Chem.* **1984**, *24*(1), 17–24.

62. Simons, J.H. Fluoropiperidine compounds. (Minnesota Mining and Manufacturing Co.) *US Patent* 2,519,983, **1950**.

63. Mitsch, R.A. Fluorimino or azomethine-containing compounds. (Minnesota Mining and Manufacturing Co.) *US Patent* 3,674,785, **1972**.

64. Huang, H.N.; Lagow, R.J.; Roesky, H. Novel synthesis of unusual classes of fluorocarbon organosulfur compounds using elemental fluorine as a reagent. *Inorg. Chem.* **1991**, *30*(4), 789–794.

65. Behr, F.E. Cyclic perfluoro aliphatic disulfonic acid anhydrides. (Minnesota Mining and Manufacturing Co., USA.) *US Patent* 4,329,478, **1982**.

66. Koshar, R.J. Cyclic perfluoroaliphatic disulfonimides. *US Patent* 57,327, **1982**.

67. Koshar, R.J. Cyclic perfluoroaliphatic disulfonimides. (Minnesota Mining and Manufacturing Co., USA.) *US Patent* 4,429,093, **1984**.

68. DesMarteau, D.D., Fluorine chemistry: a chemical gardener's paradise. In *Fluorine Chemistry at the Millennium*; Banks, R.E., (Ed.); Elsevier: **2000**; pp 179–202.

69. Koppel, I.; Koppel, J.; Maria, P.-C.; Gal, J.-F.; Notario, R.; Vlasov, V.M.; Taft, R.W. Comparison of Bronsted acidities of neutral NH-acids in gas phase, dimethyl sulfoxide and water. *Int. J. Mass Spectrom. Ion Process.* **1998**, *175*(1,2), 61–69.

70. Koppel, I.A.; Taft, R.W.; Anvia, F.; Zhu, S.-Z.; Hu, L.-Q.; Sung, K.-S.; DesMarteau, D.D.; Yagupolskii, L.M.; Yagupolskii, Y.L.; et al. The gas-phase acidities of very strong neutral Bronsted acids. *J. Am. Chem. Soc.* **1994**, *116*(7), 3047–3057.

71. Lamana, W.M.; Krause, L.J.; Summerfield, J.W. Battery containing bis (perfluoroalkylsulfonyl)imide and cyclic perfluoroalkylenedisulfonylimide. *World patent Appl.* WO 96-US13145, **1997**.

72. Zhu, S.Z.; DesMarteau, D.D. Synthesis and decomposition of benzenediazonium tris ((trifluoromethyl)sulfonyl)methanide, $C_6H_5N_2+(CF_3SO_2)3C$- and benzenediazonium bis((trifluoromethyl)sulfonyl)amide $C_6H_5N_2+(CF_3SO_2)2N$ and the cyclic analog, $C_6H_5N_2+$ cyclo-$SO_2(CF_2)3SO_2N$. *Inorg. Chem.* **1993**, *32*(2), 223–226.

73. Singh, S.; DesMarteau, D.D.; Zuberi, S.S.; Witz, M.; Huang, H.N. *N*-Fluoroperfluoroalkylsulfonimides. Remarkable new fluorination reagents. *J. Am. Chem. Soc.* **1987**, *109*(23), 7194–7196.

74. Banks, R.E.; Barlow, M.G.; Eds.; *Specialist Periodical Report: Fluorocarbon and Related Chemistry*, Burlington House: London, **1976**; Vol. 3, pp 208–209.

75. Tiers, G.D. The chemistry of perfluoro ethers. II. Ether cleavage with simultaneous replacement of α-fluorine by chlorine. *J. Am. Chem. Soc.* **1955**, *77*, 6703–6704.

76. Sokolov, S.V.; Mazalov, S.A. Ring opening and reverse cyclization of perhalogenated amines under the influence of alumium and antimony halides. *Dokl. Akad. Nauk SSSR* **1965**, *162*(5), 1071–1074.

77. Pushkina, L.N.; Kollegov, V.F.; Stepanov, A.P.; Mazalov, S.A.; Sokolov, S.V. Methods of preparation and properties of organofluorine compounds. VI. Some tertiary perfluorinated heterocyclic amines and products of their transformations. *Zh. Obshch. Khim.* **1966**, *36*(9), 1619–1625.

78. Mazalov, S.A.; Sokolov, S.V. Methods of preparation and properties of organofluorine compounds. IV. *N*-α,α-Dichlorononafluoroamylimino chloride of ω,ω,ω-trichlorohexafluorovaleric acid. *Zh. Obshch. Khim.* **1966**, *36*(7), 1330–1337.

79. Petrov, V.A.; Kunanets, V.K.; Makarov, K.N.; German, L.S. Isomerization of perfluorinated cyclic amines by antimony pentafluoride. *Izv. Akad. Nauk SSSR, Ser. Khim.* **1989** (3), 646–649.

80. Petrov, V.A.; Kunanets, V.K.; Kvasov, B.A.; Galakhov, M.V.; Makarov, K.N.; German, L.S. Cleavage of perfluorinated tertiary amines with antimony pentafluoride. *Izv. Akad. Nauk SSSR*, **1989** (1), 1616–1619.

81. Petrov, V.A.; Krespan, C.G.; Smart, B.E. Electrophilic reactions of fluorocarbons under the action of aluminum chlorofluoride, a potent Lewis acid. *J. Fluorine Chem.* **1996**, *77*(2), 139–142.

82. Banks, R.E.; Mullen, K.; Williamson, G.E. *N*-Fluoro-compounds. III. Some free-radical reactions of perfluoro-*N*-fluoropiperidine. *J. Chem. Soc. C* **1968** (21), 2608–2612.

83. Banks, R.E.; Parker, A.J.; Sharp, M.J.; Smith, G.F. *N*-Fluoro compounds. V. Photochemical reactions of perfluoro-*N*-fluoromorpholine, and its thermal decomposition; preparation and properties of perfluoromorpholin-*N*-oxyl. *J. Chem. Soc., Perkin Trans. 1* **1973**, (1), 5–13.

84. Mitsch, R.A. Organic fluoronitrogens. II. The reductive defluorination reaction. *J. Am. Chem. Soc.* **1965**, *87*(2), 328–333.

85. Banks, R.E.; Haszeldine, R.N.; Hatton, R. Synthesis of group Vb fluorides and *N*-fluoro compounds. *Tetrahedron Lett.* **1967** (41), 3993–3996.

86. Banks, R.E.; Murtagh, V.; Tsiliopoulos, E. *N*-Halo compounds. Part 12. Site-specific fluorination of carbanions with perfluoro-*N*-fluoropiperidine. *J. Fluorine Chem.* **1991**, *52*(3), 389–401.

87. Banks, R.E. *Fluorocarbons and Their Derivatives*, 2nd ed.; Macdonald Technical and Scientific: London, **1970**; pp 133–134.

88. Banks, R.E.; Hatton, R. *N*-Halo compounds. Part V. Direct conversion of perfluoro-*N*-fluoropiperidine and -morpholine into the corresponding N–H compounds. *J. Fluorine Chem.* **1978**, *11*(5), 563–565.

89. Polishchuk, V.R.; German, L.S. Interaction of perfluoropiperidine with phenols. Mechanisms of *ortho-* and *para*-fluorination. *Tetrahedron Lett.* **1972** (51), 5169–5172.

90. Polishchuk, V.R.; Medvedev, B.Y.; Bubnov, N.N.; German, L.S.; Knunyants, I.L. Direct *o*-fluorination of *N,N*-dialkylanilines. *Izv. Akad. Nauk SSSR, Ser. Khim.* **1972** (12), 2805–2807.

91. Banks, R.E.; Murtagh, V., Perfluoro-*N*-fluoropiperidine. In *Handbook of Reagents for Organic Synthesis. Fluorine-Containing Reagents*; Paquette, L.A., Paquette, L.A., (Eds.); John Willey & Sons, Inc.: 2007; pp 415–416.

92. Furin, G.G., Introduction of fluorine by N−F compounds. In *Methods of Organic Chemistry (Houben–Weyl)*; Baasner, B.; Hagemann, H.; Tatlow, J.C., Eds.; George Thieme Verlag: New York, 2000; Vol. E10a, pp 435–436.

93. Banks, R.E.; Mullen, K.; Nicholson, W.J.; Oppenheim, C.; Prakash, A. *N*-Fluoro compounds. IV. Photochemical and fluoride-initiated reactions between perfluoro-*N*-fluoropiperidine and perfluoropropene. *J. Chem. Soc., Perkin Trans. 1* **1972**, (8), 1098–1103.

94. Lal, G.S.; Pez, G.P.; Syvret, R.G. Electrophilic NF fluorinating agents. *Chem. Rev.* **1996**, *96*(5), 1737–1755.

95. Yamabe, M.; Kumai, S. Polyfluorodiacyl fluoride. (Asahi Glass Co., Ltd., Japan.) *US Patent* 4,151,200, **1979**.

96. Behr, F.E. Preparation and reactions of omega-halosulfonyl perfluoroalkane sulfonates. (Minnesota Mining and Manufacturing Co., USA.) *European Patent* EP 280392, **1988**.

97. Essers, M.; Haufe, G., Difluormaleic anhydride. In *Handbook of Reagents for Organic Synthesis. Fluorine-Containing Reagents*; Paquette, L.A., (Ed.); John Willey & Sons, Inc.: 2007; pp 206–207.

98. Pavlik, F.J. Tri(fluoroaliphatic)-alpha-fluosulfato amines and derivatives. (Minnesota Mining and Manufacturing Co., USA.) *US Patent* 3,956,293, **1976**.

99. Maeda, M.; Abe, M.; Kojima, M. Bromofluorination of double bonds using *N*-bromo imides and tetrabutylammonium fluoride as a source of fluoride. *J. Fluorine Chem.* **1987**, *34*(3–4), 337–346.

100. Young, J.A.; Durrell, W.S.; Dresdner, R.D. Fluorocarbon nitrogen compounds. VIII. Mono-, di-, tri-, and tetraacyl derivatives, oxadiazoles, and ω-bromo acyl isocyanates. *J. Am. Chem. Soc.* **1962**, *84*, 2105–2109.

101. Abe, T.; Shreeve, J.n.M. Perfluoro(tetramethylene)sulfilimine and its *N*-chloro- and *N*-bromo-derivatives. Perfluoro(tetramethylene)sulfoximine and its *N*-chloro-derivative. *J. Chem. Soc., Chem.* **1981** (5), 242–243.

102. Abe, T.; Shreeve, J.n.M. (*F*-Tetramethylene)sulfimide and (*F*-tetramethylene) sulfoxyimide. Some *N*-alkyl, *N*-(*F*-alkyl) and *N*-halo derivatives. *Inorg. Chem.* **1981**, *20*(9),2894–2899.

103. Dodman, P.; Tatlow, J.C. Highly fluorinated heterocycles. Part XII. Some reactions of the heptafluoro-*p*-dioxanyl anion. *J. Fluorine Chem.* **1976**, *8*(3), 263–274.

104. Coe, P.L.; Dodman, P.; Tatlow, J.C. Highly fluorinated heterocycles. XI. Preparation of some polyfluoro-*p*-dioxenes. Reactions of hexafluoro-*p*-dioxene. *J. Fluorine Chem.* **1975**, *6*(2), 115–128.

105. Krespan, C.G.; Dixon, D.A. Perhalodioxins and perhalodihydrodioxins. *J. Org. Chem.* **1991**, *56*(12), 3915–3923.

106. Chambers, R.D.; Vaughan, J.F.S.; Mullins, S.J. An unusual rearrangement of a diepoxide. *J. Chem. Soc., Chem. Commun.* **1995** (6), 629.

107. Chambers, R.D.; Clark, D.T.; Holmes, T.F.; Musgrave, W.K.R.; Ritchie, I. Polyfluoroheterocyclic compounds. XXIII. Monoenes and dienes derived by the fluorination of hexafluorobenzene and of perfluoro and chlorofluoro heteroaromatic compounds. Mechanism for fluorination by cobalt fluorides. *J. Chem. Soc., Perkin Trans. 1* **1974**, (1), 114–125.

108. Barnes, R.N.; Chambers, R.D.; Hercliffe, R.D.; Musgrave, W.K.R. Polyhaloaromatic compounds. Part 34. Syntheses of perfluoroaza- and -diazacyclohexadiene derivatives by fluorination of perfluoroazines and -diazines. *J. Chem. Soc., Perkin Trans. 1* **1981**, (7), 2059–2064.

109. Ulrich, H.; Kober, E.; Schroeder, H.; Raetz, R.; Grundmann, C. Synthesis of polyfluorinated heterocycles by indirect fluorination with silver fluorides. III. Fluoropiperideines, fluoropyrrolines, and fluoropyrrolenine. *J. Org. Chem.* **1962**, *27*, 2585–2589.

110. Nishida, M.; Fukaya, H.; Abe, T. A new synthetic method for perfluorocycloimines. *J. Fluorine Chem.* **1996**, *76*(1), 3.

111. Bailey, A.R.; Banks, R.E.; Barlow, M.G.; Nickkho-Amiry, M. Fluorocarbon derivatives of nitrogen. Part IV. Perfluoro-1-azacyclohex-1-ylcesium. *J. Fluorine Chem.* **1980**, *15*(4), 289–298.

112. Bailey, A.R.; Banks, R.E. *N*-Halo compounds. Part 8. Perfluoro-*N*-chloropiperidine. *J. Fluorine Chem.* **1983**, *23*(1), 87–96.

113. Nishida, M.; Ono, T.; Abe, T. Reactions of perfluorocycloimines with (polyfluoroalkoxy) trimethylsilanes and polyfluoroalkyltrifluoromethanesulfonates. *J. Fluorine Chem.* **2001**, *110*(1), 63.

114. Barnes, R.N.; Chambers, R.D.; Matthews, R.S. Reactions involving fluoride ion. Part 26. Nitranions from perfluoroazacyclohex-ene and -diene derivatives. *J. Fluorine Chem.* **1982**, *20*(3), 307–312.

115. Banks, R.E.; Oppenheim, C. Fluorocarbon derivatives of nitrogen. Part I. Some reactions of perfluoro-1-azacyclohexene. *J. Fluorine Chem.* **1978**, *12*(1), 27–34.

116. Bailey, A.R.; Banks, R.E. Studies in azide chemistry. Part X. Synthesis of perfluoro-2-azido-1-azacyclohexene. *J. Fluorine Chem.* **1984**, *24*(1), 117–124.

117. Coe, P.L.; Holton, A.G. Some reactions of diazomethane with polyfluoroazaolefins. *J. Fluorine Chem.* **1977**, *10*(6), 553–564.

118. Banks, R.E.; Besheesh, M.K.; Lawrence, N.J.; Tovell, D.J. *N*-Halo compounds. Part 21. Perfluoro(2,2,6,6-tetramethyl-*N*-fluoropiperidine) and its 2,6-dimethyl analog. *J. Fluorine Chem.* **1999**, *97*(1–2), 79–84.

119. Nishida, M.; Ono, T.; Abe, T. Reactions of alicyclic perfluoroimines with trimethyl (pentafluorophenyl)silane. *Nippon Kagaku Kaishi* **2000** (11), 817–820.

120. Nishida, M.; Fukaya, H.; Hayakawa, Y.; Ono, T.; Fujii, K.; Uekusa, H. Multiple pentafluorophenylation of 2,2,3,3,5,6,6-heptafluoro-3,6-dihydro-2*H*-1,4-oxazine with an organosilicon reagent: NMR and DFT structural analysis of oligo(perfluoroaryl) compounds. *Helv. Chim. Acta* **2006**, *89*(11), 2671–2685.

121. Nishida, M.; Ono, T.; Abe, T. Reactions of alicyclic perfluoroimines with trimethyl (trifluoromethyl)silane. *Nippon Kagaku Kaishi* **2001** (5), 281–288.

122. Nishida, M.; Fukaya, H.; Abe, T.; Okuhara, K. Polyfluoroalkylation and polyfluoroalkoxylation of perfluoro(5,6-dihydro-2*H*-1,4-oxazine) in the presence of fluoride anion. *J. Fluorine Chem.* **1998**, *91*(1), 1–3.

123. Barnes, R.N.; Chambers, R.D.; Hewit, C.D. Reactions involving fluoride ion. Part 34. A novel ring contraction. *J. Fluorine Chem.* **1986**, *34*, 59–61.

124. Barnes, R.N.; Chambers, R.D.; Hercliffe, R.D.; Middleton, R. Photochemistry of halocarbon compounds. Part 4. Photochemical conversions of some fluorinated aza- and diazacyclohexadienes. *J. Chem. Soc., Perkin Trans. 1* **1981**, (12), 3289–3291.

10

SEVEN-MEMBERED AND LARGER RING-FLUORINATED HETEROCYCLES

VIACHESLAV A. PETROV

DuPont Central Research and Development Experimental Station, PO Box 80500, Wilmington, DE 19880-0500, USA

10.1 INTRODUCTION

In this chapter, the synthesis and the chemistry of fluorinated seven and eight-membered and larger heterocyclic rings containing oxygen, nitrogen, or sulfur are reviewed with emphasis on the synthesis and chemical transformations. The chapter is organized according to compound class, proceeding from heterocycles containing oxygen atoms (one, two etc., following the substitution pattern of heterocycles from 1,2-, 1,3-, and 1,4-dioxa derivatives) to heterocycles containing both oxygen and nitrogen containing heterocycles (organized in same order: 1,2-, 1,3-, and 1,4-heteroatom pattern), and a short summary on the synthesis and chemistry of sulfur-containing heterocycles. A discussion on synthesis and chemistry of fluorinated macrocycles concludes the chapter. Since there are a limited number of heterocycles containing other heteroatoms (P, Si, Se, and Te), these classes of compounds are not included in the chapter.

Fluorinated Heterocyclic Compounds: Synthesis, Chemistry, and Applications, Edited by Viacheslav A. Petrov

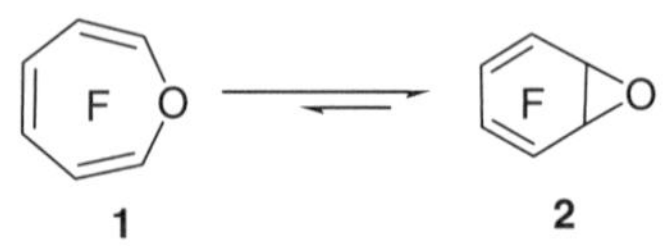

FIGURE 10.1 Equilibrium between *F*-oxepine and hexafluorobenzene oxide.

10.2 FLUORINATED OXYGEN-CONTAINING HETEROCYCLES

10.2.1 Fluorinated Derivatives of Oxepine

The number of known fluorinated (and especially perfluorinated) oxygen-containing heterocycles is limited. Parent *F*-oxepine (**1**) so far has not been isolated. It exists in rapid equilibrium with its bicyclic isomer *F*-1,2-epoxybenezene (**2**),[1,2] but in contrast to hydrocarbon analogue, the equilibrium in this case lies far on the side of **2**, with estimated equilibrium ratio **1**:**2** of 3:97 at 55°C. At this temperature, the extrapolated rate constant for formation of **1** is 6000 times smaller compared to hydrocarbon analogue (Fig. 10.1).[2]

The synthesis of oxide **2** is quite complex and involves six steps.[1,2] Two other perfluorinated derivatives of *F*-oxepine, compounds **5** and **6**, were made by thermal isomerization of bicyclic compounds *endo*-**3** and *endo*- and *exo*-**4** isomers of, prepared by reaction of Dewar-*F*-benzene with $CF_2(OF)_2$ under UV irradiation in 2% and 6% yield, respectively. At elevated temperature, both **3** and **4** undergo irreversible isomerization affording **5** and **6**, respectively (Fig. 10.2).[3,4]

The half-lives of compounds **3** and **4** are 3.8 and 6.7 min at 160°C, respectively. Low molecular weight perfluorinated polymer (MW ~1100) isolated in this reaction was shown to have an unusual structure with unsaturated seven-membered ring incorporated into the backbone of the polymer.[3]

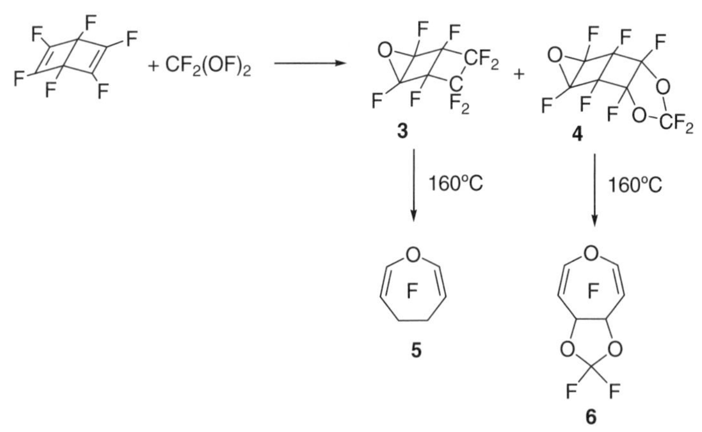

FIGURE 10.2 Preparation of compounds **5** and **6**.

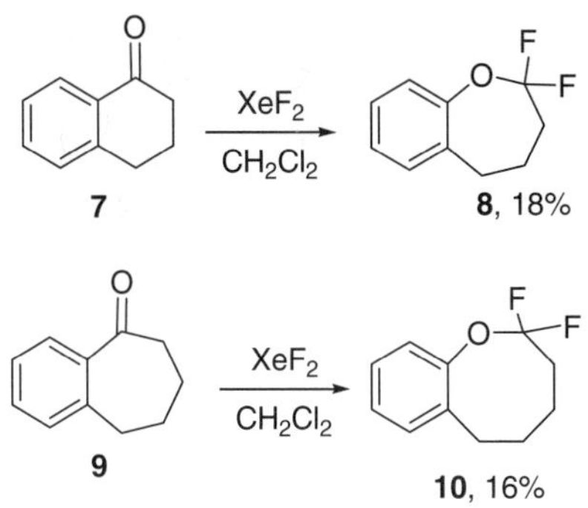

FIGURE 10.3 XeF_2-initiated ring expansion of cyclic aromatic ketones.

2,2-Difluoro-2,3,4,5-tetrahydro-2*H*-1-benzoxepine (**8**) and 2,2-difluoro-2,3,4,5-tetrahydro-2*H*-1-benzoxocin (**10**) have been prepared in low yield by oxidative ring expansion reaction of cyclic aromatic ketones **7** or **9** with XeF_2 (Fig. 10.3).[5]

Fustero and coworkers recently successfully applied a ring-closing metathesis (RCM) reaction for the stereoselective synthesis of 3-substituted -2,3,4,7- and -2,3,4,5- tetrahydrooxepines, containing a fluoroalkyl group.[6] The recent development of RCM process for the synthesis of "lightly" fluorinated carbocycles as well as oxygen- or nitrogen-containing heterocycles is summarized in an excellent review.[7]

10.2.2 Fluorinated Rings with Two Oxygen Atoms

Artemisinin is an important antimalarial drug with high activity against chloroquine-resistant *Plasmodium falciparum*. *endo*-Peroxide fragment is an important attribute of this molecule. Bonnet-Delpon and coworkers found that the introduction of CF_3 group in the molecule through the reaction of artemisinin and Ruppert–Prakash reagent [$CF_3Si(CH_3)_3$] results in a significant improvement of antimalarial activity of artemisinin.[8] Several other derivatives of artemisinin (Fig. 10.4) introduced by this group also showed promising results *in vitro* antimalarial test.[9,10]

Additional information on fluorinated derivatives of artemisinin can be found in several recent review articles[11–14] and in Chapter 12.

Synthesis of polyfluorinated 1,3-dioxepines has been developed by Burger and coworkers.[15–18] The method relies on the ring expansion of fluorinated five-membered heterocycles derived from hexafluoroacetone. For example, 4,5-dihydro-1,3-dioxepine-5-ones **12** are prepared by the reaction of 1,3-dioxalan-4-ones **11** with 1-diethylamino-1-propyne (Fig. 10.5).[15–18]

1,3-Dioxepine-5-ones undergo an interesting rearrangement at an elevated temperature. Heterocycles **12** bearing an aryl group in position 7 quantitatively rearrange into the corresponding enols **13** upon heating (Fig. 10.6).[17]

The outcome of thermal reactions depends on the reaction conditions. For example, the thermolysis of **12** ($R = H$, $R' = Ar$) at $<130°C$ in close system results in ring contraction leading to furanones **14**. When the reaction was carried out at

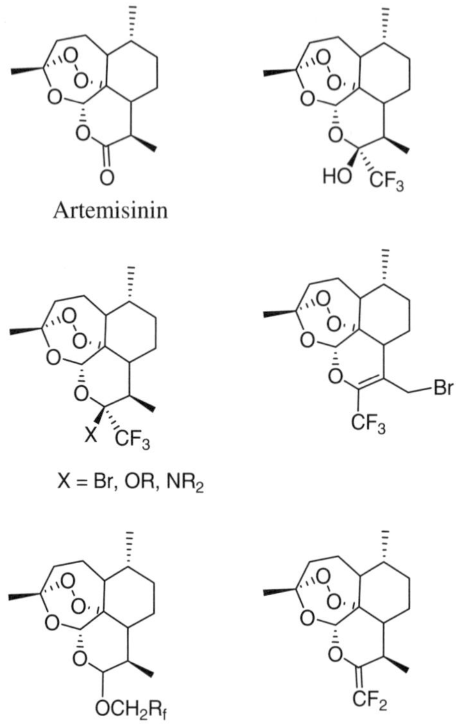

FIGURE 10.4 Artemisinin and its fluorinated derivatives.

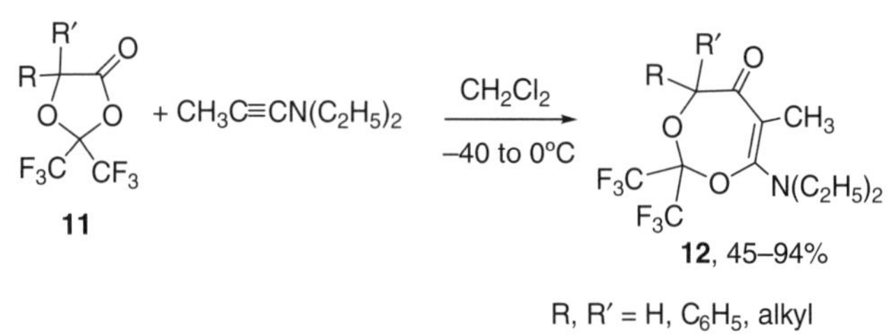

FIGURE 10.5 Synthesis of 2,2-bis(trifluoromethyl)-4,5-dihydro-1,3-dioxepine-5-ones **12**.

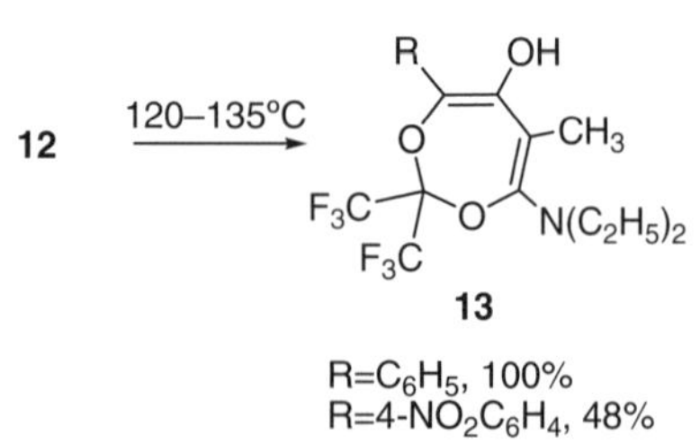

FIGURE 10.6 Rearrangement of 4,5-dihydro-2*H*-1,3-dioxepine-5-ones to enols **13**.

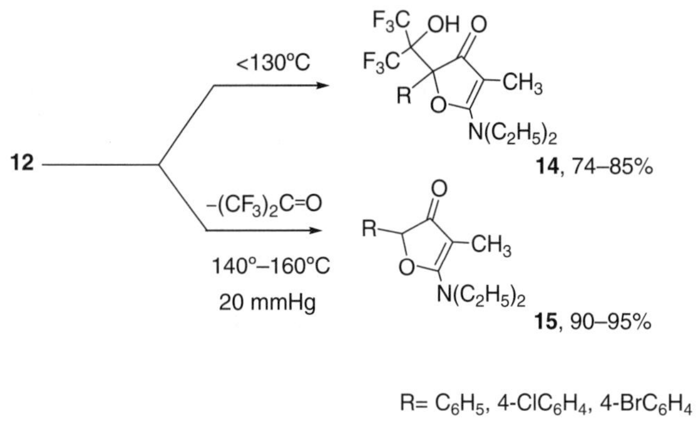

FIGURE 10.7 Thermal reactions of 4,5-dihydro-2*H*-1,3-dioxepine-5-ones **12**.

slightly higher temperature and reduced pressure, furanones **15** formed as the result of hexafluoroacetone extrusion (Fig. 10.7).

Relatively high electrophilicity of the double bond of *F*-2-methylpentene-2 (**16**; for synthesis, see Refs 19 and 20) in combination with ability of the double bond to undergo allylic shift makes this material a convenient building block for the preparation of 1,4-dioxepanes. For example, the reaction of **16** with 2,3-butanediol in the presence of triethylamine gives a mixture of 1,4-dioxepines **17** and **18**, along with dioxolane **19** (Fig. 10.8).[21]

Detailed study on the interaction of bifunctional nucleophilic reagents with **16** revealed an interesting behavior of this compound. While reaction with ethylene glycol in polar solvents (DMF, DMSO, HMTA, and acetonitrile) leads to the predominant formation of dioxolane **20**, in solvents of lower polarity predominant formation of the corresponding 1,4-dioxepine **21** as major product is observed (Fig. 10.9).[22]

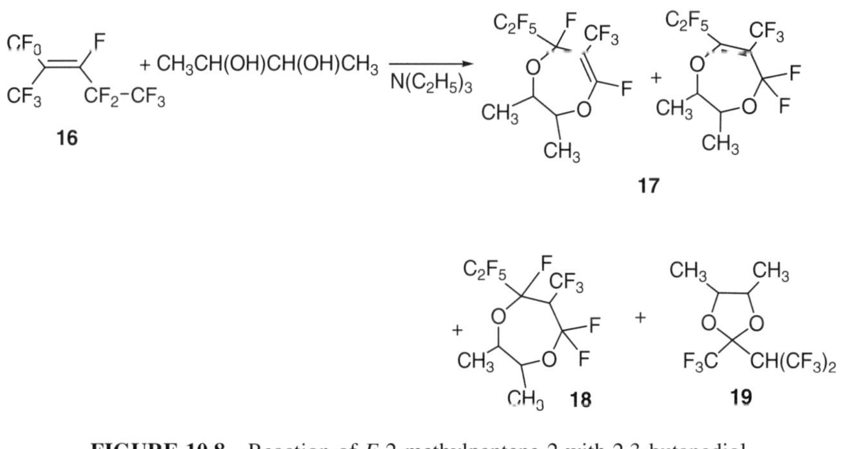

FIGURE 10.8 Reaction of *F*-2-methylpentene-2 with 2,3-butanediol.

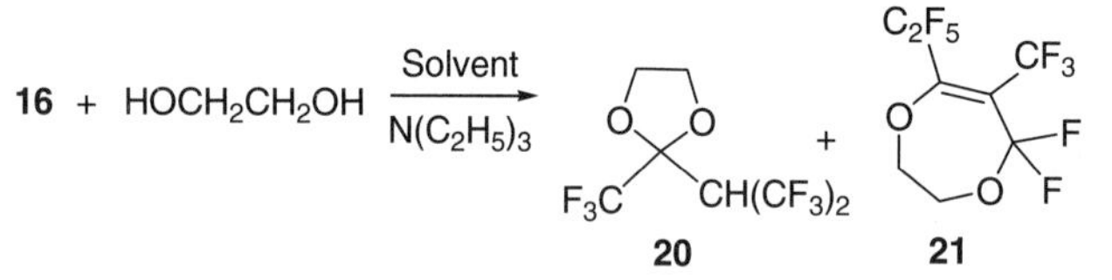

Solvent	Ratio **20:21**
DMF	81:19
Ether	40:60

FIGURE 10.9 Reaction of **16** with ethylene glycol in different solvents.

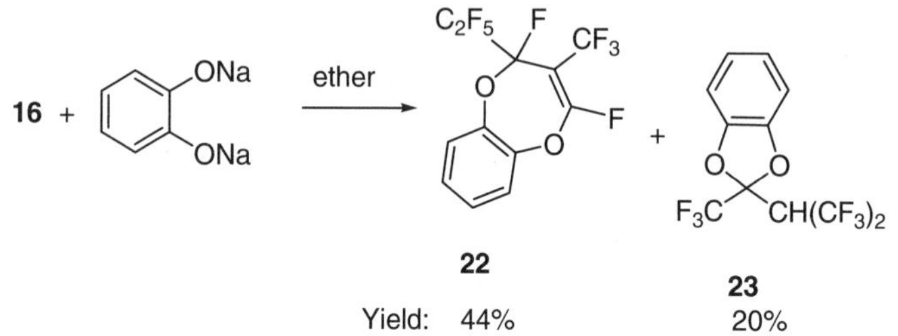

FIGURE 10.10 Reaction of **16** with catechol.

The acidic hydrolysis of a CF_2 group in **21** proceeds selectively, leading to the formation of unsaturated lactone.[22]

Reaction of **16** with the disodium salt of catechol leads to dioxepine **22** as the major product, when it is carried out in ether solvent (Fig. 10.10); however, when a $CH_3CN/N(C_2H_5)_3$ system is used, the yield of dioxolane **23** is significantly higher (62%).[23]

Reaction of olefin **24** (trimer of *F*-propene; for synthesis, see Refs [19] and [20]) with catechol in DMF affords a mixture of two isomeric benzoxazepines **25** and **26** in low to moderate yield (Fig. 10.11).[23]

Similar chemistry can be used for the construction of fluorinated heterocycles containing eight- and nine-membered rings.[24] For example, 1,5-benzodioxocin

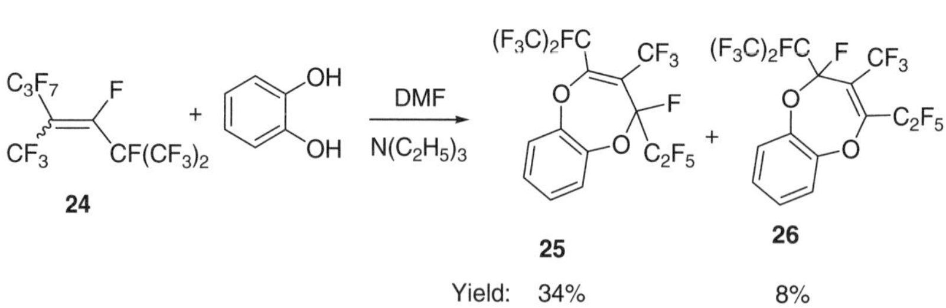

FIGURE 10.11 Reaction of **24** with catechol.

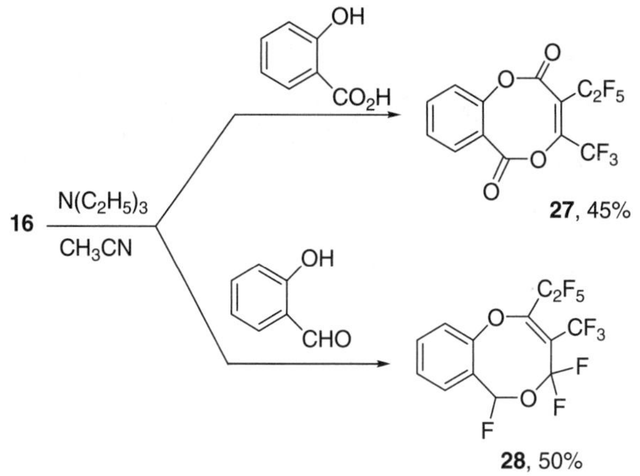

FIGURE 10.12 Construction of large oxygen-containing heterocycles using reaction of **16** with salicylic acid and salicylaldehyde.

derivatives **27** and **28** were prepared by reaction of compound **16** with salicylic acid or salicylaldehyde in the presence of triethylamine (Fig. 10.12).

Fluorinated nine-membered heterocycles **29** and **30** were synthesized by reacting **16** with phtalyl- and *o*-hydroxyphenethyl alcohols in the presence of base (Fig. 10.13).[24]

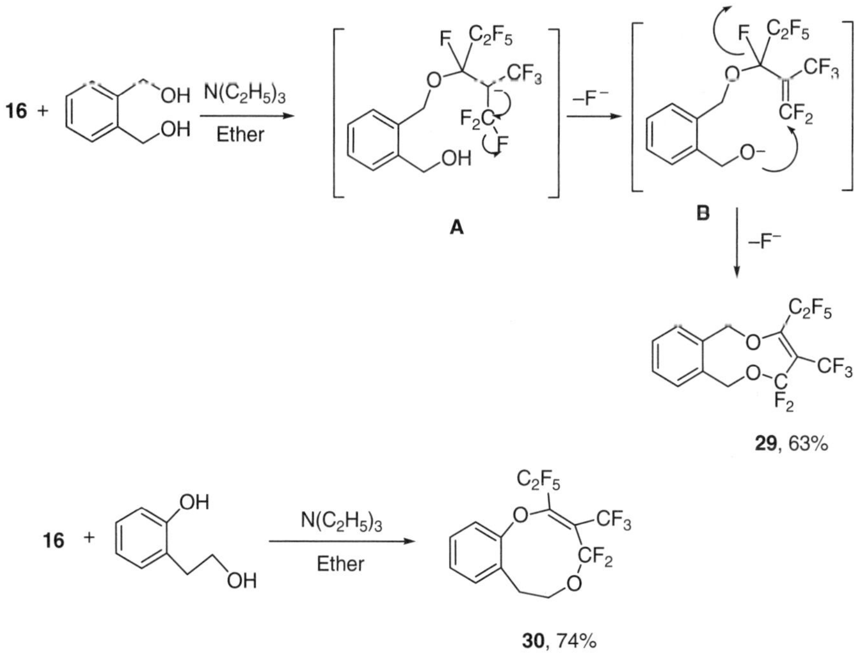

FIGURE 10.13 Synthesis of large oxygen-containing heterocycles using reaction of **16** with bifunctional nucleophiles.

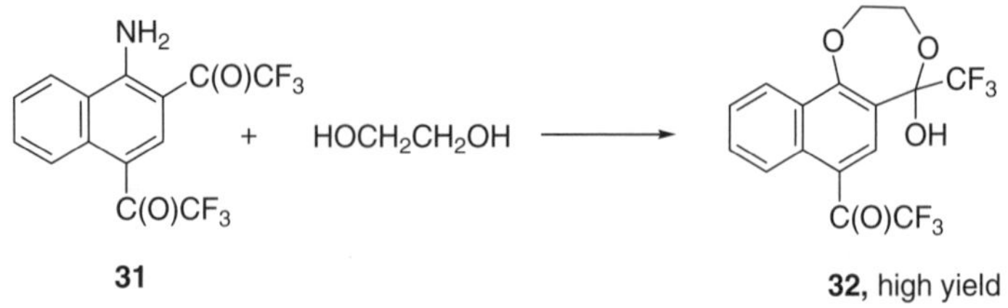

FIGURE 10.14 Preparation of naphthodioxepine **32**.

It should be pointed out that all reactions involving unsaturated fluorinated substrates and bifunctional nucleophiles rely on the attack of nucleophile on the positively charged carbon of the C=C bond (in the case of **16**, on CF=C) leading to the formation of anion **A**. The elimination of fluoride anion results in an intermediate **B** containing a highly electrophilic terminal $CF_2{=}C(CF_3)$– fragment. The final step involves cyclization through an intramolecular nucleophilic attack of an alkoxy anion with simultaneous migration of the double bond resulting in the formation of seven-membered ring (Fig. 10.13).

Naphtho-1,4-dioxepine **32** was prepared using classical cyclocondensation approach.[25] The reaction of 1-naphthylamine **31** with ethyleneglycol resulted in the high yield formation of **32** (Fig. 10.14).

10.2.3 Fluorinated Heterocycles Containing Oxygen and Nitrogen Atoms

The synthesis of perfluorinated 1,2-oxazepines **33** and **34** was carried out through unusual cyclocondensation process. The reaction of hydroxylamine with excess of **16** resulted in the formation of 1:2 adduct—1,2-oxazepine **33** in 50% yield, along with smaller amount of by-product **34** (derived from **33** as the result of HF addition across the C=C bond (Fig. 10.15), while the expected 5-fluoro-3-(pentafluoroethyl)-4-trifluoromethylisoxazole is formed in this reaction only as a minor by-product.[26]

Several 2,2-bis(trifluoromethyl)-1,3-oxazepines were prepared using an interesting reaction of 2,2-bis(trifluoromethyl)-1,3-oxazolidin-5-ones (**35**) and 1-diethylaminoprop-1-yne.[17] This reaction results in a high-yield formation of the corresponding 2-bis(trifluoromethyl) 1,3-oxazepin-5-ones **36** (Fig. 10.16).

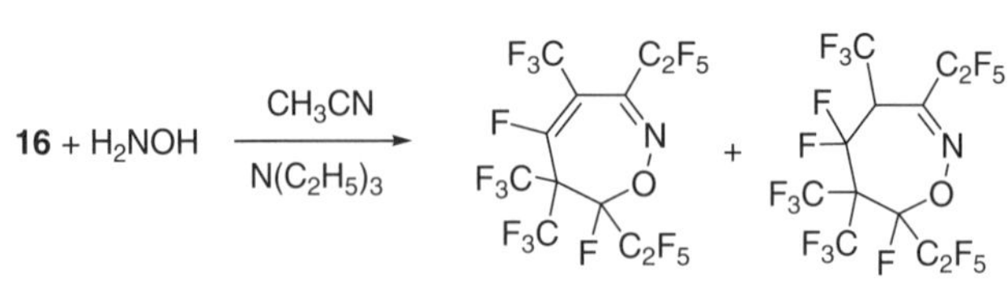

FIGURE 10.15 Synthesis of 1,2-oxazepines.

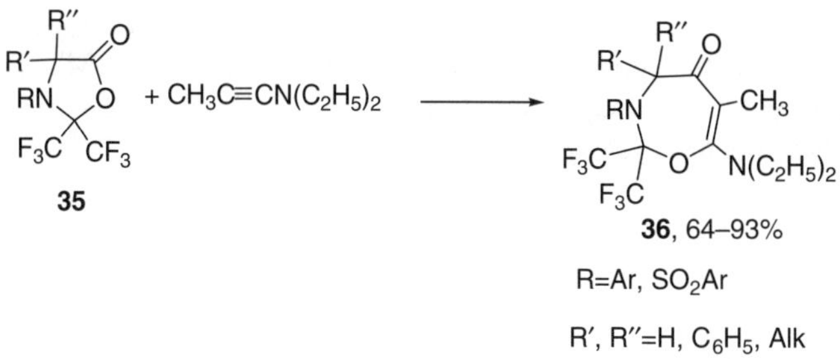

FIGURE 10.16 Synthesis of 2-bis(trifluoromethyl) 1,3-oxazepin-5-ones **36**.

1,4-Oxazepines containing fluoroalkyl substituents are often prepared using a "fluorinated building block" approach. For example, a considerable number of 1,4-oxazepines **37** (an effective inhibitors of γ-secretase) are synthesized through the condensation of a heterocyclic amine with fluorinated malonamides.[27]

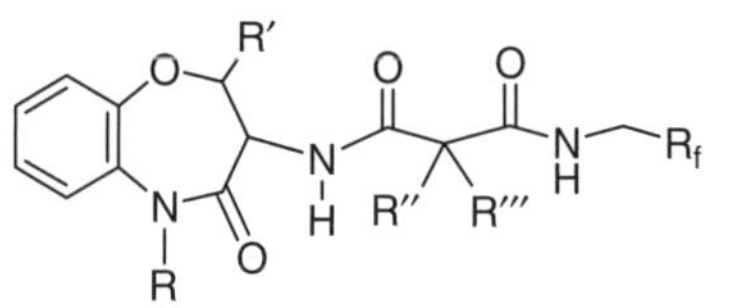

37 R_f=C_2F_5, CF_3CH_2

Trifluoromethylated 1,4-oxazepines **38** and **39** (analogues of Sustiva®, potent HIV reverse transcriptase inhibitor) were prepared by cyclocondensation of the corresponding aromatic aminoalcohols with acyl bromides RCHBrC(O)Br (Fig. 10.17).[28–31]

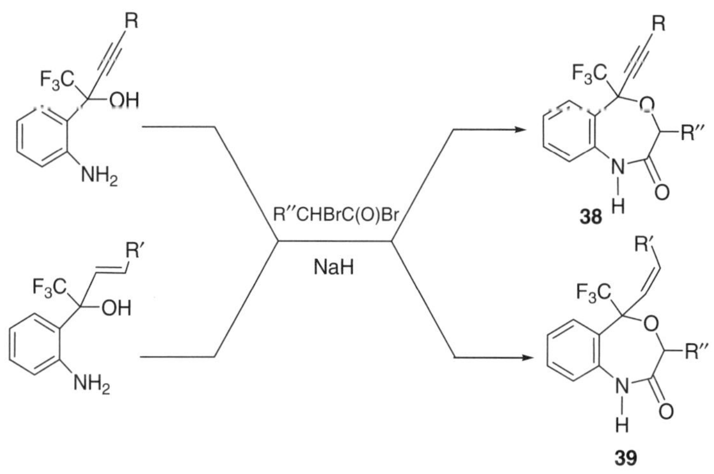

FIGURE 10.17 Preparation of 1,4-oxazepines **38** and **39**.

Several other methods have been used for the preparation of 1,4-dioxazepanes bearing trifluoromethyl groups. For instance, an intramolecular Michael addition was utilized by Abarbri and coworkers for stereoselective synthesis of seven- and eight-membered heterocycles **40** and **41**.[32]

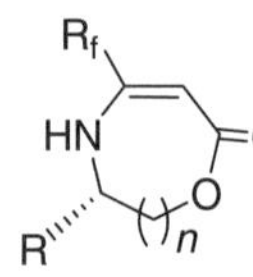

74–87%

R=alkyl, $CH_3S(CH_2)_n$

40, n=1, R_f=CF_3, C_2F_5
41, n=2, R_f=CF_3

The reaction of ethanolamines with **16** has significantly higher selectivity for the formation of seven-membered heterocycles compared to the reactions of ethyleneglycol and catechol. In this reaction, 1,4-oxazepines **42** form as a major products, along with small amount of **43**, resulting from HF addition across the double bond of **42** (Fig. 10.18).[26]

Furin and coworkers recently demonstrated that *F*-5-azanonene-4 (**44**; for preparation, see Ref. 33) is a convenient synthon for the preparation of 1,3,5-oxadiazepines **45** (Fig. 10.19).[26]

Mechanistically, this process is similar to the cyclization processess involving fluoroolefins **16** and **24**. The reaction involves the attack of nitrogen on the electron-deficient C=N double bond of **44** with the migration of the double bond, followed by a ring closure step via intramolecular nucleophilic displacement of fluorine in the CF=N- unit by alkoxy anion. The formation of **45** containing two C=N double bonds is the result of HF abstraction by triethylamine (Fig. 10.19).

The reaction of **16** with equimolar amount of *o*-aminophenols **46** produces the corresponding 1,4-oxazepines **47** (single isomer in each case) in moderate yield (Fig. 10.20).[23]

Although a similar process involving **24** is believed to proceed through an intermediate **48**, the reaction does not stop at this stage and leads to the formation

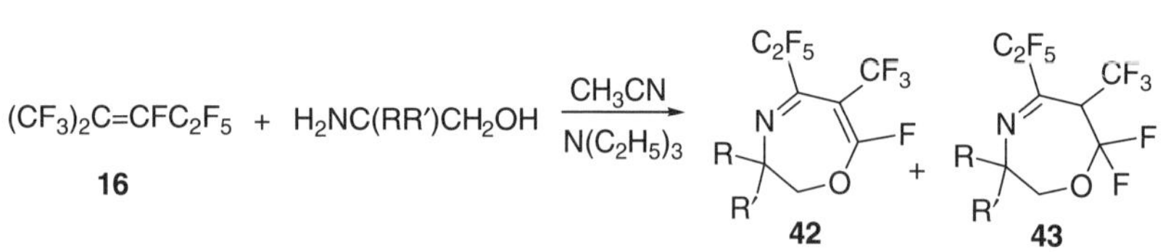

Yields: 39–70% 15–51%

R, R′=H,H; H,C_2H_5; CH_3,CH_3

FIGURE 10.18 Reaction of **16** with ethanolamines.

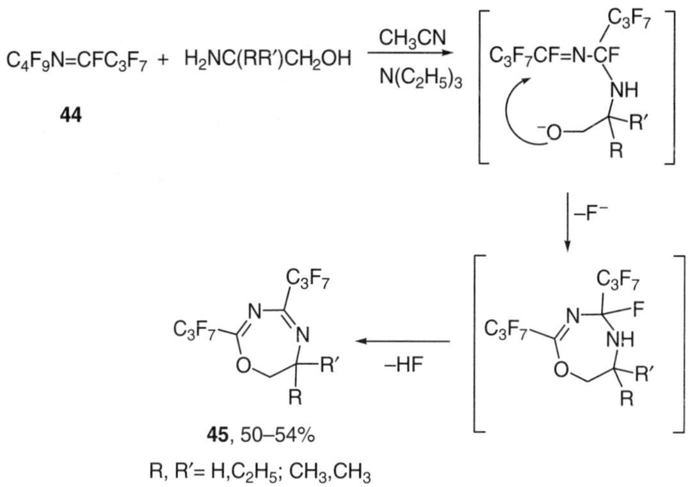

FIGURE 10.19 Use of *F*-5-azanonene-4 (**44**) for the synthesis of 1,3,5-oxadiazepines **45**.

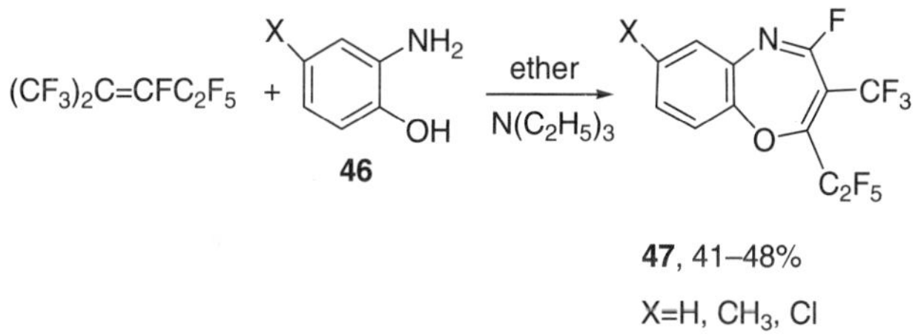

FIGURE 10.20 Reaction of **16** with *o*-aminophenols **46**.

of bicyclic oxazepines **49** through the reaction of second mole aminophenol with fluorinated C=C of intermediate **48** (Fig. 10.21).[23]

The reaction of **16** with two moles of *o*-aminobenzyl alcohol gave **50** in excellent yield (Fig. 10.22).[24]

The reaction of vinyl ketone **51** with aminophenols under microwave irradiation was shown to be an effective way for synthesizing of 1,4-oxazepin-7-ols **52** (Fig. 10.23).[34]

An interesting concept for the preparation of seven-membered heterocycles with several heteroatoms was reported in 1988 by Burger and coworkers. Base-catalyzed reaction of hexafluoroacetone imine derivatives **53** with epichlorohydrine resulted in the formation of 1,5-dioxa-3-azepines **54** and 1-oxa-3,5-diazepines **55** (Fig. 10.24).[35]

More sophisticated starting material was used in the preparation of benzo- and naphtho-1,3,5-oxadiazapine-4-ones developed by the Ukrainian research group. The synthesis is based on the reaction of isocyanate **56** with *o*-aminophenols **46**, leading to compounds **57** in 54–92% yield[36,37]. This approach is exemplified by reaction of 2-amino-1-naphtol and **56** (Fig. 10.25).

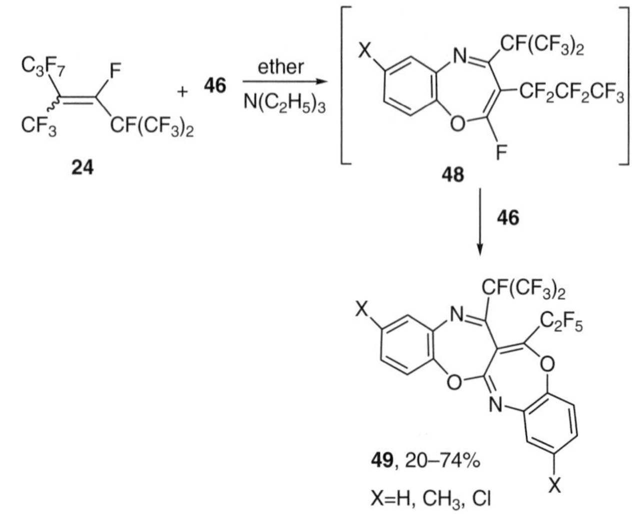

FIGURE 10.21 Preparation of bicyclic oxazepines **49** using **24**.

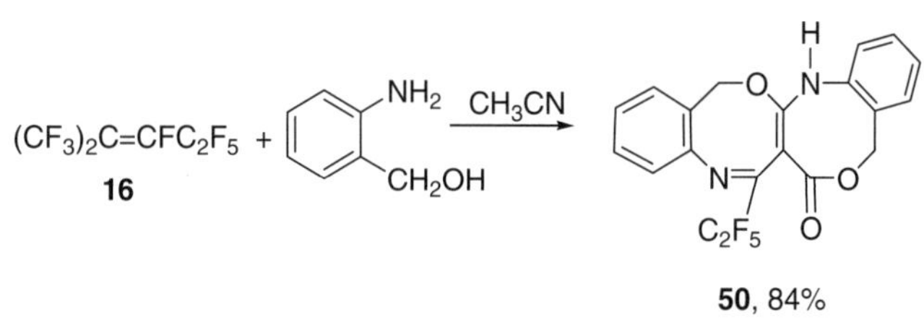

FIGURE 10.22 Reaction of **16** with *o*-aminobenzyl alcohol.

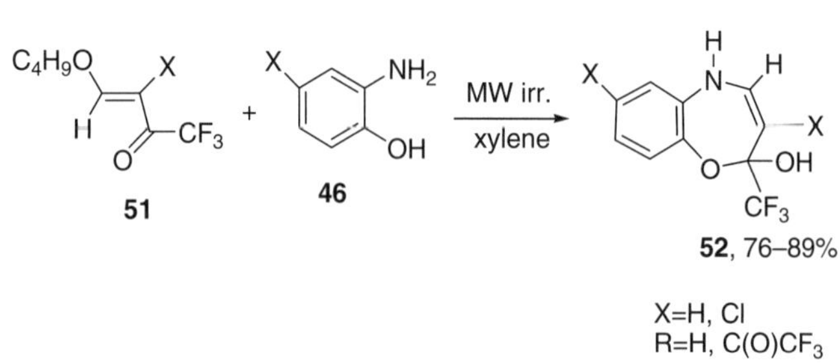

FIGURE 10.23 Microwave-assisted synthesis of fluorinated 1,4-oxazepine-7-ols.

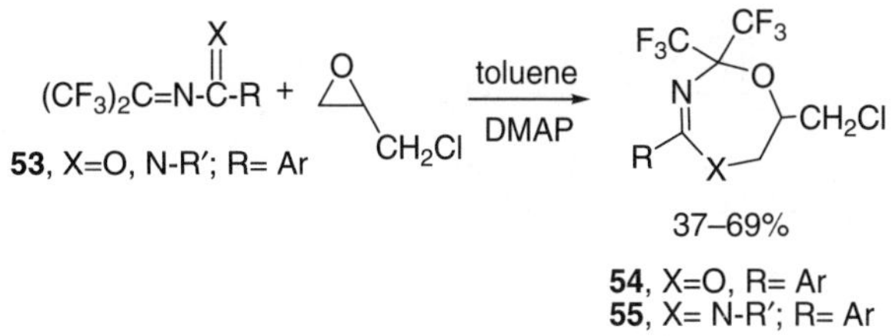

FIGURE 10.24 Synthesis of 2,2-bis(trifluoromethyl)-1,5-dioxa-3-azepines **54** and 1-oxa-3,5-diazepines **55**.

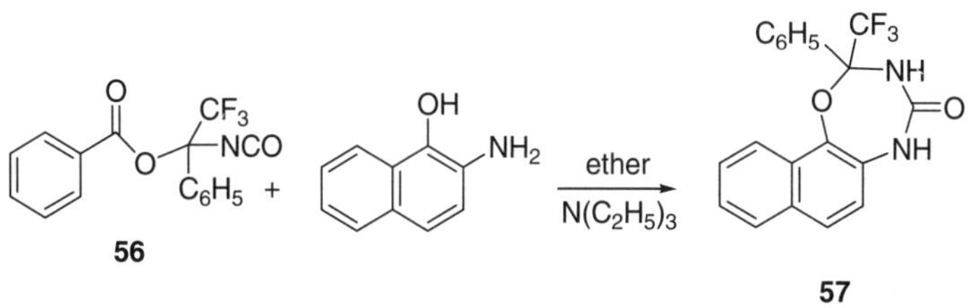

FIGURE 10.25 2-Aminonaphthol-based synthesis of **57**.

10.3 FLUORINATED NITROGEN-CONTAINING HETEROCYCLES

10.3.1 Fluorinated Nitrogen-Containing Heterocycles with One Heteroatom

Although the parent hexafluoro-1*H*-azepine is not known, several *N*-substituted hexafluoroazepines were reported. 1-Cyanohexafluoro-1*H*-azepine (**58**)[38] was prepared by thermal reaction of cyanoazide with hexafluorobenzene (Fig. 10.26).[39]

In sharp contrast to hydrocarbon analogue, which undergoes spontaneous dimerization at an ambient temperature, compound **58** is a stable solid with a melting

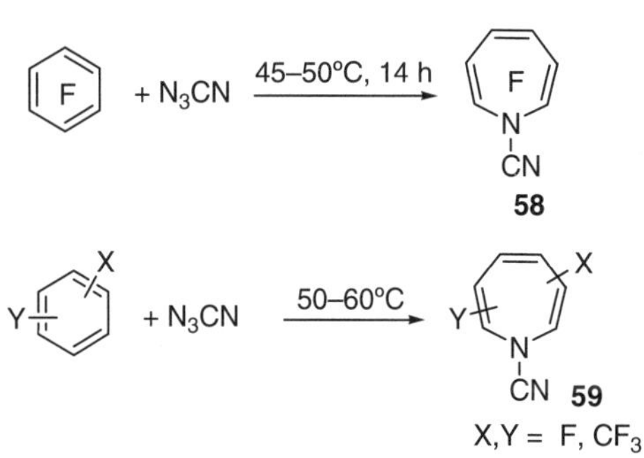

FIGURE 10.26 Preparation of fluorinated 1-cyano-1*H*-azepines **58** and **59**.

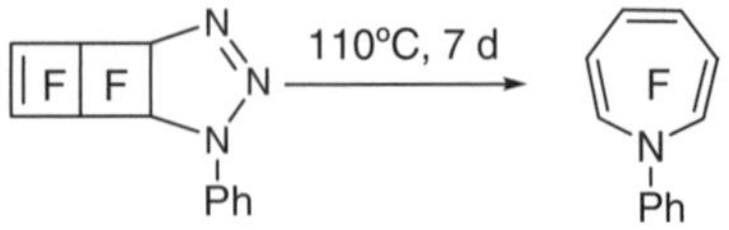

60, 93%

FIGURE 10.27 Synthesis of hexafluoro-1-phenyl-1*H*-azepine **60**.

point of 51–52°C.[38] Mono- and polyfluorinated azepines **59** (X = F or CF_3) were prepared using same method. They are more stable compared to their hydrocarbon counterparts, but undergo dimerization upon heating.[39] Among the other known hexafluoroazepines are 1-phenyl-1*H*-azepine (**60**), 1-carboethoxy-1*H*-azepine (**61**), and 1-amido-1*H*-azepine (**62**) prepared by the Haszeldine group in 1980s.[40,41] Since nitrene generated from phenylazide has a relatively low reactivity and does not react with C_6F_6, compound **60** was prepared by thermal isomerization of the cycloadduct of $C_6H_5N_3$ and hexafluoro-Dewar-benzene (Fig. 10.27).[40]

Compound **61** was synthesized through thermal reaction of C_6F_6 and ethylazidoformate, however, the azepine **62** was made by acidic hydrolysis of the nitrile group of **58** (Fig. 10.28).[41]

It should be pointed out that the formation of azepine in reaction with ethylazidoformate was observed only in case of hexafluorobenzene and the same reaction of fluorinated naphtalenes led to the exclusive formation of the corresponding aziridines as the result of [1 + 2] cycloaddition of the EtOC(O)N: to fluorinated C=C of naphthalene.[41] Azepines **60–62** undergo clean, high-yield photochemical isomerization with formation of the corresponding hexafluoro-2-azabicyclo[3.2.0]hepta-3,6-dienes.[41]

"Lightly" fluorinated seven- and eight-membered rings containing one nitrogen atom are usually prepared either by selective introduction of fluorine into the hydrocarbon ring or by using standard techniques of heterocycles synthesis. For example, 2-azepinone **63** was prepared in moderate yield through selective electrochemical fluorination of the parent hydrocarbon in the presence of $(C_2H_5)_3N{\cdot}3HF$.[42] Lactam **64** was synthesized in 52% yield by base-catalyzed reaction of caprolactam

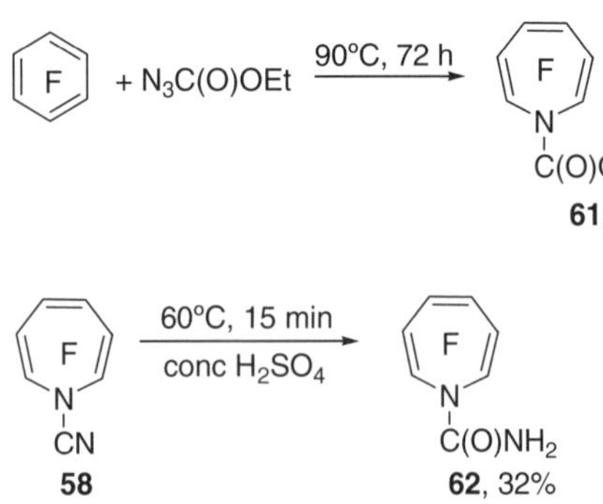

61, 21%

58

62, 32%

FIGURE 10.28 Synthesis of hexafluoroazepines **61** and **62**.

and **16**,[43] and compound **65** was isolated in low yield in the reaction of EtO(O)C $(CF_2)_4CN$ with hydrogen in the presence of H_2SO_4 and PtO catalyst.[44]

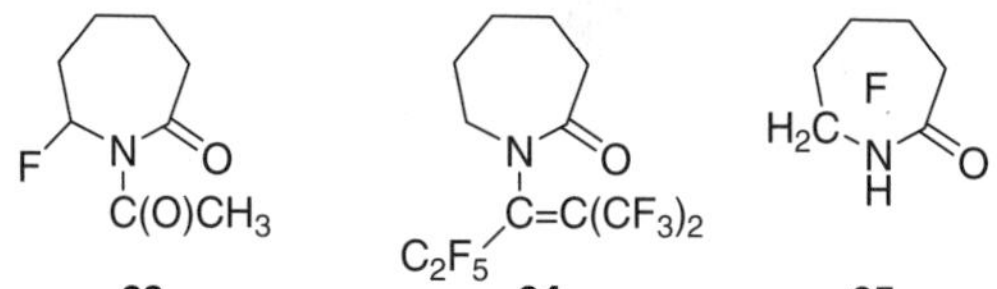

63 **64** **65**

Introduction of CF_2 group into the cyclic system can be carried out through the replacement of carbonyl group, using traditional fluorinated reagents, such as SF_4 or aminosulfuranes. Compounds **66** and **67** were prepared in 6.5–78% yield by treatment of the corresponding carbonyl-containing precursors with $(CH_3OCH_2CH_2)_2NSF_3$ in the presence of boron trifluoride catalyst.[45]

It should be pointed out that this reaction has high selectivity and only more reactive keto group is involved in the reaction with fluorinating reagent. Carbonyl-containing starting materials for these synthesis were prepared using quite popular RCM process followed by the oxidation of C=C functionality into carbonyl group.[45]

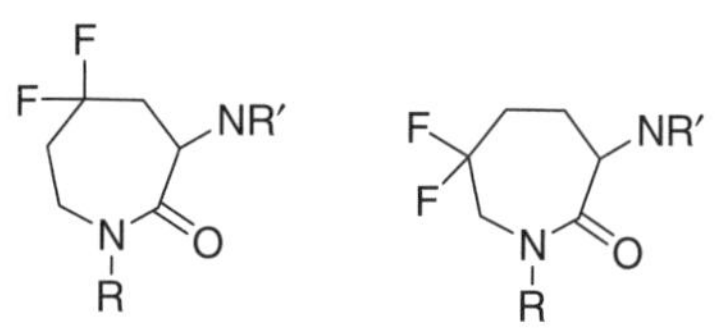

66 **67**

6.5 78%

R= ^{t}BuOC(O)–, 4-$ClC_6H_4SO_2$
R′=2,4$(CH_3O)C_6H_3CH_2$–

RCM was also successfully applied for the synthesis of wide range of nitrogen-containing heterocycles. Pioneered by Dixneuf and Osipov,[46] this approach was first applied for the synthesis of azepines **68** (Fig. 10.29). RCM reaction of 4-azanona-dienes-1,8 catalyzed by Grubs-type ruthenium catalysts results in high-yield formation of the corresponding azepines.

The position of the double bond in heterocycle can be "tuned" by right choice of starting materials. For example, isomeric azepines **69** were prepared in excellent yields from the corresponding 5-azanonadienes-1,8 (Fig. 10.29).[46]

Recently, it was shown that RCM approach, which was widely applied for the synthesis of ring-fluorinated cycloalkenes,[7,47] can be successfully used for the preparation of difluoroazepines **70** and **71** (Fig. 10.30).[7,48–50]

Review article[7] provides an overview of the synthesis of fluorinated nitrogen- and oxygen-containing heterocycles using RCM methodology.

Interesting variation of RCM process involving the double bond with vinylic fluorine substituent was recently used for the synthesis of 4-R-6-fluoro-2,3,4,7-tetrahydro-1*H*-azepines (**72**, R = H, CH_3, and Ph), prepared in 72–94% yield through the cyclization of the corresponding 2-fluoro-4-azanonadienes-1,8 (Fig. 10.31).[51]

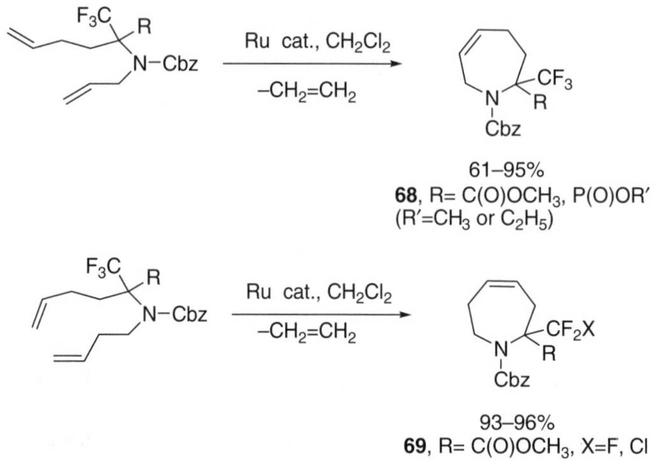

FIGURE 10.29 RCM synthesis of fluoroalkyl azepines.

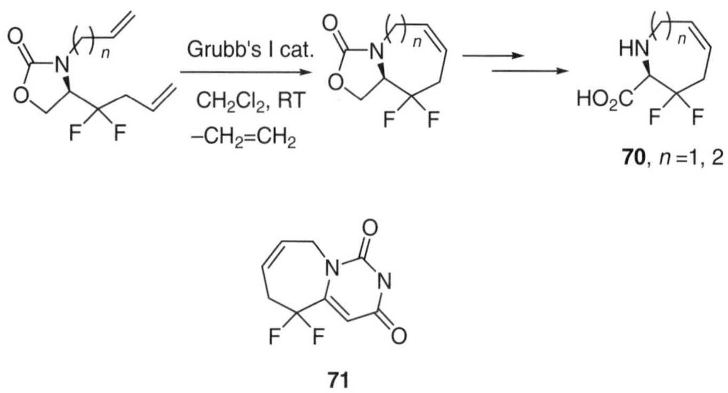

FIGURE 10.30 RCM synthesis of ring-fluorinated azepines **70** and **71**.

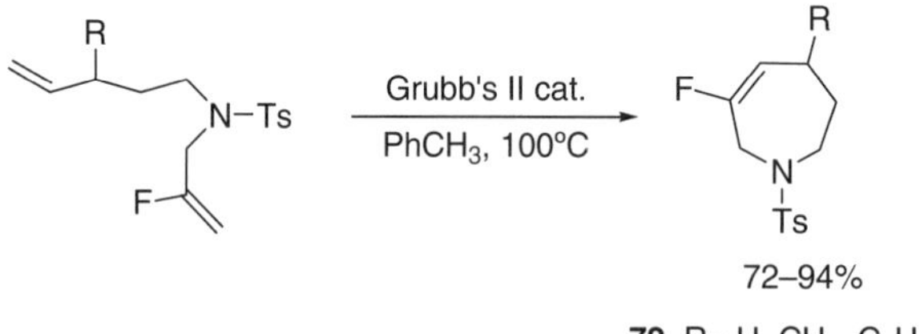

FIGURE 10.31 RCM synthesis of **72** involving fluorinated double bond.

Despite the fact that the reaction works well for the preparation of CF_3-containing five- and six-membered heterocycles, the attempt to extend this process to the preparation of CF_3-containing azepines so far was not successful.[51]

10.3.2 Fluorinated Nitrogen-Containing Heterocycles with Two or More Heteroatoms

1,2-Diazepines can also be prepared using RCM methodology. 3-Fluoro-1,2-diazepines **73** was synthesized in 64% yield through the ring closure of the corresponding unsaturated hydrazine (Fig. 10.32).[52]

Alternative synthesis of 1,2-diazepines developed by Burger and coworkers is based on retro [3 + 2] process.[53] The pyrolysis of bicyclic compounds **74** leads to high-yield formation of 5,5-bis(trifluoromethyl)-4,5-dihydro-1*H*-1,2-diazepines (**75**) through the extrusion of $CH_2{=}C(CF_3)_2$ (Fig. 10.33).

Interestingly, the reaction of **74** ($R' = C(O)R$) proceeds with the formation of isomeric 4,5-dihydro-1*H*-1,2-diazepines **76**. The reaction of compound **75** ($R = CH_3$ and $R' = C_6H_5$) with $LiAlH_4$ proceeds with selective reduction of $-N{=}C(CH_3)$ group, yielding 4,5,6,7-tetrahydro-7-methyl-3-phenyl-5,5-bis(trifluoromethyl)-1*H*-1,2-diazepine (yield 70%), but hydrogenation using H_2/Pd/C system produces isomeric 4,5,6,7-tetrahydro-3-methyl-7-phenyl-5,5-bis(trifluoromethyl)-1*H*-1,2-diazepine as the result of selective reduction of $-N{=}C(C_6H_5)-$ moiety.[53]

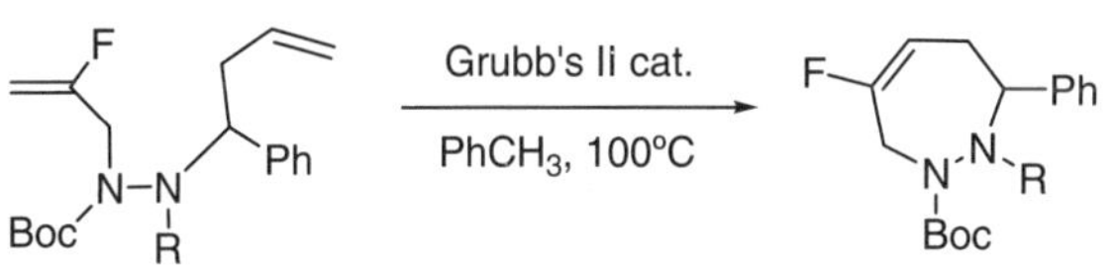

FIGURE 10.32 RCM synthesis of fluorinated 1,2-diazepines.

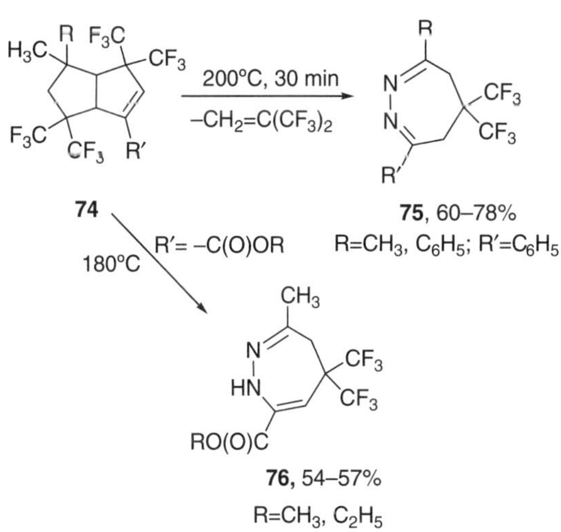

FIGURE 10.33 Synthesis of 5,5-bis(trifluoromethyl)-4,5-dihydro-1*H*-1,2-diazepines.

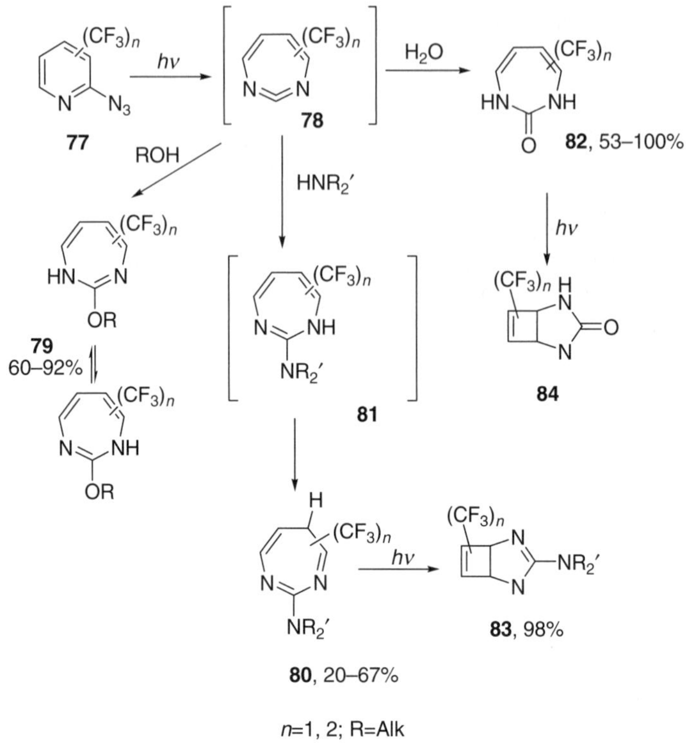

FIGURE 10.34 Photochemical synthesis of trifluoromethylated 1,3-diazepines.

Interesting synthetic approach to 1*H*- and 5*H*-1,3-diazepines based on the photolysis of 2-azido trifluoromethyl pyridines **77** was reported recently.[54,55] The photolysis of pyridines **77** (or tetrazolo[1,5-α]pyridines) results in the intermediate formation of 1,3-diazacyclohepta-1,2,4,6-tetraenes **78** (Fig. 10.34). The reaction of **78** with alcohols affords 2-alkoxy-1*H*-1,3-diazepines **79**, existing as mixture of *NH* tautomers, with free energies of activation for 1,3-proton exchange estimated to be 14–16 kcal/mol.[54] In the reaction of **77** with secondary amines, a thermodynamically stable 2-dialkylamino-5*H*-1,3-diazepines **80** form in most cases, however, kinetic isomer **81** (R = *i*-Pr; *n* = 1; 6-CF_3 isomer) was isolated and fully characterized.[54] The photolysis of **77** in the presence of water leads to the formation of the corresponding 1,3-diazapenones **82** (Fig. 10.34).

A number of chemical transformations were reported for trifluoromethylated 1,3-diazepines. The list includes acylation and catalytic hydrogenation of **79**, photochemical isomerization of **80** and **82** into 3-substituted 2,4-diazabicyclohepta-dienes (**83**) and 2,4-diazabicyclo[3.2.0]hepten-3-ones (**84**), respectively. Detailed description of chemical transformations, NMR, and single-crystal X-ray data can be found in the original article.[54]

The synthesis of 1,3,5-triazepines developed by the Burger group is based on the sequence involving cycloaddition of 2*H*-azirine (**85**) to the corresponding derivatives

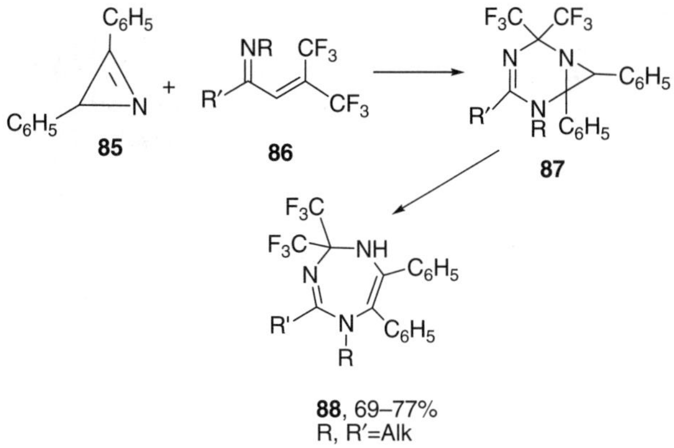

FIGURE 10.35 Hexafluoroacetone imine-based synthesis of 1,3,-triazepines.

of imine hexafluoroacetone (**86**) and acid-catalyzed ring opening of aziridine **87**, resulting in high-yield formation of 1,3,5-triazepines **88** (Fig. 10.35).[56]

The formation of nine-membered heterocycle **90** observed in the reaction of **16** and hexamethylenediamine was explained as being the result of the intramolecular cyclization of initially formed dihydroazete **89** (Fig. 10.36).[57]

1,4-Diazapines is probably the most studied group among the fluorinated seven-membered heterocycles. Although some of these materials can be prepared by the introduction of fluorinated group (e.g., CF_3CH_2–[11]) into already constructed heterocycle, the vast majority of these materials was made by the reaction of fluorinated building blocks with aliphatic or aromatic 1,4-diamines. The list of fluorinated materials used for the preparation of fluorinated 1,4-diazepines includes fluoroolefins,

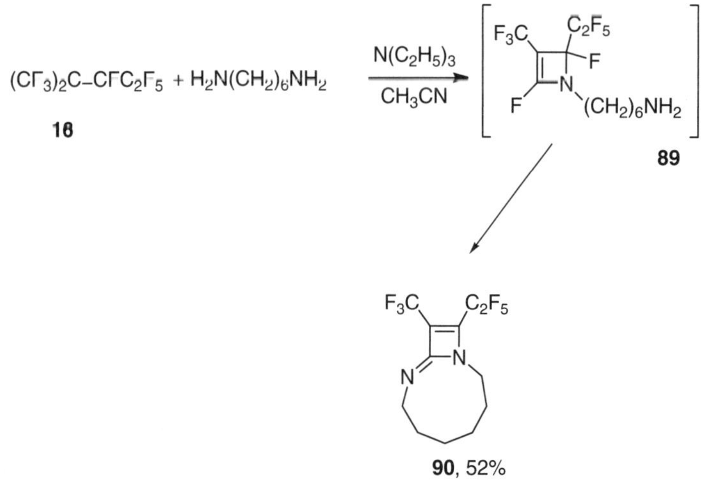

FIGURE 10.36 Synthesis of 1,3-diazepine **90**.

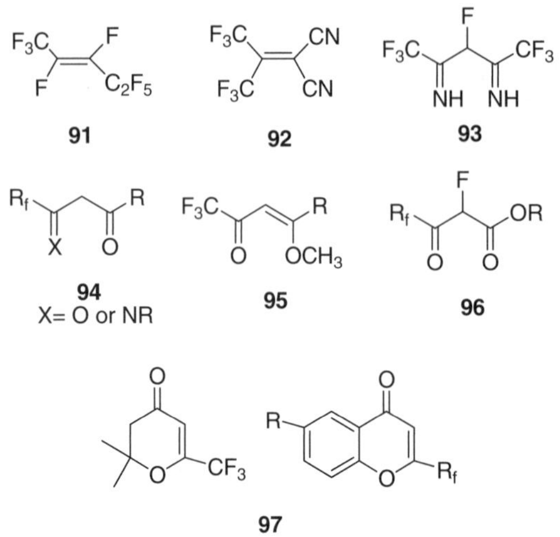

FIGURE 10.37 Major types of fluorinated synthons used in the synthesis of 1,4-diazepines.

such as **16** and **24**; readily available *F*-pentene-2 (**91**; for preparation, see Refs 58 and 59; for use in the synthesis of heterocycles, see Ref. 60); dinitrile **92** (for preparation, see Ref. 61; for use as synthon, see Refs 62 and 63); fluorinated imine **93** (for preparation, see Refs 64 and 65; for synthetic application, see Ref. 60); imidoyl fluoride **44** (for use as synthon, see Ref. 26), fluorinated β-ketones and β-iminoketones **94**[66,67]; vinyl ketones **51** (for preparation, see Ref. 68; for application in synthesis, see Refs 34 and 69); fluorinated vinyl ketones **95**; 2-fluoromalonic acid derivatives **96**[70,71]; and 2-fluoroalkyl chromones **97** (Fig. 10.37).[72,73]

Several examples demonstrating the synthesis of fluorinated 1,4-diazepines are given below. The reaction of **16** with ethylenediamine or *o*-phenylenediamine results in the formation of 1,4-diazepine **99** and **100**, but the reaction of **24** with *o*-phenylenediamines **101** under similar conditions leads to the tetracyclic compounds **102** (Fig. 10.38).[23]

Compounds **91** or **93**[60,74,75], fluorinated β-diketones, and nitrogen derivatives **94**[76] were employed in the preparation of fluorinated 1,4-diazepines **103**–**105** through the reaction with ethylenediamine **(98)** or its perchlorate complex (Fig. 10.39).

Fluorine substituent in compound **103** can be replaced by hydrogen upon treatment with more basic diethylenetriamine.[75]

The reaction of vinyl ketones **51** with **101** or **98** under microwave irradiation is a simple and effective way of construction of 1,4-diazepines (Fig. 10.40).[69]

The reaction of 1,3-diaminopyridine with **95** leads to the corresponding 3*H*-pyrido [2,3-*b*][1,4]diazepine-4-ols **107** in good yield (Fig. 10.41).[77]

Interestingly, the trichloromethyl analogue of **95** in this reaction gives only 3*H*-pyrido[2,3-*b*][1,4]diazepine-4(5*H*)-ones as the result of the hydrolysis of $-CCl_3$ group under reaction conditions.[77]

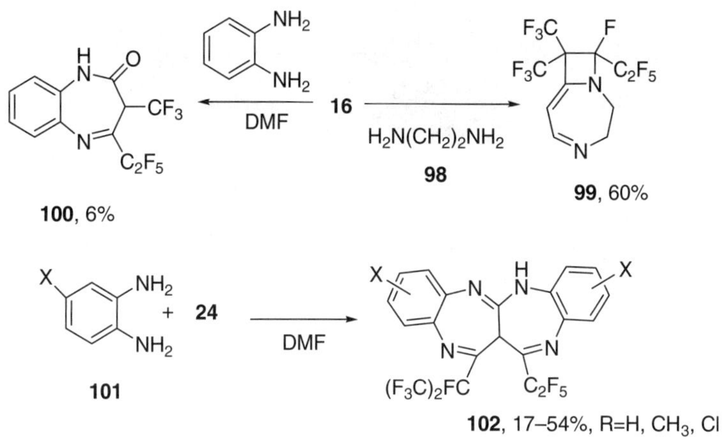

FIGURE 10.38 Synthesis of fluorinated heterocycles using *F*-propene oligomers **16** and **24**.

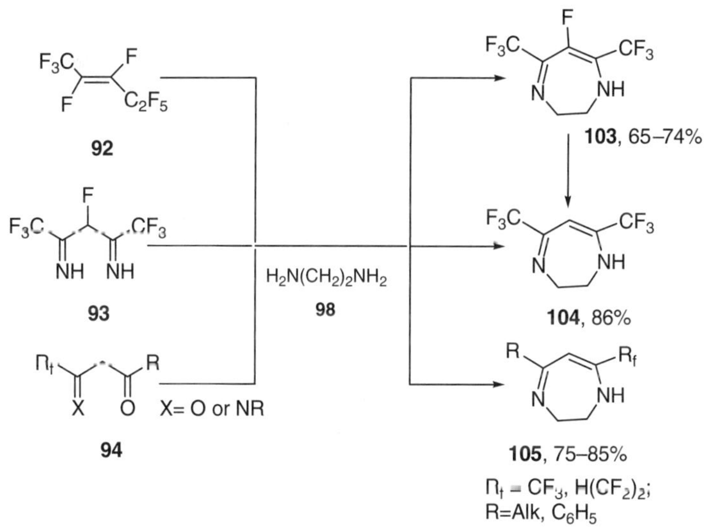

FIGURE 10.39 Synthesis of fluorinated heterocycles using ethylenediamine.

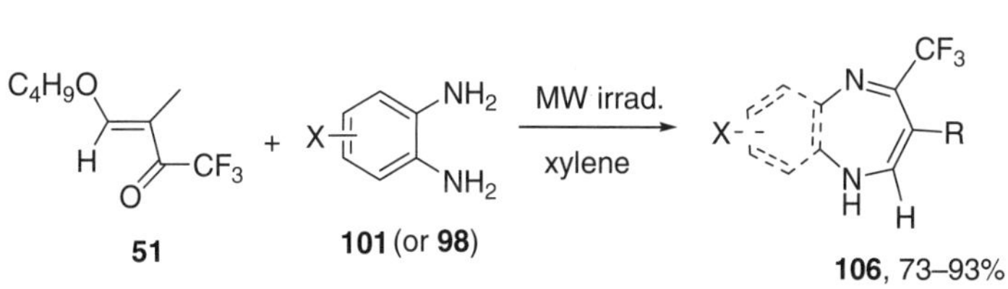

FIGURE 10.40 Microwave-assisted synthesis of fluorinated 1,4-diazepines.

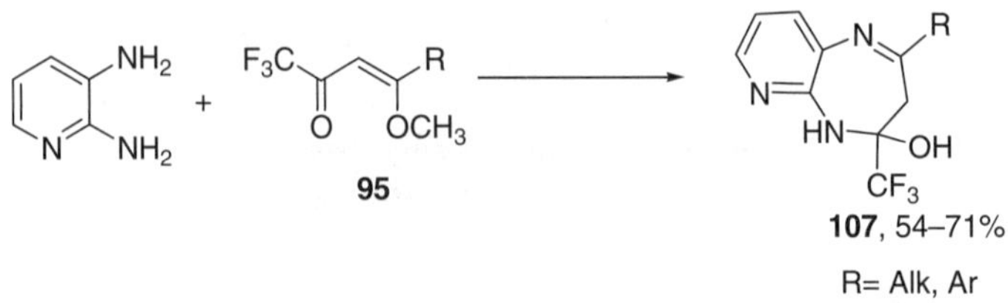

FIGURE 10.41 Synthesis of 4-trifluoromethyl-4,5-3*H*-pyrido[2,3-*b*][1,4]diazepine-4-ols.

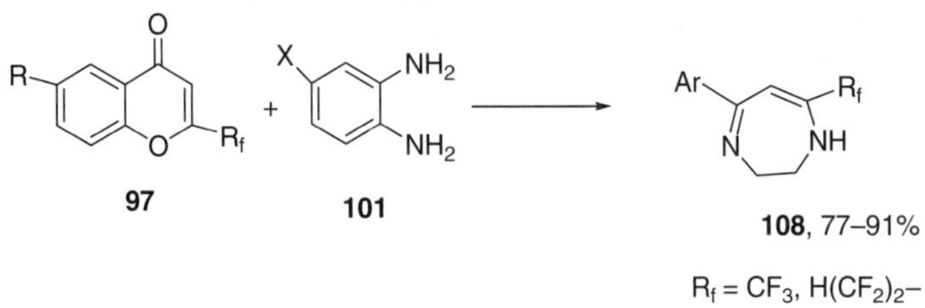

FIGURE 10.42 Chromones-based synthesis of fluorinated 1,4-diazepines.

The application of fluorinated chromones in the synthesis of 1,4-diazepines is exemplified by reaction of phenyl chromones **97** (Fig. 10.42).[72,73]

10.4 SEVEN-MEMBERED AND LARGER RING-FLUORINATED SULFUR-CONTAINING HETEROCYCLES

This group is the smallest and the least studied among fluorinated heterocycles. Parent hexafluoro-thiepine was not prepared so far, but several unsaturated eight-membered heterocycles are known. For example, compound **109** was obtained by the dimerization of the corresponding dithietene (Fig. 10.43).[78] The dimerization proceeds slowly at ambient temperature, but occurs much faster in the presence of

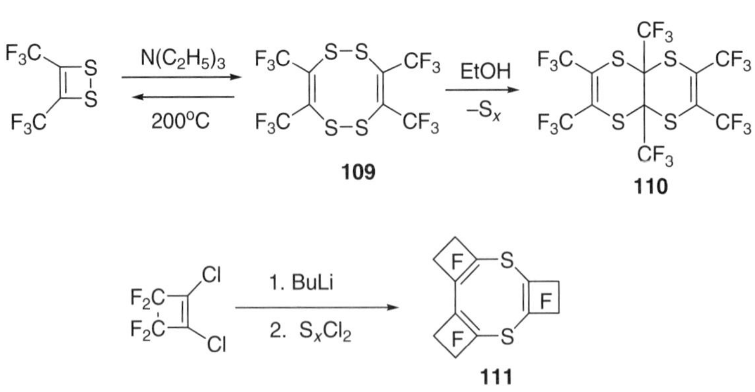

FIGURE 10.43 Synthesis of fluorinated thiocines.

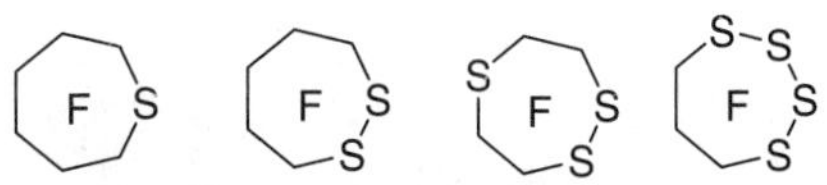

112 **113** **114** **115**

FIGURE 10.44 Perfluorinated cyclic sulfides.

catalytic amount of base. It is reversible and **109** can be converted back into dithietene upon heating.[78] The treatment of **109** with ethanol leads to formation of thermodynamic product **110**.

1,4-Dithiocin **111** was isolated in low yield along with other products in the reaction of 1,2-dichloro-*F*-cyclobutene with BuLi and sulfur chlorides.[79]

F-Thiepan (**112**) and *F*-1,2-dithiepan (**113**) were isolated from the mixture of products formed in the pyrolysis of the mixture of sulfur and tetrafluoroethylene/$F_2C{=}S$ copolymer.[80] *F*-1,2,5-Tri- (**114**) and *F*-1,2,3,4-tetra- (**115**) thiepanes were isolated in the reaction of TFE with sulfur at 300°C (Fig. 10.44).[81]

It should be pointed out that synthesis of perfluorinated sulfur(II)-containing heterocycles, requires a "custom" synthetic approach, since of these compounds cannot be prepared using oxidative methods, such as electrochemical fluorination or fluorination using elemental fluorine, due to ability of divalent sulfur to undergo oxidation.

Tetrasulfide **115** undergoes spontaneous polymerization at −40°C in acetonitrile, however, polymerization of more stable **114** requires stronger base such as triethylamine.[81] Pyrolysis of **114** leads to the formation of *F*-1,4-dithiane and sulfur.[81]

The reaction of **16** with 1,2-dithioglycol and 2-mercaptoethanol proceeds with predominant formation of five-membered heterocycles **116** and **118** along with fluorinated thiepines **117** and **119** (Fig. 10.45). However, both **117** and **119** were isolated by preparative GC, fully characterized and conformational behavior of these materials was studied using variable temperature ^{19}F NMR spectroscopy.[22]

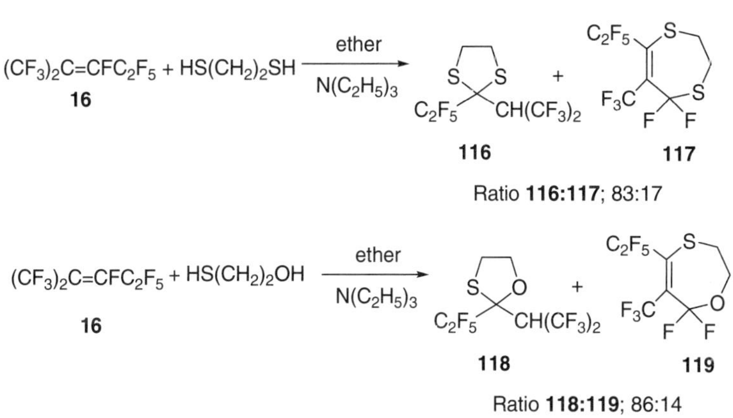

FIGURE 10.45 Synthesis of fluorinated thiepines using **16** as a building block.

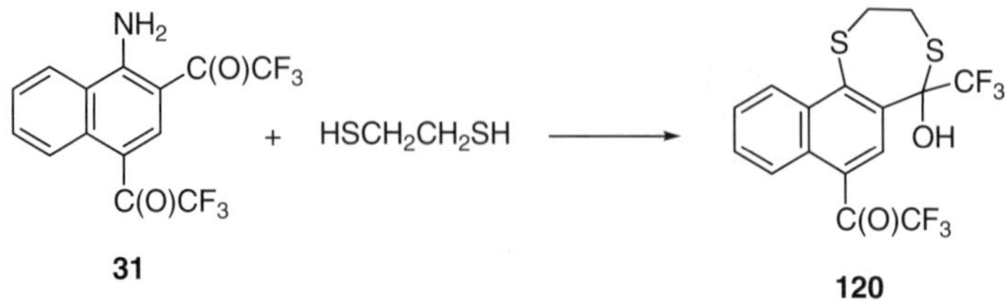

FIGURE 10.46 Preparation of trifluoromethylated 1,4-dithiepine **120**.

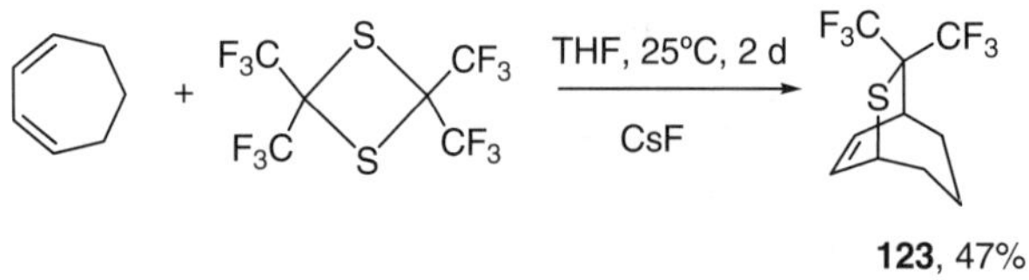

FIGURE 10.47 Synthesis of 2,2-bistrifluoromethyl thiepine **123**.

It should be pointed out that this cyclization process is extremely sensitive to the polarity of the solvent. For example, exclusive formation of **118** was observed when more polar acetonitrile was used as solvent.[22]

The reaction of **31** and 1,2-dithioglycol is highly selective affording 1,4-dithiepine **120** in excellent yield (Fig. 10.46).[25]

1,4-Oxathiepine **121** was obtained by the reaction of isocyanate **56** with thiophenol,[36] and compounds **122** was prepared by the reaction of **53** (X = S, see Fig. 10.42) and epichlorohydrin.[35]

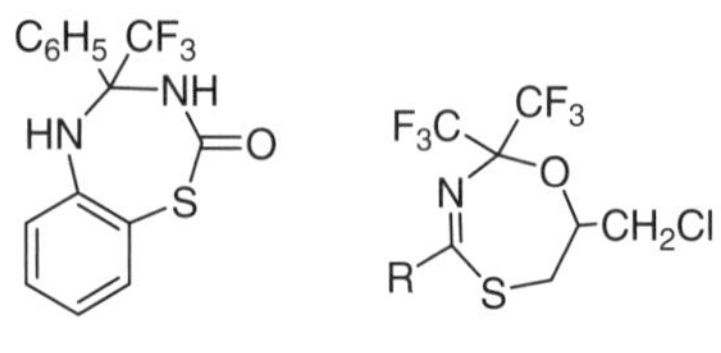

Bicyclic 2,2-bistrifluoromethyl thiepine **123** is synthesized in moderate yield by CsF-catalyzed reaction of cycloheptadiene-1,3 with the cyclic dimer of hexafluorothioacetone (Fig. 10.47).[82]

10.5 PERFLUORINATED LARGE-RING HETEROCYCLES

This class is relatively large and includes mostly oxygen- and nitrogen-containing heterocycles. Most of these materials were synthesized by exhaustive fluorination of the corresponding hydrocarbon analogues using either electrochemical fluorination (ECF) or by fluorination using elemental fluorine.

F-Oxepan **124**, for example, was isolated in low yield (3.3%) from the reaction mixture derived from ECF of 2-mehtyloxalane.[83] Under similar conditions, fluorination of hexahydroazepine mostly gives *F*-methylpiperidines; however, *F*-azepan **125** was isolated in low yield (0.6%).[84] Significantly better yields can be achieved if fluorination is carried out in cryogenic reactor using diluted F_2 gas. Using this method Lin and Lagow prepared compounds **125** and **126** in 60 and 64% yield, respectively.[85]

Yields of perfluorinated azepines bearing alkyl substituent at nitrogen prepared by ECF are higher. For example, compounds **127** were synthesized by ECF in 20–30% yield from the corresponding hydrocarbon precursors.[86]

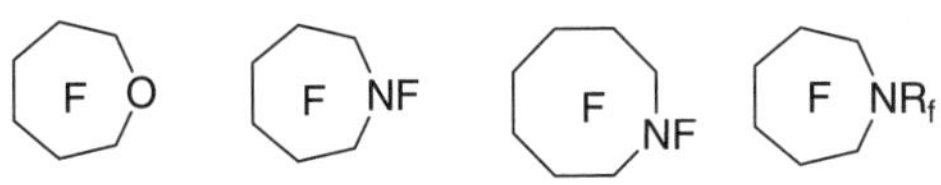

124 **125** **126** **127**

$R_f = C_2F_5$
CF_2CF_2Cl, C_3F_7
$CF_2CF_2CF_2Cl$
C_4F_9

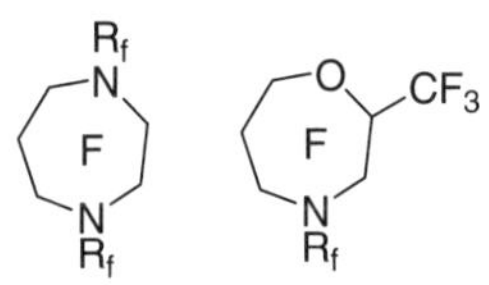

The synthesis of perfluorinated 1,4-diazepines[87] and 1,4-oxazepines[88] has been reported recently by Abe. Some reactions of compounds **127** [R_f = $-CF_2C(O)F$] involving transformation of functional group can be found in Ref. 84.

10.6 FLUORINATED MACROHETEROCYCLES

10.6.1 Polyfluorinated Macroheterocycles

Several different synthetic approaches have been used for the synthesis of fluorinated macrocycles. The simplest one relies on the introduction of a fluorinated substituent into hydrocarbon macrocycle. For example, the preparation of dibenzocrown ethers **128** is based on the reaction of the corresponding 4,4′- or 4,5′-diaminodibenzo-18-crown-6 ethers bearing amino groups on the phenyl ring with fluorinated lithium diketonate.[89]

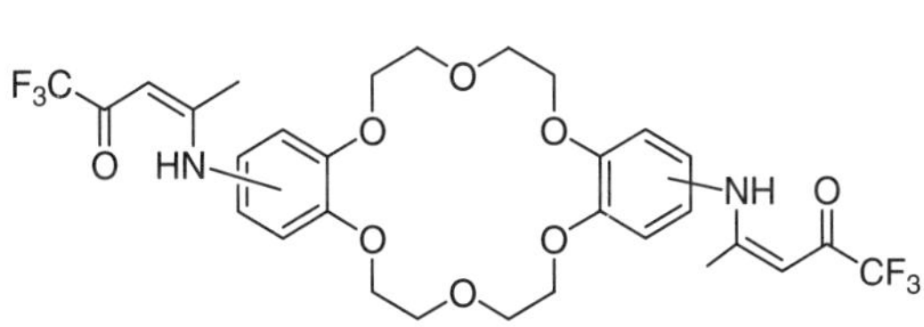

128

RCM as the final step of the synthesis was applied for the preparation of variety of fluorinated macrolactones. One of the most famous examples is the synthesis of Fludelone® (**129**)—the trifluoromethylated analogue of Epothilone®. The introduction of a CF_3 group significantly enhanced antitumor properties of this macrolide **129**.[90,91]

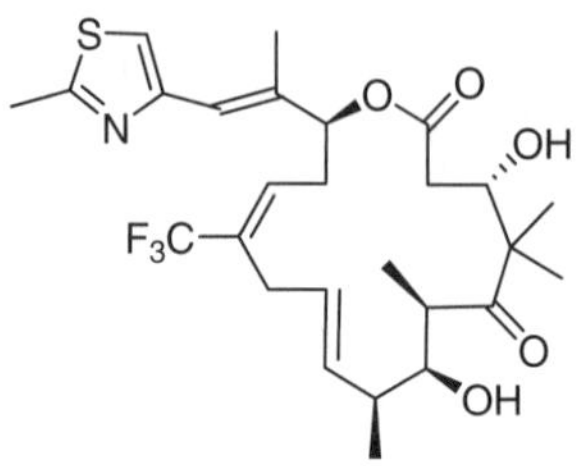

129, Fludelone

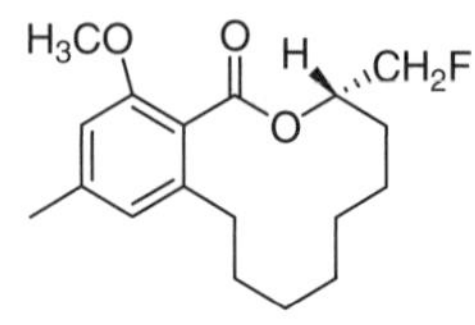

130

The Haufe group applied RCM to the preparation of both enantiomers of fluoromethyl macrolactone **130** ("Fluorolasiodiplodin"),[92] and recently Fustero and coworkers expanded this methodology for the preparation of a large number of macrocyclic difluoro- and tetrafluorolactones **131** and trifluoromethylated azalactones **132**.[93]

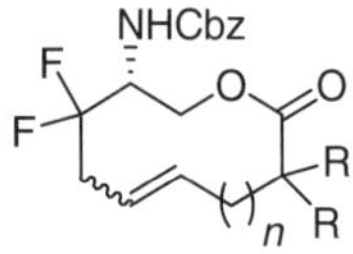

131, n=1–3, R=H or F

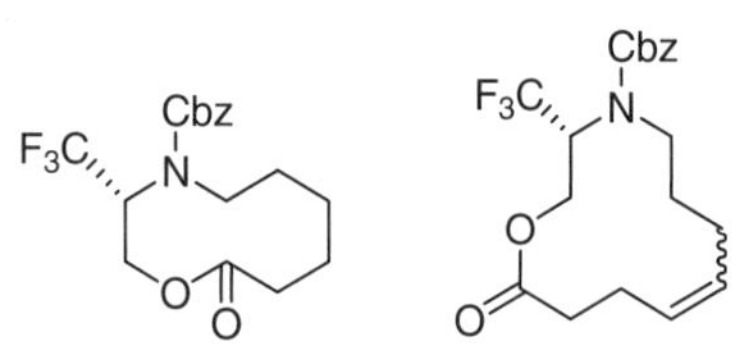

132

Interesting studies on the preparation and properties of partially fluorinated crown ethers have been published by the DuPont research group.[94] The reaction of fluorinated diene **133** and disilyl ether **134** produces the 18-membered macrocycle **135** isolated in a 50% yield. Compound **135** forms stable inclusion complex with fluoride anion, which is isolated and characterized by NMR and single-crystal X-ray diffraction.[94] Macrocycle **135** in this case behaves as a "reversed" crown ether, forming a complex with an *anion* (rather than cation), in which fluoride anion is binded to the cycle through hydrogen bonding to four different protons of $-OCH_2-$ groups of **135**. The presence of acidic C–H bonds is essential, since similarly prepared marcocycle **136** (containing $-O(CH_2)_5O-$ linkages between two tetrafluorocyclobutene units) failed to produce the complex with fluoride anion (Fig. 10.48).[94]

A similar synthetic approach was used in the synthesis of polyfluorinated oxygen- and nitrogen-containing macrocycles. Substitution reaction of fluorinated pyridines

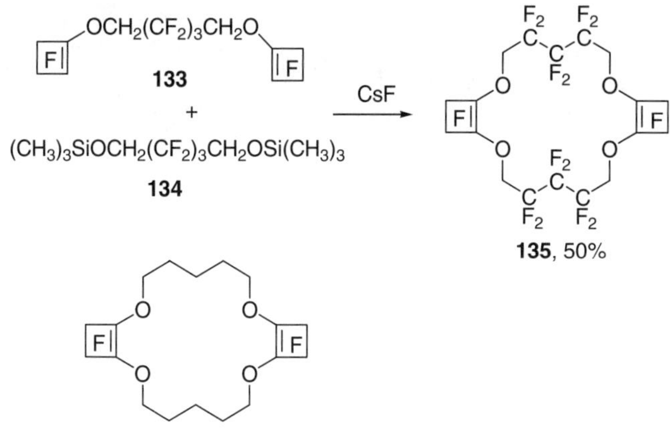

136

FIGURE 10.48 Fluorinated macrocycles **135** and **136**.

with various oxygen and nitrogen nucleophiles was used for the construction of various size fluorinated heterocycles **137**[95] (also see Chapter 8).

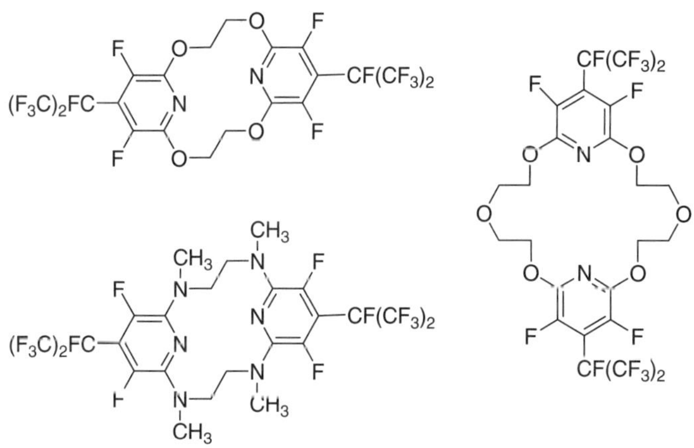

137

Interesting family of cyclic oligomers was isolated in the nucleophilic polymerization of unsaturated alcohol **138** (Fig. 10.49).[96] At a concentration of ~40 mol%, compound **138** reacts with catalytic amount of KH to give cyclic dimers **139**, which were isolated in a surprisingly high yield.

The formation of cyclic trimers, tetramers, and pentamers has also been observed in this process. Isolated dimer **139** was found to form a 2:1 complex with Bu_4NCl and the complex of unknown stoichiometry with TASF.[96]

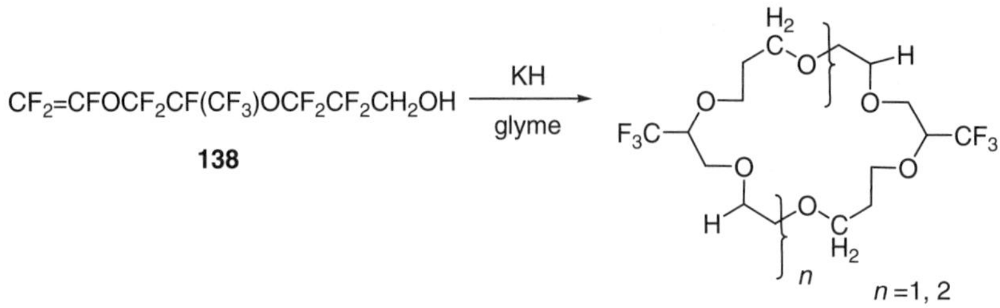

139, 55–60%

all unspecified bonds to fluorine

FIGURE 10.49 Synthesis of macrocycles **139**.

An interesting supramolecular methodology was recently applied by the Resnati group for the synthesis of "hybrid"-type macrocycles.[97] Photochemical reaction of the preformed complex leads to the formation of supramolecular ribbon **140**.

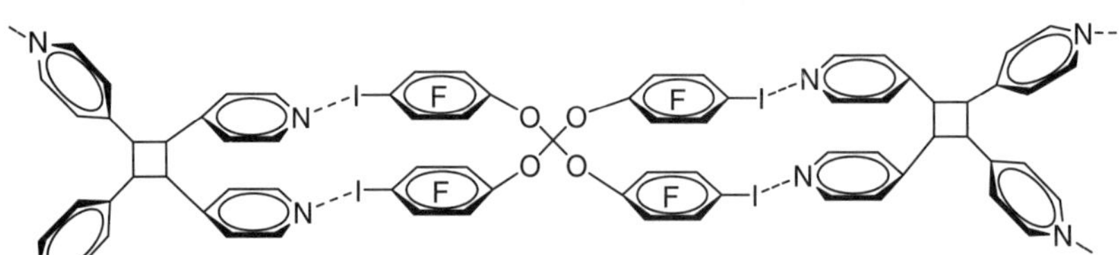

140

10.6.2 Perfluorinated Macroheterocycles

Substantial number of perfluorinated crown ethers has been prepared and characterized in the last 20 years. In 1985 the Lagow group reported synthesis of: *F*-12-crown-6, *F*-15-crown-5, and *F*-18-crown-6 (**141**) crown ethers.[98]

141

Later the same group reported synthesis of *F*-cyclohexano-, *F*-18-crown-6, *F*-24-crown-8, and *F*-[30]-crown-10 ethers, along with the largest perfluorinated crown ether known—*F*-[60]-crown-20.[99] First perfluorinated cryptand[100] and bicyclic and

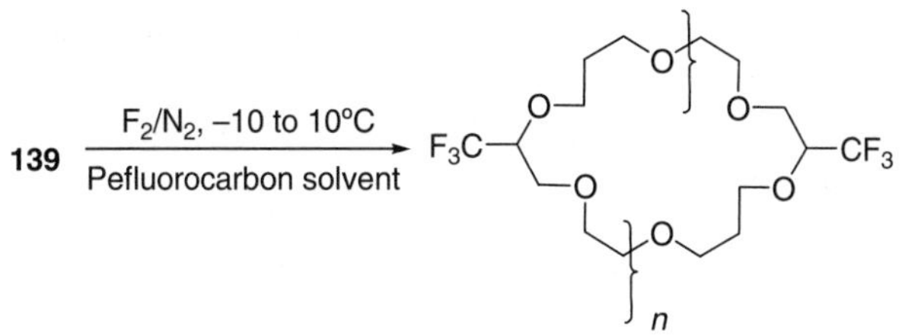

All unspecified bonds to fluorine
143, *n*=1, 82%
144, *n*=2, 85%

FIGURE 10.50 Synthesis of macrocycles **143** and **144**.

spiro ethers **142** were also prepared by exhaustive fluorination of the corresponding hydrocarbon starting materials using elemental fluorine.[98,100–104]

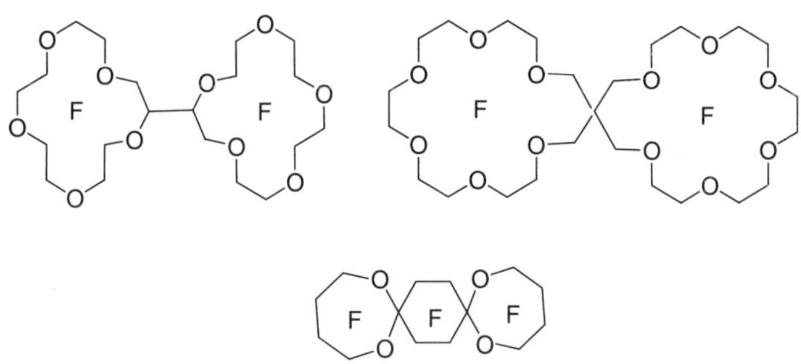

142

Additional information on the synthesis of perfluorinated linear and macrocyclic ethers can be found in two review articles.[105,106]

Partially fluorinated cyclic ethers **139** was converted to the corresponding perfluorinated macrocycles[96] using similar methodology. Since starting material **139** already contained substantial amount of fluorine, the fluorination using 25–30% mixture of F_2/N_2 can be carried out at relatively high temperature (from −10 to 10°C) using NaF as scavenger of HF and results in a high-yield formation of cyclic ethers **143** and **144** (Fig. 10.50).[96]

10.7 CONCLUSION

Substantial progress has been achieved in the synthesis and studies of chemical transformations of polyfluorinated large-ring heterocyclic compounds. A large number of new biologically fluorinated heterocyclic materials possess biological activity, and the search for new active compounds is an important factor responsible for the rapid development of synthesis of "lightly" fluorinated oxygen-, nitrogen-, and to some extent sulfur-containing heterocycles.

Polyfluorinated heterocycles containing larger rings, such as perfluorinated crown ethers or partially fluorinated "reversed" crown ethers, capable of forming complexes with anions rather than cations are another interesting and not fully investigated group of heterocyclic materials.

At this point it is clear that this branch of heterocyclic chemistry has not yet reached its maturity. While the synthetic methods and chemistry of oxygen- and nitrogen-containing seven- and eight-membered heterocycles are well developed, there is still very little known about the chemistry of larger heterocycles containing sulfur, phosphorous, and other heteroatoms.

REFERENCES

1. Takenaka, N.E.; Hamlin, R.; Lemal, D.M. Hexafluorobenzene oxide and hexafluorooxepin. *J. Am. Chem. Soc.* **1990**, *112*(18), 6715–6716.
2. Takenaka, N.E.; Lemal, D.M. The perfluorobenzene oxide–perfluorooxepin system. In *Synthetic Fluorine Chemistry*; Olah, G.A.; Chambers, R.D.; Prakash, G.K.S.,Eds.; John Wiley & Sons, Inc.: New York, 1992; pp 287–312.
3. Toy, M.S.; Stringham, R.S. Some new perfluoro ethers and copolymers from "Dewar" hexafluorobenzene. *J. Fluorine Chem.* **1979**, *13*(1), 23–39.
4. Toy, M.S.; Stringham, R.S. Thermal isomerization of perfluorobicyclo[2.2.0]hexene oxides. *J. Org. Chem.* **1979**, *44*(15), 2813–2814.
5. Zajc, B.; Zupan, M. Fluorination with xenon difluoride. 37. Room-temperature rearrangement of aryl-substituted ketones to difluoro-substituted ethers. *J. Org. Chem.* **1990**, *55*(3), 1099–1102.
6. Fustero, S.; Esteban, E.; Sanz-Cervera, J.F.; Jimenez, D.; Mojarrad, F. An enantio- and diastereoselective synthesis of fluorinated beta-aminoalkyloxepine derivatives through Mannich and ring-closing metathesis reactions. *Synthesis* **2006**, (23), 4087–4091.
7. Fustero, S.; Sanz-Cervera, J.F.; del Pozo, C.; Acena, J.L. Nitrogen-containing organofluorine compounds through metathesis reactions. *ACS Symp. Ser.* **2006**, *949* (Current Fluoroorganic Chemistry), 54–68.
8. Abouabdellah, A.; Begue, J.-P.; Bonnet-Delpon, D.; Gantier, J.-C.; Truong Thi Thanh, N.; Truong Dinh, T. Synthesis and *in vivo* antimalarial activity of 12a-trifluoromethylhydroartemisinin. *Bioorg. Med. Chem. Lett.* **1996**, *6*(22), 2717–2720.
9. Magueur, G.; Crousse, B.; Charneau, S.; Grellier, P.; Begue, J.-P.; Bonnet-Delpon, D. Fluoroartemisinin: trifluoromethyl analogues of artemether and artesunate. *J. Med. Chem.* **2004**, *47*(10), 2694–2699.
10. Magueur, G.; Crousse, B.; Ourevitch, M.; Bonnet-Delpon, D.; Begue, J.-P. Fluoroartemisinins: when a *gem*-difluoroethylene replaces a carbonyl group. *J. Fluorine Chem.* **2006**, *127*(4–5), 637–642.
11. Balsells, J.; Waters, M.; Hansen, K.; Kieczykowski, G.R.; Song, Z.J. A short and efficient synthesis of 3-[2,2,2-trifluoroethyl]hexahydro-2*H*-1,4-diazepin-2-one. *Synthesis* **2007** (18), 2779–2781.
12. Begue, J.-P.; Bonnet-Delpon, D. Fluoroartemisinins: metabolically more stable antimalarial artemisinin derivatives. *ChemMedChem* **2007**, *2*(5), 608–624.

13. Begue, J.-P.; Bonnet-Delpon, D. Recent advances (1995–2005) in fluorinated pharmaceuticals based on natural products. *J. Fluorine Chem.* **2006**, *127*(8), 992–1012.
14. Bonnet-Delpon, D.; Begue, J.-P.; Crousse, B.; Grellepois, F.; Chollet, C.; Chorki, F.; Magueur, G.; Ourevitch, M.; Nguyen, T.N.T.; Grellier, P.; Nguyen, V.H.; Truong, V.N.; Nhan, D.H.; Thu, N.T.M. New 10-trifluoromethyl monomers, dimers, and chimeras of artemisinin from a key allyl bromide precursor. *ACS Symp. Ser.* **2007**, *949* (Current Fluoroorganic Chemistry), 337–351.
15. Burger, K.; Meffert, A. Seven-membered ring heterocycles, IV. Thermolytic behavior of 2,3,4,5-tetrahydro-1,3-oxazepin-5-ones and 4,5-dihydro-2*H*-1,3-dioxepin-5-ones. *Liebigs Ann. Chem.* **1978**, (7), 1052–1059.
16. Burger, K.; Meffert, A.; Gieren, A. Seven-membered ring heterocycles, III. Synthesis of 4,5-dihydro-2*H*-1,3-dioxepin-5-ones and 2,3,4,5-tetrahydro-1,3-oxazepin-5-ones. *Liebigs Ann. Chem.* **1978**, (7), 1037–1051.
17. Burger, K.; Meffert, A.; Linder, J.; Hagn, P.; Gieren, A.; Siebels, H. Cycloaddition to five-ring heterocycles with ring expansion to seven-ring systems. *Tetrahedron Lett.* **1975**, (37), 3223–3226.
18. Gieren, A.; Siebels, H.; Burger, K.; Meffert, A. X-ray photographic structure determination of 7-diethylamino-6-methyl-4-phenyl-2,2-bis(trifluoromethyl)-4,5-dihydro-1,3-dioxepin-5-one. *Tetrahedron Lett.* **1975** (37), 3227–3230.
19. Ishikawa, N.; Maruta, M. The fluorine-19 and carbon-13 NMR studies on the structures of hexafluoropropene trimers. *Nippon Kagaku Kaishi* **1977**, (10), 1411–1415.
20. Ishikawa, N.; Sekiya, A. Preparation and isomerization of hexafluoropropene oligomers in the presence of fluoride ion. *Nippon Kagaku Kaishi* **1972**, (11), 2214–2216.
21. Sherer, K.V., Jr.; Ono, T.; Yamanouchi, K.; Yokoyama, K.Perfluoro and polyfluorinated ethers and cyclic ethers (Green Cross Corp., Japan). *European Patent* EP190393, **1986**.
22. Snegirev, V.F.; Makarov, K.N. Reaction of perfluoro-2-methyl-2-pentene with bifunctional nucleophiles. *Izv. Akad. Nauk SSSR, Ser. Khim.* **1985**, (9), 2066–2076.
23. Maruta, M.; Kubota, S.; Yoshimura, N.; Kitazume, T.; Ishikawa, N. Fluorine-containing seven-membered benzoheterocycles derived from hexafluoropropene oligomers. *J. Fluorine Chem.* **1980**, *16*(1), 75–88.
24. Ishikawa, N.; Kitazume, T.; Chino, K.; Mustafa, E.-S. Perfluoroalkylated eight- and nine-membered benzoheterocycles from perfluoro-2-methyl-2-pentene. *J. Fluorine Chem.* **1981**, *18*(4–6), 447–457.
25. Okada, E.; Tsukushi, N.; Kunihiro, N.; Tomo, Y. A facile and convenient synthetic method for fluorine-containing naphtho[1,2-*e*][1,4]diazepines, naphtho[1,2-*e*][1,4]dithiepins and naphtho[1,2-*e*][1,4]dioxepins. *Heterocycles* **1998**, *49*, 297–304.
26. Chi, K.W.; Kim, H.A.; Furin, G.G.; Zhuzhgov, E.L.; Protzuk, N. The reaction of binucleophilic reagents containing 1,4-nucleophilic centers with perfluoro-2-methylpent-2-ene and perfluoro-5-azanon-4-ene. *J. Fluorine Chem.* **2001**, *110*(1), 11–20.
27. Flohr, A.; Jakob-Roetne, R.; Wostl, W. Preparation of oxaazabenzocycloheptyl malonamides as gamma-secretase inhibitors. *US Patent Appl.* 2006/122168, **2006**.
28. Cocuzza, A.J.; Chidester, D.R.; Cordova, B.C.; Jeffrey, S.; Parsons, R.L.; Bacheler, L.T.; Erickson-Viitanen, S.; Trainor, G.L.; Ko, S.S. Synthesis and evaluation of efavirenz (Sustiva) analogs as HIV-1 reverse transcriptase inhibitors: replacement of the cyclopropylacetylene side chain. *Bioorg. Med. Chem. Lett.* **2001**, *11*(9), 1177–1179.

29. Cocuzza, A.J.; Chidester, D.R.; Cordova, B.C.; Klabe, R.M.; Jeffrey, S.; Diamond, S.; Weigelt, C.A.; Ko, S.S.; Bacheler, L.T.; Erickson-Viitanen, S.K.; Rodgers, J.D. 4, 1-Benzoxazepinone analogues of efavirenz (Sustiva) as HIV-1 reverse transcriptase inhibitors. *Bioorg. Med. Chem. Lett.* **2001**, *11*(11), 1389–1392.

30. Rodgers, J.D.; Cocuzza, A.J.1,3-Benzodiazepin-2-ones and 1,3-benzoxazepin-2-ones useful as HIV reverse transcriptase inhibitors (DuPont Pharmaceuticals Company, USA). WO 2000000479, **2000**.

31. Rodgers, J.D.; Cocuzza, A.J.; Bilder, D.M. Preparation of 1,3-benzodiazepin-2-ones and 1,3-benzoxazepin-2-ones as HIV reverse transcriptase inhibitors (DuPont Pharmaceuticals Company, USA). *US Patent* 6,204,262, **2001**.

32. Richard, S.; Prie, G.; Guignard, A.; Thibonnet, J.; Parrain, J.L.; Duchene, A.; Abarbri, M. A new synthesis of optically active 3-substituted (3*S*)-3,4-dihydro-5-(perfluoroalkyl)-2*H*-[1,4]oxazepin-7-ones. *Helv. Chim. Acta* **2003**, *86*(3), 726–732.

33. Petrov, V.A.; Belen'kii, G.G.; German, L.S. Reaction of higher perfluorinated amines with antimony pentafluoride. *Izv. Akad. Nauk SSSR, Ser. Khim.* **1985**, (8), 1934–1935.

34. Reddy, A.C.S.; Rao, P.S.; Venkataratnam, R.V. Fluoro organics: facile syntheses of novel 2- or 4-trifluoromethyl-1*H*-arylo-1,5-diazepines, oxazepines, thiazepines, 2-(1,1,1-trifluoroacetonyl)imidazoles, oxazoles and thiazoles. *Tetrahedron* **1997**, *53*(16), 5847–5854.

35. Burger, K.; Maier, G.; Kahl, T. A new concept for seven-membered ring synthesis. *Chem. Ztg.* **1988**, *112*(3), 111–112.

36. Vovk, M.V.; Dorokhov, V.I.; Samarai, L.I. Synthesis of new types of benzazepine systems. *Dokl. Akad. Nauk Ukr. SSR* **1988**, (7), 33–35.

37. Vovk, M.V.; Dorokhov, V.I.; Samarai, L.I. 1-(Acyloxy)alkyl isocyanates. IV. Reactions with *o*-aminophenols and *o*-aminothiophenols. *Zh. Org. Khim.* **1989**, *25*(11), 2394–2399.

38. Marsh, F.D.; Simmons, H.E. *N*-Cyanoazepines from cyanonitrene and aromatic compounds. *J. Am. Chem. Soc.* **1965**, *87*(15), 3529–3530.

39. Marsh, F.D. (to DuPont) *US Patent* 3,268,512, **1966**.

40. Barlow, M.G.; Culshaw, S.; Haszeldine, R.N.; Morton, W.D. Heterocyclic polyfluoro compounds. Part 41. Photochemical isomerization of 1-substituted hexafluoro-1*H*-azepines: hexafluoro-2-azabicyclo[3.2.0]hepta-3,6-dienes and their reactions. *J. Chem. Soc., Perkin Trans. 1* **1982**, (9), 2105–2109.

41. Barlow, M.G.; Harrison, G.M.; Haszeldine, R.N.; Morton, W.D.; Shaw-Luckman, P.; Ward, M.D. Heterocylic polyfluoro compounds. Part 40. Preparation of polyfluoroazepines. *J. Chem. Soc., Perkin Trans. 1* **1982**, (9), 2101–2104.

42. Sawaguchi, M.; Hara, S.; Fukuhara, T.; Yoneda, N. Electrochemical direct fluorination of lactams. *Israel J. Chem.* **1999**, *39*(2), 151–154.

43. Snegirev, V.F.; Antipin, M.Y.; Khrustalev, V.N.; Struchkov, Y.T. Reaction of perfluoro-2-methyl-2-pentene with cycloalkanone oximes. New examples of the Beckmann–Chapman rearrangement. *Izv. Akad. Nauk SSSR, Ser. Khim.* **1994**, (6), 1073–1077.

44. Rauscher, W.H.; Tucker, H. The synthesis of ethyl adipamate and 2-oxo-6,6-dihydroperfluorohexamethylenimine. *J. Am. Chem. Soc.* **1954**, *76*, 3599–3601.

45. Flohr, A.; Galley, G.; Jakob-Roetne, R.; Kitas, E.A.; Wostl, W. Preparation of fluoro substituted 2-oxo-azepan derivatives as gamma-secretase inhibitors. *US Patent Appl.* 2007/037789, **2007**.

46. Osipov, S.N.; Artyushin, O.I.; Kolomiets, A.F.; Bruneau, C.; Picquet, M.; Dixneuf, P.H. Synthesis of fluorine-containing cyclic alpha-amino acid and alpha-amino

phosphonate derivatives by alkene metathesis. *Eur. J. Org. Chem.* **2001** (20), 3891–3897.

47. Fustero, S.; Sanz-Cervera, J.F.; Piera, J.; Sanchez-Rosello, M.; Jimenez, D.; Chiva, G. Fluorinated beta-enamino esters as versatile synthetic intermediates: synthesis of fluorinated beta-amino acids and uracils. *ACS Symp. Ser.* **2005**, *911* (Fluorine-Containing Synthons), 593–610.
48. Fustero, S.; Catalan, S.; Piera, J.; Sanz-Cervera, J.F.; Fernandez, B.; Acena, J.L. Synthesis and biological evaluation of new bicyclic fluorinated uracils through ring-closing metathesis. *J. Org. Chem.* **2006**, *71*(10), 4010–4013.
49. Fustero, S.; Navarro, A.; Pina, B.; Soler, J.G.; Bartolome, A.; Asensio, A.; Simon, A.; Bravo, P.; Fronza, G.; Volonterio, A.; Zanda, M. Enantioselective synthesis of fluorinated alpha-amino acids and derivatives in combination with ring-closing metathesis: intramolecular pi-stacking interactions as a source of stereocontrol. *Org. Lett.* **2001**, *3*(17), 2621–2624.
50. Fustero, S.; Sanchez-Rosello, M.; Jimenez, D.; Sanz-Cervera, J.F.; Del Pozo, C.; Acena, J. L. Role of the *gem*-difluoro moiety in the tandem ring-closing metathesis-olefin isomerization: regioselective preparation of unsaturated lactams. *J. Org. Chem.* **2006**, *71*(7), 2706–2714.
51. De Matteis, V.; Dufay, O.; Waalboer, D.C.J.; van Delft, F.L.; Tiebes, J.; Rutjes, F.P.J.T. An improved ring-closing metathesis approach to fluorinated and trifluoromethylated nitrogen heterocycles. *Eur. J. Org. Chem.* **2007**, (16), 2667–2675.
52. De Matteis, V.; van Delft, F.L.; Tiebes, J.; Rutjes, F.P.J.T. RCM mediated synthesis of fluorinated cyclic hydrazines. *Synlett* **2007**, (3), 351–354.
53. Burger, K.; Schickaneder, H.; Zettl, C. Thermolysis of 1,5-diazabicyclo[3.3.0]oct-2-enes. 5. New simple syntheses for 4,5-dihydro-1*H*-1,2-diazepines, 5,6-dihydro-4*H*-1,2-diazepines, 4,5,6,7-tetrahydro-1*H*-1,2-diazepines, and 1-vinyl-1*H*-pyrazoles. *Liebigs Ann. Chem.* **1982**, (9), 1741–1749.
54. Reisinger, A.; Koch, R.; Bernhardt, P.V.; Wentrup, C. 1*H*-1,3-Diazepines, 5*H*-1,3-diazepines, 1,3-diazepinones, and 2,4-diazabicyclo[3.2.0]heptenes. *J. Chem. Soc., Perkin Trans. 1* **2004**, *2*(8), 1227–1238.
55. Reisinger, A.; Koch, R.; Wentrup, C. Dihydro-1,3 diazepinones and diazabicyclo[3.2.0] heptenones from pyridyl azides. *J. Chem. Soc., Perkin Trans. 1* **1998**, (15), 2247–2250.
56. Burger, K.; Kahl, T. Seven-ring synthesis. 6. 2,2-Bis(trifluormethyl)-4,5-dihydro-1,3,5-triazepine from 4,4-bis(trifluormethyl)-1,3-diazabuta-1,3-dienes. *J. Fluorine Chem.* **1989**, *42*(1), 51–58.
57. Furin, G.G.; Gatilov, Y.V.; Bagryanskaya, I.Y.; Zhuzhgov, E.L. Reaction of perfluoro-2-methylpent-2-ene with ethylenediamine and hexamethylenediamine. *Izv. Akad. Nauk SSSR, Ser. Khim.* **1999**, (8), 1578–1582.
58. Krespan, C.G.; Dixon, D.A. Fluoroolefin condensation catalyzed by aluminum chlorofluoride. *J. Fluorine Chem.* **1996**, *77*(2), 117–126.
59. Belen'kii, G.G.; Lur'e, E.P.; German, L.S. Electrophilic alkylation of fluorine-containing ethylenes by polyfluoropropenes. *Izv. Akad. Nauk SSSR, Ser. Khim.* **1975**, (12), 2728–2732.
60. Saloutin, V.I.; Skryabina, Z.E.; Burgart, Y.V.; Chupakhin, O.N.; Font-Altaba, M.; Solans, X.; Font-Bardia, M. Interaction of perfluoropent-2-ene and its 2-amino-4-imino derivative with ethylenediamine and with diethylenetriamine. *J. Fluorine Chem.* **1994**, *69*(1), 25–29.
61. Middleton, W.J. 1,1-Dicyano-2,2-bis(trifluoromethyl)ethylene. *J. Org. Chem.* **1965**, *30* (5), 1402–1407.

62. Komarov, K.V.; Chkanikov, N.D.; Galakhov, M.V.; Kolomiets, A.F.; Fokin, A.V. Reaction of 1,1-dicyano-2,2-bis(trifluoromethyl)ethylene with arylamines. *J. Fluorine Chem.* **1990**, *47*(1), 59–69.

63. Komarov, K.V.; Chkanikov, N.D.; Sereda, S.V.; Antipin, M.Y.; Struchkov, Y.T.; Kolomiets, A.F.; Fokin, A.V. Synthesis of 2-amino-3-cyano-4-bis(trifluoromethyl)-4,5-dihydro-1*H*-1,5-benzodiazepine. *Izv. Akad. Nauk SSSR, Ser. Khim.* **1988**, (8), 1920–1922.

64. Kurykin, M.A.; German, L.S.; Knunyants, I.L. Reaction of *trans*-perfluoro-2-pentene with ammonia. *Izv. Akad. Nauk SSSR, Ser. Khim.* **1980**, (12), 2827–2829.

65. Petrova, O.E.; Kurykin, M.A.; Gorlov, D.V. A convenient method for the synthesis of polyfluorinated β-diimines. *Izv. Akad. Nauk SSSR, Ser. Khim.* **1999**, *1710*–1712.

66. Pashkevich, K.I.; Filaykova, V.I.; Ratner, V.G.; Khomutov, O.G. Polyfunctional fluoroalkyl-containing carbonyl compounds in the synthesis of heterocycles. *Izv. Akad. Nauk SSSR, Ser. Khim.* **1998**, (7), 1279–1286.

67. Pashkevich, K.I.; Khomutov, O.G. Reaction of β-fluoroalkyl-α,β-dihalocarbonyl compounds with diamines. *Zh. Org. Khim.* **1999**, *35*(1), 100–105.

68. Hojo, M.; Masuda, R.; Okada, E. A convenient synthetic route to functionalized 5-trifluoroacetyl-6-trifluoromethyl-3,4-dihydro-2*H*-pyrans: hetero-Diels–Alder reaction of β,β-bis(trifluoroacetyl)vinyl ethers with electron-rich alkenes. *Synthesis* **1990**, (3), 347–349.

69. Reddy, A.C.S.; Rao, P.S.; Venkataratnam, R.V. Fluoro organics: a facile and exclusive synthesis of novel 2- or 4-trifluoromethyl(1*H*,5)arylodiazepines. *Tetrahedron Lett.* **1996**, *37*(16), 2845–2848.

70. Ishikawa, N.; Takaoka, A.; Ibrahim, M.K. Preparation of 2-fluoromalonic esters and related compounds from hexafluoropropene. *J. Fluorine Chem.* **1984**, *25*(2), 203–212.

71. Ishikawa, N.; Nakai, T.; Inouye, Y. Alkylated or alkenylated malonic acid or its derivatives having a fluorine containing aliphatic substituent and method to trap the enolate ion of the malonic acid or its derivatives. *European Patent* EP 103233, **1984**.

72. Sosnovskikh, V.Y.; Kutsenko, V.A. Reactions of 2-(polyfluoroalkyl)chromones with aliphatic diamines. *Izv. Akad. Nauk SSSR, Ser. Khim.* **1999**, (4), 817–819.

73. Sosnovskikh, V.Y.; Mel'nikov, M.Y. Reactions of regioisomeric 3,3-dimethyl- and 2,2-dimethyl-6-trifluoromethyl-2,3-dihydro-4-pyrones with *N*-nucleophiles. *Izv. Akad. Nauk SSSR, Ser. Khim.* **1999**, (5), 983–986.

74. Saloutin, V.I.; Skryabina, Z.E.; Burgart, Y.V. Novel fluorine-containing polyazaheterocycles. *Izv. Akad. Nauk SSSR, Ser. Khim.* **1992**, (9), 2170–2174.

75. Saloutin, V.I.; Skryabina, Z.E.; Burgart, Y.V.; Chupakhin, O.N.; Font-Altaba, M.; Solans, X.; Font-Bardia, M. Reaction of 2-amino-4-iminoperfluoro-2-pentene with ethylenediamine and diethylenetriamine. *Izv. Akad. Nauk SSSR, Ser. Khim.* **1993**, (10), 1773–1776.

76. Saloutin, V.I.; Skryabina, Z.E.; Burgart, Y.V. Condensation of fluoroalkyl-containing 1,3-dicarbonyl compounds with ethylenediamine. *J. Fluorine Chem.* **1992**, *56*(3), 325–334.

77. Bonacorso, H.G.; Lourega, R.V.; Deon, E.D.; Zanatta, N.; Martins, M.A.P. The first synthesis of dihydro-3*H*-pyrido[2,3-*b*][1,4]diazepinols and a new alternative approach for diazepinone analogues. *Tetrahedron Lett.* **2007**, *48*(28), 4835–4838.

78. Krespan, C.G. Bis(polyfluoroalkyl) acetylenes. IV. Fluorinated dithietenes and related heterocyclic compounds from bis(polyfluoroalkyl) acetylenes and sulfur. *J. Am. Chem. Soc.* **1961**, *83*, 3434–3437.

79. Riley, M.O.; Park, J.D. The synthesis of 1,4-dithiocin. *Tetrahedron Lett.* **1971** (30), 2871–2874.

80. Rowsell, J.R.; Wilkinson, D.G. Preparation of fluorinated cyclic sulfides and disulfides. *Chem. Commun.* **1969**, (21), 1274–1275.

81. Krespan, C.G.; Brasen, W.R. Fluorinated cyclic polysulfides and their polymers. *J. Org. Chem.* **1962**, *27*, 3995–4000.

82. Petrov, V.; Marshall, W. Hexafluorothioacetone based synthesis of fluorinated heterocycles. *J. Fluorine Chem.* **2007**, *128*(7), 729–735.

83. Abe, T.; Nagase, S. The electrochemical fluorination of 2-methyloxane, oxepane, 2-(chloromethyl)oxane and 2-(chloromethyl)oxolane. *J. Fluorine Chem.* **1979**, *13*(6), 519–530.

84. Abe, T.; Pandey, S.K.; Baba, H. Electrochemical fluorination of hexahydroazepine, methylpiperidines and methyl 1-hexahydroazepine-acetate: the preparation of *F*-(1-hexahydroazepine-acetyl fluoride) and its derivatives. *J. Fluorine Chem.* **2000**, *105* (1), 149–157.

85. Lin, W.H.; Lagow, R.J. The synthesis of perfluoro highly branched heterocyclic fluorine compounds by direct fluorination. *J. Fluorine Chem.* **1990**, *50*(1), 15–30.

86. Hayashi, E.; Abe, T.; Baba, H.; Nagase, S. The electrochemical fluorination of *N*-(ω-chloroalkyl)pipecolines and *N*-(ω-chloroalkyl)hexamethylenimines. *J. Fluorine Chem.* **1984**, *26*(4), 417–434.

87. Abe, T.; Baba, H.; Fukaya, H.; Tamura, M.; Sekiya, A. Simons electrochemical fluorination of substituted homopiperazines(hexahydro-1,4-diazepines) and piperazines. *J. Fluorine Chem.* **2003**, *119*(1), 27.

88. Abe, T.; Fukaya, H.; Ono, T.; Hayashi, E.; Soloshonok, I.; Okuhara, K. Electrochemical fluorination of *N*-containing carboxylic acids. Part 5. Fluorination of the methyl esters of *cis*-2,6-dimethylmorpholino-group substituted carboxylic acids. *J. Fluorine Chem.* **1998**, *87*(2), 193.

89. Boltachova, N.S.; Fedorova, O.V.; Ovchinnikova, I.G.; Kazheva, O.N.; Chekhlov, A.N.; Dyachenko, O.A.; Rusinov, G.L.; Filyakova, V.I.; Charushin, V.N. Fluorinated lithium 1,3-diketonates as reagents to modify podands and crown-ethers. *J. Fluorine Chem.* **2007**, *128*(7), 762–768.

90. Wilson, R.M.; Danishefsky, S.J. Small molecule natural products in the discovery of therapeutic agents: the synthesis connection. *J. Org. Chem.* **2006**, *71*(22), 8329–8351.

91. Rivkin, A.; Chou, T.-C.; Danishefsky, S.J. On the remarkable antitumor properties of fludelone: how we got there. *Angew. Chem., Int. Ed.* **2005**, *44*(19), 2838–2850.

92. Haufe, G.; Bruns, S.; Runge, M. Enantioselective ring-opening of epoxides by HF-reagents. Asymmetric synthesis of fluoro lactones. *J. Fluorine Chem.* **2001**, *112*(1), 55–61.

93. Fustero, S.; Fernandez, B.; Sanz-Cervera, J.F.; Mateu, N.; Mosulen, S.; Carbajo, R.J.; Pineda-Lucena, A.; Ramirez de Arellano, C. Asymmetric synthesis of fluorinated amino macrolactones through ring-closing metathesis. *J. Org. Chem.* **2007**, *72*(23), 8716–8723.

94. Farnham, W.B.; Roe, D.C.; Dixon, D.A.; Calabrese, J.C.; Harlow, R.L. Fluorinated macrocyclic ethers as fluoride ion hosts. Novel structures and dynamic properties. *J. Am. Chem. Soc.* **1990**, *112*(21), 7707–7718.

95. Sandford, G. Highly functionalized heterocycles and macrocycles from pentafluoropyridine. *ACS Symp. Ser.* **2005**, *911* (Fluorine-Containing Synthons), 248–261.

96. Hung, M.H.; Farnham, W.B.; Feiring, A.E.; Rozen, S. A new synthetic approach to poly- and perfluorinated polyethers. *J. Am. Chem. Soc.* **1993**, *115*(20), 8954–8959.
97. Metrangolo, P.; Meyer, F.; Resnati, G.; Ursini, M. Haloperfluorocarbons: versatile tectons in halogen bonding based crystal engineering. *ACS Symp. Ser.* **2005**, *911* (Fluorine-Containing Synthons), 514–542.
98. Lin, W.H.; Bailey, W.I., Jr.; Lagow, R.J. The first perfluoro crown ethers. *J. Chem. Soc.* **1985**, (19), 1350–1352.
99. Wei, H.-C.; Lagow, R.J. The synthesis of the largest perfluoro macrocycles; perfluoro[60]-crown-20 and perfluoro[30]-crown-10. *J. Chem. Soc.* **2000**, (21), 2139–2140.
100. Lin, T.-Y.; Lin, W.-H.; Clark, W.D.; Lagow, R.J.; Larson, S.B.; Simonsen, S.H.; Lynch, V. M.; Brodbelt, J.S.; Maleknia, S.D.; Liou, C.-C. Synthesis and chemistry of perfluoro macrocycles. *J. Am. Chem. Soc.* **1994**, *116*(12), 5172–5179.
101. Brodbelt, J.; Maleknia, S.; Liou, C.C.; Lagow, R. Selective formation of molecular oxygen/perfluoro crown ether adduct ions in the gas phase. *J. Am. Chem. Soc.* **1991**, *113* (15), 5913–5914.
102. Lagow, R.J.; Wei, H.-C. Synthesis and chemistry of the first perfluro-bis(crown ethers). In Book of Abstracts, 211th ACS National Meeting, New Orleans, LA, March 24–28, **1996**, FLUO-029.
103. Lin, T.Y.; Lagow, R.J. Synthesis of perfluorodicyclohexano-18-crown-6 ether. *J. Chem. Soc.* **1991**, (1), 12–14.
104. Lin, T.Y.; Roesky, H.W.; Lagow, R.J. The synthesis of perfluorocyclohexano-15-crown-5 ether. *Synth. Commun.* **1993**, *23*(17), 2451–2462.
105. Lagow, R.J. The discovery of successful direct fluorination syntheses: three eras of elemental fluorine reaction chemistry. In *Fluorine Chemistry at the Millennium*; Banks, R.E.,Ed.; Elsevier, 2000; pp 283–296.
106. Lagow, R.J.; Bierschenk, T.R.; Juhlke, T.J.; Kawa, H. A new synthetic procedure for the preparation and manufacture of perfluoropolyethers. In *Synthetic Fluorine Chemistry*; Olah, G.A.; Chambers, R.D.; Prakash, G.K.S., Eds.; John Wiley & Sons, Inc.: New York, 1992; pp 97–126.

PART II

APPLICATION OF FLUORINATED HETEROCYCLIC MATERIALS

11

AGRICULTURAL PRODUCTS BASED ON FLUORINATED HETEROCYCLIC COMPOUNDS

WONPYO HONG

Crop Protection Products, E. I. DuPont de Nemours and Company, Stine-Haskell Research Center, Newark, DE 19714, USA

In the search for new agricultural products having improved physical, biological, and environmental properties, one of the most generally useful chemical modifications is the introduction of fluorine atoms into lead structures. The introduction of fluorine-containing substituents in the structures of known activity has been an important strategy for optimizing the properties of agricultural and pharmaceutical products. Fluorine-containing substituents are most commonly added to carbocyclic aromatic rings, and a large number of compounds possessing fluorine-containing substituents on aryl rings have been commercialized as agricultural active ingredients.[1] On the other hand, considerably fewer compounds having fluorine-containing substituents on heterocyclic rings have been commercialized. Many of these include fluorine as a trifluoromethyl group attached to heterocyclic ring. As shown in Fig. 11.1, butafenacil possessing the trifluoromethyl moiety on the uracil ring, thifluzamide on the thiazole ring, chlorfenapyr on the pyrrole ring, fluazinam on the pyridine ring, fluacrypyrim on the pyrimidine ring, and flufenacet on the thiadiazole ring are a few examples of trifluoromethyl-containing heterocyclic agricultural active ingredients. The trifluoromethyl substituent in these compounds is usually derived from HF, trifluoroacetate, trifluoroacetic anhydride, or trifluoroacetoacetate.

The structure–activity relationships and/or biological activity profiles of these fluoro-substituted heterocyclic agricultural products and their syntheses are briefly

Fluorinated Heterocyclic Compounds: Synthesis, Chemistry, and Applications, Edited by Viacheslav A. Petrov

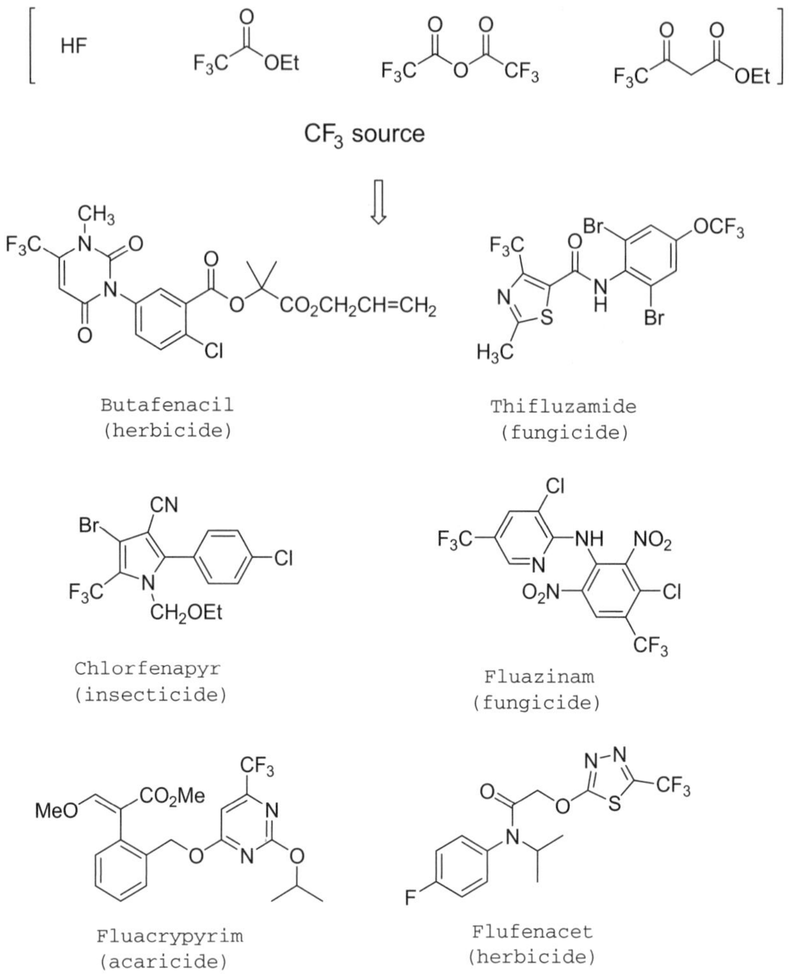

FIGURE 11.1

discussed. In particular, the synthetic methods useful for introducing the fluoro-substituted heterocyclic moieties are described. Because the processes actually used are often considered confidential by manufacturers, the syntheses described here are reasonable approaches, but are not necessarily the commercial processes.

11.1 HERBICIDES

11.1.1 Acetyl-Coenzyme A Carboxylase Inhibitors

These are chemical compounds that inhibit the acetyl-CoA carboxylase enzyme, which is responsible for catalyzing an early step in lipid and fatty acid synthesis in plants. Lipids are essential components of cell membranes, and without them,

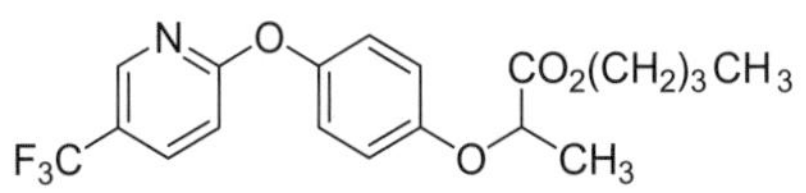

Fluazifop-butyl
(ISK/Syngenta)

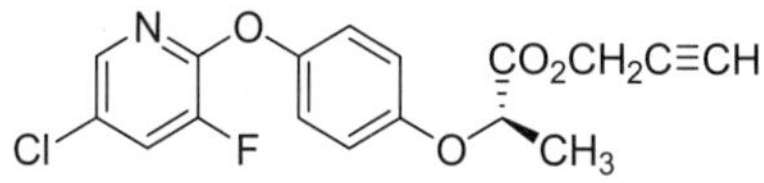

Clodinafop-propargyl
(Syngenta)

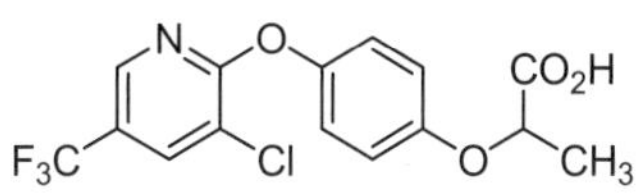

Haloxyfop
(Dow AgroSciences)

FIGURE 11.2

new cells cannot be produced. The inhibition of acetyl-coenzyme A carboxylase (ACCase) and the subsequent lack of lipid production lead to losses in cell membrane integrity, especially in regions of active growth such as meristematic tissue. Eventually, shoot and rhizome growth ceases, and shoot meristems and rhizome buds begin to die. Examples of ACCase inhibitors containing fluoro-substituted heterocyclic moieties are listed in Fig. 11.2.

Fluazifop-butyl is a herbicide discovered by Ishihara Sankyo Kaisha (ISK) and then developed jointly with ICI Plant Protection Division (now Syngenta AG). First marketed in 1983, it is now widely used in more than 50 countries for controlling annual and perennial grasses in broad-leaf crops such as cotton, soybean, and sugarbeet. The development of fluazifop-butyl, as well as flazasulfuron (herbicide), chlorfluazuron (insecticide, see Fig. 11.10), and fluazinam (fungicide, see Scheme 11.10), all by ISK, has established the utility of trifluoromethylpyridines as building blocks for agricultural active ingredients.[2] Analogues of fluazifop-butyl bearing electronically neutral substituents, the strongly electron-attracting nitro group or the electron-releasing methyl or methylthio groups instead of the trifluoromethyl at the 5-position on the pyridine ring, are all herbicidally inactive. This suggests that electronic factors are not important for activity. Besides trifluoromethyl, substituents that contribute to herbicidal activity include the halogens, suggesting that the substituent's hydrophobicity may be important. The manufacturing process developed by ISK to produce 2-chloro-5-trifluoromethylpyridine (**4**), a key intermediate to fluazifop-butyl, involves simultaneous vapor-phase chlorofluorination of 3-picoline (**1**), as shown in Scheme 11.1. Some other (chloro-)trifluoromethylpyridines such as **2**, **3**, and **5** occur as by-products in this process. The by-products **2** and **3** are reduced to **5**, which is then refed to the reactor.

11.1.2 Acetohydroxy Acid Synthase Inhibitors

These are chemical compounds that inhibit acetohydroxy acid synthase, also known as acetolactate synthase (ALS). These compounds inhibit the production of the

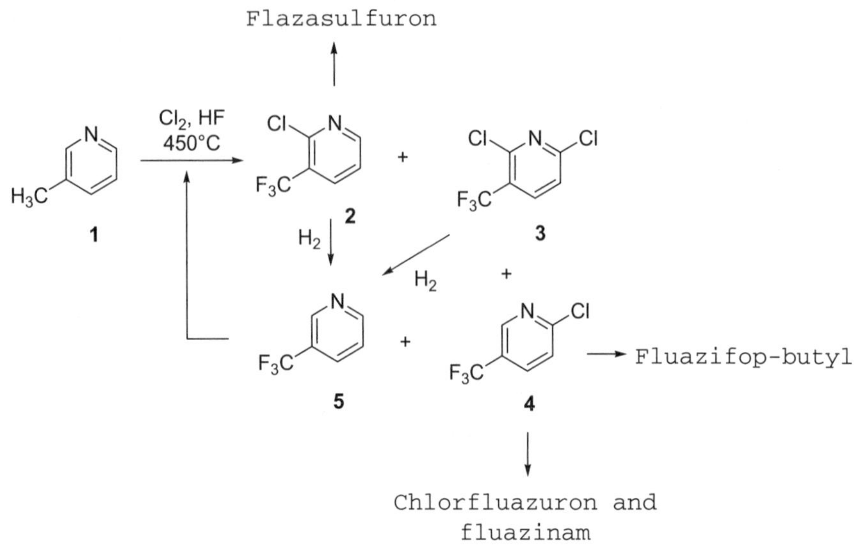

SCHEME 11.1

branched-chain aliphatic amino acids such as valine, leucine, and isoleucine, which are required for protein synthesis and cell growth. Examples of acetohydroxy acid synthase (AHAS) inhibitors containing the fluoro-substituted heterocyclic moiety are listed in Fig. 11.3.

The discovery of sulfonylurea herbicides[3,4] by DuPont in the mid-1970s began the present low-dose era of herbicide chemistry. The sulfonylureas provide a 50–100-fold greater herbicidal activity than earlier classes of active ingredients, and typically achieve crop-selective weed control at use rates of 2–75 g/ha. Such low use rates reduce the chemical load on the environment. Also, sulfonylurea herbicides are nontoxic to vertebrate and invertebrate animals and soil microorganisms, and their degradation products typically have low potential to accumulate in nontarget organisms. More than 30 sulfonylurea herbicides for use in a wide variety of crops have been commercialized or are under development. A problem that the first sulfonylurea products encountered was herbicide "carryover" to next season in some recropping situations. One of the successful strategies overcoming this problem involved introduction of an electron-withdrawing group at the 3-position of the pyridine ring of 2-pyridyl-sulfonylureas as illustrated by flazasulfuron.

As shown in Scheme 11.2, flazasulfuron, subjected to hydrolysis in neutral or acidic aqueous media, undergoes bridge contraction to give asymmetric 1,1-disubstituted urea **6**. This bridge contraction appears to involve intramolecular displacement facilitated by the pyridine nitrogen, the sulfonyl bridge, and the trifluoromethyl substituent all reducing electron density at the 2-position of the pyridine ring. In strongly alkaline solution, this rearrangement is followed by hydrolysis to give the pyridyl–pyrimidinyl amine **7**.[2] This mode of degradation, not seen with previously commercialized sulfonylurea herbicides, provides short soil half-lives (less

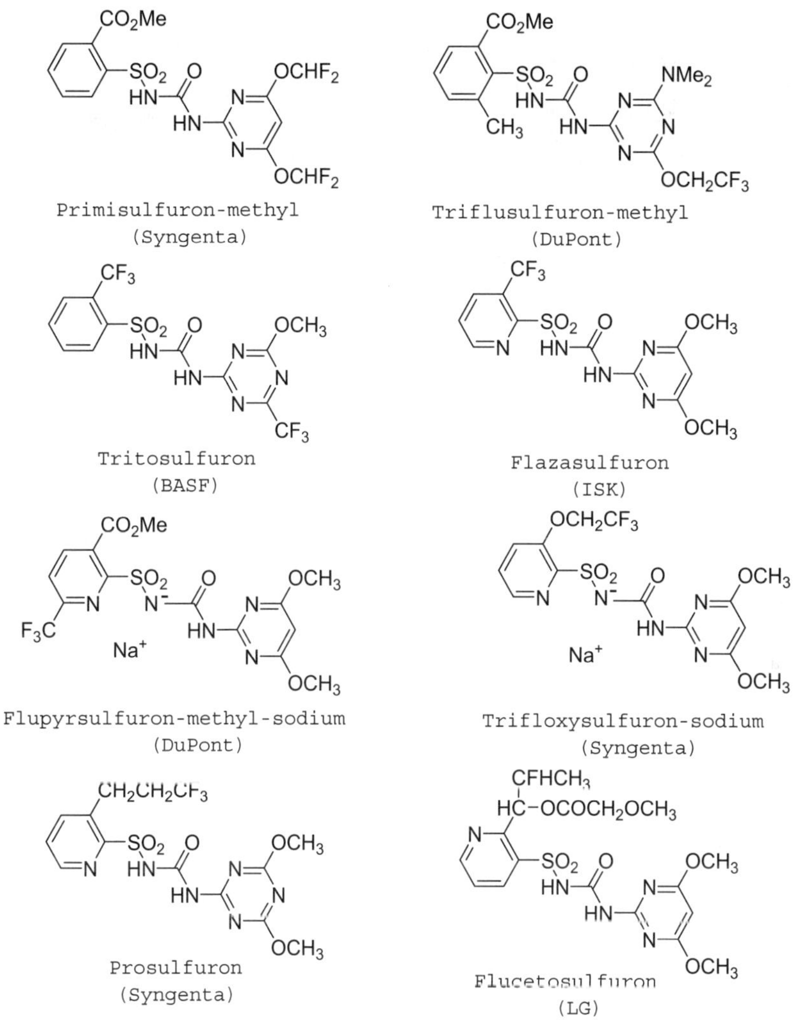

FIGURE 11.3

that 2 weeks for flazasulfuron), which minimizes carryover without relying upon the enzymes of soil microbes. Also, the trifluoromethyl group works well as an ortho-substituent, which is required for high activity in sulfonylureas. Flazasulfuron was introduced by ISK in 1989. As they had already developed a manufacturing process using trifluoromethylpyridines as intermediates, as shown in Scheme 11.1, ISK were able to use this technology. Thus, 2-chloro-3-trifluoromethylpyridine (**2**) was chosen as starting material for the synthesis of flazasulfuron, as shown in Scheme 11.3.[5] A direct thiolation using sodium hydrosulfide gives a good yield of pyridinethione **8**. The thione **8** is easily oxidized by chlorine in aqueous acetic acid to give the sulfonyl chloride **9**, which upon treatment with ammonia affords the sulfonamide **10**.

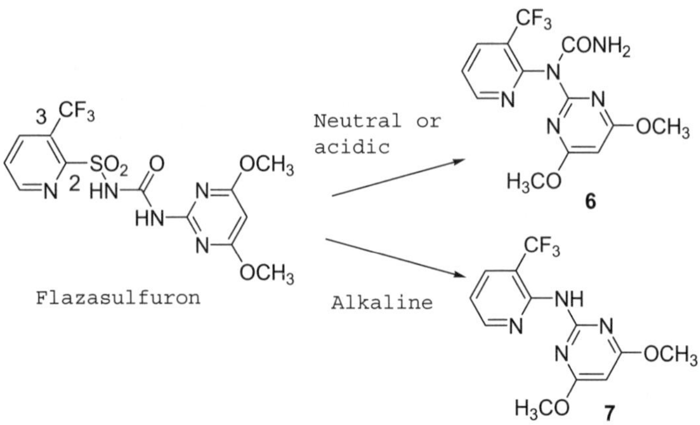

SCHEME 11.2

Many processes for forming a sulfonylurea bridge from the sulfonamide have been developed. Several processes involve phenyl carbamates as intermediates. Either the amine or the sulfonamide can be converted to the corresponding phenyl carbamate, which is then reacted with the second component to form the sulfonylurea. Flazasulfuron is prepared in good yield by both methods.

The triazolopyrimidine sulfonanilides are a class of highly active herbicides. These compounds also act by disrupting the biosynthesis of branched-chain amino acids in plants through the inhibition of AHAS. Cloransulam-methyl, diclosulam, and florasulam, as shown in Fig. 11.4, all introduced by Dow AgroSciences, are examples of the triazolopyrimidine sulfonanilide herbicides that contain a fluoro-substituted heterocyclic moiety.

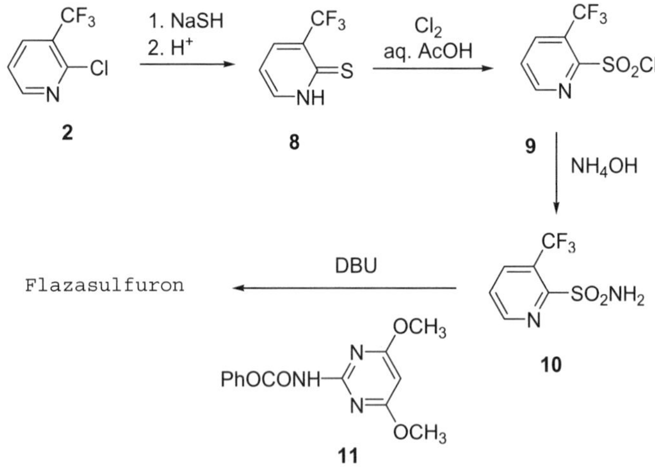

SCHEME 11.3

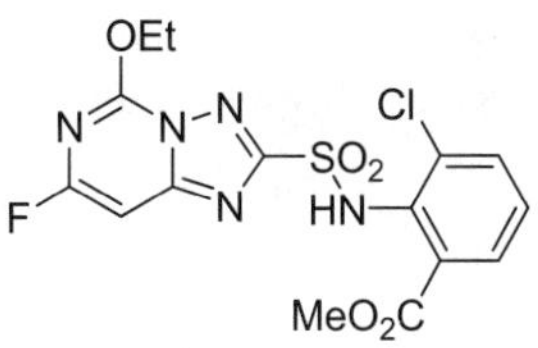

Cloransulam-methyl
(Dow)

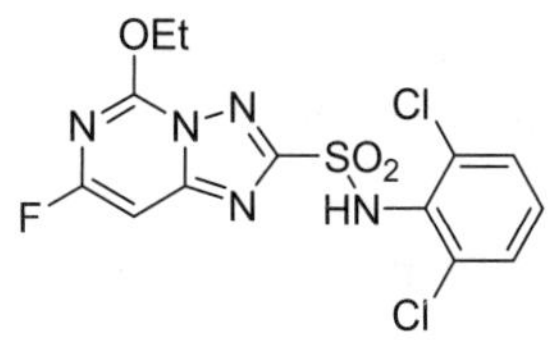

Diclosulam
(Dow)

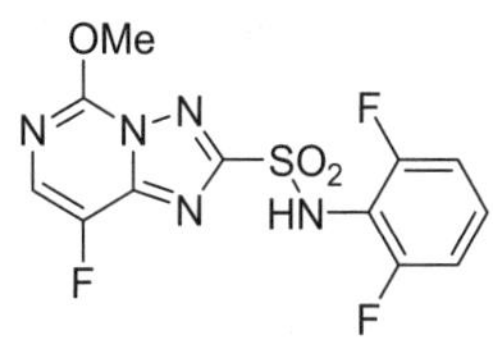

Florasulam
(Dow)

FIGURE 11.4

Both cloransulam-methyl and diclosulam share the same triazolopyrimidine ring system and are prepared from 4,6-dichloropyrimidine **12**, as shown in Scheme 11.4.[6] Treating dichloropyrimidine **12** with potassium fluoride at 144°C under reduced pressure gives 4,6-difluoropyrimidine **13**, which is isolated by distillation. Treating 4,6-difluoropyrimidine **13** with hydrazine in water generates hydrazinopyrimidine **14**, and subsequent treatment with carbon disulfide followed by benzyl chloride gives 1,2,4-triazolopyrimidine **15** via Dimroth rearrangement. Displacement of the

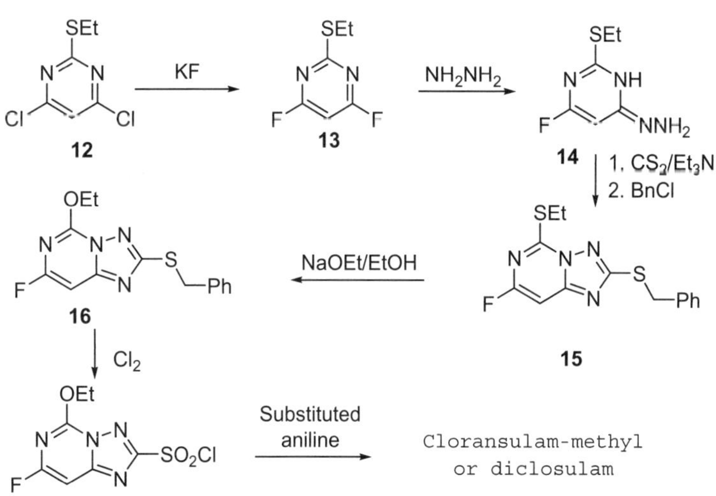

SCHEME 11.4

ethylthio group with an ethoxy group followed by oxidative chlorination and coupling with the appropriate aniline gives cloransulam-methyl or diclosulam.

11.1.3 Protoporphyrinogen Oxidase Inhibitors

These are chemical compounds that inhibit the enzyme protoporphyrinogen oxidase, quickly resulting in the formation of highly reactive compounds in plants that rupture cell membranes, causing leakage of cellular fluids. In general, the chemical structure of protoporphyrinogen oxidase (PPO) inhibitors consists of a substituted aromatic ring attached to a heterocyclic ring. A wide range of fluoro-substituted heterocycles, attached to aromatic rings having specific substitution patterns, are known to provide good biological activity. Fluoro-substituted uracil, triazolinone, and pyrazole are components of commercial products, as shown in Fig. 11.5.

Sulfentrazone was introduced in 1991 by FMC, one of the major players in this area.[7] Structure–activity studies directed toward the optimization of the 4-position of the triazolinone ring showed that optimal herbicidal activity can be obtained when the substituent is a small lipophilic group, notably difluoromethyl. As shown in Scheme 11.5, preparation of sulfentrazone involves formation of the triazolinone ring of intermediate **19**. The ring is formed in a "one-pot" reaction beginning with the condensation of the arylhydrazine **18** with acetaldehyde to form a corresponding arylhydrazone, which is reacted, without isolation, with potassium cyanate to give the triazolidine intermediate. The oxidation of the triazolidine is accomplished by the addition of an aqueous solution of sodium hypochlorite or by bubbling chlorine through the solution. Difluoromethylation at the 4-position of the triazolinone ring is accomplished by treating the triazolinone **19** with potassium carbonate and

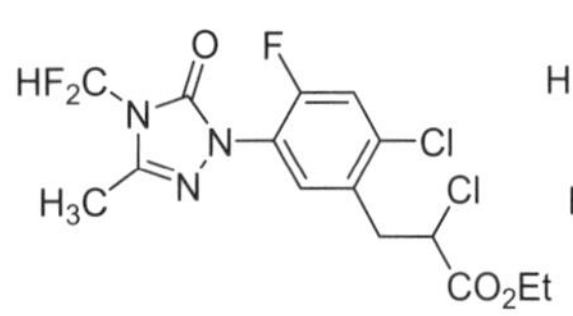

Carfentrazone-ethyl
(FMC)

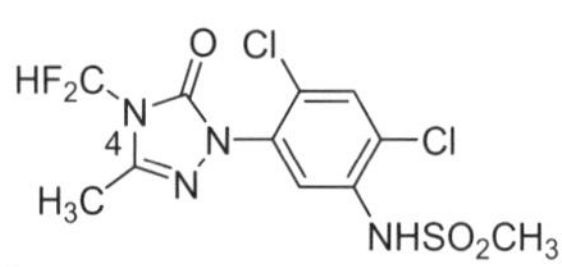

Sulfentrazone
(FMC)

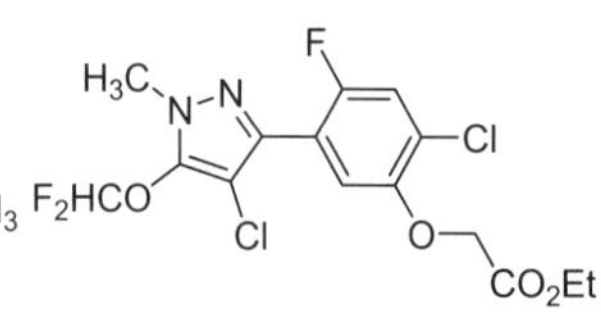

Pyraflufen-ethyl
(Nihon Nohyaku)

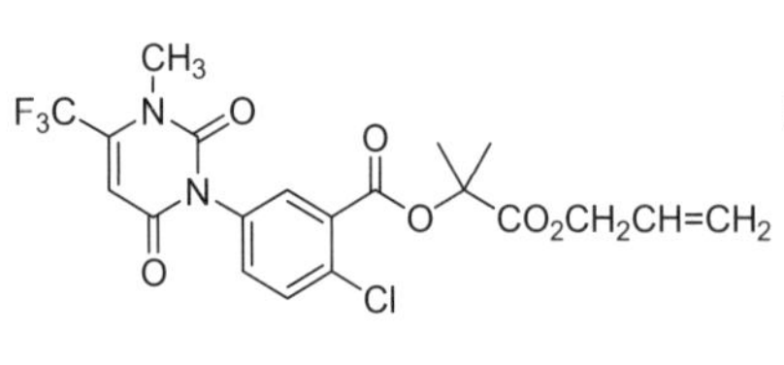

Butafenacil
(Syngenta)

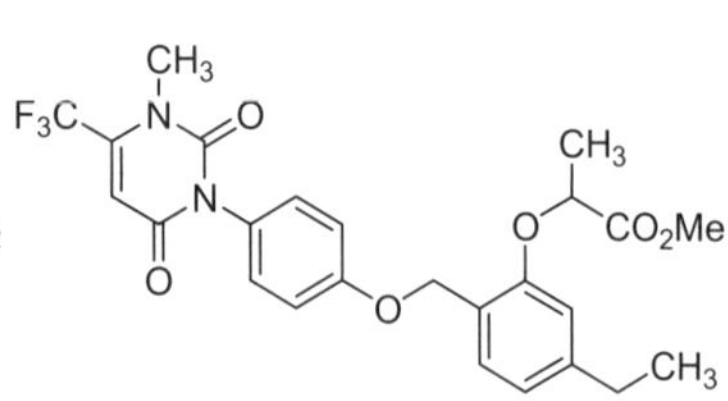

Benzfendizone
(FMC)

FIGURE 11.5

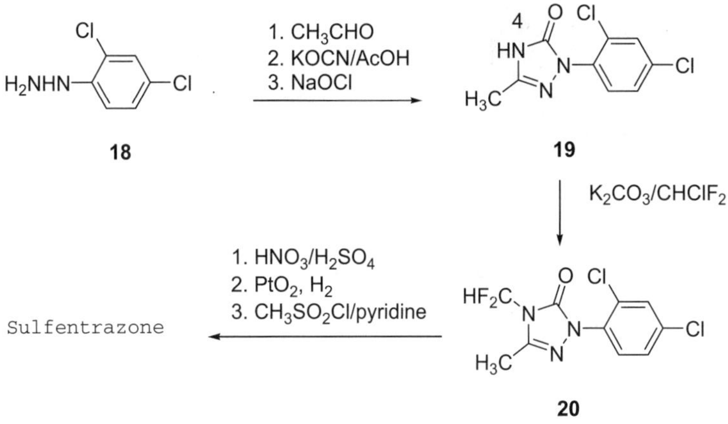

SCHEME 11.5

chlorodifluoromethane. Under optimum reaction conditions, only traces of the $-OCHF_2$ isomer are obtained, and this can be recycled back to starting material **19** by treatment with HCl. Nitration of triazolinone **20** with nitric acid in the presence of sulfuric acid followed by catalytic hydrogenation in the presence of PtO_2 in ethanol followed by treatment with methanesulfonyl chloride in the presence of pyridine gives sulfentrazone.

Replacement of the triazolinone ring with a uracil ring results in a dramatic increase in biological activity. An example of a uracil herbicide is butafenacil developed by Novartis (now Syngenta AG). Its preparation is shown in Scheme 11.6.

The uracil portion of the molecule is prepared by cyclization of the corresponding aryl isocyanate **22** with crotonate **21** in the presence of base.[8] The crotonate **21** is prepared by condensing methylamine with trifluoroacetoacetate, one of the most frequently used building blocks for the construction of trifluoromethyl-containing heterocycles.

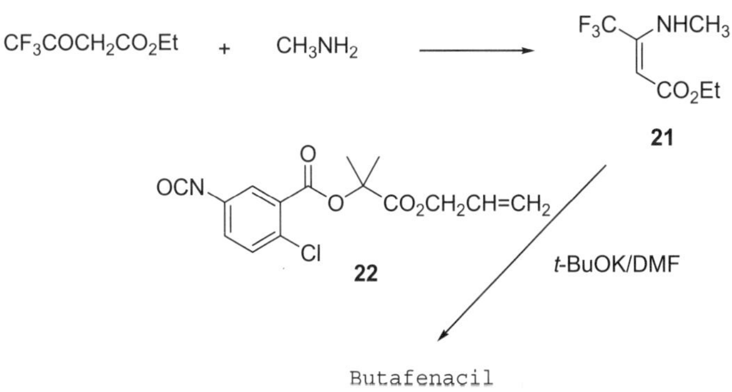

SCHEME 11.6

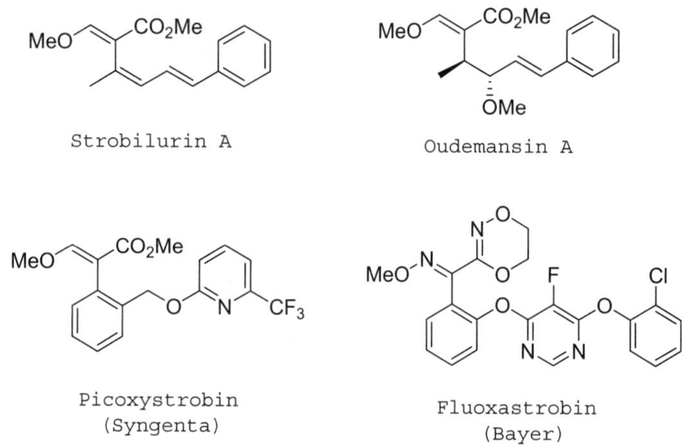

FIGURE 11.6

11.2 FUNGICIDES

11.2.1 Quinone Outside Inhibitor Fungicides

Quinone outside inhibitor (QoI) fungicides disrupt complex III mitochondrial respiration in fungi by inhibiting oxidation of ubiquinol at the "quinone outside" (Q_o) site of the cytochrome bc_1 complex, located in the inner mitochondrial membrane of fungi. Inhibiting mitochondrial respiration prevents normal fungal growth and development. The strobilurin compounds are the most important fungicides in this class.[9] Knowledge of the structures and antifungal activities of the fungicidal natural products strobilurin A and oudemansin A (see Fig. 11.6) from Basidiomycetes fungi led to the development of this class of agricultural fungicides. The earliest examples were first sold in 1996, and there are currently nine commercial strobilurin fungicides, with many others in development. Registrations have been obtained on a wide range of crops throughout the world, and indeed the strobilurins can be considered one of the most valuable classes of single-site fungicides discovered by the agrochemical industry. Picoxystrobin and fluoxastrobin,[10] shown in Fig. 11.6, are examples of strobilurin fungicides containing fluoro-substituted heterocyclic moieties.

Picoxystrobin was developed by Zeneca Agrochemicals (now Syngenta AG) and first registered in 2001.[11] Picoxystrobin is prepared as shown in Scheme 11.7.[12] Coupling the 2-hydroxypyridine **24** with corresponding benzyl chloride **25** under phase-transfer conditions followed by introduction of the methoxymethylene moiety yields picoxystrobin. The introduction of the trifluoromethyl group of **24** is achieved by chlorination of 2-picoline (**23**) followed by fluorination using HF and hydrolysis.

11.2.2 Carboxamide Fungicides

These inhibit mitochondrial function by disrupting complex II (succinate dehydrogenase) in the respiratory electron transport chain. Inhibiting respiration prevents the

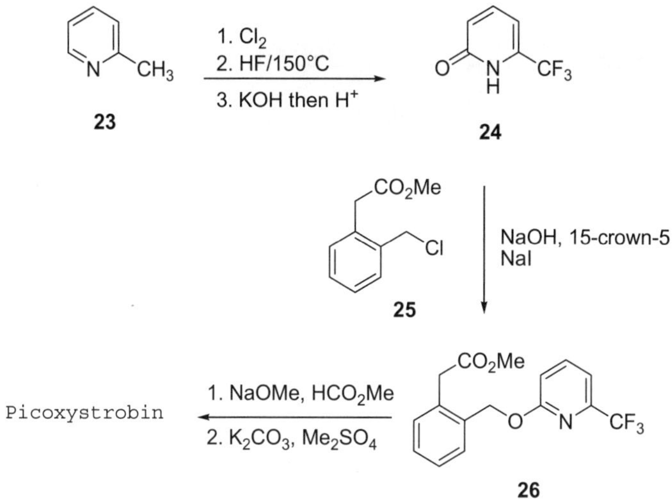

SCHEME 11.7

fungus from making ATP, and thus inhibits growth and reproduction. The first-generation compound carboxin, shown in Fig. 11.7, was developed about 40 years ago by Uniroyal Chemical Co. (now Chemtura Corp.) as a seed treatment fungicide. Thifluzamide developed in the late nineties by Monsanto (and sold to Rohm & Haas Co., now Dow AgroSciences)[13] shows higher activity than carboxin but its spectrum is not broadened. A new and more active fungicide, penthiopyrad, with the same mode of action as carboxin but a much broader antifungal spectrum, was discovered

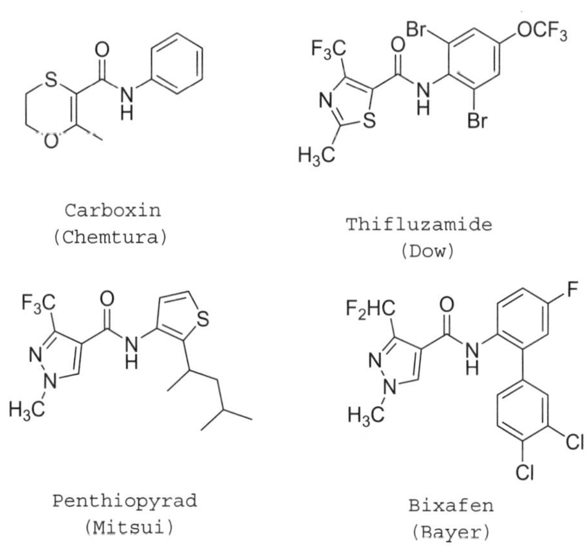

FIGURE 11.7

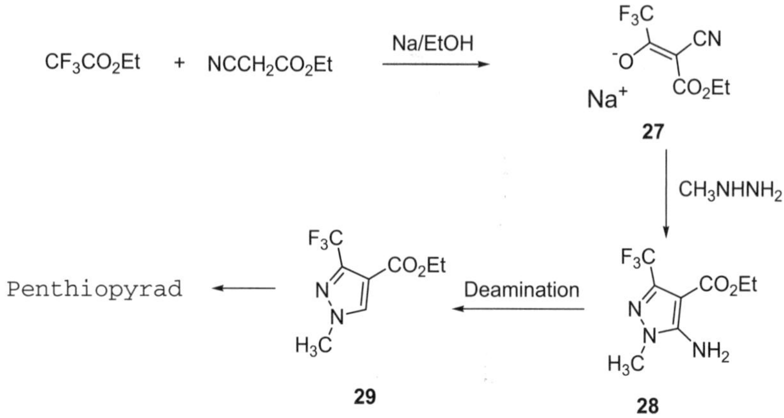

SCHEME 11.8

by Mitsui Chemicals.[14] Although carboxin is active primarily against rusts and *Rhizoctonia* disease, penthiopyrad that contains two heteroaromatic rings also shows high fungicidal activity. Branched alkyl substitution on the thiophene ring plays an important role in expanding its antifungal spectrum.

The key intermediate for the synthesis of penthiopyrad, 1-methyl-3-trifluoromethylpyrazole-4-carboxylate **29**, is prepared as shown in Scheme 11.8 by treatment of ethyl trifluoroacetate with ethyl cyanoacetate in the presence of a base such as sodium methoxide to give butenoate **27** followed by treatment of methylhydrazine sulfate in the presence of trifluoroacetic acid.[15] The deamination of **28** by diazotization–deazotization gives the key intermediate **29**. The final steps to penthiopyrad involve amination of ester **29** with the appropriate 2-alkenyl-3-aminothiophene derivative followed by hydrogenation using a noble metal catalyst.[16]

Bixafen is another pyrazole carboxamide recently developed by Bayer CropScience.[17] These active ingredients illustrate the fluoro-substituted pyrazole, thiazole, and pyrrole carboxamides as left part of the molecules that are being developed by several companies.

11.2.3 Phenylpyrrole Fungicides

These inhibit a mitogen-activated protein (MAP) kinase associated with osmotic signal transduction in fungi. Fludioxonil (see Fig. 11.8) is an example of this fungicide class.[18]

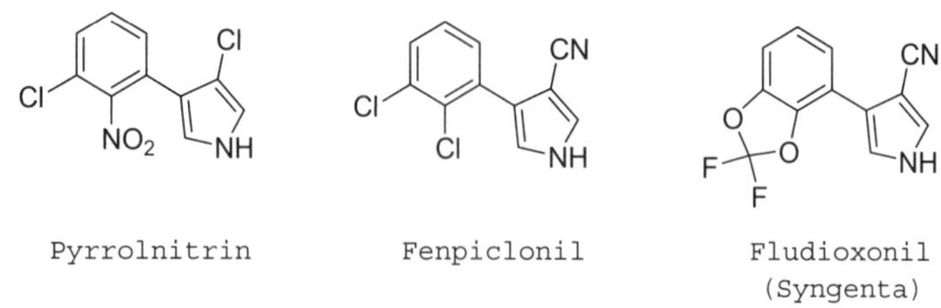

FIGURE 11.8

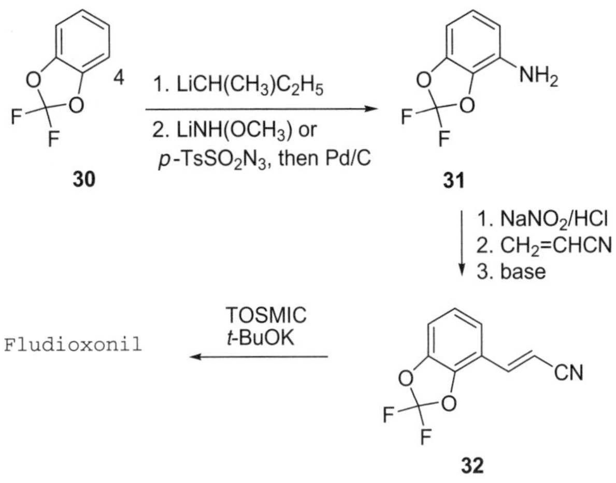

SCHEME 11.9

It was discovered by Ciba-Geigy AG (now Syngenta AG) for seed treatment[19] and foliar use.[20] Fludioxonil was first marketed in France as a cereal seed dressing in 1993 and subsequently as a foliar fungicide in 1995. Pyrrolnitrin (see Fig. 11.8), an antifungal secondary metabolite from *Pseudomonas pyrocinia*, served as the lead structure for synthesis optimization in this area. Because of its interesting antifungal activity, pyrrolnitrin was initially developed as an antimycotic for topical application in human medicine and also served as a lead structure for pharmaceutical research.[21] Pyrrolnitrin is not useful for plant protection due to rapid decomposition upon exposure to environmental conditions especially in light. Pyrrolnitrin analogues more suitable for plant protection were first described in 1969 in a Japanese patent application.[22] Stability is increased by introducing CN on the pyrrole ring and electron-withdrawing groups such as difluoro-1,3-dioxolane on the phenyl ring. Fenpiclonil was initially selected for development by Ciba-Geigy, but was superseded by fludioxonil.

The synthesis of fludioxonil is shown in Scheme 11.9. The key intermediate 3-phenyl-acrylonitrile **32** is prepared by Meerwein arylation of acrylonitrile using the diazonium salt of 2,3-(difluoromethylenedioxy)aniline (**31**). Treatment of intermediate **32** with *p*-toluenesulfonyl methylisocyanide (TOSMIC) in the presence of strong base gives fludioxonil.[23] The key intermediate **31** is available from several vendors. It is also prepared from 2,2-difluoro-1,3-benzodioxole (**30**), commercially available in bulk, by metalation at the 4-position of compound **30** followed by either direct conversion to aniline **31** using lithium methoxyamide or conversion to an azido derivative using *p*-toluenesulfonyl azide followed by reduction.[24]

11.2.4 Oxidative Phosphorylation Uncoupling Fungicides

These inhibit fungal respiration by uncoupling oxidative phosphorylation. Inhibiting respiration prevents normal fungal growth and development. Fluazinam, discovered by ISK and introduced in 1990, is a fluoro-substituted heterocycle containing fungicide in this group.[25] The best fungicidal activities were observed in this class

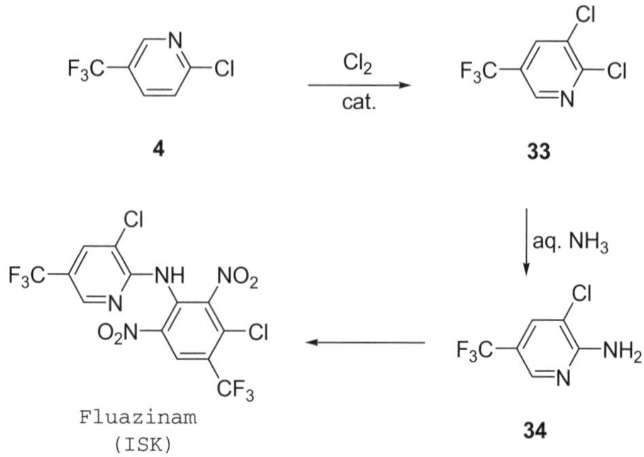

SCHEME 11.10

of compounds with *N*-phenylpyridinamines substituted with one trifluoromethyl group and one or two halogens (preferably chlorine) or two trifluoromethyl groups. Analogues substituted with cyano or nitro groups on the pyridine ring showed diminished activity, and substituted hydrophilic carboxylic acid groups or methylamino groups eliminated activity. The results can be explained as hydrophobicity and/or electronegativity effects. Both the trifluoromethyl group and chlorine are hydrophobic and have relatively small volumes. Indeed, studies involving different combinations of electronic, steric, and hydrophobic groups suggest that an electron-withdrawing effect combined with hydrophobic character is important for achieving high fungicidal activity. The importance of the trifluoromethyl substituent providing electronegativity as well as hydrophobicity on the pyridine ring of fluazinam contrasts with the role of the trifluoromethyl substituent on the pyridine rings of fluazifop-butyl (see Fig. 11.2) and chlorfluazuron (see Fig. 11.10), which improves biological activity mainly through its hydrophobic character.

The preparation of fluazinam is shown in Scheme 11.10. The key intermediate that is used to produce fluazifop-butyl, 2-chloro-5-trifluoromethylpyridine (**4**) (see Scheme 11.1), is used as a starting material for the synthesis of fluazinam as well. Further chlorination of pyridine **4** with chlorine gas in the presence of a catalyst such as acetic acid followed by amination with ammonia gives aminopyridine **34**. The synthesis of fluazinam is completed by nucleophilic reaction of **34** with 2,4-dichloro-3,5-dinitrobenzotrifluoride.

11.3 INSECTICIDES

11.3.1 Oxidative Phosphorylation Uncouplers

Dioxapyrrolomycin (Fig. 11.9) isolated from a *Streptomyces* culture was found to have moderate activity against certain insects and mites and to be a potent uncoupler

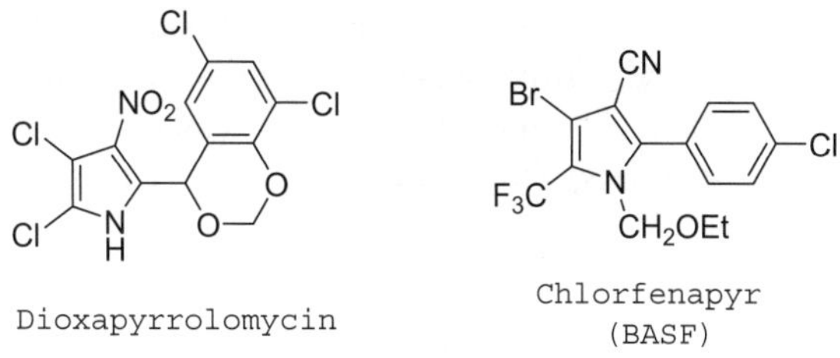

FIGURE 11.9

of oxidative phosphorylation. These findings, along with the novel structure, led to the investigation of analogues having other combinations of substituents on the pyrrole ring. Resulting from this effort was chlorfenapyr, which was developed by American Cyanamid (now BASF AG) (Fig. 11.9).[26]

Replacement of a halogen by a trifluoromethyl group in biologically active compounds often resulted in enhanced activity. In the synthesis program leading to chlorfenapyr, replacing the chlorine with a trifluoromethyl group at the 5-position of the pyrrole ring was found to improve both potency and spectrum but also cause severe phytotoxicity. This problem was solved by attaching an ethoxymethyl group on the nitrogen of pyrrole ring, which also reduces mammalian toxicity. The synthesis of chlorfenapyr is shown in Scheme 11.11. Oxazolinone **37**, the key intermediate for the synthesis of chlorfenapyr, is prepared from the corresponding amino acid derivative **36** in which the trifluoromethyl group is derived from trifluoroacetic anhydride, one of the popular trifluoromethyl sources.[27] Utilizing the thermal cyclization conditions described by Albonico[28] with a large excess of

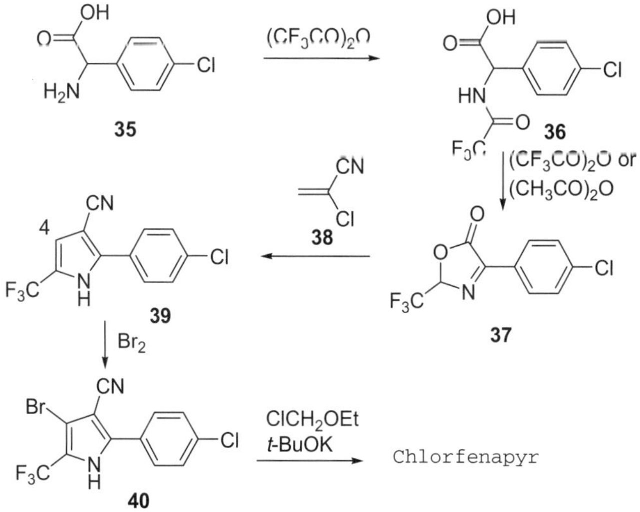

SCHEME 11.11

α-chloroacrylonitrile **38**, the pyrrole-3-carbonitrile **39** is obtained. Bromination of **39** at the 4-position of the pyrrole nucleus is accomplished using a variety of standard methods, including *N*-bromosuccinimide, bromine/acetic acid, or bromine/sodium acetate/acetic acid. Alkylation on nitrogen using chloromethyl ethyl ether then gives chlorfenapyr.

11.3.2 γ-Aminobutyric Acid-Regulated Chloride Channel Blockers

γ-Aminobutyric acid (GABA) is a major inhibitory neurotransmitter in the nervous system of animals.[29] The fast type of inhibitory neurotransmission is mediated via the ionotropic GABA receptor in postsynaptic neurons. The ionotropic GABA receptor is a chloride channel that is gated by the binding of GABA released from the terminus of presynaptic neurons. Chloride ions flow into the postsynaptic neuron through the GABA-gated channel to reverse hyperpolarization, thus suppressing nerve excitation. Blocking the inhibitory effects of GABA causes unregulated excitatory neurotransmission. The GABA receptor site that is of particular interest to agrochemists binds antagonists noncompetitively. Noncompetitive antagonists binding to this site within the channel stabilize the closed conformation of the channel and cause excitation of animal neurons. Fipronil is a noncompetitive GABA antagonist discovered by Rhone-Poulenc (now Bayer CropScience) in 1987. Fipronil[30] is a highly effective insecticide against both piercing-sucking and chewing insects, and can be effectively delivered via soil, foliar, bait, or seed treatment applications. The preparation of fipronil is shown in Scheme 11.12. Phenylpyrazole **43**, a key intermediate to prepare fipronil, can be prepared from corresponding aniline **42** and ethyl 2,3-dicyanopropionate **41** in the presence of nitrosyl sulfuric acid. A trifluoromethylthio group is introduced by treatment of phenylpyrazole **43** with trifluoromethylsulfenyl chloride in the presence of a base such as pyridine, and subsequent oxidation of resulting sulfide **44** with *m*-chloroperbenzoic acid gives fipronil.

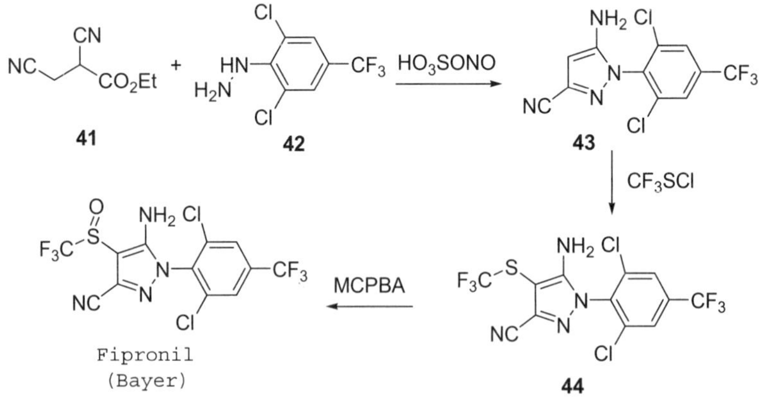

SCHEME 11.12

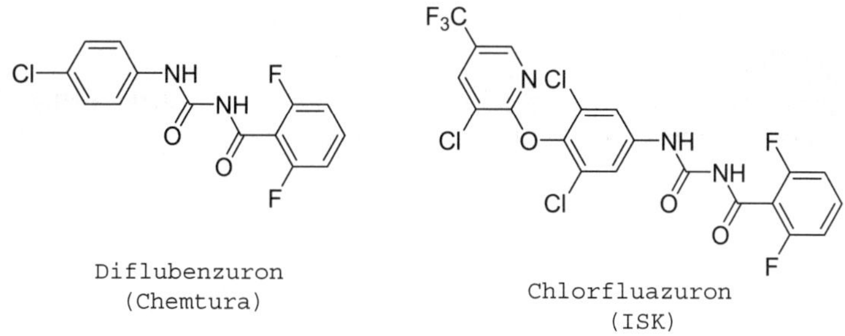

FIGURE 11.10

11.3.3 Chitin Synthesis Inhibitors

Chitin synthase is the arthropod enzyme that converts uridine diphosphoryl-*N*-acetylglucosamine into chitin, the β-1,4-linked polymer of *N*-acetylglucosamine. Chitin synthase is a member of the class of enzymes known as polymerizing glycosyltransferases, which are responsible for the synthesis of critical structural biopolymers, including cellulose as well as chitin. Inhibition of chitin synthase is widely recognized as a largely untapped, intrinsically mammalian-safe, arthropodicidal mechanism of action. Because the exoskeleton of insects consists mostly of chitin, disrupting the formation, deposition, and/or cross-linking of chitin can lead to insect death, especially during molting.[31] Modeling studies suggest the benzoylurea moiety, which is a common structural feature of chitin synthesis inhibitors, resembles diphosphoryl-*N*-acetylglucosamine. Thus, the ability of the benzoylurea containing molecules to bind to the diphosphoryl-*N*-acetylglucosamine site on chitin synthase seems plausible. Diflubenzuron (Fig. 11.10) was introduced in 1975 by Philips-Duphar B.V. (now Chemtura Corp.). Since the publication of the Philips-Duphar patent disclosing diflubenzuron, a diverse family of structures has emerged in this area[32]. Chlorfluazuron, which was introduced by ISK in 1982, was invented by adding a 3-trifluoromethylpyridinyloxy moiety to the benzoylurea skeleton of diflubenzuron, resulting in superior larvicidal activity.[2]

As shown in Scheme 11.13, to prepare chlorfluazuron, ISK utilized 2-chloro-5-trifluoromethylpyridine (**4**) as their key intermediate, which is also used as a starting

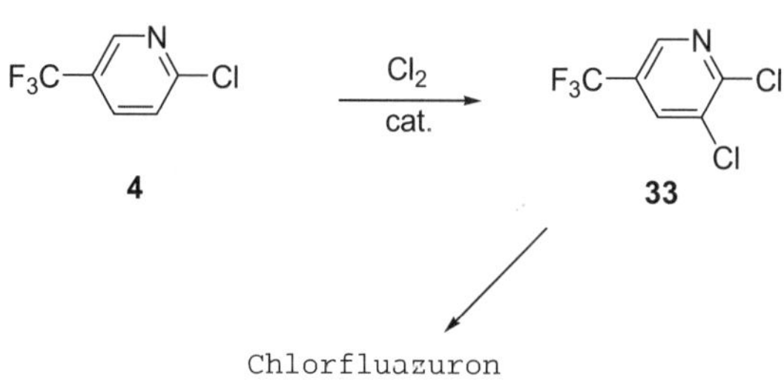

SCHEME 11.13

material for the synthesis of fluazifop-butyl (see Scheme 11.1) and fluazinam (see Scheme 11.10).

This provides another demonstration of the utility of the trifluoromethylpyridine group as a building block for agricultural products.

11.4 SUMMARY

Compounds having fluorine-containing substituents on heterocyclic rings have gained importance as active ingredients in agricultural products. Discovery programs typically explore introduction of fluorine-containing substituents, as their electronegativity and hydrophobicity can improve biological activity. Heterocycles with fluorine-containing substituents are amenable to large-scale preparation. A variety of fluorine-containing heterocyclic compounds have been commercialized.

REFERENCES

1. Tomlin, C.D.S., Ed.; *A World Compendium: The Pesticide Manual*, 14th ed.; The British Crop Protection Council: 2006.
2. Haga, T.; Tsujii, T.; Hayashi, K.; Kimura, F.; Sakashita, N.; Fujikawa, K. *Synthesis and Chemistry of Agrochemicals II, ACS Symposium Series No. 443*; Baker, D.R.; Fenyes, J.G.; Moberg, W.K., Eds.; American Chemical Society: Washington, DC, 1991; pp 107–119.
3. Levitt, G.; *Synthesis and Chemistry of Agrochemicals II, ACS Symposium Series No. 443*; Baker, D.R.; Fenyes, J.G.; Moberg, W.K., Eds.; American Chemical Society: Washington, DC, 1991; pp 17–31.
4. Brown, H.M.; Kearney, P.C. *Synthesis and Chemistry of Agrochemicals II, ACS Symposium Series No. 443*; Baker, D.R.; Fenyes, J.G.; Moberg, W.K., Eds.; American Chemical Society: Washington, DC, 1991; pp 32–49.
5. Kimura, F.; Haga, T.; Sakashita, N.; Honda, T.; Murai, S. *Japan Patent Kokai Tokkyo Koho* 88 146873, **1988**.
6. VanHeertum, J.C.; Gerwick, B.C., III; Kleschick, W.A.; Johnson, T.C. *US Patent* 5,163,995, **1992**.
7. Theodoridis, G.; Baum, J.S.; Hotzman, F.W.; Manfredi, M.C.; Maravetz, L.L.; Lyga, J.W.; Tymonko, J.M.; Wilson, K.R.; Poss, K.M.; Wyle, M.J. Synthesis and Chemistry of Agrochemicals III, *ACS Symposium Series No. 504*; Baker, D.R.; Fenyes, J.G.; Steffens, W.J., Eds.; American Chemical Society: Washington, DC, 1992; pp 134–146.
8. Sting, A.R. *European Patent* 0831091, **1997**.
9. Sauter, H.; Steglich, W.; Anke, T. *Angew. Chem., Int. Ed.* **1999**, *28*, 1328–1349.
10. Dutzmann, S.; Mauler-Machnik, A.; Kerz-Mohlendick, F.; Applegate, J.; Heinemann, U. *Proceedings of the BCPC Conference–Pest and Diseases*, 2002, *1*, 365–370.
11. Godwin, J.R.; Bartlett, D.W.; Clough, J.M.; Godfrey, C.R.A.; Harrison, E.G.; Maund, S. *Proceedings of the BCPC Conference–Pest and Diseases*, 2000, *2*, 533–540.

12. Worthington, P.A.; Munns, G.R.; Jones, R.V.; Standen, M.C.; Ritchie, D.J.; Forrester, J. *World Patent Publication* 97/01538, **1997**.
13. O'Reilly, P.; Kobayashi, S.; Yamane, S.; Phillips, W.G.; Raymond, P.; Castanho, B. *Proceedings of the BCPC Conference–Pest and Diseases*, 1992, *1*, 427–438.
14. Yanase, Y.; Yoshikawa, Y.; Kishi, J.; Katsuta, H. In *Pesticide Chemistry*, Ohkawa, H.; Miyagawa, H.; Lee, P., Eds.; Wiley-VCH Verlag GmbH & Co., 2007; pp 295–303.
15. Yoshida, K.; Matsuo, S.; Kitashima, T.; Tomiya, K.; Kodaka, K. *European Patent* 1067121, **2001**.
16. Katsuta, H.; Kitashima, T.; Sato, K. *European Patent* 1524271, **2005**.
17. Dunkel, R.; Rieck, H.; Elbe, H.; Wachendorff-Neumann, U.; Kuck, K.*World Patent publication* 03/070705, **2003**.
18. Nyfeler, R.; Ackermann, P. *Synthesis and Chemistry of Agrochemicals III, ACS Symposium Series No. 504*; Baker, D.R.; Fenyes, J.G.; Steffens, J., Eds.; American Chemical Society: Washington, DC, 1992; pp 395–404.
19. Leadbeater, A.J.; Nevill, D.J.; Steck, B.; Nordmeyer, D. *Proceedings of the BCPC Conference–Pest and Diseases*, 1990, *2*, 825–830.
20. Gehmann, K.; Nyfeler, R.; Leadbeater, A.J.; Nevill, D.; Sozzi, D. *Proceedings of the BCPC Conference–Pest and Diseases*, 1990, *2*, 399–406.
21. Umio, S.; Kariyone, K.; Tanaka, K.; Nakamura, H. *Chem. Pharm. Bull. (Tokyo)* **1969**, *17*, 559–566.
22. Nakanishi, K.; Shimizu, K.; Ouo, K.; Ueda, K. *Japan Patent Kokai Tokkyo Koho* 50 002011, **1975**.
23. Nyfeler, R.; Ehrenfreund, J. *European Patent* 0206999, **1986**.
24. Schlosser, M.; Gorecka, J.; Castagnetti, E. *Eur. J. Org. Chem.* **2003**, 452–462.
25. Nakajima, T.; Komyoji, T.; Akagi, T.; Nagatani, K.; Haga, T.; Matsuo, N.; Suzuki, K. *Synthesis and Chemistry of Agrochemicals IV, ACS Symposium Series No. 584*; Baker, D. R.; Fenyes, J.G.; Basarab, G.S., Eds.; American Chemical Society: Washington, DC, 1995; pp 443–448.
26. Addor, R.W.; Babcock, T.J.; Black, B.C.; Brown, D.G., Diehl, R.E.; Furch, J.A.; Kameswaran, V.; Kamhi, V.M.; Kremer, K.A.; Kuhn, D.G.; Lovell, J.B.; Lowen, G.T.; Miller, T.P.; Peevey, R.M.; Siddens, J.K.; Treacy, M.F.; Trotto, S.H.; Wright, D.P., Jr.; *Synthesis and Chemistry of Agrochemicals III, ACS Symposium Series No. 504*; Baker, D.R.; Fenyes, J.G.; Steffens W.J., Eds.; American Chemical Society: Washington, DC, 1992; pp 283–297.
27. Kuhn, D.G.; Kamhi, V.M.; Furch, J.A.; Diehl, R.E.; Trotto, S.H.; Lowen, G.T.; Babcock, T. J. *Synthesis and Chemistry of Agrochemicals III, ACS Symposium Series No. 504*; Baker, D.R.; Fenyes, J.G.; Steffens, W.J., Eds.; American Chemical Society: Washington, DC, 1992; pp 298–305.
28. Benages, I.; Albonico, S. *J. Org. Chem.* **1978**, *43*, 4273–4276.
29. Ozoe, Y.; Ishkawa, S.; Tomiyama, S.; Ozoe, F.; Kozaki, T.; Scott, J.G. *Synthesis and Chemistry of Agrochemicals VII, ACS Symposium Series No. 948*; Lyga, J.W.; Theodoridis, G., Eds.; American Chemical Society: Washington, DC, 2007; pp 39–50.
30. Buntain, I.G.; Hatton, L.R.; Hawkins, D.W.; Pearson, C.J.; Roberts, D.A. *European Patent* 0296117, **1988**.
31. Henrie, R.N., II; Cullen, T.; Dugan, B.; Zhang, L.; Deng, Y., Simpson, S.F.; Black, B.; Schuler, F.; Yu, S.J. *Synthesis and Chemistry of Agrochemicals VII, ACS Symposium*

Series No. 948; Lyga, J.; Theodoridis, W.G., Eds.; American Chemical Society: Washington, DC, 2007; pp 83–92.

32. Dixson, J.A.; Plummer, E.I.; Kral, R.M.; Seelye, D.E.; *Synthesis and Chemistry of Agrochemicals II, ACS Symposium Series No. 443*; Baker, D.R.; Fenyes, J.G.; Moberg, W.K., Eds.; American Chemical Society: Washington, DC, 1991; pp 459–468.

12

PHARMACEUTICALS CONTAINING FLUORINATED HETEROCYCLIC COMPOUNDS

KENJI UNEYAMA[1] AND KENJI SASAKI[2]

[1] *Department of Applied Chemistry, Faculty of Engineering, Okayama University, 3-1-1, Tsushima-Naka, Okayama 700-8605, Japan*

[2] *Faculty of Pharmaceutical Science, Okayama University, 1-1-1, Tsushima-Naka, Okayama 700-8530, Japan*

12.1 INTRODUCTION

About 10% of the total sales of pharmacuticals currently used for the medical treatment are containing fluorine atom. Over 50 years, many fluorinated medicinal and agrochemical candidates have been discovered and the interest toward development of fluorinated chemicals has been steadily increased. In particular, the strong requirement for the extremely effective drug candidates increased the interest in the unique biological properties arose from fluorine. And also the high availability of the fluorinated synthetic blocks and the effective fluorinating reagents, the widely reliable fluorination technology, and the accumulation of basic and advanced knowledge of the fluorine chemistry rapidly accelerated their developments.

A great number of examples have been reported that the introduction of a fluorine or fluorinated groups into an organic molecule induces dramatic change in its chemical, physical, and also pharmacological properties. The unique properties of fluoroorganic molecules may arise from the properties such as (i) the greatest electronegativity of fluorine, (ii) the largest strength of the carbon–fluorine bond,

Fluorinated Heterocyclic Compounds: Synthesis, Chemistry, and Applications, Edited by Viacheslav A. Petrov

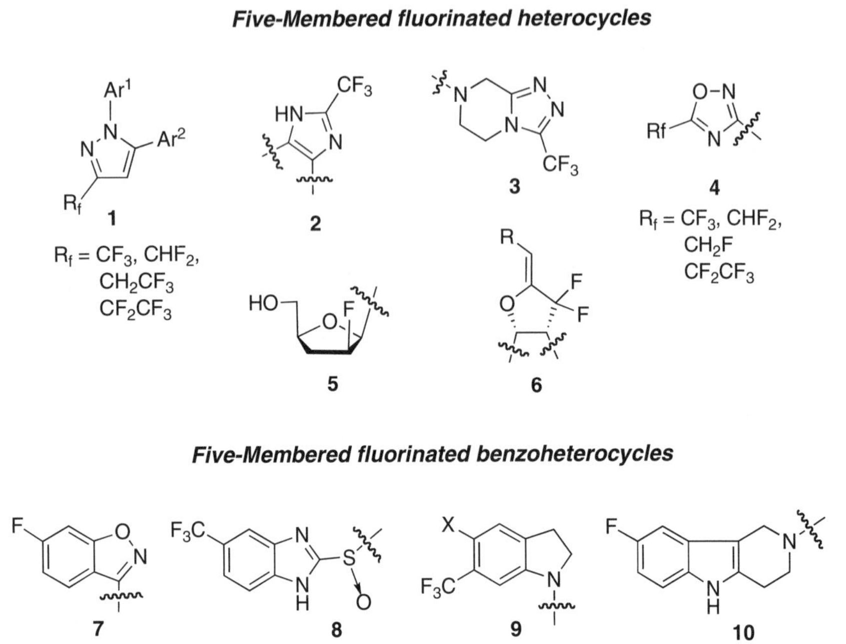

FIGURE 12.1 Five-membered fluorinated heterocyclic blocks used for therapeutic drugs.

(iii) the hardness and the low van der Waals interaction due to the low polarizability, (iv) the increased hydrophobicity, and (v) the second smallest atomic size of the fluorine atom. These factors are operative singly or sometimes cooperatively to affect the pharmacological properties of the fluorinated molecules.[1]

Fluorine and trifluoromethyl group have been used mostly in the present fluorinated medicines, and very few difluoromethyl and difluoromethylene moieties have been employed. Since the active research for the difluoromethylation and difluoromethylenation methods has taken place, the difluoro moieties will be candidates for the structural units of new drugs in near future.[2]

Fluorinated heterocycles are frequently employed as a partial unit in the synthesis of fluorinated medicines. Figures 12.1 and 12.2 summarize fluorinated heterocyclic units currently used for the synthesis of therapeutic drugs. Majority of fluorinated drugs are constructed by five- and six-membered nitrogen heterocycles containing fluorine, trifluoromethyl, difluoromethyl, fluoromethyl, 2,2,2-trifluoroethyl, and pentafluoroethyl groups. In this chapter, the syntheses of these fluorinated hererocycles related to pharmaceuticals are described, specially focusing on the methods for the construction of heterocyclic rings and the introduction of fluorine functionality. In Sections 12.2 and 12.3, syntheses of the five-membered heterocycles including nitrogen and oxygen heterocycles, and five-membered benzoheterocycles are described, respectively. In Section 12.4, six-membered fluorinated heterocycles are described. Other heterocycles are presented in Section 12.5.

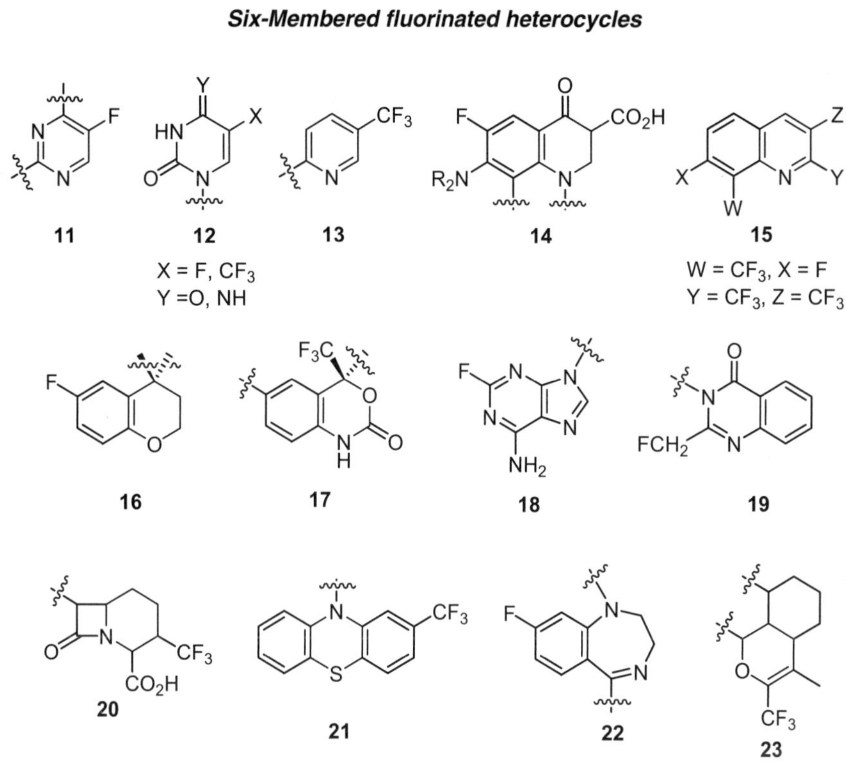

FIGURE 12.2 Six-membered fluorinated heterocyclic blocks used for therapeutic drugs.

12.2 FIVE-MEMBERED FLUORINATED HETEROCYCLES

12.2.1 1-(*N*-aryl)-3-Trifluoromethylpyrazoles

Celecoxib **24**, a highly selective COX-2 inhibitor and a nonsteroidal anti-inflammatory drug (NSAID), is used for the treatment of arthritis and pain. The core skeleton of Celecoxib **24** is a 3-trifluoromethylpyrazole ring, which has been synthesized by [2 + 3] cyclization and condensation of 1,3-diketones or their equivalents. Figure 12.3 summarizes three synthetic approaches toward 3-trifluoromethylpyrazole compounds: via condensation of 1,3-diketones or α,β-unsaturated ketones with arylhydrazine (route **A**, Schemes 12.1–12.3), via [2 + 3] cycloaddition of trifluoroacetamide with substituted ethenes or acetylene (route **B,** Schemes 12.4 and 12.5), and via cycloaddition of CF_3-alkene with C–N=N unit (route **C**, Scheme 12.6).

Scheme 12.1 shows a synthetic example via route **A**, which provides tri and difluoromethylated pyrazole analogues in reasonable yields. Condensation of 4-aryl-1,1,1-trifluoro-2,4-butanedione **26** with aryl hydrazine **27** in aqueous *N,N*-dimethylacetamide provides the desired pyrazoles **28** in good yields with an excellent regioselectivity (3-CF_3-5-aryl >99.5–99.8%). The high regioselectivity arises from

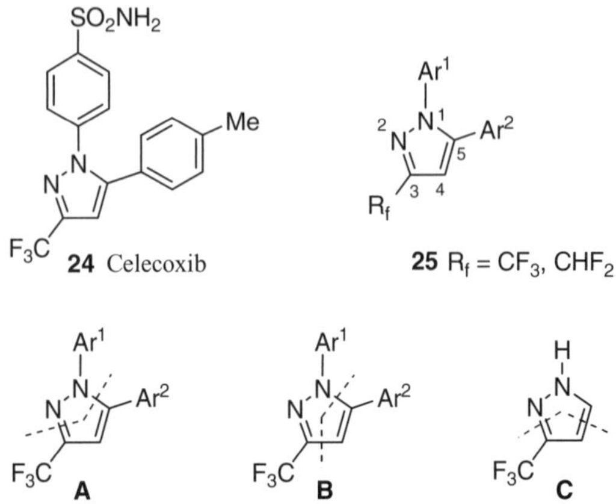

FIGURE 12.3 Construction of pyrazole skeleton involved in Celecoxib.

the exclusive nucleophilic addition of terminal amino group of aryl hydrazine to the more electron-deficient trifluoroacetyl carbonyl group rather than benzoyl carbonyl group.[3] The same protocol is applicable for the synthesis of 3-difluoromethylpyrazoles as shown in the table of Scheme 12.1.[3]

^{18}F Labeled Celecoxib **31** was synthesized by the replacement of bromine in the *N*-protected **30** by ^{18}F.[4] For the preparation of bromodifluoromethyl compound **30**, condensation of 1,3-diketone **29** with 4-aminosulfonylphenylhydrazine is applicable (Scheme 12.2).

Another construction of 1-*N*-aryl-3-trifluoromethylpyrazole skeleton via route **A** is shown in Scheme 12.3. The condensation of 4-ethoxy-3-buten-2-one **32** as a synthetic equivalent of 1,3-diketone with aryl hydrazine **33** provides a desired product **34** although the regioselectivity is poor (Scheme 12.3).[5] The reaction of **32** with hydrazine provides 3-trifluoromethylpyrazol **36**, which can further be converted to *N*-aryl-trifluoromethylpyrazole.[6]

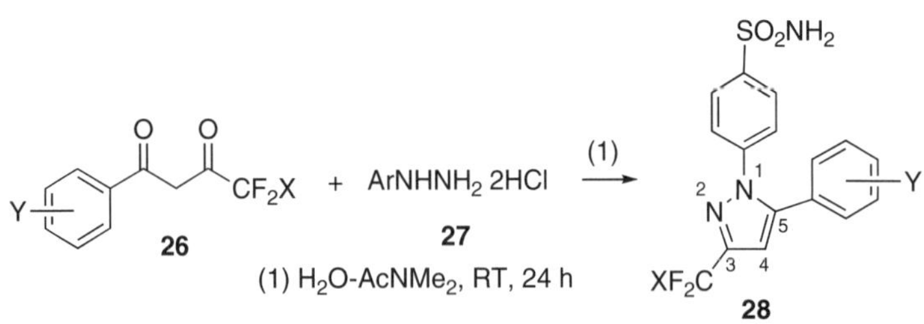

Yield of **28** (%)

Y	*p*-NO$_2$	*p*-MeO	*p*-Br	H
X = F	79	71	78	77
X = H	60	85	78	66

SCHEME 12.1

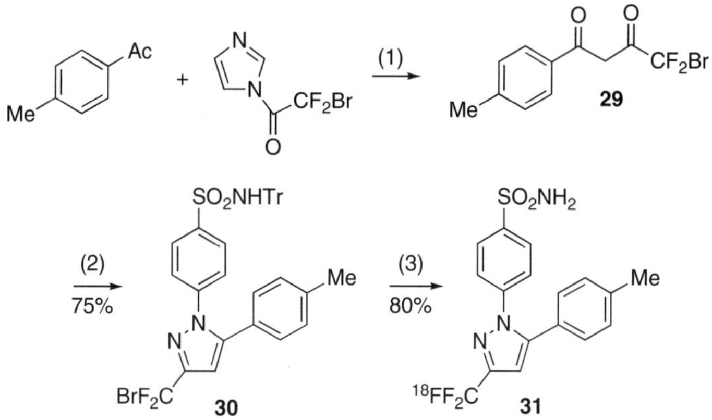

(1) $(Me_3Si)_2NNa$ / THF, –78°C; (2) (i) 4-aminosulfonylphenyl hydrazine, EtOH, reflux, 16 h, (ii) $(4\text{-}MeOC_6H_4)_2CPhCl$ - Et_3N/CH_2Cl_2; (3) (i) $Et_4N\ ^{18}F$/DMSO, 135°C, 40 min, (ii) CF_3CO_2H/CH_2Cl_2, RT, 5 min.

SCHEME 12.2

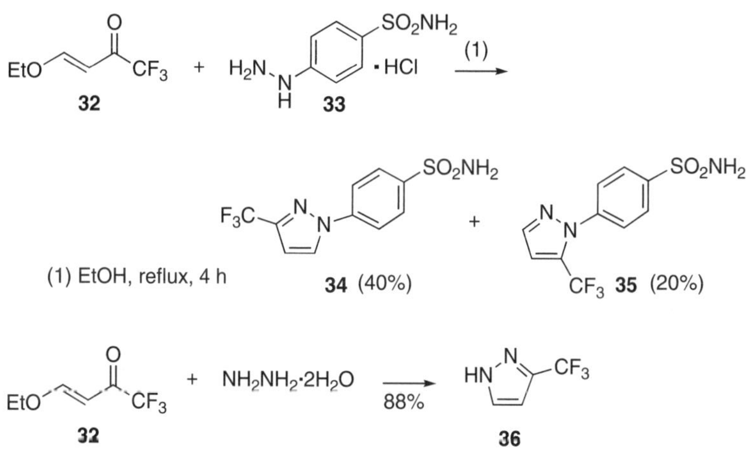

(1) EtOH, reflux, 4 h

SCHEME 12.3

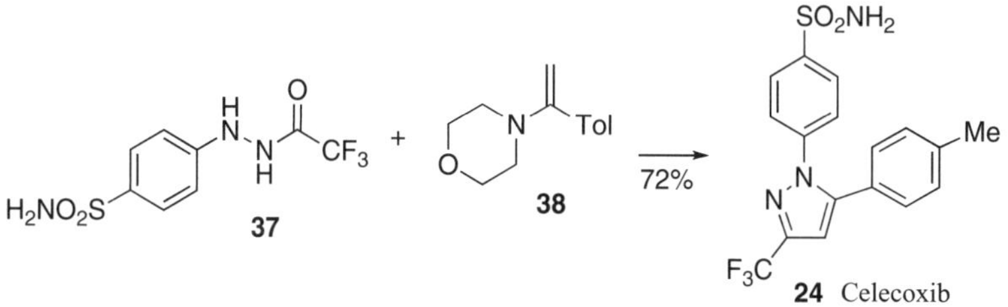

(1) (i) $PhSO_2Cl$ - *N*-Me-morpholine/AcOEt, (ii) Et_3N/THF

SCHEME 12.4

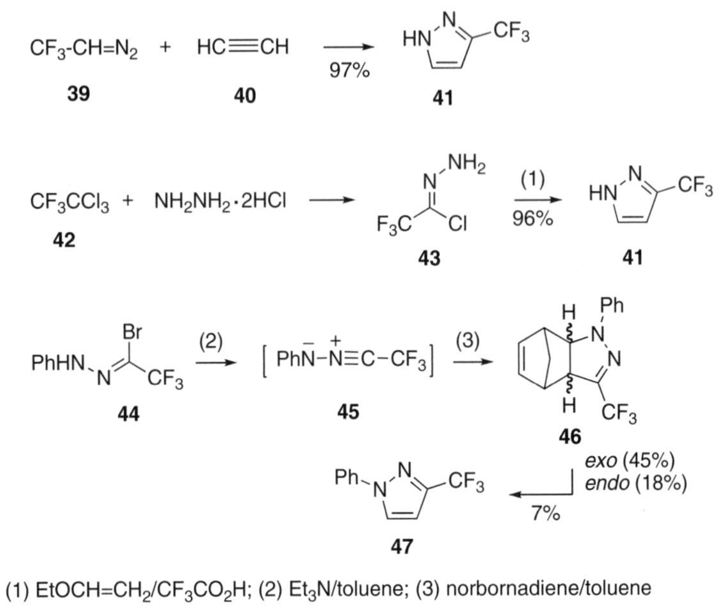

SCHEME 12.5

Scheme 12.4 shows an example of the construction of the pyrazole ring by [2 + 3] cycloaddition via route **B**. Celecoxib **24** is synthesized in 70% yield by a condensation–cyclization sequence.[7]

Three synthetic approaches toward 3-trifluoromethylpyrazole **41** are summarized in Scheme 12.5, all of which involve [2 + 3] cycloaddition. Either combinations of diazotrifluoroethane **39** with acetylene **40**[8] or trifluoroacetimidoyl chloride **43** with ethyl vinyl ether[9] leads to **41**. However, cycloaddition of a zwiterionic intermediate **45** with norbornadiene affords **47** (after removal of norbornene fragment), but only as a minor product.[10] Intermediate **45** is generated by the dehydrobromination of imidoylbromide **44** with triethylamine.

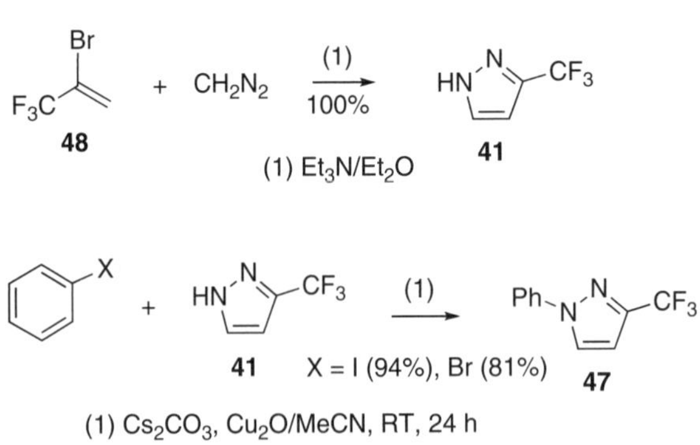

SCHEME 12.6

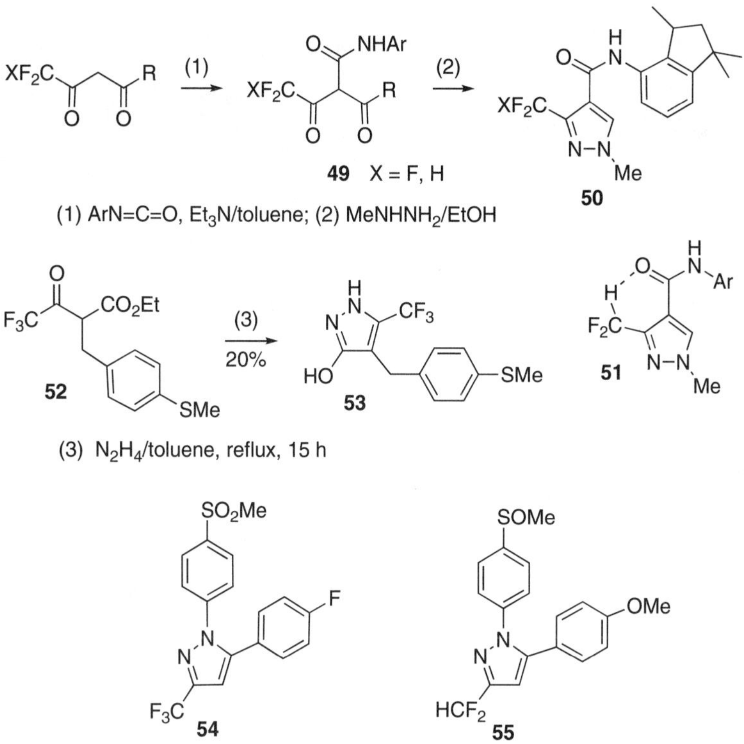

SCHEME 12.7

1,3-Dipole cycloaddition of diazomethane to 2-bromo-3,3,3-trifluoropropene **48** (Scheme 12.6) followed by dehydrobromination affords **41**, quantitatively.[11] A copper catalyzed *N*-phenylation reaction of **41** with phenyl bromide or iodide under mild conditions provides *N*-phenyl-3-trifluoromethylpyrazole **47** in excellent yields.[12]

N-Methyl-(1,1,3-trimethylindan-4-yl)pyrazole-4-carboxamides **50**, agricultural fungicides, are prepared by condensation of **49** with methylhydrazine (Scheme 12.7). Interestingly, the difluoromethyl compound **50** (X = H) is more effective as a curative fungicide against *Alternaria Solani* on tomatos than CF_3-compound **50** (X = F).[13] It is believed that the hydrogen bonding between carbonyl and hydrogen of HCF_2 group is responsible for higher biological activity of **50** (X = H).[14] The related 5-hydroxy-pyrazole **53**, a new potent anti-hyperglycemic agent, is also prepared from **52** by the similar condensation with hydrazine.[15] Structurally similar compounds **54**[16a,b] and **55**[16c] are known to be an anti-inflammatory agent.

12.2.2 2-Trifluoromethylimidazoles

Flumizole **56** (Fig. 12.4) is an immunoregulatory and anti-inflammatory agent. In animal studies, it was found to be several times more active compared with indomethacin (rat foot edema and prostaglandin synthetase tests).[17] The related

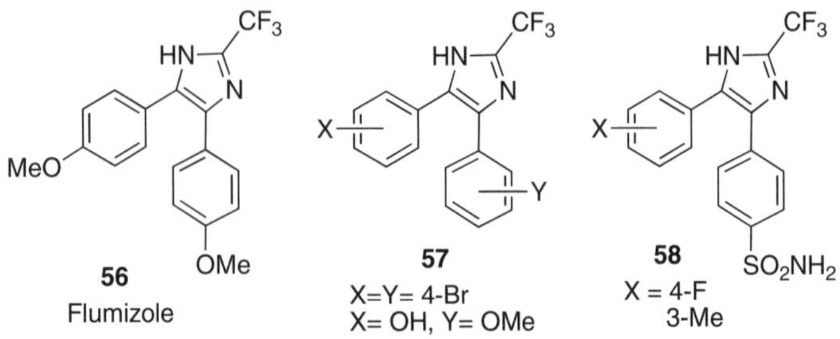

FIGURE 12.4 Flumizole and the related imidazoles.

4,5-diaryl-2-trifluoromethylimidazoles **57**[17] and **58**[18] are also known to have potent antiinflammatory effect.

Most of 4,5-diaryl-2-trifluoromethylimidazoles have been synthesized by the methods shown in Scheme 12.8. The imidazoles **60** are synthesized by the condensation of trifluoroacetaldehyde equivalent with applicable diaryl-1,2-diketones,[17] where a carbonyl group is replaced with imino group *in situ* in the presence of ammonium acetate.[19] The imidazole **63** is synthesized by the sequential substitution–condensation reactions of α-bromoketone **62** with trifluoroacetiminoamide **61**.[19]

The syntheses of 2-trifluoromethylbenzimidazoles (**65**, **68**, and **71**) are summarized in Scheme 12.9. Although condensation of phenylenediamines **64** with trifluoroacetic acid is the most convenient method[20], the reaction of phenylenediamines with either trifluoroacetimidoyl derivatives **67**[21] or **70**[22] also provides the desired benzimidazoles **68** and **71** in good yields. A direct radical trifluoromethylation of imidazole ring **72** with trifluoromethyl iodide has a poor regioselectivity.[23]

12.2.3 3-Trifluoromethyl-[1,2,4]triazolo[4,5-*a*]piperazines

Sitagliptin **75** is a potent, new oral anti-diavetic drug used for control of type 2 diabetes. Sitagliptin, discovered through the optimization of a class of β-aminoacid,

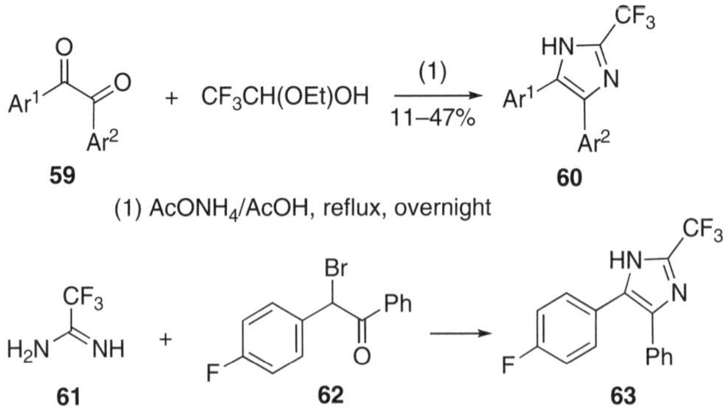

SCHEME 12.8

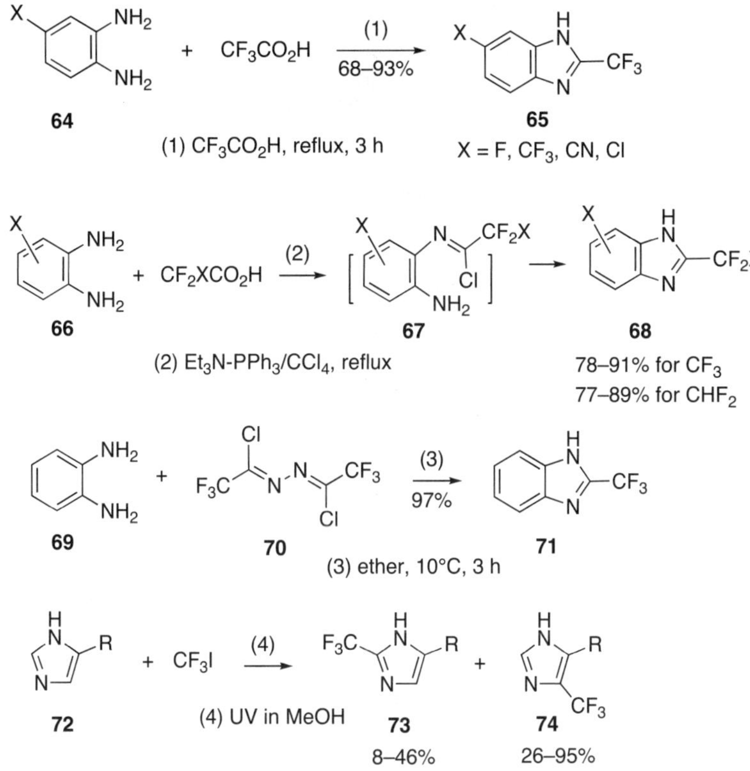

SCHEME 12.9

inhibits dipeptidyl peptidase IV (DPP-4), which deactivates the production of glucagon like peptides 1 (GLP-1), and glucose-dependent insulinotropic polypeptides (GIP)[24] and this increases glucose-dependent insulin secretion from β-cell. As for GLP-1, it also inhibits glucagons secretion. This compound **75** has been synthesized by condensation of β-amino acid **79** with 3-trifluoromethyl-[1,2,4]triazolo-[4,3-a] piperazines **80** (Scheme 12.10). The derivatives with 2,5-and 3,4-difluorophenyl and 2,4,5-trifluorophenyl groups were prepared by the same procedure.[25]

Two syntheses of triazolopiperazines **80** and **89** are shown in Schemes 12.11 and 12.12. The piperazine ring of **80** can be constructed by either intramolecular condensation–cyclization of ethylene diamine with chloroacetyl moiety (Scheme 12.11)[26] or hydrogenation of the corresponding triazolopyrazine **87** and **88** (Scheme 12.12).[27] The 3-trifluoromethyl[1,2,4]triazole ring of **80** was constructed by intramolecular dehydration of **83**. The overall yield of **80** via oxodiazole **82** is satisfactory, however, yields of 3-pentafluoroethyl and 2,2,2-trifluoroethyl **89** via 2-hydrazopyrazine are affected by low yield of **88** (Scheme 12.12).

N-Methyl and *N*-benzyl derivatives **90** ($R^2 = Me$, Bn) are also prepared by the reaction of the corresponding oxadiazole **82** with *N*-methyl and *N*-benzyl ethylene

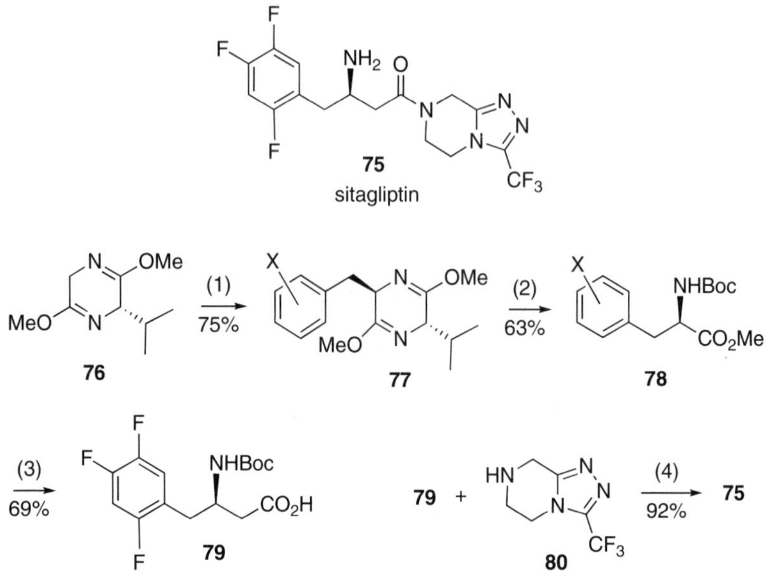

(1) (i) BuLi/THF, −78°C, (ii) $ArCH_2Br$; (2) (i) 1*N* HCl/MeCN, (ii) MeOH, (iii) $(Boc)_2O$, Et_3N/CH_2Cl_2; (3) (i) LiOH / THF-H_2O, (ii) iso-BuOCOCl, Et_2O/H_2O, (iii) CH_2N_2 (Ardnt–Eistert); (4) (i) HOBT, EDC, DIPEA/DMF, (ii) 1*N* HCl

SCHEME 12.10

diamines in 72% and 66%, respectively.[28] 6-Substituted derivatives **91** are also known.[29]

12.2.4 5-Trifluoromethyl-1-oxa-2,4-diazoles

Compound **92** (R = CH_3, WIN 61893) has a broad spectrum against picornavirus and is useful for the treatment of the common cold resulting from human rhinovirus

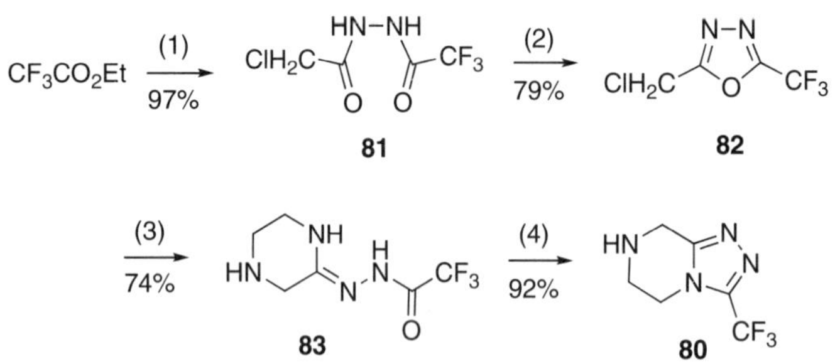

(1) (i) NH_2NH_2, 25°C, (ii) $ClCH_2COCl$, NaOH; (2) $POCl_3$, 80°C, 17 h; (3) $NH_2CH_2CH_2NH_2$/MeOH −20°C; (4) MeOH, reflux

SCHEME 12.11

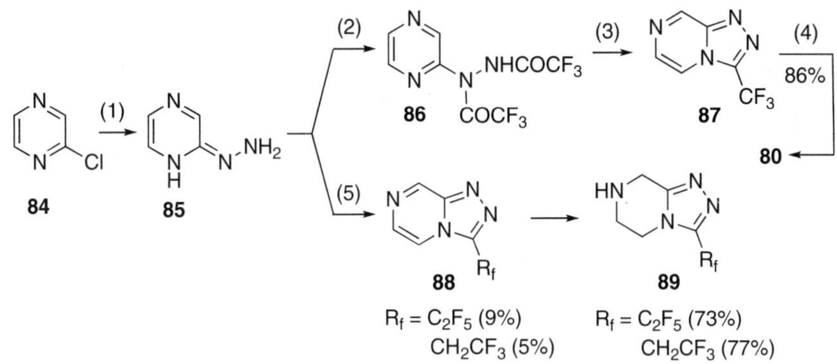

(1) NH_2NH_2/H_2O, 63–65°C, 17 h; (2) $(CF_3CO)_2O$, 20°C, 1 h; (3) (i)H_3PO_4, 75°C, 5 h, (ii) NH_4OH-H_2O, <40°C; (4) H_2 - Pd/EtOH, 45°C, 50 psi, overall yield 26% from **85**; (5) $CF_3CF_2CO_2H$/PPA, 155°C, 8 h

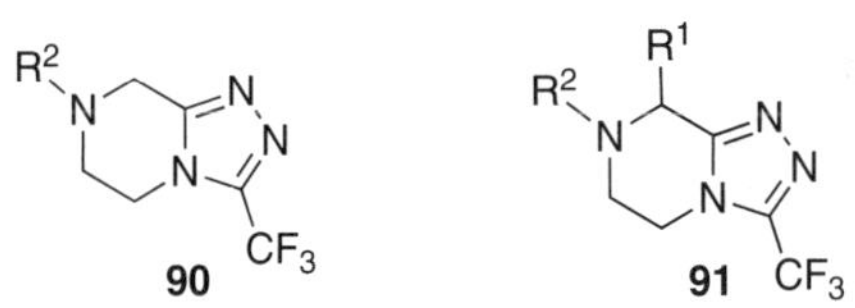

SCHEME 12.12

infections. However, the acid-sensitive oxazoline and oxadiazole rings increase lability against oral administration, and the methyl group on the oxazoline ring is easily metabolized into hydroxymethyl group. Introduction of trifluoromethyl or polyfluoroalkyl groups into the heteroatom rings could decrease the acid lability and metabolic instability of this kind of compounds.[30]

5-Trifluoromethyl-1-oxa-2,4-diazole of **92** ($R_f = CF_3$) is prepared by base-catalyzed condensation of **93**[31] with trifluoroacetic anhydride in excellent yield (Scheme 12.13). The same method is applicable for the preparation of monofluoromethyl, difluoromethyl, and 1,1-difluoroethyl compounds although their yields are low. The reaction of aminooxime **93** with 2-(trifluoroethylidene)-1,3-dithiane **95** under acid-catalyzed conditions, followed by ring-opening of the 1,3-dithiane ring and recyclization of oxadiazole ring provides 5-(2,2,2-trifluoroethyl)-1-oxa-2,4-diazole **96** in 38% yield (Scheme 12.13).[30]

Structurally similar compound **99** with trifluoromethyl group in isoxazole ring is prepared by the intramolecular cyclization of alkynyl oxime **98** (Scheme 12.14).[30]

12.2.5 2-Fluorotetrahydrofurans

Lodenosine **100** is an orally active drug candidate for treatment of a human immunodeficiency virus (HIV) and AIDS. Its structure is based on (2,3-dideoxy-2-fluoro-β-D-threo-pentofuranosyl)adenine (Fig. 12.5).[32] The related compound Dianosine **101** has been clinically used, however, it is orally less active due to the

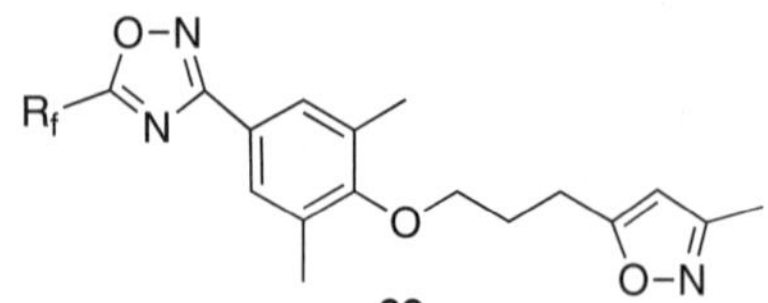

92

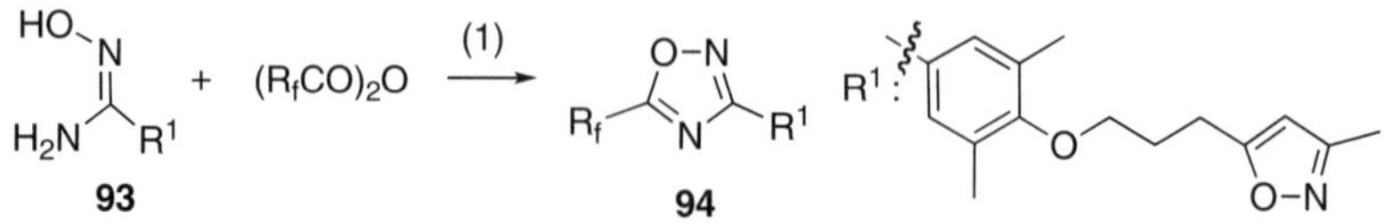

(1) Py, reflux Rf; CF_3 (87%), CH_2F (46%), CHF_2 (56%), CF_2CH_3 (40%)

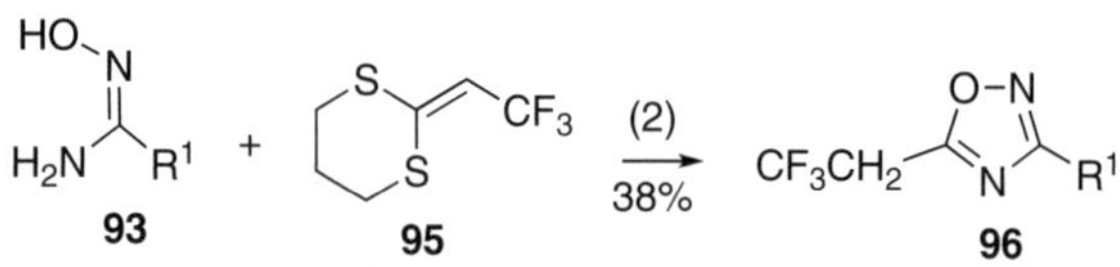

(2) CF_3SO_3Ag/THF, reflux, 22 h

SCHEME 12.13

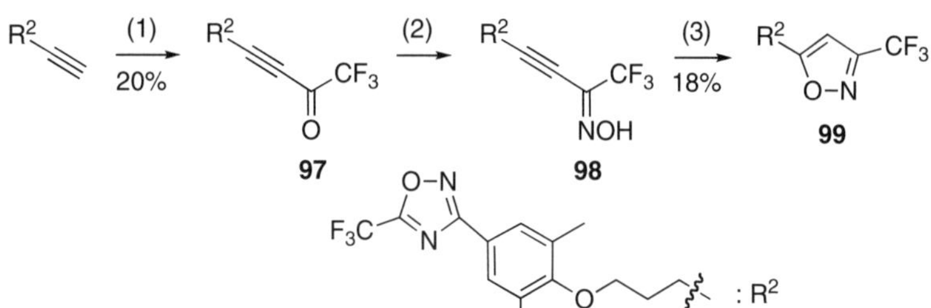

(1) (i) BuLi/THF, –78°C, (ii) $(CF_3CO)_2O$, BF_3-Et_2O; (2) NH_2OH, KOH; (3) benzene, reflux

SCHEME 12.14

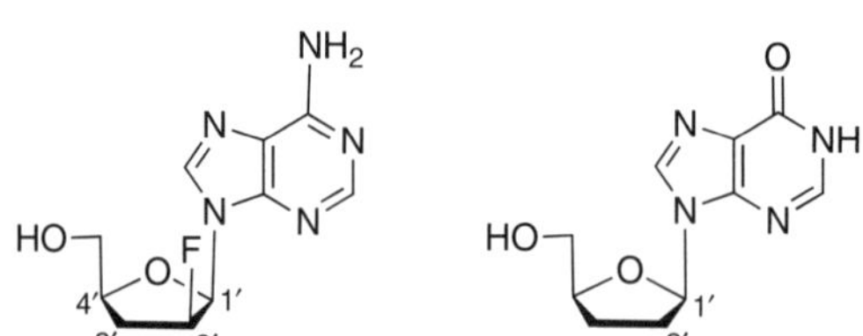

100 Lodenosine

101 Dianosine

FIGURE 12.5 Lodenosine and Dianosine.

instability under gastric acid conditions. The strong electron-withdrawing effect of fluorine atom at C_2 plays an essential role for increasing the stability of the aminal moiety involved in **100** under acidic conditions.

The synthetic strategies for **100** are divided into two categories; (A) the stereoselective preparation of 2-fluorofuranose and the coupling with an adenine derivative and (B) the stereoselective fluorination at C_2 of the furanosyl adenine.

The stereoselective introduction of 2′-fluorine in compound **101** can be achieved either by stereospecific replacement of α-2′-hydroxyl group with fluorine (Schemes 12.15–12.17) or diastereoselective catalytic hydrogenation of 2-fluoro-2,5-disubstituted-2,5-dihydrofuran moiety in derivatives **123** and **130** (Schemes 12.18 and 12.19).

Stereospecific nucleophilic replacement of α-2′-hydroxyl group can be conducted using either fluoride ion or DAST as fluorinating agents. Nucleophilic substitution of secondary hydroxyl group by fluoride ion is difficult and cannot be achieved in the reaction of the corresponding mesylate, tosylate, and even triflate derivatives **102b**. Surprisingly, compound **102c** (Scheme 12.15) undergoes exchange when reacting with KHF_2 in 2,3-butanediol at 160°C, providing **103** in 63% yield.[33] The conversion of benzoate **103** into bromide **104** and coupling of **104** with purine derivative, dehydroxylation at 3′-position of **105b** by radical hydrogenation of xanthate of **105b** ($R^2 = CS_2Me$) provided the desired compound **106**.[34]

Diethylaminosulfur trifluoride (DAST) is one of the most commonly used reagents for stereospecific replacement of hydroxyl group by fluorine, which usually proceeds with the inversion of stereochemistry. Scheme 12.16 shows two examples. The reactions of **107** and **111** with DAST provide fluorides **108**[35] and **112**[36] in moderate yields, both of which are precursors of Lodenosine **100**.

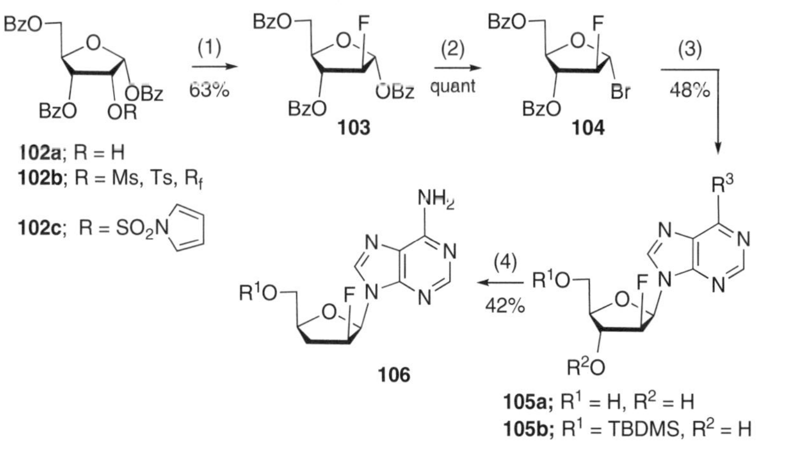

(1) KHF_2 / 2,3-butanediol, 160°C, 1 h; (2) 30% HBr/AcOH; (3) (i) TMS-purine derivative,100°C, 1 h (48%), (ii) NH_3/MeOH, 10°C, 14 h (92%), (iii)TBDMS-Cl, imidazole/DMF; (4) (i) CS_2, NaH, MeI/DMF, (ii) Bu_3SnH, AIBN/toluene, 90°C

SCHEME 12.15

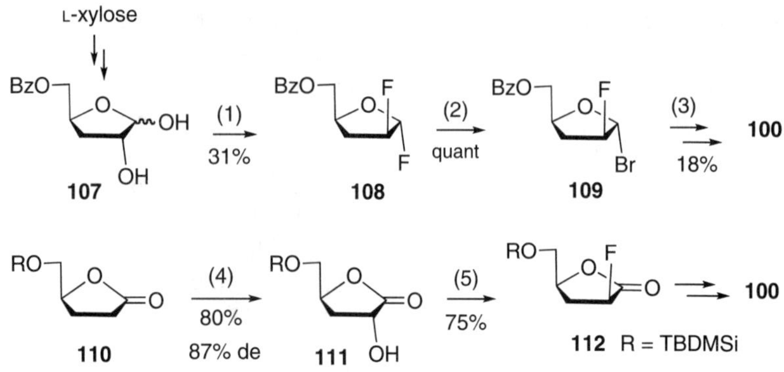

(1) DAST-DMAP/CH_2Cl_2; (2) 45% HBr/AcOH; (3) (i) 6-Cl-purine-NaH/MeCN, (ii) NH_3/MeOH, 100°C; (4) (i) Li(HMDS)/THF, –78°C, (ii) MoOPH; (5) DAST/CH_2Cl_2, –78°C to RT.

SCHEME 12.16

The fluorination using DAST was also applied for the 2′-hydroxytetrahydrofuran **114** bearing purine unit on C1′ (Scheme 12.17).[37] The presence of bromine at C3′ leads bromine migration during fluorination. When DAST reacts with hydroxyl group at the first step, the bromine next to hydroxyl group migrates to C2′, and then C3′ is attacked by fluoride ion, producing **117** in 75%. Bromine migration through conformation **119** is proposed in transition state.[38]

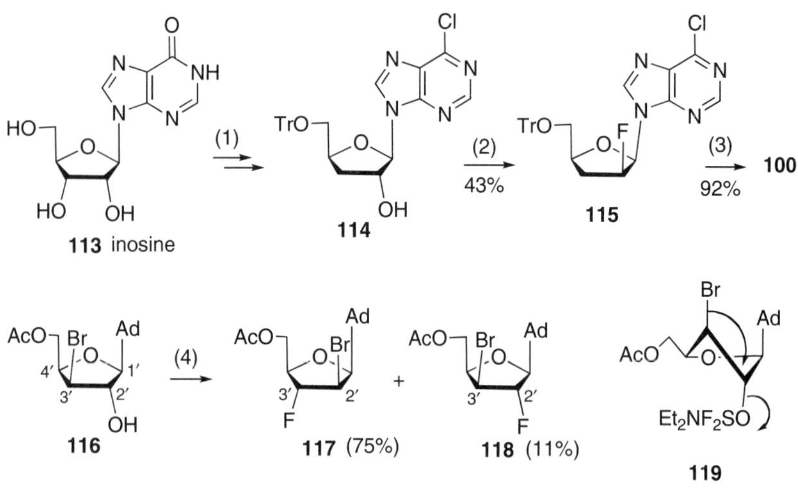

(1) 42% from inosine **108** via six steps; (2) DAST, Py/CH_2Cl_2; (3) (i) NH_3/THF, (ii) 37% HCl aq. MeOH; (4) DAST (49%), MOST (86%)/CH_2Cl_2

SCHEME 12.17

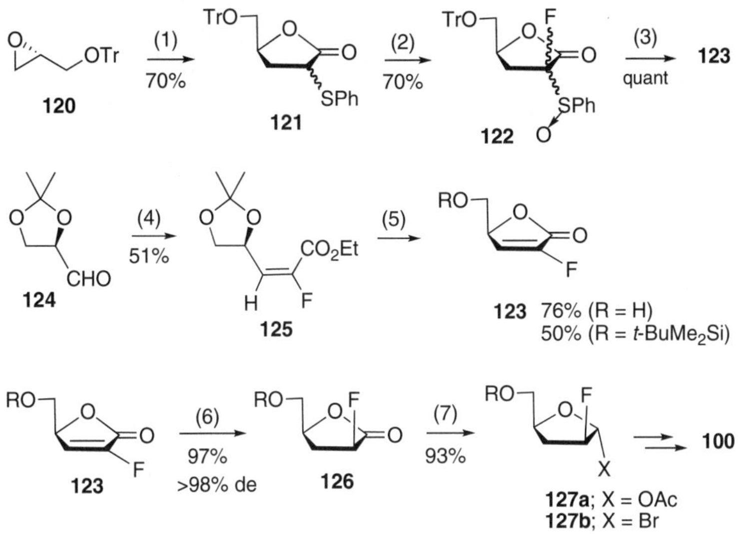

(1) Tr = CPh_3, (i) $PhSCH_2CO_2H$, LDA/THF, 0°C, 1 h, (ii) HCl, (iii) toluene, reflux; (2) (i) MTO (3 mol%), urea-H_2O_2/CH_2Cl_2, (ii) NaH/THF, 0°C, (iii) Selectfluor; (3) $NaHCO_3$/toluene, reflux; (4) *n*-BuLi, $(EtO)_2POCHFCO_2Et$/THF, –78°C, 1 h; (5) (i) TsOH, (ii) TBDMSiCl; (6) (i) R = *p*-Tol, Pd-C, H_2 (10 psi)/AcOEt; (7) (i) red-Al, 2-hydroxypyridine/THF, –30°C, (ii) Ac_2O, cat conc H_2SO_4/CH_2Cl_2, RT, (iii) 30% HBr/AcOH

SCHEME 12.18

The stereoselective fluorination at C2′ relies on the diastereoselective catalytic hydrogenation of 2-fluoro-2,5-disubstituted-2,5-dihydrofuran derivatives. The approach requires optically pure starting material **123**, which is prepared either from epoxide **120**[39] or glyceraldehyde acetal **124**[40] as shown in Scheme 12.18. Stereoselective catalytic hydrogenation occurs from the less-hindered face of **123**, providing **126** exclusively.[41] The same hydrogenation protocol is applicable for the final step (**130** → **100**) of Lodenosine synthesis (Scheme 12.19).[42]

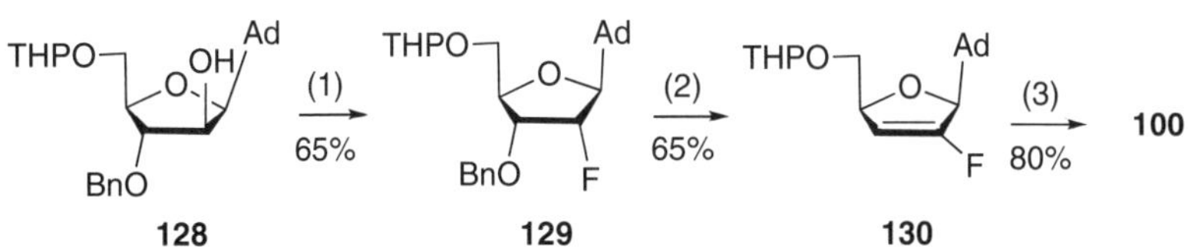

(1) (i) $R_{f2}O$, $DMAP/CH_2Cl_2$, 0°C, (ii) TBAF/THF, 0°C; (2) (i) $Pd(OH)_2/C$, cyclohexene/EtOH, reflux, (ii) Tf_2O, DMAP, Py/CH_2Cl_2, 0°C, (iii) *t*-BuOK/DMSO, RT; (3) H_2, Pd/C, THF, (ii) pyridinium DOWEX 50, EtOH, MeOH

SCHEME 12.19

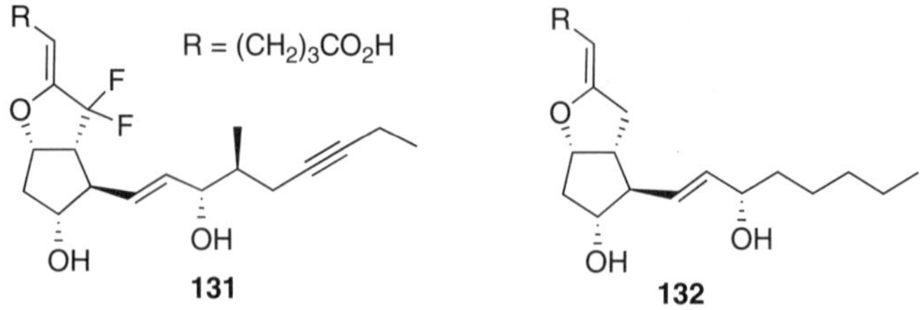

FIGURE 12.6 Difluoroprostacyclin.

12.2.6 3,3-Difluorotetrahydrofurans

7,7-Difluoroprostacyclin **131** (Fig. 12.6) is useful for prevention and treatment of ophthalmic circulatory disturbance. Prostacyclin **132** was originally found to be effective as an inhibitor of platelet aggregation and as a vasodilator in maintaining homeostatic circulation.[43] The introduction of one or two fluorines at 7- or 10-position of **132** dramatically improves the stability—no decomposition of **131** in pH 6.5 buffer solution at 25°C as observed after 30 days in contrast to facile hydrolysis of **132** (half-life = 1.5 min) under the analogous conditions.[44a]

Construction of furan ring of **131** by acid-catalyzed cyclization of **134** cannot be achieved due to low reactivity of the α,α-difluorinated double bond toward protonation. Therefore, the following sequence was employed for the total synthesis of **131** (Scheme 12.20): fluorination of lactone **133**, introduction of a double bond to **136** by Wittig olefination, and modification of R^1 part of **137**.[44] The use of manganese enolate of lactone **133** is useful for one-pot difluorination. The reaction of the lithium enolate of **133** with the conventional electrophilic fluorinating reagent such as *N*-fluorobenzenesulfonimide provides only monofluoride **135**.

12.3 FUSED FIVE-MEMBERED FLUORINATED HETEROCYCLES

12.3.1 6-Fluoro-1,2-benzisoxazoles

Risperidone **139a**[45] and the related compounds such as Iloperidone **139b**[46], Abaperidone **139c**[47], and QF-0510B **139d**[48] (Fig. 12.7) are potential atypical antipsychotic agents for the treatment of schizophrenia. All of them contain 2-[4-(6-fluoro-1,2-benzisoxazol-3-yl)-1-piperidinyl]ethyl structural block.

The 6-fluoro-1,2-benzisoxazole skeleton is synthesized by an intramolecular nucleophilic cyclization of oxime **142** in an excellent yield (Scheme 12.21). It is well known that a fluorine atom on the aromatic ring is very active toward nucleophilic substitution reactions.[49]

12.3.2 5-Trifluoromethyl and 5-Fluorobenzimidazoles

Omeprazole and its (*S*)-isomer, esomeprazole **144** (Fig. 12.8) belongs to a class of potent gastric acid secretion inhibitors and are used as antiulcer agents.[50] The skeleton

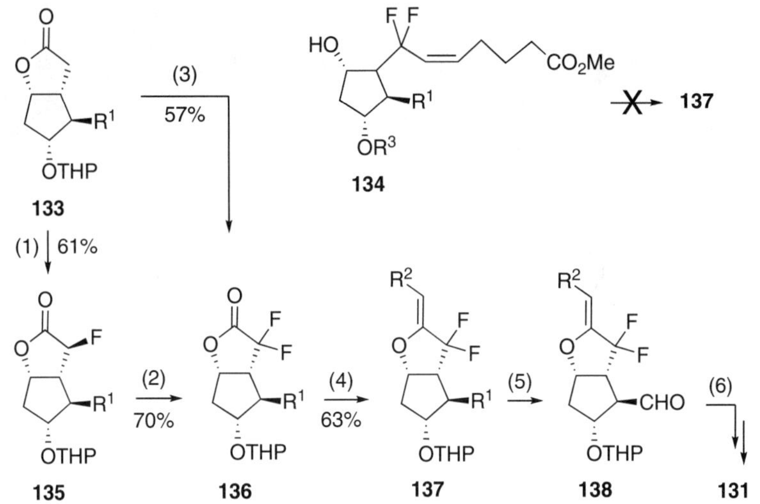

$R^1 = CH_2OSiMe_2Bu\text{-}t$ $R^2 = (CH_2)_3CO_2Me$

(1) $LiN(SiMe_3)_2$, $(PhSO_2)_2NF$/THF, –78°C; (2) $KN(SiMe_3)_2$, $ZnCl_2$, $(PhSO_2)_2NF$/THF-toluene –78°; (3) (i) $KN(SiMe_3)_2$, $MnBr_2$/THF-toluene, (ii) $(PhSO_2)_2NF$; (4) (i) $[P(Ph)_3(CH_2)_4CO_2H]Br$, $NaN(SiMe_3)_2$/THF (Z:E = 87:13), (ii) MeI, $EtNPr_2$/DMSO; (5) (i) TBAF/THF(94%), (ii) DCC, CF_3CO_2H, Py/DMSO; (6) (i) NaH, keto-phosphonate (90%), (ii) $NaBH_4$, $CeCl_3$, MeOH, (iii) TsOH/MeOH, (iv) NaOH/MeOH (30%)

SCHEME 12.20

of **144** contains a 2-[(2-pyridinyl)methylsulfinyl]benzimidazole moiety. Some fluorinated omeprazole analogues **145a**[51], **145b**[52], and **145c**[52] are shown in Fig. 12.8.

Precursors **151** are synthesized by coupling of benzimidazoly-2-thiol **147** with 2-pyridylmethyl chloride **150c** as shown in Scheme 12.22.[52,53] The thiols **147** are

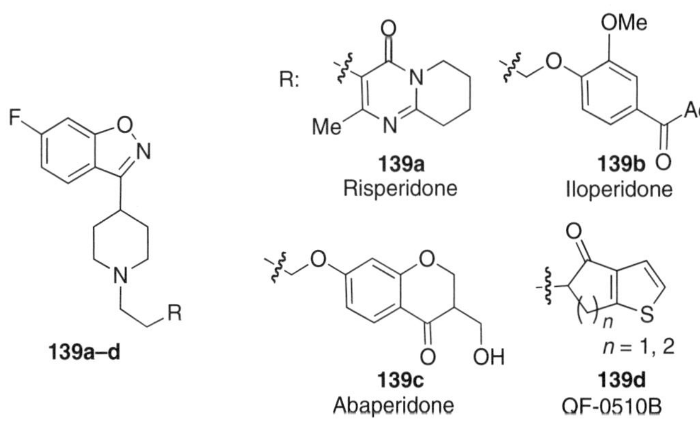

FIGURE 12.7 Risperidone and the related compounds.

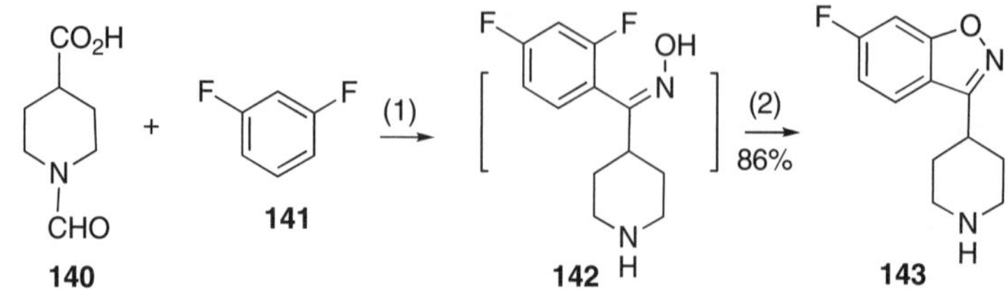

(1) (i) $SOCl_2/CH_2Cl_2$, 30–40°C, 3 h, (ii) $AlCl_3/CH_2Cl_2$, RT - 0°C, (iii) NH_2OH, H_2SO_4; (2) KOH/MeOH, reflux

SCHEME 12.21

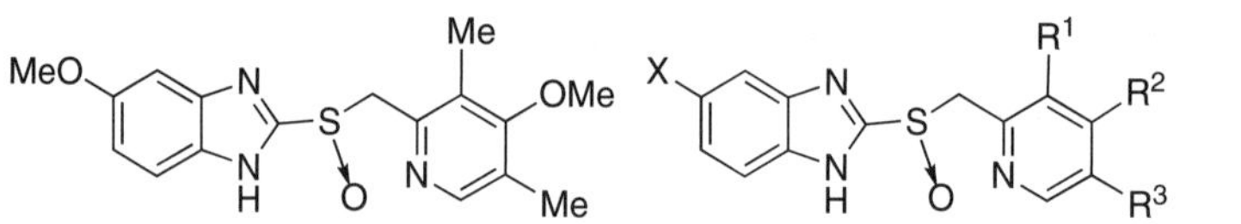

144 Omeprazole

145

145a; X = CF_3, R^1= Me, R^2 = $O(CH_2)_2OBu$, R^3 = H
145b; X = F, R^1 = Me, R^2 = OMe, R^3 = Me
145c; X = OMe, R^1 = F, R^2 = *N*-morpholinyl, R^3 = H

FIGURE 12.8 Omeprazole and the related compounds.

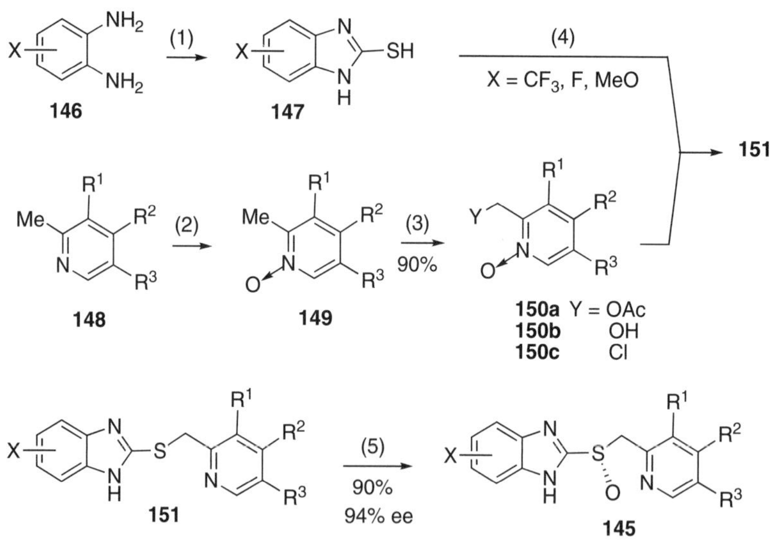

(1) KSCSOEt/EtOH-H_2O, reflux, POh; (2) *m*PCBA/CH_2Cl_2; (3) (i) Ac_2O, 100°C, 4 h, (ii) aq NaOH, 100°C, (iii) $SOCl_2/CH_2Cl_2$; (4) NaOH/MeOH-H_2O, RT or reflux (5) (i) Ti(O-*i*-Pr)$_4$/(*S,S*)-diethyltartrate/H_2O, toluene, 54°C, 1 h, (ii) Et(*i*-Pr)$_2$N, PhC(Me)$_2$OOH, 30°C, 1 h.

SCHEME 12.22

prepared by condensation of 1,2-phenylenediamine with sodium ethyl dithiocarbonate.[53] The appropriately substituted 2-pyridylmethyl chlorides **150c** are prepared by an intramolecular migration of acetoxyl group from nitrogen of pyridine ring to 2-methyl group of **149**, followed by hydrolysis and chlorination of the hydroxyl group (Scheme 12.22).[52,53] A direct chlorination of 2-methyl group with tosyl chloride via *N*-oxide **149** was also reported.[54]

Enantioselective transformation of sulfide **151** to (*S*)-sulfoxide **145** was achieved in 90% yield and 94% ee by Sharpless oxidation using cumene hydroperoxide.[55]

12.3.3 6-Trifluoromethylindolines

Appropriately substituted 1-[(3-pyridyloxy-5-pyridyl)carbamoyl]-6-trifluoromethyl-indolines **152** are selective antagonists of 5-HT2C receptor and are useful for treatment of a range of CNS disorders, in particular anxiety and depression.[56] 1-Indolyl-3-pyridylcarbamate is an important structural feature of compounds **152** (Fig. 12.9).

One of the synthetic approaches to **152** is shown in Scheme 12.23. The carbamate of **152** is formed by the coupling of 3-aminopyridine **155** with indolines **158**.[57] 6-Trifluoromethylindolines **158** (Y = CF_3) are prepared by a sequence of reactions starting from **156** (see Scheme 12.53).[58] Methyl group is introduced in **159** (X = OMe, Y = CF_3) by the sequential reactions including demethylation with TMSI (trimethylsilyl iodide) in refluxing chloroform, triflation of hydroxyl group with Tf_2O/Py, and Pd-catalyzed methylation of the aryl triflate **160** with Me_4Sn in DMF to produce **161**.[57b]

12.3.4 8-Fluoro-2,3,4,5-tetrahydro-1*H*-pyrido[4,3-b]indole; 8-fluoro-γ-carboline

While Gevotroline (WY 47384) **162** containing 8-fluoro-γ-carboline skeleton and compounds **163** are orally active atypical antipsychotic agents,[59] the corresponding β-carbolines do not show activity when administered orally (Fig. 12.10).

The 8-fluoro-γ-carboline **166** is prepared by a standard Fisher indole synthesis.[60] The condensation of 4-fluorophenylhydrazine (**164**) with **165**. 4-Phenyl compound **168** is synthesized by acid-catalyzed ring closure of **167** (Scheme 12.24).[61]

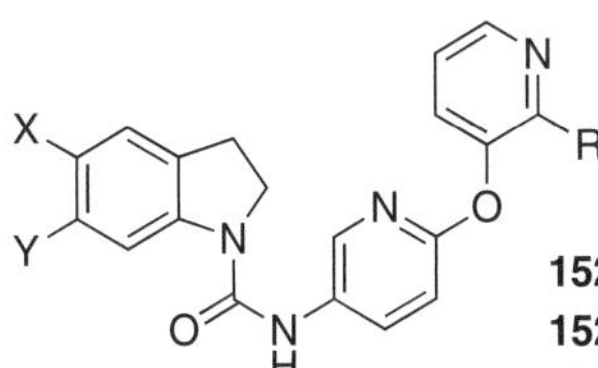

152

152a: R = H, X = OMe, Y = CF_3
152b: R = Me, X = OMe, Y = CF_3
152c: R = Me, X = Me, Y = CF_3
152d: R = Me, X = Me, Y = Cl

5-HT2C receptor antagonists

FIGURE 12.9 5-HT2C receptor antagonists.

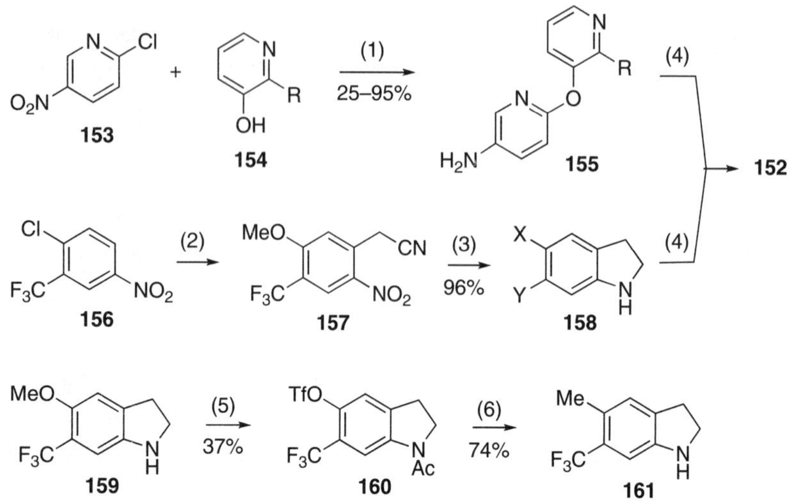

(1) (i) NaH/DMF, 0°C - RT, 18 h (25–95%), (ii) $SnCl_2$-HCl/EtOH, 50°C, 1 h, (80–100%); (2) (i) NaOR, ROH, reflux, 3 h (99%); (ii) 4-chlorophenoxyacetonitrile, KO-*t*-Bu, DMF, 10°C, 1 h (78%); (3) H_2/10% Pd-C, 50 psi, AcOH, EtOH/H_2O, RT, 0.5 h (96%); (4) (i) PhOCOCl, Et_3N, CH_2Cl_2, −20°C, 1 h, (ii) indoline, Et_3N/DMF, 100°C, 1 h, (35–85%); (5) (i) TMSI/$CDCl_3$, reflux, 65 h, (ii) Ac_2O/DCM, (iii) Tf_2O/Py, 0°C - RT, 18 h; (6) (i) $SnMe_4$, $Pd(Ph_3)_2Cl_2$, LiCl/DMF, 110°C, (ii) NaOH/MeOH-H_2O, reflux, 18 h.

SCHEME 12.23

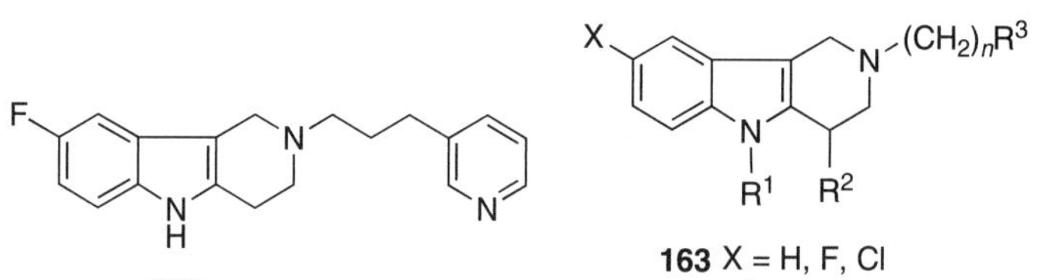

162
Gevotroline (WY 47384)

163 X = H, F, Cl
R^1 = H, 4-F-C_6H_4
R^2 = H, Ph
R^3 = 2-Py, 3-Py, 4-Py

FIGURE 12.10 Gevotroline and the related compounds.

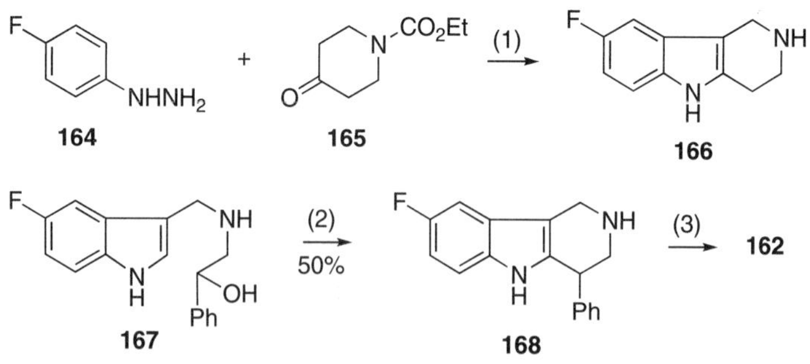

(1) (i) heating in alcohol, (ii) NaOH; (2) conc H_2SO_4, 0°C; (3) $Cl(CH_2)_nR^3$

SCHEME 12.24

12.4 SIX-MEMBERED FLUORINATED HETEROCYCLES

12.4.1 2- or 4-Substituted 5-Fluoropyrimidines

12.4.1.1 BMY 14802

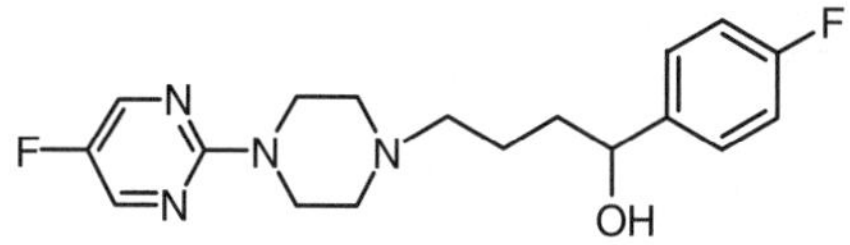

169

BMY 14802

BMY 14802 **169** emerged as the lead compound from the series of l-(pyrimidin-2-yl)-piperazine derivatives on the basis of its good activity and duration of action. Compound **169** has been found to possess *in vivo* behavioral activity indicative of possible anti-psychotic utility with minimal side effect liability. It is encouraging that several pieces of empirical evidence point toward the compound having limbic versus striatal selectivity. Since it does not bind to dopamine receptors, the demonstration of its antipsychotic activity in man could render **169** a breakthrough drug that may challenge the dopamine hypothesis of schizophrenia. Positive clinical findings would also help to confirm the viability of σ-selective ligands as useful antipsychotic drugs.[62]

The preparation of 1-(5-fluoropyrimidin-2-yl)piperazine moiety of **169** is depicted in Scheme 12.25. Heating 2-chloro-5-fluoro-4-(methylthio)pyrimidine[63] (prepared from 5-fluorouracil) with *N*-(ethoxycarbonyl)piperazine in acetonitrile afforded **170**. The latter was converted into **171** by desulfurization, followed by acid hydrolysis of the carbamate moiety.

As shown in Scheme 12.26, the alkylation of the piperazines by the ketal derivative of 4-chloro-4′-fluorobutyrophenone gave, the ketone **172** upon aqueous acidic workup. The use of unprotected ketone results in significantly lower yields. Target compound **169** was obtained in good yield via sodium borohydride reduction of its ketone precursor.

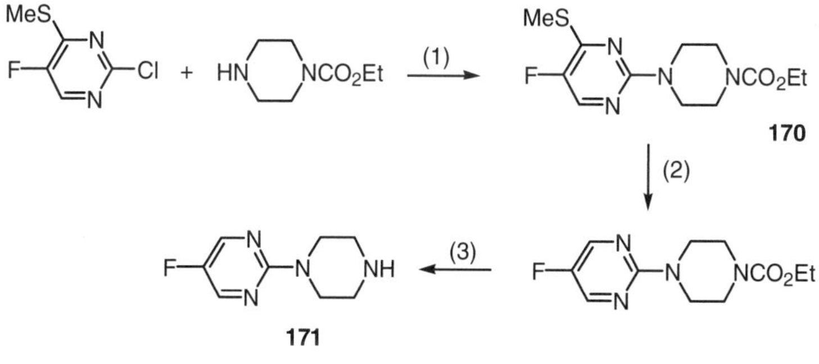

(1) K_2CO_3, MeCN, reflux; (2) Raney Ni, EtOH, reflux; (3)(i) 6N HCl, reflux, (ii) 50% NaOH.

SCHEME 12.25

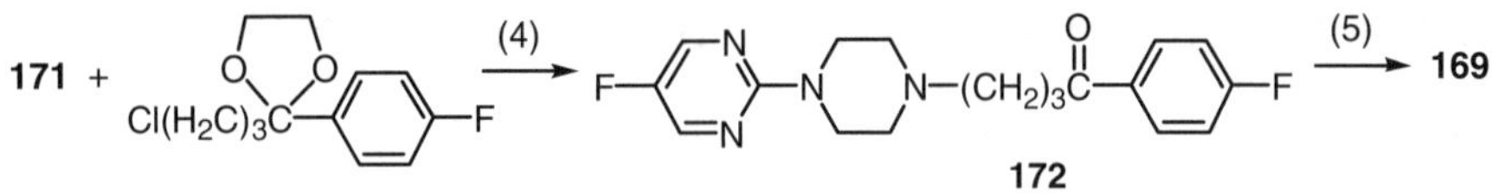

(4)(i) K_2CO_3, KI, MeCN, reflux, (ii) 3N-HCl; (5) $NaBH_4$, EtOH.

SCHEME 12.26

12.4.1.2 Voriconazole (Vfend)

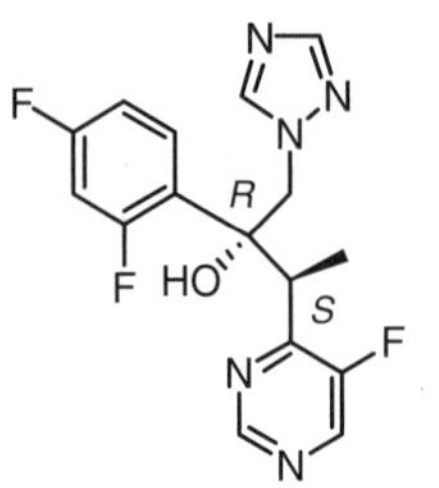

173
Voriconazole

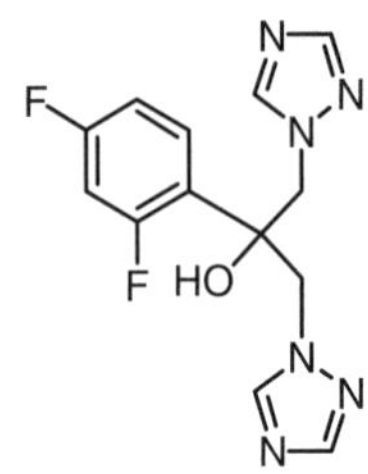

Fluconazole

Voriconazole **173**[64] and fluconazole are structurally related. Similar to other azole based antifungal agents, mode of action of **173** involves the inhibition of cytochrome P450 14α-demethylase (P45014DM).[65,66] Triazole **173** has favorable *in vitro* activity against a variety of fungi and is generally considered to be a fungistatic agent against *Candida* spp. and *Cryptococcus neoformans*.

In the synthesis of **173**, the relative stereochemistry is set in the addition of a 4-(1-metallo-ethyl)-5-fluoropyrimidine derivative to 1-(2,4-difluorophenyl)-2-(1*H*-1,2,4-triazol-1-yl)-1-ethanone **174** (Scheme 12.27). The diastereocontrol of this can be controlled by pyrimidine substitution pattern and reaction conditions of the metalation step.[67] Good diastereoselectivity (12:1) is obtained using an zinc derivative of **175**. After removal of the chlorine from the pyrimidine ring, of the desired stereoisomer of **173** is isolated via a diastereomeric resolution using salt (1*R*)-10-camphorsulfonic acid (10-CSA). Synthetic routes to the pyrimidine partner have also been evaluated. Shown in Scheme 12.28, the initial six-step route from 5-fluorouracil **177** can be replaced by a four-step process, involving fluorination of methyl 3-oxopentanoate and cyclization with formamidine acetate.[67]

Improved method for preparation of **173** is reported by Wang and Gu[68] (Scheme 12.29). It comprises of four-step synthesis of **179**, conversion of acid into *S*-2-(5-fluoropyrimidin-4-yl)propionyl chloride, Friedel–Crafts reaction to obtain ketone **180**, and the reaction with 1-methyl-1*H*-1,2,4-triazole under basic condition.

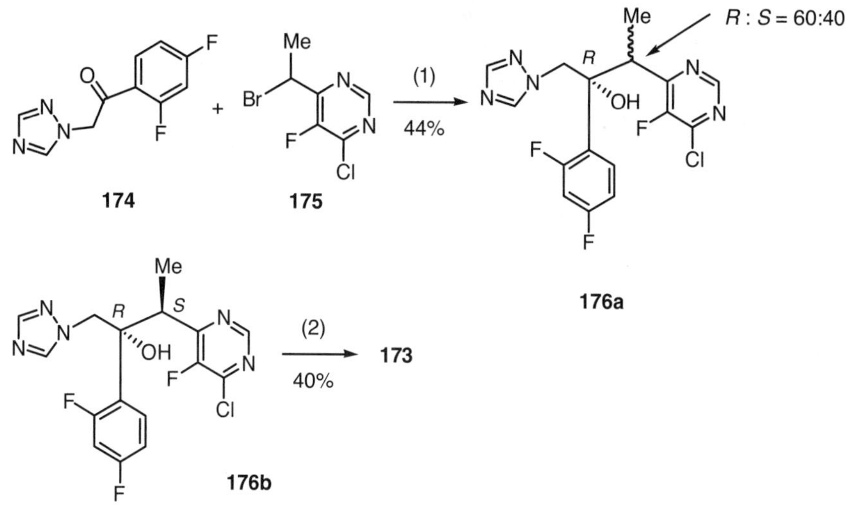

(1) (i) $LiN(i\text{-}Pr)_2$, Zn, I_2 in THF; (2) (i) 10-CSA, (ii) NaOH.

SCHEME 12.27

Six-step route

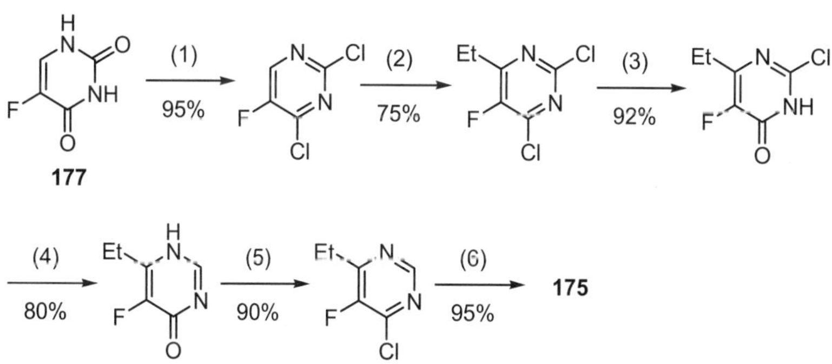

(1) (i) $POCl_3$, in $PhNMe_2$, (ii) HCl; (2) (i) EtMgBr in THF, (ii) Et_3N in THF, (iii) I_2 in THF, (iv) HCl; (3) (i) NaOH in H_2O, (ii) HCl; (4) (i) H_2/Pd in EtOH; (5) (i) $POCl_3$, Et_3N in CH_2Cl_2, (ii) HCl; (6) (i) Bromosuccinimide, AIBN in CH_2Cl_2.

Four-step route

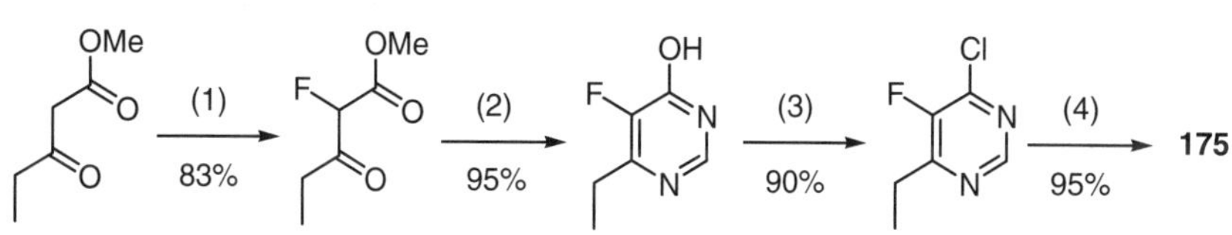

(1) F_2; (2) $NH{=}CHNH_2 \cdot AcOH$, NaOMe, (3) (i) $POCl_3$, Et_3N in CH_2Cl_2; (4) NBS, AIBN in CH_2Cl_2.

SCHEME 12.28

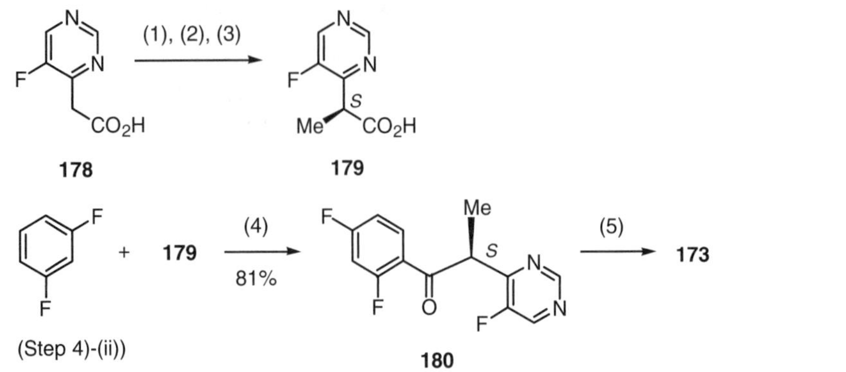

(1) H_2SO_4-EtOH, 94% yield; (2) Me_2SO_4, K_2CO_3, 86% yield; (3) HCl, resolving, 86% yield; (4) (i) $SOCl_2$, (ii) $AlCl_3$ in $PhNO_2$; (5) (i) 1-methyl-1*H*-1,2,4-triazole, BuLi, (ii) NH_4Cl.

SCHEME 12.29

12.4.2 2,4-Dioxo-5-fluoropyrimidines and 2,4-Dioxo-5-(trifluoromethyl)pyrimidines

12.4.2.1 Fluorouracil (5FU)

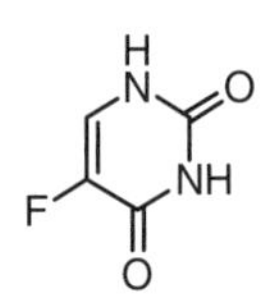

177

Fluorouracil

Introduced about 40 years ago, Fluorouracil **177** (5-fluorouracil, also called 5FU) is used for treatment of various types of cancer, including bowel, breast, stomach, and gullet (esophagus) cancer. Numerous methods of synthesis of **177** are reported.

The 2-fluoroenals, -enones, and -enesters may serve as building blocks for the synthesis of fluorinated heterocycles.[69–73] In this respect, methyl 2-fluoro-3-methoxyacrylate **181a**[72] and the corresponding acyl chloride **181c** were used for the efficacious synthesis of fluorine-bearing pyrazolones, pyrimidines (e.g., **177**), coumarines, and benzothiopyranones. Methyl ester **181a** is readily prepared by using the methyl methoxyacetate **182** derived to 1,2-dimethoxy-l-trimethylsilyloxyethene as the precursor (Scheme 12.30). Initially formed as a 1:1 mixture of stereoisomers, it can readily be converted into the pure (*Z*) component by treatment with catalytic amounts of lithium thiophenolate. The ester **181a** is saponified and the resulting acid **181b** is converted into the acyl chloride **181c** (Scheme 12.31).[74]

When Shi et al.[74] used **181a** and *O*-(trimethylsilyl)urea **183** for preparation of 5-fluorouracil **177**, the acylation product **184** was isolated in moderate yield. The

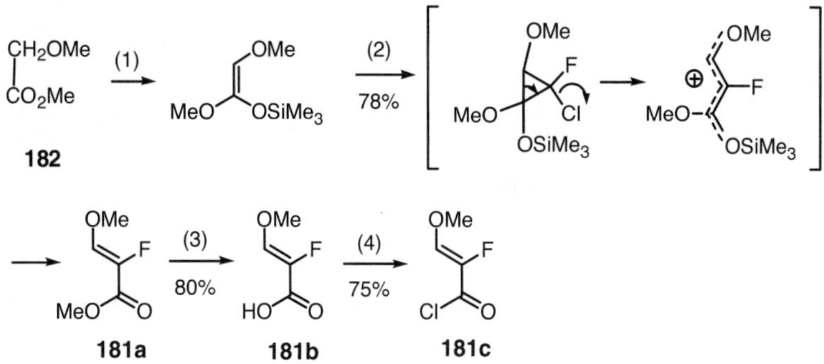

(1) LDA, Me_3SiCl, in THF; (2) (i) $FCHCl_2$, ^tBuOK, –70°C, PhSLi ; (3) Acid hydrolysis; (4) $SOCl_2$.

SCHEME 12.30

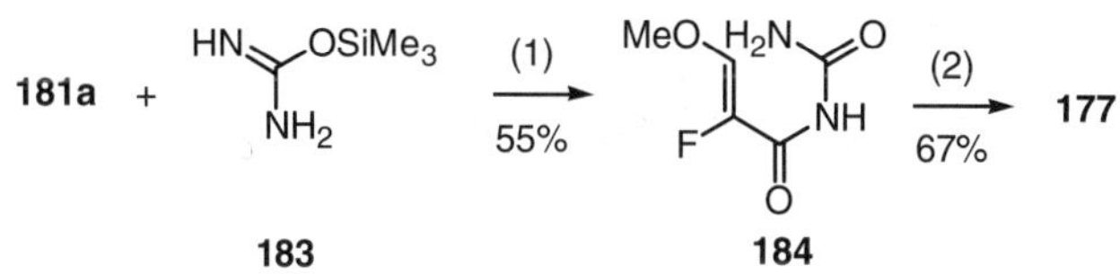

(1) MeLi, in Et_2O; (2) aq. NaOH.

SCHEME 12.31

cyclization into **177** (67%) occurred upon treatment with sodium hydroxide (Scheme 12.31). Previous syntheses of **177** using mono fluoroacetic acid as the starting material[75] or elemental fluorine,[76] or trifluoromethyl hypofluorite,[77] as the fluorinating agents had presented hazards.

As shown in Scheme 12.32, the fluorination of 5,6-unsubstituted dioxinone **185** with F_2 followed by treatment with triethylamine also affords the compound **186**, which can be converted into fluorinated heterocyclic compounds (e.g., **177**).[78]

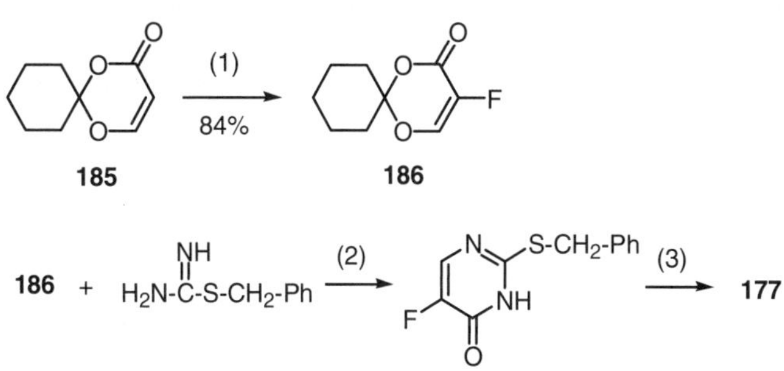

(1) (i) F_2, in MeCN, (ii) Et_3N, in CH_2Cl_2; (2) in xylene; (3) HCl.

SCHEME 12.32

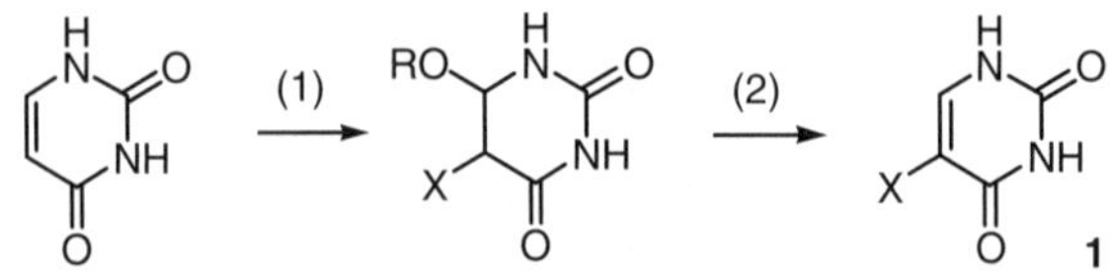

187

188; X = F
189; X = Cl
190; X = Br

177; X = F
191; X = Cl
192; X = Br

(1) X_2 or CF_3OF (X = F, Cl, Br), ROH (R = H, Ac, alkyl); (2) Δ or base.

SCHEME 12.33

Direct fluorination of uracil **187** (Scheme 12.33) can be carried out using fluorine or CF_3OF. Yield of **177** varies in small-scale experiments between 76% and 92%. It is reported to be lower in large-scale preparations (25–78% yield),[79] however, the direct fluorination of **187** by F_2 in solution is practiced on commercial scale. Fluorination of **187** goes through intermediates **188–190**. In contrast to **189** and **190**, compound **188** (X = F) is stable and removal of ROH from **188** proceeds at elevated temperature.[79]

12.4.2.2 Carmofur

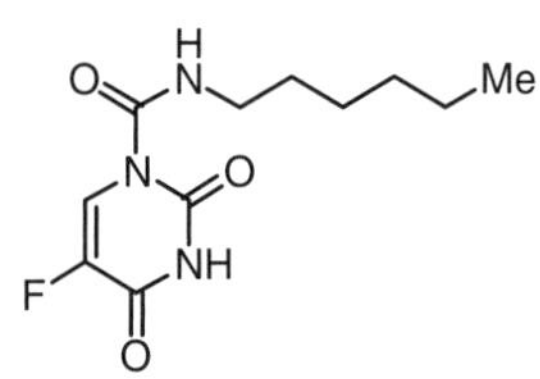

193

Carmofur

Carmofur **193** is a chemotherapy drug that is given as a treatment for some types of cancer and orally active cytostatic derivative of **177**. The introduction of carbamoyl moiety into **177** significantly affects the toxicity and tumor affinity of **177**.[80]

For preparation of compound **193**, carbamoylation of **177** by three general methods was studied extensively:[80] (i) the reaction of **177** with isocyanates (Method 1), (ii) the reaction of 1-chlorocarbonyl-5-fluorouracil **194** with amines (Method 2), and (iii) the reaction of **177** with carbamoyl chlorides (Method 3). These three methods usually give 1-carbamoyl-5-fluoruracils **195** (Scheme 12.34). 3-Carbamoyl-5-fluorouracils are not formed in these reactions.

Compounds **195** has strong internal hydrogen bonding in chloroform, a non-H-bonded structure in Me_2SO at 80°C, and mixed structures in Me_2SO at 25°C. Thirtysix compounds having structure **195** were prepared, and all showed antitumor activity. Among them, **193** appeared to be the most promising antitumor agent for oral administration. It retains well-balanced lipo- and hydro-philicity, is stable toward acids, and moreover, decomposed moderately in a tumor. It has right pharmacodynamic properties such as Tegafur described in the next section.

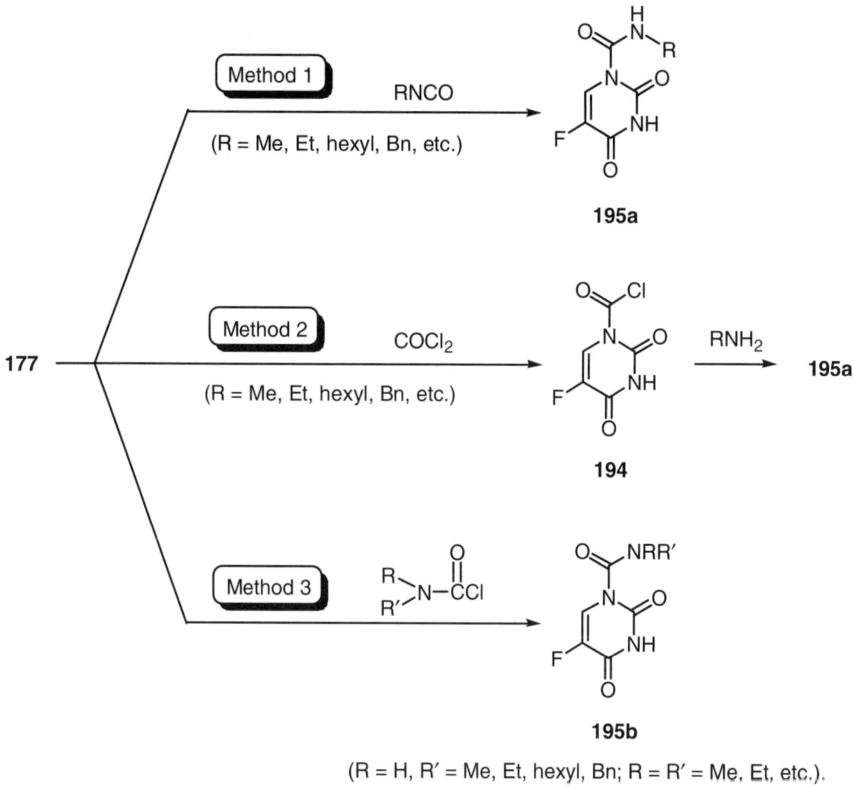

SCHEME 12.34

Wei et al.[81] have also reported the synthesis of **193** and 1-cyclohexylcarbamoyl-5-fluorouracil (CHCFU) **196** from **177** by condensation reaction with hexyl isocyanate or cyclohexyl isocyanate in the presence of 4-(dimethylamino)pyridine catalyst (Scheme 12.35). Higher yields of **193** and **196** shorter reaction time are reported for these reactions.

12.4.2.3 Tegafur

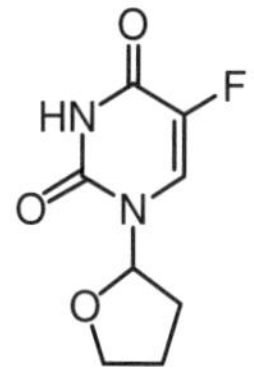

197

Tegafur

Tegafur **197** is usually used in combination with uracil **187** or **177** (Tegafur–uracil, UFT) for treatment of certain types of cancer, especially bowel cancer. When

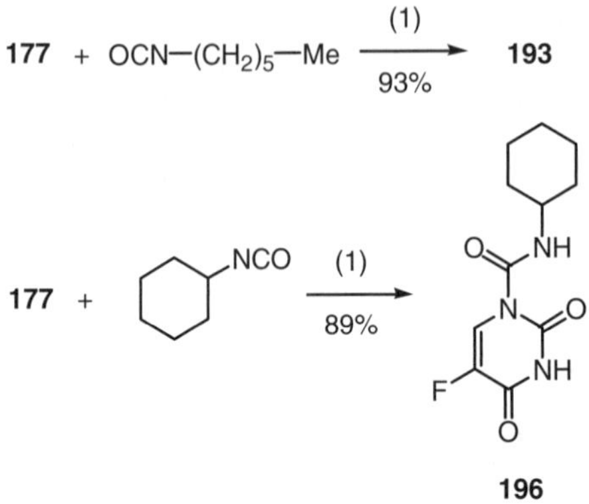

(1) 4-(Dimethylamino)pyridine, in pyridine.

SCHEME 12.35

tegafur–uracil is applied, in the cancer cells **197** is slowly converted into fluorouracil **177**. Compound **187** slows the breakdown of **177**, which means that **177** stays in the cancer cells for a longer time. Combining these two drugs leads to higher concentration of **177** in the tumor cells.[82]

A number of pyrimidine derivatives, containing modified sugar moieties instead of the naturally occurring ribose or deoxyribose, have been intensively studied.[83] Among these, 1-(tetrahydro-2-furyl)pyrimidine derivative was successfully synthesized in good yields by the reactions of trimethylsilylated fluorouracil with 2-acetoxytetrahydrofuran using CsCl as catalyst in MeCN under mild conditions (Scheme 12.36).[83] Compound **197** has also been prepared by treating **177** with 2,3-dihydrofuran in pyridine in the presence of a carboxylic acid, followed by addition of an aqueous solution of a carboxylate salt to the reaction mixture. This cost-effective method gives **197** of high purity (Scheme 12.36).[84]

Miyashita et al.[79] have also reported the direct fluorination of 1-(2-tetrahydrofuryl)-uracil **198** using fluorine or CF_3OF in several solvent systems. The yield of **197** in small-scale experiments is high although large-scale preparations seem to be less successful (Scheme 12.37).

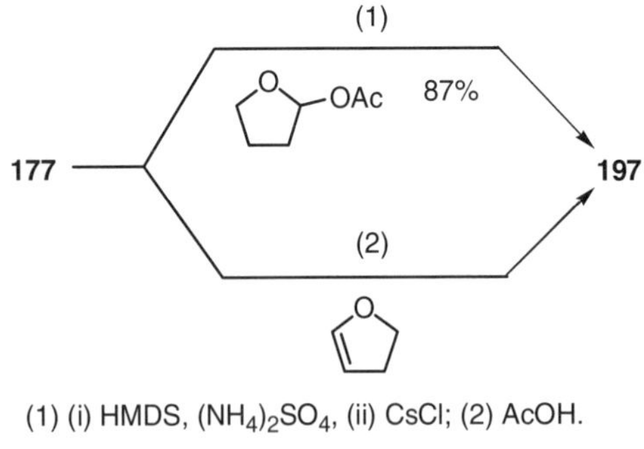

(1) (i) HMDS, $(NH_4)_2SO_4$, (ii) CsCl; (2) AcOH.

SCHEME 12.36

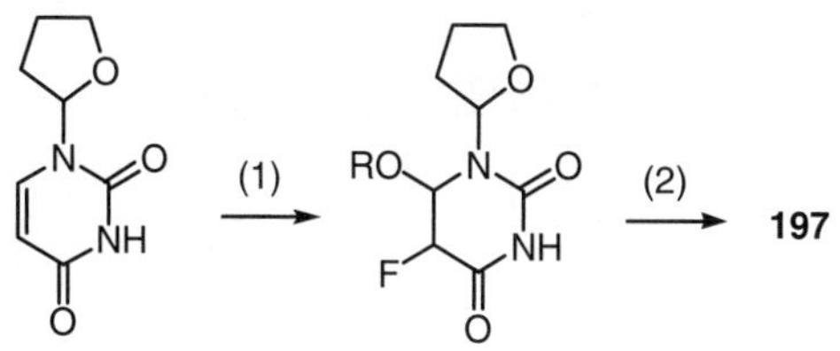

198

(1) F_2 or CF_3OF, ROH (R = H, Ac, alkyl); (2) Δ or base.

SCHEME 12.37

12.4.2.4 Doxifluridine (5′-DFUR)

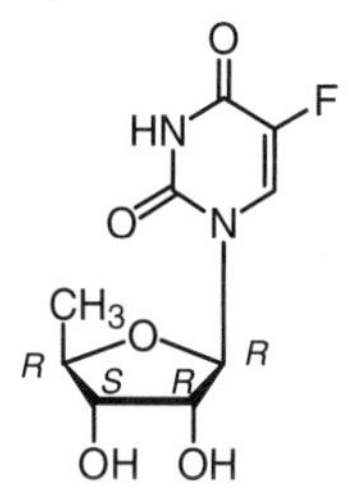

199

Doxifluridine

Doxifluridine (5′-DFUR, **199**) is an oral prodrug of 5FU (**177**). Compound **199**, designed to circumvent the rapid degradation of **177** by dihydropyrimidine dehydrogenase in the gut wall, is converted into **177** by action of pyrimidine nucleoside phosphorylase.[85]

Various processes for the production of **199** are known, the one reported by Kiss et al.[86] is particularly important. The process for production of **199** comprises the coupling reaction of a ribose derivative modified at 5′-position and activated 5-fluorouracil derivative **200**, according to the Scheme 12.38.

However, the reaction is not suitable for large-scale production. In fact, the formation of the significant amount of impurities reduces the yield and complicated the isolation and purification of the final product.

Recently, Bertolini and Frigerio[87] reported an improved method. Their preparation of **199** is performed by coupling of 5-deoxy-D-ribose derivative with silylated fluoropyrimidine in the presence of Lewis acid at subzero temperature in inert organic solvent. The one-pot reaction of **177** with chlorotrimethylsilane and hexamethyldisilazane in methylene chloride, followed by coupling with 1′,2′,3′-triacetyl-5′-deoxy-D-ribose gives pure 2′,3′-diacetyl-5′-deoxy-5-fluorouridine **201** in 90% yield (Scheme 12.39).

Compound **199** is also synthesized from **177** via a series of reactions including silylation, condensation, saponification, ketal formation, iodation, hydrogenolysis, and hydrolysis, giving the product with overall yield of 55% (Scheme 12.40).[88]

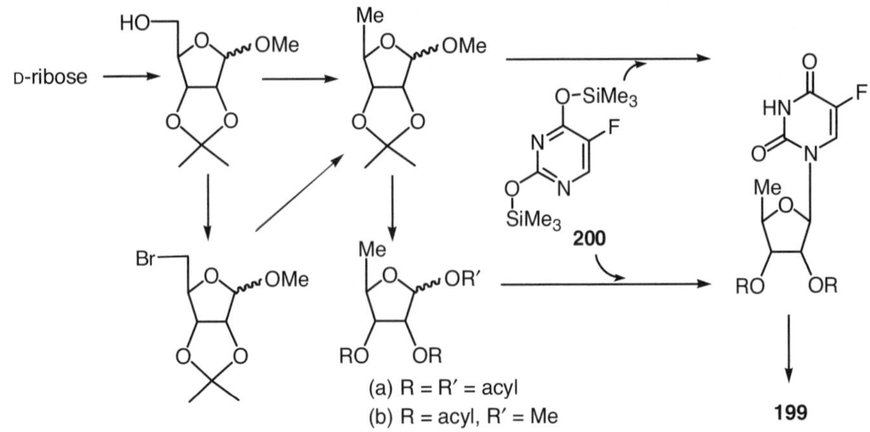

Concept of Kiss et al. Method

SCHEME 12.38

12.4.2.5 Trifluridine (5′-Trifluoromethyl-2′-deoxyuridine; α,α,α-Trifluorothymidine; TFT)

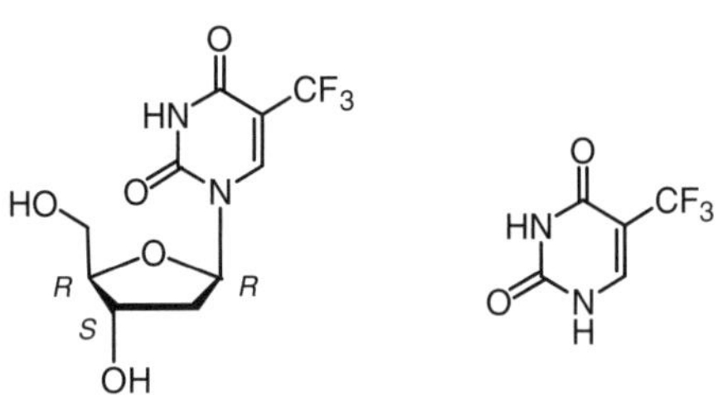

202 Trifluridine

205 5-Trifluoromethyluracil

Trifluridine **202** is an antiviral drug for topical treatment of epithelial keratitis caused by herpes simplex virus. It is a modified form of deoxyuridine and similar to be

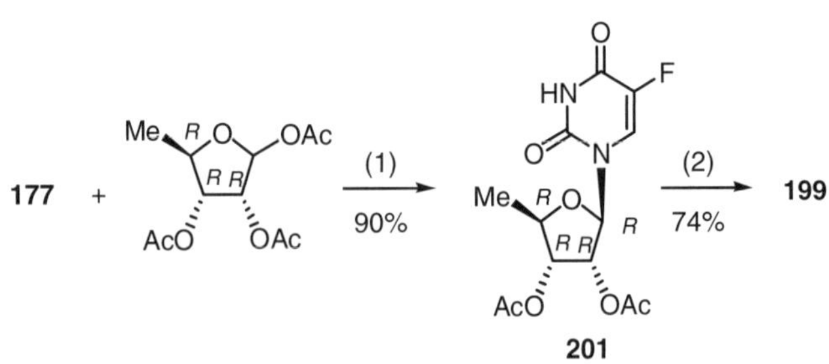

(1) (i) Me_3SiCl, $(Me_3Si)_2NH$, in CH_2Cl_2, 4 h, reflux; (ii) $SnCl_4$, cooled; (iii) HCl. (2) (i) NaOMe, in MeOH, RT, (ii) HCl, pH 4–4.2.

SCHEME 12.39

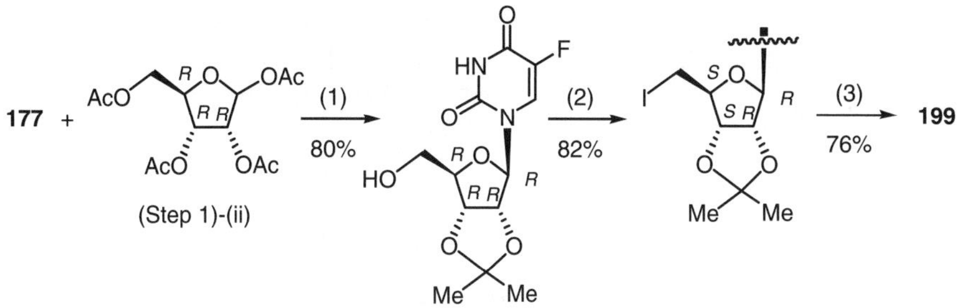

(1) (i) $(Me_3Si)_2NH$, $(NH_4)_2SO_4$, reflux; (ii) $AlCl_3$, in CH_2Cl_2, <30°C; (iii) $NaHCO_3$, in H_2O, 6.5 atm; (iv) NaOH, MeOH; (v) H_2SO_4, $CuSO_4$, in acetone, in H_2O, reflux; (vi) Na_2CO_3, pH 7. (2) (i) $[(PhO)_3PMe]^+I^-$, in DMF, 25–40°C, (ii) in MeOH. (3) (i) AcONa, H_2/Ni, in MeOH; (ii) in H_2O, in CH_2Cl_2, 1 h, 50°C; (iii) H_2SO_4, in MeOH, reflux; (iv)$CaCO_3$, in H_2O, pH 5–6; (v) in CH_2Cl_2, reflux.

SCHEME 12.40

incorporated into viral DNA replication. Compound **202** exhibits antitumor[89] and antiviral activities.[90]

This nucleoside **202** can be synthesized in a number of ways; biological transformation using nucleic base transferase,[91] trifluoromethylation of 2′-deoxyuridine or 2′-deoxy-5-halouridine,[92] or a coupling reaction of silylated 5-trifluoromethyluracil **203**[93] and 2-deoxyribose derivative **204**[94] (see Scheme 12.47). In the last case, 5-trifluoromethyluracil **205** is an important intermediate for the preparation of **202**.

Compound **205** can be synthesized by (i) the reaction of uracil-5-carboxylic acid **206** with an excess of sulfur tetrafluoride and small amount of water (0.5 mL per 45 g of sulfur tetrafluoride) (Scheme 12.41),[95] (ii) radical trifluoromethylation of **187** using bis(trifluoromethyl)mercury in the presence of azoisobutyronitrile (AIBN) (Scheme 12.41),[96] (iii) an electrolysis of a solution **187** in trifluoroacetic acid (Scheme 12.41),[97] (iv) the reaction of 5-trifluoromethyl-5,6-dihydrouracil **207**[98] with bromine in acetic acid followed by heating in dimethylformamide (DMF) solution

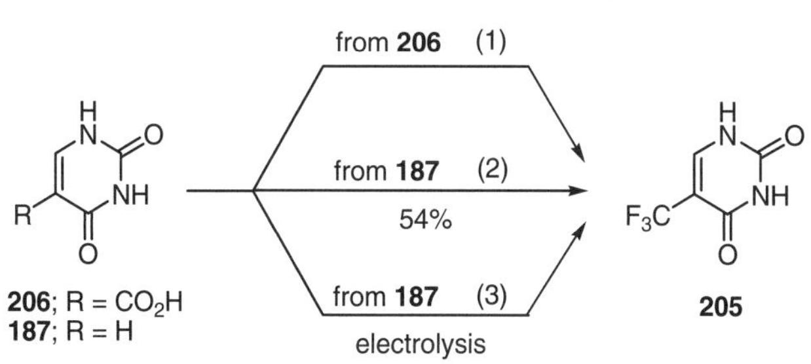

(1) SF_4, small amount of H_2O (0.5 ml per 45 g SF_4); (2) $(CF_3)_2Hg$, AIBN; (3) a nickel anode and an iron cathode, 15–18 V and 5–10 mA/cm^2, KHF_2 and CF_3CO_2H, in ethylene glycol.

SCHEME 12.41

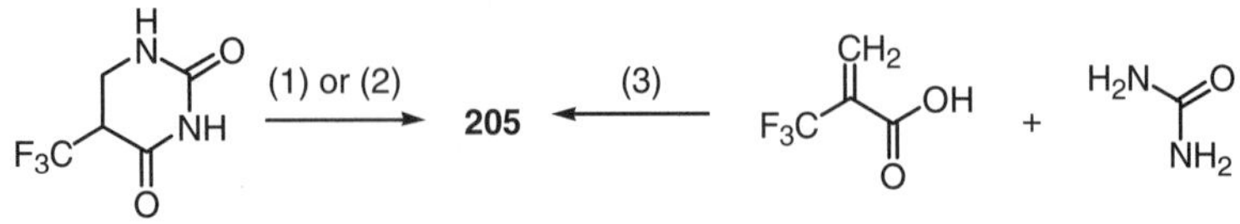

207 **208** **209**

(1) Br_2, 100%; (2) $CuBr_2$, H_2O, in DMF, in H_2O; (3) acetic anhydride.

SCHEME 12.42

(Scheme 12.42),[99] or (v) the reaction of trifluoromethylacrylic acid **208** with urea (Scheme 12.42).[100]

Andres and Marhold[93b] reported the improved method for the preparation of **205**, which employs the chlorination of thymine **210** to give pyrimidine **211**. The reaction of **211** with hydrogen fluoride yields the new 2,4-difluoro-5-(trifluoromethyl)pyrimidine **212**, and the hydrolysis gives the target compound **205** in a good yield (Scheme 12.43).

Recently, Yamakawa et al.[101] obtained **205** by the reaction of **187** with R_f–X [R_f = perfluoroalkyl; X = Cl, Br, I] in the presence of hydrogen peroxide and an iron compound (Scheme 12.44).

Umetani et al.[102] prepared **205** by the treatment of 2-(trifluoromethyl)-2-propenoic acid phenyl ester **213** with **209** and cyclization of the resulting phenyl ester **214** (Scheme 12.45).

Conventional processes for preparing **202** involve biological transformation of thymidine using nucleic base transferase[91] and trifluoromethylation on the 5-position of 2′-deoxyuridine derivatives using CF_3Cu.[93] That is, treatment of halogenated nucleoside derivative **215** with a solution of a F_3C–Cu complex obtained from CF_3I and Cu powder (HMPA, 120°C, 2.5 h) gave the trifluoromethylated compound in moderate yield, which gives **202** (Scheme 12.46) after deprotection. However, this

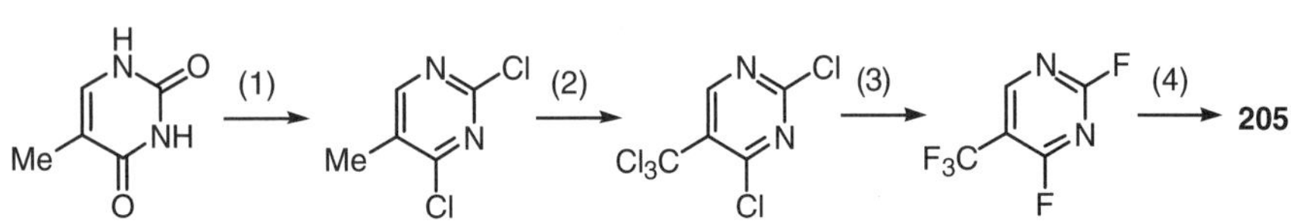

210 **211** **212**

(1) $POCl_3$; (2) Cl_2, UV irradiation; (3) HF; (4) H_2O, KF.

SCHEME 12.43

$$\mathbf{187} + CF_3I \xrightarrow{(1)} \mathbf{205}$$

(1) $FeSO_4$, H_2O_2, H_2SO_4, DMSO.

SCHEME 12.44

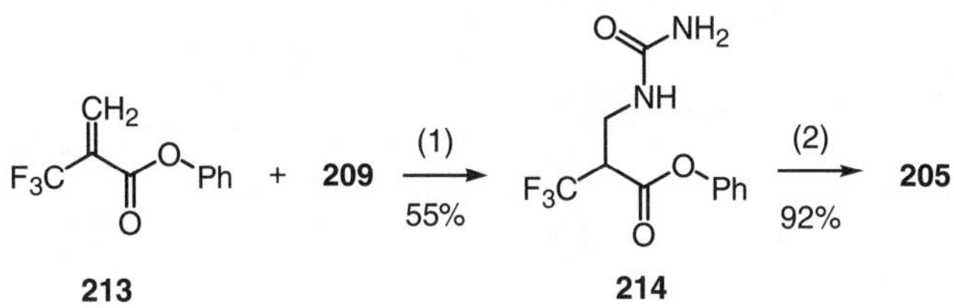

(1) 90°C for 2 h in DMF; (2) 120°C for 8 h in DMF.

SCHEME 12.45

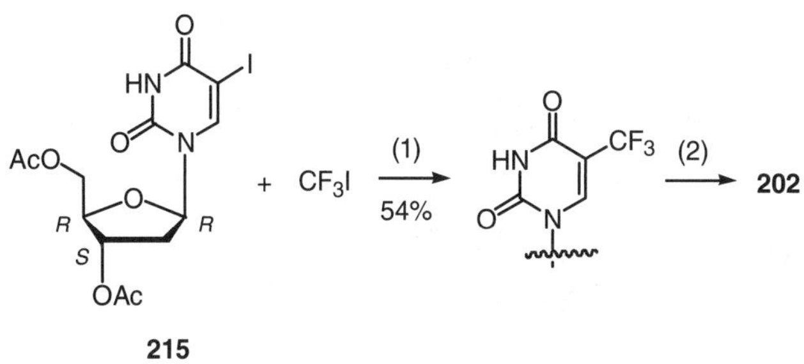

(1) Cu, P(O)(NMe$_2$)$_3$; (2) MeOH-NH$_3$, 5°C, overnight.

SCHEME 12.46

method was not cost-effective for industrial manufacturing because of low yield and the expensive reagents.

The best protocol for preparing **202** has been reported by Kawakami et al.[93a] and the key reaction is a glycosylation of 1-α-chloro-2-deoxyribose derivative **204**[94] with silylated 5-trifluoromethyluracil **203** (Scheme 12.47). However, a contamination of a significant amount of an (*R*)-anomer is a drawback of this process. To increase the stereoselectivity of the process, the following changes were made: (i) an addition of metal catalyst like $ZnCl_2$, (ii) $CHCl_3$ solvent, and (iii) large excess use of **203** (2 equiv was required for a 75:25 ratio of β-anomer:α-anomer). To meet the CGMP (current good manufacturing practices) regulation standard, the residual amount of transition metal catalysts and $CHCl_3$ in an active pharmaceutical ingredient should be strictly controlled at low level. For health and environmental reasons, $CHCl_3$ solvent should be avoided. Additionally, usage of a stoichiometric amount of **203** is preferred for an economical process. Improved synthesis by Komatsu and Umetani[103] involves a new glycosylation protocol for synthesis of **202**. Key features of the synthesis include: (i) an equimolar amount of **205**, (ii) the glycosylation is performed under high concentration, (iii) the reaction is carried out at 50°C.

Target compound **202** is also provided by reacting a halogenated perfluoroalkane (e.g., trifluoromethyl iodide) with a nucleic acid base (2′-deoxyuridine **216**) in the presence of a sulfoxide, hydrogen peroxide, and an iron (II) salt[101] (Scheme 12.48).

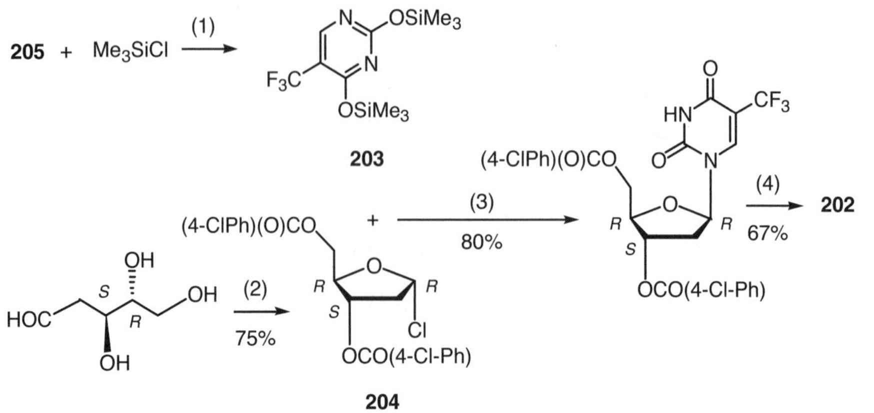

(1) Me_3SiCl, $(Me_3Si)_2NH$, in 1,2-dichloroethane, the product was used for next step without purification; (2) $4\text{-}ClC_6H_4COCl$, HCl or AcCl, in MeOH, pyridine; (3) $ZnCl_2$, in $CHCl_3$; (4) in MeOH

SCHEME 12.47

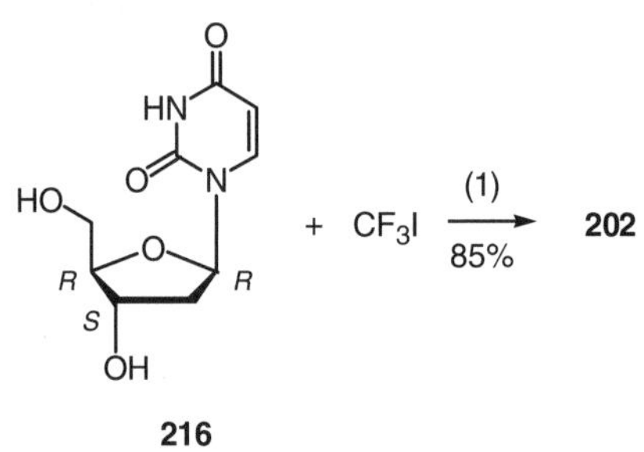

(1) $FeSO_4$, H_2O_2, H_2SO_4, DMSO.

SCHEME 12.48

12.4.3 4-Amino-5-fluoro-2-pyrimidones

12.4.3.1 Flucytosine (5-Fluorocytosine, 5-FC)

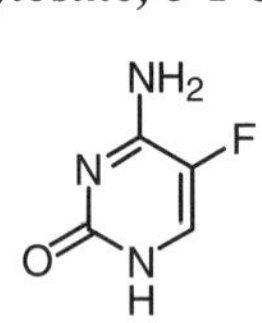

217

Flucytosine

Flucytosine[104] (5-fluorocytosine; 5-FC; 4-amino-5-fluoro-2-pyrimidone) **217** is an antimetabolite type of antifungal drug. It is activated by deamination within the fungal cells to **177**.

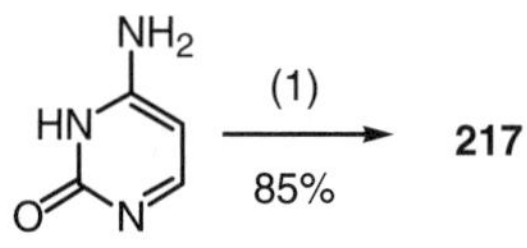

218

(1) CF_3OF in CCl_3F, Et_3N in 50% aq. MeOH.

SCHEME 12.49

Currently, the compound **217** is the only available antimetabolite drug having antifungal activity. It inhibits fungal protein synthesis by replacing **187** with **177** in fungal RNA. **217** also inhibits thymidylate synthetase via 5-fluorodeoxyuridine monophosphate and thus interferes with fungal DNA synthesis.[105]

Robins and Naik[106] have reported that the reaction of cytosine (4-amino-2-(1*H*)-pyrimidinone) **218** with CF_3OF in CCl_3F followed by treatment with triethylamine in 50% aqueous methanol gives **217** in 85% yield (Scheme 12.49).

On the other hand, the products of the reaction of F_2 or AcOF with **187** and **218** dissolved in acetic acid and/or water were studied by using ^{18}F as a tracer.[107] They have shown that the reaction of F_2 or AcOF with **187** in acetic acid and water yields, among others (**177**, **219**), both geometric isomers of the 5-F,6-OAc or 5-F,6-OH adducts (**220**, **221**). On the basis of the observation that a part of the 5-fluoropyrimidines is directly formed and the fact that both F_2 and AcOF led to the same "electrophilic substitution" products, an alternative radical cation mechanism was suggested which involves the formation of cationic intermediate **225** (Figures 12.11 and 12.12).

Takahara and Hisanaga[108] reported that the reaction of F_2 with the uracil ring systems in acetic acid proceeds via 5,6-difluoro adducts (2 cis isomers), which were

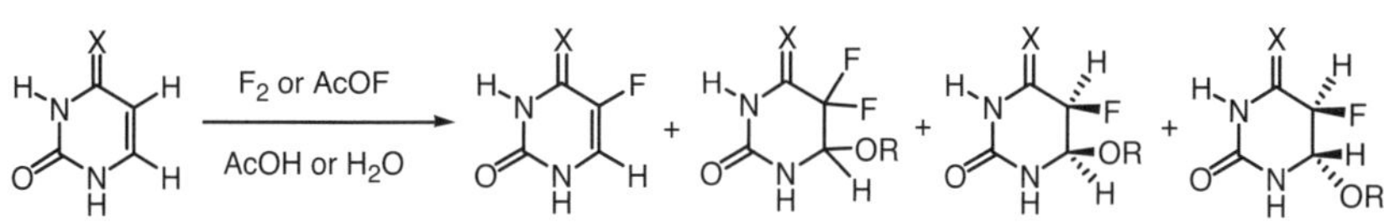

X	X	X	R	X	R	X	R
187; O	**177**; O	**219a**; O	AC	**220a**; O	Ac	**221a**; O	Ac
		b; O	H	**b**; O	H	**b**; O	H
218; NH	**217**; NH	**222a**; NH	AC	**223a**; NH	Ac	**224a**; NH	Ac
		b; NH	H	**b**; NH	H	**b**; NH	H

FIGURE 12.11 Fluoro products formed from the reaction of F_2 and AcOF with uracil (**187**) and cytosine (**218**) dissolved in acetic acid and/or water.

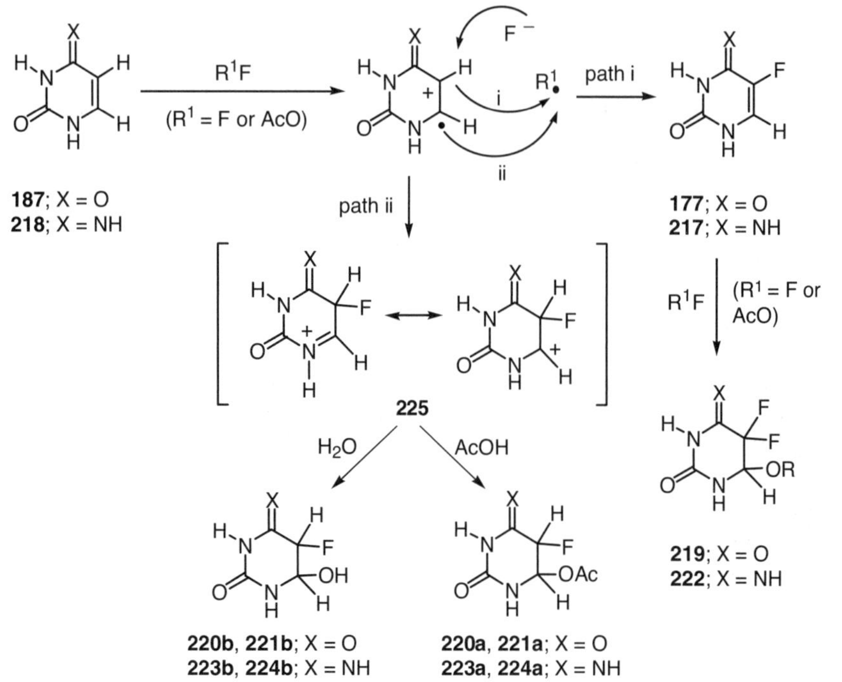

FIGURE 12.12 Proposed mechanism for the reaction of F_2 and AcOF with uracil (**187**) and cytosine (**218**).

converted to 4 isomer adducts by subsequent solvolysis. Time course adducts were traced by ^{19}F-NMR, but in the case of the cytosine ring systems, the adducts were not detected. Comparing with **187**, the reaction rate of the fluorination of **218** with F_2 is very slow. However, when HF is used as a solvent, the reaction rate of **218** becomes similar to that of **187**. They speculated that the role of HF was for increase of the electronegativity of the carbon at 5-position of the pyrimidine ring.[108] So that, **217** was obtained in good yield using HF as a solvent for the fluorination of **218** with F_2 (Scheme 12.50).

$$\mathbf{218} \xrightarrow[90\%]{(1),\ (2)} \mathbf{217}$$

(1) F_2, in HF, –5°C; (2) conc NH_3 aq.

SCHEME 12.50

12.4.3.2 *Reverset (RVT, β-D-2′,3′-Didehydro-2′,3′-dideoxy-5-fluorocytidine, D-D4FC)*

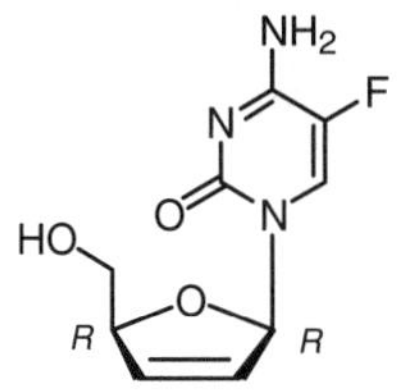

226

Reverset

Reverset **226** is an anti-HIV agent. It belongs to a group of nucleoside reverse transcriptase inhibitors (NRTIs). Compound **226** prevents HIV from entering the nucleus of healthy T-cells.

Fu and Zhu[109] have used crude preparation of *N*-deoxyribosyl transferases (NdRT-II, from *Lactobacillus helveticus*) to catalyze the transfer of a glycosyl moiety from a donor nucleoside to an acceptor base. Optimal conditions for the transglycosylation reaction to make **226** starting from β-D-2′,3′-unsatutrated thymidine (D-D4T)[110] **227** and **217** were identified after optimization of several experimental parameters including reaction time, concentration of substrate, pH, and the type of buffer. This group developed a practical procedure for enzmatic synthesis of **226** from **217** and **227** (Scheme 12.51). The method will be useful in the manufacture of important nucleoside analogues for antiviral therapy. Compound **227** was synthesized[110] as shown in Scheme 12.52, that is, thymidine **228** was dimesylated and treated with aqueous sodium hydroxide to afford **229** in 70% yield. Upon treatment with NaH in DMF, this yielded **227** in good yield.[111]

Choudhury et al.[112] employed an unprecedented palladium mediated Ferrier type rearrangement glycosidation and succeeded in synthesis of **226** (Scheme 12.53). It is noteworthy that classical Ferrier rearrangement does not work on furanoid glycals.

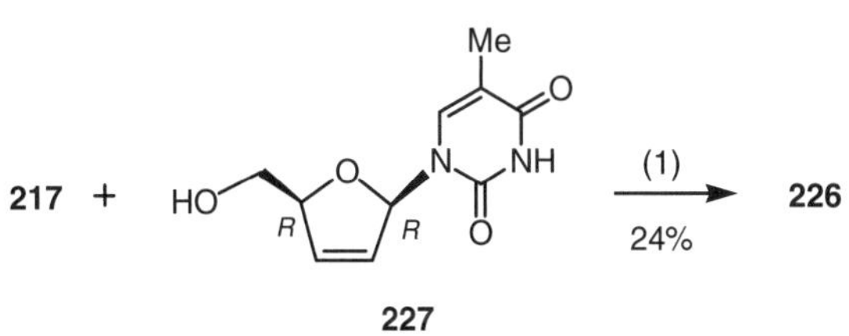

(1) NaOH, nucleoside deoxyribosyl transferase (from *Lactobacillus helveticus*).

SCHEME 12.51

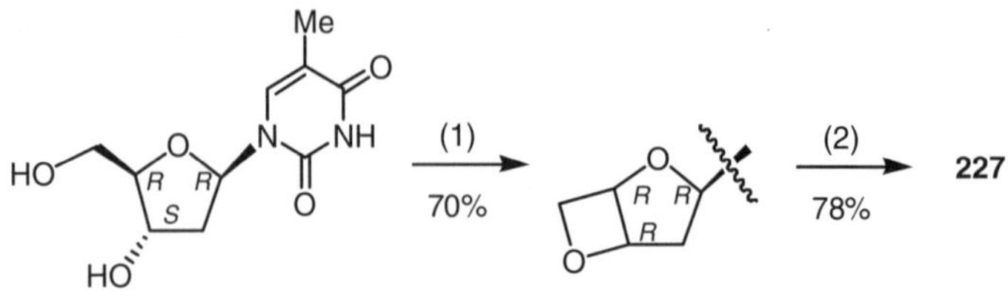

228 229

(1) (i) MsCl, pyridine; (ii) aq. NaOH. (2) NaH.

SCHEME 12.52

12.4.3.3 Emtricitabine (5-Fluoro-1-(2R,5S)-[2-(hydroxymethyl)-1,3-oxathiolan-5- yl]cytosine)

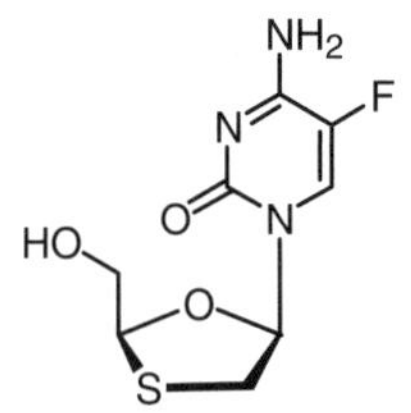

230

Emtricitabine

Emtricitabine **230** is an analogue of cytidine. The drug works by inhibiting of reverse transcriptase of human immunodeficiency virus type 1 (HIV-1), the enzyme that copies HIV RNA into new viral DNA. By interfering with this process, which is central to the replication of HIV, **230** can help to lower the amount of HIV, or "viral load," in a patient's body and can indirectly increase the number of immune system cells (called T-cells or CD4+T-cells). Both of these changes are associated with healthier immune systems and decreased likelihood of serious illness.

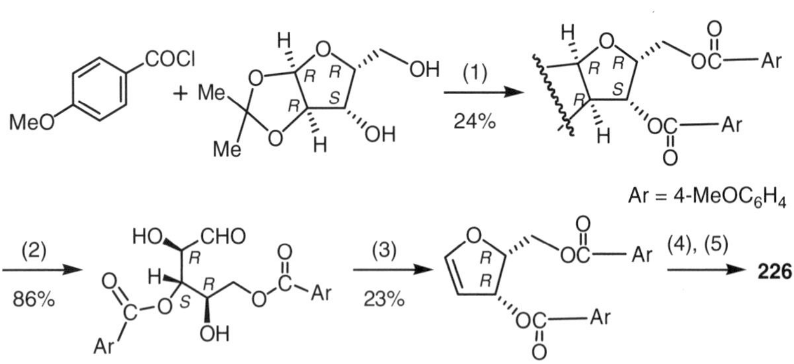

(1) pyridine. (2) H_2SO_4. (3) I_2, PPh_3, 1*H*-imidazole. (4) **217**, DBU, $Pd(PPh_3)_4$, 35%. (5) NaOMe, 92%.

SCHEME 12.53

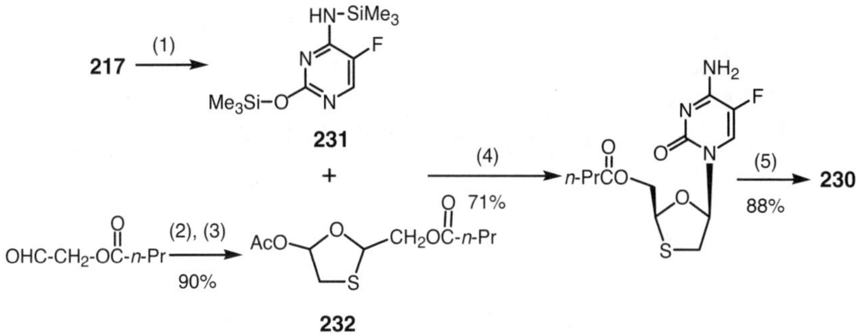

(1) $(Me_3Si)_2NH$, $(NH_4)_2SO_4$. (2) $HSCH_2CO_2H$; 90%. (3) (i) $(t\text{-BuO})_3AlH{\cdot}Li$; (ii) Ac_2O; (iii) $NaHCO_3$; 70%. (4) (i) $SnCl_2$; (ii) NH_4OH; (iii) enantioselective enzymic resolution. (5) (i) NaOMe; (ii) NH_4Cl.

SCHEME 12.54

As shown in Scheme 12.54, oxathiolane nucleoside **230**, has been prepared by the method of Schinazi and Liotta.[113] Thus, the coupling of silylated flucytosine **231** preparing from **217** with lactol **232** gave racemic compound, 2-hydroxymethyl-5-(5-fluorocytosin-1-yl)-1,3-oxathiolane [(±)-FTC], which afforded the title compound **230** after an enantioselective enzymic resolution and hydrolysis.

12.4.3.4 Capecitabine

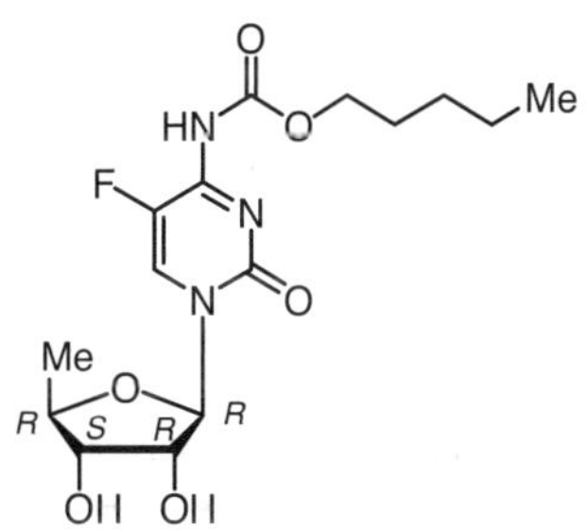

233
Capecitabine

Capecitabine **233** is a fluoropyrimidine derivative and a chemotherapy drug that is given as a treatment for some types of cancer, including advanced bowel cancer or breast cancer.

Tao and Chen[114] have reported the effective preparation of **233**. It comprises (i) adding aminopyrimidine **234** to $4\text{-}O_2NC_6H_4OCO_2(CH_2)_4Me$ and (ii) alkoxycarbonylating at the *N*-4 position under a basic condition to obtain the compound **235**. Deprotection of **235** gave 76% of **233** (Scheme 12.55).

The preparation of ^{18}F-labeled **233** has been performed by Moon et al.[115] via glycosidation from **218** and D-ribose followed by electrophilic fluorination of the resulting 5′-deoxycytidine derivative **237** with radiochemical yield of 5–15% and radiochemical purity of >95% (Scheme 12.56).

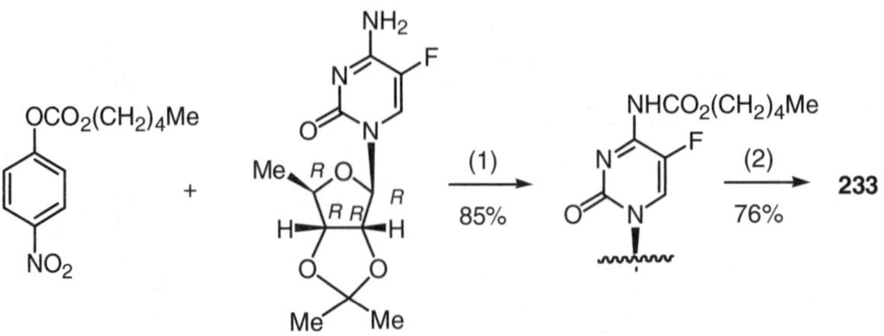

234 235

(1) K_2CO_3, in DMF, 35°C. (2) DOWEX 50, in EtOH, 40–45°C.

SCHEME 12.55

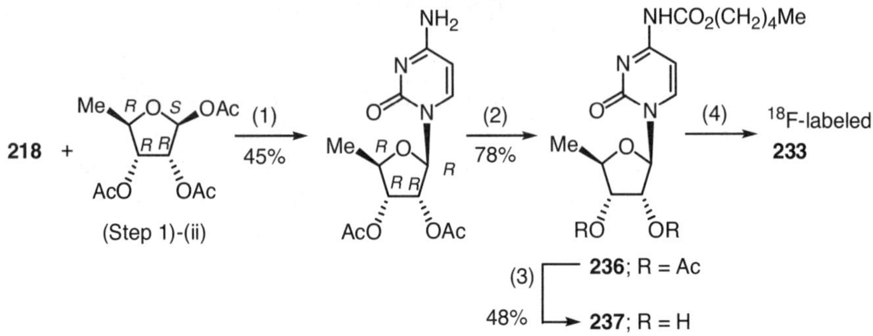

(1) (i) Me_3SiCl, $(Me_3Si)_2NH$, reflux; (ii) Me_3SiCl, NaI, in MeCN, RT. (2) $ClCO_2(CH_2)_4Me$, pyridine, in CH_2Cl_2, 0°C. (3) NaOH, in H_2O and MeOH, 0°C. (4) $^{18}F_2$, CF_3CO_2H, $MeCO_2Na$, in H_2O, RT.

SCHEME 12.56

12.4.4 2-Substituted 5-(trifluoromethyl)pyridines

12.4.4.1 Tipranavir

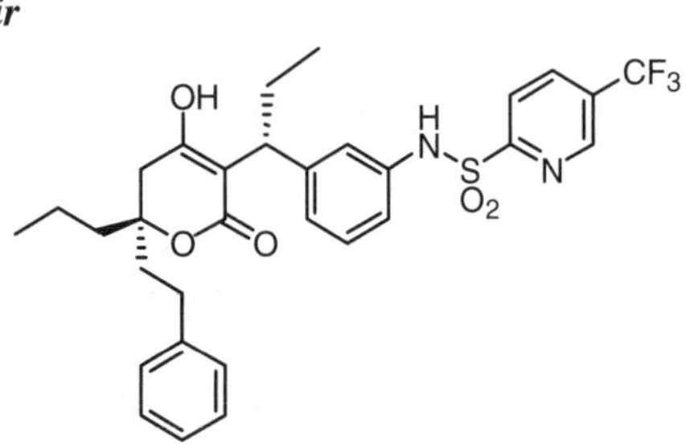

238

Tipranavir

The development of new therapeutic agents to combat the human immunodeficiency virus continues to be an intense area of pharmaceutical research.

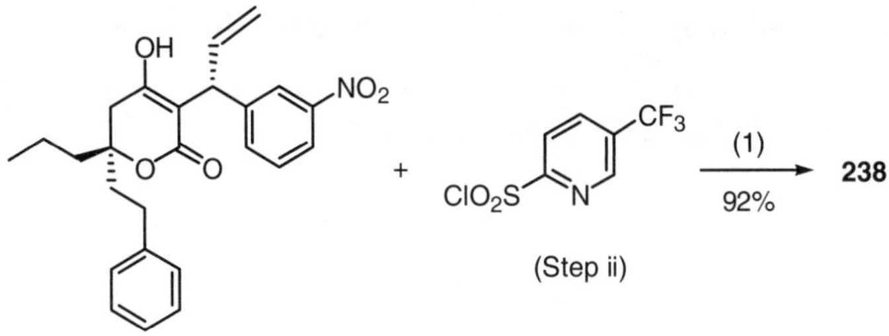

239 240

(1) (i) 5 mol % Pd/C, H_2 (1 atm), in MeOH, 25°C; (ii) Pyridine, in CH_2Cl_2, –25°C.

SCHEME 12.57

Tipranavir **238**, a unique nonpeptidic protease inhibitor (PI), has demonstrated remarkable pharmacokinetic properties and offers the significant advantage of oral bioavailability.

Target compound **238** is synthesized in 15 steps from readily available starting materials in 25% overall yield by utilizing Pd- and Mo-catalyzed asymmetric reactions to control the quaternary and tertiary stereogenic centers, respectively.[116] The fluorine containing part is introduced at the final stage by the reaction of a ketone **239** with 5-(trifluoromethyl)-2-pyridinesulfonyl chloride **240**[117] (Scheme 12.57).

For preparing the key intermediate **239**, aldol coupling of aldehyde **241** and ester **242** was performed (NaHMDS, −78°C) followed by Dess–Martin oxidation to give β-ketoester as a 1:1 mixture of C-3 epimers. After deprotection, pyrone formation furnished compound **239** in 77% yield (97% yield based on recovered starting material, Scheme 12.58).

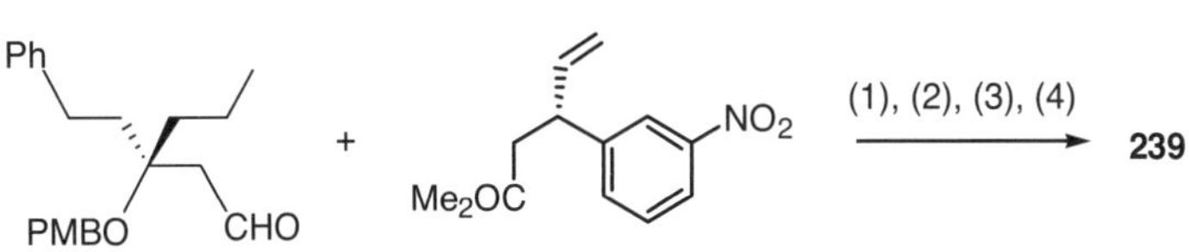

241 242

(1) NaHMDS, in THF, –78°C. (2) Dess–Martin periodinane, in CH_2Cl_2, 25°C; 89% two steps. (3) CAN, CH_3CN/H_2O; 88%. (4) NaOH, MeOH, 4°C; 77%.

SCHEME 12.58

12.4.5 6-Fluoro-1,4-dihydro-4-oxo-3-quinolinecarboxylic Acids

12.4.5.1 Norfloxacin (NOROXIN, 1-Ethyl-6-fluoro-1,4-dihydro-4-oxo-7-(1-piperazinyl)-3-quinolinecarboxylic Acid)

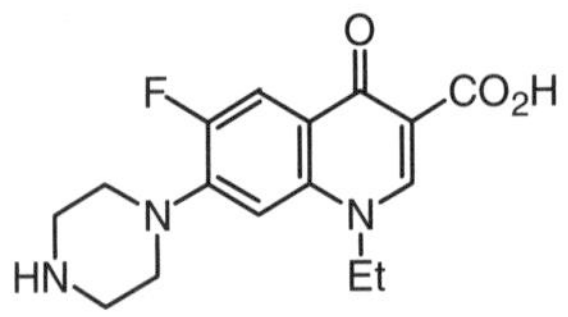

243

Norfloxacin

Fluoroquinolones are used to treat bacterial infections. These drugs are prescribed for the treatment of bone and joint, skin, ear, urinary tract infections, inflammation of the prostate and serious diseases such as bronchitis, pneumonia, tuberculosis, sexually transmitted diseases (STDs), and infections affecting people with AIDS.[118]

Norfloxacin **243** is one of oral broad-spectrum fluoroquinolone antibacterial agents used in the treatment of certain infections caused by bacteria, such as gonorrhea, prostate, and urinary tract infections. The mechanism of action of **243** involves inhibition of the A subunit of bacterial DNA gyrase, an enzyme which is essential for DNA replication.[119]

The preparation of **243** is based on condensation of piperazine **244** with chloroquinolone derivative **245** exemplified by Scheme 12.59.[120]

The typical synthesis of the intermediate **245** can be carried out by two routes: (i) preparation from 3-chloro-4-fluoroaniline **248** and malonate **249** through the formation of **250** and **251** (Scheme 12.60)[121]; (ii) preparation from 2,4-dichloro-5-fluorobenzoyl chloride **252** and enamine **253** or $CH_2(CO_2Et)_2$ (Scheme 12.61).[122]

Heravi et al.[123] have prepared **243** through borate complexes after formation of **245**. In this method, **245** was converted to borate complex, which was treated with **244** in the presence of triethylamine to afford **243** in high yields (Scheme 12.62).

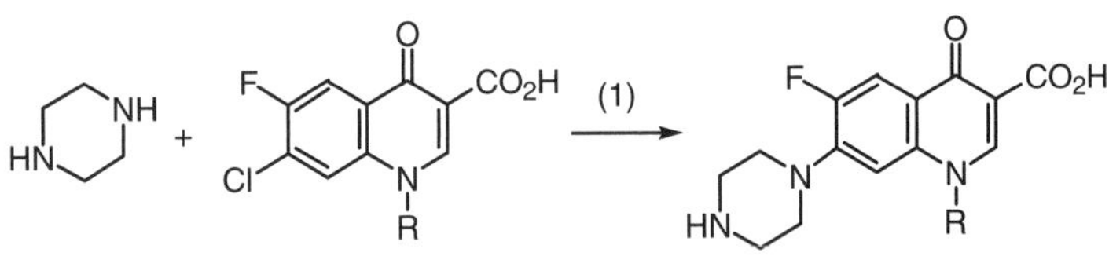

244 **245**; R = Et **243** (norfloxacin); R = Et (from **245**)

246; R = cyclopropyl **247** (ciprofloxacin); R = cyclopropyl (from **246**)

In the case of **243** from **245**
(1) BuOH, RT --> 135°C, 2 h.
In the case of **247** from **246**
(1) Et_3N, in isopentyl alcohol, RT-->130°C, 7 h; 80%.

SCHEME 12.59

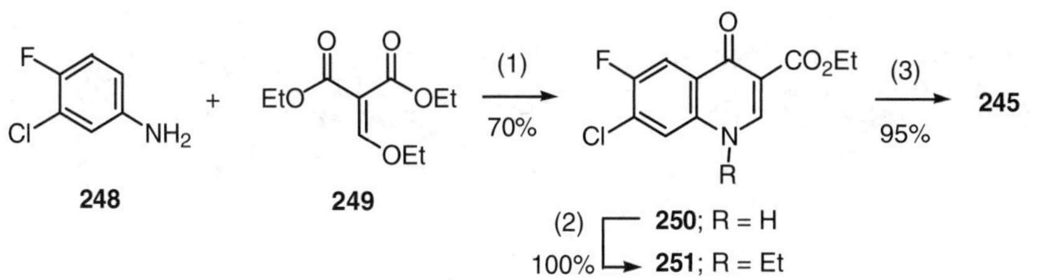

(1) In diphenyl ether. (2) EtI, K_2CO_3. (3) KOH, in H_2O.

SCHEME 12.60

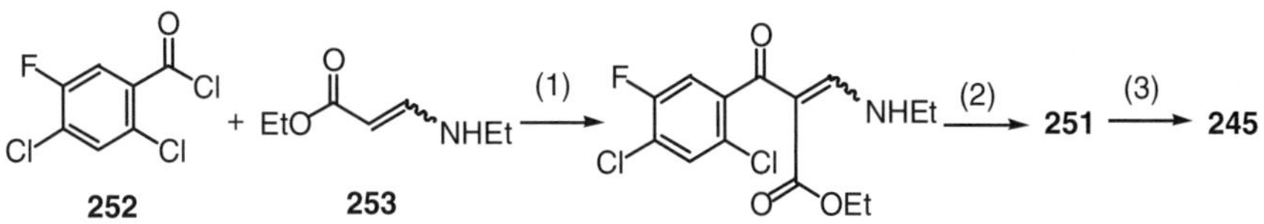

(1) Et_3N, in dioxane. (2) K_2CO_3, in DMF. (3) H_2SO_4, in H_2O.

SCHEME 12.61

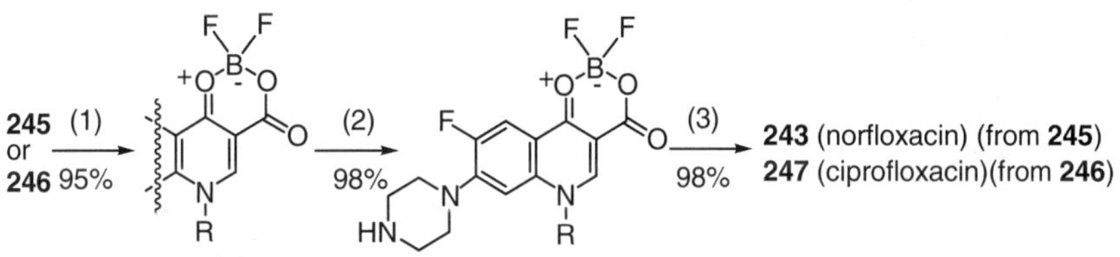

R = Et (from **245**)
R = cyclopropyl (from **246**)

In the case of **243** from **245**
(1) Et_3N, BF_3-Et_2O, in Et_2O, 5 h, reflux; 95%. (2) **244**, Et_3N, in DMSO, 3 h, RT; 98%.
(3)(i) NaOH, in H_2O, 1 h, reflux; (ii) AcOH, RT, pH 7; 98%.
In the case of **247** from **246**
(1) Et_3N, BF_3-Et_2O, in CH_2Cl_2, 5 h, reflux; 71%. (2) **244**, Et_3N, in DMSO, 3 h, RT; 96%.
(3)(i) NaOH, in H_2O, 1 h, reflux; (ii) AcOH, RT, pH 7; 90%.

SCHEME 12.62

12.4.5.2 Ciprofloxacin (Cipro, Ciproxin, and Ciprobay, 1-Cyclopropyl-6-fluoro-4-oxo-7-(1-piperazinyl)-1,4-dihydro-3-quinolinecarboxylic Acid)

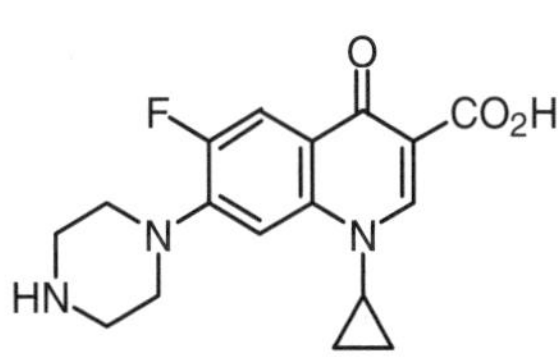

247
Ciprofloxacin

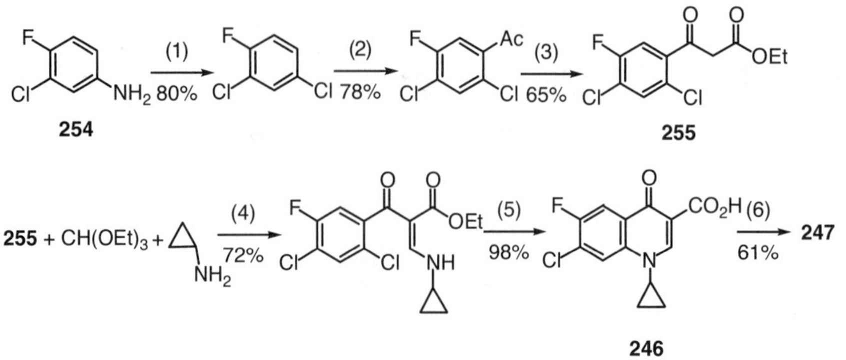

(1) $NaNO_2$, HCl, in H_2O. (2) AcCl, $AlCl_3$. (3) Diethyl carbonate, NaH. (4) (i) Ac_2O; (ii) in EtOH. (5) (i) NaH, in dioxane; (ii) KOH, in H_2O; (iii) HCl, in H_2O. (6) **244**, in DMF.

SCHEME 12.63

Ciprofloxacin **247** is used to treat or prevent certain bacterial infections. Compound **247** is also used to treat or prevent anthrax in people who may have been exposed to anthrax germs in the air. Its mode of action depends upon blocking bacterial DNA replication by binding itself to an enzyme called DNA gyrase, and causing double-stranded breaks in the bacterial chromosome.

The title compound **247** was prepared in six steps in 18% overall yield starting from 3-chloro-4-fluoroaniline **254** (Scheme 12.63).[124]

A suitable system for the realm of combinatorial chemistry as well as its application to the first library approach toward **247**, solely using microreaction technology is reported by Schwalbe et al.[125] (Scheme 12.64). A known one-pot

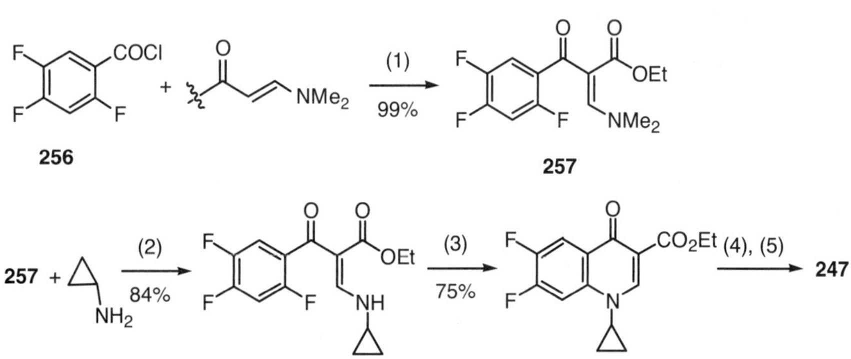

(1) Bu_3N, in toluene. (2) AcOH. (3) DBU, in 1-methyl-2-pyrrolidinone. (4) **244**, Et_3N, in *t*-BuOH; 99%. (5)(i) NaOH, in H_2O; $NaHCO_3$, in H_2O; (iii) HCl, in H_2O: 92%.

SCHEME 12.64

batch procedure for the synthesis of **247** was split in its individual reaction steps, which were successfully adapted to a continuous reaction step. In their method, particularly it was shown, the first step of the synthesis (the acylation reaction of a β-dimethylamino acrylate with trifluorobenzoic acid chloride **256**) was accessible to synthesis of high quantities without any difficulties to yield a primary building block suitable for subsequent library synthesis. In a first diversification step, the Michael addition of a set of primary amines was followed by nucleophilic ring closure providing the difluoroquinolone system, which was subjected to a second diversification step by means of a nucleophilic aromatic substitution reaction.

Compound **247** was also prepared by condensation of 7-chloro-1-cyclopropyl-6-fluoro-4-oxo-1,4-dihydroquinolin-3-carboxylic acid **246** with anhydrous **244** in up to 80% (Scheme 12.59).[126]

Same as the preparation of **243**, Heravi et al.[123] employed borate complexes as an intermediate. Thus, **246** was converted to borate complex. This compound was treated with **244** in presence of triethylamine to afford **247** in high yields (Scheme 12.62).

12.4.5.3 Levofloxacin (Levaquin, (−)-(S)-9-fluoro-3-methyl-10-(4-methyl-1-piperazinyl)-7-oxo-2,3-dihydro-7H-pyrido[1,2,3-de]-1,4-benzoxazine-6-carboxylic Acid Hemihydrate)

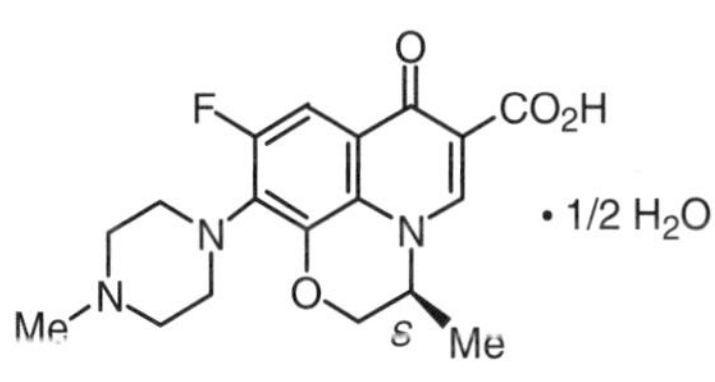

258

Levofloxacin

Levofloxacin **258** is a third generation of fluoroquinolone type antibiotic. This medicine is effective against a number of Gram-positive and Gram-negative bacteria. Because of its broad spectrum of action, **258** is frequently used for symptomatic treatment of infections of unknown etiology.

Efficient preparation of **258** via 1,2-cyclic sulfamidate has been reported.[127] (Scheme 12.65). The amidate **261** undergoes efficient and regiospecific nucleophilic cleavage with 2-bromophenol **262**, and Pd(0)-mediated amination to provide substituted and enantiomerically pure (3*S*)-3-methyl-1,4-benzoxazine **263**. This chemistry provides a short and efficient entry to **263**, a late stage intermediate in the synthesis of **258**. For selected previously reported studies on synthesis of **258**, see Ref. 128.

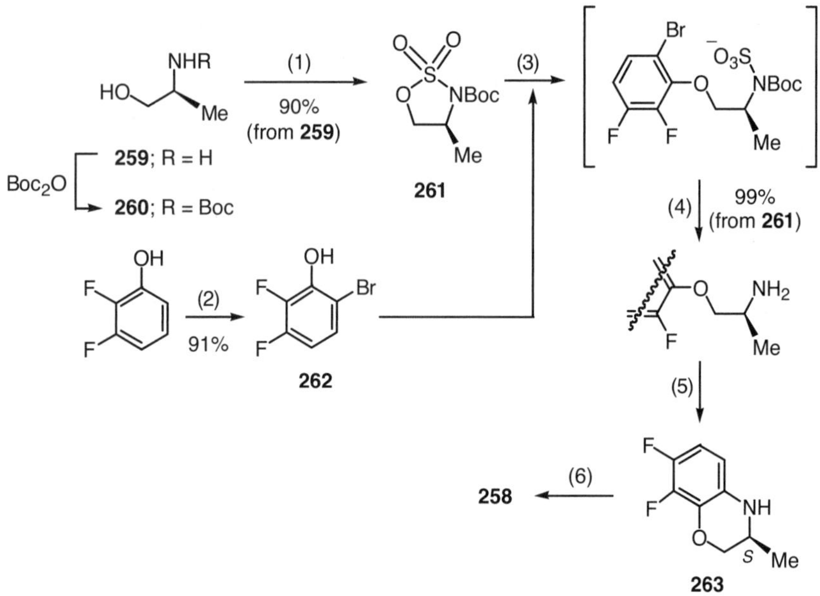

(1) (i) $SOCl_2$, (ii) $RuCl_2$, $NaIO_4$. (2) Br_2, *t*-$BuNH_2$. (3) NaH, in DMF. (4) H_3O^+. (5) $Pd(OAc)_2$ (5 mol%), xantphos (7.5 mol%), *t*-BuONa, in PhMe. (6) Ref. 128a,d.

SCHEME 12.65

12.4.6 7-Fluoro- and 2,8-Bis(trifluoromethyl) quinolines

12.4.6.1 DX-8951f (1-Amino-9-ethyl-5-fluoro-1,2,3,9,12,15-hexahydro-9-hydroxy-4-methyl-(1S,9S)-10H,13H-benzo[de]pyrano[3′,4′:6,7]indolizino[1,2-b]-quinoline-10,13-dione Methanesulfonate (1:1), Exatecan Mesylate)

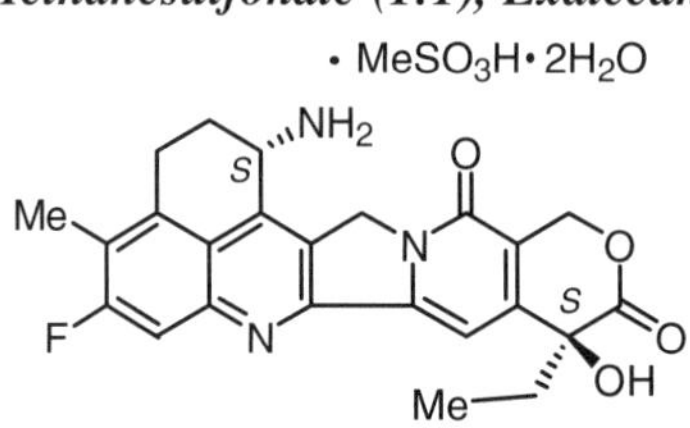

264

DX-8951f

The novel DNA topoisomerase I inhibitor, DX-8951f (exatecan mesylate, **264**) which does not require metabolic activation is fluorinated, hexacyclic derivative of camptothecin.[129] Compound **264** was reported to be 6- and 28-fold more potent than SN-38 and topotecan, respectively.[130] Compound **264** has shown efficacy in a variety of human tumor xenografts.[131]

Compound **264** has a significant antitumor effects against cell lines and xenografts in which CPT-11 was not effective.[131,132] It may also overcome P170-glycoprotein

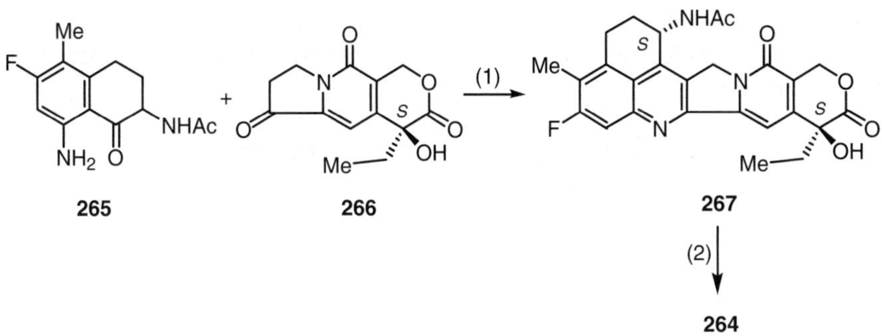

(1) Pyridine, p-$MeC_6H_4SO_3H$, in toluene. (2) $MeSO_3H$, H_2O, in toluene

SCHEME 12.66

(Pgp)-mediated multidrug resistance, as it is highly effective against cell lines overexpressing the Pgp transporter both *in vitro* and *in vivo*.[130,131] At this moment, however, little is known on the influence of the expression of the multidrug resistance-associated protein 1 (MRP1) or the lung resistance protein (LRP) on the activity of **264**. It has been already demonstrated that, unlike SN-38 and topotecan, **264** is a poor substrate for the breast cancer resistance protein (BCRP).[133]

Compound **264** (Scheme 12.66) was prepared by treating the benzopyranoindolizinoquinoline derivative **267** with compound **266** by a known procedure,[134] with methanesulfonic acid and followed by recrystallization.[135] Compound **264** is non-hydroscopic and easy to handle material.

As shown in Scheme 12.67, the process for producing the compound **265**, (a useful intermediate for production of camptothecin derivatives) involves the hydrogenation of 5-acetamido-7-fluoro-8-methyl-1-tetralone **270**,[136] to 5-acetamido-7-fluoro-8-

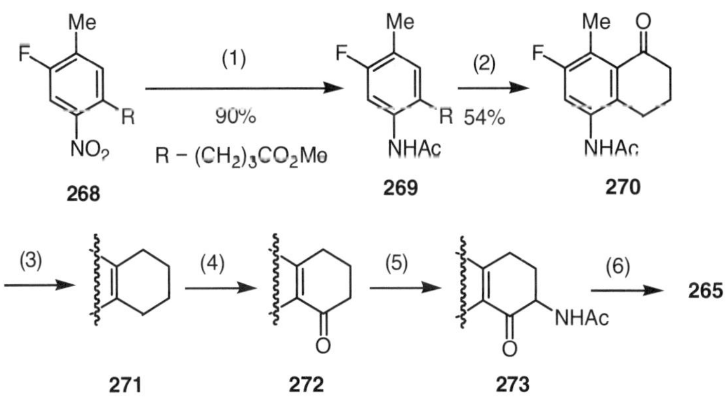

(1)(i) Ni_2B, HCl 1 M; (ii) Et_3N, Ac_2O. (2) (i) NaOH 10%; (ii) PPA, 110°C. (3) $H_2/PdCl_2$, HCl, normal hydrogen pressure, 8 h. (4) $NaHCO_3$, $KMnO_4$, in acetone. (5)(i) t-BuOK, in THF; (ii) Bu nitrite; (iii) HCl; (iv) Ac_2O, AcOH, Zn powder. (6) (i) HCl; (ii) K_2CO_3.

SCHEME 12.67

methyl-1,2,3,4-tetrahydronaphthalene **271** followed by oxidation and introduction of a protected amino group to give tetralone **273** and selective deprotection to free amino group with an acid.[137]

12.4.6.2 Mefloquine ((R,S*)-(±)-α-2-Piperidinyl-2,8-bis(trifluoromethyl)-4-quinoline-methanol Hydrochloride)*

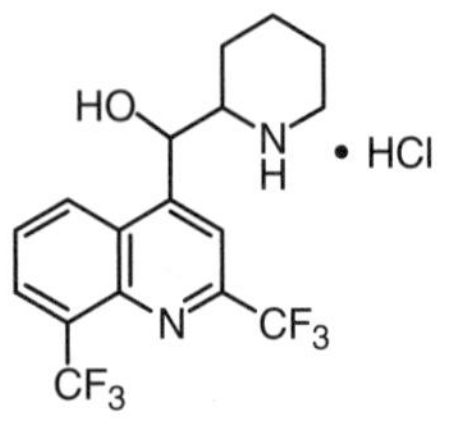

274
Mefloquine

Mefloquine **274** is a useful alternative treatment agent for *Plasmodium vivax* malaria and *Plasmodium falciparum* malaria in areas, where chloroquine is still recommended as the first-line therapeutic agent.[138] Mefloquine is a chiral molecule with two asymmetric carbon centers, and it exists as a mixture of diastereomers. The drug is currently manufactured and sold as a racemate of the (+/−) *R*,*S* enantiomers by Hoffman–LaRoche. The (+) enantiomer is more effective in treating malaria, and the (−) enantiomer specifically binds to adenosine receptors in the central nervous system, which may explain some of its psychotropic effects.[139] It is not known whether mefloquine can be transformed into its stereoisomer *in vivo*.

Many researchers have published the preparation of **274** since 1971.[140] One of the typical preparation routes for **274** has been reported by Kumarn et al.[141] As shown in Scheme 12.68, the key intermediate, 4-hydroxy-2,8-bis-(trifluoromethyl)quinoline

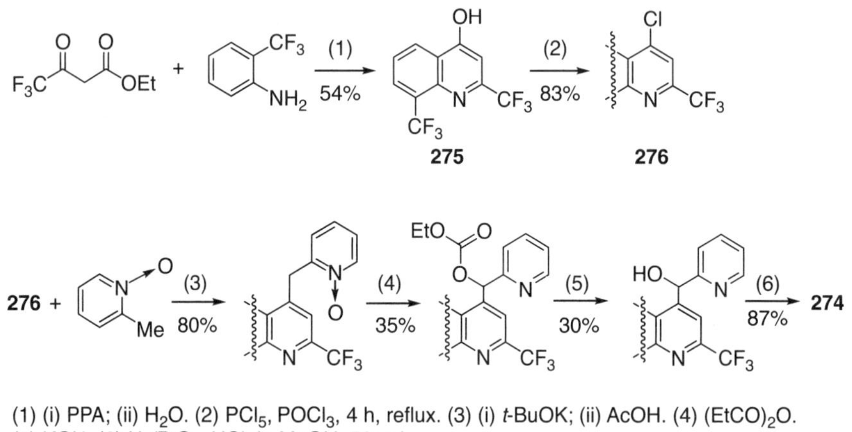

(1) (i) PPA; (ii) H_2O. (2) PCl_5, $POCl_3$, 4 h, reflux. (3) (i) *t*-BuOK; (ii) AcOH. (4) $(EtCO)_2O$. (5) KOH. (6) H_2/PtO_2, HCl, in MeOH, 50 psi.

SCHEME 12.68

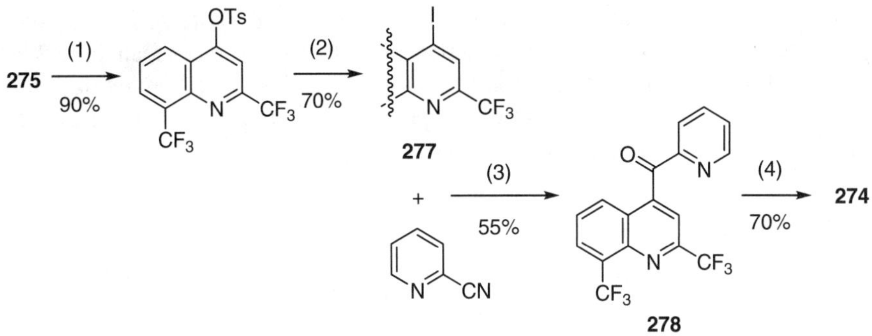

(1) (i) TsCl, Et_3N, in $CHCl_3$. (2) Phosphorus Red, I_2, in AcOH. (3) Mg, EtMgBr, in Et_2O. (4) 5% Pt/C, HCl, in EtOH, 20–24 h, RT.

SCHEME 12.69

275, was prepared via heterocyclization of ethyl 4,4,4-trifluoroacetoacetate with 2-(trifluoromethyl)aniline.

Nageswar et al.[142] employed 4-iodo-2,8-bis(trifluoromethyl)quinoline **277** as starting material for the synthesis of **278** (Scheme 12.69). It was reacted with 2-pyridinecarbonitrile to afford intermediate **278**, which was converted into racemic **274** by catalytic hydrogenation in 70% yield.

Kansal et al.[143] have reported a process for the manufacture of a mixture of racemic erythro and threo mefloquine hydrochloride (**274**) (Scheme 12.70) through the intermediacy of α-(1-acyl-α-2-piperidyl) 2,8-bis(trifluoromethyl)quinoline-4-yl ketone **279** and α-2-piperidyl 2,8-bis(trifluromethyl)quinolin-4-yl ketone hydrochloride **280** utilizing a reduction system which would provide the good yield of the erythro

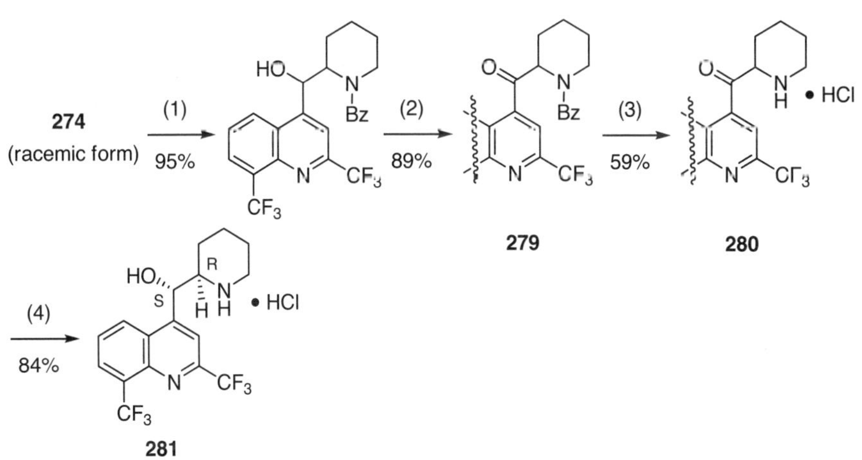

(1) BzCl, NaOH, in H_2O. (2) Jones reagent, in acetone. (3) HCl, in MeOH. (4) (i) $NaBH_4$, in MeOH, cooled, 1 h; (ii) AcOH.

SCHEME 12.70

mefloquine hydrochloride **281**, free of the undesired threo diastereoisomer. Compound **280** could be obtained by acylation of **274** followed by oxidation of **279**, deprotection step and the reduction of C=O in **280** with sodium borohydride. The process being simple and cost-effective provide for manufacture of the desired biologically active antimalarial compound.

12.4.7 6-Fluorobenzopyrans

12.4.7.1 Sorbinil ((4S)-2,3-dihydro-6-fluorospiro[4H-l-benzopyran-4,4′-imidazolidine]-2′,5′-dione)

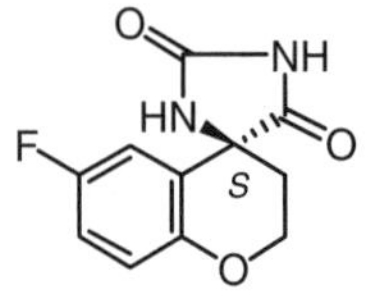

282
Sorbinil

Sorbinil **282** is a compound of potential therapeutic interest, because it prevents or alleviates the chronic complications of diabetes mellitus, due to its ability to inhibit the enzyme aldose reductase.[144] Sarges et al. prepared **282** and its enantiomer by the reaction sequence shown in Scheme 12.71, involving a brucine resolution of the racemic hydantoin precursor **284**.[145] The free base of brucine forms a crystalline complex with **282**, whereas the other enantiomer of **282** only forms a crystalline complex with brucine hydrochloride. Since this resolution technique does not work with certain congeners of sorbinil, a synthesis via an asymmetric induction sequence (Scheme 12.72) has also been developed[146] that seems generically applicable to optically active spiro hydantoins. Both methods[145,146] required 2,3-dihydro-6-fluoro-4*H*-1-benzopyran-4-one **283** and the introduction of the amino acid functionality

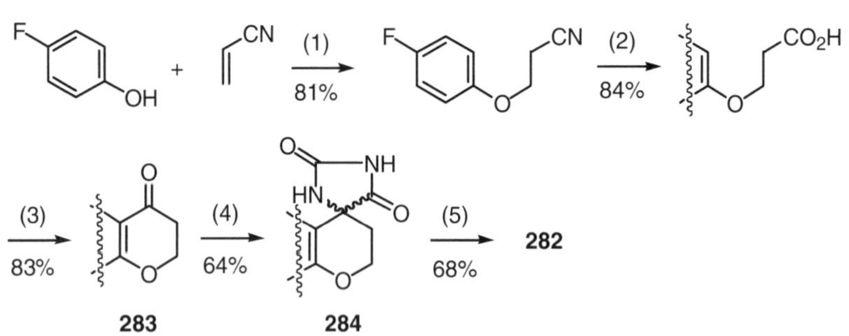

(1) Triton B. (2) HCl, HCO_2H. (3) PPA. (4) KCN, $(NH_4)_2CO_3$, $EtOH/H_2O$.
(5) (i) Brucine; (ii) HCl.

SCHEME 12.71

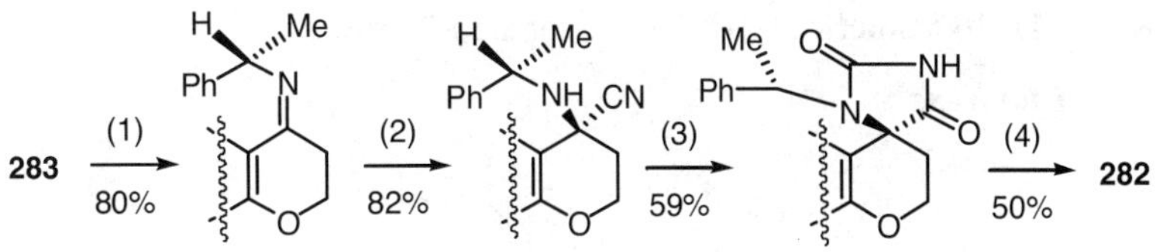

(1) (S)-Ph-CH(Me)NH_2, $TiCl_4$. (2) HCN/EtOH. (3) (i) CSI; (ii) HCl. (4) HBr, AcOH-H_2O.

SCHEME 12.72

through either Bucherer–Bergs[147] [$(NH_4)_2CO_3$, KCN] or Strecker[148] [HCN, (*S*)-2-phenethylamine] reactions.

To avoid the use of toxic cyanides in the large-scale preparation of **282**, Urban et al.[149] employed a compound **288** as a starting material. *N*-Benzoyl-α-hydroxyglycine was added to a solution of 2-(4′-fluorophenoxy)ethyl bromide in methanesulfonic acid to give *N*-benzoyl-5-fluoro-2-(2-bromoethoxy)phenylglycine **285**. The *N*-benzoyl amino acid group of **285** was dehydrated to oxazolidin-5-one **286** and it underwent subsequent intramolecular spiroalkylation to 2,3-dihydro-6-fluorospiro[4*H*-1-benzopyran-4,4′-2-phenyloxazolidin]-5-on **287** in high yield upon treatment with acetic anhydride and triethylamine (Scheme 12.73). Acidic hydrolysis of **287** provided the desired racemic spiro amino acid **288**, completing a three-step insertion of a glycine moiety. The methyl ester of **288** (***rac*-289**) was resolved by 8stereospecific hydrolysis with α-chymotrypsin and the resulting (***S***)-**289** was converted to **282** with sodium cyanate in acetic acid.

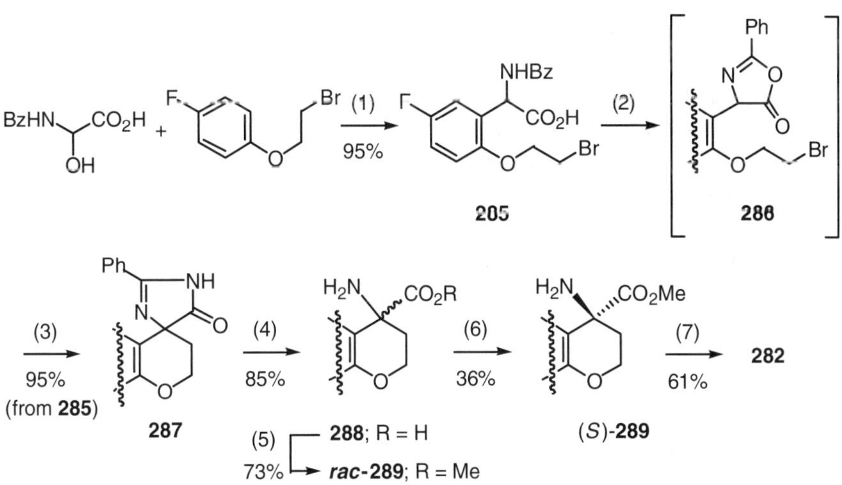

(1) Methanesulfonic acid. (2) Ac_2O, Et_3N. (3) Et_3N. (4) HCl, HCO_2H. (5) $SOCl_2$, MeOH. (6) in NaCl, 6N HCl to pH 5, ?-chymotrypsin. (7) sodium cyanate, in AcOH.

SCHEME 12.73

12.4.8 4-(Trifluoromethyl)-2*H*-3,1-benzoxazin-2-ones

12.4.8.1 Efavirenz

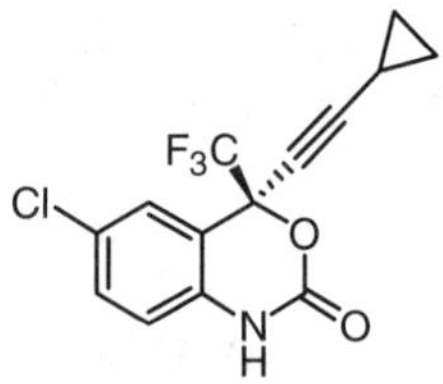

290

Efavirenz

Inhibition of human immunodeficiency virus type 1 reverse transcriptase by nucleosides such as AZT, DDC, DDI, D4T, and 3TC is a proven therapy for delaying the progression of AIDS.[150] However, the rapid viral mutation to resistant strains requires the development of new therapeutic agents.[151–153] The recent developments of both protease inhibitors and nonnucleoside reverse transcriptase inhibitors offer hope of effective treatment, especially when coadministered.[154] Efavirenz **290** is a nonnucleoside reverse transcriptase inhibitor that shows high potency against a variety of HIV-1 mutant strains.[155]

The compound **290**[150] has 1,4-dihydro-4-(trifluoromethyl)-2*H*-3,1-benzoxazin-2-one moiety as a core skeleton. The synthesis of **290** required the preparation of *p*-methoxybenzyl ketoaniline **291**, cyclopropylacetylene **292**, and (1*R*,2*S*)-*N*-pyrrolidinylnorephedrine **293** intermediate (see Fig. 12.13).

Thus, the compound **291** was prepared from 4-chloroaniline **294** in 76% overall yield as shown in Scheme 12.74.[150] More recently, Jiang et al.[156] have reported the alternative synthetic method for aminophenyl ketone derivative **295** as shown in Scheme 12.75. That is, α-(trifluoromethyl)ethenyl boronic acid **296** was conveniently prepared in 90% yield by reacting the readily available 2-bromotrifluoropropene with alkyl borate $B(OMe)_3$ and magnesium. Compound **296** underwent Suzuki coupling reactions with aryl halide **297** in THF/methanol at 70°C in the presence of 2 mol% tetrakis(triphenylphosphine)palladium and sodium carbonate to afford α-(trifluoromethyl)styrene derivative **298**. Dihydroxylation and oxidative cleavage of the alkene moiety of **298** followed by reduction of the nitro group with Raney nickel gives the aminophenyl ketone **295**, an intermediate in the synthesis of **291**, in 89% yield from **298**.

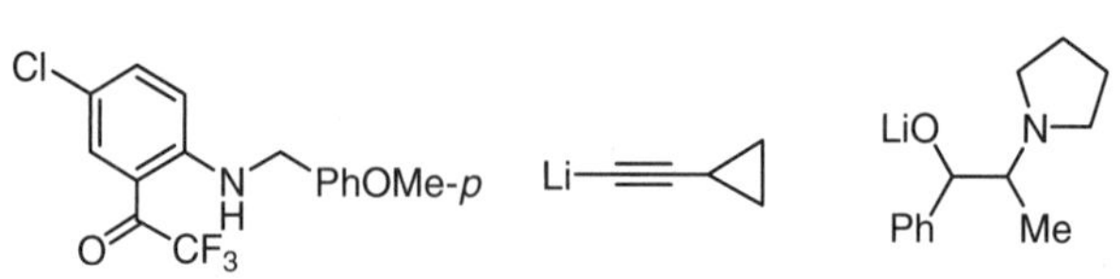

291 **292** **293**

FIGURE 12.13 Intermediates required in the synthesis of Efavirenz.

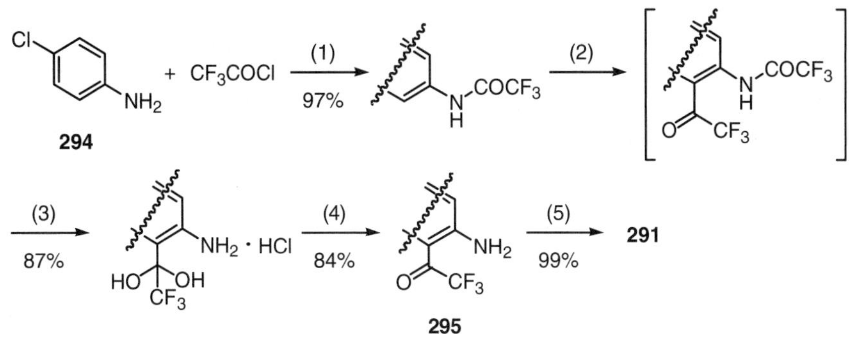

(1) aq. NaOH, MTBE. (2) (i) *n*-BuLi, TMEDA-MTBE. (3) HCl-HOAc. (4) aq. NaOAc, MTBE. (5) *p*-methoxybenzyl alcohol, cat *p*-TsOH.

SCHEME 12.74

Compound **292** is readily prepared by treating 5-chloropentyne with *n*-butyllithium or *n*-hexyllithium in cyclohexane or THF.[157] Alternative procedures from methyl cyclopropyl ketone have also been published.[158]

The preparation of **293** and its application as a ligand in asymmetric synthesis has been reported in the literature.[159,160] In which, the alkylation of (1*R*,2*S*)-norephedrine **299** with 1,4-dibromobutane using K_2CO_3 as base was reported to give **293** in only 33% yield. However, it was found that reproducibly excellent yields (97%) and product purity could be obtained using $NaHCO_3$ as base and toluene as solvent (Scheme 12.76).[150]

During the synthesis of **290**, establishment of the quarternary carbon center in an asymmetric manner presented a unique challenge. It was envisioned that the most efficient route to **290** would involve enantioselective addition of a metal

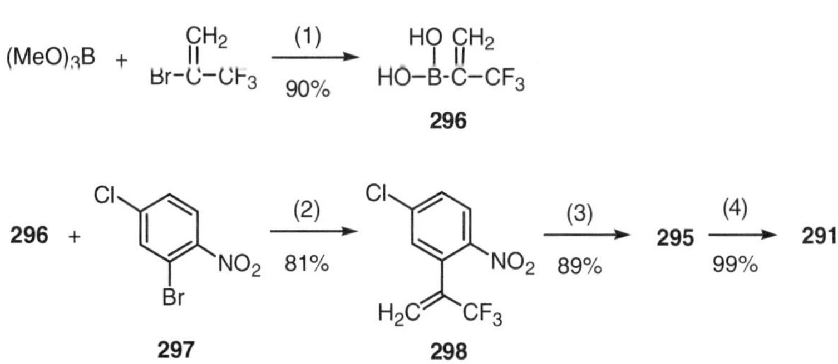

(1)(i) Mg/THF; (ii) HCl. (2) Na_2CO_3, $Pd(PPh_3)_4$. (3)(i) $NaIO_4$, OsO_4; (ii) H_2/Ni. (4) *p*-methoxybenzyl alcohol, cat *p*-TsOH (Ref. 150).

SCHEME 12.75

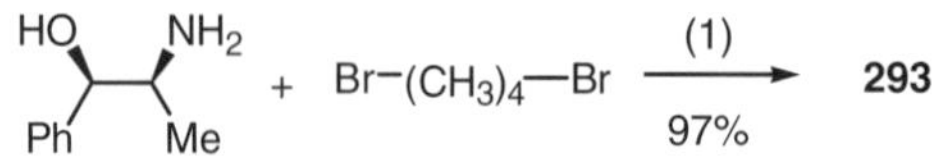

299

(1) $NaHCO_3$.

SCHEME 12.76

cyclopropylacetylide to a trifluoromethyl ketone **291** (with or without *N*-protection) mediated by a chiral additive. As shown in Scheme 12.77, the chiral complex **300** was prepared by reaction of *n*-butyllithium (or *n*-hexyllithium) with a mixture of **292** and **293** and at −10–0°C (aggregate equilibration). After addition of ketoaniline **291** the reaction was quenched with aqueous citric acid to give **301** in 91–93% isolated yield and >99.5% ee. The chiral additive **293** is easily recycled from the aqueous layer by basification with NaOH (>99% purity, 98% yield). The ligand has been recycled up to nine times in subsequent chiral addition reactions to give product **301** with typical chemical and optical purity.

Two routes were employed for conversion of the amino alcohol **301** into **290**. Initially, **301** was converted into the benzoxazinone **302**, which was then deprotected to give **290** (Scheme 12.78).[161] A more practical route to **290** involving oxidative cleavage using quinones, such as DDQ was realized by simply reversing the order of ring closure (benzoxazinone formation) and *N*-deprotection (Scheme 12.78).[150] Reaction of compound **301** with 1 equiv of DDQ (toluene, 0–10°C) proceeded quantitatively to give an 11.5:1 mixture of diastereomeric cyclic aminals **303a** and **303b**.[162] The DDHQ (dichlorodicyanohydroquinone) byproduct from this reaction is essentially insoluble in toluene, enabling removal by filtration and efficient recycling using established literature procedures.[163] The cyclic aminals **303a/303b** were then directly converted into the desired amino alcohol **304**. Reaction of **303a/303b**

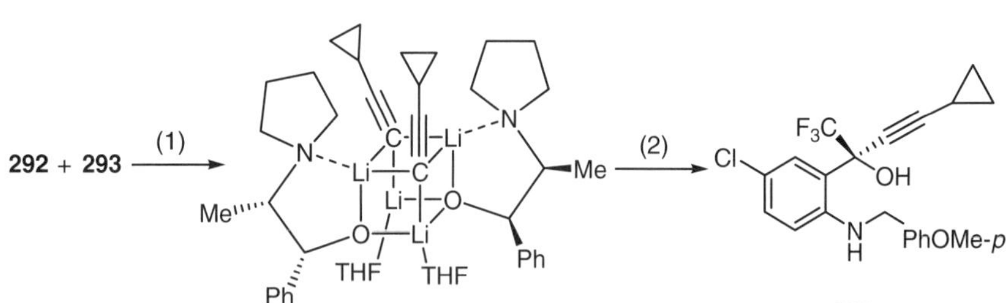

300

301

98% ee
upon crystallization
from toluene–hexane
91–93% yield

(1) *n*-BuLi, THF, −10°C to 0°C. (2) (i) **291**, −50°C; (ii) citric acid.

SCHEME 12.77

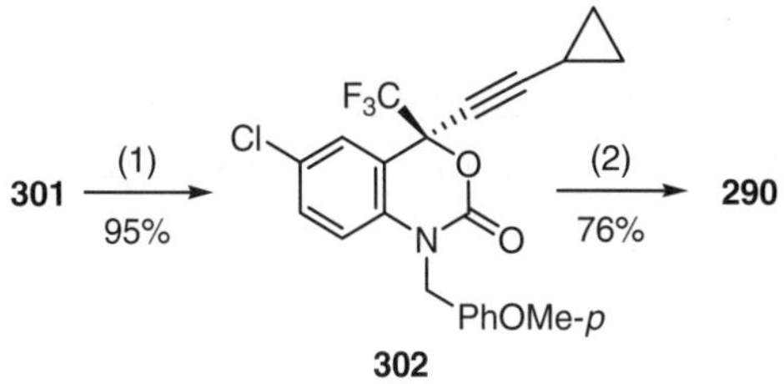

(1) $COCl_2$, Et_3N. (2) CAN.

SCHEME 12.78

with NaOH in MeOH effected clean dissociation to **304** (as its Na alkoxide) and *p*-methoxybenzyaldehyde. Since direct isolation of **304** in the presence of *p*-methoxybenzaldehyde was difficult, it was reduced in situ into *p*-methoxybenzyl alcohol using $NaBH_4$. Conversion of **304** into **290** (Scheme 12.79) was accomplished using phosgene (in the absence of base), methyl chloroformate (with K_2CO_3), or 4-nitrophenyl chloroformate (with K_2CO_3). This reaction presumably proceeds via intermediate **305** followed by ring closure. After aqueous workup (aqueous $NaHCO_3$), **290** was crystallized from THF-heptane and was isolated in 93–95%) and >99.5% ee.

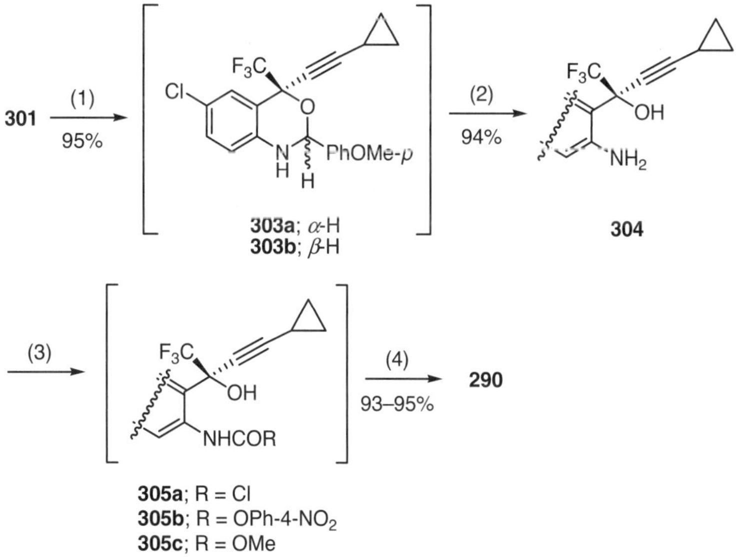

(1) DDQ, in toluene. (2) NaOH-MeOH, $NaBH_4$. (3) ClCOR. (4) Δ, or Base.

SCHEME 12.79

12.4.9 2-Fluoroadenines (2-Fluoro-6-aminopurines)

12.4.9.1 Fludarabine and Fludarabine Phosphate (9-β-D-Arabinofuranosyl-2-fluoro-adenine and 9-β-D-Arabinofuranosyl-2-fluoroadenine-5′-phosphate)

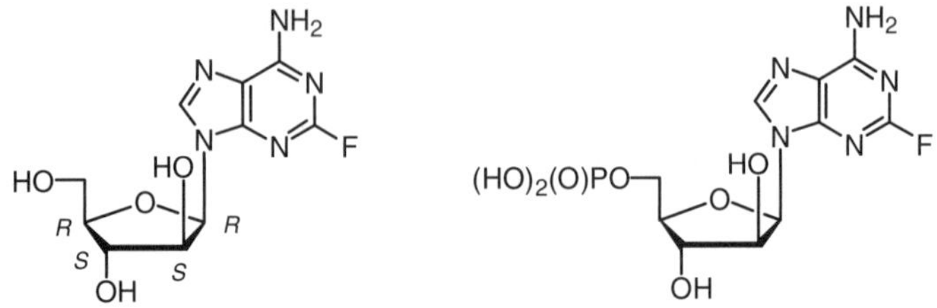

306 Fludarabine

307 Fludarabine phosphate

Fludarabine (**306**) is the anticancer agent, which is mainly used in the treatment of hematologic malignancies, and fludarabine phosphate (9-β-D-arabinofuranosyl-2-fluoroadenine-5′-phosphate, **307**) is a chemotherapeutically effective form of the drug which is converted to the parent drug **306** *in vivo*.[164]

Recently Xu et al.[165] have reported the synthesis of the title compound **306** from 2,6-diaminopurine **308** as shown in Scheme 12.80. That is, compound **308** was acylated with methoxyacetic anhydride to afford 2,6-di(methoxyacetamido)purine

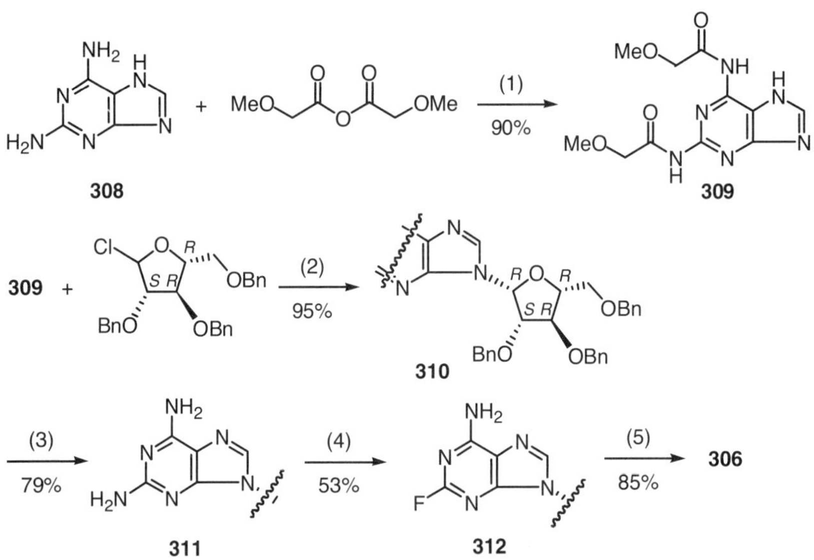

(1) in pyridine, Δ. (2) $EtN(i\text{-}Pr)_2$, in $ClCH_2CH_2Cl$, reflux, 36 h. (3) NaOMe, in MeOH, Δ. (4) (i) HBF_4, in THF; (ii) HBF_4, $NaNO_2$, in H_2O; (iii) NaOH, in H_2O and AcOEt, pH 8. (5) (i) HCl, $H_2/PdCl_2$, in H_2O and $(CH_2OMe)_2$; (ii) NH_3.

SCHEME 12.80

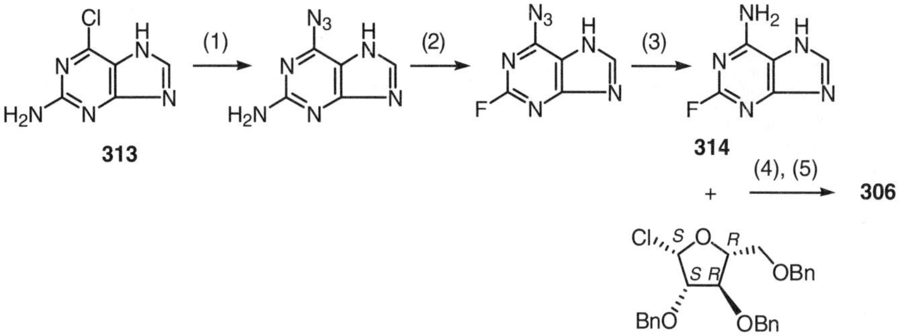

(1) (i) NaN_3, in H_2O and DMSO, Δ; (ii) H_2O, RT. (2) (i) HBF_4, $NaNO_2$, in H_2O and THF; (ii) Na_2CO_3 in H_2O. (3) H_2 in H_2O, RT. (4) Et_3N, in $ClCH_2CH_2Cl$, Δ. (5) H_2/Pd, in MeOH.

SCHEME 12.81

309, which is treated with 2,3,5-tri-*O*-benzyl-D-arabinofuranosyl chloride to form 2,6-di(acetamido)-9-(2,3,5-tri-*O*-benzyl-β-D-arabinofuranosyl)purine **310**. After removal of a protecting group on **310**, the resulting 2-amino-9-(2,3,5-tri-*O*-benzyl-β-D-arabinofuranosyl)adenine **311** was subject to diazotization and fluorination in HBF_4–tetrahydrofuran mixture to provide compound **312**. Then **306** was obtained by catalytic hydrogenation, that is, reductive debenzylation and treatment with boron trichloride in 30% overall yield.

Li et al.[166] have also reported the synthesis of **306** from 2-amino-6-chloropurine **313** and 2,3,5 tris *O* benzyl β D arabinofuranosyl chloride (Scheme 12.81). The raw material was cheap and easy to obtain, and the synthesis process was short and the operation was simple. The total yield was high to 19.6%. It was suitable for industrial production.

A facile and convenient preparation of 2-fluoroadenine **314**, was also achieved by Eaton and Denny.[167] Reaction of solid sodium nitrite with **308** in anhydrous hydrogen fluoride provided pure **314** in 22% yield. Montgomery and Hewson[168] also synthesized **314** via Baltz–Schiemann reaction, followed by column chromatography purification and crystallization to give a 0.7% yield of pure **314**. In a modified synthesis,[168] a higher yield (6%) was obtained.

For the alternative production of nucleosides, Zuffi and Monciardini[169] have reported a process by a transglycosylation reaction catalyzed by uridine phosphorylase and purine nucleoside phosphorylase. Specifically, their invention generally relates to a process for immobilizing cells and to the use of a resin for immobilizing cells, and the process is exemplified by the production of **306** from 1-β-D-arabinofuranosyluracil **315** and **314** (Scheme 12.82).[169] Similar biotransformations for the preparation of **306** and **307** from **314** and arabinofuranosyluracil have been reported by Farina et al. (using *Enterobacter aerogenes*),[170] and Hummel–Marquardt et al. (using esterases (e.g., pig liver esterase) or lipases).[171]

Phosphorylation of **306** was performed in 76% yield by treating dried **306** with $POCl_3$–Me_3PO_4.[172] An enzymatic procedure for the phosphorylation of **306** in the

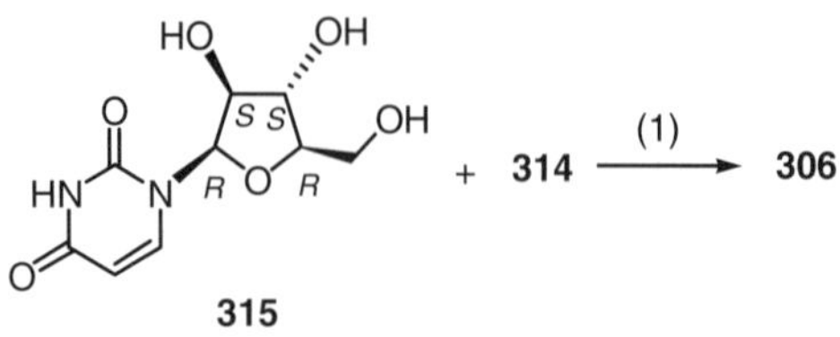

(1) (i) recombinant *Escherichia coli*, KH_2PO_4, in DMSO and H_2O, 6 days, 60°C, pH 7.

SCHEME 12.82

position 5′ is also described. The free or immobilized purified enzymes, or enzymes contained in cell pastes or cells were used.[173] Furthermore, the improved method to transform the nucleoside into monophosphates using nucleoside phosphotransferase from *Erwinia herbicola* was reported.[174]

12.4.10 2-(Fluoromethyl)quinazolin-4-one

12.4.10.1 Afloqualone [6-Amino-2-(fluoromethyl)-3-(2-methylphenyl)-quinazolin-4-one]

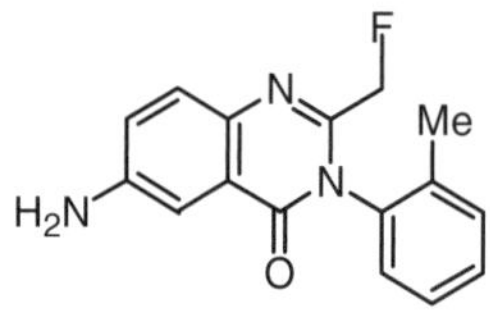

316

Afloqualone

Afloqualone (Arofuto) **316** is an analogue of methaqualone developed in the 1980s in Japan. It has sedative and muscle relaxant effects,[175] and has had some clinical use, although it causes photosensitization as a side effect that can cause skin problems such as dermatitis.[176]

The title compound **316** was prepared[177] as shown in Scheme 12.83 in three steps from the corresponding anthranilic acid derivative **317**. A fusion reaction of anilines with isatoic anhydride is well known as a general method[178] for the preparation of anthranilanilides. However, this method requires handling of phosgene in the preparation of isatoic anhydride. Tani et al.[177] reported the reaction of **317** with thionyl chloride in boiling benzene, followed by treatment with *o*-toluidine. This procedure afforded the anthranilanilides **318** in moderate yield. No effort was made to isolate the intermediates formed in the first step of this procedure. The 2-(chloromethyl)-3-tolyl-4(3*H*)-quinazolinone **319** was prepared *via* chloroacetylation of **318** with excess chloroacetyl chloride in acetic acid, followed by spontaneous cyclization

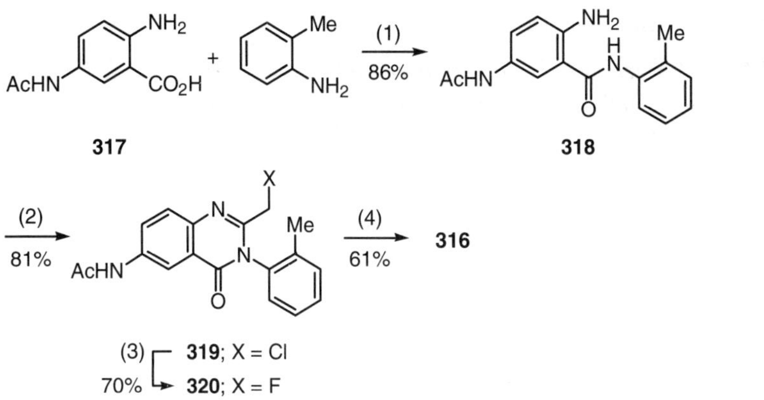

(1) $SOCl_2$, in boiling benzene. (2) $ClCH_2COCl$, in AcOH. (3) KF, in ethylene glycol. (4) 10% HCl-MeOH, RT.

SCHEME 12.83

at 110°C according to Petyunin's procedure.[179] The chlorine atom in $-CH_2Cl$ group of **319** could be readily displaced by the fluorine atom using excess of anhydrous potassium fluoride (3 molar ratio) at 160°C in ethylene glycol for 2–4 h to give the 2-(fluoromethyl)quinazolone **320** in good yields. Alternatively, the reaction using KHF_2 (5 molar ratio) in diethylene glycol under a similar condition gave a similar yield (55%).[177] The final product **316** prepared by deprotection of **320**.

As shown in Scheme 12.84, the title compound **316** is also prepared by the reduction[180] of 2-fluoromethyl-6-nitro-3-tolyl-4(3*H*)-quinazolinone **322**,[181] which is formed by treatment of 5-nitroanthranilic acid with monofluoroacetyl chloride, followed by reaction of acid **321** with *o*-toluidine. High toxicity of monofluoroacetic acid derivatives used in this synthesis is a significant drawback of this route.

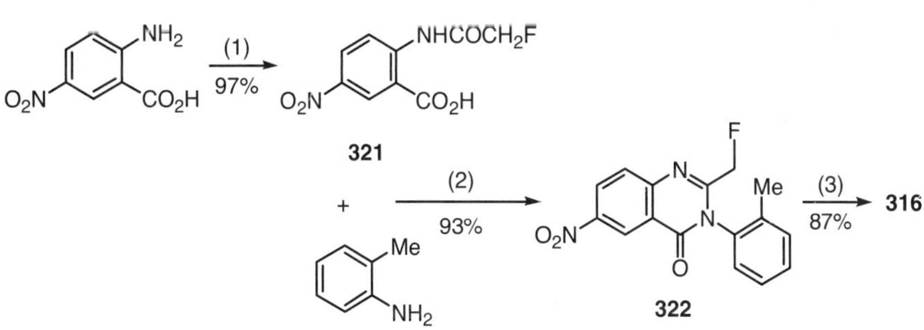

(1) FCH_2COCl, in MeCN. (2) PCl_3, in CH_2Cl_2. (3) $Na_2S_2O_4$, aq. NH_3.

SCHEME 12.84

12.4.11 3-Trifluoromethylcarbacephem

12.4.11.1 7-Phenylglycyl-3-trifluoromethylcarbacephem [(6R,7S)-7-[[(2R)-Aminophenylacetyl]amino]-8-oxo-3-(trifluoromethyl)-1-azabicyclo[4.2.0]oct-2-ene-2-carboxylic Acid]

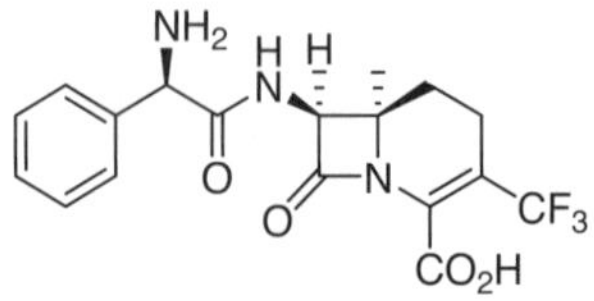

323

3-Trifluoromethylcarbacephem

Carbacephems are structurally related to cephalosporins. First reported in the early 1970s,[182] the carbacephems maintain the potent antibacterial activity characteristic of the parent cephalosporins.[183] More recently, in a comparative study cephalosporin versus carbacephem it was demonstrated carbacephem analogues possess enhanced chemical stability.[184] It was recognized that the carbacephem's equipotent biological activity concomitant with decreased chemical reactivity provided working opportunities as medicine to carbacephem.[185] D-(−)-phenylglycyl side chains attached to the C-7 position of cephalosporin antibiotics allow active transport across the intestinal barrier.[186]

Introduction of trifluoromethyl group to carbacephem moiety and derivatization to the 7-phenylglycyl-3-trifluoromethylcarbacephems **323**[187] was carried out according to Scheme 12.85. Readily available **324** was used as starting material.[187] The conversion of triflate **324** to vinyl bromide **325**[188] provided a useful substrate for the preparation of trifluoromethyl compound **326** by the coupling with *in situ* generated Burton's [CF_3Cu] reagent.[189] Displacement of **324** with LiBr in 2,6-lutidine produced the bromides as a mixture of Δ-2 and Δ-3 isomers that could be isomerized with DBU

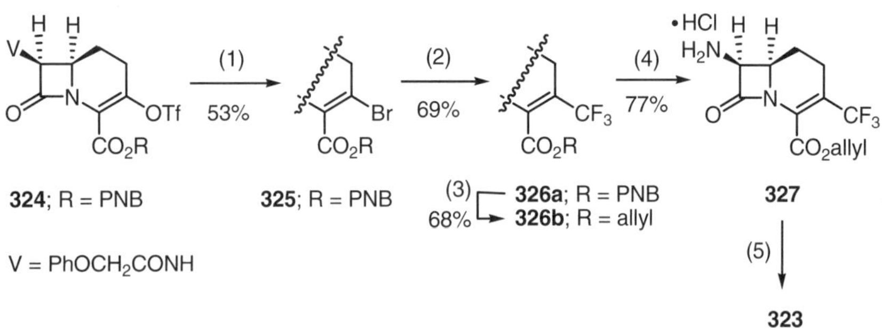

Conditions: (1) (i) LiBr, Lutidine, DMF; (ii) DBU, CH_2Cl_2. (2) Zn, CF_2Br_2, DMF, CuBr. (3) (i) Zn/HCl; (ii) allylBr, Bu_4NHSO_4. (4) PCl_5, *iso*-BuOH. (5) (i) *t*-Boc-arylglycine, 1-Chloro-3,5-dimethoxytriazine, NMM, CH_2Cl_2; (ii) $Pd(PPh_3)_4$, *N*-Bu_3SnH; (iii) TFA, Et_3SiH.

SCHEME 12.85

and crystallized to give the desired Δ-3-vinyl bromides **325**. Pregeneration of Burton's [CF_3Cu] reagent using CF_2Br_2/DMF/zinc/CuBr[190] and subsequent coupling with **325** gave the desired trifluoromethyl derivative **326a**. *p*-Nitrobenzyl (PNB) ester removal from **326a** with zinc/HC1 and esterification of the tetrabutylammonium salt with allyl bromide yielded **326b** whose phenoxyacetyl side chain was cleaved with $PC1_5$ and butanol to give **327**. The amino acid was activated with 1-chloro-3,5-dimethoxytriazine[191] and coupled to the 7-amino group. Deallylation with $(Ph_3)_4Pd$/tributyltin hydride[192] and reverse phase chromatography provided the desired **323**.

12.4.12 2-(Trifluoromethyl)phenothiazine

12.4.12.1 Fluphenazine (2-[4-[3-[2-(Trifluoromethyl)-10H-phenothiazin-10-yl]-propyl]piperazin-1-yl]ethanol, Prolixin)

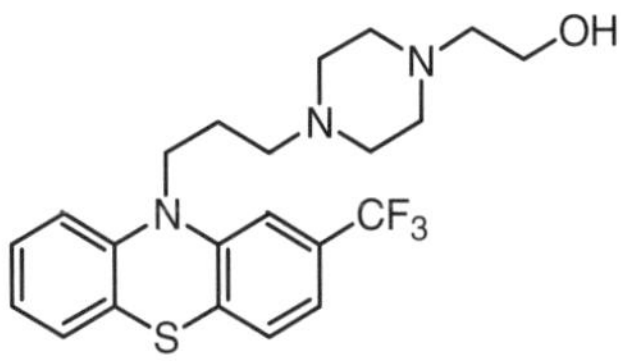

328
Fluphenazine

Fluphenazine exhibited antimutagenic activity in lymphocyte cultures, markedly decreasing genotoxic effects of standard mutagenic agents present in cell cultures. However, the strong neuroleptic effect, combined with other serious side effects on the central nervous system, limits its use.[193]

2-(Trifluoromethyl)phenothiazine **330**, is synthesized as shown in Scheme 12.86. Intermediate **329** was synthesized by coupling 3-trifluoromethylaniline with

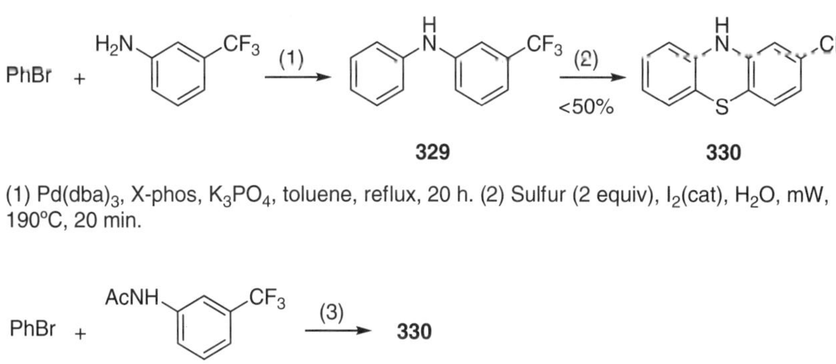

(1) $Pd(dba)_3$, X-phos, K_3PO_4, toluene, reflux, 20 h. (2) Sulfur (2 equiv), I_2(cat), H_2O, mW, 190°C, 20 min.

(3) (i) Na_2CO_3, Sorbitan monoleate, CuI; (ii) NaOH, H_2O; (iii) Sulfur (2 equiv), I_2(cat).

SCHEME 12.86

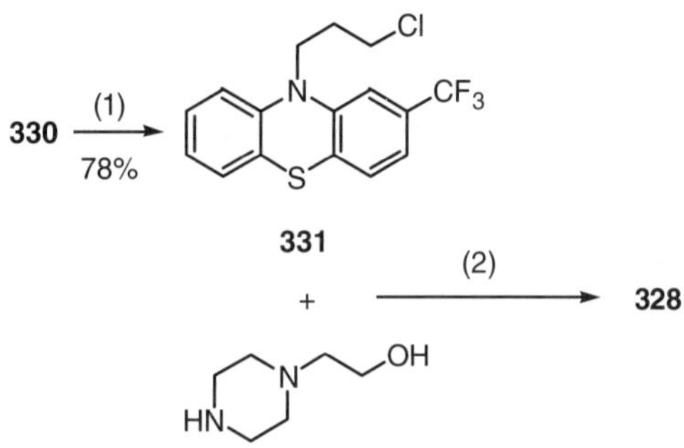

(1) $Cl(CH_2)_3Br$, NaOH, Et_4NCl, H_2O. (2) MeCOEt, KI, Na_2CO_3, reflux for 20 h.

SCHEME 12.87

bromobenzene using Buchwald palladium coupling.[194] The diphenylamine **329** was then cyclized through reaction with sulfur and catalytic amount of I_2[195] under microwave irradiation[196] to give substituted 10*H*-phenothiazine **330**. Compound **330** is also synthesized in high yield and selectivity by the *N*-phenylation of 3-trifluoromethylacetanilide with bromobenzene in the presence of anhydrous sodium carbonate, CuI, and Tween-80 [polysorbate 80, polyoxyethylene (20) sorbitan monooleate] into *N*-phenyl-*N*-(3-trifluoromethylphenyl)acetamide (Scheme 12.86). Thionylative cyclization of the *N*-(3-trifluoromethylphenyl)aniline using elemental sulfur with I_2 as the catalyst[197] lead to the formation of **330**. It has also been prepared from commercially available *N*-phenyl-3-(trifluoromethyl)benzenamine.[198]

After 3-chloropropylation of **330** (78%) under phase-transfer catalysis conditions[199] compound **328** was prepared by refluxing a mixture of compound **331** with 1-piperazine ethanol, KI, Na_2CO_3, and MeCOEt for 20 h[200] (Scheme 12.87).

Alternative syntheses of **331** were also reported by Beck et al.,[200] Schepartz,[201] or Sardessai.[202]

12.4.13 9-Fluoro-1,4-benzodiazepine

12.4.13.1 Timelotem (10-Fluoro-3-methyl-7-(2-thienyl)-1,2,3,4,4a,5-hexahydropyrazino[1,2-a][1,4]benzodiazepine)

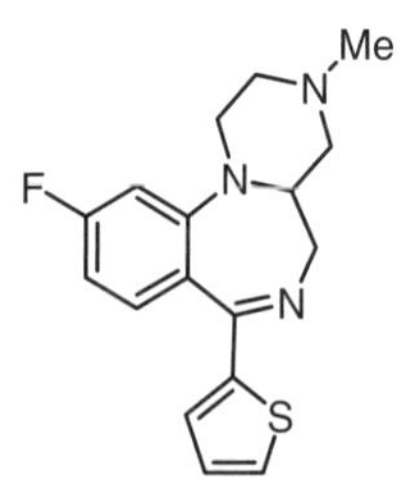

332

Timelotem

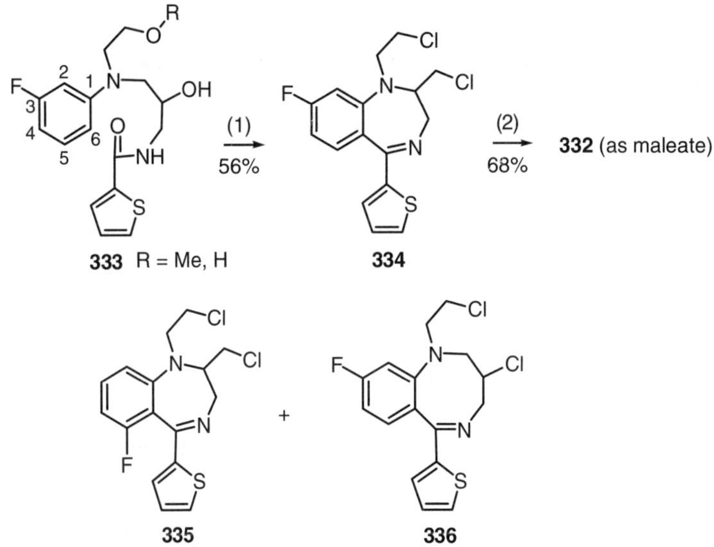

(1) $POCl_3/CHCl_3$, 140°C, 16 h; (2) $MeNH_2/MeOH\text{-}H_2O$, 90°C, 14 h.

SCHEME 12.88

Timelotem **332**, is an antipyretic.[203] One of the synthetic routes is shown in Scheme 12.88.[203] Intramolecular Friedel–Crafts type cyclization was employed as a key reaction for seven-membered ring construction. The reaction of amide group with $POCl_3$ transforms **333**[204] to the corresponding imidoyl chloride which undergoes Friedel–Crafts type ring closure, where the imino cation attacks mainly the less-hindered C_6 rather than C_2 carbon of phenyl ring (yield; **334** = 56%, **335** = 5.2%). In the ring-closure step, *N*-(2-methoxyethyl)-substituted amide **333** (R = Me) afforded a desired product **334** in 56%, meanwhile *N*-(2-hydroxyethyl)-substituted amide **333** (R = H) underwent normal Friedel–Crafts reaction, affording eight-membered, non-rearranged product **336** in 62% along with **334** (7%) and **335** (1%) by products.

12.4.14 10-(trifluoromethyl)pyrano[4,3-*j*]-1,2-benzodioxepin

12.4.14.1 10-Trifluoromethylartemisinin [(3R,5aS,6R,8aS,12R,12aR)-3,12-Epoxy-10-(trifluoromethyl)-3,6,9-trimethyl-3,4,5,5a,6,7,8,8a-octahydro-12H-pyrano[4,3-j]-1,2-benzodioxepin]

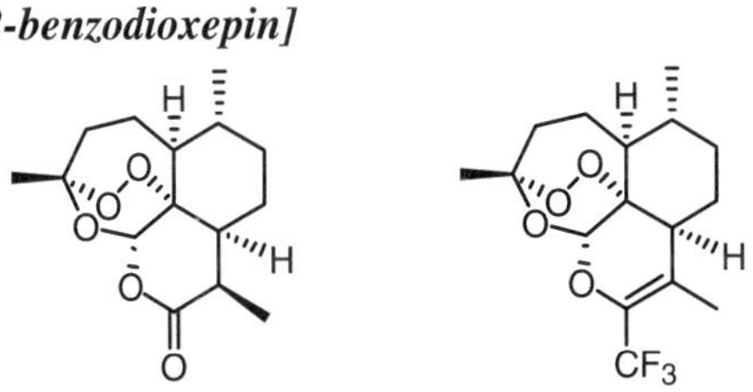

337 Artemisinin

338 10-Trifluoromethyl artemisinin

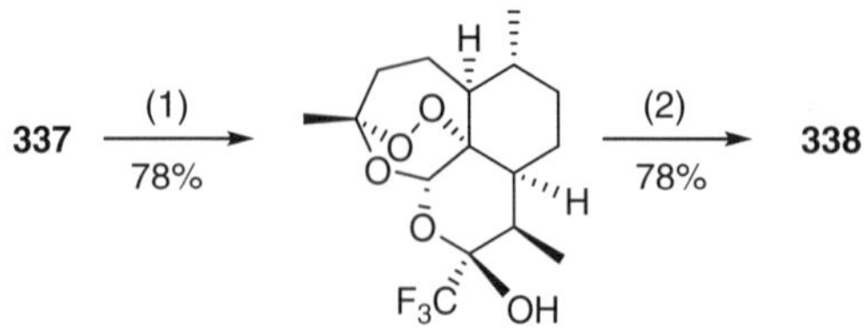

339

(1) (i) $TMSCF_3$, in THF, $TBAF \cdot 3H_2O$ (0.1 equiv); (ii) H_2O. (2) $SOCl_2$, in pyridine.

SCHEME 12.89

Artemisinin **337** is a natural efficient antimalarial drug for the treatment of multidrug-resistant forms of *P. falciparum*.[205] However, pharmacological problems associated with **337**, especially a short plasma half-life prompted scientists to search for more metabolically stable derivative. 10-Trifluoromethyl artemisinin **338** was shown to have improved stability.[206]

As shown in Scheme 12.89, hemiketal **339** was prepared from **337**[207] by treatment with $CF_3Si(CH_3)_3$[208] in the presence of TBAF-$3H_2O$ catalyst at room temperature and desilylation occurred after addition of water. The reaction is stereoselective and led to hemiketal **339** in high yield (78%). The desired compound **338** was obtained a good yield (78%) by treatment of **339** with thionyl chloride and pyridine.

REFERENCES

1. Uneyama, K. Fluorine in drug designs. *Organofluorine Chemistry*; Blackwell Publishing: Oxford, UK, 2006.
2. (a) Begue, J.-P.; Bonnet-Delpon, D. Bioorganic and medicinal chemistry of fluorine. *Chimie bioorganique et medicinale du fluor*; CNRS and EDP Science, 2005. (b) Ojima, I.; McCarthy, J.R.; Welch, J.T. Biomedical Frontiers of Fluorine Chemistry, *ACS Symposium Series 639*, ACS, 1996. (c) Kukhar, V.P.; Soloshonok, V.A., Eds.; *Fluorine-Containing Amino Acids*; John Wiley & Sons, Ltd: Chichester, 1994. (d) Kobayashi, Y.; Kumadaki, I.; Taguchi, T., Eds.; *Fluorine in Pharmacy*; Hirokawa Publishing Co.: Tokyo, 1992. (e) Welch, J.T.; Eswarakrishnan, S. *Fluorine in Bioorganic Chemistry*; John Wiley & Sons, Inc.: New York, 1991.
3. Gosselin, F.; O'Shea, P.D.; Webster, R.A.; Reamer, R.A.; Tillyer, R.D.; Grabowski, E.J.J. *Synlett* **2006**, 3267–3270.
4. Prabhakaran, J.; Underwood, M.D.; Parsey, R.V.; Arango, V.; Majo, V.J.; Simpson, N.R.; Van Heertum, R.; Mann, J.J.; Kumar, J.S.D. *Bioorg. Med. Chem.* **2007**, *15*, 1802–1807.
5. Balsamo, A.; Coletta, I.; Guglielmotti, A.; Landolfi, C.; Mancini, F.; Martinelli, A.; Milanese, C.; Minutolo, F.; Nencetti, S.; Orlandini, E.; Pinza, M.; Rapposelli, S.; Rossello, A. *Eur. J. Med. Chem.* **2003**, *38*, 157–168.
6. Pal, M.; Khanna, I.; Subramanian, V.; Padakanti, S.; Pillarisetti, S. *PCT Int. Appl.* WO 2006058201 A2 20060601, **2006**.

7. Oh, L.M. *Tetrahedron Lett.* **2006**, *47*, 7943–7946.
8. Atherton, J.H.; Fields, R. *J. Chem. Soc. C*: **1968**, 1507–1513.
9. Tang, X.-Q.; Hu, C.-M. *J. Chem. Soc., Perkin Trans. 1* **1994**, 2161–2163.
10. Tanaka, K.; Masuda, H.; Mitsuhashi, K. *Bull. Chem. Soc. Jpn.* **1986**, *59*, 3901–3904.
11. Jiang, B.; Xu, Y.; Yang, J. *J. Fluorine Chem.* **1994**, *67*, 83–85.
12. Cristau, H.-J.; Cellier, P.P.; Spindler, J.-F.; Taillefer, M. *Eur. J. Org. Chem.* **2004**, 695–709.
13. (a) Nishida, S.; Ohsumi, T.; Tsushima, K.; Matsuo, N.; Maeda, K.; Inoue, S. *PCT Int. Appl.* WO 8602641 A1 19860509, **1986**. (b) Nishida, S.; Tsushima, K.; Matsuo, N.; Osumi, T.; Maeda, K.; Inoue, S. *Jpn. Kokai Tokkyo Koho* JP 62096472 A 19870502, CAN 107:96715, **1987**.
14. Erickson, J.A.; McLoughlin, J.I. *J. Org. Chem.* **1995**, *60*, 1626–1631.
15. Kees, K.L.; Fitzgerald, J.J., Jr.; Steiner, K.E.; Mattes, J.F.; Mihan, B.; Tosi, T.; Mondoro, D.; McCaleb, M.L. *J. Med. Chem.* **1996**, *39*, 3920–3928.
16. (a) Tsuji, K.; Konishi, N.; Spears, G.W.; Ogino, T.; Nakamura, K.; Tojo, T.; Ochi, T.; Shimojo, F.; Senoh, H.; Matsuo, M. *Chem. Pharm. Bull.* **1997**, *45*, 1475–1481. (b) Matsuo, M.; Tsuji, K.; Ogino, T.; Konishi, N. *Eur. Pat. Appl.* EP 554829 A2 19930811, **1993**. (c) Penning, T.D.; Talley, J.J.; Bertenshaw, S.R.; Carter, J.S.; Collins, P.W.; Docter, S.; Graneto, M.J.; Lee, L.F.; Malecha, J.W.; Miyashiro, J.M.; Rogers, R.S.; Rogier, D.J.; Yu, S.S.; Anderson, G.D.; Burton, E.G.; Cogburn, J.N.; Gregory, S.A.; Koboldt, C.M.; Perkins, W.E.; Seibert, K.; Veenhuizen, A.W.; Zhang, Y.Y.; Isakson, P.C. *J. Med. Chem.* **1997**, *40*, 1347–1365.
17. Lombardino, J.G.; Wiseman, E.H. *J. Med. Chem.* **1974**, *17*, 1182–1188.
18. Barta, T.E.; Stealey, M.A.; Collins, P.W.; Weier, R.M. *Bioorg. Med. Chem. Lett.* **1998**, *8*, 3443–3448.
19. Watson, N.S.; Chan, C.; Ross, B.C. *Ger. Offen.* DE 3932887 A1 19900405, **1990**.
20. Navarrete-Vazquez, G.; Rojano-Vilchis, M.; Yepez-Mulia, L.; Melendez, V.; Gerena, L.; Hernandez-Campos, A.; Castillo, R.; Hernandez-Luis, F. *Eur. J. Med. Chem.* **2006**, *41*, 135–141.
21. Ge, F.; Wang, Z.; Wan, W.; Lu, W.; Hao, J. *Tetrahedron Lett.* **2007**, *48*, 3251–3254.
22. Barlow, M.G.; Bell, D.; O'Reilly, N.J.; Tipping, E. *J. Fluorine Chem.* **1983**, *23*, 293–299.
23. Kimoto, H.; Fujii, S.; Cohen, L.A. *J. Org. Chem.* **1982**, *47*, 2867–2872.
24. Drucker, D.; Easley, C.; Kirkpatrick, P. *Nature Rev.* **2007**, *6*, 109–110.
25. Biftu, T.; Feng, D.; Qian, X.; Liang, G.-B.; Kieczykowski, G.; Eiermann, G.; He, H.; Leiting, B.; Lyons, K.; Petrov, A.; Sinha-Roy, R.; Zhang, B.; Scapin, G.; Patel, S.; Gao, Y.-D.; Singh, S.; Wu, J.; Zhang, X.; Thornberry, N.A.; Weber, A.E. *Bioorg. Med. Chem. Lett.* **2007**, *17*, 49–52.
26. (a) Balsells, J.; DiMichele, L.; Liu, J.; Kubryk, M.; Hansen, K.; Armstrong, J.D., III. *Org. Lett.* **2005**, *7*, 1039–1042. (b) Xiao, Y.; Armstrong, J.D., III; Krska, S.W.; Njolito, E.; Rivera, N.R.; Sun, Y.; Rosner, T.; Clausen, A.M. *PCT Int. Appl.* WO 2006081151 A1 20060803, **2006**.
27. Hansen, K.B.; Balsells, J.; Dreher, S.; Hsiao, Y.; Kubryk, M.; Palucki, M.; Rivera, N.; Steinhuebel, D.; Armstrong, J.D., III, Askin, D., Grabowski, E.J.J. *Org. Process. Res. Dev.* **2005**, *9*, 634–639.

28. Balsells, J.; DiMichele, L.; Liu, J.; Kubryk, M.; Hansen, K.; Armstrong, J.D., III. *Org. Lett.* **2005**, *7*, 1039–1042.
29. Kim, D.; Wang, L.; Beconi, M.; Eiermann, G.J.; Fisher, M.H.; He, H.; Hickey, G.J.; Kowalchick, J.E.; Leiting, B.; Lyons, K.; Marsilio, F.; McCann, M.E.; Patel, R.A.; Petrov, A.; Scapin, G.; Patel, S.B.; Roy, R.S.; Wu, J.K.; Wyvratt, M.J.; Zhang, B.B.; Zhu, L.; Thornberry, N.A.; Weber, A.E. *J. Med. Chem.* **2005**, *48*, 141–151.
30. Diana, G.D.; Rudewicz, P.; Pevear, D.C.; Nitz, T.J.; Aldous, S.C.; Aldous, D.J.; Robinson, D.T.; Draper, T.; Dutko, F.J. *J. Med. Chem.* **1995**, *38*, 1355–1371.
31. The aminooxime **93** is prepared by the reaction of either nitrile or carboxylate with hydroxylamine in ethanol.
32. (a) Marquez, V.E.; Tseng, C.K.-H.; Kelley, J.A.; Mitsuya, H.; Broder, S.; Roth, J.S.; Driscoll, J.S. *Biochem. Pharmacol.* **1987**, *36*, 2719–2722. (b) Marquez, V.E.; Tseng, C.K.-H.; Mitsuya, H.; Aoki, S.; Kelley, J.A.; Ford, H.; Roth, J.S.; Broder, S.; Johns, D.G.; Driscoll, J.S. *J. Med. Chem.* **1990**, *33*, 978–985.
33. Tann, C.H.; Brodfuehrer, P.R.; Brundidge, S.P.; Sapino, C., Jr.; Howell, H.G. *J. Org. Chem.* **1985**, *50*, 3644–3647.
34. Barchi, J.J., Jr.; Marquez, V.E.; Driscoll, J.S.; Ford, H., Jr.; Mitsuya, H.; Shirasaka, T.; Aoki, S.; Kelley, J.A. *J. Med. Chem.* **1991**, *34*, 1647–1655.
35. Xiang, Y.; Cavalcanti, S.; Chu, C.K.; Schinazi, R.F.; Pal, S.B.; Zhu, Y.-L.; Cheng, Y.-C. *Bioorg. Med. Chem. Lett.* **1995**, *5*, 877–880.
36. Choudhury, A.; Jin, F.; Wang, D.; Wang, Z.; Xu, G.; Nguyen, D.; Castoro, J.; Pierce, M.E.; Confalone, P.N. *Tetrahedron Lett.* **2003**, *44*, 247–250.
37. (a) Takamatsu, S.; Maruyama, T.; Katayama, S.; Hirose, N.; Naito, M.; Izawa, K. *J. Org. Chem.* **2001**, *66*, 7469–7477. (b) Takamatsu, S.; Maruyama, T.; Izawa, K. *Yuki Gosei Kagaku Kyokaishi* (*J. Synth. Org. Chem.*, written in Japanese) **2005**, *63*, 864–878, CAN 144:212947.
38. Katayama, S.; Takamatsu, S.; Naito, M.; Tanji, S.; Ineyama, T.; Izawa, K. *J. Fluorine Chem.* **2006**, *127*, 524–528.
39. Caille, J.-C.; Miel, H.; Armstrong, P.; McKervey, M.A. *Tetrahedron Lett.* **2004**, *45*, 863–865.
40. Patrick, T.B.; Lanahan, M.V.; Yang, C.; Walker, J.K.; Hutchinson, C.L.; Neal, B.E. *J. Org. Chem.* **1994**, *59*, 1210–1212.
41. Jin, F.; Wang, D.; Confalone, P.N.; Pierce, M.E.; Wang, Z.; Xu, G.; Choudhury, A.; Nguyen, D. *Tetrahedron Lett.* **2001**, *42*, 4787–4789.
42. (a) Siddiqui, M.A.; Driscoll, J.S.; Marquez, V.E. *Tetrahedron Lett.* **1998**, *39*, 1657–1660. (b) Woltermann, C.J.; Lapin, Y.A.; Kunnen, K.B.; Tueting, D.R.; Sanchez, I.H. *Tetrahedron* **2004**, *60*, 3445–3449.
43. Vane, J.; O'Grady, J., Eds.; *Therapeutic Applications of Prostaglandis*; Edward Arnold: London, 1993.
44. (a) Nakano, T.; Makino, M.; Morizawa, Y.; Matsumura, Y. *Angew. Chem., Int. Ed.* **1996**, *35*, 1019–1021. (b) Matsumura, Y.; Nakano, T.; Makino, M.; Morizawa, Y. *Eur. Pat. Appl.* EP 669329 A1 19950830, **1995**. (c) Matsumura, Y.; Nakano, T.; Mori, N.; Morizawa, Y. *Chimia* **2004**, *58*, 148–152.
45. (a) Strupczewski, J.T.; Allen, R.C.; Gardner, B.A.; Schmid, B.L.; Stache, U.; Glamkowski, E.J.; Jones, M.C.; Ellis, D.B.; Huger, F.P.; Dunn, R.W. *J. Med. Chem.*

1985, *28*, 761–769; (b) Basappa; Mantelingu, K.; Sadashiva, M.P.; Rangappa, K.S. *Indian J. Chem. B* **2004**, *43B*, 1954–1957.

46. Strupczewski, J.T.; Bordeau, K.J.; Chiang, Y.; Glamkowski, E.J.; Conway, P.G.; Corbett, R.; Hartman, H.B.; Szewczak, M.R.; Wilmot, C.A.; Helsley, G.C. *J. Med. Chem.* **1995**, *38*, 1119–1131.
47. Bolos, J.; Anglada, L.; Gubert, S.; Planas, J.M.; Agut, J.; Princep, M.; De la Fuente, A.; Sacristan, A.; Ortiz, J.A. *J. Med. Chem.* **1998**, *41*, 5402–5409.
48. (a) Ravina, E.; Negreira, J.; Cid, J.; Masaguer, C.F.; Rosa, E.; Rivas, M.E.; Fontenla, J.A.; Loza, M.I.; Tristan, H.; Cadavid, M.I.; Sanz, F.; Lozoya, E.; Carotti, A.; Carrieri, A. *J. Med. Chem.* **1999**, *42*, 2774–2797. (b) Brea, J.; Masaguer, C.F.; Villazon, M.; Isabel, C. M.; Ravina, E.; Fontaine, F.; Dezi, C.; Pastor, M.; Sanz, F.; Isabel, L.M. *Eur. J. Med. Chem.* **2003**, *38*, 433–440.
49. Uneyama, K. Nucleophilic substitution on fluoroaromatic ring. *Organofluorine Chemistry*; Blackwell Publishing: Oxford, UK, 2006.
50. Lindberg, P.; Brandstrom, A.; Wallmark, B.; Mattsson, H.; Rikner, L.; Hoffman, K.-J. *Med. Res. Rev.* **1990**, *10*, 1–54.
51. Souda, S.; Ueda, N.; Miyazawa, S.; Tagami, K.; Nomoto, S.; Okita, M.; Shimomura, N.; Kaneko, T.; Fujimoto, M. *Eur. Pat. Appl.* EP 295603 A1 19881221, **1988**.
52. Ife, R.J.; Dyke, C.A.; Keeling, D.J.; Meenan, E.; Meeson, M.L.; Parsons, M.E.; Price, C. A.; Theobald, C.J.; Underwood, A.H. *J. Med. Chem.* **1989**, *32*, 1970–1977.
53. Kuhler, T.C.; Fryklund, J.; Bergman, N.; Weilitz, J.; Lee, A.; Larsson, H. *J. Med. Chem.* **1995**, *38*, 4906–4916.
54. Rane, R.A.; Pathak, R.K.; Kaushik, C.P.; Rao, K.V.V. Prasad Kumar, A. *Synth. Commun.* **2002**, *32*, 1211–1217.
55. (a) Cotton, H.; Elebring, T.; Larsson, M.; Li, L.; Sorensen, H.; von Unge, S. *Tetrahedron: Asymmetry* **2000**, *11*, 3819–3825. (b) Raju, S.V.N.; Purandhar, K.; Reddy, P.P.; Reddy, G.M.; Reddy, L.A.; Reddy, K.S.; Sreenath, K.; Mukkanti, K.; Reddy, G.S. *Org. Process. Res. Dev.* **2006**, *10*, 33–35.
56. (a) Kennett, G.A.; Lightowler, S.; De Biasi, V.; Stevens, N.C.; Wood, M.D.; Tulloch, I.F.; Blackburn, T.P. *Neuropharmacology* **1994**, *33*, 1581–1588. (b) Palvimaki, E.P.; Roth, B. L.; Majasuo, H.; Laakso, A.; Kuoppamaki, M.; Syvalahti, E.; Hietala, J. *Psychopharmacology* **1996**, *126*, 234–240. (c) Bromidge, S.M.; Duckworth, M.; Forbes, I.T.; Ham, P.; King, F.D.; Thewlis, K.M.; Blaney, F.E.; Naylor, C.B.; Blackburn, T.P.; Kennett, G.A.; Wood, M.D.; Clarke, S.E. *J. Med. Chem.* **1997**, *40*, 3494–3496.
57. (a) Bromidge, S.M.; Dabbs, S.; Davies, S.; Duckworth, D.M.; Forbes, I.T.; Jones, G.E.; Jones, J.; King, F.D.; Saunders, D.V.; Blackburn, T.P.; Holland, V.; Kennett, G.A.; Lightowler, S.; Middlemiss, D.N.; Riley, G.J.; Trail, B.; Wood, M.D. *Bioorg. Med. Chem. Lett.* **2000**, *10*, 1863–1866. (b) Bromidge, S.M.; Dabbs, S.; Davies, D.T.; Davies, S.; Duckworth, D.M.; Forbes, I.T.; Gaster, L.M.; Ham, P.; Jones, G.E.; King, F.D.; Mulholland, K.R.; Saunders, D.V.; Wyman, P.A.; Blaney, F.E.; Clarke, S.E.; Blackburn, T.P.; Holland, V.; Kennett, G.A.; Lightowler, S.; Middlemiss, D.N.; Trail, B.; Riley, G.J.; Wood, M.D. *J. Med. Chem.* **2000**, *43*, 1123–1134.
58. Bromidge, S.M.; Dabbs, S.; Davies, D.T.; Duckworth, D.M.; Forbes, I.T.; Ham, P.; Jones, G.E.; King, F.D.; Saunders, D.V.; Starr, S.; Thewlis, K.M.; Wyman, P.A.; Blaney, F.E.; Naylor, C.B.; Bailey, F.; Blackburn, T.P.; Holland, V.; Kennett, G.A.; Riley, G.J.; Wood, M.D. *J. Med. Chem.* **1998**, *41*, 1598–1612.

59. Abou-Gharbia, M.; Patel, U.R.; Webb, M.B.; Moyer, J.A.; Andree, H.; Muth, E.A. *J. Med. Chem.* **1987**, *30*, 1818–1823.
60. Walker, G.N.; Moore, M.A. *J. Org. Chem.* **1960**, *26*, 432–439.
61. Montclair, L.B.; Corraz, J. *US Patent* 3,522,262, **1970**.
62. Yevich, J.P.; New, J.S.; Lobeck, W.G.; Dextraze, P.; Bernstein, E.; Taylor, D.P.; Yocca, F. D.; Eison, M.S.; Temple, D.L., Jr. *J. Med. Chem.* **1992**, *35*, 4516–4525.
63. (a) Uchytilova, V.; Holy, A.; Cech, D.; Gut, J. *Collect. Czech. Chem. Commun.* **1975**, *40*, 2347–2352. (b) Ueda, T.; Fox, J.J. *J. Med. Chem.* **1963**, *6*, 697–701.
64. Doctor fungus: http://www.doctorfungus.org/Thedrugs/Voriconazole.htm.
65. Odds, F.C.; Cheesman, S.L.; Abbott, A.B. *J. Antimicrob. Chemother.* **1986**, *18*, 473–478.
66. Patterson, T.F. *J. Chemother.* **1999**, *11*, 504–512.
67. Butters, M.; Ebbs, J.; Green, S.P.; MacRae, J.; Morland, M.C.; Murtiashaw, C.W.; Pettman, A. *Org. Process. Res. Dev.* **2001**, *5*, 28–36, and references therein.
68. Wang, Y.; Gu, S. *Faming Zhuanli Shenqing Gongkai Shuomingshu* CN 100999518 A 20070718, **2007**.
69. Schlosser, M. *J. Fluorine Chem.* **1991**, *54*, 192–193.
70. Shi, G.-q.; Schlosser, M. *J. Fluorine Chem.* **1992**, *58*, 135.
71. Shi, G.-q.; Schlosser, M. *Tetrahedron* **1993**, *49*, 1445–1456.
72. Shi, G.-q.; Takagishi, S.; Schlosser, M. *Tetrahedron* **1994**, *50*, 1129–1134.
73. Ghosh, S.N.; Schiosser, M. *J. Fluorine Chem.* **1994**, *67*, 53–56.
74. Shi, G.-q.; Wang, Q.; Schlosser, M. *Tetrahedron* **1996**, *52*, 4403–4410.
75. Duschinsky, R.; Pleven, E.; Heidelberger, C. *J. Am. Chem. Soc.* **1957**, *79*, 4559–4560.
76. (a) Cech, D.; Holy, A. *Collect. Czech. Chem. Commun.* **1976**, *41*, 3335–3342. (b) Hillers, S.; Lazdin, A.; Veinbergs, A.; Snikeris, D.; Knunyants, I.L.; German, L.S.; Kazmina, N.B. *US Patent* 3,846,429 19741105, **1974**. (c) Nakamura, J.; Takeuchi, T.; Sakuma, K.; Tomioka, H. *Jpn. Kokai Tokkyo Koho* JP 53144584 A 19781215, CAN 91:20535, **1978**; (d) Kobayashi, Y.; Kumadaki, I.; Nakazato, A. *Tetrahedron Lett.* **1980**, *21*, 4605–4606; (e) Haas, A.; Kortmann, D. *Chem. Ber.* **1981**, *114*, 1176–1179; see also an indirect method: Lal, G.S.; Pastore, W.; Pesaresi, R. *J. Org. Chem.* **1995**, *60*, 7340–7342.
77. Robins, M.J.; Naik, S.R. *J. Am. Chem. Soc.* **1971**, *93*, 5277–5278.
78. Sato, M.; Kaneko, C.; Iwaoka, T.; Kobayashi, Y.; Iida, T. *J. Chem. Soc., Chem. Commun.* **1991**, *10*, 699–700.
79. Miyashita, O.; Matsumura, K.; Shimadzu, H.; Hashimoto, N. *Chem. Pharm. Bull.* **1981**, *29*, 3181–3190, and references therein.
80. Ozaki, S.; Ike, Y.; Mizuno, H.; Ishikawa, K.; Mori, H. *Bull. Chem. Soc. Jpn.* **1977**, *50*, 2406–2412.
81. Wei, R.; Yang, T. *Zhongguo Yaowu Huaxue Zazhi* **2001**, *11*, 49–50.
82. Cancerbackup: http://www.cancerbackup.org.uk/Home.
83. Lee, C.H.; Kim, J.Y.; Kim, W.J.; Kim, Y.H.; *Heterocycles* **1990**, *31*, 211–214, and references therein.
84. Umemoto, T.; Ogura, E.; Mukono, T. *Jpn. Kokai Tokkyo Koho* JP 61112073 A 19860530, CAN 106:18602, **1986**.
85. National Cancer Institute home page: http://www.nci.nih.gov/.

86. Kiss, J.; D'Souza, R.; Van Koeveringe, J.A.; Arnold, W. *Helv. Chim. Acta* **1982**, *65*, 1522–1537.
87. Bertolini, G.; Frigerio, M. *PCT Int. Appl.* WO 2005040184 A1 20050506, **2005**.
88. Dong, H.; Qian, H. *Zhongguo Yiyao Gongye Zazhi* **2002**, *33*, 108–110.
89. (a) Umeda, M.; Heidelberger, C. *Cancer Res.* **1968**, *28*, 2529–2538. (b) Yamashita, J.; Yasumoto, M.; Takeda, S.; Matsumoto, H.; Unemi, N. *J. Med. Chem.* **1989**, *32*, 136–139.
90. (a) Kaufman, H.E.; Heidelberger, C. *Science* **1964**, *145*, 585–586. (b) Wingard, J.R.; Stuart, R.K.; Saral, R.; Burns, W.H. *Antimicrob. Agents Chemother.* **1981**, *20*, 286–290. (c) Carmine, A.A.; Brogden, R.N.; Heel, R.C.; Speight, T.M.; Avery, G.S. *Drugs* **1982**, *23*, 329–353.
91. (a) Stout, M.G.; Hoard, D.E.; Holman, M.J.; Wu, E.S.; Siegel, J.M. *Methods Carbohydr. Chem.* **1976**, *7*, 19–24. (b) Heidelberger, C.; Parsons, D.G.; Remy, D.C. *J. Med. Chem.* **1964**, *7*, 1–5. (c) Heidelberger, C.; Parsons, D.G.; Remy, D.C. *J. Am. Chem. Soc.* **1962**, *84*, 3597–3598.
92. (a) Schwarz, B.; Cech, D.; Reefschlaeger, J. *J. Prakt. Chem.* **1984**, *326*, 985–993;(b) Hein, L.; Cech, D.; Liebenthal, C. *Ger. (East)*, **1976**, GEXXA8 DD 119423 19760420.
93. (a) Kawakami, H.; Ebata, T.; Koseki, K.; Matsushita, H.; Naoi, Y.; Itoh, K.; Mizutani, N. *Heterocycles* **1990**, *31*, 569–574. (b) Kobayashi, Y.; Yamamoto, K.; Asai, T.; Nakano, M.; Kumadaki, I. *J. Chem. Soc., Perkin Trans. 1* **1980**, 2755–2762. (c) Andres, P.; Marhold, A. *J. Fluorine Chem.* **1996**, *77*, 93–95, and references therein.
94. (a) Xu, S.; Meng, J.; Li, Jinliang; J., Q.; Huang, F.; Guo, H. *Faming Zhuanli Shenqing Gongkai Shuomingshu* CN 1789274 A 20060621, **2006**. (b) Wang, Z.-X.; Duan, W.; Wiebe, L.I.; Balzarini, J.; De Clercq, E.; Knaus, E.E. *Nucleosides Nucleotides Nucleic Acids* **2001**, *20*, 11–40.
95. Mertes, M.P.; Saheb, S.E. *J. Pharm. Sci.* **1963**, *52*, 508–509.
96. Schwarz, B.; Cech, D.; Reefschlager, J. *J. Prakt. Chem.* **1984**, *326*, 985–993.
97. Hein, L.; Cech, D. *Z. Chem.* **1977**, *17*, 415–416.
98. Fuchikami, T.; Ojima, I. *Tetrahedron Lett.* **1982**, *23*, 4099–4100.
99. Fuchikami, T.; Yamanouchi, A. *Chem. Lett.* **1984**, 1595–1598.
100. Fuchikami, T.; Yamanouchi, A.; Ojima, I. *Synthesis* **1984**, 766–768.
101. Yamakawa, T.; Yamamoto, K.; Uraguchi, D.; Tokuhisa, K. *PCT Int. Appl.* WO 2007055170 A1 20070518, **2007**.
102. Umetani, H.; Fukazawa, N.; Komatsu, H. *Jpn. Kokai Tokkyo Koho* JP 2001302611 A 20011031, CAN 135:331437, **2001**.
103. Komatsu, H.; Umetani, H. *Org. Process. Res. Dev.* **2002**, *6*, 847–850.
104. Doctor Fungus: http://www.doctorfungus.org/thedrugs/Flucytosine.htm.
105. (a) Bennett, J.E. *Ann. Intern. Med.* **1977**, *86*, 319–322. (b) Scholer, H.J. Flucytosine. In *Antifungal Chemotherapy*; Speller, D.C.E., Eds.; John Wiley & Sons, Ltd: Chichester, 1977; pp 35–106.
106. Robins, M.J.; Naik, S.R. *J. Chem. Soc., Chem. Commun.* **1972**, *1*, 18–19.
107. Visser, G.W.M.; Boele, S.; Van Halteren, B.W.; Knops, G.H.J.N.; Herscheid, J.D.M.; Brinkman, G.A.; Hoekstra, A. *J. Org. Chem.* **1986**, *51*, 1466–1471.
108. Takahara, T.; Hisanaga, Y. *Nippon Kagaku Kaishi* **1985**, *10*, 2034–2039.

109. Fu, S.-J.; Zhu, L.-M. *Biocat. Biotransform.* **2006**, *24*, 253–256.
110. Chen, S.-W.; Wang, Y.-H.; Jin, Y.; Tian, X.; Zheng, Y.-T.; Luo, D.-Q.; Tu, Y.-Q. *Bioorg. Med. Chem. Lett.* **2007**, *17*, 2091–2095.
111. (a) McGuigan, C.; Pathirana, R.N.; Snoeck, R.; Andrei, G.; De Clercq, E.; Balzarini, J. *J. Med. Chem.* **2004**, *47*, 1847–1851. (b) Horwitz, J.P.; Chua, J.; Da Rooge, M.A.; Noel, M.; Klundt, I.L. *J. Org. Chem.* **1966**, *31*, 205–211.
112. Choudhury, A.; Pierce, M.E.; Nguyen, D.; Storace, L.; Confalone, P.N. *Tetrahedron Lett.* **2005**, *46*, 8099–8102.
113. Schinazi, R.F.; Liotta, D.C.; Choi, W.-B. *Aust. Pat. Appl.* AU 2004200957 A1 20040401, **2004**.
114. Tao, G.; Chen, Y. *Faming Zhuanli Shenqing Gongkai Shuomingshu* CN 1935828 A 20070328, **2007**.
115. Moon, B.S.; Shim, A.Y.; Lee, K.C.; Lee, H.J.; Lee, B.S.; An, G. I.I.; Yang, S.D.; Chi, D.Y.; Choi, C.W.; Lim, S.M.; Chun, K.S. *Bull. Korean Chem. Soc.* **2005**, *26*, 1865–1868.
116. Trost, B.M.; Andersen, N.G. *J. Am. Chem. Soc.* **2002**, *124*, 14320–14321, and references therein.
117. (a) Judge, T.M.; Phillips, G.; Morris, J.K.; Lovasz, K.D.; Romines, K.R.; Luke, G.P.; Tulinsky, L.J.; Tustin, J.M.; Chruschel, R.A.; Dolak, L.A.; Mizsak, S.A.; Watt, M.W.; Morris, J.; Vander Velde, S.L.; Strohbach, J.W.; Gammill, R.B. *J. Am. Chem. Soc.* **1997**, *119*, 3627–3628. (b) Fors, K.S.; Gage, J.R.; Heier, R.F.; Kelly, R.C.; Perrault, W.R.; Wicnienski, N. *J. Org. Chem.* **1998**, *63*, 7348–7356.
118. Healthatoz: http://www.healthatoz.com/healthatoz/Atoz/.
119. Goldstein, E.J.C. *Am. J. Med.* **1987**, *82*, 3–17.
120. Lulla, A.; Kankan, Rajendra N.R. *Indian Pat. Appl.* IN 2002MU00447 A 20040228, **2004**.
121. For example, see Coll, A.P.; Morte, S.S. *Afinidad* **2000**, *57*, 447–450.
122. For example, see Grohe, K.; Heitzer, H. *Liebigs Annalen der Chemie,* **1987**, *1987*, 29–37.
123. Heravi, M.M.; Jaddi, Z.S.; Oskooie, H.A.; Khaleghi, S.; Ghassemzadeh, M. *J. Chem. Res.* **2005**, *9*, 578–579.
124. Dai, G.; Shi, D.; Zhou, L.; Mei, S. *Zhongguo Yiyao Gongye Zazhi* **1992**, *23*, 153–154.
125. Schwalbe, T.; Kadzimirsz, D.; Jas, G. *QSAR Comb. Sci.* **2005**, *24*, 758–768.
126. Huang, Q.; Liu, G.; Li, J.; Ban, C.; Ren, B.; Luo, T. *Jingxi Shiyou Huagong* **2005**, *2*, 20–21.
127. Bower, J.F.; Szeto, P.; Gallagher, T. *Org. Lett.* **2007**, *9*, 3283–3286.
128. (a) Hayakawa, I.; Atarashi, S.; Imamura, M.; Yokohama, S.; Higashihashi, N.; Sakano, K.; Ohshima, M. *Eur. Pat. Appl.* EP 206283 A2 19861230, **1986**. (b) Atarashi, S.; Yokohama, S.; Yamazaki, K.; Sakano, K.; Imamura, M.; Hayakawa, I. *Chem. Pharm. Bull.* **1987**, *35*, 1896–1902. (c) Sakano, K.; Yokohoma, S.; Hayakawa, I.; Atarashi, S.; Kadoya, S. *Agric. Biol. Chem.* **1987**, *51*, 1265–1270. (d) Mitscher, L.A.; Sharma, P.N.; Chu, D.T.W.; Shen, L.L.; Pernet, A.G. *J. Med. Chem.* **1987**, *30*, 2283–2286. (e) Atarashi, S.; Tsurumi, H.; Fujiwara, T.; Hayakawa, I. *J. Heterocycl. Chem.* **1991**, *28*, 329–331. (f) Kang, S.B.; Ahn, E.J.; Kim, Y.; Kim, Y.H. *Tetrahedron Lett.* **1996**, *37*, 9317–9320. (g) Satoh, K.; Inenaga, M.; Kanai, K. *Tetrahedron: Asymmetry* **1998**, *9*, 2657–2662. (h) Adrio, J.; Carretero, J.C.; Ruano, J.L.G.; Pallares, A.; Vicioso, M. *Heterocycles* **1999**, *51*, 1563–1572.

129. van Hattum, A.H.; Pinedo, H.M.; Schluper, H.M.M.; Erkelens, C.A.M.; Tohgo, A.; Boven, E. *Biochem. Pharmacol.* **2002**, *64*, 1267–1277.

130. Mitsui, I.; Kumazawa, E.; Hirota, Y.; Aonuma, M.; Sugimori, M.; Ohsuki, S.; Uoto, K.; Ejima, A.; Terasawa, H.; Sato, K. *Jpn. J. Cancer Res.* **1995**, *86*, 776–782.

131. Kumazawa, E.; Jimbo, T.; Ochi, Y.; Tohgo, A. *Cancer Chemother. Pharmacol.* **1998**, *42*, 210–220.

132. Takiguchi, S.; Kumazawa, E.; Shimazoe, T.; Tohgo, A.; Kono, A. *Jpn. J. Cancer Res.* **1997**, *88*, 760–769.

133. Boven, E.; van Hattum, A.H.; Hoogsteen, I.; Schlüper, H.M.M.; Pinedo, H.M. *Ann. NY Acad. Sci.* **2000**, *922*, 175–177.

134. (a) Terasawa, H.; Sato, K.; Mitsui, I. *Jpn. Kokai Tokkyo Koho* JP 06087746 A 19940329, CAN 121:134541, **1994**. (b) Terasawa, H.; Ejima, A.; Ohsuki, S.; Uoto, K. *Eur. Pat. Appl.* EP 495432 A1 19920722, **1992**.

135. Kamihara, S.; Kanai, K.; Noguchi, S.; Terasawa, H.; Kitaoka, H. *Eur. Pat. Appl.* EP 737686 A1 19961016, **1996**.

136. Cartoni, A.; Madami, A.; Palomba, D.; Marras, M.; Berettoni, M.; Olivieri, L.; Ettorre, A.; Cipollone, A.; Animati, F.; Maggi, C.A.; Monteagudo, E. *Tetrahedron* **2003**, *59*, 1309–1316.

137. Kamihara, S.; Kanai, K.; Noguchi, S. *PCT Int. Appl.* WO 9626181 A1 19960829, **1996**.

138. Maguire, J.D.; Krisin; Marwoto, H.; Richie, T.L.; Fryauff, D.J.; Baird, J.K. *Clin. Infect. Dis.* **2006**, *42*, 1067–1072.

139. Shepherd, R. *PCT Int. Appl.* WO 9839003 A1 19980911, **1998**.

140. Ohnmacht, C.J.; Patel, A.R.; Lutz, R.E. *J. Med. Chem.* **1971**, *14*, 926–928.

141. Kumar, A.; Chaterjee, S.K. *Indian* IN 178385 A1 19970419, **1997**.

142. Nageswar, Y.V.D.; Meshram, H.M.; Sattur, P.B.R. *Indian* IN 170910 A1 19920613, **1992**.

143. Kansal, V.K.; Maniyan, P.P.; Deshmukh, S.S.; Gupta, N.L. *Indian* IN 185394 A1 20010113, **2001**.

144. (a) Sarges, R.; Belletire, J.L.; Schnur, R.C.; Peterson, M.J. *ACS/CSJ Chemical Congress*, Medicinal Chemistry Section, April 24, 1979, Honolulu, HI, Abstract No. 16. (b) Peterson, M.J.; Sarges, R.; Aldinger, C.E.; McDonald, D.P. *Metab. Clin. Exp. 28* (Suppl. 1), **1979**, 456–461.

145. (a) Sarges, R. *US Patent* US 4130714 19781219, **1978**. (b) Sarges, R.; Bordner, J.; Dominy, B.W.; Peterson, M.J.; Whipple, E.B. *J. Med. Chem.* **1985**, *28*, 1716–1720.

146. Sarges, R.; Howard, H.R.; Kelbaugh, P.R. *J. Org. Chem.* **1982**, *47*, 4081–4085.

147. Ware, E. *Chem. Rev.* **1950**, *46*, 403–470.

148. (a) Patel, M.S.; Worsley, M. *Can. J. Chem.* **1970**, *48*, 1881–1884. (b) Harada, K.; Okawara, T. *J. Org. Chem.* **1973**, *38*, 707–710.

149. Dirlam, N.L.; Moore, B.S.; Urban, F.J. *J. Org. Chem.* **1987**, *52*, 3587–3591.

150. Pierce, M.E.; Parsons, R.L., Jr.; Radesca, L.A.; Lo, Y.S.; Silverman, S.; Moore, J.R.; Islam, Q.; Choudhury, A.; Fortunak, J.M.D.; Nguyen, D.; Luo, C.; Morgan, S.J.; Davis, W. P.; Confalone, P.N.; Chen, C.-y.; Tillyer, R.D.; Frey, L.; Tan, L.; Xu, F.; Zhao, D.; Thompson, A.S.; Corley, E.G.; Grabowski, E.J.J.; Reamer, R.; Reider, P.J. *J. Org. Chem.* **1998**, *63*, 8536–8543.

151. Romero, D.L. *Ann. Rep. Med. Chem.* **1994**, *29*, 123–132.

152. Tucker, T.J.; Lyle, T.A.; Wiscount, C.M.; Britcher, S.F.; Young, S.D.; Sanders, W.M.; Lumma, W.C.; Goldman, M.E.; O'Brien, J.A.; Ball, R.G.; Homnick, C.F.; Schleif, W.A.; Emini, E.A.; Huff, J.R.; Anderson, P.S. *J. Med. Chem.* **1994**, *37*, 2437–2444.
153. Quinn, T.C. *Lancet* **1996**, *348*, 99–106.
154. De Clercq, E. *J. Med. Chem.* **1995**, *38*, 2491–2517.
155. Young, S.D.; Britcher, S.F.; Tran, L.O.; Payne, L.S.; Lumma, W.C.; Lyle, T.A.; Huff, J.R.; Anderson, P.S.; Olsen, D.B.; Carroll, S.S.; Pettibone, D.J.; O'Brien, J.A.; Ball, R.G.; Balani, S.K.; Lin, J.H.; Chen, I.-W.; Schleif, W.A.; Sardana, V.V.; Long, W.J.; Byrnes, V.W.; Emini, E.A. *Antimicrob. Agents Chemother.* **1995**, *39*, 2602–2605.
156. Jiang, B.; Wang, Q.-F.; Yang, C.-G.; Xu, M. *Tetrahedron Lett.* **2001**, *42*, 4083–4085.
157. Corley, E.G.; Thompson, A.S.; Huntington, M. *Org. Synth.* **2000**, *77*, 231–235.
158. (a) Schoberth, W.; Hanack, M. *Synthesis* **1972**, *12*, 703–703. (b) Scott, L.T.; Cooney, M.J.; Otte, C.; Puls, C.; Haumann, T.; Boese, R.; Carroll, P.J.; Smith, A.B., III; de Meijere, A. *J. Am. Chem. Soc.* **1994**, *116*, 10275–10283. (c) Lambert, J.B.; Papay, J.J.; Mark, H.W. *J. Org. Chem.* **1975**, *40*, 633–636.
159. Niwa, S.; Soai, K. *J. Chem. Soc., Perkin Trans. 1* **1990**, *4*, 937–943.
160. (a) Soai, K.; Yokoyama, S.; Hayasaka, T. *J. Org. Chem.* **1991**, *56*, 4264–4268. (b) Moffett, R.B. *J. Org. Chem.* **1949**, *14*, 862–867.
161. Thompason, A.S.; Corley, E.G.; Huntington, M.F.; Grabowski, E.J.J. *Tetrahedron Lett.* **1995**, *36*, 8937–8940.
162. The structure of **299b** was confirmed using heteronuclear NOE (HOESY) NMR experiments. Yu, C.; Levy, G.C. *J. Am. Chem. Soc.* **1984**, *106*, 6533.
163. Newman, M.S.; Khanna, V.K. *Org. Prep. Proced. Int.* **1985**, *17*, 422–423.
164. Bauman, J.G.; Wirsching, R.C. *PCT Int. Appl.* WO 9412514 A1 19940609, **1994**.
165. Xu, X.-B.; Wang, Y.-Q.; Shen, Y.-J. *Jingxi Yu Zhuanyong Huaxuepin* **2006**, *14*, 16–18.
166. Li, Y.-Q.; Lu, M.; Lv, C.-X.; Lu, H.-F. *Jiangsu Huagong* **2005**, *33*, 121–123.
167. Eaton, C.N.; Denny, G.H., Jr. *J. Org. Chem.* **1969**, *34*, 747–748.
168. Montgomery, J.A.; Hewson, K. *J. Am. Chem. Soc.* **1960**, *83*, 463–468.
169. (a) Zuffi, G.; Monciardini, S. *Eur. Pat. Appl.* EP 1835035 A1 20070919, **2007**. (b) Zuffi, G.; Ghisotti, D.; Oliva, I.; Capra, E.; Frascotti, G.; Tonon, G.; Orsini, G. *Biocat. Biotransform.* **2004**, *22*, 25–33.
170. Farina, P.; Petrucciani, L.; Colombo, P.; Caprioli, G. *Eur. Pat. Appl.* EP 1464708 A1 20041006, **2004**.
171. Hummel-Marquardt, H.; Kennecke, M.; Weber, A.; Schmitz, T.; Tilstam, U.; Nickisch, K. *Ger. Offen.* DE 4418474 A1 19951123, **1995**.
172. Blumbergs, P.; Khan, M.S.; Kalamas, R.L.; Patel, A.; Lamontagne, M.P. *PCT Int. Appl.* WO 9200312 A1 19920109, **1992**.
173. Cotticelli, G.; Salvetti, R. *PCT Int. Appl.* WO 2007144168 A1 20071221, **2007**.
174. Barai, V.N.; Kvach, S.V.; Zinchenko, A.I.; Mikhailopulo, I.A. *Biotechnol. Lett.* **2004**, *26*, 1847–1850.
175. Ochiai, T.; Ishida, R. *Jap. J. Pharmacol.* **1982**, *32*, 427–438.
176. Ishikawa, T.; Kamide, R.; Niimura, M. *J. Dermatol.* **1994**, *21*, 430–433.

177. Tani, J.; Yamada, Y.; Oine, T.; Ochiai, T.; Ishida, R.; Inoue, I. *J. Med. Chem.* **1979**, *22*, 95–99.

178. Hickinbottom, W.J.; In *The Chemistry of Carbon Compounds*; Vol. IIIA, Rodd, E.H., Ed.; Elsevier: New York-Amsterdam, 1954; pp 580.

179. Petyunin, P.A.; Kozhevnikov, Yu.V.; *Khim. Geterotsikl. Soedin.* **1967**, 415–418.

180. Kamifuji, T.; Okatake, M. *Jpn. Kokai Tokkyo Koho* JP 05043556 A 19930223, CAN 119:72623, **1993**.

181. Kamifuji, T.; Okatake, M. *Jpn. Kokai Tokkyo Koho* JP 05097788 A 19930420, CAN 119:139264, **1993**.

182. Cook, G.K.; Hornback, W.J.; McDonald, J.H., III; Munroe, J.E.; Snyder, N.J. *Bioorg. Med. Chem. Lett.* **1998**, *8*, 1261–1266.

183. Hornback, W.J.; Munroe, J.E. *US Patent* US 5099015 A 19920324, **1992**.

184. Blaszczak, L.C.; Brown, R.F.; Cook, G.K.; Hornback, W.J.; Hoying, R.C.; Indelicato, J. M.; Jordan, C.L.; Katner, A.S.; Kinnick, M.D.; McDonald, J.H., III; Morin, J.M.; Munroe, J.E.; Pasini, C.E. *J. Med. Chem.* **1990**, *33*, 1656–1662.

185. (a) Cook, G.K.; Hornback, W.J.; Jordan, C.L.; McDonald, J.H., III; Munroe, J.E. *J. Org. Chem.* **1989**, *54*, 5828–5830. (b) Crowell, T.A.; Halliday, B.D.; McDonald, J.H., III; Indelicato, J.M.; Pasini, C.E.; Wu, E.C.Y. *J. Med. Chem.* **1989**, *32*, 2436–2442. (c) Cook, G.K.; McDonald, J.H., III; Alborn, W., Jr.; Boyd, D.B.; Eudaly, J.A.; Indelicato, J.M.; Johnson, R.; Kasher, J.S.; Pasini, C.E.; Preston, D.A.; Wu, E.C.Y. *J. Med. Chem.* **1989**, *32*, 2442–2450.

186. (a) Dantzig, A.H.; Duckworth, D.C.; Tabas, L.B. *Biochim. Biophys. Acta* **1994**, *1191*, 7–13. (b) Tsuji, A.; Nakashima, E.; Kagami, I.; Yamana, T. *J. Pharm. Sci.* **1981**, *70*, 768–772.

187. (a) Evans, D.A.; Sjorgen, E.B. *Tetrahedron Lett.* **1985**, *26*, 3787–3790. (b) Bodurow, C. C.; Boyer, B.D.; Brennan, J.; Bunnell, C.A.; Burks, J.E.; Carr, M.A.; Doecke, C.W.; Eckrich, T.M.; Fisher, J.W.; Gardner, J.P.; Graves, B.J.; Hines, P.; Hoying, R.C.; Jackson, B.G.; Kinnick, M.D.; Kochert, C.D.; Lewis, J.S.; Luke, W.D.; Moore, L.L.; Morin, J.M., Jr.; Nist, R.L.; Prather, D.E.; Sparks, D.L.; Vladuchick, W.C. *Tetrahedron Lett.* **1989**, *30*, 2321–2324.

188. Blaszczak, L.C.; Munroe, J.E.; Spry, D.O. *Eur. Pat. Appl.* EP 301877 A1 19890201, **1989**.

189. Kobayashi, Y.; Yamamoto, K.; Kumadaki, I. *Tetrahedron Lett.* **1979**, *42*, 4071–4072.

190. Burton, D.J.; Wiemers, D.M. *J. Am. Chem. Soc.* **1985**, *107*, 5014–5015.

191. Kaminski, Z. *Tetrahedron Lett.* **1985**, *26*, 2901–2904.

192. Dangles, O.; Guibe, F.; Balavoine, G.; Lavielle, S.; Marquet, A. *J. Org. Chem.* **1987**, *52*, 4984–4993.

193. Gasiorowski, K.; Malinka, W.; Swiatek, P.; Jaszczyszyn, A. *Cell. Mol. Biol. Lett.* **2003**, *8*, 927–942.

194. Madrid, P.B.; Polgar, W.E.; Toll, L.; Tanga, M.J. *Bioorg. Med. Chem. Lett.* **2007**, *17*, 3014–3017.

195. Yale, H.L.; Sowinski, F.; Bernstein, J. *J. Am. Chem. Soc.* **1957**, *79*, 4375–4379.

196. Mayer, M.; Lang, P.T.; Gerber, S.; Madrid, P.B.; Pinto, I.G.; Guy, R.K.; James, T.L. *Chem. Biol.* **2006**, *13*, 993–1000.

197. Nikitenkova, L.P., Artamonova, N.N., Shner, V.F., Salov, B.V., Khokhlova, V.M.; Shkurko, E.N. *Russ.* RU 2167156 C1 20010520, **2001**.

198. (a) Yale, H.L.; Sowinski, F.; Bernstein, J. *J. Am. Chem. Soc.* **1957**, *79*, 4375–4379. (b) Smith, N.L. *J. Org. Chem.* **1950**, *15*, 1125–1130.

199. Shner, V.F.; Artamonova, N.N.; Salov, B.V.; Maksimova, N.F.; Khokhlova, V.M. *Russ.* RU 2167157 C1 20010520, **2001**.

200. Beck, I.; Dietz, A.; Halmos, J.; Jakfalvi, E.; Ladanyi, L.; Rozsa, L.; Simonyi, I.; Toreki, J. *Hung. Teljes* HU 38343 A2 19860528, **1986**.

201. Schepartz, A.; Cuenoud, B. *J. Am. Chem. Soc.* **1990**, *112*, 3247–3249.

202. Sardessai, M.S.; Brander, M.J.; Midha, K.K.; Hawes, E.M. *J. Label. Compd. Radiopharm.* **1986**, *23*, 317–327.

203. Heitmann, W.; Liepmann, H.; Maetzel, U.; Zeugner, H.; Fuchs, A.M.; Kraehling, H.; Ruhland, M.; Mol, F.; Tulp, M.T.M. *Eur. J. Med. Chem.* **1988**, *23*, 249–256.

204. Liepmann, H.; Milkowski, W.; Zeugner, H. *Eur. J. Med. Chem.* **1976**, *11*, 501–507.

205. Klayman, D.L. *Science* **1985**, *228*, 1049–1055.

206. Magueur, G.; Crousse, B.; Ourevitch, M.; Begue, J.-P.; Bonnet-Delpon, D. *J. Org. Chem.* **2003**, *68*, 9763–9766.

207. (a) Ye, B.; Wu, Y. *Faming Zhuanli Shenqing Gongkai Shuomingshu* CN 1053237 A 19910724, **1991**. (b) Bharel, S.; Gulati, A.; Abdin, M.Z.; Srivastava, P.S.; Vishwakarma, R.A.; Jain, S.K. *J. Nat. Prod.* **1998**, *61*, 633–636. (c) Orabi, K.Y.; Galal, A.M.; Ibrahim, A.-R.S.; El-Feraly, F.S.; Khalifa, S.I.; El-Sohly, H.N. *Phytochem.* **1999**, *51*, 257–261. (d) Ma, J.; Weiss, E.; Kyle, D.E.; Ziffer, H. *Bioorg. Med. Chem. Lett.* **2000**, *10*, 1601–1603; (e) Sy, L.-K.; Brown, G.D. *Tetrahedron* **2002**, *58*, 897–908, and references therein. (f) Sy, L.-K.; Brown, G.D. *Tetrahedron* **2002**, *58*, 909–923. (g) Khalifa, S.I. *Bull. Fac. Pharm. (Cairo Univ.)* **2000**, *38*, 143–146. (h) Yadav, J.S.; Babu, R. S.; Sabitha, G. *Tetrahedron Lett.* **2003**, *44*, 387–389, and references therein; (i) Yadav, J. S.; Babu, R.S.; Sabitha, G. *ARKIVOC* **2003**, *3*, 125–139. (j) Bhattacharya, A.K.; Siddique, M.S.; Bhakuni, R.S.; Jain, D.C.; Sharma, R.P. *Indian* IN 184682 A1 20000923, **2000**. (k) Zhang, Y.; Ye, H.; Li, G. *Yingyong Yu Huanjing Shengwu Xuebao* **2003**, *9*, 616–618. (l) Weathers, P.J.; DeJesus-Gonzalez, L.; Kim, Y.J.; Souret, F.F.; Towler, M.J. *Plant Cell Rep.* **2004**, *23*, 414–418. (m) Singh, C.; Tiwari, P. *Indian* IN 191269 A1 20031018, **2003**. (n) Reiling, K.K.; Renninger, N.S.; Mcphee, D.J.; Fisher, K.J.; Ockey, D.A. *PCT Int. Appl.* WO 2006128126 A1 20061130, **2006**. (o) Tatineni, R.; Doddapaneni, K.K.; Dalavayi, S.; Kulkarni, S.M.; Mangamoori, L.N. *Proc. Biochem.* **2006**, *41*, 2464–2467. (p) Liu, Y.; Xiao, W.; Wong, M.-K.; Che, C.-M. *Org. Lett.* **2007**, *9*, 4107–4110.

208. (a) Prakash, G.K.S.; Yudin, A.K. *Chem. Rev.* **1997**, *97*, 757–786. (b) Singh, R.P.; Shreeve, J.M. *Tetrahedron* **2000**, *56*, 7613–7632.

13

PRACTICAL USES OF FLUORINATED HETEROCYCLES

PAUL R. RESNICK

FluoroScience LLC, 104 Rivergreen Court, Cary, NC 27518, USA

Fluorine-containing heterocycles are not just laboratory curiosities. They have many practical uses. The objective of this chapter is to emphasize how these materials have been used in present day commercial applications. Many fluorine-containing heterocycles and their derivatives have been synthesized, which may be found in the catalogs of laboratory supply houses. In addition, there is a vast body of work in the literature about the chemistry of these heterocycles and the derivatives. It is the intent of this chapter to discuss those fluorinated heterocycles that have been used as products and as starting materials for nonpharmaceutical applications that have commercial and practical uses. The focus of this chapter is on heterocycles in which a fluorine atom is directly bonded to the hetero-ring system.

The fluorinated heterocyclic compound with the greatest commercial use has been perfluoropropylene oxide, (HFPO, I). It has been used as both a monomer and as an intermediate.[1] There have been many syntheses of HFPO, but virtually all of them consist of the oxidation of hexafluoropropylene (Eq. 13.1).

(13.1)

HFPO, I

The major routes are oxidation with molecular oxygen or alkaline hydrogen peroxide or electrochemically. The separation of HFPO, bp −27.4°C, from starting hexafluoropropylene, bp −29.4°C, is performed by extractive distillation.[2]

Fluorinated Heterocyclic Compounds: Synthesis, Chemistry, and Applications, Edited by Viacheslav A. Petrov

HFPO readily reacts with nucleophiles. Reaction occurs primarily at the center carbon to give an alkoxide intermediate that is in equilibrium with fluoride ion and an acyl fluoride (Eq. 13.2).

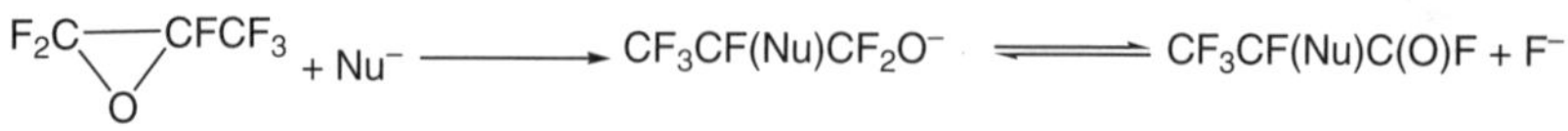

(13.2)

The alkoxide itself is a nucleophile and may react with additional HFPO to form a second alkoxide, and so on. A commercially important example is the reaction of HFPO with fluoride ion (Eq. 13.3). Each of the alkoxide intermediates is in equilibrium with an acid fluoride (Eq. 13.4). By proper choice of cation, solvent, reaction temperature, and pressure, the number of HFPO units in the reaction product may be varied from one, that is, CF_3CF_2COF, to more than 100.

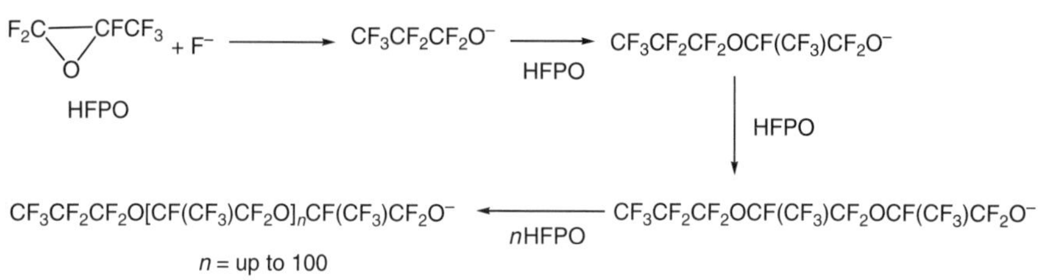

(13.3)

$$CF_3CF_2CF_2O^- \rightleftharpoons CF_3CF_2C(O)F + F^- \tag{13.4}$$

The higher molecular weight acid fluoride products of the oligomerization may be decarbonylated by a variety of means to give fluoro-endcapped inert materials known as Krytox® (II) (Eq. 13.5). Different grades of Krytox® oils are obtained by fractionation. Krytox® oils are highly lubricious liquids with low volatility that are compatible with oxygen

$$\begin{array}{c} CF_3CF_2CF_2O[CF(CF_3)CF_2O]_nCF(CF_3)C(O)F \rightarrow \\ CF_3CF_2CF_2O[CF(CF_3)CF_2O]_nCF_2CF_3 \\ \text{II, Krytox}^{\circledR} \end{array} \tag{13.5}$$

and are resistant to aggressive chemicals.[3] Krytox® greases are prepared by the addition of low molecular weight polytetrafluoroethylene to Krytox® oil. Krytox® is thermally stable, nonflammable even in liquid oxygen, is nontoxic and is insoluble in water, acids, bases, and most organic solvents. It has very low vapor pressure and can be used between about −60°C and 399°C. Krytox® is highly resistant to ionizing radiation and can withstand extreme pressure and high mechanical stress. The presence of the strong electron-withdrawing fluorine groups renders the ether oxygen atoms inert resulting in a structure whose chemical properties are very similar to polytetrafluoroethylene. The carbon–oxygen bonds of perfluoroethers often are more stable than the carbon–carbon bonds of perfluorocarbons. Both perfluorinated ethers and perfluorinated amines are nonbasic and do not form complexes with protic acids or boron trifluoride. The presence of oxygen in the oligomeric backbone plus the trifluoromethyl side chains results in the elimination of any crystallinity in these inert liquid materials.

A more important use of HFPO is as an intermediate for the syntheses of a wide range of trifluorovinyl ether monomers that are used as comonomers in fluorinated plastics and elastomers. The general synthesis of this class of monomers is exemplified by the synthesis of perfluoromethyl vinyl ether (IV, PMVE) (Eq. 13.6).

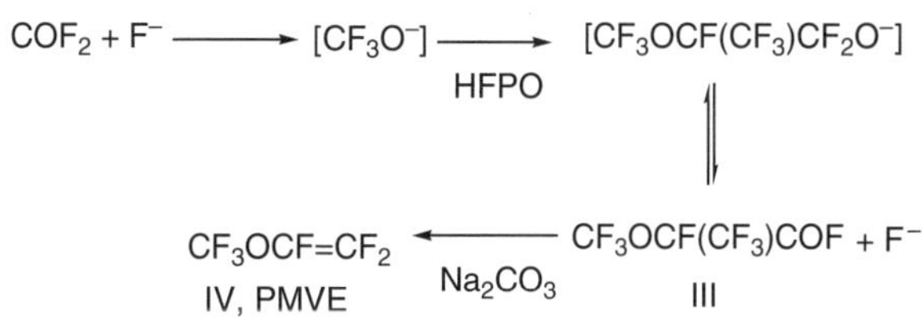

(13.6)

Reaction of an acyl fluoride, via its alkoxide with HFPO can be made to yield a 1:1 adduct (III) that may be decarboxylated to the desired trifluorovinyl ether (IV). Reaction conditions have been optimized to give high yields of the desired reaction product between the acyl fluoride and HFPO whether they be the 1:1 or 1:2 or higher adducts. The trifluorovinyl ethers are difficult to homopolymerize, but copolymerize readily with other monomers such as tetrafluoroethylene, vinylidene fluoride, ethylene, and so on, to give fluoroplastics sold as high performance grades of Teflon® PFA, Hyflon®, Neoflon®, Fluon®, Dyneon®, and so on. These ethers are vital components in modern high-performance fluoroelastomers such as Kalrez®, Dai-el®, and Tecnoflon®. HFPO (I) is also a key starting material for the synthesis of the monomers used in commercial perfluorinated ion-exchange membranes such as Nafion® and Flemion®. Table 13.1 lists some of the monomers that arise from HFPO.

TABLE 13.1 Some Commercial Trifluorovinyl Ether Monomers Based on HFPO

Monomer	Starting Material	Use
$CF_2{=}CFOCF_3$	COF_2	Elastomer
$CF_2{=}CFOCF_2CF_3$	CF_3COF	Plastic
$CF_2{=}CFOCF_2CF_2CF_3$	CF_3CF_2COF (made from HFPO)	Plastic
$CF_2{=}CFOCF_2CF(CF_3)OCF_2CF_2SO_2F$	FSO_2CF_2COF	Ionomer
$CF_2{=}CFOCF_2CF(CF_3)OCF_2CF_2COOCH_3$	CH_3OOCCF_2COF	Ionomer
$CF_2{=}CFOCF_2CF_2CF_2CF_2SO_2F$	$FSO_2CF_2CF_2CF_2COF$	Ionomer
$CF_2{=}CFOCF_2CF_2CF_2COOCH_3$	$CH_3OOCCF_2CF_2COF$	Ionomer
$CF_2{=}CFOCF_2CF(CF_3)OC_6F_5$	$C_6F_5O^- M^+$	Elastomer
$CF_2{=}CFOCF_2CF(CF_3)OCF_2CF_2CN$	CH_3OOCCF_2COF	Elastomer

The same acid fluoride intermediates may also be used to synthesize stable fluids. If the intermediates are decarboxylated in a protic solvent such as water at elevated temperature the carboxyl groups are replaced by a proton (Eq. 13.7).

$$CF_3CF_2CF_2O[CF(CF_3)CF_2O]_nCF(CF_3)C(O)F \xrightarrow[NaOH/H_2O]{} \underset{V, n = 1-7}{CF_3CF_2CF_2O\,[CF(CF_3)CF_2O]_nCFHCF_3} \quad (13.7)$$

These "hydro" endcapped liquids ($V, n = 1$–7) were commercialized by DuPont as the Freon® E series fluorocarbons for use as stable fluids, heat-transfer liquids, and dielectric coolants. The acyl fluorides were also converted into inert materials by dimerization with loss of the carbonyl group either electrochemically or photochemically (Eq 13.8). Both processes were used to make dimers and mixed dimers to tailor the boiling point of the final products.

$$C_3F_7O[CF(CF_3)CF_2O]_nCF(CF_3)C(O)F \rightarrow C_3F_7O[CF(CF_3)CF_2O]_n(CF_3)CF\text{-}CF(CF_3)[OCF_2CF(CF_3)]_nOC_3F_7 \quad (13.8)$$

The materials are virtually chemically inert and were commercialized for a short time as Hostinert® fluids for use as stable liquids with excellent dielectric properties and compatibility with most metals, plastics, and elastomers.

HFPO oligomers, especially the trimer, have been converted to their corresponding ammonium salts to give excellent surfactants (Eq 13.9). However, their use has been curtailed due to their negative toxicological and environmental properties.

$$CF_3CF_2CF_2OCF(CF_3)CF_2OCF(CF_3)COF \xrightarrow[(1)\ H_2O\ (2)\ NH_3]{} CF_3CF_2CF_2OCFCF_2OCF(CF_3)COO^-NH_4^+ \quad (13.9)$$

HFPO undergoes an electrophilic rearrangement catalyzed by Lewis acids to give hexafluoroacetone (HFA, VII), in virtual quantitative yield[4] (Eq. 13.10). This reaction is used in a commercial process to make HFA and competes favorably with the preparation of HFA

(13.10)

that starts from hexachloroacetone. Thus, the whole family of products prepared from HFA may be considered as arising from HFPO. These include, hexafluoroisopropanol, (VIII), Sevoflurane®, (IX), bisphenol-AF, (X), and hexafluoroisopropylidene bisphthalic dianhydride, (XI), used as a comonomer in high performance polymers.[5]

$(CF_3)_2CHOH$ $(CF_3)_2CHOCH_2F$
VIII XI

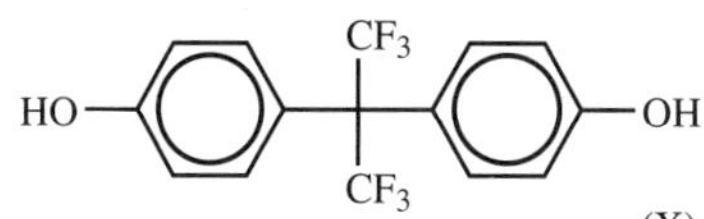
(X)

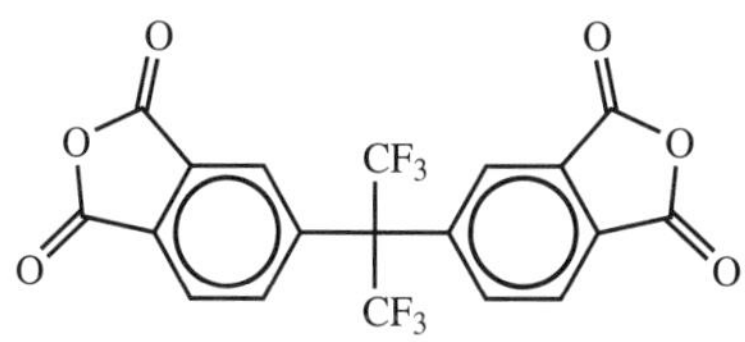
(XI)

HFPO also serves as the starting material for trifluoropyruvic acid derivatives. For example, reaction of HFPO with methanol yields methyl 3,3,3-trifluoro-2-methoxy propionate (XII) that in turn yields methyl trifluoropyruvate (XIII) when treated with antimony pentafluoride.[6,7] The yields of both these reactions are well over 90% (Eq 13.11).

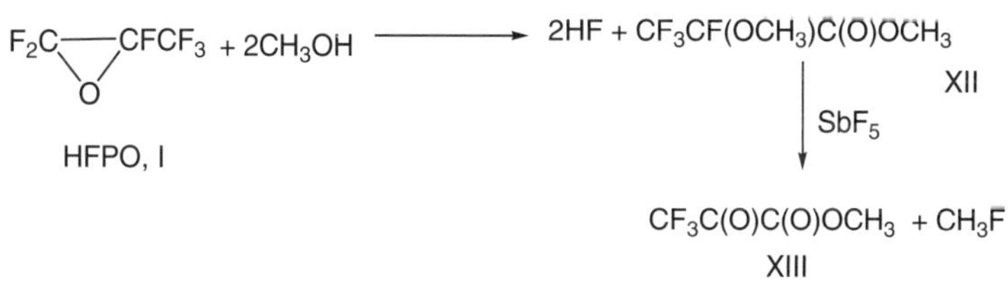

HFPO is a useful source of difluorocarbene. When heated above 160°, HFPO decomposes to difluorocarbene and trifluoroacetyl fluoride[8,9] (Eq 13.12). This reaction has been utilized to prepare thin perfluorinated conformal coatings on a wide variety of substrates using a hot-wire chemical vapor deposition process commercialized by GVD Corporation.[10,11,12]

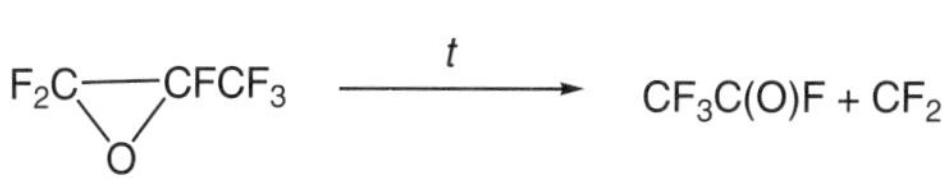
(13.11)

HFPO, I

Although large number of fluorinated epoxides has been synthesized, to date, HFPO is the only compound that has commercial interest.[13] Tetrafluoroethylene oxide, TFEO, has been prepared and is extremely reactive, isomerizing easily to trifluoroacetyl fluoride[14,15] (Eq 13.13). Although the chemistry of TFEO has been described, this facile isomerization so far has precluded its use in the preparation of commercial products.[16,17,18]

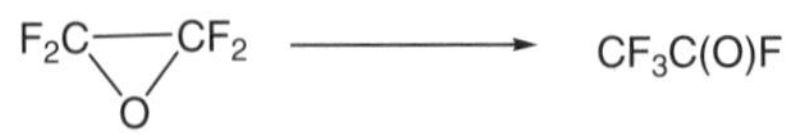

TFEO

(13.12)

A practical synthesis of 2,2-bis(trifluoromethyl)oxirane (XIV) has been described[19] (Eq 13.14). The compound has been used to introduce the -$C(CF_3)_2OH$ group into the polymer backbone of photopolymers highly transparent at 157 nm, but

XIV

(13.13)

commercial exploitation of this chemistry has yet to be developed.[20] Many additional fluorinated heterocycles have been prepared for use in microlithography, but to date none have been produced as components of commercial polymers and products.

An oxetane, 2,2,3,3-tetrafluorooxetane (XV), has been utilized by the Daikin Corporation to prepare a large number of derivatives as well as a perfluorinated polyether oil, Demnum®.[21] The compound was first prepared by the reaction of TFE, anhydrous hydrogen fluoride and formaldehyde (Eq 13.15a) and later by the high temperature oxidation of 1,1,2,2-tetrafluoropropane at elevated temperature[22,23] (Eq 13.15b).

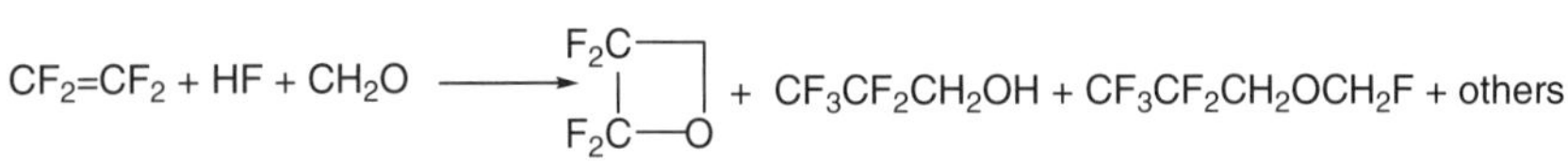

XV

(13.14)

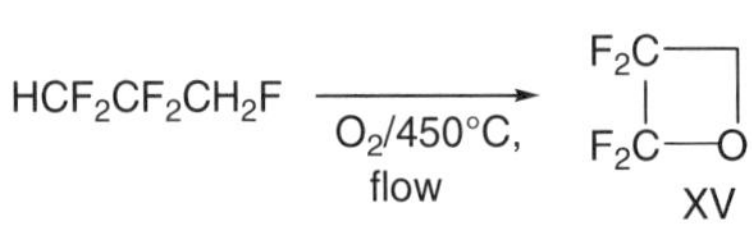

XV

(13.15)

The chemistry of XV is similar to that of HFPO. It can be isomerized to an acyl fluoride with Lewis acids (Eq 13.16) but more importantly, it will react with nucleophiles to open the oxetane ring. Reaction with NaI yields an acyl fluoride that may be dehalogenated and esterified to give useful α-fluoroacrylate monomers (Eq 13.17).

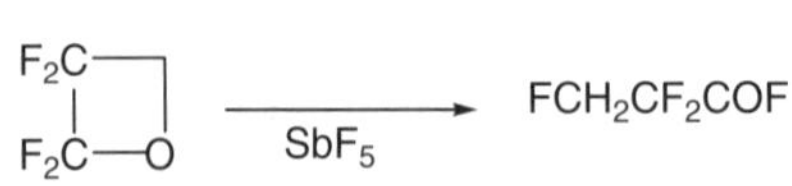

XV

(13.16)

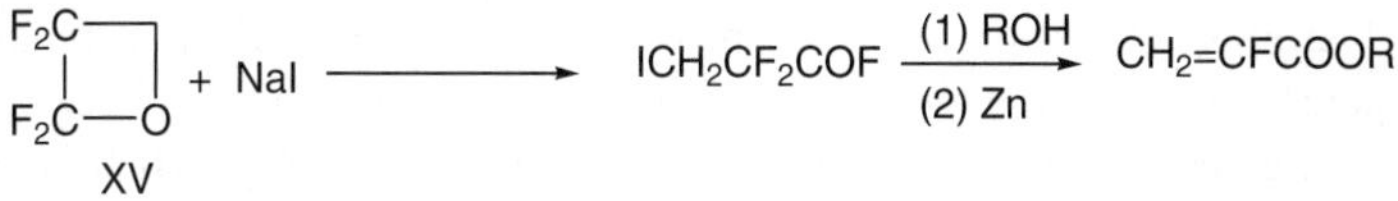

(13.17)

The most important use of XV is its telomerization with fluoride ion to give a linear hydrofluoroacyl fluoride (XVI), that is fluorinated with elemental fluorine to give a linear perfluorinated polyether (XVIII), which is sold by Daikin as Demnum® (Eq 13.18). Note that Demnum® and Krytox® are isomeric materials. In general, their properties are quite similar, but there are subtle differences in physical properties and in chemical stability in tribological applications.

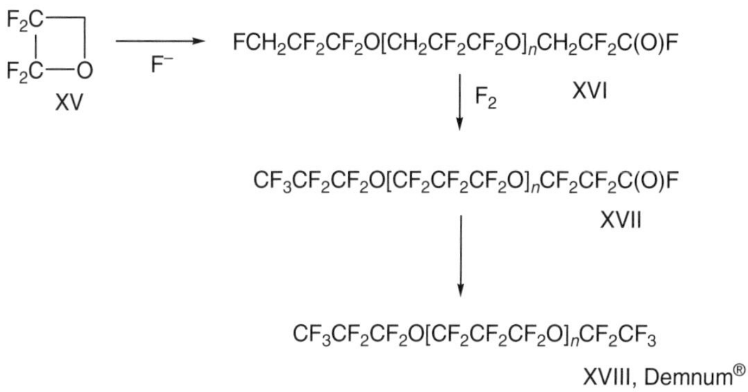

(13.18)

The intermediate acyl fluoride, (XVII), can be converted using classical chemical reactions to acids, alcohols, vinyl ethers, and iodides that in turn may be used as surfactants and intermediates for monomers and polymers. These are the same reactions that have been exploited in the chemistry of HFPO oligomers.

Perfluorinated dioxole monomers have been used to prepare a series of amorphous fluoropolymers such as Teflon® AF and Hyflon® AD. A third amorphous fluoropolymer, Cytop® contains perfluorotetrahydrofuran and perfluorotetrahydropyran rings, but is prepared in a cyclopolymerization process from an acyclic monomer. These amorphous fluoropolymers retain the outstanding chemical, thermal, and surface properties associated with perfluorinated polymers while also having unique electrical, optical, and solubility characteristics.

Teflon® AF, sold by DuPont, is a copolymer of 2,2-bistrifluoromethyl-4,5-difluoro-1,3-dioxole (PDD, XIX) and tetrafluoroethylene.[24]

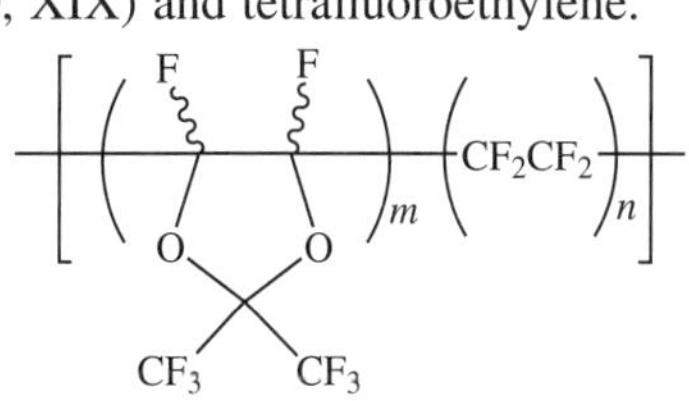

Teflon® AF

PDD monomer is synthesized in four steps starting from hexafluoroacetone (Eq 13.19). The glass transition temperature of the copolymers, T_g, is a function of the mole percent PDD in the polymer and increases with amount of PDD in the polymer while the density of the polymers decrease with increasing PDD content.

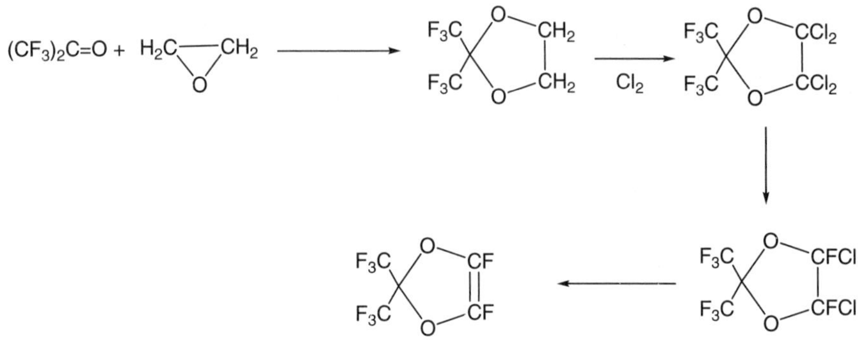

PDD, XIX

(13.19)

It is also interesting to note that the T_g may vary greatly with small changes in the structure of the dioxole. For example the homopolymer of PDD has a T_g of 335°C while the homopolymer of the isomeric 2-pentafluoroethyl-4,5-difluoro-1,3-dioxole, (XX), has a T_g of 150°C.

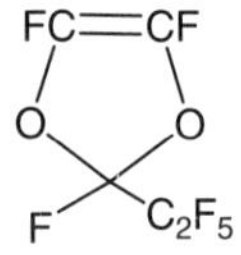

XX

It should be noted that these dioxoles are extremely reactive in free radical polymerizations. Dioxoles are the first fluorinated monomers containing an internal olefinic structure that homopolymerize and possess reactivity similar to tetrafluoroethylene. This high reactivity is believed to a result of the steric accessibility of the double bond.

The other commercial dioxole monomer is 4-trifluoromethoxy-2,2,5-trifluoro-1,3-dioxole, (TDD, XXI). Copolymers of TDD and tetrafluoroethylene are produced and sold by Solvay-Solexis as Hyflon® AD amorphous fluoropolymer (XXII).[25] Monomer XXII (TDD) has been synthesized by two related routes via perfluorohypofluorite intermediates (Eq 13.20). In the first route, trifluoromethylhypofluorite formed from either COF_2 or CO and fluorine is added

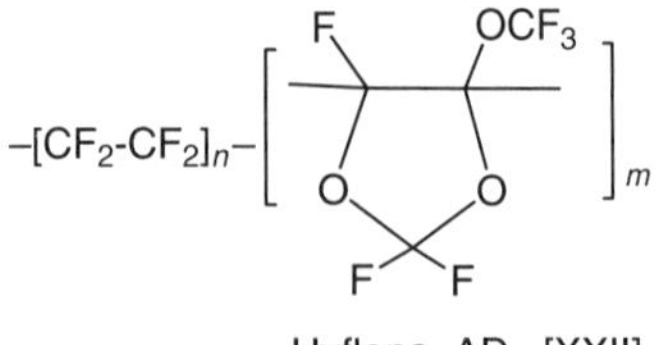

Hyflon® AD [XXII]

tetrachloroethylene followed by dechlorination to give 1,1,2-trichloro-trifluoromethoxyethylene. This in turn is reacted with $CF_2(OF)_2$ formed from carbon dioxide and fluorine to give 2,2,4-trifluoro-4,5-dichloro-5-trifluoromethoxy-1,3-dioxolane, which is dechlorinated with zinc to give TDD monomer. The second route uses the same steps and same reagents but in a different order. The heterocycle is first formed by the reaction of $CF_2(OF)_2$ with tetrachloroethylene followed by dechlorination and subsequent addition of trifuoromethylhypofluorite. The last step, dechlorination, is the same in both routes.

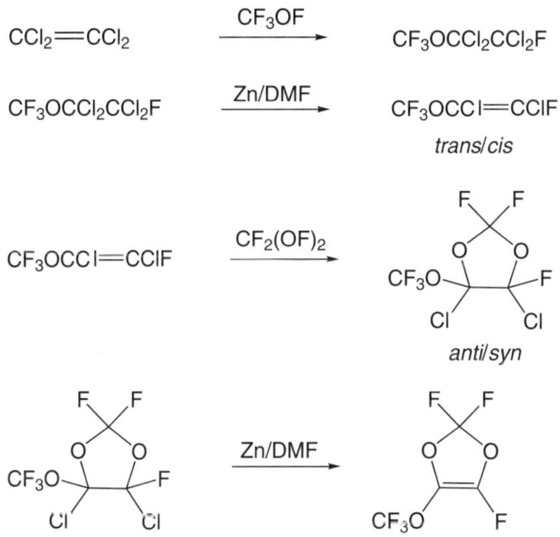

(13.20)

TDD, XXI

Hyflon® AD amorphous fluoropolymer is used in optical devices, pellicles in semiconductor manufacture, as a dielectric and as a separation membrane. Small amounts of TDD have been used as a modifier in ethylene-chlorotrifluoroethylene polymers to increase stress crack resistance. Minute amounts of TDD are used also as a modifier in polytetrafluoroethylene to improve elastic modulus, reduce creep and permeability and increase transparency. It has been suggested that the much higher reactivity of TDD and other fluorinated dioxoles relative to other modifiers gives a more uniform distribution of the modifier in the polymer chain that results in a greater increase in the desired properties at lower concentration of modifier in the polymer.

The third amorphous fluoropolymer, Cytop®[26,27] (XXIII), is produced by Asahi Glass from an acyclic monomer by means of a cyclopolymerization. The resulting polymer contains a perfluorinated tetrahydrofuran ring and has a T_g of 108°. The polymerization of

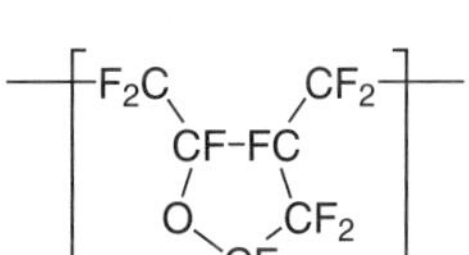

Cytop®, XXIII

the monomer, $CF_2{=}CFOCF_2CF_2CF{=}CF_2$ (XXIV) results primarily in cyclization to a five-membered heterocyclic ring. The ratio of five- to six-membered rings has been shown to be 84:16 and does not change during the course of the polymerization.[28] Cytop® is used in the preparation of Lucina® optical fiber,[29] in antireflective coatings, as a conformal coating and other specialty applications. The monomer has been prepared using two synthetic routes. The initial route starting with chlorotrifluoroethylene is shown below (Eq 13.21).

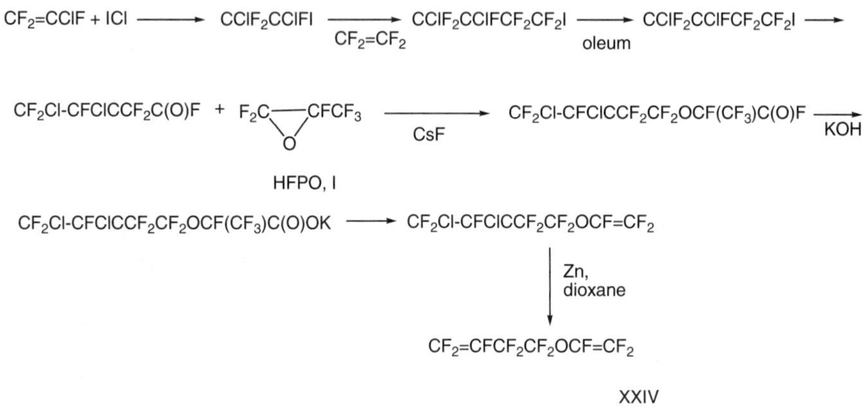

(13.21)

A second route utilizes the fluorination of a hydrocarbon as the key synthetic step.[30] (Eq 13.22) Both are multistep processes that result in a high-cost product.

$$CH_2ClCHClCH_2CH_2OCH(CH_3)CH_2OH \xrightarrow{R_fC(O)F} CH_2ClCHClCH_2CH_2OCH(CH_3)CH_2OC(O)CRf \xrightarrow{F_2}$$

$$CF_2Cl\text{-}CFClCF_2CF_2OCF(CF_3)CF_2OCRf \xrightarrow{F^-} R_fC(O)F + CF_2Cl\text{-}CFClCF_2CF_2OCF(CF_3)COF \xrightarrow{Na_2CO_3}$$

$$CF_2Cl\text{-}CFClCF_2CF_2OCF{=}CF_2 \longrightarrow CF_2{=}CFCF_2CF_2OCF{=}CF_2 \qquad (13.22)$$

All three commercial amorphous fluoropolymers, Teflon® AF, Hyflon® AD, and Cytop® posses a unique set of properties. All dissolve in fluorinated solvents and thus may be spin coated to produce thin films and coatings. The polymers may also be extruded and molded using traditional polymer processing techniques. Note that the polymers are not soluble in hydrocarbon solvents or water and retain the chemical and thermal stability of perfluorinated polymers such as Teflon®. These polymers have lower density than the well-known semicrystalline perfluorinated polymers such as pTFE that results in lower refractive index, lower thermal conductivity, higher gas permeability, and lower dielectric constant. The polymers are transparent and have excellent mechanical properties below their T_g due to their amorphous character. The presence of a heterocyclic ring in the polymer backbone of these materials is key to these properties. Small changes in monomer structure lead to changes in the microstructure of the polymers and large changes in the physical properties of the polymers. For example, Teflon® AF may have a density of less than 1.7, a refractive index of less than 1.3, and one of the highest gas permeability of any known polymer. The commercial usefulness of these amorphous polymers depends on the utilization of a combination of these properties.

As previously noted perfluorinated ethers and amines are virtually chemically inert. This property has been the rationale for the commercial production of a large number of perfluorinated inert fluids. Although most of these materials were acyclic fluorocarbons, amines, and ethers, a number of commercial products were fluorinated heterocycles. These materials were all manufactured using the Simons process in which all the hydrogen atoms in an organic compound are replaced with fluorine atoms by electrolysis in anhydrous hydrogen fluoride. Hydrocarbon amines and ethers were particularly useful in this process due to their high solubility in anhydrous HF. By far the largest volume of inert fluid fluorinated heterocycle produced was a mixture of perfluoro-2-butyltetrahydrofuran and perfluoro-2-propyltetrahydropyran, $C_8F_{16}O$ (XXV), known as 3M's Fluorinert® FC-75 boiling at 102°. This material was made in the electrochemical fluorination of octanoic acid derivatives along with perfluorooctanoyl fluoride in the process used to prepare perfluorooctanoic acid (Eq 13.23).

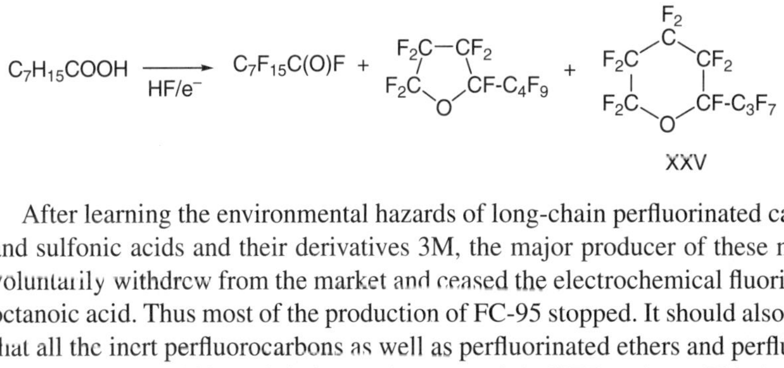

(13.23)

XXV

After learning the environmental hazards of long-chain perfluorinated carboxylic and sulfonic acids and their derivatives 3M, the major producer of these materials, voluntarily withdrew from the market and ceased the electrochemical fluorination of octanoic acid. Thus most of the production of FC-95 stopped. It should also be noted that all the inert perfluorocarbons as well as perfluorinated ethers and perfluorinated amines have very high global warming potential, GWP, values. This has further diminished their commercial usefulness. Since all the perfluorinated ethers and amines are chemically inert and have similar properties, their primary distinguishing characteristic was boiling point. Hence, a large number of these "inert fluorocarbon" materials were synthesized and commercialized. Most were acyclic compounds along with some cyclic compounds containing only carbon and fluorine. A few fluorinated heterocycles such as perfluoro-*N*-methylmorpholine, bp 50°C, perfluoro-*N*-ethylmorpholine, bp 72°C, and perfluoro-*N*-isopropylmorpholine, bp 97°C (XXVI) were also produced, but they never achieved wide spread use.

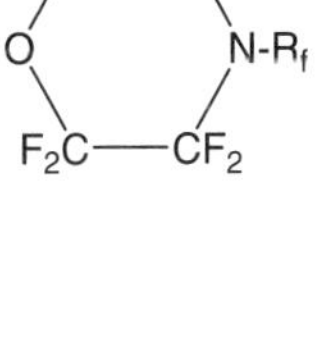

R_f=CF_3, C_2F_5, $CF(CF_3)_2$, XXVI

A fluorinated heterocycle, 3,3,4,4-tetrafluoro-1,2-oxathietane-2,2-dioxide more commonly called tetrafluoroethylene sultone (XXVII), is the starting material for the synthesis of Nafion® perfluorinated ion exchange resins. This compound was synthesized independently at DuPont[31] and the Russian Academy of Sciences[32] by the reaction of tetrafluoroethylene and sulfur trioxide (Eq 13.24).

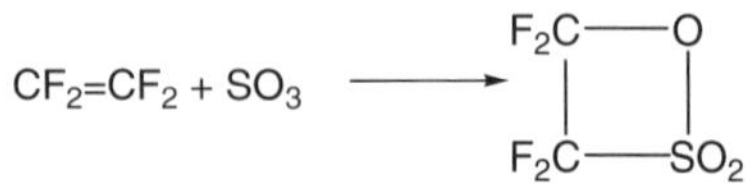

XXVII

(13.24)

The key Nafion® monomer, is prepared from the sultone by rearrangement to fluorosulfonyldifluoroacetyl fluoride followed by reaction with two moles of another fluorinated heterocycle, HFPO, and decarboxylation with sodium carbonate[33] (Eq 13.25).

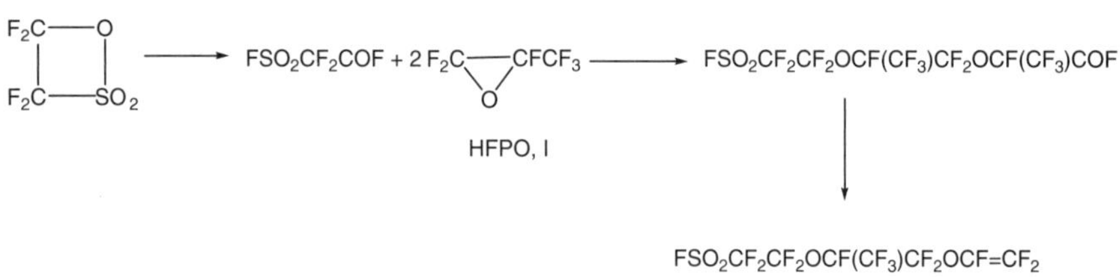

XXVIII

(13.25)

The monomer XXVIII is copolymerized with tetrafluoroethylene to give a polymer-containing pendant fluorosulfonyl groups that are then hydrolyzed and acid exchanged to produce Nafion® (XXIX) The resulting polymer combines the chemical, thermal, and oxidative stability of perfluorinated polymers such as polytetrafluoroethylene with the properties of a highly acidic fluorinated sulfonic acid. Nafion® is used in a variety of electrochemical applications such as the synthesis of chlorine and caustic and as the conductive membrane of many modern fuel cells. It has also been used in water electrolysis and as an acid catalyst in many proprietary commercial processes.

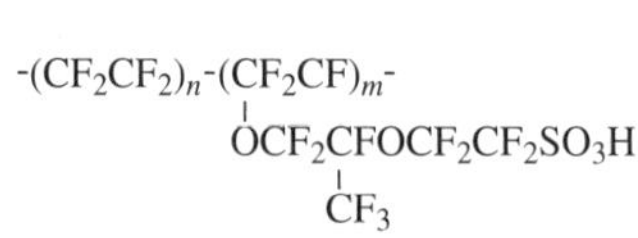

Nafion®, XXIX

The recent commercial phase out of long-chain perfluorinated sulfonic and carboxylic acids and materials that may degrade them in the environments has spawned a search for environmental-friendly substitutes for these surfactants and surface protection agents. Many approaches have and are being investigated and substitutes are beginning to reach the marketplace. One such family of substitutes is based on oligomers and polymers made from fluoroalkoxyoxetanes[34] (XXX, Eq. 13.26). Although these products and their intermediates do not involve fluorine directly bonded to the heterocycle, they should be noted since they exploit the properties associated with multiple carbon–fluorine bonds.[35] The final products are environmental-friendly and have good surfactant and surface protection properties. They are sold by Omnova Corporation as PolyFox®.

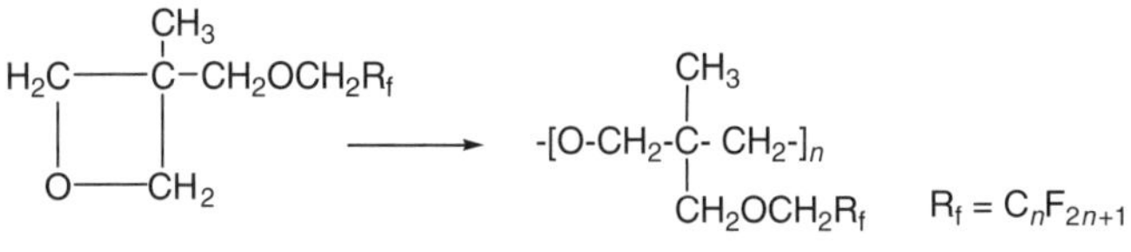

XXX

(13.26)

The future will see many more applications of fluorine-containing heterocycles. The bulk of these applications will be considered as building blocks to add fluorine to primarily hydrocarbon molecules and polymers to utilize the unique and desired properties that relatively small amounts of fluorine can provide.

REFERENCES

1. Millauer, H.; Schwertfeger, W.; Siegemund, G. *Angew. Chem., Int. Ed.* **1985**, *24*, 161.
2. (a) Oda, O.; Uchida, K.; Morikawa, S. *US Patent* 4,134,796, **1979**. (b) Sulzbach, R. *US Patent* 4,358,348, **1982**.
3. See www.2.dupont.com/Lubricants/en_US/assets/downloads/H-58505-1.pdf.
4. (a) Squire, E.N. *US Patent* 4,302,608, **1981**.(b) Morin, D.E. *US Patent* 3,213,134, **1965**.
5. Appel, W.K.; Blech, B.A.; Strobbe, M. Monomers and polymers from hexafluoroacetone. In *Organofluorine Chemistry;* Banks, R.E.; Smart, B.E.; Tatlow, J.C.,Eds.; Plenum: New York, NY, 1994; 413, 0-306-44610-3.
6. Sianesi, D.; Pasetti, A.; Tarli, F. *J. Org. Chem.* **1966**, *31*, 2312
7. Anderson, D.G.; England, D.C.; Milian, A.S. *US Patent* 4,357,282, **1982**.
8. Kennedy, R.C.; Levy, J.B. *J. Fluorine Chem.* **1976**, *7*, 101
9. Mahler, W.; Resnick, P.R. *J. Fluorine Chem.* **1974**, *3*, 451.
10. Cobianu, C.P.; Shiffer, S.R.; Serban, B.C.; Bradley, A.D.; Mihaila, M.N. *US Patent* 7,318,351, **2008**.
11. Lau, K.K.S.; Mao, Y.; Lewis, H.G.P.; Murthy, S.K.; Olsen, B.D.; Loo, L.S.; Gleason, K.K. *Thin Solid Films* **2006**, *501*, 211.
12. Lau, K.K.S.; Murthy, S.K.; Lewis, H.G.P.; Caulfield, J.A.; Gleason, K.K. *J. Fluorine Chem.* **2003**, *122*, 93.
13. Resnick, P.R. Perfluoroepoxides. In *Fluorine Chemistry: A Comprehensive Treatment;* Howe-Grant, M., Ed.; John Wiley & Sons, Inc.: New York, NY, 1995; 294, 0-471-12031-6.
14. Gibbs, H.H.; Warnell, J.L. *French Patent* 1322597, **1963**.

15. Caglioti, V.; Lenzi, M.; Mele, A. *Nature* **1964**, *201*, 4919.
16. Lenzi, M.; Mele, A. *J. Chem. Phys.* **1965**, *43*, 1974.
17. Warnell, J.L. *US Patent* 3,250,806, **1966**.
18. Warnell, J.L. *German Patent* DE 1191576, **1965**.
19. Petrov, V.A.; Marshall, W.J.; Krespan, C.G.; Cherstkov, V.F.; Avetisian, E.A. *J. Fluorine Chem.* **2004**, *125*, 99.
20. Crawford, M.K.; Tran, H.V.; Schadt, F.L.; Zumsteg, F.C.; Feiring, A.E.; Fryd, M. *US Patent Application* 207413, **2007**.
21. (a) Ohsaka, Y. Perfluoropolyether fluids (Demnum®) based on oxetanes. In *Organofluorine Chemistry*; Banks, R.E.; Smart, B.E.; Tatlow, J.C., Ed.; Plenum: New York, NY, 1994; 463, 0-306-44610-3.(b) Ohsaka, Y.; Tohzuka, T; Takaki, S. *US Patent* 4,845,268, 1989.
22. (a) Weinmayr, V. *J. Org. Chem.* **1963**, *28*, 492. (b) Ohsaka, Y.; Tozuka, T.; Takaki, S. *US Patent* 4,845,268, **1986**. (c) Ohsaka, Y.; Takaki, S.; Sakai, H. *US Patent* 4,886,581, **1989**. (d) Ikeda, S.; Sonoi, T. *US Patent Application* 191617, **2007**.
23. (a) Henne, A.L. Abstracts of the 1st International Symposium on Fluorine Chemistry **1959**, 24. (b) Aisaka, Y.; Kono, S. *Japanese Patent* 04068314B, **1992**.
24. Resnick, P.R.; Buck, W.H. Teflon® AF amorphous fluoropolymers. In *Modern Fluoropolymers;* Scheirs, J., Ed.; John Wiley & Sons, Inc.: New York, 1997; 397, 0-471-97055-7.
25. Arcella, V.; Ghielmi, A.; Tommasi, G. *Ann. NY Acad. Sci.* **2003**, *984*, 226.
26. Sugiyama, N. Perfluoropolymers obtained by cyclopolymerization and their applications. In *Modern Fluoropolymers;* Scheirs, J.,Ed.; John Wiley & Sons, Inc.: New York, 1997; 541, 0-471-97055-7.
27. See www.agc.co.jp/english/chemicals/shinsei/cytop/cytop.htm.
28. Yamamoto, K.; Ogawa, G. *J. Fluorine Chem.* **2005**, *126*, 1403.
29. See www.lucina.jp/eg_lucina/indexeng.htm.
30. (a) Iwaya, M.; Okamoto, H.; Oharu, K. *US Patent* 7,102,031, **2006**. (b) Ito, T.; Kaneko, Y. *US Patent* 7,247,757, **2007**.
31. (a) England, D.C. *US Patent* 2,852,554, **1958**. (b) England, D.C.; Dietrich, M.A.; Lindsey, R.V. *J. Am. Chem. Soc.* **1960**, *82*, 6181.
32. Knunyants, I.L.; Sokolski, G.A. *Angew. Chem., Int. Ed.* **1972**, *11*, 583.
33. Connolly, D.J.; Gresham, W.F. *US Patent* 3,282,875, **1966**.
34. Mansouri, H.R.; Thomas, R.R.; Garnier, S.; Pizzi, A. *J. Appl. Polym. Sci.* **2007**, *106*, 1683.
35. (a) Malik, A.A.; Archilbald, T.G. *US Patent* 5,654,450, **1997**. (b) Malik, A.A.; Archilbald, T.G. *US Patent* 5,807,977, **1998**.

INDEX

Fluorinated Heterocyclic Compounds: Synthesis, Chemistry, and Applications, Edited by Viacheslav A. Petrov